Spray Drying Encapsulation of Bioactive Materials

Advances in Drying Science and Technology

Series Editor

Arun S. Mujumdar
McGill University, Quebec, Canada

Handbook of Industrial Drying, Fourth Edition
Arun S. Mujumdar

Advances in Heat Pump-Assisted Drying Technology
Vasile Minea

Computational Fluid Dynamics Simulation of Spray Dryers: An Engineer's Guide
Meng Wai Woo

Handbook of Drying of Vegetables and Vegetable Products
Min Zhang, Bhesh Bhandari, and Zhongxiang Fang

Intermittent and Nonstationary Drying Technologies: Principles and Applications
Azharul Karim and Chung-Lim Law

Thermal and Nonthermal Encapsulation Methods
Magdalini Krokida

Industrial Heat Pump-Assisted Wood Drying
Vasile Minea

Intelligent Control in Drying
Alex Martynenko and Andreas Bück

Drying of Biomass, Biosolids, and Coal: For Efficient Energy Supply and Environmental Benefits
Shusheng Pang, Sankar Bhattacharya, Junjie Yan

Drying and Roasting of Cocoa and Coffee
Ching Lik Hii and Flavio Meira Borem

Heat and Mass Transfer in Drying of Porous Media
Peng Xu, Agus P. Sasmito, and Arun S. Mujumdar

Freeze Drying of Pharmaceutical Products
Davide Fissore, Roberto Pisano, and Antonello Barresi

Frontiers in Spray Drying
Nan Fu, Jie Xiao, Meng Wai Woo, Xiao Dong Chen

Drying in the Dairy Industry
Cécile Le Floch-Fouere, Pierre Schuck, Gaëlle Tanguy, Luca Lanotte, Romain Jeantet

Spray Drying Encapsulation of Bioactive Materials
Seid Mahdi Jafari and Ali Rashidinejad

For more information about this series, please visit: www.crcpress.com/Advances-in-Drying-Science-and-Technology/book-series/CRCADVSCITEC

Spray Drying Encapsulation of Bioactive Materials

Edited by
Seid Mahdi Jafari
Ali Rashidinejad

CRC Press
Taylor & Francis Group
Boca Raton London New York

CRC Press is an imprint of the
Taylor & Francis Group, an **informa** business

First edition published 2021
by CRC Press
6000 Broken Sound Parkway NW, Suite 300, Boca Raton, FL 33487-2742

and by CRC Press
2 Park Square, Milton Park, Abingdon, Oxon, OX14 4RN

© 2022 Taylor & Francis Group, LLC
CRC Press is an imprint of Taylor & Francis Group, LLC

ISBN: 978-0-367-36646-9 (hbk)
ISBN: 978-1-032-04417-0 (pbk)
ISBN: 978-0-429-35546-2 (ebk)

Typeset in Times
by SPi Technologies India Pvt Ltd (Straive)

Advances in Drying Science and Technology

Series Editor: Dr. Arun S. Mujumdar

It is well known that the unit operation of drying is a highly energy-intensive operation encountered in diverse industrial sectors ranging from agricultural processing, ceramics, chemicals, minerals processing, pulp and paper, pharmaceuticals, coal polymer, food, forest products industries as well as waste management. Drying also determines the quality of the final dried products. The need to make drying technologies sustainable and cost effective via application of modern scientific techniques is the goal of academic as well as industrial R&D activities around the world.

Drying is a truly multi- and interdisciplinary area. Over the last four decades the scientific and technical literature on drying has seen exponential growth. The continuously rising interest in this field is also evident from the success of numerous international conferences devoted to drying science and technology.

The establishment of this new series of books entitled Advances in Drying Science and Technology is designed to provide authoritative and critical reviews and monographs focusing on current developments as well as future needs. It is expected that books in this series will be valuable to academic researchers as well as industry personnel involved in any aspect of drying and dewatering.

The series will also encompass themes and topics closely associated with drying operations, e.g., mechanical dewatering, energy savings in drying, environmental aspects, life cycle analysis, techno-economics of drying, electrotechnologies, control and safety aspects, and so on.

About the Series Editor

Dr. Arun S. Mujumdar is an internationally acclaimed expert in drying science and technologies. He is the Founding Chair in 1978 of the International Drying Symposium (IDS) series and Editor-in-Chief of Drying Technology: An International Journal since 1988. The 4th enhanced edition of his Handbook of Industrial Drying published by CRC Press has just appeared. He is recipient of numerous international awards including honorary doctorates from Lodz Technical University, Poland and University of Lyon, France.

Please visit www.arunmujumdar.com for further details.

Contents

Preface

Bioactive materials are health-promoting ingredients naturally present in plant or animal sources, and are highly demanded by both consumers and manufacturers. These ingredients can be used for the manufacture of functional foods, pharmaceuticals, cosmetics, and several other products. Some common examples of bioactive ingredients include phenolics, carotenoids, essential fatty acids, peptides, prebiotics, essential oils, phytosterols, and bioactive live organisms (probiotics). The main challenge of the incorporation of these valuable ingredients into functional products is their low stability under environmental/process conditions, as well as their low bioavailability within the body. Thus, the encapsulation of bioactive materials within different carriers is a suitable and efficient strategy for overcoming such challenges.

One of the most popular and common technologies for the encapsulation of bioactive ingredients and nutraceuticals is spray drying. This process results in highly stable bioactive-loaded powders that can easily be handled, stored, and applied in different formulations. This book will guide the readers on how to produce such encapsulated bioactive powders using spray drying technology. Recent findings obtained by international experts in this field are provided to make a framework for the expansion of relevant studies in similar fields such as the spray drying encapsulation of various food ingredients and nutraceuticals using various biopolymers.

The overall aim of the *Spray Drying Encapsulation of Bioactive Materials* is to bring science and industrial applications together on encapsulated ingredients, with an emphasis on the spray dried bioactive compounds that enable novel/enhanced properties or functions. This book covers recent and applied research in all types of bioactives processed by spray drying. The original results related to the experimental, theoretical, formulation, and/or applications of bioactive-loaded powders for different purposes are provided in enough depth in different chapters. After presenting a brief overview of bioactive ingredients in Chapter 1, the process of spray drying and its application for bioactive compounds have been explained in Chapter 2. Then, spray drying encapsulation of categorized bioactives have been covered in the following chapters; i.e., phenolic compounds (Chapter 3), anthocyanins (Chapter 4), carotenoids (Chapter 5), vitamins (Chapter 6), minerals (Chapter 7), essential oils (Chapter 8), essential fatty acids and functional oils (Chapter 9), proteins and bioactive peptides (Chapter 10), and probiotics (Chapter 11). Chapter 12 has been devoted to the nanospray drying of bioactive ingredients, as a new and emerging process. Other important topics including the advances in the spray drying process for encapsulation of bioactives, and characterization of bioactive-loaded powders have been explained in Chapters 13 and 14, respectively. Finally, Chapters 15 and 16 deal with the application of spray dried encapsulated bioactives in the food and pharmaceutical industries, respectively.

All who are engaged in either micro- or nanoencapsulation of food, nutraceutical, and pharmaceutical bioactive ingredients worldwide can use this book as either a textbook or a reference, which will give the readers quality and recent knowledge on the potentials of spray drying encapsulation, as well as its novel applications in developing bioactive delivery systems. We hope this book will stimulate further research in this rapidly growing area, and will enable scientists and manufacturers to use the spray drying process as a common, effective, and economic technique for the encapsulation of various bioactive ingredients.

The editors appreciate the great cooperation of all contributors for taking time from their busy schedules and helping with the formation of this great piece of work. Also, it is necessary to express our sincere gratitude to the editorial staff at CRC Press (Taylor and Francis) for their help and support throughout this project. Finally, a special acknowledgment goes to our families for their understanding and encouragement during the edition of this great project.

Seid Mahdi Jafari

Ali Rashidinejad

Editor Biographies

Prof. Seid Mahdi Jafari received his PhD in Food Process Engineering from the University of Queensland (Australia), in 2006. He has been working on the nanoemulsification and nanoencapsulation of food bioactive ingredients for the past 15 years. Now, as a full Professor, he is an academic member of GUASNR (Iran) and Adjunct Prof. in UVigo (Spain). He has published more than 250 papers in top-ranked International Food Science Journals (h-index=57 in Scopus) and 60 book chapters, along with editing 36 books, with Elsevier, Springer, and Taylor & Francis.

In November 2015, he was awarded as one of the top 1% world scientists by Thomson Reuters (Essential Scientific Indicators) in the field of Biological Sciences. In December 2017, he was selected as one of the top national researchers by the Iranian Ministry of Science, Research, and Technology. He has been awarded as one of the world's highly cited researchers by Clarivate Analytics (Web of Science), in November 2018, 2019 and 2020, and a top reviewer in the field of agricultural and biological sciences by Publons (September 2018 and 2019).

Dr. Ali Rashidinejad received his PhD in Food Science from the University of Otago (New Zealand). For the last decade, he has been working on the protection and delivery of bioactive materials and their incorporation into functional food products.

He is currently a research scientist and an academic member of Riddet Institute, Massey University (New Zealand).

In the field of encapsulation and delivery of bioactive materials, Ali has published numerous works in top-ranked international journals and reference books. Dr. Rashidinejad is the lead inventor of 'FlavoPlus™', a new method for the delivery of hydrophobic flavonoids in functional foods. He is specifically interested in the behavior of bioactive compounds in the food matrix and gastrointestinal tract, with a particular focus on antioxidant-enriched functional foods.

List of Contributors

Aida Firdaus MN Azmi
Faculty of Applied Sciences
Universiti Teknologi MARA
Selangor, Malaysia

Aleksandra Jedlińska
Institute of Food Sciences
Warsaw University of Life Science SGGW
Warsaw, Poland

Ali Rashidinejad
Riddet Institute
Massey University
Palmerston North, New Zealand

Amir Pouya Ghandehari Yazdi
Department of Food Research and
   Development
Zar Research, and Industrial Development Group
Alborz, Iran

Ana Isabel Bourbon
INL – International Iberian Nanotechnology
   Laboratory
Braga, Portugal

Arlete Maria Lima Marques
INL – International Iberian Nanotechnology
   Laboratory
Braga, Portugal

and

CEB – Centre of Biological Engineering
University of Minho
Campus de Gualtar, Portugal

Arup Nag
Riddet Institute
Massey University
Palmerston North, New Zealand

Asli Can Karaca
Karaca Department of Food Engineering, Faculty
   of Chemical and Metallurgical Engineering
Istanbul Technical University
Istanbul, Turkey

Ayhan Topuz
Faculty of Engineering, Department of Food
   Engineering
Akdeniz University
Antalya, Turkey

Beraat Ozcelik
Department of Food Engineering
Faculty of Chemical and Metallurgical
   Engineering
Istanbul Technical University
Istanbul, Turkey

Clive A. Prestidge
Clinical and Health Sciences
University of South Australia
Adelaide, Australia

and

ARC Centre of Excellence in Convergent
   Bio-Nano Science and Technology
University of South Australia
Adelaide, Australia

Cordin Arpagaus
Eastern Switzerland University of Applied
   Sciences
Institute for Energy Systems
Buch, Switzerland

Dorota Nowak
Department of Food Engineering and
   Process Management, Institute of Food
   Sciences
Warsaw University of Life Sciences (SGGW)
Warsaw, Poland

Dorota Witrowa-Rajchert
Department of Food Engineering and
   Process Management, Institute of Food
   Sciences
Warsaw University of Life Sciences (SGGW)
Warsaw, Poland

Elham Assadpour
Department of Food Materials and Process
  Design Engineering
Gorgan University of Agricultural Sciences and
  Natural Resources
Gorgan, Iran

Emilia Janiszewska-Turak
Department of Food Engineering and Process
  Management, Institute of Food Sciences
Warsaw University of Life Sciences (SGGW)
Warsaw, Poland

Emrah Eroglu
Food Safety and Agricultural Research Center
Akdeniz University
Antalya, Turkey

and

Faculty of Engineering, Department of Food
  Engineering
Akdeniz University
Antalya, Turkey

Esra Capanoglu
Department of Food Engineering,
  Faculty of Chemical and Metallurgical
  Engineering
Istanbul Technical Universit
Istanbul, Turkey

Handan Basunal Gulmez
Faculty of Engineering, Department of Food
  Engineering
Akdeniz University
Antalya, Turkey

Haroldo Cesar Beserra Paula
Federal University of Ceará
Ceará, Brazil

Hayley B. Schultz
Clinical and Health Sciences
University of South Australia
Adelaide, Australia

and

ARC Centre of Excellence in Convergent
  Bio-Nano Science and Technology
University of South Australia
Adelaide, Australia

Irisvan Da Silva Ribeiro
Federal University of Ceará
Ceará, Brazil

Jose Antonio Couto Teixeira
CEB - Centre of Biological Engineering
University of Minho
Braga, Portugal

Katarzyna Samborska
Institute of Food Sciences
Warsaw University of Life Sciences (SGGW)
Warsaw, Poland

Khadijeh Khoshtinat
Department of Food Technology Research,
  Faculty of Nutrition Sciences and Food
  Technology
National Nutrition and Food Research Institute
Shahid Beheshti University of Medical Science
Tehran, Iran

Khashayar Sarabandi
Department of Nutrition and Food Sciences,
  School of Medicine
Zahedan University of Medical Sciences
Zahedan, Iran

Leila Kamali Rousta
Department of Food Research and
  Development
Zar Research, and Industrial Development Group
Alborz, Iran

Lorenzo Pastrana
INL – International Iberian Nanotechnology
  Laboratory
Braga, Portugal

Mansoureh Geranpour
Department of Food Materials and Process
  Design Engineering
Gorgan University of Agricultural Sciences and
  Natural Resources
Gorgan, Iran

Marco Faieta
Faculty of Bioscience and Technology for
  Food, Agriculture and Environment
University of Teramo
Teramo, Italy

Merve Tomas
Department of Food Engineering, Faculty of
  Engineering and Natural Sciences
Istanbul Sabahattin Zaim University
Istanbul, Turkey

Miguel Angelo Cerqueira
INL – International Iberian Nanotechnology
  Laboratory
Braga, Portugal

Paola Pittia
Faculty of Bioscience and Technology for
  Food, Agriculture and Environment
University of Teramo
Teramo, Italy

Paul Joyce
Clinical and Health Sciences
University of South Australia
Adelaide, Australia

and

ARC Centre of Excellence in Convergent
  Bio-Nano Science and Technology
University of South Australia
Adelaide, Australia

Pouria Gharehbeglou
Faculty of Food Science and Technology
Gorgan University of Agricultural Sciences and
  Natural Resources
Gorgan, Iran

Raseetha Vani Siva Manikam
Faculty of Applied Sciences
Universiti Teknologi MARA
Selangor, Malaysia

Regina Celia Monteiro De Paula
Federal University of Ceará
Ceará, Brazil

Ruba Almasri
Clinical and Health Sciences
University of South Australia
Adelaide, Australia

and

ARC Centre of Excellence in Convergent
  Bio-Nano Science and Technology
University of South Australia
Adelaide, Australia

Saeed Mirarab Razi
Kalleh Research and Development Center
Amol, Iran

Seid Mahdi Jafari
Faculty of Food Science & Technology
Gorgan University of Agricultural Sciences and
  Natural Resources
Gorgan, Iran

Selene Maia De Morais
State University of Ceará
Ceará, Brazil

Siew Young Quek
School of Chemical Sciences
The University of Auckland
Auckland, New Zealand.

and

Centre of Research Excellence in Food Research
Riddet Institute
Palmerston North, New Zealand

Surajit Sarkar
Keventer Agro Limited
West Bengal, India

Tahlia R. Meola
Clinical and Health Sciences
University of South Australia
Adelaide, Australia

and

ARC Centre of Excellence in Convergent
  Bio-Nano Science and Technology
University of South Australia
Adelaide, Australia

Takeshi Furuta
Department of Biotechnology, Graduate School
  of Engineering
Tottori University
Tottori, Japan

Tze Loon Neoh
Centre for Macroalgal Resources & Biotechnology
College of Science and Engineering, James
    Cook University
Queensland, Australia

Yongchao Zhu
School of Chemical Sciences
The University of Auckland
Auckland, New Zealand

Zahra Akbarbaglu
Department of Food Science and Technology,
    Faculty of Agriculture
University of Tabriz
Tabriz, Iran

Zahra Beig Mohammadi
Department of Food Science and Technology
North Tehran Branch
Islamic Azad University
Tehran, Iran

Zeynep Saliha Gunes
Department of Food Engineering
Faculty of Engineering and Natural Sciences
Istanbul Sabahattin Zaim University
Istanbul, Turkey

1 Bioactive Compounds
Chemistry, Structure, and Functionality

Saeed Mirarab Razi
Kalleh Research and Development Center, Iran
Ali Rashidinejad
Massey University, New Zealand

CONTENTS

1.1 INTRODUCTION

Bioactive compounds are non-nutritional ingredients that are found in small amounts in various foods. Overall, these compounds are present in many species of plants, animals, microorganisms, and marine organisms (Kris-Etherton et al., 2002). Each living body processes various chemical compounds for its subsistence and survival. Generally speaking, all compounds of the biological system could be classified into two groups; (a) primary metabolites such as proteins, lipids, amino acids, and carbohydrates that have a role in growth and development; and (b) secondary metabolites that help living organisms to control local challenges and enhance their ability to survive by interacting with their surrounding (Azmir et al., 2013; Harborne, 1993). On the other hand, the secondary metabolites have no role in the growth and productivity of some restricted taxonomic groups of microorganisms and possess uncommon chemical structures (Smith & Berry, 1978). Thus, bioactive compounds in plants can be defined as the secondary metabolites with elicit toxicological or pharmacological properties in animals and human (Bernhoft, 2010).

Bioactive compounds have important roles in living plants. A large variety of vegetables and fruits provide various nutrients and a wide variety of bioactive compounds, including phytochemicals (e.g., flavonoids, carotenoids phenolic acids, and alkaloids), vitamins (e.g., folate, vitamin C, and provitamin A), minerals (e.g., calcium, potassium, and magnesium), and fibers (Liu, 2004). Nowadays, bioactive compounds have numerous applications in the food, pharmaceutical, and medical industries. For example, in the food industry, bioactive compounds are used in the production of functional foods (i.e., enhancing functionality), increasing shelf-life, improving nutritional quality, and enhancing consumer acceptance (Tajkarimi et al., 2010). These compounds have functional properties such as antimicrobial (essential oils), antioxidant (phenolics), probiotic, and flavoring properties.

Beside their numerous advantages, bioactive compounds might impart off-flavors in food products, as a result of rapid degradation and combination with other ingredients in food systems, which ultimately lead to a reduction in the functional quality of food components (Quirós-Sauceda et al., 2014). Encapsulation is a suitable way for improving or preserving the functionality of bioactive compounds. Moreover, the encapsulation of bioactive compounds can modify the physical properties of food materials and overcome solubility incompatibilities between food ingredients.

Different categories for bioactive compounds are defined. Croteau et al. (2000) categorized the derived bioactive compounds of plants into three main groups: (1) terpenes and terpenoids, with about 25,000 types; (2) alkaloids, with about 12,000 types; and (3) phenolic compounds, with about 8000 types. Most bioactive compounds have specific structural properties due to different biosynthesis pathways (Azmir et al., 2013). In this chapter, various bioactive compounds (as well as bioactive live organisms), their chemistry, structure, and functional properties are discussed, while their encapsulation by spray drying technique will be covered in the following chapters of this book.

1.2 PHENOLIC COMPOUNDS (PHENOLICS; POLYPHENOLS; POLYPHENOLIC COMPOUNDS)

1.2.1 BACKGROUND

Phenolics are compounds with one or more aromatic rings and hydroxyl groups in their structures (Liu, 2013). Dietary polyphenols have received growing attention lately due to the potential antioxidant activities. Several health benefits have been shown for polyphenols from various sources; e.g., anti-carcinogenic and anti-cardiovascular activities. Generally, these compounds are the products of the secondary metabolism in plants that are classified into flavonoids, phenolic acids, stilbenes, coumarins, and tannins, and contribute to the growth and color of the corresponding plants. In addition, phenolic compounds can protect plants against parasites, pathogens, predators, and UV irradiation (Liu, 2013). The phenolic contents of various fruits and vegetables are shown in Figure 1.1. Among the common fruits, blueberry, blackberry, and pomegranate contain higher phenolic contents (Liu, 2013; Wolfe et al., 2008). Similarly, among vegetables spinach possesses the greatest amount of phenolic contents.

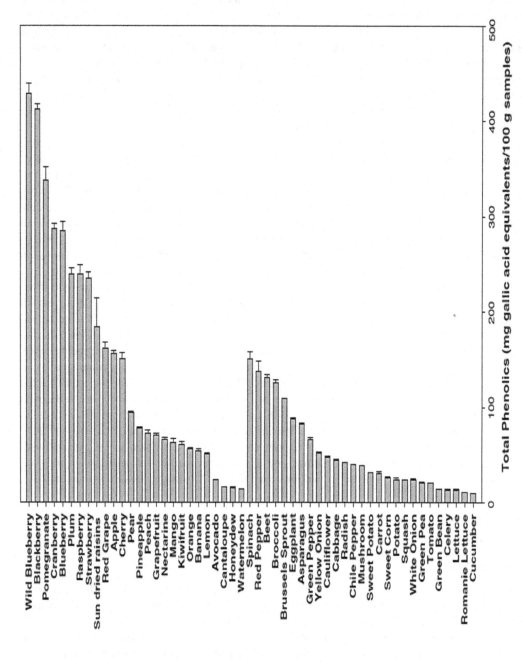

FIGURE 1.1   The concentration of phenolic contents in different fruits and vegetables (Liu, 2013).

1.2.2 Classification, Chemistry, and Structure

Generally, more than 8000 kinds of polyphenols have been reported with a diversity of chemical structures (Bravo, 1998). They exist within a wide group of natural plant phenolic compounds, which can range from simple molecules (phenolic acids) to highly polymerized compounds (tannins). These are produced through two main metabolic mechanisms: the shikimate pathway and the acetate pathway (Harborne, 1993). Polyphenolic compounds can occur in conjugated forms and can be linked to one or more sugar residues (monosaccharides, disaccharides, and oligosaccharides) via hydroxyl groups (Bravo, 1998). This occurs via direct linkages of an aromatic carbon atom to the sugar units, although the association with other molecules such as organic and carboxylic acids, lipids, amines, and even other phenols, can also occur (El, 2009; Spencer et al., 1988).

The most common method of classifying polyphenols is based on the number of carbons. Polyphenols may also be categorized into different types depending upon the number of phenolic rings in the structure and/or the structural elements that bind these rings together (Manach et al., 2004). The most important group of polyphenols is flavonoids, which include more than 5000 compounds (Harborne, 1993). Flavonoids themselves can be sub-divided into six main sub-categories (according to the type of heterocyclic moieties involved): flavonols, flavanones, flavones, isoflavones, flavanols (catechins and proanthocyanidins), and anthocyanidins (Manach et al., 2004). The concentration of phenolic contents in various plant sources (fruits and vegetables) is shown in Figure 1.1 (Liu, 2013).

1.2.3 Phenolic Acids

Phenolic acids are classified into two main groups, including hydroxycinnamic acid and hydroxybenzoic acid derivatives (Table 1.1). Hydroxycinnamic acid derivatives contain p-coumaric, ferulic, caffeic, and sinapic acids. They are linked to cell wall components, including lignin, cellulose, and proteins through ester bonds (Liu, 2007). Hydroxybenzoic acid derivatives contain p-hydroxybenzoic, protocatechuic, gallic, vanillic, and syringic acids. These compounds commonly exist in the bonded form in foods, and are part of a complex structure such as lignins and hydrolysable tannins. These components are also adhered to fibers, proteins, and sugars and exist in sugar derivatives and organic acids in foods originating from plants (Liu, 2013). Phenolic acids can be hydrolyzed by enzymes or upon alkaline or acid hydrolysis (Tsao, 2010).

1.2.4 Flavonoids

Flavonoids are one of the most important and largest groups of phenolic compounds in plant foods. These components have a typical structure, including two aromatic rings (A and B) linked by three carbons to oxygenated heterocyclic ring, or Ring C (Table 1.1). According to the differences in the structure of Ring C, they can be categorized as flavonols (e.g., quercetin, myricetin, kaempferol, and galangin), flavones (e.g., luteolin, chrysin, and apigenin), anthocyanidins (e.g., cyanidin, peonidin, malvidin, pelargonidin, and delphinidin), flavanols (e.g., catechins), flavanones, and isoflavonoids (e.g., genistein, glycitein, daidzein, and formononetin) (Liu, 2013). In nature, flavonoids mostly exist as conjugates in glycosylated or esterified forms; however, they might be converted to aglycones due to the conditions during food processing (Tsao, 2010).

1.2.4.1 Isoflavones, Neoflavonoids, and Chalcones

Isoflavones mostly exist in the leguminous family of plants. These components have their Ring B linked to the C3 position of Ring C (Table 1.1). This type of flavonoids exists in legumes such as soybeans, fava beans, chickpeas, peanuts, pistachios, and nuts. Dalbergin is the most common neoflavonoid found in the plant kingdom. Fruits like apples and hops (or beers) are the common sources of open-ring chalcones (Mazur et al., 1998). They also exist in tomatoes, strawberries, pears, and certain wheat products.

TABLE 1.1

Some Common Flavonoid Compounds and Their Structure

Flavonoid Compound(s)	Structure
Benzoic acid	
Cinnamic acid	
Flavonoid backbone	
Isoflavones	
Neoflavonoids	
Dalbergin	
Flavones	 Apigenin R = H Luteolin R = OH Tangeretin R = H Nobiletin R = OCH₃
Flavonols	 Kaempferol R_1 = H, R_2 = H Quercetin R_1 = H, R_2 = OH Myricetin R_1 = OH, R_2 = OH Isorhamnetin R_1 = OCH₃ R_2 = H
Flavanones	 Naringenin R_1 = H, R_2 = OH Hesperetin R_1 = OH, R_2 = OCH₃

Flavonoid Compound(s)	Structure

Flavanonols

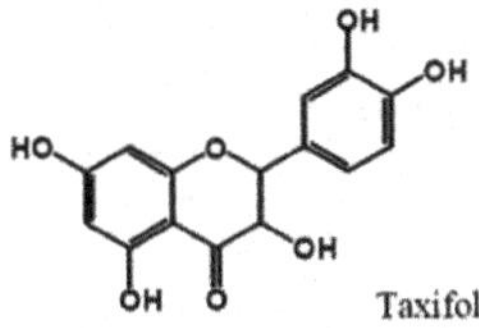

Taxifolin

Catechins

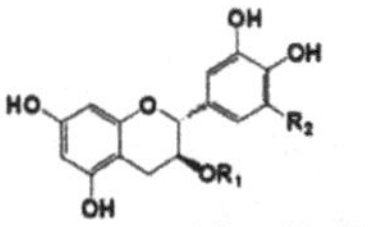

(+)-Catechin: $R_1 = R_2 = H$;
(+)-Catechin gallate: $R_1 = $ gallyl, $R_2 = H$;
(+)-Gallocatechin: $R_1 = H, R_2 = OH$;
(+)-Gallocatechin gallate: $R_1 = $ gallyl, $R_2 = OH$;

Epicatechins

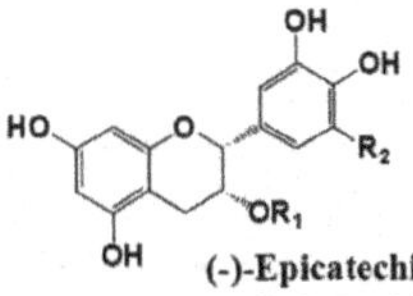

(-)-Epicatechin: $R_1 = R_2 = H$;
(-)-Epicatechin gallate: $R_1 = $ gallyl, $R_2 = H$;
(-)-Epigallocatechin: $R_1 = H, R_2 = OH$;
(-)-Epiallocatechin gallate: $R_1 = $ gallyl, $R_2 = OH$;

Procyanidins

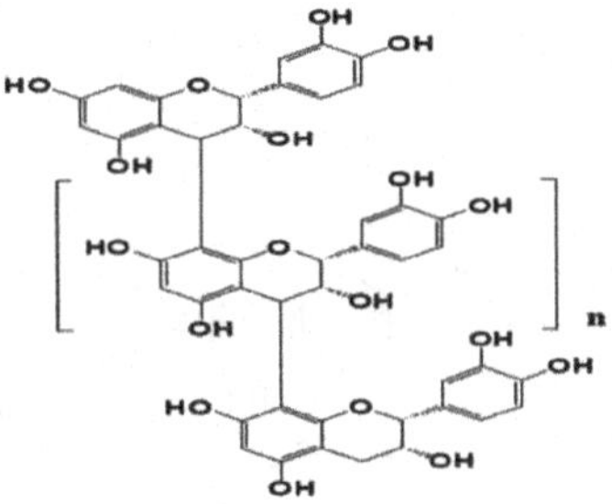

Procyanidins: $n > 0$;
Oligomeric procyanidins: $n = 0-7$

Procyanidin B1

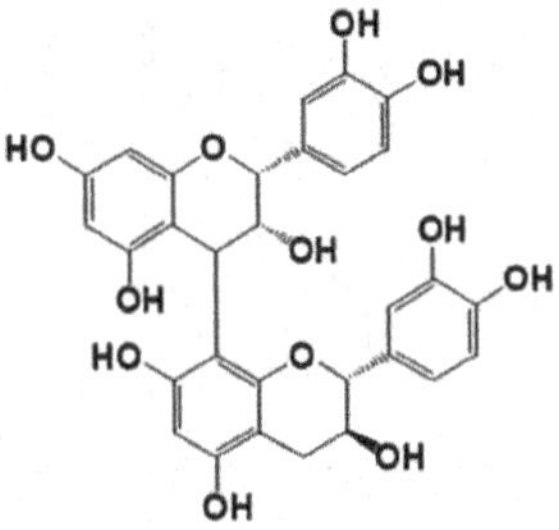

Procyanidin B2

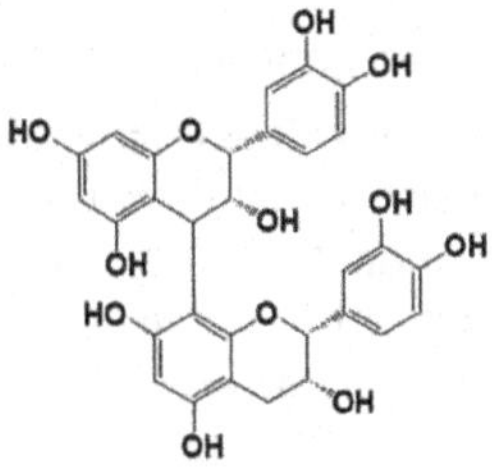

(Continued)

TABLE 1.1
(*Continued*)

Flavonoid Compound(s)	Structure
Procyanidin A2	
Procyanidin C1	
Theaflavin	
Avenanthramides	 Avenanthramide A: R = H; Avenanthramide B: R = OCH$_3$; Avenanthramide C: R = OH
Capsaicinoids	 Capsaicin Dihydrocapsaicin

1.2.4.2 Flavones, Flavonols, Flavanones, and Flavanonols

Among all polyphenols, flavones and their 3-hydroxy derivatives flavonols, containing their methoxides, glycosides, and other acylated products on all three rings, are the largest subgroup (Table 1.1). Taxifolin from citrus is a common flavanonol. Some of the flavanonols have typical substitution patterns, such as furan flavanones, prenylated flavanones, and benzylated flavanones. Flavones are present in high quantities in flowers, leaves, and fruits. Red peppers, celery, chamomile, parsley, and mint are the rich sources of flavones. Flavonols are building blocks of proanthocyanins and they are a subgroup of flavonoids with a ketone group. These compounds are widely present in fruits and vegetables. Lettuce, onions, apples, kale, tomatoes, grapes, berries, and tea are the main sources of flavonols.

Flavanones are another subgroup of flavonoids which exist in all citrus fruits, such as lemons, oranges, and grapes. Naringenin, hesperitin, hesperidin, and eriodictyol are some examples of this class of flavonoids. Flavonones have several health benefits, because they have the ability of scavenging free radicals. Flavanonols or dihydroflavonols are the 3-hydroxy derivatives of flavanones. They are commonly found in bananas, blueberries, apples, peaches, and pears.

1.2.4.3 Flavanols and Proanthocyanidins

Flavan-3-ols or flavanols are generally called catechins. Unlike most flavonoids, there is no C4 carbonyl in Ring C and a double bond between C2 and C3 of flavanols (Table 1.2). Catechin and epicatechin are examples of *trans-* and *cis-* configuration of these flavonoids, respectively. Both these configurations have two stereoisomers such as (−)-catechin, (+)-catechin, (−)-epicatechin, and (+)-epicatechin. The two common isomers in plants are (+)-catechin and (−)-epicatechin (Table 1.1).

TABLE 1.2

The Concentration of Some Anthocyanins in the Selected Fruits and Their By-Products (Ramos et al., 2014)

Fruit/Sub-Product	Anthocyanin (mg/100 g)
Apple (Red Delicious)	1.7
Bilberry	300–98
Black currant	130–476
Black olives	42–228
Blackberry	82.5–325.9
Blueberry	25–495
Cherry	2–450
Chokeberry	410–1480
Cranberry	67–140
Crowberry	360
Elderberry	200–1816
Gooseberry	2–43.3
Peach	4.2
Pomegranate (juice)	600–765
Port wine	14–110
Purple corn	1642
Raspberry	20–687
Red apple	1.3–12
Red grape	30–75
Red wine	16.4–35
Saskatoon berry	234
Strawberry	19–55

Many fruits contain flavanols; especially, the skins of grapes, blueberries, and apples. Monomeric flavanols (epicatechin and catechin) and their derivatives are the main flavonoids in cacao bean and tea leaves. Catechin and epicatechin can form polymers which can be referred to as proanthocyanidins, because an acid-catalyzed cleavage of the polymeric chain forms anthocyanidins (Tsao et al., 2003).

Proanthocyanidins can be considered as condensed tannins. Flavanols and oligomers are strong antioxidants which have numerous health benefits. When fermented, tea flavanols might form unique dimers such as theaflavin. Different structures of proanthocyanidins (dependent on interflavanic linkages) are presented in Table 1.1.

1.2.4.4 Polyphenolic Amides

Several polyphenols might possess N-containing functional substituents. Avenanthramides in oats and capsaicinoids in chili peppers are the two common polyphenolic amides in foods (Table 1.1).

1.2.4.5 Anthocyanins

Anthocyanins are the natural plant pigments imparting blue, red, and purple colors to flowers, fruits, leaves, and some vegetables. These pigments have some applications in beverages and foods as natural colorants. Anthocyanins are classified as flavonoids and their basic structure is containing flavylium cation (C6-C3-C6) that can be linked to various hydroxyl or methoxyl groups or sugars.

Different sugars are attached to anthocyanins; glucose is the most common sugar followed by rhamnose, galactose, xylose, arabinose, and rutinose. With a variation of the attached sugars, anthocyanins can be mono-, di-, or tri-glycosides. Moreover, the sugar residues might be acylated by aliphatic/aromatic acids. Anthocyanidins are known as aglycones part of anthocyanins and among 19 anthocyanidins, six of which are considered as major anthocyanins; cyanidin, malvidin, delphinidin, peonidin, pelargonidin, and petunidin (Figure 1.2 A). Different parameters such as structure, oxygen, pH, light, temperature, and intermolecular and intramolecular association with other compounds (e.g., sugars, ascorbic acids, metal ions, and proteins) can affect the color stability of anthocyanins (Burton-Freeman et al., 2016). With pH changing from acidic to basic, the anthocyanin colors change reversely from red to blue. The structures of anthocyanins at different pH levels are presented in Figure 1.2 B. Anthocyanins can donate a hydrogen atom to scavenge free radicals (Burton-Freeman et al., 2016).

Anthocyanins with different concentrations are present in fruits and their by-products, some of examples of which are presented in Table 1.2.

The consumption of anthocyanins has various health benefits; e.g., the prevention of diabetes, cancer, obesity, cardiovascular diseases, as well as neuronal, and visual impairment (Ramos et al., 2014). Moreover, anthocyanins are natural antioxidants that possess inflammatory ability. The stability of anthocyanins is dependent on their structure as follows: (a) co-pigmentation influence, during which the molecular or complex associations are formed by the pigments and other colorless organic compounds or metallic ions. Inter and intermolecular co-pigmentations, metal complexion, and self-association can improve anthocyanin stability (Figure 1.3); (b) temperature influence, where an increase in the temperature decreases the stability of anthocyanin; (c) pH influence, where the stability is higher at the lower pH compared to the higher pH. At the pH range of 1, 2–4, and 5–6, flavylium cation (purple and red colors), quinoidal (blue), carbinol pseudo-base, and chalcone are predominant, respectively. In alkaline pH, the degradation of anthocyanins can occur; (d) oxygen influence, where the degradation of anthocyanin is faster in the presence of oxygen; (e) solid content and water activity influence, where a faster degradation of anthocyanins can occur in more concentrated liquid systems. Furthermore, the stability of anthocyanins is significantly higher when stored in dry crystalline form (Sui et al., 2018).

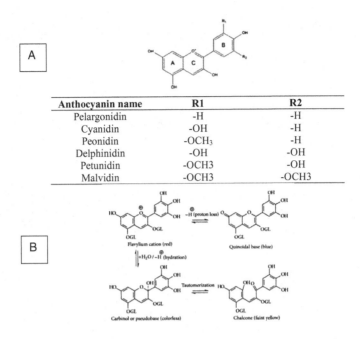

FIGURE 1.2 Structures of the main anthocyanidins (A) and their structures with a change in pH levels in aqueous solution (R1 = H or glycoside; R2 and R3 = H or a methyl group) (B).

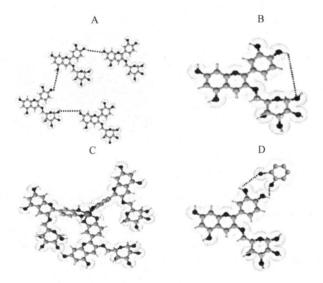

FIGURE 1.3 The co-pigmentation (A), self-association (B), intramolecular co-pigmentation (C), metal complexation, and (D) intermolecular co-pigmentation of anthocyanin molecules.

1.2.5 OTHER POLYPHENOLS

There are many non-flavonoid polyphenols in foods that play an important role in human health. Among these, the following are the most important compounds; resveratrol (found in grapes and red wine), ellagic acid and its derivatives (exist in berry fruits like strawberries), lignans (found in the bound forms in sesame, flax, and many grains), curcumin (widely found in turmeric), rosmarinic

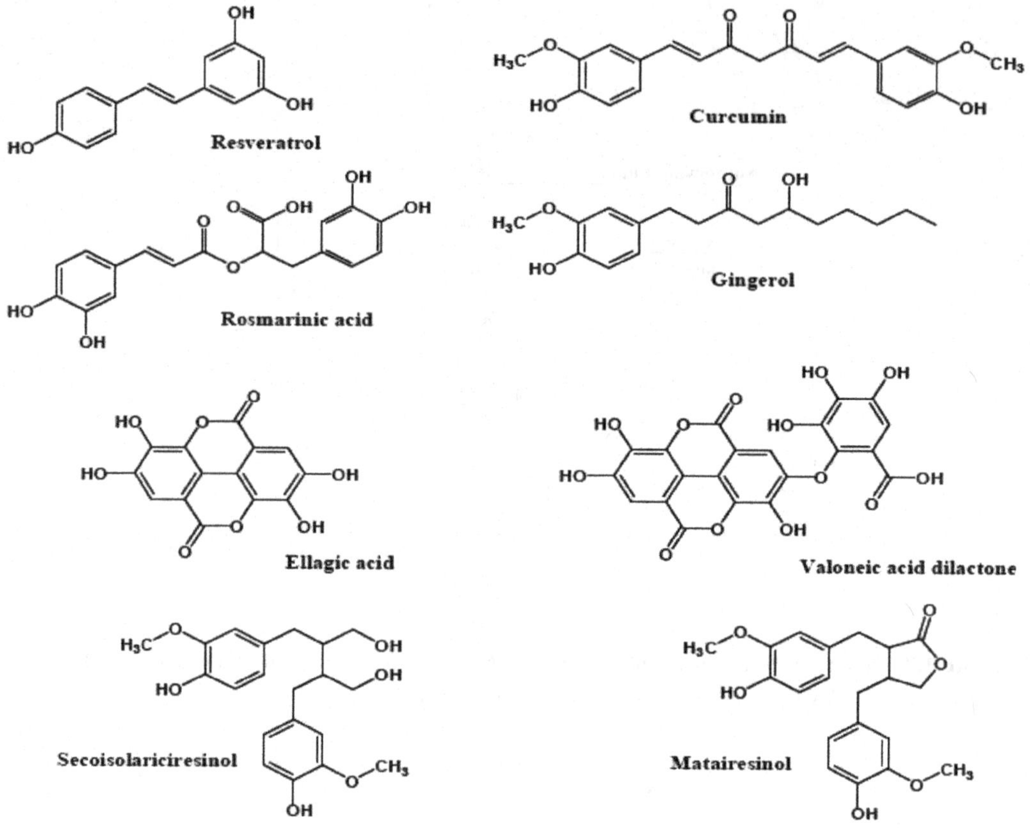

FIGURE 1.4 Other important polyphenols that are found in natural sources such as fruits, vegetables, herbs, and spices (Tsao, 2010).

acid, ellagic acid (a dimer of caffeic acid), and gallic acid (Tsao, 2010). The structures of these components are presented in Figure 1.4.

1.2.6 FUNCTIONALITY AND HEALTH EFFECTS OF POLYPHENOLS

Several polyphenols have anti-inflammatory and antioxidant properties. These can have therapeutic effects on cancer, neurodegenerative disorders, cardiovascular disease, and obesity (Table 1.3). Some studies have been carried out on the protective effect of polyphenols against oxidative stress and the results have shown that these compounds can produce hydrogen peroxide (H_2O_2) which can help regulate immune response actions namely, cellular growth.

1.3 CAROTENOIDS

1.3.1 BACKGROUND

The quality of food and its deterioration can be determined by measuring its color. Chlorophylls, carotenoids, and anthocyanins, which are the main pigment families in fruits and vegetables, are responsible for green, red-yellow, and blue-violet colorations, respectively (Fernández-García et al., 2012). These compounds are mostly insoluble or poorly soluble in water, meaning that they are hydrophobic, lipophilic, and soluble in solvents like acetone, chloroform, and alcohol. Carotenoids have been used as restoring colorants in the food industry for a very long time, especially in the

TABLE 1.3

Some of the Potential Health Effects of Polyphenols in Humans

Type of Disease/ Disorder	Phenolic Compound(s)	Proven Health Effects	References
Neurodegenerative diseases	Curcumin, catechins, and resveratrol Curcumin, ginsenosides, myricetin, and ginkgetin	- Antioxidant and immunomodulatory properties that might prevent Alzheimer's disease and dementia. - The iron-chelating ability that prevents neurotoxicity, and results in a neuroprotective effect against neurodegenerative diseases.	Jiang et al. (2017)
Inflammation	Phenolic compounds including phenolic acids, proanthocyanidins, and flavonoids	- Inhibit systemic and localized inflammations using restoring the redox balance to decrease oxidative stress.	Zhang and Tsao (2016)
Cancer	Flavonoids such as anthocyanins, flavanols, catechins, flavones, isoflavones, and flavanones	- Neutralize free radicals and reduce cancer risk by preventing cellular growth in tumors. - A positive effect on specific types of cancers such as prostate, colon, and endometrial.	Cory et al. (2018)
Cardiovascular disease	Various flavonoids	- Improve ventricular health, enzymatic modulation, declined platelet activity, anti-inflammatory effects, and decrease blood pressure.	Wang et al. (2014)
Type 2 diabetes	Various polyphenolic compounds	- Inhibit Type 2 diabetes by the protection of beta cells from glucose toxicity, antioxidant, and anti-inflammatory effects, regulation of the transport of glucose, and deceleration of starch digestion that results in better glycemic control.	Xiao and Hogger (2015)
Obesity	Resveratrol, curcumin, catechins, and genistein (an isoflavone)	- Anti-obesogenic effects through the inhibition of lipogenesis, adipocyte oxidation, and the enhancement of energy expenditure, which results in improved weight loss and maintenance. - Protein-binding properties that can prevent lipid, starch, and protein digestion in the gastrointestinal tract by interacting with digestive enzymes.	Wang et al. (2014) Barrett et al. (2018)

products that lose part of their natural color (e.g., fruit juices, sausages, pasta, beverages, candies, and margarines) (Mezzomo & Ferreira, 2016).

1.3.2 CLASSIFICATION, CHEMISTRY, AND STRUCTURE

Attention to carotenoid consumption has increased since Peto et al. (1981) reported that β-carotene consumption protected the body against some types of cancer such as lung and stomach cancer. Generally speaking, there are two broad classifications of carotenoids that include carotenes and xanthophylls, with the difference that carotenes are hydrocarbons and do not contain oxygen while xanthophylls contain oxygen. Two types of carotenoids are available in nature: (a) β-carotenes, which are consisted of the linear hydrocarbons that can be cyclized either at one end or at both ends of molecules; and (b) the oxygenated derivatives of carotenes like lutein, neoxanthin, violaxanthin, and zeaxanthin, known as xanthophylls (Mezzomo & Ferreira, 2016).

Although more than 700 carotenoids have been found in nature, most of them are not present in our diet and only about 40 carotenoids can be absorbed, metabolized, and/or used in our bodies

(Fernández-García et al., 2012). Yet only six carotenoids profile (including α- and β-carotene, lycopene, β-cryptoxanthin, zeaxanthin, and lutein) are detected in human blood plasma. The content of these carotenoids in some foods and their structures are presented in Table 1.4, and their chemical structure is also represented in Figure 1.5.

According to the data shown in Table 1.5, lutein, β-carotene, and β-cryptoxanthin are the most abundant carotenoids in green vegetables. Carrots and pumpkins are the main sources of α-carotene, while fruits and vegetables such as oranges, carrots, red bell peppers, broccoli, potatoes, and green vegetables are the best sources of β-carotene. β-Cryptoxanthin can be found in ripe red peppers and some tropical fruits like papaya. Lycopene is the main pigment of tomato (and its derived products such as sauces), pink grapefruit, and watermelon. Green vegetables, such as spinach, Brussels sprouts, broccoli, and peas, are the main source of lutein, while egg yolks and corn are rich sources of zeaxanthin (Fernández-García et al., 2012). About 10% of carotenoids are the main structural form of vitamin A precursors, such as a β-type ring in β-carotene and β-cryptoxanthin. Only the carotenoids with at least one β-type ring with one polyene chain including at least 11 carbon atoms that have no oxygenated functional groups are potential precursors of vitamin A (Fernández-García et al., 2012).

Provitamin A activity of the main carotenoids (relative to β-carotene) is shown in Table 1.5. Fruits and vegetables are the main sources of these carotenoids in our diet. Carotenoids with provitamin activity can be converted into retinol with an oxidation process and central fracture of the molecule. β,β-carotene 15,15′-monooxygenase (EC 1.13.11.21) is an enzyme that can catalyze this metabolic process (Goodman et al., 1966).

1.3.3 FUNCTIONALITY AND HEALTH EFFECTS

Carotenoids have various applications in the food, cosmetics, pharmaceutical, and animal feed industries, which are related to their coloring properties. These pigments can be used for food fortification due to their possible provitamin A activity, different biological functions, and health benefits; e.g., decreasing the risk of degenerative diseases, increasing the strength of the immune system, antioxidant properties, and anti-obesity activities (Mezzomo & Ferreira, 2016). Some of the carotenoids are precursors of provitamin A and this vitamin has a significant role in growth, development, reproduction, survival of epithelial tissues, immune system, and visual cycle acting in the regeneration of photoreceptors. Its deficiencies can result in changes in the skin, night blindness, and corneal ulcers, as well as disorders in growth and learning in childhood.

Dark green and yellow-orange leafy vegetables are the rich sources of provitamin A carotenes, and the body can convert these compounds to vitamin A as needed (Mezzomo & Ferreira, 2016). Carotenoids that are either precursors of vitamin A or nonprecursors of this vitamin can protect the body against cancer. Various parameters, such as light, heat, and acids, can cause trans-isomerization of carotenoids (the best stable form in nature), enhancing the loss of color and provitamin activity. Depending on their structure, carotenoids are susceptible to enzymatic or nonenzymatic (e.g., metals, oxygen availability, temperature, light, antioxidants, and prooxidants) oxidation (Mezzomo & Ferreira, 2016).

1.4 MINERALS

1.4.1 BACKGROUND

Minerals have an important role in the functionality of every organism on earth. Specification of the minerals in mineralized prehistoric human remains shows food habits and dietary, living, and environmental conditions of the populations. Comprehension of feeding habits discovered by our ancestors helps us to clarify the evolution of the species (De la Guardia & Garrigues, 2015). In the body, teeth and bones are considered as the indicators for the exposure of humans to minerals. The separation and characterization of the bones constituent minerals are somehow difficult, but they are

TABLE 1.4

Carotenoid Content of Some Common Food Products (mg/100 g). Modified From Fernández-García et al. (2012)

Sources	α-Carotene	β-Carotene	β-Cryptoxanthin	Lutein or Zeaxanthin	Lycopene
Green vegetables					
Lettuce	-	1272	-	2635	-
Spinach	-	5597	-	11,398	-
Brussels Sprouts	6	450	-	1590	-
Vegetables/tubers					
Beans	147	408	-	-	-
Broccoli	1	779	-	2	-
Pepper	59	2379	2205	-	-
Pumpkin	4795	6940	-	-	-
Potato	-	6	-	-	-
Tomato	112	393	-	130	3025
Carrot	4649	8836	-	-	-
Onion	6	-	-	-	-
Fruits					
Pineapple	30	-	-	-	-
Banana	5	21	-	-	-
Grape	5	603	12	13	-
Mango	17	445	11	-	-
Melon	27	1595	-	40	-
Orange	16	51	122	187	-
Watermelon	-	295	103	17	4868
Pear	6	27	-	-	-
Cereals					
Corn	33	30	-	884	-
Wheat	-	100	-	35	-
Vegetable oils					
Olive	-	219	30	5990	-
Palm	24	38	-	-	-

- Not reported/defined

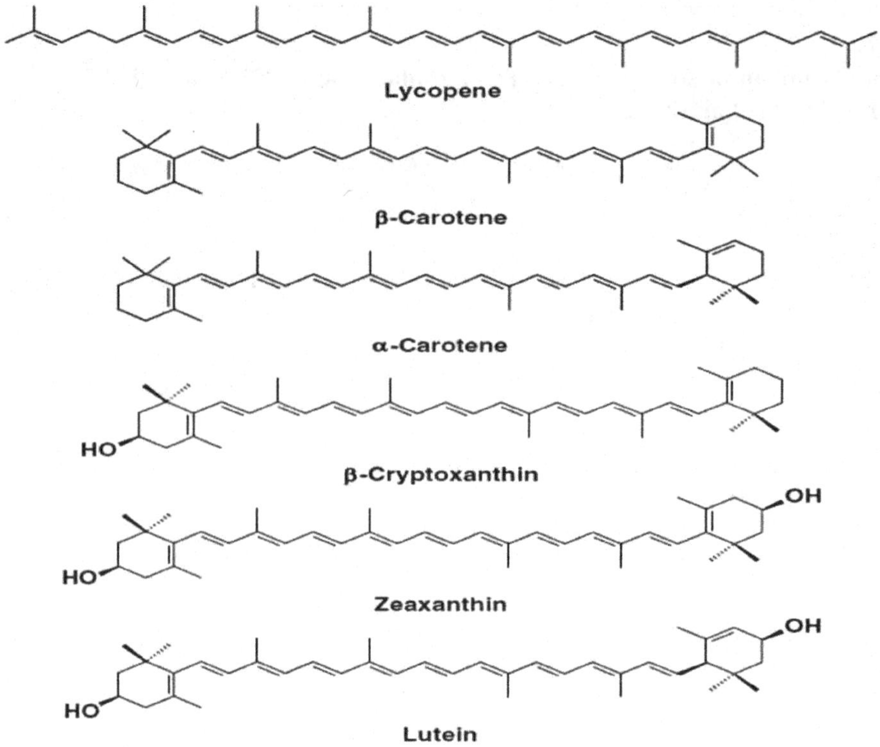

FIGURE 1.5   Chemical structures of typical carotenoids in food.

TABLE 1.5
Provitamin A Activity (Relative to B-Carotene) of the Main Carotenoids. Modified From Fernández-García et al. (2012)

Carotenoid	Percent activity
Trans-β-Carotene	100
9-*cis*-β-Carotene	38
13-*cis*-β-Carotene	53
Trans-α-Carotene	53
9-*cis*-α-Carotene	13
13-*cis*-α-Carotene	16
Trans-β-Cryptoxanthin	57
9-*cis*-β-Cryptoxanthin	27
15-*cis*-β-Cryptoxanthin	42
β-Carotene-5,6-epoxide	21
Mutatochrome	50
Y-Carotene	42–50
β-Zeacarotene	20–40

the object of numerous researches to specify the form of feeding practiced by our ancestors (De la Guardia & Garrigues, 2015). To discriminate between omnivorous and herbivorous food patterns, the concentration of trace components like strontium (Sr) and barium (Ba) has been applied, where a low amount of Sr. shows low consumption of food based on animal origin. The consumption of diets containing minerals such as magnesium (Mg), copper (Cu), iron (Fe), and zinc (Zn) has decreased in recent years, which can be due to a sedentary lifestyle. Furthermore, the downward trend in the concentration of minerals in foods can be related to the intensive farming practices that cause depletion of the mineral content of the soil (De la Guardia & Garrigues, 2015).

1.4.2 CLASSIFICATION, CHEMISTRY, AND STRUCTURE

The chemical and physical properties of a mineral depend on its internal structure and chemical composition. Some minerals have complex chemical constituents, while some are pure metals or salts. Generally, they might classify into three groups: macro (major), micro (trace), and ultra-trace elements. More than 49 nutrients are required for the organic metabolic of humans. Of these, for the most important physiological and biochemical activities, 23 mineral elements are needed. However, in comparison with proteins, carbohydrates, and lipids, small quantities of minerals are needed.

Minerals are essential for the functioning of the organism, so they should be consumed regularly through diet (Prentice et al., 2012). The macrominerals exist at higher proportions in the body tissues, and present in greater amounts in the diet. Although microminerals are needed in a smaller amount, they are essentially the same as macrominerals. A third group is present that of the essential trace components or oligo-elements, considered thus when the daily demand is very small (De la Guardia & Garrigues, 2015). The macrominerals include calcium, phosphorus, magnesium, sodium, potassium, chloride, and sulfur. The microelements include iron, copper, cobalt, magnesium, potassium, zinc, iodine, manganese, molybdenum, fluorine, selenium, chromium, and sulfur. The macrominerals are needed in amounts of more than 100 mg/day, while the microminerals are needed in the amounts less than 100 mg/day. The ultratrace elements include silicon, boron, arsenic, and nickel, which are available in some animals and are assumed to be essential for these animals.

The previous studies showed that minerals had an important role in the activity of enzymes, either in their active sites or as cofactors. Macrominerals (sodium (Na), chloride (Cl), and potassium (K)) are the key intracellular components. These macrominerals are responsible for the formation of the environment around almost all cells, across which gases and metabolites pass from one side to another side (De la Guardia & Garrigues, 2015). Although the microminerals are essential elements of biological structures, they can be toxic at a higher concentration than that required for the physiological actions. Furthermore, this toxicity can be extended to other components that are not considered essential nutrients: by having similar atomic properties, they can imitate the reactivity of a micromineral (Fraga, 2005).

The dietary reference intakes (DRIs) set recommendations for minerals for apparently healthy individuals and are defined experimentally under controlled conditions. The DRIs of nutrients for humans during various stages of life are shown in Table 1.6. These include the highest and lowest amount of nutrients necessary for the normal activity of the organisms, defined by recommended dietary allowances (RDA) for the safe daily intake.

1.4.3 ESSENTIAL, TRACE, AND TOXIC ELEMENTS IN FOODS

Dietary minerals are inorganic compounds containing essential and essential trace components. Essential components occur in our body in mg/kg quantities. These essential elements include Ca, Na, Mg, Fe, K, Zn, as well as Fl and Cl ions. Essential trace components are needed in milligram and sub-milligram quantities. Cu, Co, Mn, Se, Cr, Mob, Ni, and I are considered as essential trace elements. The nutritional imbalance spectrum has two aspects, including: (a) high intake causes potential toxicity; and (b) low intake causes deficiencies. The deficiency of essential elements causes structural and physiological abnormalities in humans.

The non-essential components are food contaminants with cumulative features; Ar, Ni, Bar, Cad, Pb, Sb, Hg, and Al. These elements are considered dangerous for humans. The presences of these elements in the food are due to: (a) natural sources like raw materials; and (b) process that causes contamination of food throughout the supply chain. Table 1.7 shows the major functions of trace elements in the body. Metabolic functions of biochemical modes of action of micro and macrominerals are also presented in Table 1.8.

1.4.3.1 Iron

This element is necessary for oxygen transportation and cellular energy generation. It is also a functional cofactor in more than 200 metalloenzymes. Iron can be toxic when it is consumed at a higher dose. This is due to Fenton chemistry-derived free radicals, because iron homeostasis is highly regulated. In young children, the deficiency of iron results in psychomotor and cognitive development, which cannot reverse even with enough intake of this element later in life.

1.4.3.2 Calcium

This element is necessary for the growth and maintenance of the structure, muscular function, normal blood clotting, and nerve conduction. Furthermore, it is a cofactor in enzyme reactions. In addition, calcium is one of the main minerals in the body, and is present at about 2% of adult body weight and 99% of this element in the body is present in bone and the remaining 1% is present in teeth, extracellular fluid, connective tissue, and blood. In plasma, calcium is necessary for neuromuscular activity throughout the body, especially for the heart and nervous system.

The World Health Organization (WHO) states that the lack of calcium in the body is an important problem in the world, especially in young children up to two years old, pregnant women, children, and adolescents. There is rare calcium toxicity due to excess dietary intake and it can occur in people who use a high amount of calcium or vitamin D supplements. Toxicity might result in cardiac arrhythmias, nephrolithiasis, constipation, delirium, and in extreme cases death (Zand et al., 2015).

1.4.3.3 Zinc

Zinc is a cofactor in many enzymes and possesses a structural role in zinc finger proteins for transcription factors. This element has antioxidant and anti-inflammatory properties. Zinc is required for the function of cells that are involved in neutrophils, innate immunity, and natural killer cells. Moreover, it is necessary for RNA transcription, DNA synthesis, cell division, and activation. The lack of zinc in our body can lead to retardation in growth, rough skin, male hypogonadism, delayed wound healing, poor appetite, abnormal neurosensory changes, and cell-mediated immune dysfunction (Zand et al., 2015).

1.4.4 Functionality and Health Effects of Minerals

Minerals have important roles in the functionality of cells of the body. Some minerals such as Ca, Mg, Cl, P, Na, and K (i.e., macrominerals) are needed in large quantities in the body. These minerals are necessary for bone, heart, brain, and muscle development and functions. On the other hand, some minerals, such as Cr, Cu, Fl, I, Fe, Zn, Se, Mo, and Mg (i.e., trace minerals), are required in small quantities. Except for chromium, which helps the body keep blood sugar levels normal, all trace minerals are incorporated into hormones or enzymes needed in body processes (metabolism).

1.5 VITAMINS

1.5.1 Background

The crucial period for the identification of vitamins and their biological, chemical, and physical properties was in the early 20th century. From 1912, Frederick Hopkins (1861–1947) reported that the presence of the organic substance in food is necessary for health. After that, the Polish biochemist, Casimir Funk (1884–1967), suggested the term 'vitamins' for these components.

1.5.2 Classification, Chemistry, and Structure

1.5.2.1 Water-soluble Vitamins

Water-soluble vitamins include vitamin B groups and vitamin C. These components have different roles in our body and their deficiencies might have undesirable effects. For example, the deficiency of vitamin B_1 causes disorder of the nerves and heart muscle wasting; the deficiency of vitamin B_2, results in the inflammation of the tongue, skin, and lips, ocular disturbances, nervous problems; vitamin B_3 deficiency is associated with skin lesions, gastrointestinal disorders, and nervous problems; in the case of vitamin B5, its deficiency leads to gastrointestinal disorders, nervous problems, weakness, fatigue, sleep disorders, restlessness, and nausea; vitamin B_6 deficiency can cause confusion, dermatitis, convulsions, anemia, and mental depression; B_8 deficiency results in neurological disorders, dermatitis, hair loss, and conjunctivitis; B_9 deficiency may result in a disorder in the formation of red blood cells, headache, weakness, irritability, and palpitations; for vitamin B_{12}, its deficiency is linked with gastrointestinal disorders, smoothness of the tongue, and nervous problems; lastly, vitamin C deficiency causes anemia, weak wound healing, swollen and bleeding gums, and soreness and hardness of the joints, and, at the lower extremities, bleeding under the skin and in deep tissues.

1.5.2.2 Fat-soluble Vitamins

This group of vitamins includes A, D, E, K, and their deficiencies result in different symptoms. For example, ocular disorders that result in blindness, dry skin, growth retardation, diarrhea, and vulnerability to infection are the symptoms of vitamin A deficiency. Incomplete bone growth in children and soft bones in adults are the symptoms of vitamin D deficiency. Peripheral neuropathy and dissociation of red blood cells are symptoms of vitamin E deficiency. Disorders such as the blood clotting and internal bleeding are the symptoms of vitamin K deficiency. The structures, food sources, and functionality of vitamins are presented in Table 1.9.

1.5.3 Functionality and Health Effects

Vitamins have various biological roles in the human body and most water-soluble vitamins are part of the enzyme required for the metabolism of energy. Vitamins B_1, B_3, and B_{12} are important for nerve function. Vitamin C is an antioxidant and has an assistant role in iron absorption. Fat-soluble vitamins cannot be easily excreted, compared to water-soluble vitamins. Vitamin A has numerous functions in the body such as tooth and bone growth and healthy skin, and it is needed for the vision. Vitamin D is stored in bones and is needed for the appropriate absorption of calcium. Vitamin E is an antioxidant and protects cell walls. Vitamin K is required for blood clotting. More information about these vitamins is presented in Table 1.9.

1.6 ESSENTIAL OILS

1.6.1 Background

Essential oils are natural, liquid, volatile, and mixtures of compounds with low molecular weight, which are formed by aromatic plants. The strong odor is a prominent feature of essential oils. Hydrodistillation and steam and solvent extraction, are two common processes for the extraction of essential oils. The first information about essential oils was obtained in the 16th century, derived from the Quinta essential drug. The name of 'essential oil' is associated with the flammability of these components. A large number of researchers have tried to define essential oils. According to Agence Française de Normalisation (AFNOR), essential oils are components of vegetable or raw

TABLE 1.6

Dietary Reference Intakes (DRIs) of Nutrients for Humans During Various Stages of Life

Nutrients	Males (9 to >70 Years)		Female (9 to >70 Years)			
Macronutrients	*Min.*	*Max.*	*Min.*	*Max.*	*Pregnancy*	*Lactation*
Carbohydrates (g)*	100	100	100	100	135	160
Lipids (% of energy)*	20	35	20	35	20–35	20–35
Linoleic acid* (% of energy)	5	10	5	10	5–10	5–10
Linolenic acid* (% of energy) Protein (g)†	0.6	1.2	0.6	1.2	0.6–1.2	0.6–1.2
Protein (g)†	0.66	0.87	0.66	0.76	0.88	1.05
Histidine (mg/kg daily)	11	13	11	12	15	15
Isoleucine (mg/kg daily)	15	18	15	17	20	24
Leucine (mg/kg daily)	34	40	34	38	45	50
Lysine (mg/kg daily)	31	37	31	35	41	42
Methionine + cysteine (mg/kg daily)	15	18	15	17	20	21
Phenylalanine + tyrosine (mg/kg daily)	27	33	27	31	36	41
Threonine (mg/kg daily)	16	19	16	18	21	24
Tryptophan (mg/kg daily)	4	5	4	5	5	7
Valine (mg/kg daily)	19	23	19	22	25	28
Water (L/day)‡	2.4	3.7	2.1	2.7	3	3.8
Micronutrients						
Macroelements						
Calcium (Ca) (mg/day)†	800	1100	800	1100	800–1000	800–1000
Magnesium (Mg) (mg/day)†	200	350	200	300	290–335	255–300
Phosphorus (P) (mg/day)†	580	1055	580	1055	580–1055	580–1055
Sodium (Na) (g/day)‡	1.2	1.5	1.2	1.5	1.5	1.5
Potassium (K) (g/day)‡	4.5	4.7	4.5	4.7	4.7	5.1
Chloride (Cl) (g/day)‡	1.8	2.3	1.8	2.3	2.3	2.3
Microelements						
Iron (Fe) (mg/day)†	5.9	6	5	8.1	22–23	6.5–7
Zinc (Zn) (mg/day)†	7	9.4	6.8	7.3	9.5–10.5	10.4–10.9
Iodine (I) (µg/day)†	73	95	73	95	785–800	985–1000
Copper (Cu) (µg/day)†	540	700	540	700	785–800	985–1000
Selenium (Se) (µg/day)†	35	45	35	45	49	59
Molybdenum (Mo) (µg/day)†	26	34	26	34	40	35–36
Chromium (Cr) (µg/day)‡	25	35	21	25	29–30	44–45
Fluoride (F) (mg/day)‡	2	4	2	3	3	3
Manganese (Mn) (mg/day)‡	1.9	2.3	1.6	1.8	2	2.6

(Continued)

Nutrients	Males (9 to >70 Years)		Female (9 to >70 Years)			
Macronutrients	Min.	Max.	Min.	Max.	Pregnancy	Lactation
Vitamins and Choline						
Vitamin A (µg/day)[†]	445	630	420	500	530	885–900
Vitamin E (mg/day)[†]	9	12	9	12	12	16
Vitamin C (mg/day)[†]	39	75	39	60	66–70	96–100
Thiamine (mg/day)[†]	0.7	1	0.7	1	1.2	1.2
Riboflavin (mg/day)[†]	0.8	1.1	0.8	0.9	1.2	1.3
Vitamin B_6 (mg/day)[†]	0.8	1.4	0.8	1.3	1.6	1.7
Niacin (mg/day)[†]	9	12	9	11	14	13
Folate (µg/day)[†]	250	320	250	320	520	450
Vitamin B_{12} (µg/day)[†]	1.5	2	1.5	2	2.2	2.4
Vitamin D (µg/day)[†]	10	10	10	10	10	10
Vitamin K (µg/day)[‡]	60	120	60	90	75–90	75–90
Pantothenic acid (mg/day)[‡]	4	5	4	5	6	7
Biotin (µg/day)[‡]	20	30	20	30	30	35
Choline (mg/day)[‡]	375	550	375	425	450	550

* AMDR: acceptable macronutrient distribution ranges.

[†] EAR: estimated average requirement.

[‡] AI: adequate intake.

TABLE 1.7
The Major Functions of Trace Elements in the Body (Al-Fartusie & Mohssan, 2017; Pais & Jones Jr., 1997)

Element	RNI*/day	Functions	Deficiency
Calcium	525 mg	Healthy teeth and bones, blood clotting, enzyme activation, and regulation of muscle contraction.	Irritability, rickets, tremors in newborn babies, and jitteriness.
Magnesium	75 mg	Glycolysis, synthesis of RNA, replication of DNA, steady heart rate, and metabolism of vitamin D.	Sleep disorders, poor nail, irritability, and muscle weakness.
Iron	7.8 mg	Forms the main part of hemoglobin and has a function in cognitive development.	Pale skin, mental problems, anemia, and tiredness.
Zinc	5 µg	Immune and growth functions, site-specific antioxidants, synthesis, and activation of proteins and enzymes.	Growth retardation, night blindness, loss of appetite, deficiency of the immune.
Copper	0.3 mg	Connective tissue synthesis, catalyst in the mobilization of iron.	Reduced skin tone, declined plasma iron, skeletal demineralization
Selenium	10 µg	Interaction with heavy metals, antioxidants, maintaining a healthy immune system.	Vulnerability to infection, and asthma.

* The recommended (UK) nutrient intake for an infant (6–9-month of age).

TABLE 1.8

Biochemical Modes and Metabolic Functions of Macrominerals and Microminerals. Modified From De la Guardia and Garrigues (2015)

Minerals	Metabolic Functions	Biochemical Modes of Action
Microminerals		
Chromium	Help regulate normal blood glucose levels; carbohydrate metabolism.	Lipoprotein metabolism, potentiates insulin action, gene expression.
Arsenic	Not needed for humans	-
Boron	Its function is not clear.	-
Cobalt	Exists in vitamin cyanocobalamin.	-
Copper	Synthesis of elastin and collagen; the formation of red cells, hemoglobin, and enzymes; acts as an antioxidant.	Part of many metalloenzymes (Zn/Cu superoxide dismutase, lysyl oxidase) associated with the respiratory chain and metabolism of iron
Fluoride	Teeth and bone mineralization; stimulates new bone formation.	Calcium fluorapatite
Iodide	Inhibits goiter and cretinism; energy metabolism.	Ingredient of triiodothyronine and thyroid hormones thyroxine.
Iron	Electron and oxygen transport; inhibits microcytic hypochromic anemia; cellular metabolism.	Part of myo/hemoglobin and a wide range of enzymes.
Manganese	Not clear; some antioxidants have a role in the formation of bone.	Part of the enzymes involved in cholesterol, amino acid, and carbohydrate metabolism.
Molybdenum	Cofactor for enzymes; catabolism of pyrimidine, sulfur amino acids, and purines.	An agent of electron transfer for enzymes (xanthine oxidase and sulfite oxidase) and involved as a cofactor
Nickel	Its biological role in humans is not clear; might have the role of a cofactor of metalloenzymes and simplify iron adsorption or microorganism metabolism.	-
Selenium	Control of thyroid hormone action; cellular antioxidant, oxidation, and reduction of vitamin C and other molecules.	Exists in the structure of tyrosine deiodinase, glutathione peroxidase, and T lymphocyte receptor expression.
Silicon	Its biological function is not clear in humans; might have a role in bone function.	
Vanadium	Its biological function is not clear in humans.	
Zinc	Control of differentiation; synthesis of protein; the immune system; acts in several enzymes involved in macronutrient metabolism and sexual maturation.	Part of multiple proteins and enzymes, enzyme cofactor involved in the control of gene expression.
Macrominerals		
Calcium	Components of teeth and bones; have roles in intracellular communication and blood coagulation.	Contractility and excitability of striated muscles, retain the permeability of cell membranes.
Chlorine	Regulate the fluid volume of cells with sodium and normal cell function; food digestion.	Regulate acid–base, salt, and water balance.

Minerals	Metabolic Functions	Biochemical Modes of Action
Magnesium	Component of teeth and bones; have a role in the relaxation of muscle; regulate acid–base, salt and water balance; enzyme cofactor.	Transportation of conduction, which has a role in the enzymatic process for energy production.
Phosphorus	Component of teeth, bones, and some lipids; buffering capacity, transfer, and storage of energy, synthesis of nucleotide.	Energy reserve as ATP and CP.
Potassium	Reduces markers of bone turnover; recurrence of kidney stones; excess sodium intake causes it acts to blunt the increasing blood pressure.	Nerve impulse generation and transmission
Sodium	Regulate the fluid volume of cells with sodium and normal cell function.	Regulate of acid–base, salt, and water balance, nerve impulse generation, and transmission.
Sulfur (as inorganic sulfate)	When sulfur-containing compounds are required it provides sulfur	Needed for 3′-phosphoadenosine-5′-phosphate biosynthesis

TABLE 1.9
Vitamins; Their Food Sources, and Their Effect on the Human Body. Modified From Galanakis (2016)

Vitamin	Food Sources	Health Effects	Structure
Vitamin B1 (Thiamin)	Watermelon, spinach, tomato, soy milk, ham, pork chops, and sunflower seeds.	Supports nerve function and energy metabolism.	
Vitamin B2 (Riboflavin)	Eggs, liver, broccoli, fish, meat, mushroom, spinach, milk.	Supports normal vision, energy metabolism, and skin health.	
Vitamin B3 (Niacin)	Grains, potatoes, dairy products, nuts, meat, poultry, spinach, tomato, and fish.	Improves food digestion, nerve function, and promotes lipoprotein profiles and skin health.	
Vitamin B5 (Pantothenic acid)	Mushrooms, whole grains, avocados, tomato, chicken, and broccoli.	Supports energy metabolism.	
Vitamin B6 (Pyridoxine)	Beans, eggs, nuts, red meat, spinach, fish, legumes, banana, potatoes, and watermelon.	Have a significant role in the metabolism of amino acids and fatty acids, as well as the production of red blood cells, decrease in homocysteine levels, and risk of heart diseases.	
Vitamin B8 (Biotin)	Egg yolks, grains, soybeans, and fish.	Improves healthy bones and hair, and converts food into energy, fat, and glycogen synthesis.	

(Continued)

TABLE 1.9
(*Continued*)

Vitamin	Food Sources	Health Effects	Structure
Vitamin B9 (Folic acid)	Orange juice, broccoli, tomato, spinach, and legumes.	Supports the formation of DNA synthesis and new cell, and declines homocysteine levels.	
Vitamin B12 (Cyanocobalamin)	Meat, fish, eggs, milk, cheese, poultry, and fortified soy milk.	Reduces homocysteine levels, reduces the risk of heart diseases, helps in new cell production, break down fatty and amino acids, and supports nerve cell.	
Vitamin C (Ascorbic acid)	Citrus fruits, kiwis, strawberry, potatoes, broccoli, spinach, snow peas, tomatoes, and bell peppers.	Take parts in new cell synthesis, hydrolysis of fatty acids and amino acids, supports nerve cell preservation, acts as an antioxidant, and enhances the immune system.	
Vitamin A (Retinol)	Beef, spinach, liver, eggs, fish, shrimp, fortified milk, cheese, carrots, mangoes, squash, pumpkin, turnip greens, and broccoli.	Supports the growth of skin, vision, bone, and tooth, immunity, and reproduction, and acts as an antioxidant.	
Vitamin D	Fortified milk and margarine, egg yolk, fatty fish, liver, and self-synthesis via sunlight.	Improves bones mineralization, preserve normal blood levels of calcium and phosphorus to strengthen bones.	Vitamin (D3)
Vitamin E	Polyunsaturated plant oils, avocados, margarine, sunflower seeds, whole grains, nuts, wheat, tofu, and shrimp.	Antioxidant, protects vitamin A, includes lipids and antioxidant enzymes, supports cell membrane stabilization, regulate oxidation reactions.	
Vitamin K	Cabbage, broccoli, spinach, eggs, liver, milk, and Brussels sprouts.	The synthesis of blood-clotting proteins and regulation blood calcium.	

material, which are produced by steam distillation or by mechanical processes from the epicarp of citrus, or dry distillation. Afterward, these components are separated from the aqueous phase using physical methods. Essential oils are insoluble in water and soluble in ether, alcohol, and fixed oils. These volatile oils at room temperature are colorless and liquid. Unless a few cases such as cinnamon, vetiver, and sassafras, they have a density less than unity. They have a distinctive refractive index, a characteristic odor, and a high optical activity.

Essential oils may be present in all parts of aromatic plants: flowers (e.g., orange, lavender, pink, and the (clove) flower bud), leaves (e.g., eucalyptus, mint, thyme, bay leaf, savory, sage, pine), needles, underground organs/roots, rhizomes (e.g., ginger, sweet flag), seeds (e.g., coriander, carvi), fruits (e.g., anise, citrus epicarps), and bark (e.g., cinnamon, rosewood, and sandalwood). These components should be used at a limited amount because of the toxicity problems and their intense aroma. These oils can change the organoleptic properties of the food.

Essential oils can form in special or groups of cells; particular organs such as flower calyces, leaves, fruit, and roots are the sources of these components. These components have various applications in different fields such as cosmetics, perfumes, spices, nutrition aromatherapy, and phytotherapy (Lahlou, 2004).

1.6.2 Extraction and Composition

There are some extraction methods, such as microwave and solvent extraction, or hydrodistillation, which have limited the extraction of essential oils. Supercritical carbon dioxide is an appropriate method in which the labile compounds are maintained. Furthermore, the time process is decreased, and the presence of toxic solvent residues is avoided. Although using liquid carbon dioxide for the extraction of essential oils at high pressure and low temperature produces a more natural organoleptic profile, it is an expensive process. It appears that the most common method for the extraction of essential oils is steam distillation (Moyler, 1998). Antimicrobial activity and composition of essential oils are dependent on the extraction method. Moreover, harvesting season, ripeness, and geographical source all have an effect on essential oil composition.

Essential oils contain 85%–99% volatile components, which are mixtures of low molecular weight components such as terpenes, terpenoids, and other aromatic and aliphatic constituents. 1%–15% of essential oils are non-volatile components. The main components of some essential oils and the method for determination of their antioxidant properties are presented in Table 1.10 (Sánchez-González et al., 2011). These components have various applications in the food industry (as flavoring, antioxidant, antibacterial, and antifungal agents). Some of these applications are presented in Table 1.10. Essential oils have been used in meat, fish, fruit, vegetables, and dairy products as antimicrobial and antioxidants components (Sánchez-González et al., 2011).

1.6.3 Classification, Chemistry, and Structure

Among about 3000 essential oils known to date, only 300 have commercial importance for the use in agronomic, pharmaceutical, sanitary, perfume, cosmetic, and food industries. These essential oils contain phenylpropanoids, terpenoids, and other components such as aliphatic and aromatic constituents. Monoterpenes, sesquiterpenes, and oxygenated derivatives can be found in the terpenes group, all of which have a low molecular weight (Raut & Karuppayil, 2014). The presences of bioactive volatile components are necessary for biological activities and antimicrobial properties of essential oils. The major components of the essential oils are terpenoids (Chávez-González et al., 2016).

Organic volatile compounds with low molecular weight are present in organic oils. These volatile compounds belong to different chemical groups such as alcohols, ketones, ketones, ethers or oxides, esters, amines, amides, phenols, heterocycles, and mainly the terpenes. The structures of some terpenes are presented in Figure 1.6. A large number of compounds in essential oils that have been

TABLE 1.10

The Plant Source and the Main Components of Some Essential Oils and Their Tested Antioxidant and Antimicrobial Properties (Sánchez-González et al., 2011)

The Common Name of The Essential Oil(s)	The Latin Name of Plant Source	Major Components	Antioxidant Assay Used
Bergamot	*Citrus bergamia*	Limonene Linalool	ABTS[a]
Cinnamon	*Cinnamomum zeylandicum*	Trans-cinnamaldehyde	ABTS[a]
Coriander	*Coriandrum sativum* (seeds)	Linalool	DPPH[b]
Clove	*Syzygium aromaticum*	Eugenol Eugenyl acetate	DPPH[b]
Eucalyptus	*Eucalyptus globulus*	Eucalyptol	DPPH[b] Thiobarbituric acid
Lemon	*Citrus limon*	Limonene Valencene Ocimene	ABTS[a]
Oregano	*Origanum vulgare*	Carvacrol Thymol Y-Terpinene ρ – Cymene	DPPH[b] Thiobarbituric acid
Rosemary	*Rosemarinus officinalis*	Bomyl acetate Camphor 1,8-Cineole α-Pinene	ABTS[a] DPPH[b]
Sage	Salvia officinalis *L.*	Camphor α – Pinene β – Pinene 1,8-Cineole α – Thujone	ABTS[a]
Thyme	Thymus vulgaris	Borneol Viridiflorol Thymol Carvacrol γ – Terpinene ρ – Cymene	Aldehyde Carboxylic acid ABTS[a] DPPH[b]
Tea Tree	*Melaleuca alternifolia*	Terpinen-4-ol α – Terpinene 1,8-Cineole γ – Terpinene	DPPH[b]

[a] 2,2-Azino-bis(3-ethylbenzthiazoline-6-sulphonic acid)
[b] 2,2-Diphenyl-1-picrylhydrazyl 2

recognized so far belong to the family of terpenes, including functionalized derivatives of alcohols (geraniol, α-bisabolol), esters (γ-tepinyl acetate, cedrylacetate), ketones (menthone, p-vetivone) of aldehydes (citronellal, sinensal), and phenols (thymol). These compounds include non-terpenic compounds that are bio-generated through the phenylpropanoids pathway; e.g., eugenol, cinnamaldehyde, and safrole (Dhifi et al., 2016).

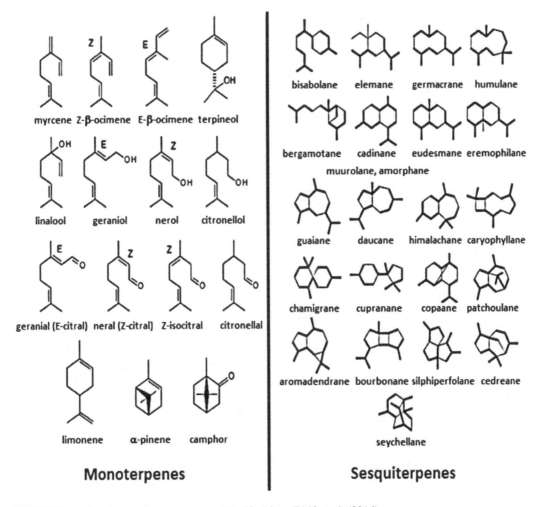

Monoterpenes

Sesquiterpenes

FIGURE 1.6 Structures of some terpenes. Modified from Dhifi et al. (2016).

1.6.4 FUNCTIONALITY AND HEALTH EFFECTS

Essential oils have various applications in pharmaceutical and food industries. Many health benefits, such as their effect on chronic, infectious, and acute diseases, have been reported. They have various functional properties, some of which are discussed below:

1.6.4.1 Antibacterial Activity

The hydrophobicity is one of the most important properties of essential oils, which help them partition into lipids of the cell membrane and disrupt the structure of bacteria, which makes it more permeable and cause the leakage of ions and other molecules. The leakage of bacterial cells at a certain amount can be tolerated without disruption, but more leakage of cell contents can lead to cell death. The highest antibacterial activity has been observed in the essential oils containing aldehydes or phenols (e.g., cinnamaldehyde, carvacrol, citral, eugenol, or thymol), followed by the essential oil containing terpene alcohols. The weaker activity has been observed in the essential oils containing esters or ketones (e.g., β-myrcene, α-thujone, or geranyl acetate), while volatile oils including terpene hydrocarbons have been reported to be typically inactive (Dhifi et al., 2016; Dorman & Deans, 2000; Sikkema et al., 1994). In general, essential oils specified with a high concentration of phenolic compounds, such as thymol, carvacrol, and eugenol, have significant antibacterial activities. These compounds participate in the functions such as disruption of the cytoplasmic membrane, the driving force of protons, active transport, electron flow, and coagulation of cell contents (Dhifi et al., 2016; Dorman & Deans, 2000; Sikkema et al., 1994).

1.6.4.2 Antioxidant Activity

Phenolic compounds and secondary metabolites with conjugated double bonds have antioxidative properties. The essential oils of cinnamon, nutmeg, basil, clove, oregano, parsley, and thyme are specified by the greatest important antioxidant properties (Aruoma, 1998). The most active compounds are carvacrol and thymol, due to their phenolic structures. The redox properties of these phenolic compounds have a significant role in neutralizing free radicals. Moreover, certain alcohols, ketones, ethers, aldehydes, and monoterpenes (e.g., linalool, 1,8-Cineole, geranial/neral, citronellal, isomenthone, and menthone) can be responsible for the antioxidant activity of essential oils (Aruoma, 1998). These components that possess a great scavenging capacity of free radicals might lessen the risk of some disease such as brain dysfunction, heart disease, cancer, and immune system decline (Aruoma, 1998; Burt, 2004; Dhifi et al., 2016).

1.6.4.3 Anti-inflammatory Activity

Essential oils can be effective in inflammatory diseases such as rheumatism, allergies, or arthritis. They can inhibit histamine release or decrease the formation of the inflammation mediators (Maruyama et al., 2005). The anti-inflammatory activity of these compounds might be due to their antioxidant activities and their interactions with signaling cascades involving cytokines and regulatory transcription factors, and on the expression of pro-inflammatory genes (Dhifi et al., 2016). Cancer chemoprotective activity, cytotoxicity, allelopathic activity, repellent, and insecticidal activity, are other biological activities of essential oils (Dhifi et al., 2016).

1.7 ESSENTIAL FATTY ACIDS

1.7.1 Background

Essential fatty acids (EFAs) are polyunsaturated fatty acids that are necessary for health, because they cannot be synthesized in the body and should be provided by foods. Omega-3 (ω-3) and omega-6 (ω-6) are the two families of EFAs. Omega-3 and Omega-6 fatty acids have a final carbon–carbon double bond in ω-3 and ω-6 positions, respectively. Designations of ω-3 and ω-6 have great importance, and the double bonds are in the *cis*-configuration in ω-3 fatty acids. In these components, two hydrogen atoms are on the same side of the double bond. Eicosapentaenoic acid (EPA), α-linolenic acid (ALA), and docosahexaenoic acid (DHA) have different properties for which they could be classified as functional ingredients (Kaur et al., 2014).

1.7.2 Sources of Essential Fatty Acids and Their Status in Diet

Flaxseed is one of the richest sources of ALA, while it can also be found in walnut, hemp, soybean, and canola oil. It is also commonly present in the chloroplast of green leafy vegetables. Fish and fish oil are the main sources of EPA, with a content ranging between 39% and 50% for both saltwater and fresh fish, respectively. DHA, which is the main brain ω-3 fatty acid (65% of the brain is made of fat and 50% of this fat is DHA), can be found in fish oil and red-brown algae. Oxygen, light, and heat can destruct ALA and it might be toxic if not protected. It can be destroyed faster than linoleic acid. ω-3 fatty acids are essential nutrients, because within 95%–99% of ω-3 fatty acids are less than what is required for good health, so this is the most therapeutic of essential nutrients (Kaur et al., 2014).

1.7.3 Sources of ω-6 Fatty Acids and their Status

1.7.3.1 Linoleic Acid

The fatty acid composition (linoleic and α-linolenic) of various essential fatty acids in different oils is presented in Table 1.11. Linoleic acid is a polyunsaturated omega-6 fatty acid and must receive through the human diet. Linoleic acid is composed of an 18-carbon chain with *cis*-configuration of double bonds at the 9th and 12th carbons from the carbonyl functional group (C18:2 ($\Delta^{9,12}$)). It is insoluble in water and very soluble in acetone, diethyl ether, benzene, and ethanol. The consumption of this fatty acid is necessary for maintaining proper health. It is proven that the deficiency of linoleate (the salt form of the acid) in rates can result in poor wound healing, skin scaling, and hair loss. The richest sources of linoleic acid are safflower, sunflower, and corn. It is also found in medium amounts in soybean, almond, and sesame, and in small quantities in canola, olive oils, and peanut. Coconut and palm kernel are poor sources of this fatty acid (Table 1.11). Depending upon the need, the body can convert linoleic acid into other fatty acids.

1.7.3.2 Gamma-linolenic Acid

γ-linolenic acid is categorized as 18:3 (n−6), meaning that this fatty acid is a carboxylic acid with a 18-carbon chain and contains three *cis* double bonds at positions 6, 9, and 12 (18:3 ($\Delta^{6,9,12}$)). γ-linolenic acid is the first intermediate in the metabolic conversion of linoleic acid to arachidonic

TABLE 1.11

The Average Content of Linoleic and A-Linolenic (Ala 18:3 N−3) Acids in Various Oils (g/100 g Fat). Modified From Kaur et al. (2014)

Oil	Linoleic Acid 18:2(N-6)	Ala 18:3(N−3)	Total Unsaturated Fatty Acids
Soybean	50.8	6.8	80.7
Cotton seed	50.3	0.4	69.6
Corn	57.3	0.8	82.8
Safflower	73	0.5	86.3
Sunflower	66.4	0.3	88.5
Sesame	40	0.5	80.5
Olive	8.2	0.7	81.4
Peanut	31	1.2	77.8
Rapeseed (Zero erucic acid)	22.2	11	88
Rapeseed (High erucic acid)	12.8	8.6	88.4
Cocoa butter	2.8	0.2	36
Coconut	1.8	-	7.9
Palm kernel	1.5	-	12.9
Palm	9	0.3	47.7
Almond	18.2	0.5	87.7
Cashew	17	0.4	73.8
Chestnut	35	4	75.5
Walnut	61	6.7	86.5
Butter	2.3	1.4	32.6

acid. Evening primrose oil (7%–10%), black current (15–20 g/100 g), and borage (18–26 g/100 g) are the main sources of γ-linolenic acid. It is present in small quantities in human milk and organ meats (Table 1.11).

1.7.3.3 Dihomo-γ-linolenic Acid

Dihomo-γ-linolenic acid (DGLA) is an n−6 polyunsaturated fatty acid created by desaturation and a carbon chain elongation reaction from linoleic acid, then further converted into arachidonic acid. In low quantities, this fatty acid has been shown to have an antithrombotic action; however, it generates an inflammatory or thrombotic action at higher concentrations, converting into arachidonic acid (Dihomo-γ-linolenic acid levels and obesity in patients with type 2 diabetes). This 20-caron fatty acid (C 20:3 ($\Delta^{8,11,14}$)) is synthesized from GLA.

1.7.3.4 Arachidonic Acid

This fatty acid that contains 20 carbons and three double bonds (C 20:4 ($\Delta^{5,8,11,14}$)) is present in meat, eggs, and dairy products. Linoleic acid is present in high quantities in human diets. The intake of linoleic acid has been increased during the past century because of the use of corn and safflower in higher quantities. The presence of linoleic acid at high quantities and ω-3 fatty acids at low quantities in the diet might cause hypertension, chronic inflammation, and a tendency to blood clotting, which increases the risk of heart attack and stroke. An increment in the quantity of linoleic acid slows down the metabolism of EPA, ALA, and DHA by preventing Δ^6 desaturase that might also decline with age.

1.7.4 Functionality and Health Effects

ω−3 fatty acids have significant importance in human health because these fatty acids can prevent some diseases, such as myocardial infarction and cardiovascular diseases, bowel disease, psoriasis, mental illnesses, cancer, and bronchial asthma. These fatty acids have roles in the improvement of several conditions related to human health, including increased energy levels, stamina, and performance; enhancing learning, concentration, calmness, behavior, and IQ; reducing cardiovascular risk factors; preventing cancer growth and metastasis; enhancing insulin sensitivity; speeding the healing of wounds because of accidental injury, physical exertion, and surgery; lowering inflammation and joint pain; increasing bone mineral metabolism; and helping weight management, improving fat burning, and decreasing fat production (Kaur et al., 2014).

1.8 BIOACTIVE ENZYMES

1.8.1 Background

The concept of enzymes as biocatalysts was initially introduced in 1883 by the discovery of catalyzing starch into sugars by diastase. In the 20th century, the main landmarks were advanced in the route of enzyme isolation and purification, the perception that enzymes are proteins with biochemical activity. Catalyzing reactions in different biological processes in all living cells perform using enzymes. They are extremely efficient catalysts due to the acceleration of various reactions. The enzyme activities have been dependent on their structures, for more than 100 years; the hypotheses of 'induced-fit' and 'lock-and-key' have proposed that the structural interactions between substrates and enzymes have a great role in enzyme catalysis. However, these hypotheses are incomplete, because they fail to explain the detailed mechanism of the enhancement of reaction rates achieved by enzymes (Agarwal, 2006).

TABLE 1.12

The Main Enzymes From Animals, Plants, and Bacterial Sources Used in the Food Industry

Enzyme	Source	Action In Food	Food Applications
Alpha-Amylase	Cereal seeds such as barley, wheat	Hydrolysis of starch to oligosaccharides	Breadmaking, brewing, and malting
Beta-Amylase	Sweet potato	Hydrolysis of starch to maltose	Making of high malt syrups
Papain	Latex of unripe papaya fruit	Hydrolysis of protein in food and beverage	Meat tenderization, chill haze inhibition in beer
Bromelain	Pineapple	Hydrolysis of protein in muscle and connective tissue	Meat tenderization
Ficin	Fig	As bromelain	Same as papain and bromelain, but it is expensive
Trypsin	Bovine/porcine pancreas	Hydrolysis of food proteins	Making hydrolysates for food flavoring
Chymosin (rennet)	Calf abomasum	Hydrolysis of kappa-casein	Coagulation of milk in cheesemaking
Pepsin	Bovine abomasum	As chymosin. Also, more general casein hydrolysis in cheese	Generally, exist with chymosin as part of rennet
Lipase/esterase	The gullet of goat and lamb; pig pancreas; calf abomasum	Hydrolysis of triglyceride	Flavor enhancement, modification of fat by interesterification
Lipoxygenase	Soybean	Oxidation of unsaturated fatty acids in flour	Improvement of bread dough
Lysozyme	Hen egg white	Bacterial cell wall hydrolysis	Inhibition of late blowing defects in cheese using spore-forming bacteria
Lactoperoxidase	Cheese whey and bovine colostrum	Oxidation of thiocyanate ion to bactericidal hypothiocyanate	Cold sterilization of milk
Alpha-Amylase	*Bacillus spp. Aspergillus spp. Microbacterium imperial*	Hydrolysis of wheat starch	Dough softening, enhancing bread volume, helping with the production of sugars for yeast fermentation
Alpha-Acetolactate	*Bacillus subtilis*	Production of acetoin from acetolactate	Decrease in wine maturation time
Amyloglucosidase	*Aspergillus niger Rhizopus spp.*	Converts starch dextrin to glucose	Have a role in the production of high fructose corn syrup, and lite beers production
Aminopeptidase	*Aspergillus spp.*; *Lactococcus lactis*; *Rhizopus oryzae*	Releases free amino acids from proteins and peptides	Accelerating cheese maturation
Catalase	*Aspergillus niger Micrococcus luteus*	Convert hydrogen peroxide to oxygen and water	Oxygen removal technology combined with glucose oxidase
Cellulase	*Aspergillus niger Trichoderma spp*	Kappa-casein hydrolyses	Coagulation of milk for cheese production

(Continued)

TABLE 1.12
(Continued)

Enzyme	Source	Action In Food	Food Applications
Chymosin	*Aspergillus awamori Kluyveromyces Lactis*		
Beta-Galactosidase (lactase)	*Aspergillus spp. Kluyveromyces spp.*	Converts milk lactose to galactose and glucose	Sweetening whey and milk, decrease the crystallization in ice cream including whey, enhancing whey protein concentrates functionality, producing lactulose and products for lactose-intolerant individuals
Glucose isomerase	*Bacillus coagulans Actinplanes missouriensis Streptomyces lividans Streptomyces rubiginosus*	Converts glucose to fructose	Making high fructose corn syrup
Protease(proteinase)	*Aspergillus spp. Rhizomucor miehei Cryphonectria parasitica Penicillium citrinum Rhizopus niveus Bacillus spp.*	Hydrolysis of kappacasein, hydrolysis of animal and vegetable food proteins, hydrolysis of wheat glutens	Coagulation of milk or cheese production, and bread dough improvement

1.8.2 CLASSIFICATION AND CHEMISTRY

The main enzymes with different sources and their action in food are presented in Table 1.12.

Enzymes can be classified into six different groups, based on the kind of catalyze of chemical reaction:

- Oxidoreductases: catalyze reduction/oxidation reactions that commonly involve the transfer of electrons. Oxidases or dehydrogenases belong to this group.
- Transferases: these enzymes can transfer a functional group like phosphate or methyl group and generally have roles in the transfer of a radical. Transmethylases (e.g., of a methyl group), Transacetylases (e.g., of an acetyl group), Transglycosidases (e.g., monosaccharides), phosphomutases and transphosphorylases (e.g., of a phosphate group), and transaminases (of an amino group) are some of the examples of this group.
- Hydrolases: catalyze the hydrolysis of different bonds. The removal or addition of water is involved in the hydrolase reaction. Hydrolases include carbohydrates, esterases, deaminases amidases, nucleases, and proteases. Hydrases include aconitase, fumarase, enolase, and carbonic anhydrase.
- Lyases: split different bonds using reactions other than oxidation and hydrolysis. This reaction involves the cleaving or forming a C-C bond. An example is the group of desmolases.
- Isomerases: catalyze isomerization in a single molecule and involves changing the structure or geometry of a molecule. An example is glucose-isomerase.
- Ligases: connect two molecules using covalent bonds; e.g., DNA lygase responsible for linking two fragments of DNA.

Enzymes have different functions inside living organisms. They are necessary for cell regulation and signal transduction, as well as movement generation. In general, the metabolic pathways in a cell are specified by the amount and types of enzymes that exist in the cell. Enzymes have a great role in the digestive systems of mammals and other animals. Enzymes like amylases and proteases can break down large starch and protein molecules, respectively. Smaller fragments can be formed as a result of breakdown reactions and these fragments can be absorbed easier using the intestines of animals.

Enzymes are commonly globular proteins, possessing a size range from over 60 to further than 2500 amino acids, i.e., molecular weight (MW) of ± 6000–250.000. The three-dimensional structure of enzymes determines their activity. The majority of enzymes have a bigger size than the substrates they act on, and, accordingly, only a small section of enzymes is directly involved in catalysis. This small part is known as the active site and this place usually includes few amino acids (3–4) that are directly involved in the catalytic process. An enzyme can bound with the substrate close to, or even in the active site (van Oort, 2010).

1.8.3 FUNCTIONALITY AND HEALTH EFFECTS

A large number of enzymes can affect the functional behavior of proteins directly or indirectly. These effects might be detrimental or beneficial in the food product(s). Generally, the majority of enzymes that affect protein functionality belong to hydrolases. Proteases have an important role in characterizing the functionalities of a great variety of proteins by acting directly on these biopolymers. However, a low number of hydrolases might have an indirect effect on protein functionality, as the role adenosine triphosphatase plays in the physical structure of meat proteins. The second group of enzymes that can change the functionally of proteins is oxidoreductases. They can change such a functionality indirectly by a product of the enzyme reaction. For example, lipoxygenase produces lipid hydroperoxides, which can oxidize protein sulfhydryl groups to disulfides thus change the structure of the protein. There is no clear demonstration of other groups of enzymes (e.g., isomerases, lyases, and ligases) regarding their effect on the functionality of proteins in foods, although it is conceivable that they do.

1.9 BIOACTIVE PEPTIDES

1.9.1 BACKGROUND

Bioactive peptides are a source of nutrition and have several potential physicochemical functions in our body. These peptides are protein-derived fragments and enzymatic proteolysis during gastrointestinal digestion can release these peptides. *In vitro*, peptides can be released in foods during the processing of foods such as ripening, cooking, and fermentation, as well as using food-grade proteolytic enzymes, small fragments of bioactive peptides might be attained in hydrolysates. The degree of hydrolysis during isolation affects the bioavailability of isolated peptides (Sánchez & Vázquez, 2017).

1.9.2 FUNCTIONALITY AND HEALTH BENEFITS

The activity of a peptide is related to its structure such as the type of *C*- and *N*-terminal amino acid, the amino acid composition, the length of the peptide chain, the hydrophilic/hydrophobic properties of the amino acid chain, the charge of amino acids which forms peptide. For example, peptides with a higher angiotensin-converting enzyme (ACE) inhibitory activity generally have basic *N*-terminal or aromatic amino acids, a higher amount of positively and hydrophobic charged amino acids in *C*-terminal. The amino acid sequence of some peptide and their corresponding activity are reported in Table 1.13.

Peptides derived from milk proteins can prevent peroxidation of essential fatty acids with their antioxidant properties. Antimicrobial peptides are a plentiful and diverse group of molecules

TABLE 1.13

Examples of Some Bioactive Peptides and Their Activity

Sequence	Origin activity	Reference
FSDKKIAK	Antimicrobial	Capriotti et al. (2016)
EQLTK	Antimicrobial	Wada and Lönnerdal (2014)
LKP	Antihypertensive, ACE	Erdmann et al. (2008)
VPPIPP	Antihypertensive	Korhonen (2009)
IBYW	Hipoglucemiante	Guerin et al. (2016)
GGGGYPMYPLR	Inmunpmodulatory	Agyei et al. (2016)
GLLVDLL	Anticancer	Sanjukta and Rai (2016)
YGLF	Opioids	Meisel and Bockelmann (1999)
LDAVNR	Anti-inflammatory	Fan et al. (2014)
NIGK	Platelet-activating inhibition	Fan et al. (2014)

produced by many cell types and tissues in a variety of plant, invertebrate, and animal species. Their amino acid composition, cationic charge, amphipathicity, and size cause them attach to and insert into membrane bilayers. These peptides can prevent cell growth of different detrimental microorganisms such as fungi and bacteria. An abundant number of peptides of animal and plant origin possess cytotoxic effects. Peptides are endogenous ligands for the reception of opioids. Antihypertensive peptides, which are derived from corn, milk, and fish protein sources, are known as ACE inhibitors. These compounds are necessary for the regulation of blood pressure. Peptides which are derived from milk, egg soy, and plant sources have shown anti-inflammatory properties. Immunomodulatory peptides obtained from hydrolysates of soybean and rice proteins act to stimulate reactive oxygen species, which triggers non-specific immune defense systems (Sánchez & Vázquez, 2017).

1.10 BIOACTIVE LIVE ORGANISMS (PROBIOTICS)

1.10.1 BACKGROUND

Probiotics are nonpathogenic living microorganisms that can include a broad category, but mostly refer to the commensal bacterial flora which have beneficial effects on human health and disease prevention, on condition that they are administered/consumed in adequate amounts. Probiotics are frequently, but not necessarily, commensal bacteria. Élie Metchnikoff was the first researcher to introduce probiotic bacteria as bioactive living organisms (Metchnikoff, 1908). This researcher believed that the daily ingestion of probiotics from different products such as soured milk could decrease the effects of harmful intestinal organisms, which produce life-shortening autotoxins. The Food and Agriculture Organization (FAO) and the World Health Organization (WHO) have defined probiotics as live microorganisms, which have some benefits for their host (when administered/consumed in adequate amounts).

1.10.2 CLASSIFICATION AND STRUCTURE

Although, Lactobacillus and Bifidobacterium are the main probiotic microorganisms, Streptococcus, Lactococcus, Enterococcus, Propionibacterium, and Saccharomyces yeasts are widely present in various foods (Azad et al., 2018). Some examples of the probiotic strains are given in Table 1.14. These microorganisms are used in various functional foods, including dairy products (e.g., milk,

TABLE 1.14

Some Examples of Probiotic Microorganisms (Chugh & Kamal-Eldin, 2020)

Lactobacillus	Bifidobacterium	Others
L. acidophilus NCFM,	*B. adolescentis*,	*Enterococcus faecium*,
L. bulgaricus,	*B. animalis*,	*Pediococcus pentosaceus*, and
L. casei,	*B. breve*,	*Saccharomyces boulardii*,
L. delbrueckii,	*B. bifidum*,	
L. kefranofaciens M1,	*B. lactis*,	
L. paraplantarum,	*B. longum*, and	
L. paracasei,	*B. pseudocatenulatum*,	
L. plantarum,		
L. reuteri,		
L. rhamnosus, and		
L. salivarus		

acidified milks, cheeses, creams, yogurts, and ice cream) or non-dairy products (e.g., bread and cereal fiber snacks, meats and meat products, and juices and other fruit products). For the evident effect of probiotics in food, the amount of viable probiotic cells should be about 10^8–10^9 colony forming units (cfu) *per* day, depending on the bacterial and the physiological properties of the host (Chugh & Kamal-Eldin, 2020; Terpou et al., 2019). Probiotics with various mechanisms involving a great range of bioactive compounds can confer benefits health. These mechanisms include: (a) changing the pH and reducing the oxygen availability by probiotic microorganisms, which decrease favorite conditions of intestinal and prevent multiplication of pathogenic bacteria; (b) producing bacteriocins; (c) synthesis of essential micronutrients like amino acids, vitamins, and enzymes and increasing the bioavailability of dietary nutrients; (d) motivation of the host immune system; (e) increasing the metabolic activity of carbohydrates (Chugh & Kamal-Eldin, 2020; Schepper et al., 2017).

1.10.3 FUNCTIONALITY AND HEALTH EFFECTS

Probiotic bacteria have various beneficial effects, delivered through a range of mechanisms which are briefly presented in Table 1.15.

1.10.4 PRODUCTION OF VARIOUS COMPOUNDS BY BIOACTIVE LIVE ORGANISMS

One of the most important aspects of probiotics in regards with their bioactivity is that they can produce different compounds as follows:

1.10.4.1 Production of Bacteriocins

Probiotics have different antimicrobial activities, which can fight against harmful bacteria. They act with non-specific mechanisms and produce a short-chain fatty acid that results in the modulation of the hydrogen peroxide and pH, or the production of particular toxins with limited killing ranges. Bacteriocins are ribosomally-synthesized peptides made by specified bacterial strains to fight against harmful bacteria through the use of antibiotic-like effects. Bacteriocins and their

TABLE 1.15

Mechanisms of Action of Probiotic Bacteria. Modified From Chugh and Kamal-Eldin (2020)

Beneficial Effect(s)	Mechanisms
Act against harmful bacterial in the gastrointestinal tract.	Indirect prevention of growth of harmful bacteria using changing the competition and pH for nutrients and oxygen, as well as the direct prevention secretion of bacteriocins.
Increase the nutritional value and metabolic activities of the food.	Production of co-factors and vitamins Metabolism of dietary components
Anti-inflammatory, antioxidant, and immune-modulating effects.	Promotion of anti-inflammatory cytokine production Increase in T-cell numbers and activity
Enhancement in gut function and decrease in allergy.	Suppression of hypersensitivity Reduction of irritable bowel syndrome symptoms
Decrease the risk of cardiovascular diseases.	Cholesterol decrement using deconjugation of bile salts Anti-hypertensive peptides production
Decrease the risk of different types of cancer.	Carcinogenic metabolites detoxification

TABLE 1.16

Bacteriocins Produced by Lactic Acid Bacteria and Their Classification, Properties, and Examples (Chugh & Kamal-Eldin, 2020)

Class	Properties	Examples	Typical producing bacteria
I	Very small peptides (<5 KDa) lanthionine or methyllanthionine-containing bacteriocins called lantibiotics.	Lactocin, nisin, mersacidin	*Lactobacillus lactis subsp. lactis*
IIa	Non-modified, heat-stable, small (<10 KDa), cationic hydrophobic peptides have a double glycine leader peptide active against *Listeria*	Pediocin PA1, leucocin A, Sakicin A	*Leuconostoc gelidum*
IIb	Non-modified, heat-stable, small (<10 KDa), cationic hydrophobic and amphiphilic peptides need the synergy of two complementary peptides at nano- or picomolar range	Lactococcin G, enterocin X, plantaricin A	*Enterococcus faecium*
IIc	Non-modified, heat-stable, small (<10 KDa), cationic hydrophobic peptides Affect cell wall formation and membrane permeability	Acidocin B, reuterin 6 enterocin P	*Lactobacillus acidophilus*
III	Sensitive to heat, large peptides (>30 kDa),	Lysostaphin, helveticin J enterolysin A	*Lactobacillus helveticus*

classification, which are produced by lactic acid bacteria, are shown in Table 1.16. These components are small, heat-stable, helical, amphiphilic cationic molecules, which vary in terms of mode of action and their molecular attributes. Probiotic bacteria secreted the antimicrobial peptides to simplify the competition with host-microbiota as colonizing peptides. The antimicrobial activity of bacteriocins might be due to their acting as membrane perturbs or pore-forming agents or through the prevention of cellular division processes. Bacteriocins also have antiviral, anticancer, and plant protection properties.

Probiotic microorganisms, especially lactic acid bacteria and bifidobacteria, have various metabolic activities catalyzed by lipases, proteases, esterases, and amylases that may be lacking in the host. These bacteria can produce lactase and β-galactosidase enzymes and reduce lactose intolerance.

1.10.4.2 Production of Amino Acids and Peptides

De novo biosynthesis pathway is used by gastrointestinal bacteria to produce several amino acids, which play a key role in the metabolism of carbohydrates and lipids and the formation of short-chain fatty acids. Furthermore, probiotic bacteria can ferment proteins and produce amino acids, small peptides, lactones, phenols, and indoles, which maintain energy balance and contribute to anti-inflammatory and antioxidant effects. The aromatic amino acids have a substrate role in the production of different metabolites, which has the main role in decreasing some malfunctions.

1.10.4.3 Production of Short-chain Fatty Acids (SCFAs)

Probiotic bacteria can produce carbohydrate-digesting enzymes in small and large intestines, which produce hexose sugars from indigestible carbohydrates. Probiotic bacteria can use this product to produce SCFAs as follows:

$$59\ C_6H_{12}O_6 + 38\ H_2O \rightarrow 60\ CH_3COOH + 18\ CH_3CH_2CH_2COOH$$
$$+ 22\ CH_3CH_2COOH + 96\ CO_2 + 134\ H_2 + Heat\ (27) \tag{1.1}$$

The SCFAs mainly include acetate, propionate, and butyrate. This is in addition to the organic acids (acetic acid and lactic acid), branched-chain fatty acids (e.g., isovaleric, isobutyric, and 2-methyl butyric acids) heat, and gases (22; 28; 29) that are produced this way (Rowland et al., 2018).

1.10.4.4 Production of Vitamins

Probiotic bacteria can produce growth factors and critical nutrients, vitamin B12 (not synthesized by plants), and vitamin K. *Propionibacterium shermani* is a probiotic bacterium, which is able to produce propionic acid and vitamin B12. Moreover, probiotic bacteria can produce a large variety of vitamins belonging to the B-complex group (e.g., thiamine) (Chugh & Kamal-Eldin, 2020).

1.10.4.5 Production of Antioxidant Compounds

A large number of Bifidobacterium and Lactobacillus species such as *L. acidophilus*, *L. delbrueckii subsp. bulgaricus, L. casei, L. paracasei, L. rhamnosus, L. plantarum, L. fermentum, Streptococcus thermophilus, B. breve, B. animalis, and B. longum* possess antioxidant activity. This is because, probiotic bacterial enzymes can hydrolyze proteins and produce peptides, which have antioxidant, immune-modulating, and anti-inflammatory properties (Chugh & Kamal-Eldin, 2020; Mishra et al., 2015).

1.10.4.6 Production of Anti-inflammatory and Immune-modulating Compounds

These bacteria can regulate lymphocytes, antibodies, interleukins, and cytokines, and chemokine production, which play significant roles in protecting the intestinal epithelial cells against disorders like inflammation. These bacteria can interact with dendritic cells and the toll-like receptors of the epithelial and with monocytes/macrophages and B- and T-lymphocytes. Mature dendritic cells interact with naïve T cells, depending on the produced cytokines which further are differentiated into T-helper or regulatory. Toll-like receptors control the maturation of dendritic cells responsible for the differentiation of T-helper 0 to T-regulatory which regulates the T-helper 1, T-helper 2, and T-helper 17 inflammatory responses.

1.10.4.7 Production of Exopolysaccharides

Exopolysaccharides are long-chain linear biopolymers with side chains of homopolysaccharide or heteropolysaccharide carbohydrate units that are high molecular weight compounds linked with α- and β-glycosidic bonds. Many probiotic bacteria can produce exopolysaccharides from sugar nucleotide precursors using glycosyltransferase and glycantransferase enzymes (Chugh & Kamal-Eldin, 2020).

1.11 TECHNIQUES FOR EXTRACTION OF BIOACTIVE COMPOUNDS

Extraction methods have an important role in the outcome of any medicinal plant study (Azmir et al., 2013). There are various factors such as solvent, pressure, temperature, pH, and properties of plant matrix that can change the extraction efficiency/outcome (Hernández et al., 2009). Several extraction methods can be used for the extraction of plant bioactives. Non-conventional methods have high environmental-friendly properties because of the decrease in the use of synthetic and organic chemicals and operational time, as well as the improvement in quality and yield of extract. To improve the selectivity of bioactive components from plant materials and also to increase the overall yield, the effects of various parameters such as ultrasound, pulsed electric field, enzyme digestion, extrusion, microwave heating, Ohmic heating, supercritical fluids, and accelerated solvents have been investigated as non-conventional methods (Azmir et al., 2013). On the other hand, conventional extraction methods like Soxhlet and maceration can be used as a reference to evaluate new extraction methods.

Generally, all of the extraction methods for bioactive compounds from plant sources have the same purpose: (a) to enhance the selectivity of methods; (b) to extract specific bioactive compounds; (c) to enhance the sensitivity of bioassay with increment the concentration of targeted compounds; (d) to change bioactive compounds form to a suitable state for detection and separation; and (e) to present a considerable and reproducible method that is independent of change in the sample matrix (Smith, 2003).

1.11.1 CONVENTIONAL EXTRACTION TECHNIQUES

Different extraction techniques can be used for the extraction of bioactive compounds. Most of these techniques are dependent on the extraction power of the solvents and using mixing or heating processes. Conventional extraction techniques are classified into three groups: (1) Soxhlet extraction; (2) Maceration; and, (3) Hydrodistillation (Azmir et al., 2013). A brief explanation of these methods is provided below, while the reader can read more information about these techniques in the report by Azmir et al. (2013).

1.11.1.1 Soxhlet Extraction

Soxhlet extraction was first designed for lipid extraction but nowadays it is used for the extraction of various compounds. In particular, it has been used for the extraction of bioactive compounds from different natural sources. Soxhlet is used as a model for the comparison of a new extraction method. In this method a small quantity of the dry sample is put in a thimble and transferred in the distillation flask containing a specific solvent. During this process, the solvent carries extracted solutes into the bulk liquid. The solvent passes back to the solid bed of the sample and the solute has remained in the distillation flask.

1.11.1.2 Maceration Extraction

Maceration is known as an inexpensive and popular way to extract bioactive compounds, especially essential oils. The maceration process on a small scale includes several steps. Grinding of the plant material is the first step to enhance the surface area for better mixing with the solvent. Next, a suitable solvent (named menstruum) is added in a closed vessel. Afterward, while the liquid is strained off, the solid residue of the extraction process (marc) is pressed to recover large quantities of the occluded solutions. Impurities are separated by filtration. Extraction can be simplified by occasional shaking that enhances diffusion and eliminates the concentrated solution from the surface of the sample for bringing new solvent to the menstruum.

1.11.1.3 Hydrodistillation Extraction

Hydrodistillation is a common method for the extraction of bioactive compounds from plants. It could be carried out before dehydration of the plant material and organic solvents are not involved in this process. Hydrodistillation techniques are classified into three groups; direct steam distillation, water distillation, and water and steam distillation. In hydrodistillation, after packing the plant material, enough amount of water is added and brought to boil. Alternatively, the plant sample is exposed to direct steam. Bioactive compounds of the plant tissue are released by acting steam and hot water as the main influential parameters. Vapor mixture of oil and water condenses using indirect cooling by water and then bioactive compounds and oil separate from water using a specific separator.

1.11.2 Non-conventional Extraction Techniques

There are some challenges for the extraction of bioactive compounds using conventional techniques; long extraction time, expensive solvents, the need for high purity solvents, low extraction selectivity, and thermal disintegrate unstable compounds (De Castro & García-Ayuso, 1998). Therefore, non-conventional extraction methods are applied to overcome such limitations. Ultrasound-assisted extraction, microwave-assisted extraction, enzyme-assisted extraction, pulsed electric field assisted extraction, pressurized liquid extraction, and supercritical fluid extraction are some of the newest extraction methods for extraction of bioactive compounds from plant materials (Azmir et al., 2013). These methods are advantageous in terms of using safer chemicals and solvents auxiliaries, the design for energy performance, consumption of renewable feedstock, decrease in derivatives and catalyzers, and the design for limiting the possible degradation of the bioactive compounds. They are also more cost-effective and produce less pollution (Azmir et al., 2013). A brief explanation of two of the most popular non-conventional techniques is given in the following sections. More information about different non-conventional extraction techniques is presented by Azmir et al. (2013).

1.11.2.1 Ultrasound-assisted Extraction

In ultrasound-assisted extraction, ultrasound energy simplifies inorganic and organic compounds leaching from the plant matrix. The main effect of ultrasound is the increment of mass transfer and accelerated access of solvent to plant cell materials. Two major types of physical phenomena are involved in ultrasound extraction: a) rinsing the substances of the cell after dissociation of the walls; b) diffusion across the cell wall. Different parameters, such as particle size, moisture content of the sample, solvent properties, and milling degree, have a significant effect on the extraction efficiency. Pressure, temperature, time, and frequency of the sonication are other substantial parameters for the action of ultrasound. The main advantages of ultrasound-assisted extraction are the decrease in extraction energy, time, and use of solvent. Ultrasound waves pass through a medium by forming compression and expansion. In this process, bubbles form, grow, and collapse, resulting in the production of high quantities of energy. The temperature of the bubbles is about 5000 K, pressure 1000 atm with a heating and cooling rate above 1010 K/s.

1.11.2.2 Pulsed-electric Field Extraction

Pulsed-electric field extraction has been identified as a helpful technique for improving the extraction, pressing, drying, and diffusion processes during the past decade. In this process, for increasing the extraction efficiency, the structure of the cell membrane is destroyed. Electric potential passes through the membrane of the cell, when living cells are suspended in the electric field. According to the dipole nature of the membrane, molecules are separated according to their charge in the cell membrane. When the transmembrane potential is exceeding a critical value (about 1 V), repulsion happens between the charge-carrying molecules, which causes the formation of the pores in weak areas of the membrane and give rise to permeability.

1.12 STABILITY AND BIOAVAILABILITY OF BIOACTIVES AND THE IMPORTANCE OF ENCAPSULATION

The potential beneficial effects of bioactives not only depend upon the amount consumed, but also their stability in food (during processing and storage) and the gastrointestinal tract (GIT), as well as their bioavailability.

1.12.1 STABILITY

Almost all bioactives can be degraded during the manufacture, processing, and storage of food products or the functional foods containing these ingredients, as well as in the biological fluids (e.g., gastric juice) that can reduce their health-promoting effects. The auto-oxidation and epimerization are the two main mechanisms known regarding the instability of bioactive compounds such as polyphenols (Ananingsih et al., 2013). Auto-oxidation can occur rapidly by losing the hydrogen atoms and forming semiquinone radical intermediates, quinone oxidized, and superoxide products. This type of degradation usually causes a color transformation of the aqueous solution from transparent to brown which indicates the formation of large molecular weight compounds through the polymerization of the compounds. The instability mechanisms can be affected by pH, temperature, initial concentration of bioactives, antioxidants, the existence of oxygen, ionic strength of the medium, metal ions, and other active components. For example, polyphenols such as catechins are greatly stable at pH < 4, while they become gradually unstable as the pH increases from 4 to 8, with the highest instability at pH values above 8 (Ananingsih et al., 2013; Tang et al., 2019). Generally, factors such as high temperature can decrease the antioxidant activity of bioactive compounds due to increasing the rate of oxidation, polymerization, and epimerization (Fang et al., 2019). The high oxygen levels of the environment and low concentration of antioxidants increase the rate of bioactive oxidation.

Bioactive live organisms (i.e., probiotics) also face stability challenges in food (processing and storage) and through the GIT, because of enzymatic reactions (proteases), pH variations, and bile salts, which negatively affect their bioavailability. Moreover, probiotics face some crucial situations when designing probiotic-based functional foods; e.g., relative humidity (RH), water activity (aw), temperature, and pH, all of which are crucial for probiotic survivability and long-term storability.

1.12.2 BIOAVAILABILITY

Within the human body, bioactive compounds encounter different pH values from acidic in the stomach to alkaline in the duodenum. Since the absorption of these compounds mostly occurs in the duodenum and considering the neutral and alkaline condition of the GIT, which are suitable for auto-oxidation of bioactives, their low stability through the GIT can be one of the reasons for their low bioavailability (Krupkova et al., 2016). Therefore, the bioavailability of bioactives has been reported to be very uncertain. To obtain knowledge about bioactive metabolism and bioavailability, evaluation of the biological activity within target tissues is vital (Manach et al., 2004).

Most bioactive compounds (e.g., flavonoids) can be identified in the portal vein of rats after oral administration (Okushio et al., 1996), which can indicate their intestinal absorption. However, it appears that very little has been published on the distribution of bioactives in tissues in humans; whereas, there are some encouraging data available from animal studies. For instance, in the study where rats were given 0.6% green tea catechins in their drinking water for 28 days (Kim et al. (2000), some catechins were found in remarkable concentrations in some of the rats' organs such as the kidneys, large intestine, lung, prostate, oesophagus, and bladder. It was also reported that a higher level of epigallocatechin gallate (EGCG) could appear in the large intestine and oesophagus than other rats' organs. This was associated with a probable poor systemic absorption of EGCG. It was also observed that tea catechins could be rapidly and extensively metabolized (Kim et al., 2000).

In the case of bioactive live organisms, there is no doubt about the importance of intestinal microflora for the maintenance of human's health in terms of the major physiological functions such as the maintenance of mucosal health, production of nutrients, barrier function restoration, stimulation of bowel motility, secretion of antimicrobial substances, immune stimulation, and improved bioavailability (Holzapfel et al., 1998; Shanahan, 2002).

Several *in vivo* and *in vitro* techniques have been employed for assessing the rate and degree of digestion, as well as the absorption of food and relative release of bioactive compounds such as phenolic compounds from food matrices. The *in vivo* methods are considered to be the most reliable and accurate assessments for measuring the recovery from blood, urine, or feces (Olthof et al., 2003; Rios et al., 2002). Nonetheless, such techniques can be expensive, time-consuming, and difficult to carry out. Thus, many researchers have developed functional models to mimic the action of the human digestive system. One example is the complex simulated gastrointestinal computer-controlled system known as TIM (TNO-Intestinal Models), which was developed by the Netherlands Organization for Applied Scientific Research. This machine can obtain information about interactions between nutritional and functional food constituents, nutrient absorption, the effect of food processing on nutritional and functional properties of food, measuring the recovery of bioactive compounds such as polyphenols, and the effectiveness of specific compounds within a digestive system. In such a system, which consists of two parts that simulate the function of the stomach and small intestine, as well as the function of the colon, parameters such as temperature, pH, digestive juice flow, and the release of digestive enzymes and bile, are simulated and controlled. Several simulated models for digesting different food matrices on a small scale have been designed (Arkbåge, 2003; Avantaggiato et al., 2004; Lamothe et al., 2014; Martinez et al., 2013; Sarkar et al., 2009). Conditions such as correct pH control, maximum contact of the food with the digestive fluids and dialysis membrane, and movement of the food matrix during the digestive process should mimic human *in vivo* conditions as closely as possible.

1.12.3 The Importance of Encapsulation

A possible solution to solve the above-mentioned challenges in the case of both bioactive compounds and bioactive live organisms (and their successful incorporation into functional foods) is their encapsulation using different systems. Controlled micro- and nanodelivery systems can be used to enhance the stability of bioactive compounds within food products and during the storage period, guarantee sustainable and prolong release profile, increase the absorption rate in the small intestine, and enhance the bioavailability and bioaccessibility of bioactive food ingredients (Fernandez, 2009; Frascareli et al., 2012; Rashidinejad et al., 2014; Rashidinejad et al., 2016).

Encapsulation can also enhance the functional activities of these ingredients. For example, it is reported that the encapsulation of *L. fermentum* strain UCO-979C can increase its inhibitory effects on *Helicobacter pylori*, a principal pathogenic risk factor for gastric carcinoma (Vega-Sagardía et al., 2018). The importance of encapsulation systems in probiotic-based functional foods can be mainly attributed to their probiotic-protecting functions against food matrices and interactions with food elements, food processing conditions, and gastrointestinal fluids. For instance, an encapsulation system based on the extrusion of alginate-milk microspheres was used to improve the survivability of *L. bulgaricus* and its resistance against adverse environmental situations such as acidic conditions (i.e., pH = 2.0–2.5), high concentration of bile salt (i.e., 1.0%–2.0%) and long-term storage of 30 days (Shi et al., 2013). Moreover, it has been reported that the regular consumption of probiotic-based functional/synbiotic foods should be ~100 g per day (~10^9 viable cells), which could provide a daily dose of 10^6–10^8 cfu.mL^{-1}/cfu.g^{-1} (Hill et al., 2014). Therefore, the protection activity of encapsulation systems should be able to provide these minimum concentrations of the viable probiotic cells into the colon. In fact, encapsulation systems could potentially reduce regular daily consumption of such functional foods with the same daily delivery of viable probiotic cells, if they are well-engineered for evading probiotic-losing situations.

The materials used for the encapsulation of bioactives (known as wall materials/encapsulants/coating materials) should meet three most important requirements in order to be used in a food-grade nanodelivery system: (i) be made up of safe and food-grade components; (ii) be stable against degradation during processing and storage situations; (iii) not have adverse effects on the desired organoleptic properties of foods (Babazadeh et al., 2019; Bahrami et al., 2019; Ghanbarzadeh et al., 2016). In addition, they should meet the criteria in terms of the acceptable sensory attributes such as acceptable flavor and appearance, as well as optimum rheological properties (Ahmadi et al., 2019). Nevertheless, it appears to be a significant conflict between the data from *in vitro* investigations and those from *in vivo* studies, mostly due to the erratic bioavailability of bioactives. In addition, the stability of bioactive materials, as well as the safety of their delivery particles, remain the substantial challenges concerning the delivery of the encapsulated bioactives and their incorporation into the corresponding functional food products. Therefore, there is a need for some systematic clinical trials with reference to the administration dose, duration and interval practices, and the frequency of consumption.

1.13 CONCLUDING REMARKS

Bioactive materials are composed of two categories; bioactive compounds and bioactive live organisms. Bioactive compounds are the secondary metabolites of the plants that can improve their overall ability to survive environmental challenges. Furthermore, these compounds show health-promoting effects in humans. A wide variety of herbs, vegetables, and fruits provide a range of bioactive compounds; phenolic compounds (e.g., flavonoids, isoflavones, neoflavonoids, chalcones, flavones, flavonols, flavanones, flavanonols, flavanols, proanthocyanidins, and anthocyanins), carotenoids, vitamins (e.g., folate, vitamin C, and provitamin A), minerals (e.g., calcium, potassium, magnesium), essential oils, essential fatty acids, peptides, amino acids, and enzymes. Nowadays, bioactive compounds are widely being used in the medical, pharmaceutical, and food industries. These compounds can be incorporated into functional food products, to improve their nutritional quality, shelf life, and consumer acceptance. Moreover, these compounds possess antimicrobial, antioxidant, and flavoring properties. Most bioactive compounds have low water solubility that leads to their poor absorption in the gastrointestinal tract, and accordingly, a limited bioavailability. These compounds are also susceptible to environmental factors such as light and temperature. Therefore, encapsulation techniques such as spray drying can be very beneficial for the protection of various bioactive compounds against the above-mentioned factors, as well as their targeted delivery. Encapsulation technologies can also enhance the solubility of bioactive compounds, as well as controlling their targeted release in both food and the gastrointestinal tract.

Bioactive live organisms (i.e., probiotics) are nonpathogenic microorganisms (commensal bacterial flora) with beneficial effects on human health and disease prevention. Several mechanisms of action of these bioactive live organisms are associated with the relative prevention and/or treatment of diseases such as inflammatory bowel diseases. These effects are linked to the properties such as immunomodulation and initiation of an immune response, antimicrobial activity and suppression of bacterial growth, suppression of human T-cell proliferation, and the enhancement of barrier activity. Like bioactive compounds, bioactive live organisms are also susceptible to the environmental conditions (as well as the conditions of the upper part of the gastrointestinal tract), meaning that techniques such as spray drying encapsulation can be beneficial for the protection of their livability and efficacy.

REFERENCES

Agarwal, P. K. (2006). Enzymes: an integrated view of structure, dynamics and function. *Microbial Cell Factories*, 5(1), 2.

Agyei, D., Ongkudon, C. M., Wei, C. Y., Chan, A. S., & Danquah, M. K. (2016). Bioprocess challenges to the isolation and purification of bioactive peptides. *Food and Bioproducts Processing*, 98, 244–256.

Ahmadi, P., Tabibiazar, M., Roufegarinejad, L., & Babazadeh, A. (2019). Development of behenic acid-ethyl cellulose oleogel stabilized Pickering emulsions as low calorie fat replacer. *International Journal of Biological Macromolecules*. 150, 974–981.

Al-Fartusie, F. S., & Mohssan, S. N. (2017). Essential trace elements and their vital roles in human body. *Indian Journal of Advances in Chemical Science*, 5(3), 127–136.

Ananingsih, V. K., Sharma, A., & Zhou, W. (2013). Green tea catechins during food processing and storage: a review on stability and detection. *Food Research International*, 50(2), 469–479.

Arkbåge, K. (2003). Vitamin B12, folate and folate-binding proteins in dairy products, Doctoral thesis, Swedish University of Agricultural Sciences, Agraria, 1401–6249, 430.

Aruoma, O. I. (1998). Free radicals, oxidative stress, and antioxidants in human health and disease. *Journal of the American Oil Chemists' Society*, 75(2), 199–212.

Avantaggiato, G., Havenaar, R., & Visconti, A. (2004). Evaluation of the intestinal absorption of deoxynivalenol and nivalenol by an in vitro gastrointestinal model, and the binding efficacy of activated carbon and other adsorbent materials. *Food and Chemical Toxicology*, 42(5), 817–824.

Azad, M., Kalam, A., Sarker, M., Li, T., & Yin, J. (2018). Probiotic species in the modulation of gut microbiota: an overview. *BioMed Research International*, 2018, 1–8.

Azmir, J., Zaidul, I., Rahman, M., Sharif, K., Mohamed, A., Sahena, F., & Omar, A. (2013). Techniques for extraction of bioactive compounds from plant materials: a review. *Journal of Food Engineering*, 117(4), 426–436.

Babazadeh, A., Tabibiazar, M., Hamishehkar, H., & Shi, B. (2019). Zein-CMC-PEG multiple nanocolloidal systems as a novel approach for nutra-pharmaceutical applications. *Advanced Pharmaceutical Bulletin*, 9(2), 262.

Bahrami, A., Delshadi, R., Jafari, S. M., & Williams, L. (2019). Nanoencapsulated nisin: an engineered natural antimicrobial system for the food industry. *Trends in Food Science & Technology*, 94, 20–31.

Barrett, A. H., Farhadi, N. F., & Smith, T. J. (2018). Slowing starch digestion and inhibiting digestive enzyme activity using plant flavanols/tannins—a review of efficacy and mechanisms. *LWT*, 87, 394–399.

Bernhoft, A. (2010). A brief review on bioactive compounds in plants. *Bioactive Compounds in Plants-benefits and Risks for Man and Animals*, 50, 11–17.

Bravo, L. (1998). Polyphenols: chemistry, dietary sources, metabolism, and nutritional significance. *Nutrition Reviews*, 56(11), 317–333.

Burt, S. (2004). Essential oils: their antibacterial properties and potential applications in foods—a review. *International Journal of Food Microbiology*, 94(3), 223–253.

Burton-Freeman, B., Sandhu, A., & Edirisinghe, I. (2016). *Anthocyanins nutraceuticals* (pp. 489–500), Elsevier.

Capriotti, A. L., Cavaliere, C., Piovesana, S., Samperi, R., & Laganà, A. (2016). Recent trends in the analysis of bioactive peptides in milk and dairy products. *Analytical and Bioanalytical Chemistry*, 408(11), 2677–2685.

Chávez-González, M., Rodríguez-Herrera, R., Aguilar, C., Kon, K., & Rai, M. (2016). *Essential oils: A natural alternative to combat antibiotics resistance* (pp. 227–237), Elsevier Academic Press, London Wall, London, UK.

Chugh, B., & Kamal-Eldin, A. (2020). Bioactive compounds produced by probiotics in food products. *Current Opinion in Food Science*, 32, 76–82.

Cory, H., Passarelli, S., Szeto, J., Tamez, M., & Mattei, J. (2018). The role of polyphenols in human health and food systems: a mini-review. *Frontiers in Nutrition*, 5, 87.

Croteau, R., Kutchan, T. M., & Lewis, N. G. (2000). Natural products (secondary metabolites). *Biochemistry and Molecular Biology of Plants*, 24, 1250–1319.

De Castro, M. L., & Garcia-Ayuso, L. (1998). Soxhlet extraction of solid materials: an outdated technique with a promising innovative future. *Analytica Chimica Acta*, 369(1–2), 1–10.

De la Guardia, M., & Garrigues, S. (2015). *Handbook of mineral elements in food*. John Wiley & Sons.

Dhifi, W., Bellili, S., Jazi, S., Bahloul, N., & Mnif, W. (2016). Essential oils' chemical characterization and investigation of some biological activities: a critical review. *Medicine*, 3(4), 25.

Dorman, H., & Deans, S. G. (2000). Antimicrobial agents from plants: antibacterial activity of plant volatile oils. *Journal of Applied Microbiology*, 88(2), 308–316.

El Gharras, H. (2009). Polyphenols: food sources, properties and applications–a review. *International Journal of Food Science & Technology*, 44(12), 2512–2518.

Erdmann, K., Cheung, B. W., & Schröder, H. (2008). The possible roles of food-derived bioactive peptides in reducing the risk of cardiovascular disease. *The Journal of Nutritional Biochemistry*, 19(10), 643–654.

Fan, X., Bai, L., Zhu, L., Yang, L., & Zhang, X. (2014). Marine algae-derived bioactive peptides for human nutrition and health. *Journal of Agricultural and Food Chemistry*, 62(38), 9211–9222.

Fang, J., Sureda, A., Silva, A. S., Khan, F., Xu, S., & Nabavi, S. M. (2019). Trends of tea in cardiovascular health and disease: a critical review. *Trends in Food Science & Technology*, 88, 385–396.

Fernandez, A. T.-G., Sergio, Lagaron, Jose, Maria. (2009). Novel route to stabilization of bioactive antioxidants by encapsulation in electrospun fibers of zein prolamine. *Food Hydrocolloids*, 23(5), 1427–1432.

Fernández-García, E., Carvajal-Lérida, I., Jarén-Galán, M., Garrido-Fernández, J., Pérez-Gálvez, A., & Hornero-Méndez, D. (2012). Carotenoids bioavailability from foods: from plant pigments to efficient biological activities. *Food Research International*, 46(2), 438–450.

Fraga, C. G. (2005). Relevance, essentiality and toxicity of trace elements in human health. *Molecular Aspects of Medicine*, 26(4–5), 235–244.

Frascareli, E. C., Silva, V. M., Tonon, R. V., & Hubinger, M. D. (2012). Effect of eprocess conditions on the microencapsulation of coffee oil by spray drying. *Food and Bioproducts Processing*, 90(3), 413–424.

Galanakis, C. M. (2016). *Nutraceutical and functional food components: Effects of innovative processing techniques*. Academic Press.

Ghanbarzadeh, B., Babazadeh, A., & Hamishehkar, H. (2016). Nano-phytosome as a potential food-grade delivery system. *Food Bioscience*, 15, 126–135.

Goodman, D. S., Huang, H. S., & Shiratori, T. (1966). Mechanism of the biosynthesis of vitamin A from β-carotene. *Journal of Biological Chemistry*, 241(9), 1929–1932.

Guerin, J., Kriznik, A., Ramalanjaona, N., Le Roux, Y., & Girardet, J.-M. (2016). Interaction between dietary bioactive peptides of short length and bile salts in submicellar or micellar state. *Food Chemistry*, 209, 114–122.

Harborne, J. (1993). *Introduction to ecological biochemistry*, 4th ed.: Academic Press, Elsevier, London.

Hernández, Y., Lobo, M. G., & González, M. (2009). Factors affecting sample extraction in the liquid chromatographic determination of organic acids in papaya and pineapple. *Food Chemistry*, 114(2), 734–741.

Hill, C., Guarner, F., Reid, G., Gibson, G. R., Merenstein, D. J., Pot, B., & Salminen, S. (2014). Expert consensus document: the international scientific association for probiotics and prebiotics consensus statement on the scope and appropriate use of the term probiotic. *Nature Reviews Gastroenterology & Hepatology*, 11(8), 506.

Holzapfel, W. H., Haberer, P., Snel, J., & Schillinger, U. (1998). Overview of gut flora and probiotics. *International Journal of Food Microbiology*, 41(2), 85–101.

Jiang, H., Wang, J., Rogers, J., & Xie, J. (2017). Brain iron metabolism dysfunction in Parkinson's disease. *Molecular Neurobiology*, 54(4), 3078–3101.

Kaur, N., Chugh, V., & Gupta, A. K. (2014). Essential fatty acids as functional components of foods–A review. *Journal of Food Science and Technology*, 51(10), 2289–2303.

Kim, S., Lee, M. J., Hong, J., Li, C., Smith, T. J., Yang, G. Y., & Yang, C. S. (2000). Plasma and tissue levels of tea catechins in rats and mice during chronic consumption of green tea polyphenols. *Nutrition and Cancer*, 37(1), 41–48.

Korhonen, H. (2009). Milk-derived bioactive peptides: from science to applications. *Journal of Functional Foods*, 1(2), 177–187.

Kris-Etherton, P. M., Hecker, K. D., Bonanome, A., Coval, S. M., Binkoski, A. E., Hilpert, K. F.,… Etherton, T. D. (2002). Bioactive compounds in foods: their role in the prevention of cardiovascular disease and cancer. *The American Journal of Medicine*, 113(9), 71–88.

Krupkova, O., Ferguson, S. J., & Wuertz-Kozak, K. (2016). Stability of (–)-epigallocatechin gallate and its activity in liquid formulations and delivery systems. *The Journal of Nutritional Biochemistry*, 37, 1–12.

Lahlou, M. (2004). Essential oils and fragrance compounds: bioactivity and mechanisms of action. *Flavour and Fragrance Journal*, 19(2), 159–165.

Lamothe, S., Azimy, N., Bazinet, L., Couillard, C., & Britten, M. (2014). Interaction of green tea polyphenols with dairy matrices in a simulated gastrointestinal environment. *Food & Function*, 5(10), 2621–2631.

Liu, R. H. (2004). Potential synergy of phytochemicals in cancer prevention: mechanism of action. *The Journal of Nutrition*, 134(12), 3479S–3485S.

Liu, R. H. (2007). Whole grain phytochemicals and health. *Journal of Cereal Science*, 46(3), 207–219.

Liu, R. H. (2013). Dietary bioactive compounds and their health implications. *Journal of Food Science*, 78(s1), A18–A25.

Manach, C., Scalbert, A., Morand, C., Rémésy, C., & Jiménez, L. (2004). Polyphenols: food sources and bioavailability. *The American Journal of Clinical Nutrition*, 79(5), 727–747.

Martinez, R. C., Cardarelli, H. R., Borst, W., Albrecht, S., Schols, H., Gutiérrez, O. P., & Zoetendal, E. G. (2013). Effect of galactooligosaccharides and Bifidobacterium animalis Bb-12 on growth of Lactobacillus amylovorus DSM 16698, microbial community structure, and metabolite production in an in vitro colonic model set up with human or pig microbiota. *FEMS Microbiology Ecology*, 84(1), 110–123.

Maruyama, N., Sekimoto, Y., Ishibashi, H., Inouye, S., Oshima, H., Yamaguchi, H., & Abe, S. (2005). Suppression of neutrophil accumulation in mice by cutaneous application of geranium essential oil. *Journal of Inflammation*, 2(1), 1.

Mazur, W. M., Duke, J. A., Wähälä, K., Rasku, S., & Adlercreutz, H. (1998). Isoflavonoids and lignans in legumes: nutritional and health aspects in humans. *The Journal of Nutritional Biochemistry*, 9(4), 193–200.

Meisel, H., & Bockelmann, W. (1999). Bioactive peptides encrypted in milk proteins: proteolytic activation and thropho-functional properties. *Antonie van Leeuwenhoek*, 76(1), 207–215.

Metchnikoff, E. (1908). *The prolongation of life*. Putnam.

Mezzomo, N., & Ferreira, S. R. (2016). Carotenoids functionality, sources, and processing by supercritical technology: a review. *Journal of Chemistry*, 2016, 1–16.

Mishra, V., Shah, C., Mokashe, N., Chavan, R., Yadav, H., & Prajapati, J. (2015). Probiotics as potential antioxidants: a systematic review. *Journal of Agricultural and Food Chemistry*, 63(14), 3615–3626.

Moyler, D. (1998). *CO2 extraction and other new technologies: an update on commercial adoption*. Paper presented at the *International Federation of Essential Oils and Aroma Trades—21st International Conference on Essential Oils and Aroma's*. IFEAT, London.

Okushio, K., Matsumoto, N., Kohri, T., Suzuki, M., Nanjo, F., & Hara, Y. (1996). Absorption of tea catechins into rat portal vein. *Biological & Pharmaceutical Bulletin*, 19(2), 326–329.

Olthof, M. R., Hollman, P. C., Buijsman, M. N., Van Amelsvoort, J. M., & Katan, M. B. (2003). Chlorogenic acid, quercetin-3-rutinoside and black tea phenols are extensively metabolized in humans. *The Journal of Nutrition*, 133(6), 1806–1814.

van Oort, M. (2010). Enzymes in food technology–introduction. *Enzymes in Food Technology*, 2.

Pais, I., & Jones Jr., J. B. (1997). *The handbook of trace elements*. CRC Press.

Peto, R., Doll, R., Buckley, J. D., & Sporn, M. (1981). Can dietary beta-carotene materially reduce human cancer rates? *Nature*, 290(5803), 201–208.

Prentice, A., Dibba, B., Sawo, Y., & Cole, T. J. (2012). The effect of prepubertal calcium carbonate supplementation on the age of peak height velocity in Gambian adolescents. *The American Journal of Clinical Nutrition*, 96(5), 1042–1050.

Quirós-Sauceda, A. E., Ayala-Zavala, J. F., Olivas, G. I., & González-Aguilar, G. A. (2014). Edible coatings as encapsulating matrices for bioactive compounds: a review. *Journal of Food Science and Technology*, 51(9), 1674–1685.

Ramos, P., Herrera, R., & Moya-Leon, M. A. (2014). Anthocyanins: food sources and benefits to consumer's health. *Handbook of anthocyanins*. Hauppauge, NY: Nova Science Publishers, 373–394.

Rashidinejad, A., Birch, E. J., Sun-Waterhouse, D., & Everett, D. W. (2014). Delivery of green tea catechin and epigallocatechin gallate in liposomes incorporated into low-fat hard cheese. *Food Chemistry*, 156, 176–183.

Rashidinejad, A., Birch, E. John, Sun-Waterhouse, Dongxiao, & Everett, D. W. (2016). Effect of liposomal encapsulation on the recovery and antioxidant properties of green tea catechins incorporated into a hard low-fat cheese following in vitro simulated gastrointestinal digestion. *Food and Bioproducts Processing*, 100, 238–245.

Raut, J. S., & Karuppayil, S. M. (2014). A status review on the medicinal properties of essential oils. *Industrial Crops and Products*, 62, 250–264.

Rios, L. Y., Bennett, R. N., Lazarus, S. A., Rémésy, C., Scalbert, A., & Williamson, G. (2002). Cocoa procyanidins are stable during gastric transit in humans. *The American Journal of Clinical Nutrition*, 76(5), 1106–1110.

Rowland, I., Gibson, G., Heinken, A., Scott, K., Swann, J., Thiele, I., & Tuohy, K. (2018). Gut microbiota functions: metabolism of nutrients and other food components. *European Journal of Nutrition*, 57(1), 1–24.

Sánchez, A., & Vázquez, A. (2017). Bioactive peptides: a review. *Food Quality and Safety*, 1(1), 29–46.

Sánchez-González, L., Vargas, M., González-Martínez, C., Chiralt, A., & Chafer, M. (2011). Use of essential oils in bioactive edible coatings: a review. *Food Engineering Reviews*, 3(1), 1–16.

Sanjukta, S., & Rai, A. K. (2016). Production of bioactive peptides during soybean fermentation and their potential health benefits. *Trends in Food Science & Technology*, 50, 1–10.

Sarkar, A., Goh, K., Singh, R., & Singh, H. (2009). Behaviour of an oil-in-water emulsion stabilized by β-lactoglobulin in an in vitro gastric model. *Food Hydrocolloids*, 23(6), 1563–1569.

Schepper, J. D., Irwin, R., Kang, J., Dagenais, K., Lemon, T., Shinouskis, A., & McCabe, L. R. (2017). *Probiotics in gut-bone signaling understanding the gut-bone signaling axis* (pp. 225–247): Springer.

Shanahan, F. (2002). The host–microbe interface within the gut. *Best Practice & Research Clinical Gastroenterology*, 16(6), 915–931.

Shi, L. E., Li, Z. H., Li, D. T., Xu, M., Chen, H. Y., Zhang, Z. L., & Tang, Z. X. (2013). Encapsulation of probiotic Lactobacillus bulgaricus in alginate–milk microspheres and evaluation of the survival in simulated gastrointestinal conditions. *Journal of Food Engineering*, 117(1), 99–104.

Sikkema, J., de Bont, J. A., & Poolman, B. (1994). Interactions of cyclic hydrocarbons with biological membranes. *Journal of Biological Chemistry*, 269(11), 8022–8028.

Smith, J. E., & Berry, D. R. (1978). *Developmental mycology* (Vol. 3): Wiley.

Smith, R. M. (2003). Before the injection—modern methods of sample preparation for separation techniques. *Journal of Chromatography A*, 1000(1–2), 3–27.

Spencer, C. M., Cai, Y., Martin, R., Gaffney, S. H., Goulding, P. N., Magnolato, D., ... Haslam, E. (1988). Polyphenol complexation—some thoughts and observations. *Phytochemistry*, 27(8), 2397–2409.

Sui, X., Zhang, Y., Jiang, L., & Zhou, W. (2018). Anthocyanins in food, *Encyclopedia of Food Chemistry*, 10–17.

Tajkarimi, M., Ibrahim, S. A., & Cliver, D. (2010). Antimicrobial herb and spice compounds in food. *Food Control*, 21(9), 1199–1218.

Tang, G.-Y., Meng, X., Gan, R.-Y., Zhao, C.-N., Liu, Q., Feng, Y.-B., ... Corke, H. (2019). Health functions and related molecular mechanisms of tea components: an update review. *International Journal of Molecular Sciences*, 20(24), 6196.

Terpou, A., Papadaki, A., Lappa, I. K., Kachrimanidou, V., Bosnea, L. A., & Kopsahelis, N. (2019). Probiotics in food systems: significance and emerging strategies towards improved viability and delivery of enhanced beneficial value. *Nutrients*, 11(7), 1591.

Tsao, R. (2010). Chemistry and biochemistry of dietary polyphenols. *Nutrients*, 2(12), 1231–1246.

Tsao, R., Yang, R., Young, J. C., & Zhu, H. (2003). Polyphenolic profiles in eight apple cultivars using high-performance liquid chromatography (HPLC). *Journal of Agricultural and Food Chemistry*, 51(21), 6347–6353.

Vega-Sagardía, M., Rocha, J., Sáez, K., Smith, C. T., Gutierrez-Zamorano, C., & García-Cancino, A. (2018). Encapsulation, with and without oil, of biofilm forming *Lactobacillus fermentum* UCO-979C strain in alginate-xanthan gum and its anti-*Helicobacter pylori* effect. *Journal of Functional Foods*, 46, 504–513.

Wada, Y., & Lönnerdal, B. (2014). Bioactive peptides derived from human milk proteins—mechanisms of action. *The Journal of Nutritional Biochemistry*, 25(5), 503–514.

Wang, S., Sun, Z., Dong, S., Liu, Y., & Liu, Y. (2014). Molecular interactions between (−)-epigallocatechin gallate analogs and pancreatic lipase. *PLoS One*, 9(11), e111143.

Wolfe, K. L., Kang, X., He, X., Dong, M., Zhang, Q., & Liu, R. H. (2008). Cellular antioxidant activity of common fruits. *Journal of Agricultural and Food Chemistry*, 56(18), 8418–8426.

Xiao, J., & Hogger, P. (2015). Dietary polyphenols and type 2 diabetes: current insights and future perspectives. *Current Medicinal Chemistry*, 22(1), 23–38.

Zand, N., Christides, T., & Loughrill, E. (2015). Dietary intake of minerals. *Handbook of mineral elements in food*, John Wiley & Sons, Ltd., Oxford, 23–39.

Zhang, H., & Tsao, R. (2016). Dietary polyphenols, oxidative stress and antioxidant and anti-inflammatory effects. *Current Opinion in Food Science*, 8, 33–42.

2 Spray Drying Encapsulation of Bioactive Ingredients

An Overview of Various Aspects

Takeshi Furuta
Tottori University, Japan

Tze Loon Neoh
James Cook University, Australia

CONTENTS

2.1 INTRODUCTION

Encapsulation technology has been advanced significantly over recent decades and penetrated our daily lives through widespread application in the food and nutraceutical industries. A great number of food products can be found to rely on encapsulation technology to provide their respective unique attributes. Encapsulation is the process of enveloping active materials within a capsule made of other immiscible materials. It is further classified into subgroups based on the resultant capsule size: microencapsulation produces core material-loaded particles within the size range of 1–800 μm while nanoencapsulation produces particles between 1 nm and 1 μm (Ruiz Canizales, Velderrain

Rodríguez, Domínguez et al., 2019). The substances being encapsulated are generally active or functional compounds, which are referred to as core materials, active agents, or fills. By contrast, the capsules that form a barrier between the core materials and the outer atmosphere are normally called wall materials, carrier materials, shells, or matrices.

Microencapsulation techniques provide several beneficial and desirable functionalities such as improving the handling of active ingredients (by converting liquid to solid), stabilizing labile active ingredients throughout processing, prolonging shelf-life of final products, and masking flavors and tastes. As many food active ingredients are either in the liquid form, heat-labile, volatile, chemically reactive, or of a combination of these characteristics, microencapsulation aids in facilitating processing and formulating functional, healthy, and tasty foods (Barrow, Wang, Adhikari, & Liu, 2013). Microencapsulation technology is also employed to improve the stability, solubility, and bioavailability and mask the bitterness of bioactives (e.g., polyphenols), which are sensitive to environmental conditions to facilitate the enrichment in food products (Aizpurua-Olaizola et al., 2016; Sun-Waterhouse, Wadhwa, & Waterhouse, 2013).

In food application settings, microencapsulation of active ingredients is commonly achieved by various techniques such as spray drying, freeze drying, extrusion, complex coacervation, inclusion complexation, etc. Of all these techniques, spray drying is the most common method that has been employed for the encapsulation of food bioactive ingredients. This technique generally involves the dispersion of core material in a solution of wall material, which is then sprayed into a hot chamber. Figure 2.1 illustrates a typical process for the production of microcapsules loaded with hydrophobic bioactive compound by spray drying. Basically, a spray dryer is made up of several components, namely feed pump, atomizer or spray nozzle, air heater, air dispenser, drying chamber, and systems for exhaust air cleaning and powder recovery (Furuta, Soottitantawat, Neoh, & Yoshii, 2011). During a spray drying process, the feed is atomized or sprayed into the drying chamber where a hot stream of gas flows through. As the solvent (commonly water) evaporates rapidly, the minuscule atomized droplets transform into solid particles with the core material entrapped within the wall material matrix (Figure 2.1). The size and morphology of the atomized droplets govern the final characteristics of the powder products. The recent developments of nozzle systems are reviewed and discussed in Section 2.2. The morphology of spray dried particles is predominantly affected by the drying rate and temperature of the spray dried particles, which is another topic covered in Section 2.2.

Since the majority of food core materials are bioactive substances extracted from living organisms, they are generally hydrophobic, making it difficult to obtain a stable dispersion without phase separation throughout the spray drying operation by simply mixing these materials in the aqueous feed solutions. Hence, emulsification of the core materials in the feeds using surface-active agents

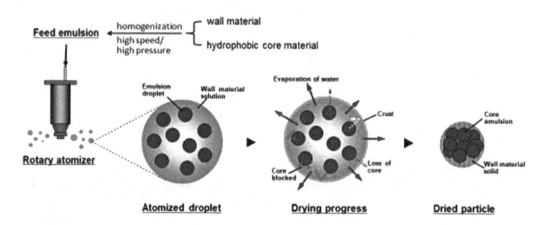

FIGURE 2.1 Spray drying microencapsulation of a model core material in an oil in water emulsion, of which the wall material is dissolved in the aqueous dispersed phase.

(i.e., emulsifiers) is normally used to kinetically stabilize the droplets of core materials by reducing the interfacial tension at the oil/water interface. In common, the more conventional techniques such as high-speed homogenization and high-pressure homogenization involve the use of a combination of mechanical forces such as turbulence, cavitation, impact, and high shear to break up larger droplets into smaller ones (Figure 2.1). There are two basic types of emulsions; oil in water (O/W) and water in oil (W/O) emulsions, as well as two other more complex types of double emulsions; water in oil in water (W/O/W) and oil in water in oil (O/W/O) emulsions. O/W emulsion refers to a system in which the hydrophobic component of the system is dispersed in the aqueous continuous phase; by contrast, in the case of W/O emulsion the aqueous phase is dispersed in the hydrophobic continuous phase. In addition, there is also another type of emulsion called 'multi-layered emulsion', which can be constructed by depositing polyelectrolyte onto the surface of emulsion droplets via the formation of covalent complexes with the oppositely charged emulsifier. These topics are discussed in Section 2.3.

The selection of wall material is an important part of the microencapsulation process. For food applications, the wall materials are restricted to food-grade materials approved by authorities such as the US Food and Drug Administration (FDA) and European Food Safety Authority (EFSA), or certified as generally recognized as safe (GRAS). Depending on the physicochemical characteristics of the core material and the purpose of encapsulation, the wall material of choice for spray drying microencapsulation should at least fulfill these criteria: i) excellent protection barrier; ii) high aqueous solubility, and; iii) good film-forming property (Section 2.4). Ubbink and Krüger (2006) proposed an alternative approach for the selection of encapsulation materials, based on the fundamental scientific understanding of the target application over the more established method of trial and error. The wall materials known to be suitable for spray drying microencapsulation for food applications are predominantly carbohydrate polymers (e.g., maltodextrins, octenyl succinic anhydride-modified starches, gum Arabic, carrageenans, carboxymethyl cellulose, soluble soybean polysaccharide, and chitosan) and proteins (e.g., gluten, whey proteins, and gelatin) (Wandrey, Bartkowiak, & Harding, 2010).

Spray dried particles generally have the matrix type morphology, meaning that the core material is spread over the matrix of wall material either in the form of multiple small droplets or at a smaller, more homogenous, or even the molecular level (Figure 2.1). The state in which the core ingredients are present inside the particles is essentially important for food manufacturing and this will be discussed in Section 2.5. Concerning the structure of microcapsules, the storage stability and release of microencapsulated core materials are also important aspects of microencapsulation. Numerous studies have reported the oxidative stability and release of core ingredients into fluids (including simulated gastrointestinal fluids) and the atmosphere. The oxidative stability of encapsulated oil(s) seems to be closely related to the proportion of core material present near the outer surface of the particles. The techniques and protocols for the quantification of these surface oils are discussed in Section 2.6.

2.2 ATOMIZATION OF FEED LIQUID AND DRYING MECHANISM OF ATOMIZED DROPLETS

2.2.1 Atomization Techniques

Atomization is the key process in spray drying that determines the particle size distributions of the droplets and hence the powders, the structure of microparticles, and the encapsulation efficiency (EE). EE is generally defined as the percentage of total added core material encapsulated within the spray dried microcapsules. Atomization of feed liquid, which can be a solution, dispersion, or emulsion, is accomplished by applying energy onto the bulk of feed to transform it into fine droplets. The common types of atomizers used in spray drying for food applications are rotary atomizers and high-pressure spray nozzles. Ultrasonic nozzles, although relatively less common, have also been reported as a viable atomization technique for spray drying of food products (Turan, Cengiz, & Kahyaoglu, 2015; Turan, Cengiz, Sandıkçı, Dervisoglua, & Kahyaoglu, 2016).

Liu, Chen, and Selomulya (2015) covered in their review the production of uniform microparticles using an in-house developed microfluidic jet spray drying technique. The microparticles generated had tightly-controlled characteristics and sizes. The technique was claimed to be useful for correlating the effects of formulations and spray drying conditions with the physicochemical properties and functionalities of spray dried microparticles. The reviewed studies demonstrated that the understanding of the precursor's formulations and properties and the spray drying conditions could be leveraged to manipulate the particle size, morphology, and microstructure of spray dried microparticles. Alternatively, in the development of bench-top spray dryers by manufacturers such as Büchi (Switzerland), two-fluid nozzles are commonly used for the atomization of feed liquid into fine droplets due to the short residence time of the sprayed droplets inside the drying chamber.

Turan et al. (2015) utilized an ultrasonic nozzle in the production of spray dried microspheres of blueberry juice and extract. The size distributions of the microspheres and the characteristics of the powders were compared with those prepared by spray drying with a conventional pressure, rotary, or two-fluid nozzle. The ultrasonic atomization produced powders with relatively higher total phenolic content, antioxidant activity, and anthocyanin content than the spray dried powders obtained by the conventional atomization techniques. Furthermore, the microspheres were significantly uniform in size with a smooth surface. The ultrasonic atomization technique was proven beneficial for spray drying microencapsulation of blueberry extract. Kondo, Niwa, and Danjo (2014) developed a three-fluid nozzle with a three-layered concentric structure consisted of inner and outer liquid passages and an outermost gas passage. The nozzle was applied for the production of a powder that has a sustained-release property. The model drug supplied via the center nozzle was successfully encapsulated in ethylcellulose supplied through the concentric peripheral nozzle, resulting in microcapsules with the sustained-release property. The authors concluded that the three-fluid nozzle would be useful as a novel microencapsulation method, thanks to the unique structure of the nozzle that allowed separate supplies of core and wall materials. This type of three-fluid nozzle has also been studied for the preparations of therapeutic microparticles by Sunderland, Kelly, and Ramtoola (2015) and Wan, Maltesen, Andersen, et al. (2014a, 2014b).

Kašpar, Jakubec, and Štěpánek (2013) attempted to produce chitosan microparticles cross-linked by tri-polyphosphate (TPP) using two- and three-fluid nozzles. A novel *in situ* cross-linking method was employed, by which chitosan and TPP were fed through the different channels of the three-fluid nozzle, and cross-linking occurred within the atomized droplets. The resultant particles were suitable as a microencapsulation vehicle as they permitted the use of higher chitosan concentrations with favorable stability in aqueous media. Liu, Chen, Cheng, and Selomulya (2016) used a pilot-scale spray dryer equipped with a microfluidic aerosol nozzle for microencapsulation of curcumin in whey protein isolate (WPI) to enhance the solubility of curcumin. They managed to produce uniform spray dried particles of whey protein isolate-curcumin complexes, achieving an 11,355-fold increase in curcumin solubility compared with the raw crystals. The monodispersed nozzle was further applied for the microencapsulation of fish oil in WPI. Thermal pre-treatment for cross-linking the WPI before spray drying demonstrated an improvement in particle properties; e.g., high docosahexaenoic acid (DHA) oil content, better antioxidation property, and slow release of encapsulated oil in simulated gastric fluids (SGF) (Wang, Liu, Chen, & Selomulya, 2016).

Munoz-Ibanez, Azagoh, Dubey, Dumoulin, and Turchiuli (2015) investigated the breakup of dispersed droplets during the atomization of O/W emulsions to identify the parameters governing droplet breakup in emulsions. They found that the most significant parameters were the rotational speed of the wheel of the rotary atomizer and the gas flow rate of the two-fluid nozzle. These researchers also indicated that the median diameters of the atomized droplets could be expressed as a function of the Capillary number and that the dispersed droplet breakup occurred only above critical Capillary numbers for different atomization techniques; the capillary number (Ca) is defined as $\mu V/\sigma$ (μ: viscosity, V: velocity, σ: surface tension), which is the ratio of the inertia of the fluid to its surface tension and can be applied to estimate the criterion of the droplet breakup in atomizing the emulsion systems. Lin and Phan (2013) studied a numerical simulation of the air-liquid two-phase flows both inside and around a rotating disk atomizer. The atomizer was installed with multiple spray nozzles.

They found that some of the nozzles could reach flooding conditions at low rotating disk speeds and feed flow rates when feed liquid was ununiformly distributed through the distributors.

2.2.2 Understanding the Drying Mechanism of Droplets

Different approaches have been adopted by researchers in the field of spray drying to elucidate the drying mechanism of sprayed droplets. Single droplet drying technique was used to monitor the transient change in shape, temperature, and mass of a droplet during drying. The researchers at Monash University in Australia have developed a new bench-top apparatus for drying of single suspended droplets, and have conducted extensive studies on the drying process with the equipment (Fu, Woo, Lin, Zhou, & Chen, 2011). These researchers (Khem, Woo, Small, Chen, & May, 2015) employed the same technique to determine the protective mechanism of WPI, trehalose, lactose, and skim milk on *Lactobacillus plantarum* A17 by monitoring cell survival, droplet temperature, and corresponding mass changes. Fu, Woo, and Chen (2011) adopted the glass-filament single droplet drying technique to study the drying kinetics of a single milk droplet and extended the technique for observing the dissolution behavior of the dried single particle. Material migration within the droplet during the drying process was studied based on the observed wetting and dissolution behavior. The technique has also been used to determine the changes in temperature, moisture content, and diameter of skim milk droplets with 50 wt% initial solids to study the drying kinetics (Fu, Woo, Selomulya, et al., 2012). The correlation of the empirical data with the master activation-energy curve using the Reaction Engineering Approach (REA) was found to give a good description of the drying histories, supporting the suitability of REA for the interpretation of the drying behavior of skim milk with high initial solid concentrations.

The same technique was further employed by Wang, Che, Selomulya, and Chen (2014) to investigate the droplet drying behavior of DHA-containing emulsions with different core concentrations and types of wall materials during convective drying. The authors proved that REA could be successfully applied to describe the drying time course (mass history) of the droplet for different formulation systems. The model resulted from the correlation of the data (changes in droplet size, moisture content, and droplet/particle temperature) by REA has been demonstrated to accurately describe the drying behavior of different systems. Chew et al. (2014) conducted a study to understand the drying and rehydration behaviors of milk protein concentrate by the same technique and approach. Chew and co-worker (Chew, Fu, Gengenbach, Chen, & Selomulya, 2015) later conducted a similar study on emulsions modeled after compositions of human milk at various lactation periods. Surface composition was determined qualitatively by X-ray photoelectron spectroscopy (XPS) for comparison between the particles produced by the glass-filament single droplet drying and the spray dried powders obtained by single-stream drying using a microfluidic spray dryer to determine the effects of process conditions. While the REA was capable of describing the droplet drying kinetics of many materials, they identified the inadequacy of the approach to predict the particle formation process during drying. Both studies (Chew et al., 2014) pointed to the critical influence of bulk and surface compositions on particle formation throughout drying.

Tian, Fu, Wu, et al. (2014) adopted the single droplet drying approach and dissolution tests of spray dried milk powders produced with a microfluidic jet spray dryer to gain insight into the drying mechanisms. They investigated the process of shell formation on milk particles during drying and the wettability of the particles obtained by spray drying of milk feeds with different initial solid contents. The addition of surfactants (Tween 80 or lecithin) was shown to limit the effectiveness in improving the wettability of spray dried milk powders from high-solid milk feeds (>20 wt%), presumably due to rapid shell formation and high feed viscosity that could hinder the redistribution of free surfactant molecules on the milk droplet surface.

Besides glass-filament single droplet drying, a single droplet can also be suspended by acoustic levitation for the characterization of droplet behavior during drying. In 1978, a group of researchers from Kyoto University (Japan) developed a novel technique for floating a single droplet in the air to study the drying behavior of the levitated droplet in order to avoid the deformation that occurs in filament-suspended droplets near the end of drying (Toie, Okazaki, & Furuta, 1978). They succeeded for the first time in obtaining the drying data of skim milk and aqueous NaCl throughout the entire drying process. Subsequently, they carried on with the accurate measurement of droplet temperature and the establishment of a correlation for the heat transfer coefficient around the droplet using the same technique (Toei, Furuta, & Okazaki, 1982). Briefly, these researchers devised an acoustic levitator using a cylindrical hollow ultrasonic lead titanate-zirconate transducer and a piezoelectric element to generate a field of the standing wave inside the hollow of the transducer. The droplet (approximately 2 mm in diameter) was placed at the node of the standing wave and floated stably in the air against gravity. A carbon dioxide laser was used to supply the heat for drying the levitated droplet. The changes in the morphology of the droplet during drying were captured with a high-speed camera and the temperature was recorded with a combination of an infrared thermometer and a high-speed recorder. They successfully recorded the contraction and expansion of the droplet during the drying process, from which the drying rate was determined. The major limitation of this particular technique was that drying could only be performed within a narrow temperature range. Zaitone and Lamprecht (2013) adopted the technique of acoustic levitation single droplet drying to simulate the full-scale spray drying microencapsulation process for investigating the drying kinetics and the particle formation mechanism during drying. This technique offered easier adjustment and better control of formulation variables and process parameters.

Nijdam, Kachel, Scharfer, Schabel, and Kind (2015) measured the water content profiles within thin aqueous films of lactose and bovine serum albumin (BSA) during drying using inverse microscope Raman spectroscopy to understand component diffusion within droplets during spray drying. The study elucidated the underlying mechanisms of BSA accumulation that occurred only within the depth of 10 nm from surfaces of dried films but not in the bulk of the film - a phenomenon which has also been observed in spray dried particles (Faldt & Bergenstahl, 1994). No segregation of lactose and BSA was observed in the films dried at 80°C even though the solute-fixed coordinate system drying model predicted that there were significant water content gradients within the films. The findings cast doubt on elucidating solute segregation by diffusion mechanism, based on which the authors proposed that BSA accumulation was instead driven by the surface activity of BSA.

Ivey and Vehring (2010) developed two models based on theoretical descriptions to accelerate the formulation and process development. The first one was a thermodynamic model based on the global energy and mass balance of the spray drying process for the prediction of the outlet air temperature and the residual water content of powders. The second one utilized a stochastic model based on a Monte-Carlo simulator to predict the aerodynamic size distributions of complex spray dried microparticles containing a discrete solid phase. The measured mass median aerodynamic diameters were well simulated by using the stochastic drying model. Both models allowed rank-ordering of formulation and process variables by the relative importance and were proven to aid the understanding of the spray drying unit operation. Boraey and Vehring (2014) modeled drying processes with large Péclet numbers, in which the droplet surface recession rate was much faster than the diffusion in the droplets; the Péclet number, written as *Pé* (*LV*/α for heat transfer and *LV*/*D* for mass transfer, *L*: characteristic length, *V*: velocity, α: thermal diffusivity, *D*: diffusivity) is a dimensionless number used to estimate the significance between the axial dispersion and the bulk flow of heat and/or species. The predicted dried-particle diameter, particle density, and aerodynamic diameter by the model were in good agreement with numerical and experimental results found in the literature.

2.3 EMULSION TYPES

In encapsulation of hydrophobic core materials such as fish and plant oils by spray drying, the core material requires to be emulsified in the aqueous solution of wall material, usually a carbohydrate and protein solution as if the core material is 'dissolved' homogeneously in the liquid. As illustrated in Figure 2.2, there are various types of emulsions: primary emulsions (O/W and W/O) and double

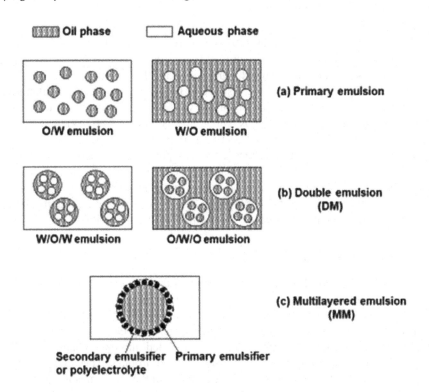

FIGURE 2.2 Schematic illustrations of different types of emulsions. (a) Primary oil-in-water (O/W) and water-in-oil (W/O) emulsions, (b) Double water-in-oil-in-water (W/O/W) and oil-in-water-in-oil (O/W/O) emulsions, and (c) Multi-layered (layer-by-layer) emulsion.

emulsions (W/O/W and O/W/O). The emulsifier layer formed at the interface between the dispersed and continuous phases may have either a single- or a multi-layer structure, with the single layer more common. The emulsions of which the dispersed droplets are surrounded by multi-layered emulsifiers are called 'multi-layered emulsions'. Cross-linking and thermal treatment of emulsifiers are alternative methods for increasing the emulsifying ability. Besides the structure of emulsion, the size distribution of droplets also affects the characteristics of the resultant microparticle, particularly the recovery of the core material during spray drying and storage, as well as its release characteristics during the gastrointestinal digestion tract.

2.3.1 W/O/W Emulsions

The preparation of W/O/W emulsions involves a two-step process: first, an aqueous core is dispersed and emulsified with a hydrophobic emulsifier to obtain a primary W/O emulsion, and then this W/O emulsion is further dispersed in a second aqueous phase (Figure 2.2). Colloid mills, rotor-stator systems, and high-pressure homogenizers are often used as emulsification techniques.

Several groups of researchers have used W/O/W double emulsions to prepare microparticles encapsulating different core materials. Maisuthisakul and Gordon (2012), Lee, Ganesan, Baharin, and Kwak (2015), and Berendsen, Güell, and Ferrando (2015) conducted spray drying of W/O/W double emulsions of extracts of mango, peanut sprout, and procyanidin, respectively. The primary W/O emulsions were prepared in soybean oil, medium-chain triglycerides (MCT), and sunflower oil, respectively, with polyglycerol polyricinoleate (PGPR) as a hydrophobic emulsifier. The outer phases were combinations of aqueous gum Arabic (GA), maltodextrin (MD), and sodium alginate solutions (Maisuthisakul & Gordon, 2012), aqueous whey protein concentrate (WPC) or MD solution (Lee et al., 2015), and WPI combined with carboxymethylcellulose (CMC), GA or chitosan (Berendsen et al., 2015). Berendsen et al. (2015) prepared the secondary emulsions by membrane emulsification and extensively measured the droplet sizes at different stages. Response surface methodology (RSM) was applied by Maisuthisakul and Gordon (2012) for the optimization of the composition of the outer (or continuous) phase. Esfanjani, Jafari, Assadpoor, and Adeleh Mohammadi (2015) prepared a primary W/O emulsion of saffron water extract in sunflower oil using Span 80, from which two kinds of W/O/W emulsions (single-layered stabilized by WPC alone and double-layered stabilized by complex of WPC and pectin) were prepared. Drusch, Hamann, Berger, et al. (2012) measured the accumulation of surface-active compounds on the surface of the droplet in a time interval relevant for the possible change of surface composition. They modified a commercial contact angle meter to enable monitoring of the change in surface tension within an extremely short time comparable to droplet drying in a spray dryer. The hydrolyzed milk proteins possessed high surface activity which could alter the surface composition of the spray dried particles.

2.3.2 MULTI-LAYERED EMULSION

Jiménez-Martín et al. (2015a, 2015b, and 2016) studied the microencapsulation of fish oil using both double- and multi-layered emulsions. For the preparation of the multi-layered emulsion, a primary O/W emulsion was first prepared by mixing fish oil and lecithin (emulsifier) overnight, adding an acetic acid solution to the mixture, and then homogenizing the mixture. To obtain the multi-layered emulsion, chitosan in acetic acid solution was blended into the prepared primary emulsion to form a lecithin-chitosan complex on the surface of the oil droplets. They evaluated the effect of the two different types of emulsions on the fatty acid profile of the microencapsulated fish oil during microencapsulation and storage (Jiménez-Martín et al., 2016) and the physicochemical properties and oxidative stability of the microcapsules (Jiménez-Martín et al., 2015a). In a separate study, they compared the influence of monolayered and multi-layered fish oil emulsions and also investigated the impact of chitosan concentration in the multi-layered emulsion on storage stability, microencapsulation yield, and EE (Jiménez-Martín et al., 2015b).

A similar study was conducted by Carvalho, Silva, and Hubinger (2014) on multi-layered emulsion of green coffee oil stabilized by lecithin and chitosan. Multi-layered emulsion of Miglyol 812 N stabilized by pea protein isolate (PPI)/pectin complex was applied for the encapsulation of the MCT and compared with that stabilized by PPI alone, in regard to the enhancement of emulsion stability during spray drying (Gharsallaoui et al., 2010) and improvement of flavor protective properties (Gharsallaoui et al., 2012). Lim and Roos (2016) examined the differences between single-layered and multi-layered emulsions in the ability to protect spray drying microencapsulated carotenoids during storage. Serfert, Schröder, Mescher, et al. (2013) investigated the impacts of atomization and drying on the functionality of spray dried β-lactoglobulin single-layered emulsion and multi-layered emulsion with bilayer interface consisting of the globular β-lactoglobulin and pectin with different degrees of methoxylation. Single droplet drying by acoustic levitation revealed that the emulsions had similar drying behavior regardless of how they were stabilized. The relatively more pronounced upshift in the droplet size distribution of reconstituted single-layered emulsion indicated a better stabilization effect of the bilayer interface against droplet coalescence during spray drying. The β-lactoglobulin-low methoxylated pectin bilayer rendered the best protection against oxidation to the fish oil in both liquid and spray dried states.

Aberkane, Roudaut, and Saurel (2014) also evaluated the ability of PPI/high methoxyl pectin bilayer interface to improve the stability of polyunsaturated fatty acid (PUFA)-rich oil emulsion during spray drying and the oxidative stability of the spray dried emulsion during storage in comparison with that of PPI-single layer interface. In contrast to the results reported by Serfert et al. (2013), a more distinct increase in droplet size of the reconstituted double-layered emulsion indicated that oil droplet coalescence was more likely to occur in the double-layered emulsion during spray drying. However, better protection of spray drying microencapsulated PUFA-rich fish oil emulsion against oxidation by the PPI/pectin bilayer interface was observed, which was in agreement with that observed by Serfert et al. (2013).

2.3.3 Cross-linking of Emulsifiers

Cross-linking of protein-based emulsifiers by biochemical enzyme reaction or heat treatment provides another means to enhance the emulsifying ability of proteins. Bao et al. (2011) achieved cross-linking of sodium caseinate (NaCas) by microbial transglutaminase (MTGase). They reported that the emulsifying properties of NaCas could be improved by cross-linking, but extensive cross-linking was found to decrease the emulsifying capacity. Similarly, bovine and caprine caseins were cross-linked using MTGase by Mora-Gutierrez, Attaie, Kirven, and Farrell Jr (2014) for the micro-encapsulation of algae oil, resulting in the extended oxidative stability of the spray dried algae oil emulsions. The effects of modification of protein-based interfacial layer through enzymatic cross-linking on various microencapsulated lipophilic core materials (e.g., sunflower oil (Damerau et al., 2014), squalene oil (Ghani, Matsumura, Yamauchi, et al., 2016), and fish oil (Pourashouri et al., 2014)) have been studied. In contrast to Bao et al. (2011) and Mora-Gutierrez et al. (2014), these studies reported that the cross-linking of interfacial protein layer (NaCas (Damerau et al., 2014, Ghani et al., 2016) and fish gelatin (Pourashouri et al., 2014)) showed either minor or insignificant improvement in oil stability (Damerau et al., 2014), retention (Ghani et al., 2016), EE, and surface oil content (Pourashouri et al., 2014).

Kašpar et al. (2013) proposed a new technique for the preparation of cross-linked chitosan microparticles by TPP anions using a three-fluid nozzle, in which two different *ex-situ* and *in-situ* cross-linking methods have been proposed as described in Section 2.2. Ribeiro, Laurentino, Alves, et al. (2015) investigated the cross-linking modification of GA with different concentrations of sodium trimetaphosphate (STMP) to obtain an efficient emulsifier. A decrease in viscosity was observed in the modified GA with the increase of STMP concentration. The EE of the studied essential oil was increased to 97% by cross-linking of GA with 6% STMP from the 85% observed in the microcapsules of unmodified GA.

2.3.4 Thermal Treatment for Cross-linking and/or Maillard Reaction

Mezzenga and Ulrich (2010) reported a novel method for the production of solid oil powders with high oil content, by which the protein (β-lactoglobulin) monolayer that adsorbed at the oil–water interface in a primary protein-stabilized O/W emulsion was cross-linked by heat treatment. Unlike ordinary spray drying microencapsulation, the emulsion was spray dried without the addition of hydrocolloids as wall material and the entrapment of oil was achieved solely through the physical strength of the cross-linked protein monolayer. Analogous to the aforementioned study, Wang et al. (2016) employed a thermal treatment for cross-linking of WPI to produce a protein-stabilized oil–water interface before spray drying and successfully produced microcapsules with ultra-high oil content (83%–90%). Serfert, Lamprecht, Tan, et al. (2014) also applied a thermal treatment at 90°C under acidic conditions to β-lactoglobulin from WPI to obtain fibrillar β-lactoglobulin for microencapsulation of fish oil. The fibrillar β-lactoglobulin showed improved emulsifying activity, EE, and oxidative protection compared with native WPI.

Li et al. (2015) and Tang and Li (2013) prepared soy protein isolate (SPI)-GA and SPI-lactose conjugates, respectively, as novel emulsifiers using a dry-heated Maillard reaction. The SPI-GA and SPI-lactose conjugates exhibited higher emulsifying properties. The emulsion of lycopene prepared with the SPI-GA conjugates was of smaller sizes than those prepared with the mixture of SPI and GA (Li et al., 2015). The thermal pre-treatment of SPI with lactose improved the retention efficiency and thermal stability of the spray dried soy oil emulsion (Tang & Li, 2013). Augustin et al. (2015) showed that a combination of pH increase (pH 6.4 → pH 7.5) and high-heat treatment (90°C) improved the encapsulating performance of buttermilk to produce omega-3 powders. The enhancement can presumably be partially attributed to the formation of Maillard reaction products in the protein-carbohydrate-based encapsulating material (buttermilk and glucose syrup), since the Maillard reaction has been shown to improve the oxidative stability of omega-3 oil powders (Sanguasri & Augustin, 2001).

2.3.5 MISCELLANEOUS METHODS FOR THE PREPARATION OF EMULSIONS

Nakagawa and Fujii (2015) conducted a study on spray drying of lipid cores that dissolve β-carotene. Electrostatic complexation of SPI and GA was induced on the surface of the oil droplets by a freeze-treatment below the isoelectric point of the SPI. The slow freezing conditions were shown to enhance the EE while reducing the surface oil content. In addition, the freezing treatment also increased the mass transfer resistance of the shell matrix, resulting in delayed release of the β-carotene dissolved in the lipid core. While high-pressure homogenizers are often used as major tools in homogenization processes, Penbunditkul et al. (2012) reported that larger oil droplets were observed in a bergamot oil emulsion stabilized by octenyl succinic anhydride-modified starch (OSA-modified starch) after the emulsion was passed through a high-pressure homogenizer and the carboxylate and ester carbonyl groups of OSA were lost with the increasing number of homogenization cycles. For stable atomization, the viscosity of feed emulsion directly affects the shape and size of the emulsion droplets during atomization. Laine, Toppinen, Kivelä, et al. (2011) prepared stable and low-viscosity emulsions using acid-modified oat bran (MOB). The influence of homogenization pressure and time and the concentration of MOB on the stability, viscosity, and oil droplet size of the emulsions were studied. The concentration of MOB was found to significantly affect the mentioned emulsion properties. While the influences of homogenization pressure and time were insignificant, homogenization pressure did decrease the emulsion viscosity.

2.4 WALL MATERIALS

There have been several published works reporting on new wall materials with excellent encapsulation ability, when compared with conventional choices such as MD, GA, and OSA-modified starches. As an alternative to GA as wall material, Sarkar and Singhal (2011) developed novel wall materials by esterification of guar gum hydrolysate (GGH) with oleic acid and OSA. With a good emulsifying property (measured by turbidity of emulsions), the OSA ester of GGH was particularly regarded as a promising alternative for GA for microencapsulation. They further applied these OSA and oleate esters of GGH as wall materials to encapsulate mint oil and compared the encapsulation ability against GA and GA-OSA (Sarkar, Gupta, Variyar, Sharma, & Singhal, 2013). Both of the modified GGH showed favorable flavor encapsulation ability with emulsifying activity. Meanwhile, the authors also investigated the blending of radiation or enzymatically depolymerized guar gum (GG) with GA as wall materials for the microencapsulation of mint oil (Sarkar, Gupta, Variyar, Sharma, & Singhal, 2012). As the high viscosity of aqueous GG solution presents a major hindrance to its application in flavor encapsulation, irradiation was employed to partially depolymerize GG to decrease the viscosity. Substituting one-tenth of GA with radiation-depolymerized GG exhibited increased stability of the encapsulated mint oil with no significant alteration in emulsion characteristics as opposed to the decreased mint oil stability for the substitution with commercial partially enzymatically depolymerized GG. Irradiation of GG powder was comparable to that performed on aqueous GG solution although the former required 10 times as high an irradiation dose.

Wang, Di, Liu, and Wu (2013) extracted the polysaccharides from *Mactra veneriformis* (MVPS) and applied it as an encapsulation carrier of metformin hydrochloride. Smooth and spherical microparticles were obtained by adding chitosan to MVPS. The size of the microspheres was about 1–10 μm, which has been suggested to be appropriate for nasal pharmaceutical applications. Tatar, Tunç, Dervisoglu, Cekmecelioglu, and Kahyaoglu (2014) developed a hemicellulose-based coating isolated from economical corn wastes and examined the effectiveness of the new material for microencapsulation of fish oil. Pai, Vangala, Ng, Ng, and Tan (2015) investigated the potential use of resistant maltodextrin (RMD) as an encapsulating shell material for naringin, the flavonoid found in grapefruit. Aqueous solubility of naringin increased at higher RMD: naringin ratios and spray drying temperatures. The solubility enhancement was attributed to the stabilized dispersion of smaller amorphous particles of naringin in the RMD matrix. Inulin, a prebiotic material, has been evaluated as a wall material for its potential to encapsulate and control the release of oregano essential oil (Costa, Duarte, Bourbon, et al., 2013). The authors ascribed the difference in the release profile of microencapsulated oregano oil to the dissimilar properties of the wall matrix of inulin.

2.5 STRUCTURE OF SPRAY DRIED PARTICLES

The microstructure of spray dried particles is, without doubt, a critical factor that dictates the release characteristics of microencapsulated core materials and the storage stability and solubility of the microcapsules. Several advanced technologies such as scanning electron microscopy (SEM), confocal laser scanning microscopy (CLSM, including confocal Raman microscopy), XPS, atomic force microscopy (AFM), and positron annihilation lifetime spectroscopy are typically used to determine the external and internal structure of spray dried microparticles.

Porras-Saavedra, Palacios-González, Lartundo-Rojas, et al. (2015) prepared spray dried powders using blends of SPI, GA, and MD at various inlet air temperatures. They studied the relationship between microstructural developments, the distribution of elemental components, and characteristics related to the functional properties of the powders. The elemental atomic composition of the microparticle's surface revealed greater compositions of nitrogen in the external surfaces of microparticles of SPI, MD/GA, SPI/GA, and SPI/MD/GA. The physical characteristics of spray dried particles of NaCas and casein hydrolysate encapsulating fish oil were reported by Drusch, Serfert, Berger, et al. (2012). The spray dried carrier matrix was characterized by helium pycnometry (true density of spray dried particles), nitrogen displacement (surface area), XPS (the surface composition of wall material matrix of spray dried particles), oscillating drop tensiometry (surface/interfacial elasticity), and positron annihilation lifetime spectroscopy (free volume elements). They found that the surface coverage by protein was always higher in the particles containing hydrolyzed casein compared with the particles containing NaCas. In addition, with increasing NaCas and hydrolyzed casein content, the size of free volume increased, facilitating the autoxidation of the encapsulated core material.

Peres et al. (2011) produced nanoparticles of epigallocatechin gallate (EGCG) with a blend of GA and MD, and investigated the internal structure and density of the particles by mercury porosimetry and helium pycnometry, respectively. The pore size distribution of the spray dried particles showed that the pore diameters ranged from 10 to 100 nm, with an average diameter of 24 nm. The AFM images of the EGCG-loaded particles (EGCG/P) showed that the spherically shaped particles measured approximately 370 nm in diameter. The dynamic light scattering (DLS) measurements of the EGCG/P revealed that the average size of the emulsion was 400 nm, which was comparable with the size observed by AFM. To elucidate the conservation of properties of protein entrapped within carbohydrate matrices by spray drying, Peres et al. (2012) investigated the interactions between collagen hydrolysate (CH) and carbohydrates (MD and GA) in the spray dried particles. The CH-loaded particles (CH/P), as observed under a scanning electron microscope, had a smooth surface with no visible dents differing from the blank particles without CH, which were spherical and extensively dented. In addition, the SEM micrographs also showed that the CH/P particles had a hollow core and the particle size was larger with a thinner wall as the content of CH increased. Diffusion coefficients in solid-state measured with the nuclear magnetic resonance (NMR) technique revealed that the diffusion coefficient of CH encapsulated in carbohydrates was 10%–20% smaller than that of free CH.

Maher, Auty, Roos, Zychowski, and Fenelon (2015) examined lactose crystallization behavior in spray dried powders of coarse (conventional) and fine (nano-sized) emulsions of sunflower oil emulsified with the mixture of lactose and NaCas. Lactose in the powder particles of the nanoemulsions was found to crystallize more rapidly than that of the conventional emulsions. The isotherms and rates of lactose crystallization were successfully calculated with the Avrami equation, indicating three-dimensional crystal formation; the Avrami equation is applied essentially to express the kinetics of crystallization process. The equation is also often applied to describe or simulate the changes in materials, like chemical reaction rates. The transformation described by the Avrami equation has a characteristic s-shaped (or sigmoidal) profile. Polarized light microscopy, CLSM, and cryo-SEM were employed to inspect the internal structure of the particles and the micrographs showed that the rate of lactose crystal formation for the nanoemulsion powders was higher and the small fat globules were distributed more evenly inside the powder particles of spray dried nanoemulsions. Vicente, Pinto, Menezes, and Gaspar (2013) proposed a rational methodology to control the properties of spray dried particles, including size and morphology. To predict the particle size, they proposed an estimation model that encompassed a mechanistic atomization model and a constant evaporation rate model. The experimental particle sizes were compared with the predicted ones and strong correlations were obtained. Besides, the spray dried particles with different morphologies (e.g., shriveled, spherical, or inflated shapes) could also be obtained by manipulating the solids concentration, outlet temperature, and droplet size. The proposed methodology has allowed the production of particles with different sizes and morphologies in a controlled manner.

Sadek, Li, Schuck, et al. (2014) studied the extent to which compositional differences in protein (ratio of WPI: native phosphocaseinate (NPC)) independently influenced the morphology and functionality of the resultant spray dried powders. They used two types of drying techniques; monodispersed droplets atomized in a concurrent airflow (inlet and outlet temperatures of 193.5°C and 108.4°C, respectively) by a piezoelectric ceramic nozzle (MDSD), and a single pendant droplet deposited onto a plate surface and dried in the still air of a low temperature and humidity. The WPI particles obtained by MDSD were spherical with a smooth outer surface and had an open hollow structure. By contrast, the NPC particles were very wrinkled and dense. The sphericity and compactness indexes of the particles and the bulk density and compressibility of the powders were investigated at different WPI: NPC ratios. The particles were classified as having three well-defined types of morphology including: 1) smooth, spherical, and hollow; 2) wrinkled and dense; and, 3) hybrid-shaped, which were governed by the type of protein rather than the drying kinetics.

O/W emulsion of sunflower oil was spray dried to encapsulate the oil within a matrix of MD and GA and the properties of the spray dried particles were examined using various techniques such as low-vacuum SEM (LVSEM) for morphology, confocal Raman microscopy for distribution of components in the spray dried particles, and XPS for particle surface chemistry and surface oil coverage (Munoz-Ibanez et al., 2016). Besides, the amount of surface oil was quantified by extraction using petroleum ether. Munoz-Ibanez et al. (2015) previously reported on the occurrence of oil droplet breakup in emulsions, changing the oil droplet size distribution as a result of high shear stress during atomization. Through observation of the internal distribution of components by confocal Raman microscopy, oil droplets of different sizes resulting from oil droplet breakup due to high-speed atomization of the coarse emulsion could be seen distributed at or close to the surface of the spray dried particles. By contrast, more evenly sized oil droplets were found distributed below a well-defined layer of GA in the particles from the coarse emulsion atomized at low speed. The protecting layer of GA was also observed in the particles from the fine emulsion. The study showed that oil droplet breakup did influence the amount of surface oil in the spray dried particles by pushing the oil droplets in an atomized emulsion towards the surface.

Nuzzo, Overgaard, Bergenståhl, and Millqvist-Fureby (2017) employed dryers of varying scales (from single particle to laboratory scale to pilot scale to full scale) to produce whole milk powders for elucidating the influences of spray drying scale-up on the morphology and composition of the whole milk particles using LVSEM, confocal Raman microscopy, and XPS. The surface composition of the particles determined by XPS revealed that the surface coverages by protein and lactose increased in the order of full-scale dryer<pilot plant dryer<laboratory spray dryer; by contrast, an exact opposite trend was observed for fat surface coverage. The different surface composition in the

particles obtained with the single-particle dryer may be attributed to the absence of shear stress, the relatively long drying time, and the vibration of the droplet caused by the acoustic wave.

2.6 STORAGE STABILITY OF MICROENCAPSULATED INGREDIENTS AND SURFACE OIL

2.6.1 STORAGE STABILITY OF ENCAPSULATED CORE INGREDIENTS

Tonon, Brabet, and Hubinger (2010) evaluated the effects of temperature, water activity (a_w), and type of carrier agent on the storage stability of anthocyanins and antioxidant activity of açai juice encapsulated in MDs, GA, and tapioca starch. A two-step reduction was observed in anthocyanin degradation; the first step occurred at a higher reaction rate and the second step at a lower rate. The antioxidant activity decreased over time with increasing a_w during storage, which was attributed to the higher degradation of anthocyanin at higher a_w. Liang, Huang, Ma, Shoemaker, and Zhong (2013) produced spray dried powders of β-carotene nanoemulsions with OSA-modified starches known as HI-CAP, CAPSUL, and CAPSUL TA. The degradation profiles of the encapsulated β-carotene over time were observed at a different relative humidity (RH) and analyzed with the Weibull model. The degradation rate constant (k) was plotted as a function of the temperature difference between the storage temperature (T) and the glass transition temperature (T_g) of the carrier solid. With increasing $T - T_g$, the mobility of the reactants including oxygen that diffused through the wall material matrices increased, and k increased, having a maximum value 2.3 at $T - T_g \approx 0°C$ for HI-CAP due to the transformation of the matrix from glassy to a rubbery state near T_g, while in the case of CAPSUL and CAPSUL TA, the maximum appeared at $T - T_g \approx -30°C$.

Ng, Jessie, Tan, Long, and Nyam (2013) investigated the improvement in quality and protection against degradation of kenaf seed oil via spray drying microencapsulation. NaCas and MD were used as encapsulation agents and soy lecithin as an emulsifier. The microencapsulated kenaf seed oil was then stored under an accelerated storage condition at 65°C over a period of 24 days. Compared with bulk oil, microencapsulation has been shown to protect against the oxidation of kenaf seed oil and preventing the degradation and loss of bioactive compounds in the oil (Ng et al., 2013). Furthermore, Ng, Choong, Tan, Long, and Nyam (2014) have also studied the impact of the total solid content in feed liquid emulsions on the oxidative stability of encapsulated kenaf seed oil. During the accelerated storage, increasing total solid content (TSC) suppressed the increases in peroxide value (POV), p-anisidine value (p-AV), total oxidation value, and free fatty acid value. The microcapsules from the feed liquid emulsion with the lowest TSC had the highest levels of the oxidation compounds, which have been attributed to the smallest particle size and the largest emulsion droplet size. As the particle size increased at higher TSC, oxidation was slowed down due to a reduction in specific surface area. Similar work was done on the antioxidant properties of the microencapsulated kenaf seed oil (Razmkhah, Tan, Long, & Nyam, 2013).

Ghani et al. (2016) examined the effects of emulsion droplet diameter on the stability of squalene oil (SQ) encapsulated in spray dried microcapsules using MD and NaCas or polymerized NaCas as wall materials. The SQ droplet diameter, which was significantly dependent on the homogenization pressure, affected the stability of the oil in the spray dried powders. The degradation kinetics of SQ powders were measured under the accelerated condition at 105°C and could be described by the Avrami equation. The natural logarithm of degradation rate constant k correlates linearly with the logarithmized average oil droplet diameter of reconstituted emulsion. Ko, Kim, and Park (2012) encapsulated ally isothiocyanate (AITC) using GA and chitosan as wall materials and investigated the availability of the microcapsules as a natural preservative for extension of shelf-life and quality improvement of Kimchi. The number of *Leuconostoc* and *Lactobacillus* species in Kimchi decreased with the application of the AITC powder. Damerau et al. (2014) investigated the oxidative stability of sunflower oil encapsulated in MD from two aspects: the effect of humidity response of the carrier matrix; and the effect of modification of emulsion interface through cross-linking.

Increasing RH improved the oxidative stability of the spray dried sunflower emulsions (POV and α-tocopherol content), due to aggregation and collapse of the matrix at higher RHs, which resulted in limited oxygen availability. The cross-linking of the protein slightly improved oxidative stability (Damerau et al., 2014).

The storage stability of fish oil encapsulated in a blend of sugar beet pectin (SBP) and glucose syrup was examined with the addition of extra virgin olive oil (EVOO) (Polavarapu, Oliver, Ajlouni, & Augustin, 2011) and EVOO plus ethylenediaminetetraacetic acid (EDTA) (Polavarapu, Oliver, Ajlouni, & Augustin, 2012). The accelerated oxidation method at 80°C and 0.5 bar (Polavarapu et al., 2011) or 5 bar (Polavarapu et al., 2012) oxygen were employed. These studies showed that SBP functioned poorly as a wall material because of its residual metal ions (Polavarapu et al., 2011) and the microcapsules with EDTA had longer induction periods (a 35% increase in induction period in the microcapsules of fish oil added with EVOO and a 55% increase in those of fish oil alone) with also slower oxygen consumption relative to their respective counterparts without EDTA (Polavarapu et al., 2012). Significant increases in oil droplet size, as well as the presence of bimodal distributions, were observed in the reconstituted emulsions of the powders after 2- and 3-month storage (Polavarapu et al., 2011, 2012).

The effect of dextrose equivalent (DE) of MD on the oxidative stability of encapsulated fish oil was investigated by Ghani, Adachi, Shiga, et al. (2017). MDs with three different DE values (11, 19, and 25) were applied for the production of fish oil microcapsules using NaCas as an emulsifier. During 15-day storage at 60°C, the microcapsules prepared with MD of DE = 11 had the highest POV. Since the particles of DE = 11 had the largest vacuoles, the emulsions inside the shell of the particle presumably moved closer to the particle surface, that, in turn, resulted in increasing surface oil. As an accelerated oxidation test, the Rancimat technique was applied for estimating the storage stability of linseed oil (Gallardo, Guida, Martinez, et al., 2013) and pitaya seed oil (Lim, Tan, Bakar, & Ng, 2012). With regard to the Rancimat method, although some researchers questioned the changes in the mechanisms of lipid oxidation due to the high temperature used in the method, some studies have shown high correlations between the oxidation stability at room temperature and that determined by the Rancimat test. The induction periods were measured as an index for the oxidation stability of these microencapsulated oils (Gallardo et al., 2013; Lim et al., 2012).

The storage stability of anthocyanins from barberry encapsulated under different temperatures and RHs were investigated (Mahdavi, Jafari, Assadpour, & Ghorbani, 2015). Among the wall materials tested, the blend of GA and MD led to higher EE and best anthocyanin stability (Mahdavi et al., 2015). First-order reaction models have been employed for describing the storage stabilities of encapsulated cores by several researchers for astaxanthin (Karaca, Low, & Nickerson, 2013), anthocyanins (Bicudo, Jó, Oliveira, et al., 2015; Burin, Rossa, Ferreira-Lima, Hillmann, & Boirdignon-Luiz, 2011; Ferrari, Germer, Alvim, & Aguirre, 2013; Mahdavi et al., 2015), rice bran oil (Charoen, Jangchud, Jangchud, Harnsilawat, & McClements, 2015), sulforaphane (Wu, Zou, Mao, Huang, & Liu, 2014), and sulforaphene (Tian et al., 2015). The activation energy of the degradation reaction of astaxanthin was determined by an Arrhenius plot of the reaction rate constant (Pu, Bankston, & Sathivel, 2011). Jiménez-Martín et al. (2015b) produced microcapsules of primary and multi-layered emulsions of fish oil and measured oxidative stability. Using thiobarbituric acid reactive substance (TBARS) values as the oxidative index, they reported significantly higher TBARS values for the microcapsules of primary emulsion compared with those of multi-layered emulsion. They also reported that the microcapsules of the multi-layered emulsion prepared with lecithin-chitosan had higher resistance against lipid oxidation during storage. Zhou, Yuan, Zhao, Zhao, and Wang (2013) introduced the measurement of the oxygen pressure (OPR) in microcapsules to assess the oxygen permeability of a wall material as a means to quantify the quality of the wall material. To estimate the OPR inside the powder, a linear relation between OPR and the oxidation rate constants (first-order reaction rate) was assumed, which were determined from the oxidation rates under vacuum and atmospheric conditions. As an application of OPR method, crocetin was microencapsulated in three different types of wall materials, including β-cyclodextrin (CD), MD, and GA. The OPRs were larger in the order of β-CD > MD > GA, meaning that crocetin degraded more readily in the order of β-CD, MD, and GA. The authors also found that the liquid: solid ratio of the feed emulsions was an important factor affecting the oxygen transmission ratio as the OPR also increased with the increase in liquid: solid ratio from 5:1 to 15:1, although the increase in OPR was statistically insignificant. Kuang, Zhang, Bajaj, et al. (2015) studied the storage stability of lutein microencapsulated in mixture wall materials of MD and sucrose. The increased lutein degradation at higher storage temperatures and oxygen transmission rates suggested that lutein was highly unstable and susceptible to thermal and oxidative degradations. The enthalpy relaxation kinetics of the lutein microcapsules was determined by fitting the Kohlrausch–Williams–Watts model to the empirical moisture adsorption isotherms of the microcapsules. The thermodynamic analysis of the minimum integral entropy theory was also applied to explain the optimum storage conditions of encapsulated carotenoids (Guadarrama-Lezama et al., 2014), allspice essential oil (Sánchez-Sáenz et al., 2011), paprika oleoresin (Rascón et al., 2015), and canola oil (Bonilla, Azuara, Beristain, & Vernon-Carter, 2010). The optimum storage conditions were established using the minimum integral entropy in conjunction with the adsorption isotherm modeled by the Guggenheim-Anderson-de-Boer (GAB) equation.

Costa, Silva, Toledo, Azevedo, and Borges (2015) microencapsulated Swiss cheese bioaroma in the mixture of MD and CAPSUL by spray drying and determined the T_g of the spray dried microparticles at various values of a_w. The increase in the equilibrium moisture content of the microparticles caused a significant reduction in the T_g (Costa et al., 2015). The antioxidant activity of natural beetroot juice powder (Pitalua, Jimenez, Vernon-Carter, & Beristain, 2010) and eugenol-rich clove extract powder (Chatterjee & Bhattacharjee, 2013) were evaluated during storage. The storage stability of betalains in the beetroot juice powder at 30°C depended significantly on the a_w with the greatest stability observed when $a_w < 0.521$ (Pitalua et al., 2010). After 30 days of storage, no significant difference was observed between the nonencapsulated and encapsulated clove extracts in terms of antioxidant activity in soybean oil. The release of antioxidants from the encapsulated clove powder was observed to have occurred in a controlled manner (Chatterjee & Bhattacharjee, 2013).

2.6.2 Surface Oil: Definition and Impacts of Spray Drying Conditions

2.6.2.1 Surface Oil Content and its Impact on the Stability of Encapsulated Oil

As previously mentioned in Section 2.1, most functional substances extracted from agricultural products and seafood are generally poorly water-soluble hydrophobic substances, which are mostly unstable upon exposure to light, heat, and oxygen. Therefore, they are mixed with emulsifiers and solutions of wall materials and turned into O/W emulsions and are eventually spray dried to improve their storage stability. The storage stability of spray dried particles is dependent on the oxidative stability of the oil droplets dispersed within the matrix of the spray dried particles. The oil droplets that are present adjacent to the surface oxidize rapidly due to the penetration of oxygen from the surroundings, resulting in harmful oxidation products and off-flavor. Surface oil can be quantified by mixing the spray dried particles with hexane or petroleum ether, followed by filtering the suspension and determining the amount of oil that dissolves in the filtrate. Although XPS can be employed for elemental analysis of particle surface, this method determines only the percentage surface coverage of elements but not their absolute contents.

2.6.2.2 Parameters Controlling the Amount of Surface Oil

For quantification of surface oil in spray dried powder, the powder is first subjected to extraction by hexane or petroleum ether. Briefly, the solvent is added to the powder at the ratio of 5 mL of solvent to 0.3–1.0 g of powder. The mixture is stirred for 15 min and then filtered. The oil dissolved in the filtrate is typically quantified gravimetrically or by chromatography as the amount of surface oil (Drusch & Berg, 2008; Ghani, Adachi, Sato, et al., 2017). Ghani, Adachi, Sato, et al. (2017) investigated various aspects in regard to the relation between surface oil ratio s (defined as the weight ratio of surface oil to the total oil in a powder) and oil droplet size of emulsion. The powders obtained by

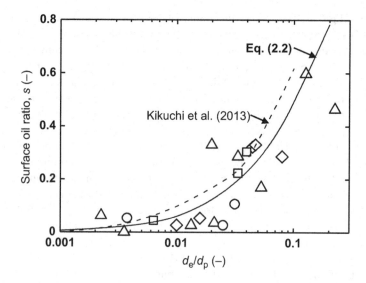

FIGURE 2.3 Changes of the surface oil ratio *s* as a function of the mean diameter ratio of reconstituted oil droplets to spray dried particles d_e/d_p. MD (DE = 25) concentration: ○, 30 wt%; △, 40 wt%; □, 50 wt%; ◇,60 wt%. Oil: fish oil. Modified from Ghani, Adachi, Sato, et al. (2017), with permission.

spray drying of fish oil emulsions in the solutions of mixture wall material of MD (DE = 25) and NaCas (oil load: 30–60 wt% of total solid) were analyzed by thin-layer chromatography for surface oil ratio *s*, which is presented as a function of the average diameter ratio of reconstituted oil droplets to spray dried particles $z \ (= d_e/d_p)$ in Figure 2.3. The surface oil ratio *s* was found increasing abruptly with the increase in *z*. The changes in *s* with *z* could be described by a simple emulsion dispersion model. If the concentration of oil in the particles is C_0 [kg/m³] and suppose the surface oil dissolved in hexane is extracted from oil droplets dispersed within a thin layer (thickness = δ) close to the surface of the particles, *s* can be described by:

$$s = 1 - \frac{\pi}{6}\left(d_p - 2\delta\right)^3 C_0 \Big/ \frac{\pi}{6} d_p^3 C_0 = 1 - \frac{\left(d_p - 2\delta\right)^3}{d_p^3} = 6\frac{\delta}{d_p} - 12\left(\frac{\delta}{d_p}\right)^2 + 8\left(\frac{\delta}{d_p}\right)^3 \qquad (2.1)$$

And if δ is assumed to be in the same order of magnitude as the average diameter of the reconstituted oil droplet d_e, then $\delta/d_p \approx z$ and:

$$s = 6z - 12z^2 + 8z^3 = 1 - \left(1 - 2z\right)^3 \qquad (2.2)$$

The solid line presented in Figure 2.3 is the estimation of *s* by Equation 2.2, which was close to the experimental data. Mathematical simulations on the surface oil content of microcapsules based on the percolation theory have been extensively studied by Kikuchi, Yamamoto, Shiga, Yoshii, and Adachi (2013). The simulations suggested that smaller droplets were more favorable for the production of microcapsules wherein the oil had higher resistance against oxidation.

The amount of surface oil typically increases at higher spray drying temperatures, which has been attributed to the formation of the vacuole in atomized droplets due to the high temperatures (Drusch & Berg, 2008). It is presumed that this phenomenon was caused by the formation of the vacuole, which reduces the volume of the particle shell and pushes the oil droplets entrapped within the shell towards the particle surface. Ghani, Adachi, Sato, et al. (2017) and Ghani, Adachi, Shiga, et al. (2017) used MDs of different DEs to produce spray dried particles and subsequently measured

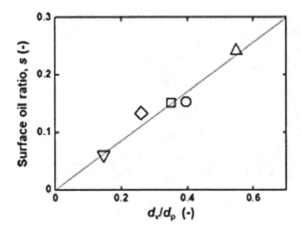

FIGURE 2.4 A linear relationship between surface oil ratio s and mean diameter ratio of vacuoles to spray dried particles d_v/d_p. DE value of MDs: $\triangle$, 8; $\bigcirc$, 11; $\square$, 19; $\lozenge$, 24; ∇, 40. Modified from Ghani, Adachi, Sato, et al. (2017), with permission.

the vacuole diameter d_v and particle diameter d_p of 400 particles by CLSM. Surface oil ratio s has been shown to have a linear correlation with d_v/d_p (Figure 2.4) (Ghani, Adachi, Sato, et al., 2017).

2.6.2.3 A New and Innovative Spray Drying Method of One-step Double Encapsulation

The technique of fluidized bed coating has been practically adopted for the coating of spray dried particles with materials such as starch to prevent the contact between the surface oil of the particles and oxygen. Normally, this is achieved via two separate processes: particle formation by spray drying, and fluidized bed coating of the particles. Schaffner (2004) devised a novel atomizer capable of simultaneously dispersing starch powder and atomizing feed emulsion into the spray tower, through which both of the aforesaid processes could be performed in one single step (one-step double encapsulation). However, there was no discussion about the morphology of the resultant particles in the patent. Following a similar concept, Takashige, Iwamoto, Shiga, et al. (2019) manufactured starch-coated spray dried fish oil emulsion using MD or sucrose as wall material. The starch powder was dispersed onto the surface of the atomized droplets of fish oil emulsion through a feed pipe, the outlet of which was located at the same horizontal level of the rotary atomizer. The starch powder was found distributed evenly across the surface of the spray dried particles irrespective of the type of wall material (MD or sucrose) (Figure 2.5). Owing to the low T_g of sucrose and thus slower film formation at the surface of atomized droplets, the spray dried particles of sucrose are known to be prone to adhesion on the internal surfaces of the spray dryer, resulting in a significant reduction in powder recovery. The simultaneous spray drying microencapsulation of fish oil in sucrose and starch coating has offered the advantage of improving the powder recovery.

2.7 RELEASE OF MICROENCAPSULATED CORE INGREDIENTS DURING STORAGE

Research on the release of encapsulated core substances can generally be classified into two types by the medium into which the substances are released; either into the atmosphere or a liquid phase (commonly, in simulated gastric and intestinal fluids (SGF and SIF, respectively)). The former is important for examining the long-term storage of the microcapsules while the latter is essential for understanding the gastrointestinal absorption of the encapsulated medicinal chemicals and nutritional components.

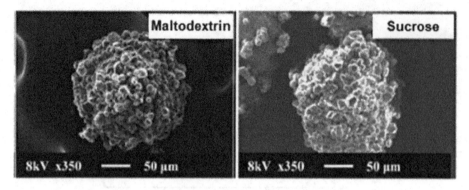

FIGURE 2.5 Scanning electron micrographs of one-step starch-coated microparticles of fish oil emulsion suspended in aqueous maltodextrin solution (left) and in aqueous sucrose solution (right). Starch feed ratio = 90 g/min. Modified from Takashige et al. (2019), with permission.

In vitro digestion profiles under simulated gastrointestinal conditions (in SGF and SIF) or water are particularly of interest across a wide array of applications (Davidov-Pardo, Arozarena, & Marín-Arroyo, 2013; Estevinho, Carlan, Blaga, & Rocha, 2016; Flores, Singh, Kerr, Phillips, & Kong, 2015; Lauro et al., 2015; Lee et al., 2015; Li et al., 2015; Subpuch, Huang, & Suwannaporn, 2016). The release profiles of core materials were described using diffusion models (Costa, Duarte, Bourbon, et al., 2012) and empirical models (Dima, Pătraşcu, Cantaragiu, Alexe, & Dima, 2016; García, Vega, Jimenez, Santos, & Robert, 2013; Robert, García, Reyes, Chávez, & Santos, 2012). The release profiles of collagen hydrolysate in ultrapure water and cellular medium have been characterized (Peres et al., 2012) and the effects of heating of SGF and SIF have also been investigated (Goyal et al., 2016).

Panyoyai, Bannikova, Small, Shanks, and Kasapis (2016) analyzed the diffusion pattern of nicotinic acid (vitamin B_3) encapsulated in WPI. The diffusivity was estimated by Fick's second law from the relative mass change of nicotinic acid over time. The diffusivity of nicotinic acid in WPI was nearly constant up to the T_g of the matrix, after which a linear increase in diffusivity with temperature was observed until it hit a maximum and then started decreasing thereafter. They also examined the relationship between the free volume of the whey protein matrix and the diffusion coefficient of nicotinic acid. Estevinho and co-workers (Estevinho et al., 2016) studied the encapsulation of vitamins (i.e., C, B_{12}, and A) in various wall materials. The release time courses in water for the vitamins C and B_{12} encapsulated with three different wall materials (i.e., chitosan, modified chitosan, and sodium alginate) revealed that both of the vitamins had highly controlled release profiles when encapsulated within the native chitosan wall matrix. The kinetic model that best fitted the experimental release results was the Weibull equation with the highest value of the correlation coefficient (Estevinho & Rocha, 2017). The investigation on microencapsulation of vitamin B_{12} has been further conducted with modified chitosan (Carlan, Estevinho, & Rocha, 2017) and a carbohydrate polymer from a marine unicellular cyanobacterium (Estevinho et al., 2019). The *in vitro* release profiles of the microencapsulated vitamin B_{12} in SGF were well described by the Weibull equation.

The releases of encapsulated primary and multi-layered emulsions of ethyl hexanoate under various RHs were extensively studied by Gharsallaoui et al. (2012). In higher RH environments, damage of the matrix structure due to a higher moisture content resulted in higher diffusion of volatile molecules from the oil droplets to the hydrated matrix, causing higher release from the powder. The release profiles were analyzed with Avrami's equation to obtain the release kinetic rates. The lowest flavor release rate constant was observed in the range of 33%–52% RH, then it increased dramatically with increasing a_w, reflecting an increase of flavor diffusion through the wall matrix. The variations of release rates were more pronounced for the multi-layered emulsion particles.

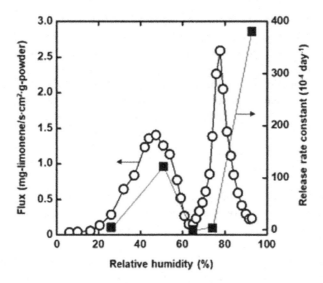

FIGURE 2.6 Comparison of release rates of *d*-limonene encapsulated in maltodextrin/gum Arabic mixture under different storage conditions. Storage conditions; o: ramping relative humidity, and ■: static relative humidity.

Kim, Han, Na, et al. (2013) developed insect-repellent films (repelling for Indian meal moth) coated with microencapsulated cinnamon oil (CO, an insect-repelling substance). The CO was microencapsulated in GA, a mixture of WPI and MD, and polyvinyl alcohol, which was mixed with polypropylene solution and spread on the surface of a low-density polyethylene film. The release rate of CO from the coated film was significantly lower for microencapsulated CO. Moth invasion was prevented by the insect-repellent films, demonstrating the potential use of the films as insect-repellent packaging materials. A similar study was also conducted with thyme oil as the insect-repelling substance (Chung et al., 2013).

It is well known that the release rate of a core material encapsulated in spray dried particles exhibits a strong dependency on environment humidity. The common release experiment using a humidity-controlled chamber is laborious and time-consuming. Further, the technique may overlook changes in the release rate caused by phase and structural changes that have taken place in the matrix. Yamamoto, Neoh, Honbou, Furuta, et al. (2012) and Yamamoto, Neoh, Honbou, Kimura, et al. (2012) developed an experimental setup for rapid evaluation of the release of *d*-limonene from spray dried powders under linearly ramped humidity at a constant temperature using an in-house built dynamic vapor sorption (DVS) system. The release fluxes (release rates) of *d*-limonene were monitored in real time by auto-sampling gas chromatography. Figure 2.6 is an example of the release flux profile of *d*-limonene encapsulated in a mixture of GA and MD. The profile as a function of a_w has two maximum peaks appearing at the similar a_w, at which the release rate constant also peaked in the static-RH release experiment (humidity-controlled chamber). The release flux profiles under ramping humidity were modeled by applying an extended Arrhenius equation. The kinetic parameters such as activation energy and frequency factor of release were estimated (Yamamoto, Neoh, Honbou, Kimura, et al., 2012).

Using a similar DVS-GC system, Takashige, Ariyanto, Adachi, and Yoshii (2020) measured the on-set RH of the *d*-limonene release from the spray dried microparticles with MD, lactose, and sucrose as wall materials. The on-set RH correlated linearly with the T_g of the wall materials. Solid amorphous glasses are known to transform into an amorphous rubbery state by gaining moisture during exposure to high-RH environments, resulting in the lowering of T_g. During storage or application, the microcapsules may undergo transient changes in moisture distribution internally. In the case of humidification, the particles may have a moisture gradient from the surface to the center, causing a gradient in T_g within a particle. This implies that the location-specific T_g would be different from the spatially averaged T_g measured by conventional DSC. Sleeuwen, Zhang, and Normand (2012) focused on the prediction of spatial T_g profiles in a microcapsule placed in a humidifying environment. The prediction of T_g consisted of a numerical simulation of moisture distribution in the particle and the calculation of the local T_g at the predicted moisture content. The predicted T_g were compared with experimental data using temperature-modulated DSC and standard DSC. The non-equilibrated (non-uniform moisture distribution inside the particle) microcapsules have broader distributions of T_g than the equilibrated (initial and final) particles.

2.8 CONCLUDING REMARKS

In recent years, functional foods are becoming more important for maintaining a healthy life in different societies. Functional foods contain various functional/bioactive ingredients, most of which are supplied as powders containing these ingredients. Most food powders are produced by spray drying technology and the quality of these powders is especially influenced by atomization and drying of the atomized droplets into solid particles. Since most of the functional core materials such as liquid lipids and flavors are fat-soluble, it is necessary to emulsify these hydrophobic core materials with an emulsifier before atomization for encapsulation in the matrix of wall material such as aqueous carbohydrate. More complex emulsion systems such as multi-layered emulsion systems have contributed to improving the stability of core materials during spray drying and storage. Atomization of the raw material solution is an important process that determines the physicochemical properties of the spray dried powders. Advancement in spraying technology has enabled the production of the droplets that are significantly uniform in terms of size. Uniform particles demonstrate the potential for application in functional food products, due to the improvement in wettability and possibly apparent solubility of such spray dried powders. Although it has not been covered in this chapter, very fine dry powders such as nano-sized powders can also be produced by spray drying. The application of nano-powders is expected to aid the development of food powders with enhanced solubility and bioavailability of the encapsulated functional ingredients.

The internal structure of spray dried particles and the state of entrapped functional core material within the particles are fundamental factors that affect the storage stability of spray dried powders and the diffusional characteristics of core material into the environment. Especially in the case of microencapsulation of liquid lipids, storage stability is of great importance. The presence of surface oil (and their concentration) is an important factor that governs this property, although few studies are available on the effect of this factor. It is also known that the dispersion and release of core materials into the surrounding environment are, to a great extent, influenced by the T_g of dry particles. These mass transfer phenomena should supposedly be analyzed based on moisture sorption-related changes in T_g of the wall material matrix. Nonetheless, these phenomena are still being analyzed on a trial and error basis using empirical diffusion models such as the Avrami equation, which are not theoretically related to the dispersion and release of encapsulated core materials. The relationship between T_g and the mass transfer rate is yet to be theoretically established. Theoretical modeling of the structure and characteristics of dry powder, including T_g, is necessary for the elucidation of the mass transfer mechanisms. Transient methods for measuring the relaxation phenomena in food powders such as the release of encapsulated core materials are useful for analyzing various factors that influence the transient phenomena.

Even though we have not included the studies on the microencapsulation of probiotics in the scope of this chapter, we found that a vast number of studies were published within a rather short time frame. This seems to be partly attributed to the development of the simplified bench-top spray dryers, manufactured by companies such as Büchi, Labmaq, and LabPlant. These spray dryers are very compact and user-friendly so that one can manufacture spray dried powders from even small quantities (down to a few ten milliliters) of the feed liquid. However, the results obtained with such small-scale dryers may not directly be reflective of what can be expected from spray dryers of larger scales.

REFERENCES

Aberkane, L., G. Roudaut, and R. Saurel. 2014. Encapsulation and oxidative stability of PUFA-rich oil micro-encapsulated by spray drying using pea protein and pectin. *Food Bioprocess Technol* 7: 1505–1517.

Aizpurua-Olaizola, O., P. Navarro, A. Vallejo, M. Olivares, N. Etxebarria, and A. Usobiaga. 2016. Microencapsulation and storage stability of polyphenols from *Vitis vinifera* grape wastes. *Food Chem* 190: 614–621.

Augustin, M. A., S. Bhail, L. J. Cheng, Z. Shen, S. Øiseth, and L. Sanguansri. 2015. Use of whole buttermilk for microencapsulation of omega-3 oils. *J Funct Foods* 19: 859–867.

Bao, S. S., X. C. Hu, K. Zhang, X. K. Xu, H. M. Zhang, and H. Huang. 2011. Characterization of spray dried microalgal oil encapsulated in cross-linked sodium caseinate matrix induced by microbial transglutaminase. *J Food Sci* 76: E112–E118.

Barrow, C. J., B. Wang, B. Adhikari, and H. Liu. 2013. Spray drying and encapsulation of omega-3 oil. In *Food Enrichment with Omega-3 Fatty Acids*, ed. C. Jacobsen, N. S. Nielsen, A. F. Horn, and A. D. M. Sørensen, 194–225. Cambridge: Woodhead Publishing.

Berendsen, R., C. Güell, and M. Ferrando. 2015. Spray dried double emulsions containing procyanidin-rich extracts produced by premix membrane emulsification: Effect of interfacial composition. *Food Chem* 178: 251–258.

Bicudo, M. O. P., J. Jó, G. A. Oliveira et al. 2015. Microencapsulation of juçara (*Euterpe edulis* M.) pulp by spray drying using different carriers and drying temperatures. *Drying Technol* 33: 153–161.

Bonilla, E., E. Azuara, C. I. Beristain, and E. J. Vernon-Carter. 2010. Predicting suitable storage conditions for spray dried microcapsules formed with different biopolymer matrices. *Food Hydrocoll* 24: 633–640.

Boraey, M. A., and R. Vehring. 2014. Diffusion controlled formation of microparticles. *J Aerosol Sci* 67: 131–143.

Burin, V. M., P. N. Rossa, N. E. Ferreira-Lima, M. C. R. Hillmann, and M. T. Boirdignon-Luiz. 2011. Anthocyanins: Optimisation of extraction from Cabernet Sauvignon grapes, microcapsulation and stability in soft drink. *Int J Food Sci Technol* 46: 186–193.

Carlan, I. C., B. N. Estevinho, and F. Rocha. 2017. Study of microencapsulation and controlled release of modified chitosan microparticles containing vitamin B12. *Powder Technol* 318: 162–169.

Carvalho, A. G. S., V. M. Silva, and M. D. Hubinger. 2014. Microencapsulation by spray drying of emulsified green coffee oil with two-layered membranes. *Food Res Int* 61: 236–245

Charoen, R., A. Jangchud, K. Jangchud, T. Harnsilawat, and D. J. McClements. 2015. The physical characterization and sorption isotherm of rice bran oil powders stabilized by food-grade biopolymers. *Drying Technol* 33: 479–492.

Chatterjee, D., and P. Bhattacharjee. 2013. Comparative evaluation of the antioxidant efficacy of encapsulated and un-encapsulated eugenol-rich clove extracts in soybean oil: Shelf-life and frying stability of soybean oil. *J Food Eng* 117: 545–550.

Chew, J. H., N. Fu, T. Gengenbach, X. D. Chen, and C. Selomulya. 2015. The compositional effects of high solids model emulsions on drying behaviour and particle formation processes. *J Food Eng* 157: 33–40

Chew, J. H., W. Liu, N. Fu, T. Gengenbach, X. D. Chen, and C. Selomulya. 2014. Exploring the drying behaviour and particle formation of high solids milk protein concentrate. *J Food Eng* 143: 186–194.

Chung, S. K., J. Y. Seo, J. H. Lim, H. H. Park, M. J. Yea, and H. J. Park. 2013. Microencapsulation of essential oil for insect repellent in food packaging system. *J Food Sci* 78: E709–E714.

Costa, J. M. G., E. K. Silva, A. A. C. Toledo, H. V. M. Azevedo, and S. V. Borges. 2015. Physical and thermal stability of spray dried Swiss cheese bioaroma powder. *Drying Technol* 33: 346–354.

Costa, S. B., C. Duarte, A. I. Bourbon et al. 2012. Effect of the matrix system in the delivery and in vitro bioactivity of microencapsulated oregano essential oil. *J Food Eng* 110: 190–199.

Costa, S. B., C. Duarte, A. I. Bourbon et al. 2013. Inulin potential for encapsulation and controlled delivery of oregano essential oil. *Food Hydrocoll* 33: 199–206.

Damerau, A., T. Moisio, R. Partanen, P. Forssell, A. M. Lampi, and V. Piironen. 2014. Interfacial protein engineering for spray dried emulsions – Part II: Oxidative stability. *Food Chem* 144: 57–64.

Davidov-Pardo, G., I. Arozarena, M. R. Marín-Arroyo. 2013. Optimization of a wall material formulation to microencapsulate a grape seed extract using a mixture design of experiments. *Food Bioprocess Technol* 6: 941–951.

Dima, C., L. Pătraşcu, A. Cantaragiu, P. Alexe, and S. Dima. 2016. The kinetics of the swelling process and the release mechanisms of *Coriandrum sativum L.* essential oil from chitosan/alginate/inulin microcapsules. *Food Chem* 195: 39–48.

Drusch, S., and S. Berg. 2008. Extractable oil in microcapsules prepared by spray-drying: Localisation, determination and impact on oxidative stability. *Food Chem* 109: 17–24.

Drusch, S., S. Hamann, A. Berger, Y. Serfert, and K. Schwarz. 2012. Surface accumulation of milk proteins and milk protein hydrolysates at the air–water interface on a time-scale relevant for spray-drying. *Food Res Int* 47: 140–145.

Drusch, S., Y. Serfert, A. Berger et al. 2012. New insights into the microencapsulation properties of sodium caseinate and hydrolyzed casein. *Food Hydrocoll* 27: 332–338.

Esfanjani, A. F., S. M. Jafari, E. Assadpoor, and A. Adeleh Mohammadi. 2015. Nano-encapsulation of saffron extract through double-layered multiple emulsions of pectin and whey protein concentrate. *J Food Eng* 165: 149–155.

Estevinho, B. N., I. Carlan, A. Blaga, and F. Rocha. 2016. Soluble vitamins (vitamin B12 and vitamin C) microencapsulated with different biopolymers by a spray drying process. *Powder Technol* 289: 71–78.

Estevinho, B. N., R. Mota, J. P. Leite, P. Tamagnini, L. Gales, and F. Rocha. 2019. Application of a cyanobacterial extracellular polymeric substance in the microencapsulation of vitamin B12. *Powder Technol* 343: 644–651.

Estevinho, B. N., and F. Rocha. 2017. Kinetic models applied to soluble vitamins delivery systems prepared by spray drying. *Drying Technol* 35: 1249–1257.

Faldt, P., and B. Bergenstahl. 1994. The surface composition of spray dried protein–lactose powders. *Colloids Surf A Physicochem Eng Asp* 90: 183–190.

Ferrari, C. C., S. P. M. Germer, I. D. Alvim, and J. M. Aguirre. 2013. Storage stability of spray dried blackberry powder, produced with maltodextrin or gum Arabic. *Drying Technol* 31: 470–478.

Flores, F. P., R. K. Singh, W. L. Kerr, D. R. Phillips, and F. Kong. 2015. *In vitro* release properties of encapsulated blueberry (*Vaccinium ashei*) extracts. *Food Chem* 168: 225–232.

Fu, N., M. W. Woo, and X. D. Chen. 2011. Colloidal transport phenomena of milk components during convective droplet drying. *Colloids Surf B Biointerfaces* 87: 255–266.

Fu, N., M. W. Woo, S. X. Q. Lin, Z. Zhou, and X. D. Chen. 2011. Reaction Engineering Approach (REA) to model the drying kinetics of droplets with different initial sizes-experiments and analyses. *Chem Eng Sci* 66: 1738–1747.

Fu, N., M. W. Woo, C. Selomulya et al. 2012. Drying kinetics of skim milk with 50 wt.% initial solids. *J Food Eng* 109: 701–711.

Furuta, T., A. Soottitantawat, T-L. Neoh, and H. Yoshii. 2011. Effect of Microencapsulation on food flavors and their releases. In *Physicochemical Aspects of Food Engineering and Processing*, ed. S. Devahastin, 1–40. Boca Raton, FL: CRC Press.

Gallardo, G., L. Guida, V. Martinez et al. 2013. Microencapsulation of linseed oil by spray drying for functional food application. *Food Res Int* 52: 473–482.

García, P., J. Vega, P. Jimenez, J. Santos, and P. Robert. 2013. Alpha-tocopherol microspheres with cross-linked and acetylated inulin and their release profile in a hydrophilic model. *Eur J Lipid Sci Technol* 115: 811–819.

Ghani, A., S. Adachi, K. Sato et al. 2017. Effect of oil-droplet diameter, and dextrose equivalent of maltodextrin on the surface-oil ratio of microencapsulated fish oil by spray drying. *J Chem Eng Jpn* 50: 799–806.

Ghani, A., S. Adachi, H. Shiga et al. 2017. Effect of different dextrose equivalents of maltodextrin on oxidation stabilityin encapsulated fish oil by spray drying. *Biosci Biotechnol Biochem* 81: 705–711.

Ghani, A. A., K. Matsumura, A. Yamauchi et al. 2016. Effects of oil-droplet diameter on the stability of squalene oil in spray dried powder. *Drying Technol* 34: 1726–1734.

Gharsallaoui, A., G. Roudaut, L. Beney, O. Chambin, A. Voilley, and R. Saurel. 2012. Properties of spray dried food flavours microencapsulated with two-layered membranes: Roles of interfacial interactions and water. *Food Chem* 132: 1713–1720.

Gharsallaoui, A., R. Saurel, O. Chambin, E. Cases, A. Voilley, and P. Cayot. 2010. Utilisation of pectin coating to enhance spray-dry stability of pea protein-stabilised oil-in-water emulsions. *Food Chem* 122: 447–454.

Goyal, A., V. Sharma, M. K. Sihag, S. Arora, A. K. Singh, and L. Sabikhi. 2016. Effect of microencapsulation and spray drying on oxidative stability of flaxseed oil and its release behavior under simulated gastrointestinal conditions. *Drying Technol* 34: 810–821.

Guadarrama-Lezama, A. Y., E. Jaramillo-Flores, G. F. Gutiérrez-López, C. Pérez-Alonso, L. Dorantes-Álvarez, and L. Alamilla-Beltrán. 2014. Effects of storage temperature and water activity on the degradation of carotenoids contained in microencapsulated chili extract. *Drying Technol* 32: 1435–1447.

Ivey, J. W., and R. Vehring. 2010. The use of modeling in spray drying of emulsions and suspensions accelerates formulation and process development. *Comput Chem Eng* 34: 1036–1040.

Jiménez-Martín, E., A. Gharsallaoui, T. Pérez-Palacios, J. R. Carrascal, and T. A. Rojas. 2015a. Volatile compounds and physicochemical characteristics during storage of microcapsules from different fish oil emulsions. *Food Bioprod Process* 96: 52–64.

Jiménez-Martín, E., A. Gharsallaoui, T. Pérez-Palacios, J. R. Carrascal, and T. A. Rojas. 2015b. Suitability of using monolayered and multilayered emulsions for microencapsulation of ω-3 fatty acids by spray drying: Effect of storage at different temperatures. *Food Bioprocess Technol* 8: 100–111.

Jiménez-Martín, E., T. A. Rojas, A. Gharsallaoui, J. R. Carrascal, and T. Pérez-Palacios. 2016. Fatty acid composition in double and multilayered microcapsules of ω-3 as affected by storage conditions and type of emulsions. *Food Chem* 194: 476–486.

Karaca, A. C., N. Low, and M. Nickerson. 2013. Encapsulation of flaxseed oil using a benchtop spray dryer for legume protein–maltodextrin microcapsule preparation. *J Agric Food Chem* 61: 5148–5155.

Kašpar, O., M. Jakubec, and F. Štěpánek. 2013. Characterization of spray dried chitosan-TPP microparticles formed by two- and three-fluid nozzles. *Powder Technol* 240: 31–34.

Khem, S., M. W. Woo, D. M. Small, X. D. Chen, and B. K. May. 2015. Agent selection and protective effects during single droplet drying of bacteria. *Food Chem* 166: 206–214.

Kikuchi, K., S. Yamamoto, H. Shiga, H. Yoshii, and S. Adachi. 2013. Surface oil content of microcapsules containing various oil fraction and oil-droplet size. *Jpn J Food Eng* 14: 169–173.

Kim, I. H., J. Han, J. H. Na et al. 2013. Insect-resistant food packaging film development using cinnamon oil and microencapsulation technologies. *J Food Sci* 78: E229–E237.

Ko, J. K., W. Y. Kim, and H. J. Park. 2012. Effects of microencapsulated Allyl isothiocyanate (AITC) on the extension of the shelf-life of Kimchi. *Int J Food Microbiol* 153: 92–98.

Kondo, K., T. Niwa, and K. Danjo. 2014. Preparation of sustained-release coated particles by novel microencapsulation method using three-fluid nozzle spray drying technique. *Eur J Pharm Sci* 51: 11–19.

Kuang, P., H. Zhang, P. R. Bajaj et al. 2015. Physicochemical properties and storage stability of lutein microcapsules prepared with maltodextrins and sucrose by spray drying. *J Food Sci* 80: E359–E369.

Laine, P., E. Toppinen, R. Kivelä et al. 2011. Emulsion preparation with modified oat bran: Optimization of the emulsification process for microencapsulation purposes. *J Food Eng* 104: 538–547.

Lauro, M. R., L. Crasci, C. Carbone, R. P. Aquino, A. M. Panico, and C. Puglisi. 2015. Encapsulation of a citrus by-product extract: Development, characterization and stability studies of a nutraceutical with antioxidant and metalloproteinases inhibitory activity. *LWT* 62: 169–176.

Lee, Y. K., P. Ganesan, B. S. Baharin, and H. S. Kwak. 2015. Characteristics, stability, and release of peanut sprout extracts in powdered microcapsules by spray drying. *Drying Technol* 33: 1991–2001.

Li, C., J. Wang, J. Shi, X. Huang, Q. Peng, and F. Xue. 2015. Encapsulation of tomato oleoresin using soy protein isolate-gum aracia conjugates as emulsifier and coating materials. *Food Hydrocoll* 45: 301–308.

Liang, R., Q. Huang, J. Ma, C. F. Shoemaker, and F. Zhong. 2013. Effect of relative humidity on the store stability of spray dried beta-carotene nanoemulsions. *Food Hydrocoll* 33: 225–233.

Lim, A. S. L., and Y. H. Roos. 2016. Spray drying of high hydrophilic solids emulsions with layered interface and trehalose-maltodextrin as glass formers for carotenoids stabilization. *J Food Eng* 171: 174–184.

Lim, H. K., C. P. Tan, J. Bakar, and S. P. Ng. 2012. Effects of different wall materials on the physicochemical properties and oxidative stability of spray dried microencapsulated red-fleshed pitaya (*Hylocereus polyrhizus*) seed oil. *Food Bioprocess Technol* 5: 1220–1227.

Lin, C. X., and L. Phan. 2013. A numerical study of both internal and external two-phase flows of a rotating disk atomizer. *Drying Technol* 31: 605–613.

Liu, W., X. D. Chen, Z. Cheng, and C. Selomulya. 2016. On enhancing the solubility of curcumin by microencapsulation in whey protein isolate via spray drying. *J Food Eng* 169: 189–195.

Liu, W., X. D. Chen, and C. Selomulya. 2015. On the spray drying of uniform functional microparticles. *Particuology* 22: 1–12.

Mahdavi, S. A., S. M. Jafari, E. Assadpour, and M. Ghorbani. 2015. Storage stability of encapsulated barberry's anthocyanin and its application in jelly formulation. *J Food Eng* 181: 59–66.

Maher, P. G., M. A. E. Auty, Y. H. Roos, L. M. Zychowski, and M. A. Fenelon. 2015. Microstructure and lactose crystallization properties in spray dried nanoemulsion. *Food Struct* 3: 1–11.

Maisuthisakul, P., and M. H. Gordon. 2012. Influence of polysaccharides and storage during processing on the properties of mango seed kernel extract (microencapsulation). *Food Chem* 134: 1453–1460.

Mezzenga, R., and S. Ulrich. 2010. Spray dried oil powder with ultrahigh oil content. *Langmuir* 26: 16658–16661.

Mora-Gutierrez, A., R. Attaie, J. M. Kirven, and H. M. Farrell Jr. 2014. Cross-linking of bovine and caprine caseins by microbial transglutaminase and their use as microencapsulating agents for *n*-3 fatty acids. *Int J Food Sci Technol* 49: 1530–1543.

Munoz-Ibanez, M., C. Azagoh, B. N. Dubey, E. Dumoulin, and C. Turchiuli. 2015. Changes in oil-in-water emulsion size distribution during the atomization step in spray-drying encapsulation. *J Food Eng* 167: 122–132.

Munoz-Ibanez, M., M. Nuzzo, C. Turchiuli, B. Bergenståhl, E. Dumoulin, and A. Millqvist-Fureby. 2016. The microstructure and component distribution in spray dried emulsion particles. *Food Struct* 8: 16–24.

Nakagawa, K., and Y. Fujii. 2015. Protein-based microencapsulation with freeze pretreatment: Spray dried oil in water emulsion stabilized by the soy protein isolate–gum acacia complex. *Drying Technol* 33: 1541–1549.

Ng, S. K., Y. H. Choong, C. P. Tan, K. Long, and K. L. Nyam. 2014. Effect of total solids content in feed emulsion on the physical properties and oxidative stability of microencapsulated kenaf seed oil. *LWT* 58: 627–632.

Ng, S. K., L. Y. L. Jessie, C. P. Tan, K. Long, and K. I. Nyam. 2013. Effect of accelerated storage on microencapsulated kenaf seed oil. *J Am Oil Chem Soc* 90: 1023–1029.

Nijdam, J., S. Kachel, P. Scharfer, W. Schabel, and M. Kind. 2015. Effect of diffusion on component segregation during drying of aqueous solutions containing protein and sugar. *Drying Technol* 33: 288–300.

Nuzzo, M., J. S. Overgaard, B. Bergenståhl, and A. Millqvist-Fureby. 2017. The morphology and internal composition of dried particles from whole milk—From single droplet to full scale drying. *Food Struct* 13: 35–44.

Pai, D., V. R. Vangala, J. W. Ng, W. K. Ng, and R. B. H. Tan. 2015. Resistant maltodextrin as a shell material for encapsulation of naringin: Production and physicochemical characterization. *J Food Eng* 161: 68–74.

Panyoyai, N., A. Bannikova, D. M. Small, R. A. Shanks, and S. Kasapis. 2016. Diffusion of nicotinic acid in spray dried capsules of whey protein isolate. *Food Hydrocoll* 52: 811–819.

Penbunditkul, P., H. Yoshii, U. Ruktanonchai, T. Charinpanitkul, S. Assabumrungrat, and A. Soottitantawat. 2012. The loss of OSA-modified starch emulsifier property during the high-pressure homogeniser and encapsulation of multi-flavour bergamot oil by spray drying. *Int J Food Sci Technol* 47: 2325–2333.

Peres, I., S. Rocha, J. Gomes, S. Morais, M. C. Pereira, and M. Coelho. 2011. Preservation of catechin antioxidant properties loaded in carbohydrate nanoparticles. *Carbohydr Polym* 86: 147–153.

Peres, I., S. Rocha, J. A. Loureiro, M. C. Pereira, G Ivanova, and M. Coelho. 2012.Carbohydrate particles as protein carriers and scaffolds: Physico-chemical characterization and collagen stability, *J Nanopart Res* 14: 1144–1155.

Pitalua, E., M. Jimenez, E. J. Vernon-Carter, and C. I. Beristain. 2010. Antioxidative activity of microcapsules with beetroot juice using gum Arabic as wall material. *Food Bioprod Process* 88: 253–258.

Polavarapu, S., C. M. Oliver, S. Ajlouni, and M. A. Augustin. 2011. Physicochemical characterisation and oxidative stability of fish oil and fish oil–extra virgin olive oil microencapsulated by sugar beet pectin. *Food Chem* 127: 1694–1705.

Polavarapu, S., C. M. Oliver, S. Ajlouni, and M. A. Augustin. 2012. Impact of extra virgin olive oil and ethylenediaminetetraacetic acid (EDTA) on the oxidative stability of fish oil emulsions and spray dried microcapsules stabilized by sugar beet pectin. *J Agric Food Chem* 60: 44–450.

Porras-Saavedra, J., E. Palacios-González, L. Lartundo-Rojas et al. 2015. Microstructural properties and distribution of components in microparticles obtained by spray-drying. *J Food Eng* 152: 105–112.

Pourashouri, P., B. Shabanpour, S. H. Razavi, S. M. Jafari, A. Shabani, and S. P. Aubourg. 2014. Impact of wall materials on physicochemical properties of microencapsulated fish oil by spray drying. *Food Bioprocess Technol* 7: 2354–2365.

Pu, J., J. D. Bankston, and S. Sathivel. 2011. Production of microencapsulated crawfish (*Procambarus clarkii*) astaxanthin in oil by spray drying technology. *Drying Technol* 29: 1150–1160.

Rascón, M. P., E. Bonilla, H. S. García, M. A. Salgado, M. T. González-Arnao, and C. I. Beristain. 2015. T_g and a_w as criteria for the oxidative stability of spray-dried encapsulated paprika oleoresin. *Eur Food Res Technol* 241: 217–225.

Razmkhah, S., C. P. Tan, K. Long, and K. L. Nyam. 2013. Quality changes and antioxidant properties of microencapsulated kenaf (*Hibiscus cannabinus L.*) seed oil during accelerated storage. *J Am Oil Chem Soc* 90: 1859–1867.

Ribeiro, F. W. M., L. S. Laurentino, C. R. Alves et al. 2015. Chemical modification of gum Arabic and its application in the encapsulation of *Cymbopogon citratus* essential oil. *J Appl Polym Sci* 132: DOI: 10.1002/app.41519

Robert, P., P. García, N. Reyes, J. Chávez, and J. Santos. 2012. Acetylated starch and inulin as encapsulating agents of gallic acid and their release behaviour in a hydrophilic system. *Food Chem* 134: 1–8.

Ruiz Canizales, J., G. R. Velderrain Rodríguez, J. A. Domínguez Avila et al. 2019. Encapsulation to protect different bioactives to be used as nutraceuticals and food ingredients. In *Bioactive Molecules in Food*, ed. J. M. Mérillon, and K. G. Ramawat, 2164–2198. Cham, Switzerland: Springer Nature.

Sadek, C., H. Li, P. Schuck et al. 2014. To what extent do whey and casein micelle proteins influence the morphology and properties of the resulting powder? *Drying Technol* 32: 1540–1551.

Sánchez-Sáenz, E. O., C. Pérez-Alonso, J. Cruz-Olivares, A. Román-Guerrero, J. G. Baéz-González, and M. E. Rodríguez-Huezo,. 2011. Establishing the most suitable storage conditions for microencapsulated allspice essential oil entrapped in blended biopolymers matrices. *Drying Technol* 29: 863–872.

Sanguasri, L., and M. A. Augustin. 2001. Encapsulation of food ingredients, WO200174175-A1

Sarkar, S., S. Gupta, P. S. Variyar, A. Sharma, and R. S. Singhal. 2012. Irradiation depolymerized guar gum as partial replacement of gum Arabic for microencapsulation of mint oil. *Carbohydr Polym* 90: 1685–1694.

Sarkar, S., S. Gupta, P. S. Variyar, A. Sharma, and R. S. Singhal. 2013. Hydrophobic derivatives of guar gum hydrolyzate and gum Arabic as matrices for microencapsulation of mint oil. *Carbohydr Polym* 95: 177–182.

Sarkar, S., and R. S. Singhal. 2011. Esterification of guar gum hydrolysate and gum Arabic with *n*-octenyl succinic anhydride and oleic acid and its evaluation as wall material in microencapsulation. *Carbohydr Polym* 86: 1723–1731.

Schaffner, D. 2004. Process for the manufacture of powders preparations of fat-soluble substance. *US Patent* WO2004/062382.

Serfert, Y., C. Lamprecht, C.-P. Tan et al. 2014. Characterisation and use of β-lactoglobulin fibrils for microencapsulation of lipophilic ingredients and oxidative stability thereof. *J Food Eng* 143: 53–61

Serfert, Y., J. Schröder, A. Mescher et al. 2013. Spray drying behaviour and functionality of emulsions with β-lactoglobulin/pectin interfacial complexes. *Food Hydrocoll* 31: 438–445.

Sleeuwen, R. M. T., S. Zhang, and V. Normand. 2012. Spatial glass transition temperature variations in polymer glass: Application to a maltodextrin–water system. *Biomacromolecules* 13: 787–797.

Subpuch, N., T. C. Huang, and P. Suwannaporn. 2016. Enzymatic digestible starch from pyrodextrinization to control the release of tocopheryl acetate microencapsulation in simulated gut model. *Food Hydrocoll* 53: 277–283.

Sunderland, T., J. G. Kelly, and Z. Ramtoola. 2015. Application of a novel 3-fluid nozzle spray drying process for the microencapsulation of therapeutic agents using incompatible drug-polymer solutions. *Arch Pharm Res* 38: 566–573.

Sun-Waterhouse, D., S. S. Wadhwa, and G. I. Waterhouse. 2013. Spray-drying microencapsulation of polyphenol bioactives: A comparative study using different natural fibre polymers as encapsulants. *Food Bioprocess Technol* 6: 2376–2388.

Takashige, S., H. D. Ariyanto, S. Adachi, and H. Yoshii. 2020. Flavor release from spray dried powders with various wall materials. *ChemEngineering* 4: 1; doi:10.3390/chemengineering4010001.

Takashige, S., S. Iwamoto, H. Shiga et al. 2019. Stability of fish oil encapsulated in spray dried powders coated with starch particles. *Food Sci Technol Res* 25: 363–371.

Tang, C., and X. R. Li. 2013. Microencapsulation properties of soy protein isolate: Influence of preheating and/or blending with lactose. *J Food Eng* 117: 281–290.

Tatar, F., M. T. Tunç, D. Dervisoglu, D. Cekmecelioglu, and T. Kahyaoglu. 2014. Evaluation of hemicellulose as a coating material with gum arabic for food microencapsulation. *Food Res Int* 57: 168–175.

Tian, G., Y. Li, Q. Yuan, L. Cheng, P. Kuang, and P. Tang. 2015. The stability and degradation kinetics of Sulforaphene in microcapsules based on several biopolymers via spray drying. *Carbohydr Polym* 122: 5–10.

Tian, Y., N. Fu, W. D. Wu et al. 2014. Effects of co-spray drying of surfactants with high solids milk on milk powder wettability. *Food Bioprocess Technol* 7: 3121–3135

Toei, R., T. Furuta, and M. Okazaki. 1982. Drying of a droplet in a non-supported state. *AIChE Symp Ser* 78: 111–117.

Toie, R., M. Okazaki, and T. Furuta. 1978. *Drying mechanism of a non-supported state*. In *Proceedings of 1st International Drying Symposium*, ed. A. S. Mujumdar, 53–59. Princeton: Science Press.

Tonon, R. V., C. Brabet, and M. D. Hubinger. 2010. Anthocyanin stability and antioxidant activity of pray-dried açai (*Euterpe oleracea* Mart.) juice produced with different carrier agents. *Food Res Int* 43: 907–914.

Turan, F. T., A. Cengiz, and T. Kahyaoglu. 2015. Evaluation of ultrasonic nozzle with spray-drying as a novel method for the microencapsulation of blueberry's bioactive compounds. *Innov Food Sci Emerg Technol* 32: 136–145.

Turan, F. T., A. Cengiz, D. Sandıkçı, M. Dervisoglua, and T. Kahyaoglu. 2016. Influence of an ultrasonic nozzle in spray-drying and storage on the properties of blueberry powder and microcapsules. *J Sci Food Agric* 96: 4062–4076.

Ubbink, J., and J. Krüger. 2006. Physical approaches for the delivery of active ingredients in foods. *Trends Food Sci Technol* 17: 244–254.

Vicente, J., J. Pinto, J. Menezes, and F. Gaspar. 2013. Fundamental analysis of particle formation in spray drying. *Powder Technol* 247: 1–7.

Wan, F., M. J. Maltesen, S. K. Andersen et al. 2014a. One-step production of protein-loaded PLGA microparticles via spray drying using 3-fluid nozzle. *Pharm Res* 31: 1967–1977.

Wan, F., M. J. Maltesen, S. K. Andersen et al. 2014b. Modulating protein release profiles by incorporating hyaluronic acid into PLGA microparticles via a spray dryer equipped with a 3-fluid nozzle. *Pharm Res* 31: 2940–2951.

Wandrey, C., A. Bartkowiak, and S. E. Harding. 2010. Materials for encapsulation. In *Encapsulation Technologies for Active Food Ingredients and Food Processing*, ed. N. J. Zuidam, and V. Nedovic, 31–100. New York: Springer.

Wang, L. C., L. Q. Di, R. Liu, and H. Wu. 2013. Characterizations and microsphere formulation of polysaccharide from the marine clam (*Mactra veneriformis*). *Carbohydr Polym* 92: 106–113.

Wang, Y., L. Che, C. Selomulya, and X. D. Chen. 2014. Droplet drying behaviour of docosahexaenoic acid (DHA)-containing emulsion. *Chem Eng Sci* 106: 181–189.

Wang, Y., W. Liu, X. D. Chen, and C. Selomulya. 2016. Micro-encapsulation and stabilization of DHA containing fish oil in protein-based emulsion through mono-disperse droplet spray dryer. *J Food Eng* 175: 74–84.

Wu, Y., L. Zou, J. Mao, J. Huang, and S. Liu. 2014. Stability and encapsulation efficiency of sulforaphane microencapsulated by spray drying. *Carbohydr Polym* 102: 497–503.

Yamamoto, C., T. L. Neoh, Y. Honbou, Y. Furuta, S. Kimura, and H. Yoshii. 2012. Evaluation of flavor release from spray dried powder by ramping with dynamic vapor sorption–gas chromatography. *Drying Technol* 30: 1045–1050.

Yamamoto, C., T. L. Neoh, Y. Honbou, S. Kimura, H. Yoshii, and T. Furuta. 2012. Kinetic analysis and evaluation of controlled release of *d*-limonene encapsulated in spray dried cyclodextrin powder under linearly ramped humidity. *Drying Technol* 30: 1283–1291.

Zaitone, B. A., and A. Lamprecht. 2013. Single droplet drying step characterization in microsphere preparation. *Colloids Surf B Biointerfaces* 105: 328–334.

Zhou, H., X. Yuan, Q. Zhao, B. Zhao, and X. Wang. 2013. Determination of oxygen transmission barrier of microcapsule wall by crocetin deterioration kinetics. *Eur Food Res Technol* 237: 639–646.

3 Spray Drying Encapsulation of Phenolic Compounds

Raseetha Vani Siva Manikam and Aida Firdaus MN Azmi
Universiti Teknologi MARA, Malaysia

CONTENTS

3.1 INTRODUCTION

Phenolic compounds may either contain one or more aromatic rings combined with hydroxyl groups. According to Dai and Mumper (2010), phenolic compounds can be found in plants. In addition, phenolic compounds can also be found in fruits, vegetables, cereals, and beverages such as coffee and wine (Ozdal et al., 2016). Phenolic compounds can range from simple molecules (phenolic acids) to highly polymerized substances (tannins). Variations in phenolic compounds occur due to the differences in hydroxylation, methoxylation, prenylation, or glycosylation (Dai & Mumper, 2010). Phenolic compounds possess health benefits such as minimizing and alleviating cancer and cardiovascular diseases, as well as being anticarcinogenic, antiatherogenic, antiulcer, antithrombotic, anti-inflammatory, antiallergenic, anticoagulant, immune-modulating, antimicrobial, vasodilatory, and having analgesic activities (Ozdal et al., 2016). Table 3.1 shows examples of phenolic compounds present in our daily diet.

TABLE 3.1

Classes of Phenolic Compounds with Representative Members and Dietary Sources (Huang, Cai, & Zhang, 2009; Tulipani et al., 2008)

Phenolic Compounds	Representative Compounds	Chemical Structure	Molecular Weight (g·mol⁻¹)	Chemical Formula	Dietary Sources
Flavones	Apigenin		270.24	$C_{15}H_{10}O_5$	Oranges, lemons, apricots, apples, blackcurrants, bananas, potatoes, spinach, onions, lettuce, beans, cereals
Flavonols	Quercetin		302.23	$C_{15}H_{10}O_7$	Honey, red grapes, citrus fruits, cherries, and raspberries
Flavan-3-ols	Epigallocatechin gallate (EGCG)		458.37	$C_{22}H_{18}O_{11}$	Green/black tea
Isoflavones	Genistein		270.24	$C_{15}H_{10}O_5$	Soy milk, tofu
Anthocyanins	Cyanidin		287.24	$C_{15}H_{11}O_6^+$	Plums, grapes, elderberries, cherries

Proantho-cyanidins	ProanthocyanidinB		578.50	$C_{30}H_{26}O_{12}$	Cranberries, grapes, walnuts, rice
Hydroxy-cinnamic acids	Caffeic acid phenethyl ester		284.31	$C_{17}H_{16}O_4$	Artichoke, oregano, thyme, basil, coffee, mushrooms, medicinal plants
Ellagitannins (ellagic acid derivatives)	Punicalagin		1084.71	$C_{48}H_{28}O_{30}$	Berry fruits, pomegranate
Stilbenes	Resveratrol		228.24	$C_{14}H_{12}O_3$	Grapes, mulberries

Bioactive ingredients such as phenolic compounds can degrade easily, especially during food handling, transportation, storage, and processing, due to the chemical and enzymatic reactions and further loss of bioactivity. Thus, the encapsulation process is highly needed. Phenolic compounds are encapsulated in order to make them more stable in functional foods and gut while improving their bioavailability and more. Besides, encapsulation can mask any unwanted flavor, smell, and taste in either food and/or pharmaceutical products. The bioavailability of encapsulated phenolic compounds highly depends on their dispersion and disintegration in the target area. Differences in physical properties of the initial material and emulsion properties could affect the encapsulation efficiency (Consoli, Hubinger, & Dragosavac, 2020). By enhancing the encapsulation efficiency, stability, and bioavailability of phenolic compounds, it could benefit consumers by preventing or alleviating aging, inflammation, cancer, and neurodegenerative diseases. According to Botelho, Canas, and Lameiras (2017), encapsulation refers to retaining the phenolic compounds (referred to as the core, fill, active, internal, or payload phase) within the wall material (referred to as coating, membrane, shell, capsule, carrier material, external phase, or matrix).

Dias, Ferreira, and Barreiro (2015) stated that the encapsulation process can extend the shelf-life by reducing disintegration or breakdown of phenolic compounds from the core into the surrounding membrane, and hence, results in targeted delivery. This situation also protects the phenolic compounds with better probabilities for solubility and flow to maintain the bioavailability. Spray drying encapsulation can protect the phenolic compounds from chemical degradation (e.g. oxidation or hydrolysis) and maintain them as functional as possible until being released in the gastrointestinal tract (GIT).

3.2   USE OF PHENOLIC COMPOUNDS IN FOOD

During plant growth, phenolic compounds play a major role against ultraviolet radiation or as a self-defense mechanism against pathogens, parasites, and predators. Upon harvest, phenolic compounds can contribute to the color of the produce and when it is incorporated into food products. For instance, phenolic compounds such as anthocyanins can cause color changes in wine when they come in contact with oxygen. Color changes are crucial to preserving and maturing wine (Dai & Mumper, 2010). In addition, phenolic compounds also lead to bitter and astringent flavors due to the interaction between phenolic compounds, mainly procyanidin, and the glycoproteins in saliva. However, astringency and bitterness are not commonly preferred by consumers due to their unpleasant taste (Leyva-Jiménez, Lozano-Sánchez, Cádiz-Gurrea, Arráez-Román, & Segura-Carretero, 2019). This might be due to the concentration of the phenolic compounds themselves. Besides, phenolic compounds can be used to prevent and alleviate health problems that are caused by the overproduction of radical species (Vulić et al., 2019). Once the imbalance between antioxidant and excessive production of free radical species occurs in humans, and consequently, it causes substantial damage or injury to major cells such as proteins, lipids, and nucleic acids that can increase the risk of cardiovascular disease, Alzheimer, cancer, stroke, neurodegenerative disorders, and other possible diseases.

Phenolic compounds have poor bioavailability, mainly due to their low water solubility and their chemical instability. For instance, maximum plasma concentration for absorbing aglycones takes up to two hours after ingestion. However, most phenolic compounds found in food systems are in the form of esters, glycosides, large molecules, or macromolecules that cannot be absorbed before being broken into smaller units (Zhang et al., 2013). These substances must be broken down by intestinal enzymes for easier absorption.

Leyva-Jiménez et al. (2019) and Kumar Singh et al. (2019) found that 5%–10% of the total dietary phenolic compounds could be absorbed in the small intestine. Thus, only a small percentage of phenolic compounds is being absorbed, and they are not fully utilized. The bioavailability of phenolic compounds depends on various factors such as the food matrix, water-solubility, technological processes, and interactions in the GIT (Leyva-Jiménez et al., 2019). Yinbin et al. (2018) reported

TABLE 3.2

Previous Studies on the Spray Drying Encapsulation of Phenolic Compounds

Sample	Results	References
Khasi mandarin orange, watermelon, carambola, pineapple (fruit juice)	- The encapsulation yield for mandarin orange was 85.2% - Total phenolic content was higher in mandarin orange compared with other fruits tested	Saikia, Mahnot, and Mahanta (2015)
Purple sweet potato (flour)	- Total phenolic content was 57.23 g/kg when encapsulated with 10 g/kg ascorbic acid and 30 g/kg maltodextrin.	Ahmed, Akter, Lee, and Eun (2010)
Cistus L herb	- Sterilization caused a 24% loss of total phenolic compounds using 1.5% carboxymethyl chitosan	Ammendola et al. (2020)
Resveratrol	- Encapsulation efficiency (97%) obtained with low shear membrane emulsification method was used	Consoli et al. (2020)
Thyme essential oil	- Encapsulation efficiency using casein and maltodextrin was 88.9% and also exhibited antimicrobial activity	Radünz et al. (2020)
Capsicum	- Total phenolic content extracted using ethanol 50% and water was insignificant	de Sá Mendes et al. (2020)
Carob pulp	- Encapsulation efficiency reduced from 95.2% to 87% after two cycles of emulsification	Wang et al. (2020)
Curcumin	- Curcumin retention was 97.9% in a mixture of soy protein isolate: soy soluble polysaccharide: maltodextrin (1:1:2)	Chen, Liu, and Tang (2020)
Pineapple waste	- Total polyphenol compounds were lower (18 mg/g) when 10% maltodextrin was used compared with 2.5% maltodextrin	de, Siacor, and Taboada (2020)
Bay leaf	- Total phenolic content was higher (430 mg/L GAE) when gum Arabic was used as wall material with an encapsulation efficiency of 96.1%	Chaumun, Goëlo, Ribeiro, Rocha, and Estevinho (2020)
Rosemary leaf	- Rosmarinic acid encapsulated in an optimized proliposome resulted in high retention (100%)	Bankole, Osungunna, Souza, Salvador, and Oliveira (2020)

that the encapsulation process retained phenolic compounds derived from plum with minimum degradation caused by reaction with oxygen, heat, and light. Commonly used carrier materials for the encapsulation of polyphenols are carbohydrate, protein isolates, and lipids.

The spray drying technology encapsulates the liquid feed into a powder form and it is the most widely used encapsulation method in the food industry (Ray, Raychaudhuri, & Chakraborty, 2016). Nowadays, the encapsulation of bioactive compounds like phenolic compounds has been receiving more interest to provide useful applications in the food industry (Ray et al., 2016). Table 3.2 indicates previous studies that have used spray drying for the encapsulation of extracted phenolic compounds. The spray drying technique is becoming even more popular nowadays for the incorporation of polyphenols in food products (Aguiar, Estevinho, & Santos, 2016; Arpagaus, Collenberg, Rutti, Assadpour, & Jafari, 2018; Jia, Dumont, & Orsat, 2016; Simões et al., 2017). Spray drying encapsulation is described as pumping the targeted active ingredients as a solution in the atomizer, then spraying it in the form of mist or nebulized droplets and drying it in the drying chamber. Next, the remaining residual solvents are evaporated using hot air, hence converting liquid drops into solid particles. Finally, the solid particles are collected in the filter compartment (Filkovà & Mujumdar, 1995).

3.3 PARAMETERS FOR EFFICIENT ENCAPSULATION OF PHENOLIC COMPOUNDS

3.3.1 INITIAL STAGE

3.3.1.1 Properties of Wall Material(s)

Selecting appropriate outer wall material is the initial step in encapsulating phenolic compounds; it should be a film-forming biopolymer that comes from a natural or synthetic polymer, and depends on the active agent and the final characteristic for final microcapsules (King, 1995). As an example, for flavor encapsulation, a suitable outer wall material must possess strong emulsifying and film formation capability. Meanwhile, the rheological properties of wall materials should also be considered; e.g., low viscosity, low hygroscopicity, and cost-effectiveness (Wu, Chai, & Chen, 2005). The chosen material in coating the active agent(s) determines the encapsulation efficiency and stability of the active agent(s) (Ray et al., 2016). An ideal outer wall material should have the following characteristics (Desai & Park, 2005):

(1) easy to work on (desirable rheological properties even at high concentration);
(2) suitable emulsification characteristics;
(3) chemically inert towards the active agent;
(4) can seal and hold the active agent during processing and storage conditions;
(5) can remove the solvent (used during the encapsulation process) completely;
(6) can retain the active agent against surrounding conditions like light, moisture, heat, and humidity;
(7) the solvent used should be acceptable according to the industry (food or pharmaceutical).

It has been reported that gums are capable to encapsulate phenolic compounds better than other coating materials (Busch et al., 2007; Dag, Kilercioglu, & Oztop, 2017). Generally, gums are used in combination with other types of carrier agents (e.g., maltodextrin or pectins), to enhance their functionalities in encapsulating active agents (Shishir, Xie, Sun, Zheng, & Chen, 2018). Lima, Madalão, Benincá, Saraiva, and Silva (2019) found that the materials suitable for encapsulating polyphenols by spray drying were maltodextrin and gum Arabic as they efficiently controlled the release of the target compound. Maltodextrin with a dextrose equivalence (DE) within 10–20 is a suitable wall material for anthocyanins and phenolic acids. Maltodextrin at DE = 20 to 23 resulted in a higher anthocyanin content (ANC) at the end of the drying process, compared with DE = 28–31 and DE = 10 (Ersus & Yurdagel, 2007).

Gum Arabic, modified starches, and hydrolyzed starches are usually used as coating materials for spray drying of phenolic compounds; these materials have good emulsifying properties and can retain volatile compounds during the drying processes. Besides, gum Arabic has good solubility, consists of prebiotic effects, and is highly resistant under GIT conditions (Badreldin, Ziada, & Blunden, 2009). The emulsion stability before drying depends on the choice of outer wall material for the encapsulation of phenolic compounds as it influences the ability to flow, mechanical stability, and shelf-life after drying. As an example, Zhang, Zhang, Chen, and Quek (2020) reported that phenolic compounds extracted from cranberries were mixed with a different mixture of outer wall materials using gum Arabic and maltodextrin at different DE values. From the findings, the phenolic content achieved the highest recovery with maltodextrin powder of DE=10 to 13. Another study by Paini et al. (2015) indicated that for the encapsulation of phenolic compounds from the olive pomace, maltodextrin was selected as a coating agent as it was easily soluble, less viscous at high solid content, and

TABLE 3.3

Previous Studies Using Different Types of Wall Materials for the Encapsulation of Phenolic Compounds

Wall Materials	Sample	Results	References
Maltodextrin and inulin	Japanese quince (JQ) fruit	- Phenolic content was higher when maltodextrin was used, followed by inulin: maltodextrin (2:1) and inulin: maltodextrin (1:2). - Flavan-3-ol, phenolic acid, and polymeric procyanidin contents were the highest for encapsulation with inulin	Turkiewicz et al. (2020)
Soy protein isolate, 1:1 soy protein isolate and soy soluble polysaccharide, 1:1 soy protein isolate with 2% maltodextrin, 1:1:2 soy protein isolate, soy soluble polysaccharide, and maltodextrin	Curcumin	- Curcumin retention was 97.9% when used 1: 1: 2 soy protein isolate, soy soluble polysaccharide, and maltodextrin - Surface morphology in 1:1:2 soy protein isolate, soy soluble polysaccharide, and maltodextrin was very smooth and fewer wrinkles	Chen et al. (2020)
Maltodextrin	Olive pomace	- Optimum inlet and outlet temperatures were 130°C with 94% yield and 76% efficiency - UV light affected 66% loss of polyphenols after 48 hours of exposure	Paini et al. (2015)
Maltodextrin and gum Arabic	Olive pomace	- The combined ratio of maltodextrin and gum Arabic (60: 40) resulted in 87.3% yield compared with other ratios (0:100, 20: 80, 40: 60, 60: 40, 80: 20, and 100:0%). - Total polyphenol yield for all ratio combinations was insignificant.	Aliakbarian, Paini, Casazza, and Perego (2015)
Maltodextrin, gum Arabic, lactose (as drying carrier), polysorbate 80, poloxamer 188 (as surfactant)	Clove	- Eugenol and eugenyl acetate lower when lactose was used as drying carrier compared with maltodextrin or mixture of maltodextrin: gum Arabic (60%) - Poloxamer 188 indicated increased eugenol and eugenyl acetate content compared with polysorbate 80	Cortés-Rojas, Souza, and Oliveira (2014)

(Continued)

TABLE 3.3
(Continued)

Wall Materials	Sample	Results	References
Modified chitosan, sodium alginate, and gum Arabic	Bay leaf	- Release pattern in order water> gum Arabic (6 min)> modified chitosan (9 min)>sodium alginate (15 min) - Microparticles with modified chitosan had smoother surface compared with others	Chaumun et al. (2020)
Casein and maltodextrin (1:4)	Thyme essential oil	- Encapsulation efficiency was 88.9% - Spherical and irregular shape compared with without sphere formation in controls (without active ingredients)	Radünz et al. (2020)
12 variations of using oleuropein/ sodium alginate ratios and inlet temperatures	Olive leaf extract	- Oleuropein/sodium alginate (1: 1.6) with inlet air 135°C resulted in 51.5% yield and 60.8% efficiency - Encapsulation efficiency was higher when sodium alginate concentration was higher	Gonzalez et al. (2019)
Colloidal silicone dioxide (as drying agent), maltodextrin, and starch	Soybean extract	- Inlet temperature was varied at 80°C, 115°C, and 150°C - Total polyphenol for soybean extract added with colloidal silicone dioxide was higher (11.05 mg/g), compared with maltodextrin (10.8 mg/g) and starch (9.87 mg/g) - Increased inlet temperature resulted in lower polyphenol content	Georgetti, Casagrande, Souza, Oliveira, and Fonseca (2008)

able to function at a given temperature. The use of pure starch as the carrier agent was not feasible, because of the hydrophilic nature and less emulsifying characteristics (Fathi, Martín, & McClements, 2014). Thus, modified starches were used as the encapsulated wall material, because these starches have superior functionalities and improved commercial applicability (Shishir et al., 2018).

Additionally, cyclodextrins have inner hydrophobic properties and exhibit external hydrophilic surface, hence, permit encapsulation with poor water-soluble molecules (e.g., polyphenol) and increase the solubility (Duchêne & Bochot, 2016). Chitosan nanoparticles prevented the degradation of tea polyphenols in the GIT and improved their absorption in intestinal epithelial cells (Liang et al., 2017). Furthermore, gelatins can also be considered as potential outer wall material for encapsulation; nonetheless, they are soluble in water and can release active ingredients quickly in an aqueous solution (Dang et al., 2017). In addition, zein (extracted from cereals such as maize) is insoluble in water and capable to be considered as a carrier material to encapsulate phenolic compounds and develop the delivery system in both food and pharmaceutical industries (Dai et al., 2017; Donsì, Voudouris, Veen, & Velikov, 2017; Yang et al., 2017). Lipids are good wall materials for the encapsulation of bioactive agents such as phenolic compounds as most phenolics are hydrophobic. Polar fat compounds such as phospholipids are suitable to be used as wall materials since they allow controlled release of active ingredients (Đorđević et al., 2016). Table 3.3 summarizes previous studies that used various wall materials to encapsulate phenolic compounds.

3.3.1.2 Spray Drying Conditions

An ideal spray drying equipment consists of a series of components, including a system for controlling the drying temperature, a system to control and pump in-feed flow of the material, and a system to collect the dried material (Silveira, Perrone, Junior, & Carvalho, 2013). Spray drying efficiency depends on the total contact surface area of the material intended to be dried in the drying chamber (Figure 3.1). Therefore, sufficient drying temperatures are needed to initiate the removal of the solvent and vaporize the liquid samples (Sokhansanj & Jayas, 1995). In the spray drying procedure, the inlet and outlet temperature should be adequate and controlled, in order to prevent minimal loss of the compounds that are sensitive to heat (Ray et al., 2016). Moreover, the dissolved solvent cannot

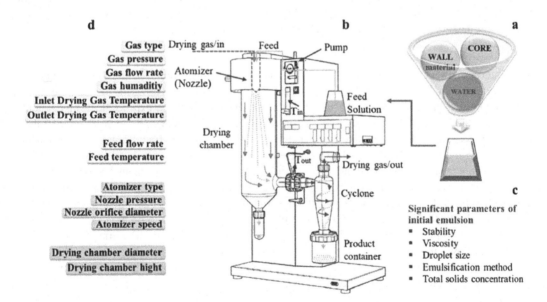

FIGURE 3.1 A schematic illustration of the process of spray drying including emulsification (a) and directed to drying step (b) with parameters to consider in the initial emulsion (c) and during the spray drying process (d). T stands for temperature. Modified from Geranpour, Assadpour, and Jafari (2020).

evaporate adequately in the expected time. Consequently, the resulting powder will still be damp, with residual moisture content and resulting in lower encapsulation yield, if spray drying is carried out with low inlet temperature. Additionally, when higher inlet temperature is used, microcapsules may crack (Kha, Nguyen, Roach, & Stathopoulos, 2014). Particle size changes along the spray drying depending on operating drying temperature. When samples are spray dried at a lower inlet/outlet temperature (e.g., about 110/74°C), scanning electron microscopy (SEM) images show shrinking of the particles compared with other those dried at the higher inlet/outlet temperature (e.g., 170/145°C and 200/173°C) (Alamilla-Beltràn, Chanona-Pérez, Jiménez-Aparicio, & Guitérrez-López, 2005).

3.3.1.3 Atomizer

Spraying the product into the chamber is one of the critical points in the spray drying of phenolic compounds. This is because the particle size of the final product is affected by the size and speed at which the droplets are being formed from the initial solution. The final droplets should be homogenous and uniform in size (Moreira et al., 2009). It is important that the atomization process is optimized for maximum volatile retention. The main role of the atomizer is to ensure samples are transformed into tiny droplets and evenly distributed in the drying chamber for mixing with heated air or gas. Atomizer must produce tiny droplets that are not large (to avoid incomplete drying) or not so small products (to prevent difficulties in recovery). Additionally, the atomizer should be able to disperse the sample materials at a certain rate into the drying chamber (Patel, Patel, & Suthar, 2009).

The form of sprayer used is also a vital factor as it affects the distribution and size of the particles being formed. There are three types of atomizers, including pressure nozzle, double-piston, and centrifugal atomizers (Vissotto, Montenegro, Santos, & Oliveira, 2006). A pressure atomizer forces the sample down a narrow orifice and pumps it into the nozzle opening at high pressure. It is used for its cost-effectiveness, but the downside effects include the clogging of the orifice (Finney, Buffo, & Reineccius, 2002). The pneumatic fluid atomizer is usually used as an atomizer for phenolic encapsulation. It is an atomizer equipped with a pneumatic pump to allow sample introduction at low pressure. Speed differences between dryer and samples cause shear stress and lead to drying of samples.

This arrangement ensures proper control of the outcome of the product in terms of the size distribution of the droplets (Samantha et al., 2015). Finney et al. (2002) found that the nozzle atomization produced a larger particle size compared with wheel (centrifugal) atomization. The selection of a proper atomization type is necessary to determine the final particle size.

3.3.1.4 The Direction of the Input

The materials can be introduced into the spray dryer in either concurrent, counter-current, or mixed flow processes. Counter-current flow deals with the flow of a liquid sample in the opposite direction with the drying air, causing the material that is almost dry to be exposed to a higher temperature, resulting in a completely dried outcome (Silveira et al., 2013). However, the mixed flow process involves both concurrent and counter-flow directions, whereby liquid samples are atomized in increasing the direction with airflow in decreasing direction being circulated inside the drying chamber. This explains that the liquid samples initiate the ascending movement, but will change to a descending movement when the air drags the material down. The hot air and liquid material leave the equipment in the lower area. This method is only used with a smaller-sized drying chamber (Finney et al., 2002; Patel et al., 2009). Phenolic encapsulation has been optimized by co-current flow where the liquid samples and drying air (or gas) are introduced in the same direction. This direction of flow causes high exposure of the droplets to high temperatures that cause high evaporation. Therefore, this direction of flow results in the final products that have lower compactness and hollow assemblies (Sokhansanj & Jayas, 1995).

3.3.1.5 Temperature

The success of polyphenol encapsulation depends on the encapsulating outer wall agent, proper temperature selection, and drying conditions (Murugesan & Orsat, 2011). Usually, certain organoleptic characteristics in sample material would degrade due to the process conditions such as high temperatures and air when entering the drying chamber. The outflow temperature of the products is caused by the combination of parameters such as the speed of liquid sample introduction, the rate of aspiration, and the dryer temperature. Both inflow and outflow temperatures of the final product determine its moisture content, where high variation between inflow and outflow temperature can cause a high amount of residual moisture content that causes instability of the phenolic compound (Moreira et al., 2009). Table 3.4 summarizes various operational conditions used by different researchers for spray drying encapsulation of phenolic compounds. The inflow

TABLE 3.4
Operational Conditions of Spray Dryer for Encapsulation of Phenolic Compounds

Parameters	Samples	Findings	References
Atomization speed (25000–35000) and maltodextrin concentration (12.5%–15%) was optimized using Response Surface Methodology	Pineapple juice	- Increased maltodextrin concentration and atomization speed decreased moisture content. - -True density was lower at high atomization speed.	Abadio, Domingues, Borges, and Olieveira (2004)
Optimised the inlet (125.6°C, 130°C, 140°C, 150°C, 154.1°C) and outlet (72.9°C, 75°C, 80°C, 85°C, 87.1°C) temperatures carried out using Response Surface Methodology	Bitter melon	- Optimum inlet and outlet - temperatures were 140°C and 80°C, respectively. - - Encapsulated material was more stable at −20°C and 10°C, compared with 30°C, until 150 days.	Tan, Kha, Parks, Stathopoulos, and Roach (2015)
Inlet temperature was varied between 140°C to 160°C. Constant parameters (outlet temperature: 80°C, maltodextrin 25%, feed flow at 10 mL/min).	Capsicum	- Lower inlet temperature resulted in a smoother morphology surface without fissures. - Higher inlet temperature resulted in spherical shape and shrinkage.	de Sa Mendes, Coimbra, et al. (2020)
12 variations of inlet temperature (135°C–160°C) with varying ratios of olive leaf extract: inulin	Olive leaf extract	- About 87.1% encapsulation efficiency and 71% yield were found with 136°C inlet temperature and olive leaf extract: inulin ratio of 1: 1.8.	Urzúa et al. (2017)
Inlet temperature varied (140°C–160°C), and maltodextrin: juice ratio also varied (30: 70, 40: 60, and 50: 50)	Tomato powder	- Increased maltodextrin: juice ratio (50:50) had higher recovery (9.13%). - Lower recovery with high inlet temperature was seen.	Sidhu, Singh, and Kaur (2019)
Maltodextrin concentration (12%–31.7%), gelatin (0.6%–6%) and emulsification method (high shear homogenization and sonication) was varied Emulsion with 20 g maltodextrin and 0.6 g gelatin/100 g was subjected to drying temperature (124°C–190°C), atomization flow (275°C–536 L/h) and emulsion flow (1.4–8.6 mL/min)	Turmeric	- Combinations with lower gelatin such as 26% maltodextrin/0.6% gelatin and 31.7% maltodextrin/0.6% gelatin exhibited lower values of storage (G′) and loss (G″) modulus. - Inlet temperature of 160 °C, atomization airflow of 418 L/h, and emulsion flow of 5 mL/min resulted in high encapsulation efficiency (77.2%).	Ferreira, Piovanni, Malacrida, and Nicoletti (2019)

temperature is the same as the evaporation of the liquid sprayed. Spray drying efficiency depends on the remaining water content and characteristics of the final product. High inlet temperatures (i.e., 170°C–200°C) cause the fast formation of droplet surface that could potentially be penetrated and cause morphological damage to phenolic compounds (Alamilla-Beltràn et al., 2005).

Polyphenols and other antioxidants are usually degraded upon thermal treatment (Suhag & Nanda, 2016). Studies conducted by Sambroska et al. (2019) found that increasing the inlet temperature for encapsulating phenolic compounds from honey has lowered the total phenolic content (TPC) in honey. Michael (1993) claimed that a higher inflow temperature for air or gas resulted in rapid moisture and solvent evaporation; however, final products that were susceptible to high heat might have changes in their physical and chemical properties. The outflow air temperature is the most important parameter in the spray drying process that ensures the success of encapsulation. The higher outflow of air temperature would produce a larger size of powder recovery (Maury, Murphy, Kumar, Shi, & Lee, 2005). Ferrari, Ribeiro, and Aguirre (2012) reported that increased drying temperature caused reduced moisture content and water activity from the encapsulation of black cherry pulp using maltodextrin. If inlet air is kept constant, decreased introduction of liquid samples would increase the temperature of the outflow material (Oliveira & Petrovick, 2009).

Preliminary studies by Samantha et al. (2015) indicated that the inlet temperature of < 130°C formed a wet layer on the walls of drying material; however, polyphenol degradation occurred if the inlet temperature was set at 160°C or higher. Sivetz and Foote (1963) found that decreasing the feed temperature for phenolic encapsulation by coffee solid extracts before drying could improve the coffee flavor. From studies conducted by Paini et al. (2015), they found that an increase in inlet temperature by 160 °C caused a decrease in bulk density of the final products and degradation of the spherical structure of the particle which caused higher porosity in the product.

3.3.2 During the Drying Process

3.3.2.1 Initial Emulsion

The retention of phenolic compounds in the final spray dried powder depends on the preparation of the initial emulsion. Thus, there are several parameters that must be taken into consideration to ensure high retention of phenolic compounds during encapsulation: the size of droplet formation, stability, viscosity, and total solid contents of the emulsion (Jafari, Assadpoor, He, & Bhandari, 2008). Retention of active ingredients and the encapsulation of phenolic compounds during spray drying is dependent on the total solid content in the emulsion. Reduced time to form a semipermeable or wet membrane at the surface of the drying material is achievable by increasing total solid content. Besides, high total solids increase emulsion stability; thus, this situation prevents the circulating movement inside the droplet (Ré, 1998).

Although there is always a good side to higher total solids, optimum infeed solid content is necessary for spray drying encapsulation of phenolic compounds (Bhandari, 2007). One of the two main reasons why optimum infeed total solid is needed is because increased wall materials that exceed their solubility can lead to undissolved wall materials. The next reason is due to the relationship between the initial viscosity of the initial emulsion and total solid content. The use of core material contributes to the infeed solid content. The emulsion viscosity also determines the volatiles retention, where it influences the loss of volatiles to the point that the surface of drying samples becomes semipermeable. Theoretically, increased viscosity for the emulsion to optimum level would decrease internal oscillation and inner flow of droplets; thus, improving the retention. Higher viscosity in emulsion could result in a larger droplet (Ré, 1998). Maltodextrin provides low viscosity and higher content of total solids for the encapsulation of phenolic compounds from cranberry juice where it recovers a high amount of TPC after spray drying. Therefore, the viscosity of infeed emulsion depends on the selection of outer wall material, to prevent the loss of more phenolic compounds.

3.3.2.2 Particle Size

Parameters and physical properties of the infeed emulsion as described previously (i.e., viscosity and total solid content) affect the particle size of the manufactured powder. Besides, the atomization parameters (rotational speed and wheel diameter) also contribute to the powder particle size (Jafari et al., 2008). As an example, higher pressure and smaller orifice result in smaller particles. Initial raw material and atomization characteristics define the properties of the final powder (Silveira et al., 2013). A high concentration of wall material reduces bulk density and increases the particle size of the powder. Studies conducted by Paini et al. (2015) indicated that increased concentration of maltodextrin resulted in a higher size of the powder. Jafari et al. (2008) stated that medium-sized particles had the highest encapsulation efficiency, whereas, smaller particles tended to disperse poorly (formed lumps on a liquid surface such as cold water).

3.4 METHOD OF EMULSIFICATION

Emulsification can be used to encapsulate compounds exhibiting either hydrophilic or hydrophobic properties (Gumus, Decker, & McClements, 2017; Jia et al., 2016). Homogenization is necessary to alter two immiscible liquids into an emulsion. This technique includes single emulsion (oil in water, O/W, and water in oil, W/O) and double emulsion (water in oil in water, W/O/W, and oil in water in oil, O/W/O). Preparation of emulsions is commonly applied to encapsulate active ingredients in an aqueous solution to form powder by drying technology (e.g., spray drying) (Fang & Bhandari, 2010). Nanoemulsions are to be used in the liquid form; however, these emulsions are more stable during storage once they are spray dried.

The emulsion prepared for spray drying (core: wall material ratio of 1: 4) is usually homogenized before being introduced to the spray dryer (Gibbs, Kermasha, Alli, & Mulligan, 1999). Homogenization of emulsions can be carried out using either a high-energy approach or a low-energy approach (Shishir et al., 2018). Emulsification using high energy may need high-end equipment such as a high-shear mixer, a high-speed homogenizer, an ultrasonicator, a microfluidizer, a membrane homogenizer, and a microchannel homogenizer, to yield micro- or nanoemulsions (Simões et al., 2017). The selection of a high-energy instrument determines the characteristics of the micro or nanoemulsion produced (Shishir et al., 2018). Increased pressure during homogenization could produce lycopene nanodispersions with a thin poly-dispersity index (PDI) that was stable for the incorporation into beverages, as reported by Shariffa et al. (2017). Compared with the high-energy emulsification technique, low-energy emulsification technique that allows a thin range of oil and surfactant needs more surfactants to produce an emulsion (Joye & McClements, 2014). Furthermore, emulsions fabricated using low energy technique are typically not stable at elevated temperatures (Saberi, Fang, & McClements, 2016).

3.5 EMULSION STABILITY

The emulsification of phenolic compounds with the chosen carrier agents could contribute to the retention and stability during spray drying. Several studies reported favorable encapsulation efficiency of various phenolic compounds. Based on Zhang et al. (2020), phenolic compounds from cranberry juice extract were encapsulated using different types of wall materials, including gum Arabic (GA), maltodextrin (M1, 10–13 DE, and M3, 17–20 DE), and a combination of GA and M1. Higher phenolic compounds and anthocyanin retention was shown by blended wall materials of GA and M1 (215.6% and 83.3%, respectively). The wall materials with the lowest phenolic compounds and anthocyanin retention were maltodextrin with 17–20 DE, which were 137.8% and 58.8%, respectively. Moreover, the particle size of the manufactured powder was lower than 10 μm in the case of all wall materials used. Individual phenolic compounds that were identified in the cranberry juice extract were: anthocyanins (cyanidin 3-galactoside, cyanidin 3-gluctoside, cyanidin 3-arabinoside, peonidin 3-galactoside, peonidin 3-glucoside, peonidin 3-arabinoside); proanthocyanins (PAC B1, catechin, PAC B2, epicatechin, PAC A2); phenolic acids (protocatechuic acid and chlorogenic acid); and, flavonols (myricetin-3-galactoside, myricetin-3-glucoside, quercetin-3-galactoside, quercetin-3-rhamnoside, phloridzin, and quercetin).

Moreover, Gonzalez et al. (2019) reported on the spray drying encapsulation of active ingredients from olive leaf extract with the carrier agent of sodium alginate (SA). Individual phenolic compounds that were identified in the olive leaf extract were: secoiridoids (oleuropein, oleuropein diglucoside, and ligstroside); simple phenols (hydroxytyrosol glucoside and verbascoside); elenolic acids (oleoside 11-methyl ester); oleosides (oleoside isomer a and b); and, flavonoids (rutin, luteolin rutinoside, luteolin glucoside, and chrysoeryol-7-O-glucoside). Total phenolic compounds of the encapsulated olive leaf extract were 30.4 mg/g and the highest phenolic compound identified was oleuropein (24.3 mg/g). The study indicated decreased phenolic content after drying and subsequently forming the manufactured powder of about 16.7 mg/g microparticles. The particle size of the manufactured powder ranged from 0.25 μm to 20 μm. Moreno, Cocero, and Rodríguez-Rojo (2018) reported previously that the encapsulation of active ingredients from red grape marc extract with three types of carrier agents such as maltodextrin (MD), whey protein isolate (WPI) (90% protein), and pea protein isolate (PPI) (75% protein). The study reported that WPI was efficient and produced high drying capability, with a yield of 69.3% for the carrier: extract (0.5: 1). The same carrier agent also showed higher phenolic compounds where ANZ retention of the TPC and total ANZ among other samples were 159 ± 1.8 mg/g and 2.73 ± 0.16 mg/g, respectively. The sample with the highest TPC and total ANZ was WPI (0.3: 1) with the value of 167 ± 9.7 mg/g and MD (0.5: 1) with the value of 3.21 ± 0.05 mg/g, respectively; however, it had lower drying yield (%) than WPI (0.5: 1) to achieve the 50% yield benchmark. The particle size of the manufactured powder with carrier agent of MD, WPI, and PPI with the ratio of 1: 1, carrier: extract solids was 16.2 ± 0.3 μm, 15.0 ± 0.8 μm, and 62.7 ± 9.5μm, respectively.

A previous study from Pieczykolan and Kurek (2019) reported on the encapsulation of anthocyanins from black chokeberry extract with the combination of polysaccharides with dietary fibers, where the polysaccharide was maltodextrin and the dietary fiber included gum Arabic, inulin, pectin, guar gum, and beta-glucan. The highest anthocyanin retention was using the carrier agent of maltodextrin and beta-glucan with the value of 3055.65 mg/100 g, while the lowest was maltodextrin and gum Arabic with the value of 1940.22 mg/100 g. When the outer coating material was highly soluble, it allowed the development of a hollow space in which the active ingredients could stay encapsulated (Tonon, Brabet, & Hubinger, 2010). The particle size of the manufactured powder with maltodextrin + gum Arabic, maltodextrin + inulin, maltodextrin + beta-glucan, maltodextrin + pectin, and maltodextrin + guar gum was 53.09, 17.66, 25.61, 17.35, and 16.29 μm, respectively. Encapsulation of phenolic compounds extracted from the spent coffee grounds was evaluated by Ballesteros, Ramirez, Orrego, Teixeira, and Mussatto (2017) with the outer wall materials of maltodextrin and gum Arabic. The retention of phenolic compounds and flavonoids was high with the combined maltodextrin and gum Arabic with a value of 204.86 mg/100 ml and with the carrier agent using maltodextrin with the value of 7.88 mg/100 mL, respectively. The particle size was < 20 μm for all the samples. This showed that maltodextrin had good encapsulation capability as a protective layer and with the combination with another carrier agent, increased its functionality.

Cortés-Rojas et al. (2014) incorporated eugenol and eugenyl acetate extracted from clove, encapsulated by a solid lipid structure system followed by spray drying. The incorporations were tested using few formulations containing glyceryl dibehenate and stearic acid (acting as solid lipids), polysorbate 80, and Poloxamer 188 (acting as surfactants), lactose, maltodextrin, and gum Arabic. Higher retention of eugenol and eugenyl acetate was reported when maltodextrin DE 10 and Poloxamer 188 were used. Both eugenol and eugenyl acetate showed retention of nearly 60% with this formulation, which was higher than other formulations. The particle size of this formulation was the largest (5.005 μm (d_{10}), 22.49 μm (d_{20}), and 48.17 μm (d_{10}).

3.6   STABILITY OF SPRAY DRIED PHENOLIC COMPOUNDS

Previous studies that explored the stability of spray dried phenolic compounds are presented in Table 3.5. Many studies have explored the stability of manufactured powder at different temperatures. Zhang et al. (2020) studied the encapsulated phenolic compounds of cranberry juice extracts by spray drying with various types of outer wall materials, including gum Arabic and maltodextrin. The manufactured powder was stored for 12 weeks at different temperatures: 4°C, 25°C, and 45°C. The total contents of individual phenolics measured during the storage were anthocyanins, proanthocyanins, phenolic acids, and flavonols. From the study, all samples with different carrier agents

TABLE 3.5
Selected Studies on the Storage Stability of Encapsulated Phenolic Compounds

Tested Parameters	Sample	Results	References
- Wall materials: gum Arabic, maltodextrin DE 10–13, DE 17–20, gum Arabic, and maltodextrin DE 10–13 blend. - Stored at 4°C, 25°C, and 45°C until 12 weeks.	Cranberry juice	- Blend wall material retained phenolic content. - Quercetin increased about 16 times after eight weeks of storage.	Zhang et al. (2020)
- Wall material: maltodextrin DE 10 - Water activity varied at 0.11, 0.22, 0.33, and 0.44. - Stored at 5°C, 25°C, and 40°C for six months.	Bayberry juice	- Phenolic compounds retention efficiency with a water activity of 0.33 was observed when stored at less than 25°C. - Gallic acid, cyanidin 3-glucoside, quercetin 3-galactoside, and quercetin 3-glucoside were retained (about 93%–97%) after spray drying. - Phenolic compounds decreased at 6%–8% at 5°C, 6%–9% at 25°C, and 7%–37% at 40°C after six months of storage.	Fang and Bhandari (2011)
- Wall materials: maltodextrin, whey protein isolate, and pea protein isolate. - Stored at 40 and 60 °C for two weeks. - Stored at 25°C for six months.	Grape marc	- Phenolic compounds encapsulated with whey protein isolate were less stable but had a controlled release rate. - Whey protein isolate was effective to obtain 50 yield with a 0.5: 1 (carrier: extract) ratio.	Moreno et al. (2018)
- Wall material: whey protein isolate. - Stored at 37 °C for two weeks.	Blueberry pomace	- Phenolic compounds increased two-fold during storage. - Each microcapsule consisted of 1.32 mg cyanidin-3-o-glucoside, 2.83 mg gallic acid equivalents (GAE).	Flores, Singh, and Kong (2014)
- Wall material: maltodextrin (DE 16.5–19.5). - Stored at 5°C, 25°C, and 45°C, at three light conditions (dark, sunlight, and artificial light) for 70 days.	Olive pomace	- Phenolic compounds reduced 21% (stored at 25°C) and 34% (stored at 45°C) after 70 days.	Paini et al. (2015)

showed fluctuating trends with storage time. There are two stages regarding the retention trend of TPC and ANC. The first stage (stored up to six weeks) indicated a linear trend. In the following stage (stored between 6 and 12 weeks), the TPC and ANC in the manufactured powder were the highest at eight weeks, ANC at 10 weeks, and proanthocyanin content at week 12, respectively; by contrast, TPC increased by about 140%. This was agreeable with a previous study from Fang and Bhandari (2011) that showed the degradation of phenolic content in bayberry juice with increased temperature (40°C) compared with storage at 5°C and 25°C, respectively. The retention of TPC was 96% after the spray drying process. Quercetin 3-glucoside was identified in bayberry for the first time and had not been reported before. This is due to the morphology of the powder formed where the powder was smoother when coated with maltodextrin, with a decreased surface area and increased oxidation stability.

The stability of encapsulated phenolic compounds from red grape marc extracts was evaluated for a short period (i.e., two weeks at 40°C and 60°C) by Moreno et al. (2018). The tests showed that there was no loss of phenolic compounds detected under either temperature, and microparticles with WPI as a carrier agent showed an increased TPC after two weeks. Furthermore, for the long-term test, the manufactured powder was stored for six months at 25°C with daylight exposure for all encapsulated materials. Phenolic compounds availability was not affected by the daylight exposure. Thus, long-term storage was feasible for phenolic compounds for up to six months. Phenolic compounds from blueberry pomace increased up to two-fold during storage at 37°C. As for the *in vitro* release test, the manufactured powder with different carriers was tested in the gastric and intestinal fluids model at 37°C for 3h (Flores et al., 2014). The slower release rate could be due to the morphology of the microcapsules, where the smoother surface allowed the disintegration of active ingredients at a lower rate (Pierucci, Andrade, Farina, Pedrosa, & Rocha-Leao, 2007).

Morphology of the microcapsules prepared with PPI showed that lower maltodextrins DE microcapsules were shallow with rough surfaces (Gharsallaoui, Saurel, Chambin, & Voilley, 2012). Additionally, pH selection for emulsion preparation is highly important as proteins precipitate at their isoelectric point and limit the release of active ingredients (Assadpour, Jafari, & Maghsoudlou, 2017). Paini et al. (2015) studied the stability of phenolic compounds extracted from olive pomace with maltodextrin (DE 16.5–19.5) as outer wall material followed by spray drying. The best sample in terms of microencapsulation efficiency was selected for stability tests, which were a sample with maltodextrin 100 g/L and were tested for storage stability (5°C, 25°C, and 45°C) and light exposures (dark, sunlight, and artificial light, 2.1 klux) for 70 days. Additionally, encapsulated material was also stored under UV light for 48h. Total polyphenols were measured during the storage. About 94% of phenolic compounds were still present in microparticles after storage, even after undergoing high temperatures for spray drying such as inlet temperature 130°C and feed flow 10 mL/min. During storage, encapsulated materials were stable at 25°C until 28 days; however, the stability declined at 45°C after 14 days. Microparticles stored at 5°C were relatively stable even after 70 days. Similarly, phenolic compounds were stable in dark storage conditions. About 44% of phenolic compounds degraded after 7 days, when kept in strong light (2.1 klux). Eight hours of UV light exposure on microparticles resulted in a 66% degradation of the phenolic compounds.

3.7 BIOAVAILABILITY/BIOACCESSIBILITY OF ENCAPSULATED PHENOLIC COMPOUNDS BY SPRAY DRYING

Before spray drying encapsulation, the selection of outer wall material is very important to ensure efficient retention of active ingredients. Phenolic compounds need to be stable upon the processing parameters such as high temperatures, while the solubility of the wall material also needs to be considered. Similarly, several aspects of microparticles need to be evaluated; microencapsulation yield, rate of release for phenolic compounds, and solubility for further application either in food or pharmaceutical purposes. Bioavailability solely depends on the chemical structure of phenolic compounds, polarity, solidity, and absorption into GIT. Meanwhile, bioaccessibility determines how the active ingredients are being released during digestion in the targeted area. The limitations for bioavailability are subjected to oxidative stability due to the elevated temperature for spray drying and storage conditions (Ozdal et al., 2016).

Ozdal et al. (2016) reported that due to the specific structures, most phenolic compounds pass through the small intestine and settle in the colon. The human intestine contains a complex system of bacteria, referred to as gut microbiota, to maintain homeostasis. Hence, phenolic compounds would be converted as metabolites by gut microbiota. These metabolites may have higher biological function compared with initial precursor structures. Furthermore, phenolic compounds lead to the inhibition of pathogenic bacteria. This causes increased good bacteria in the gut microbiota, which gives a positive impact on human health. Leyva-Jiménez et al. (2019) stated that the human gut is comprised of around 10^{13}–10^{14} bacterial cells, which were 10 times of human somatic cells. Maintaining a good gut microbiota is crucial to create a symbiosis between intestinal homeostasis and health benefits.

Faria, Fernandes, Norberto, Mateus, and Calhau (2014), Leyva-Jiménez et al. (2019), and Kumar Singh et al. (2019) stated that up to 10% of phenolic compounds are absorbed in the small intestine. The remaining phenolic compounds accumulate in the large intestine together with bile conjugates and are visible to the gut microbial. Phenolic compounds need to be transformed into low weight metabolite to enhance health benefits. The metabolites from phenolic compounds in the gut microbiota could create variations in the bacteria community. According to Kumar Singh et al. (2019), phenolic compounds can produce radical species to alter membrane permeability and exhibit anti-inflammatory properties to give better health benefits. Figure 3.2 summarizes the possibility of spray drying encapsulation of phenolic compounds, considering each step involved and their actions in gut microbiota for the benefits of human health.

Meanwhile, Figure 3.3 shows the interactions of phenolic compounds in GIT. In short, (a) phenolic compounds (blue balls) are transferred into the small intestine by binding to fiber materials in the lumen. Phenolic compounds are also transported to the large intestine to facilitate the absorption of bacteria. However, phenolic compounds could limit the digestion of starch and protein by (b) binding to these compounds, constraining either (c) digestive enzymes or (d) nutrient transporters such as glucose transporter (GLUT), sodium-glucose co-transporter (SGLT) and intracellular cholesterol transporter (NPC1L1). In addition, (e) cholesterol absorption is delayed when phenolic compounds are displaced from the micelle and enable (f) phenolic compounds to bind to the micelle membrane to reduce ω3/ω6 fatty acids or α-tocopherol.

Previously, Gonzalez et al. (2019) studied the stability of the spray dried encapsulated phenolic compounds from olive leaf extract (OLE) with a carrier agent of sodium alginate (SA) upon *in vitro* gastrointestinal digestion. The encapsulated microparticles were tested in simulated gastric, intestinal, and colonic digestion solutions. Oleuropein content was about 85%, identified as the main phenolic compounds in OLE and about 80% when encapsulated with SA. Higher oleuropein and hydroxytyrosol (HT) were found in OLE-SA and minimal degradation of oleuropein within gastric digestion allowed gradual release in intestinal digestion. This gradual release was due to the swelling capability of SA microparticles, enabling easier digestion (George & Abraham, 2006). Similarly, increased levels of oleoside 11-methyl ester and oleoside were found in intestinal digestion up to 120 min. Hazas et al. (2016) reported that Winstar rats fed with oleuropein diet had better digestion in GIT after 21 days compared with HT and secoiridoids. Oleuropein absorption was stable and did not undergo degradation in the stomach/gastric level; by contrast, secoiridoids was fully degraded. Bioavailability of oleuropein was reflected in urine secretion and indicated that resistance in the action of β-glucosidase enzymes in the small intestine resulted in higher biological activity. The *in vitro* colon metabolism pathway revealed that phenolic compounds such as HT and phenylpropionic acid were found in human feces (Mosele et al., 2014). Hence, oleuropein opens the possibility for it to be applied in pharmaceutical industries to treat colon or GIT-related diseases (Carrera-González, Ramírez-Expósito, Mayas, & Martínez-Martos, 2013;).

One of the important concerns of spray drying is the rate at which the active ingredients can be released in the GIT. The release of active ingredients should not occur before reaching GIT

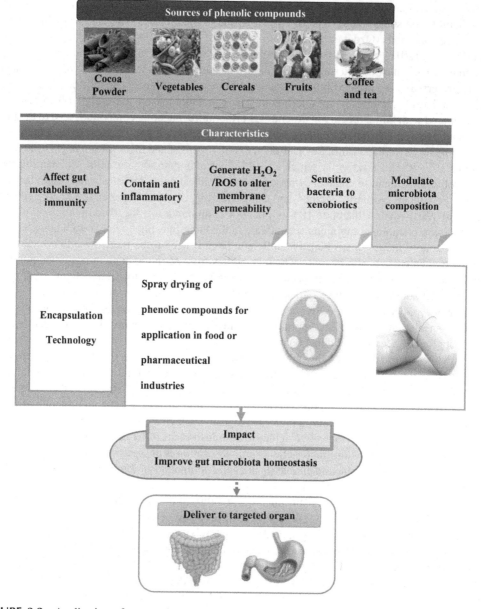

FIGURE 3.2 Application of encapsulated phenolic compounds into functional foods and an example of major dietary sources of phenolic compounds and the potential gut microbiota that are associated with the benefits on human health.

(Jia et al., 2016). Nevertheless, some outer wall material cannot achieve this target in simulated gastrointestinal fluids (SGF). For example, curcumin encapsulated using carrageenan–chitosan dual hydrogel bovine serum albumin indicated a rapid release in SGF, probably due to lower hydration of hydrogels at pH= 1.2 (Pascalau et al., 2016). Curcumin easily degrades at neutral pH. Nonetheless, swelling of proteins allows a higher amount of phenolic compounds (Kimpel & Schmitt, 2015). This problem can be solved by using emulsions. Curcumin in sodium caseinate emulsions was released at 5% and 16% in SGF and simulated intestinal fluids (SIF), respectively (Kumar et al., 2016). Proteins can easily undergo proteolysis, due to the presence of protease pepsin in gastric fluid (Kumar et al., 2016).

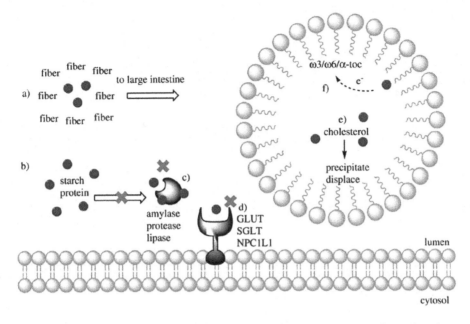

FIGURE 3.3 The gastrointestinal interactions of phenolic compounds with other nutrients, digestive enzymes, and nutrient transporters. Modified from Domínguez-Avila et al. (2017).

3.8   CONCLUDING REMARKS

Spray drying encapsulation of phenolic compounds has a wide range of applications in both the food and pharmaceutical industries. Limitations occur due to the high-energy requirement and degradation of heat-sensitive compounds during spray drying, which affects the encapsulation efficiency. The encapsulation efficiency of phenolic compounds ranges between 50%–90%, depending on outer wall material and operating conditions. Selecting the outer wall material based on active ingredients and optimizing the process condition for selective active ingredients are highly crucial for targeted application.

Encapsulated phenolic compounds can be retained effectively, especially with the use of wall materials such as maltodextrin, which is commonly used in the food industry. Several excellent functionalities of maltodextrin that act as the protective layer for bioactive compounds make it a suitable option as a carrier material. Other than that, the combination of carrier agents increases the encapsulation efficiency; e.g., maltodextrin alone has poor stability when used under high temperatures. Whey protein isolate also shows effective encapsulation of phenolic compounds but is highly sensitive to high temperatures. Each carrier material has its own property to encapsulate the selected phenolic compound; thus, the carrier agent chosen must be suitable for the targeted compounds. Moreover, the suitable storage temperature for spray dried powder loaded with phenolic compounds is about 25°C and below, in order to reduce the loss of bioactive compounds over a long storage period. Therefore, spray drying encapsulation is an effective way of retaining the phenolic compounds to be formulated in food products. The rate at which the phenolic compounds are released can be altered and their absorption in GIT can be improvised. Clinical studies have shown that phenolic compounds act as prebiotics and indicate antimicrobial activities against pathogenic bacteria in the gut. Phenolic compounds have a lower bioavailability and it is difficult to deliver them into the targeted area in the human body (i.e., cells, GIT, or tissues). Hence, encapsulation can be a useful technique for the transport and transfer of phenolic compounds into targeted regions. Due to the vast diversity of phenolic compounds from different sources, their spray drying encapsulation still requires extensive research and there are numerous questions to be answered. Additionally, the incorporation of the various phenolic compounds into different food products is yet to be systematically researched and guided.

REFERENCES

Abadio, F., Domingues, A. M., Borges, S. V., & Olieveira, V. M. (2004). Physical properties of powdered pineapple (*Ananas comosus*) juice-effect of malt dextrin concentration and atomization speed. *Journal of Food Engineering*, 64(3), 285–287.

Aguiar, J., Estevinho, B. N., & Santos, L. (2016). Microencapsulation of natural antioxidants for food application – The specific case of coffee antioxidants – A review. *Trends in Food Science & Technology*, 58, 21–39.

Ahmed, M., Akter, M. S., Lee, J.-C., & Eun, J.-E. (2010). Encapsulation by spray drying of bioactive components, physicochemical and morphological properties from purple sweet potato. *LWT- Food Science and Technology*, 43(9), 1307–1312.

Alamilla-Beltràn, L., Chanona-Pérez, J. J., Jiménez-Aparicio, A. R., & Guitérrez-López, G. F. (2005). Description of morphological changes of particles along spray drying. *Journal of Food Engineering*, 67(1–2), 179–184.

Aliakbarian, B., Paini, M., Casazza, A. A., & Perego, P. (2015). Effect of encapsulating agent on physical–chemical characteristics of olive pomace polyphenols-rich extracts. *Chemical Engineering Transactions*, 43, 97–102.

Ammendola, M., Haponska, M., Balik, K., Modrakowska, P., Matulewicz, K., Kazmierski, L., Lis, A., Kozlowska, J., Garcia-Valls, R., Giamberini, M., Bajek, A., & Tylkowski, B. (2020). Stability and anti-proliferative properties of biologically active compounds extracted from *Cistus L.* after sterilization treatments. *Scientific Reports*, 10(1), 6521.

Arpagaus, C., Collenberg, A., Rutti, D., Assadpour, E., & Jafari, S. M. (2018). Nano spray drying for encapsulation of pharmaceuticals. *International Journal of Pharmaceutics*, 546(1–2), 194–214.

Assadpour, E., Jafari, S. M., & Maghsoudlou, Y. (2017). Evaluation of folic acid release from spray dried powder particles of pectin-whey protein nano-capsules. *International Journal of Biological Macromolecules*, 95, 238–247.

Badreldin, H. A., Ziada, A., & Blunden, G. (2009). Biological effects of Gum Arabic: A review of some recent research. *Food and Chemical Toxicology: An International Journal Published for the British Industrial Biological Research Association*, 47(1), 1–8.

Ballesteros, L. F., Ramirez, M. J., Orrego, C. E., Teixeira, J. A., & Mussatto, S. I. (2017). Encapsulation of antioxidant phenolic compounds extracted from spent coffee grounds by freeze-drying and spray-drying using different coating materials. *Food Chemistry*, 237, 623–631.

Bankole, V. O., Osungunna, M. O., Souza, C. R. F., Salvador, S. L., & Oliveira, W. P. (2020). Spray-dried proliposomes: An innovative method for encapsulation of *Rosmarinus officinalis L.* polyphenols. *AAPS PharmSciTech*, 21(5), 143.

Bhandari, B. R. (2007). Spray drying and food properties. In C. C. Y. H. Hui, M. M. Farid, O. O. Fasina, A. Noohorm, & J. Welti-Chanes (Eds.), *Food drying science and technology* (pp. 215–248). DEStech Publications, Inc.

Botelho, G., Canas, S., & Lameiras, J. (2017). Development of phenolic compounds encapsulation techniques as a major challenge for food industry and for health and nutrition fields. In A. M. Grumezescu (Ed), *Nutrient Delivery: Nanotechnology in the Agri-Food Industry*, Volume 5, (pp. 535–586). Academic Press.

Busch, V. M., Pereyra-Gonzalez, A., Šegatin, N., Santagapita, P. R., Ulrih, N. P., & Buera, M. P. (2007). Propolis encapsulation by spray drying: Characterization and stability. *LWT- Food Science and Technology*, 75, 227–235.

Carrera-González, M. P., Ramírez-Expósito, M. J., Mayas, M. D., & Martínez-Martos, J. M. (2013). Protective role of oleuropein and its metabolite hydroxytyrosol on cancer. *Trends in Food Science & Technology*, 31(2), 92–99.

Chaumun, M., Goëlo, V., Ribeiro, A. M., Rocha, F. A., & Estevinho, B. N. (2020). In vitro evaluation of microparticles with *Laurus nobilis L.* extract prepared by spray-drying for application in food and pharmaceutical products. *Food and Bioproducts Processing*, 122, 124–135.

Chen, F.-P., Liu, L.-L., & Tang, C.-H. (2020). Spray-drying microencapsulation of curcumin nanocomplexes with soy protein isolate: Encapsulation, water dispersion, bioaccessibility and bioactivities of curcumin. *Food Hydrocolloids*, 105, 105821.

Consoli, L., Hubinger, M. D., & Dragosavac, M. M. (2020). Encapsulation of resveratrol using Maillard conjugates and membrane emulsification. *Food Research International*, 137, 109359.

Cortés-Rojas, D. F., Souza, C. R. F., & Oliveira, W. P. (2014). Encapsulation of eugenol rich clove extract in solid lipid carriers. *Journal of Food Engineering*, 127, 34–42.

Dag, D., Kilercioglu, M., & Oztop, M. H. (2017). Physical and chemical characteristics of encapsulated gold-enberry (*Physalis peruviana L.*) juice powder. *LWT- Food Science and Technology*, 83, 86–94.

Dai, J., & Mumper, R. J. (2010). Plant phenolic: Extraction, analysis and their antioxidant and anticancer properties. *Molecules*, 15(10), 7313–7352.

Dai, L., Sun, C., Li, R., Mao, L., Liu, F., & Gao, Y. (2017). Structural characterization, formation mechanism and stability of curcumin in zein-lecithin composite nanoparticles fabricated by antisolvent co-precipitation. *Food Chemistry*, 237, 1163–1171.

Dang, X., Yang, M., Shan, Z., Mansouri, S., May, B. K., Chen, X., Chen, H., & Woo, M. W. (2017). On spray drying of oxidized corn starch cross-linked gelatin microcapsules for drug release. *Materials Science and Engineering*, 74, 493–500.

Desai, K. G. H., & Park, H. J. (2005). Recent developments in microencapsulation of food ingredients. *Drying Technology*, 23, 1361–1394.

Dias, M. I., Ferreira, I. C. R. R., & Barreiro, M. F. (2015). Microencapsulation of bioactives for food application. *Food & Function*, 6(4), 1035–1052.

Domínguez-Avila, J. A., Wall-Medrano, A., Velderrain-Rodríguez, G. R., Chen, C.-Y. O., Salazar-López, N. J., Robles-Sánchez, M., & González-Aguilar, G. A. (2017). Gastrointestinal interactions, absorption, splanchnic metabolism and pharmacokinetics of orally ingested phenolic compounds. *Food & Function*, 8(1), 15–38.

Donsì, F., Voudouris, P., Veen, S. J., & Velikov, K. P. (2017). Zein-based colloidal particles for encapsulation and delivery of epigallocatechin gallate. *Food Hydrocolloids*, 63, 508–517.

Đorđević, V., Paraskevopoulou, A., Mantzouridou, F., Lalou, S., Panti, C. M., & Bugarski, B. (2016). Encapsulation technologies for food industry. In P. R. V. Nedovic, J. Levic, V. T. Saponjac, & G. V. Barbosa-Canovas (Ed.), *Emerging and traditional technologies for safe, healthy and quality food* (pp. 329–380). Springer International Publishing.

Duchêne, D., & Bochot, A. (2016). Thirty years with cyclodextrins. *International Journal of Pharmaceutics*, 514(1), 58–72.

Ersus, S., & Yurdagel, U. (2007). Microencapsulation of anthocyanin pigments of black carrot (*Daucus carota L.*) by spray drier. *Journal of Food Engineering*, 80, 805–812.

Fang, Z., & Bhandari, B. (2010). Encapsulation of polyphenols – A review. *Trends in Food Science & Technology*, 21(10), 510–523.

Fang, Z., & Bhandari, B. (2011). Effect of spray drying and storage on the stability of bayberry polyphenols. *Food Chemistry*, 129(3), 1139–1147.

Faria, A., Fernandes, I., Norberto, S., Mateus, N., & Calhau, C. (2014). Interplay between anthocyanins and gut microbiota. *Journal of Agricultural and Food Chemistry*, 62(29), 6898–6902.

Fathi, M., Martín, Á., & McClements, D. J. (2014). Nanoencapsulation of food ingredients using carbohydrate based delivery systems. *Trends in Food Science & Technology*, 39(1), 18–39.

Ferrari, C. C., Ribeiro, C. P., & Aguirre, J. M. D. (2012). Spray drying of blackberry pulp using maltodextrin as carrier agent. *Brazilian Journal of Food Technology*, 15(2), 157–165.

Ferreira, S., Piovanni, G. M. O., Malacrida, C. R., & Nicoletti, V. R. (2019). Influence of emulsification methods and spray drying parameters on the microencapsulation of turmeric oleoresin. *Emirates Journal of Food and Agriculture*, 31(7), 491–500.

Filkovà, I., & Mujumdar, A. S. (1995). Industrial spray drying system. In A. S. Mujumdar (Ed), *Handbook of Industrial Drying*, 2nd Ed, (pp. 15–25). Dekker.

Finney, J., Buffo, B., & Reineccius, C. (2002). Effects of type of spray drying and processing temperatures on the physical properties and stability of spray-dried flavors. *Journal of Food Science*, 67(3), 1108–1114.

Flores, F. P., Singh, R. K., & Kong, F. (2014). Physical and storage properties of spray-dried blueberry pomace extract with whey protein isolate as wall material. *Journal of Food Engineering*, 137, 1–6.

George, M., & Abraham, T. E. (2006). Polyionic hydrocolloids for the intestinal delivery of protein drugs: Alginate and chitosan – A review. *Journal of Controlled Release*, 114(1), 1–14.

Georgetti, S. R., Casagrande, R., Souza, C., Oliveira, W. P., & Fonseca, M. (2008). Spray drying of the soybean extract: Effects on chemical properties and antioxidant activity. *LWT- Food Science and Technology*, 41(8), 1521–1527.

Geranpour, M., Assadpour, E., & Jafari, S. M. (2020). Recent advantages in the spray drying encapsulation of essential fatty acids and functional oils. *Trends in Food Science & Technology*, 102, 71–90.

Gharsallaoui, A., Saurel, R., Chambin, O., & Voilley, A. (2012). Pea (*Pisum sativum, L.*) protein isolate stabilized emulsions: A novel system for microencapsulation of lipophilic ingredients by spray drying. *Food and Bioprocess Technology*, 5(6), 2211–2221.

Gibbs, B. F., Kermasha, S., Alli, I., & Mulligan, C. N. (1999). Encapsulation in the food industry: A review. *International Journal of Food Science and Nutrition*, 50, 213–224.

Gonzalez, E., Gomez-Caravaca, A. M., Gimenez, B., Cebrian, R., Maqueda, M., Martinez-Ferez, A., Segura-Carretero, A., & Robert, P. (2019). Evolution of the phenolic compounds profile of olive leaf extract encapsulated by spray-drying during in vitro gastrointestinal digestion. *Food Chemistry*, 279, 40–48.

Gumus, C. E., Decker, E. A., & McClements, D. J. (2017). Gastrointestinal fate of emulsion-based ω-3 oil delivery systems stabilized by plant proteins: Lentil, pea, and faba bean proteins. *Journal of Food Engineering*, 207, 90–98.

Hazas, M.-C. L. D. L., Piñol, C. L. D. L., Macià, A. L. D. L., Romero, M.-P. L. D. L., Pedret, A. L. D. L., Solà, R. L. D. L., Rubió, L. L. D. L., & Motilva, M. J. L. D. L. (2016). Differential absorption and metabolism of hydroxytyrosol and its precursors oleuropein and secoiridoids. *Journal of Functional Foods*, 22, 52–63.

Huang, W.-Y., Cai, Y.-Z., & Zhang, Y. (2009). Natural phenolic compounds from medicinal herbs and dietary plants: Potential use for cancer prevention. *Nutrition and Cancer*, 62(1), 1–20.

Jafari, S. M., Assadpoor, E., He, V., & Bhandari, B. (2008). Encapsulation efficiency of food flavours and oils during spray drying. *Drying Technology*, 26, 815–835.

Jia, Z., Dumont, M.-J., & Orsat, V. (2016). Encapsulation of phenolic compounds present in plants using protein matrices. *Food Bioscience*, 15, 87–104.

Joye, I. J., & McClements, D. J. (2014). Biopolymer-based nanoparticles and microparticles: Fabrication, characterization, and application. *Colloid & Interface Science*, 19(5), 417–427.

Kha, T. C., Nguyen, M. H., Roach, P. D., & Stathopoulos, C. E. (2014). Micro-encapsulation of gac oil: Optimisation of spray drying conditions using response surface methodology. *Powder Technology*, 264, 298–309.

Kimpel, F., & Schmitt, J. J. (2015). Review: Milk proteins as nanocarrier systems for hydrophobic nutraceuticals. *Journal of Food Science*, 80(11), 2361–2366.

King, C. J. (1995). Spray-drying – Retention of volatile compounds revisited. *Drying Technology*, 13(5–7), 1221–1240.

Kumar, D. D., Mann, B., Pothuraju, R., Sharma, R., Bajaj, R., & Minaxi, M. (2016). Formulation and characterization of nanoencapsulated curcumin using sodium caseinate and its incorporation in ice cream. *Food & Function*, 7(1), 417–424.

Kumar Singh, A., Cabral, C., Kumar, R., Ganguly, R., Rana, H. K., Gupta, A., Rosaria Lauro, M., Carbone, C., Reis, F., & Pandey, A. K. (2019). Beneficial effects of dietary polyphenols on gut microbiota and strategies to improve delivery efficiency. *Nutrients*, 11(9), 216.

Leyva-Jiménez, F. J., Lozano-Sánchez, J., Cádiz-Gurrea, M. D. L. L., Arráez-Román, D., & Segura-Carretero, A. (2019). Functional ingredients based on nutritional phenolics. A case study against inflammation: Lippia Genus. *Nutrients*, 11(7), 1646.

Liang, J., Yan, H., Puligundla, P., Gao, X., Zhou, Y., & Wan, X. (2017). Applications of chitosan nanoparticles to enhance absorption and bioavailability of tea polyphenols: A review. *Food Hydrocolloids*, 69, 286–292.

Lima, E. M., Madalão, M. C., Benincá, D. B., Saraiva, S. H., & Silva, P. I. (2019). Effect of encapsulating agent and drying air temperature on the characteristics of microcapsules of anthocyanins and polyphenols from juçara (*Euterpe edulis*). *International Food Research Journal*, 26(2), 607–617.

Maury, M., Murphy, K., Kumar, S., Shi, L., & Lee, G. (2005). Effects of process variables on the powder yield of spray-dried trehalose on a laboratory spray dryer. *European Journal Pharmaceutical and Biopharmaceutical*, 59(3), 565–573.

Michael, J. K. (1993). Spray drying and spray congealing of pharmaceuticals. In J. Swarbrick & J. C. Boylan (Eds), *Encyclopedia of pharmaceutical technology* (Vol. 14, pp. 207–221). Marcel Dekker INC.

Moreira, G. É. G., Costa, M. G. M., Souza, A. C. R. D., Brito, E. S. D., Medelros, M. D. F. D. D., & Azeredo, H. M. C. D. (2009). Physical properties of spray dried acerola pomace extract as affected by temperature and drying AIDS. *Food Science and Technology*, 42(2), 641–645.

Moreno, T., Cocero, M. J., & Rodríguez-Rojo, S. (2018). Storage stability and simulated gastrointestinal release of spray dried grape marc phenolics. *Food and Bioproducts Processing*, 112, 96–107.

Mosele, J. I., Martín-Peláez, S., Macià, A., Farràs, M., Valls, R. M., Catalán, U., & Motilva, M. J. (2014). Faecal microbial metabolism of olive oil phenolic compounds: In vitro and in vivo approaches. *Molecular Nutrition & Food Research*, 58(9), 1809–1819.

Murugesan, R., & Orsat, V. (2011). Spray drying for the production of nutraceutical ingredients—A review. *Food and Bioprocess Technology*, 5(1), 3–14.

de Sá Mendes, N., Coimbra, P. P. S., Santos, M. C. B., Cameron, L. C., Ferreira, M. S. L., Buera, M. P., & Gonçalves, E. C. B. A. (2020). *Capsicum pubescens* as a functional ingredient: Microencapsulation and phenolic profilling by UPLC-MSE. *Food Research International*, 135, 109292.

Oliveira, O. W., & Petrovick, P. R. (2009). Spray drying of plant extracts: Bases and applications. *Brazilian Journal of Pharmacognosy*, 20(4), 641–650.

Ozdal, T., Sela, D. A., Xiao, J., Boyacioglu, D., Chen, F., & Capanoglu, E. (2016). The reciprocal interactions between polyphenols and gut microbiota and effects on bioaccessibility. *Nutrients*, 8(2), 78.

Paini, M., Aliakbarian, B., Casazza, A. A., Lagazzo, A., Botter, R., & Perego, P. (2015). Microencapsulation of phenolic compounds from olive pomace using spray drying: A study of operative parameters. *LWT- Food Science and Technology*, 62(1), 177–186.

Pascalau, V., Soritau, O., Popa, F., Pavel, C., Coman, V., Perhaita, I., Borodi, G., Dirzu, N., Tabaran, F., & Popa, C. (2016). Curcumin delivered through bovine serum albumin/polysaccharides multilayered microcapsules. *Journal of Biomaterials Applications*, 30(6). 857–872.

Patel, R. P., Patel, M. P., & Suthar, A. M. (2009). Spray drying technology: An overview. *Indian Journal of Science and Technology*, 2(10), 44–47.

Pieczykolan, E., & Kurek, M. A. (2019). Use of guar gum, gum arabic, pectin, beta-glucan and inulin for microencapsulation of anthocyanins from chokeberry. *International Journal of Biological Macromolecules*, 129, 665–671.

Pierucci, A. P., Andrade, L. R., Farina, M., Pedrosa, C., & Rocha-Leao, M. H. (2007). Comparison of alpha-tocopherol microparticles produced with different wall materials: Pea protein a new interesting alternative. *Journal of Microencapsulation*, 24(3), 201–213.

de Ramos, R. M. Q., Siacor, F. D. C., & Taboada, E. B. (2020). Effect of maltodextrin content and inlet temperature on the powder qualities of spray-dried Pineapple (*Ananas comosus*) waste extract. *Waste and Biomass Valorization*, 11(7), 3247–3255.

Radünz, M., dos Santos Hackbart, H. C., Camargo, T. M., Nunes, C. F. P., de Barros, F. A. P., Dal Magro, J., Filho, P. J. S., Gandra, E. A., Radünz, A. L., & da Rosa Zavareze, E. (2020). Antimicrobial potential of spray drying encapsulated thyme (*Thymus vulgaris*) essential oil on the conservation of hamburger-like meat products. *International Journal of Food Microbiology*, 330, 108696.

Ray, S., Raychaudhuri, U., & Chakraborty, R. (2016). An overview of encapsulation of active compounds used in food products by drying technology. *Food Bioscience*, 13, 76–83.

Ré, M. I. (1998). Microencapsulation by spray drying. *Drying Technology: An International Journal*, 16(6), 1195–1236.

Saberi, A. H., Fang, Y., & McClements, D. J. (2016). Influence of surfactant type and thermal cycling on formation and stability of flavor oil emulsions fabricated by spontaneous emulsification. *Food Research International*, 89, 296–301.

Şahin, S., & Bilgin, M. (2018). Olive tree (*Olea europaea L.*) leaf as a waste by-product of table olive and olive oil industry: A review. *Journal of the Science of Food and Agriculture*, 98(4), 1271–1279.

Saikia, S., Mahnot, N. K., & Mahanta, C. L. (2015). Effect of spray drying of four fruit juices on physicochemical, phytochemical and antioxidant properties. *Journal of Food Processing and Preservation*, 39(6), 1656–1664.

Samantha, S. C., Bruna, A. S. M., Adriana, R. M., Fabio, B. R., Sandro, A. C. A., & Aline, R. (2015). Drying by spray drying in the food industry: Micro-encapsulation, process parameters and main carriers used. *African Journal of Food Science*, 9(9), 462–470.

Sambroska, K., Jedlińska, A., Wiktor, A., Derewiaka, D., Welosiak, R., Matwijczuk, A., & Witrowa-Rajchert, D. (2019). The effect of low-temperature spray drying with dehumidified air on phenolic compounds, antioxidant activity, and aroma compounds of rapeseed honey powders. *Food and Bioprocess Technology*, 12(6), 919–932.

Shariffa, Y. N., Tan, T. B., Uthumporn, U., Abas, F., Mirhosseini, H., Nehdi, I. A., Wang, Y. H., & Tan, C. P. (2017). Producing a lycopene nanodispersion: Formulation development and the effects of high pressure homogenization. *Food Research International*, 101, 165–172.

Shishir, M. R. I., Xie, L., Sun, C., Zheng, X., & Chen, W. (2018). Advances in micro and nano-encapsulation of bioactive compounds using biopolymer and lipid-based transporters. *Trends in Food Science & Technology*, 78, 34–60).

Sidhu, G. K., Singh, M., & Kaur, P. (2019). Effect of operational parameters on physicochemical quality and recovery of spray-dried tomato powder. *Journal of Food Processing and Preservation*, 42(10), e14120.

Silveira, A. C. P., Perrone, Í. T., Junior, P. H. R., & Carvalho, A. F. D. (2013). Spray drying: A review. *Journal of Candido Tostes Dairy Institute*, 68(391), 51–58.

Simões, L. D., Madalena, D. A., Pinheiro, A. C., Teixeira, J. A., Vicente, A. A., & Ramos, Ó. L. (2017). Micro- and nano bio-based delivery systems for food applications: In vitro behavior. *Advances in Colloid and Interface Science*, 243, 23–46.

Sivetz, M., & Foote, H. E. (1963). *Coffee processing technology*. AVI Publishing Co.

Sokhansanj, S., & Jayas, D. (1995). Drying of foodstuff. In A. S. Mujumdar (Ed), *Handbook of Indutrial Drying* (pp. 589–626). Dekker, CRC Press.

Suhag, Y., & Nanda, V. (2016). Optimization for spray drying process parameters of nutritionally rich honey powder using response surface methodology. *Cogent Food and Argriculture*, 2(1), 1176631.

Tan, S. P., Kha, T. C., Parks, S. E., Stathopoulos, C. E., & Roach, P. D. (2015). Effects of the spray-drying temperatures on the physiochemical properties of an encapsulated bitter melon aqueous extract powder. *Powder Technology*, 281, 65–75.

Tonon, R. V., Brabet, C., & Hubinger, M. D. (2010). Anthocyanin stability and antioxidant activity of spray-dried açai (*Euterpe oleracea Mart.*) juice produced with different carrier agents. *Food Research International*, 43(3), 907–914.

Tulipani, S., Mezzetti, B., Capocasa, F., Bompadre, S., Beekwilder, J., De Vos, C. R., Capanoglu, E., Bovy, A., & Battino, M. (2008). Antioxidants, phenolic compounds, and nutritional quality of different strawberry genotypes. *Journal of Agricultural and Food Chemistry*, 56(3), 696–704.

Turkiewicz, I. P., Wojdyło, A., Tkacz, K., Lech, K., Michalska-Ciechanowska, A., & Nowicka, P. (2020). The influence of different carrier agents and drying techniques on physical and chemical characterization of Japanese quince (*Chaenomeles japonica*) microencapsulation powder. *Food Chemistry*, 323, 126830.

Urzúa, C., González, E., Dueik, V., Bouchon, P., Giménez, B., & Robert, P. (2017). Olive leaves extract encapsulated by spray-drying in vacuum fried starch-gluten doughs. *Food and Bioproducts Processing*, 106, 171–180.

Vissotto, F. Z., Montenegro, F. M., Santos, J. M., & Oliveira, S. J. R. (2006). Evaluation of the influence of lecithination and agglomeration on the physical properties of a cocoa powder beverage (cocoa powder beverage lecithination and agglomeration). *Food Science and Technology*, 26(3), 666–671.

Vulić, J., Šeregelj, V., Kalušević, A., Lević, S., Nedović, V., Šaponjac, V. T., & Ćetković, G. (2019). Bioavailability and bioactivity of encapsulated phenolics and carotenoids isolated from red pepper waste. *Molecules*, 24(15), 2837.

Wang, J., Martínez-Hernández, A., de Lamo-Castellví, S., Romero, M.-P., Kaade, W., Ferrando, M., & Güell, C. (2020). Low-energy membrane-based processes to concentrate and encapsulate polyphenols from carob pulp. *Journal of Food Engineering*, 281, 109996.

Wu, K. G., Chai, X. H., & Chen, Y. (2005). Microencapsulation of fish oil by simple coacervation of hydroxy-propyl methylcellulose. *Chinese Journal of Chemistry*, 23(11), 1569–1572.

Yang, H., Feng, K., Wen, P., Zong, M.-H., Lou, W.-Y., & Wu, H. (2017). Enhancing oxidative stability of encapsulated fish oil by incorporation of ferulic acid into electrospun zein mat. *LWT- Food Science and Technology*, 84, 82–90.

Yinbin, L., Wu, L., Weng, M., Tang, B., Lai, P., & Chen, J. (2018). Effect of different encapsulating agent combinations on physicochemical properties and stability of microcapsules loaded with phenolics of plum (Prunus salicina lindl.). *Powder Technology*, 340, 459–464.

Zhang, H., Yu, D., Sun, J., Liu, X., Jiang, L., Guo, H., & Ren, F. (2013). Interaction of plant phenols with food macronutrients: Characterisation and nutritional–Physiological consequences. *Nutrition Research Reviews*, 27(1), 115.

Zhang, J., Zhang, C., Chen, X., & Quek, S. Y. (2020). Effect of spray drying on phenolic compounds of cranberry juice and their stability during storage. *Journal of Food Engineering*, 269, 109744.

4 Spray Drying Encapsulation of Anthocyanins

Katarzyna Samborska and Aleksandra Jedlińska
Warsaw University of Life Sciences SGGW, , Poland

Seid Mahdi Jafari
Gorgan University of Agricultural Sciences and Natural Resources, Iran

CONTENTS

4.1 INTRODUCTION

Anthocyanins (ANCs) belong to the group of phenolic compounds, the main group of biologically active compounds in plants (including fruits, vegetables, and herbs). Phenolic compounds are plant secondary metabolites with a diverse structure, molecular weight, physical, chemical, and biological properties (Ataie-Jafari, Karimi, & Pajouhi, 2008; Daneshzad, Shab-Bidar, Mohammadpour, & Djafarian, 2019; Duthie et al., 2006). ANCs (apart from betanins and carotenoids) are responsible for the non-green coloration of plants. They can be found in different parts: flowers, fruits, seeds, leaves, bracts, stems, roots, and tubers. These pigments, in addition to providing basic functions for plants (attracting pollinating insects, protecting plants against adverse environmental conditions), provide an attractive, aesthetic appearance for the edible parts of plants, and bring benefits to human health after their consumption. Thus, they are used as natural colorants in the food, cosmetics, and pharmaceutical industries (Boldt, Meyer, & Erwin, 2014; Cazzonelli, 2011; Delgado-Vargum Arabics, Jiménez, & Paredes-López, 2000; Johnson, 2002).

Over the past few years, there has been substantial growth in the food colorants market. At the same time, further growth of this market is expected, which has been estimated at 10% to 15% per year (Cortez, Luna-Vital, Margulis, & Gonzalez de Mejia, 2017). Colorants are added to improve the appearance of food or drinks and to restore the color(s) lost during food processing. The trend in the food colorant industry is to replace synthetic dyes in food and beverages with natural pigments. This is caused by the awareness of the environmental hazards of by-products from colorants synthesis and the 'clean label' consumer trend. The main assumptions of this trend include the use of natural ingredients and minimizing the number of ingredients; i.e., using only those which are necessary to produce the product. Additionally, increasing the use of natural colorants is associated with their impact on reducing the incidence of, for example, cancer, diabetes, or obesity. Synthetic colorants can be allergenic, toxic, and even carcinogenic. Increased ecological and sustainability awareness increase the use of natural colorants, and preferably those made from bio-waste. An example would be the use of berry pomace for the production of colorants - berry pomace represents, on average, 20% of raw materials.

However, the disadvantage of natural colorants compared with synthetic ones is their low stability (Carochom, Barreiro, Morales, & Ferreira, 2014; Cooperstone & Schwartz, 2016; Cortez et al., 2017; Gomes et al., 2019; Li et al., 2016; Phan et al., 2020; Rodriguez, Vidallon, Mendoza, & Reyes, 2016). Thus, techniques such as microencapsulation, with the application of straightforward and easily scalable methods such as spray drying, can be very beneficial in this regard. Thus, this chapter aimed to present a summary of the research carried out during the last decade in the field of encapsulation of ANCs by spray drying, which is a unique technology for their stabilization and preservation, as well as their transformation into an attractive powdered form. It should be noted that spray drying of other phenolic compounds and carotenoids has been covered in Chapters 3 and 5, respectively.

4.2 TYPES, SOURCES, AND PHYSICOCHEMICAL PROPERTIES OF ACNS

The name "anthocyanin" comes from the Greek language, and is a composite of two words: anthos–flower, kyanos–blue. This name was first used in 1835 by Marquart to describe the blue pigments of *Centaurea* flowers. ANCs are the largest group of water-soluble pigments in plants, which give the red, pink, blue, purple colors of their various parts. ANCs occur in various parts of plants: flowers (hibiscus), leaves (red cabbage), shoots (rhubarb), fruits (grapes), and even roots (radish, red onion). In most cases, ANCs accumulate in the epidermis part of the plant; in cells, they are located within the vacuoles. However, they do not occur in mushrooms and algae. So far, over 600 ANCs have been recognized and described. ANCs are commonly found in the human diet, in particular by eating violet, blue and red fruits, and vegetables, where their content is in the range of 0.1–1% of solids. Daily human ANCs intake is quite changeable and highly dependent on eating habits. Wu et al. (2006) found that the average consumption of ANCs for US adults was about 12.5 mg/day. Zamora-Ros, Knaze, and Luj'an-Barroso (2011); Zamora-Ros, Knaze, and Luján-Barroso (2011) determined the total daily consumption of ANCs by Europeans: for men, it ranged from 19.8 (Netherlands) to 64.9 mg/day (Italy), while for women, it was from 18.4 (Spain) to 44.1 mg/day (Italy). The high levels of ANCs consumption in Italy can be attributed to the Mediterranean diet, which is rich in red and purple vegetables and red wine (Cavalcanti, Santos, & Meireles, 2011; Gomes et al., 2019; Piątkowska, Kopeć, & Leszczyńska, 2011; Silva, Freitas, Maçanitab, & Quinaa, 2016; Smeriglio, Barreca, Bellocco, & Trombetta, 2016).

ANCs belong to the polyphenolic organic compounds. These are glycosides with an anthocyanidin (which belongs to flavonoids and has a C6–C3–C6 carbon skeleton). The basic structure of ANCs is presented in Figure 4.1. Anthocyanidins (aglycons) consist of an aromatic Ring A connected to a heterocyclic Ring C, which contains oxygen, connected by a carbon–carbon bond, with the third aromatic Ring B. When anthocyanidins occur in the form of glycosides (bonded to sugar moiety), then they are called ACNs. Sugar residues are attached in the form of mono-, di- and

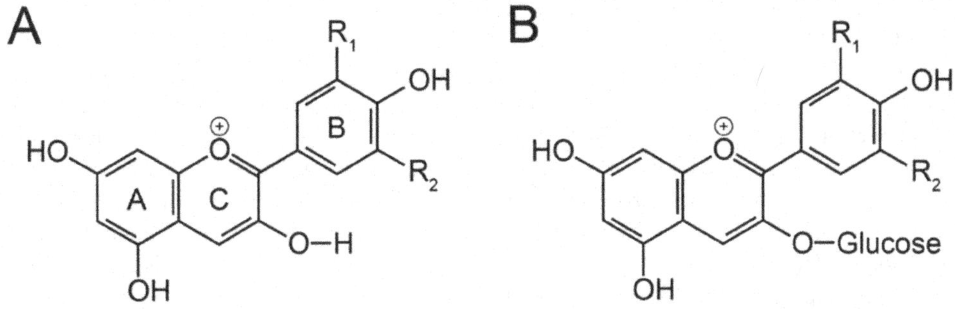

R_1	R_2	Anthocyanidin	Anthocyanin
H	H	Pelargonidin	Pelargonidin-3-glucoside
OH	H	Cyanidin	Cyanidin-3- glucoside
OH	OH	Delphinidin	Delphinidin-3- glucoside
OCH_3	H	Peonidin	Peonidin-3- glucoside
OCH_3	OH	Petunidin	Petunidin-3- glucoside
OCH_3	OCH_3	Malvidin	Malvidin-3- glucoside

FIGURE 4.1 Anthocyanidin (A) and anthocyanin (B) skeleton, types, and positions of substituents in six common anthocyanins (ANCs). Based on Castañeda, Hernandez, Páez, and Galán-Vidal (2009) and Clifford (2000).

trisaccharides, which can additionally be acylated (having a substituted organic acid) (Castañeda et al., 2009; Clifford, 2000; Konczak & Zhang, 2004).

The differences in ANCs are related to (Clifford, 2000):

- basic anthocyanidin skeleton – number and position of hydroxyl and methoxyl substituents (Figure 4.1 shows six common types);
- type, number, and position(s) of sugars attached to the skeleton; and,
- sugar acylation range and type of acylating agent(s).

The most common sugar residues are glucose, galactose, rhamnose, and arabinose; they usually occur as 3-glycosides or 3,5-diglycosides. The most popular acylating agents include cinnamic acids (feluric, sinapic, p-coumaric, caffeic) and aliphatic acids (malic, acetic, oxalic, malonic, succinic). 22 types of ANCs (aglycons) have been identified, but only six of them are widespread in food: pelarginidin, cyanidin, delphinidin, peonidine, petunidine, and malvidin, as shown in Figure 4.1. The names of these aglycons come from the plants in which they were identified. Generally, the most common ones are derivatives of cyanidin, then pelargonidin, and delphinidin. Cyanidin-3-glucoside is the most common anthocyanin found in fruits. In turn, malvidin glycosides are the most characteristic ANCs in red grapes and their products (e.g., wine and juice), while 3-glycoside pelargonidin is the predominant pigment in strawberries (Clifford, 2000; Gomes et al., 2019;

Pascual-Teresa & Sanchez-Ballesta, 2008; Sinopoli, Calogero, & Bartolotta, 2019; Xie, Su, Sun, Zheng, & Chen, 2018).

The color that ANCs give to the plant depends on the pH of the cell juice and its chemical form. In an acidic environment, they usually turn red, while in alkaline conditions blue is the dominant color. In a strongly acidic aqueous solution (pH < 1), ANCs have a red form of flavylium cations. As the pH increases, the intensity of the color decreases, due to the decrease in the concentration of flavylium cations, which changes into the form of hemiketal (colorless). Then, at pH = 5–6, the hemiketal easily tautomerizes to a form of pale yellow cis-chalcone. At the acidic, neutral, and alkaline pH values, there is a deprotonation reaction of the flavylium cations, which is associated with the formation of a purple color. The color of plants also depends on the type of ANCs contained, on their connections with metals, as well as the presence of other plant pigments (yellow flavones and orange carotenoids) (Cortez et al., 2017; Fernandes et al., 2019).

Anthocyanin synthesis is a photochemical process; therefore, fruits (e.g., apricots, peaches, apples, and pears) ripening on trees and directly exposed to solar radiation have a darker color, compared with fruits picked earlier and ripened in storehouses. Leaf discoloration occurs due to the production of ANCs, which are defensive compounds against negative environmental conditions (e.g., cold or strong ultraviolet radiation) (Gould & Lee, 2002; Landi, Tattini, & Gould, 2015; Piątkowska et al., 2011; Silva et al., 2016).

The addition of ANCs (or their extracts) into food products is valuable for two reasons: in terms of technology, ANCs shape the color of the product, which is the most important factor in terms of the perception of the product by the consumer, and in terms of health benefits, ANCs can help in the prevention or treatment of several diseases. Flowers and fruits are the richest parts of plants in terms of ANCs, but they also exist in stems, leaves, and roots. The richest sources of ANCs are colorful fruits such as grapes, cherries, berries, peaches, pomegranates, plums, and dark-colored vegetables (e.g., red radishes, black beans, eggplants, red onions, purple corn, red cabbage, and purple sweet potatoes). ANCs are present not only in raw materials but also in processed products, like wines, juices, beverages, jams, yogurts (Delgado-Vargas & Paredes-Lopez, 2003; Piątkowska et al., 2011; Smeriglio et al., 2016; Wu et al., 2006).

Grapes are the main source for ANCs isolation in the world (Piątkowska et al., 2011). They constitute one-fourth of the world's harvest of fruit, and more than half of them are red grapes. Nowadays the direction of grape processing is obtaining grape concentrate with high coloring properties. The other ANCs rich raw material is blueberries (*Vaccinium myrtillus* L.). Decoctions of these fruits are used as an anti-inflammatory, anti-diarrheal and anti-ulcer agent, as well as for cystitis and urinary inflammation. Blueberry fruits are used in the food industry as a colorant for wines, juices, and for making jams. Dried fruits are also used in herbal mixtures. Another source of ANCs is red cabbage, which is deep purple in color. This is used for food coloration, while water extracts from the leaves can be used as an indicator of pH; they change color from dark purple through shades of red to almost yellow. Moreover, numerous flowers are rich in ANCs, and after drying they are used as mixtures of herbs. Their medicinal properties are largely associated with their high ANCs content. One example of this is the flower of *Centaurea cyanus* L., which is characterized by a beautiful sapphire color, used for herbal mixtures. Cornflower is a honey plant, which is included in incense and is used in the colorant industry. It is used in herbal medicine in liver diseases and abnormal metabolism. This is characterized by an anti-inflammatory effect and is used as a diuretic in diseases of the kidneys and the urinary tract. The black mallow (*Malve hortensis*) flower containing malvidin is one of the richest anthocyanin raw materials. Dried flowers are used to dye wines, liqueurs, food formulations, and fabrics. In medicine, this material is used for curing the inflammation of the mouth and throat (Piątkowska et al., 2011; Smeriglio et al., 2016).

The health benefits of ANCs are widely known. The most important of these are its antioxidant, anti-inflammatory, anti-atherosclerotic, and anti-cancer effects. ANCs, like other polyphenolic compounds, can neutralize free radicals (atoms or molecules having an unpaired electron on the last electron shell). The antiradical activity of ANCs increases with the higher number of hydroxyl groups in Ring B and the acylation of sugar residues with phenolic acids. In addition, ANCs, due to the presence of hydroxyl groups in Ring C, show the ability to chelate metal ions (e.g., iron and copper). An important feature is also the inhibition of lipid peroxidation or self-oxidation of fatty acids. Numerous studies prove that ANCs have a much higher antioxidant potential than the best-known reference antioxidants, such as vitamin E, β-carotene, or vitamin C (Fukumoto & Mazza, 2000; Kähkönen & Heinonen, 2003; Lila, 2004; Miguel, 2011; Wang, Cao, & Prior, 1997; Wang & Stoner, 2008; Yang, Koo, Song, & Chun, 2011). Delphinidin and cyanidin anthocyanidins proved to be the most active compounds, followed by malvidin, peonidine pelarginidin, and petunidine (Lucioli, 2012).

The positive effect of ANCs on the cardiovascular system is associated with their anti-inflammatory effect, their ability to strengthen blood vessels or the inhibition of platelet aggregation. By neutralizing the free radicals, ANCs limit the formation of atherosclerotic plaques in the arteries (Lucioli, 2012).

Studies from recent years show that a diet rich in ANCs has a positive effect on the plasma lipid profile (the amount of cholesterol, triglycerides, and LDL fraction decreases) and the carbohydrate metabolism of the body by lowering elevated blood glucose and inhibiting the absorption of sugars in the small intestine, which can prevent obesity and diabetes (Smeriglio et al., 2016).

Anthocyanosides present in blueberries, strawberries, and cherries can help reduce the incidence of age-related macular degeneration (AMD) and improve visual acuity. ANCs improve night vision and adaptation to darkness, which is probably associated with an increase in the rate of rhodopsin regeneration. In addition, they have been shown to protect the retina from damage by modifying the activity of some enzymes and limit the development of glaucoma by stabilizing collagen in the trabecular mesh (Ghosh & Konishi, 2007).

Researchers are increasingly interested in the possibility of using ANCs in the prevention and treatment of various types of cancer. *In vitro* studies conducted on various cancer cell lines proved that ANCs have anti- proliferating activity. Cell divisions are inhibited at various stages of the cell cycle, either by acting on regulatory proteins or by blocking enzymes from the kinase group (Hou, Ose, & Lin, 2003; Roobha, Marappan, Aravindhan, & Devi, 2011). It has also been shown that ANCs can induce apoptosis (programmed cell death). Anti-cancer activity of ANCs has been demonstrated, among others, in relation to leukemia cell lines (Feng, Ni, Wang, et al., 2007) or cancerous colon. Noteworthy is the fact that ANCs act proapoptotically only on cancer cells, without showing such effects on healthy cells (Hou et al., 2003).

Foods rich in ANCs may also have beneficial effects against age-related neurodegeneration (Rendeiro et al., 2013; Tsuda, 2012). Shukitt-Hale, Carey, Jenkins, Rabin, and Joseph (2007) observed that ANCs from blackberries and plums could delay the onset of the decline of the nervous system, including delaying a decline in cognitive function and motor skills, through inhibiting the inflammation of nerves and modulating neural signals.

ANCs are relatively unstable and sensitive to temperature, pH, light, and UV radiation, which may result in the disappearance or modification of the expected color. One way to increase the stability of ANCs is through the addition of metal cations. It was found that mainly cyanidin, delphinidin, and petunidin are capable of metal chelating. The oxidation state of metal ions plays a key role in the formation of the anthocyanin-metal complexes. The most common metals used in anthocyanin complexes are tin, copper, iron, aluminum, magnesium, and potassium (Cavalcanti et al., 2011; Clifford, 2000; Cortez et al., 2017; Tachibana, Kimura, & Ohno, 2014).

Another way to increase the stability of ANCs is microencapsulation, which will be discussed in the following section.

4.3 ENCAPSULATION OF ACNS BY SPRAY DRYING

Microencapsulation is the incorporation of sensitive or functional core substances in a shell of wall material for protection and ease of handling (Mahdavi, Jafari, Assadpour, & Ghorbani, 2016; Edris et al., 2016). The principle of the encapsulation process is the formation of a multicomponent structure in the form of particles consisting in general of two substances: preserved (active core material) and protective (wall material/carrier). Encapsulation of bioactive compounds aims to protect it from light, oxygen, moisture, free radical degradation, and other deteriorating conditions, to improve its stability. Additionally, the change of the form from a liquid into a dry powder creates better handling properties (Robert et al., 2010; Silva, Stringheta, Teofilo, & Nolasco de Oliveira, 2013). Besides, the encapsulation of bioactives can be used to improve their bioavailability and modify the time and/or place of their release upon administration (Nedović, Kalušević, Manojlović, Petrović, & Bugarski, 2013).

There are different methods for encapsulation in the food industry; extrusion, coacervation, freeze, and spray drying are typical examples of such techniques (Edris et al. 2016). The selection of a microencapsulation method depends on the specific applications and parameters, such as required particle size, physicochemical properties, release mechanisms, process cost, etc. (Mahdavi, Jafari, Ghorbani, & Assadpoor, 2014). Among different methods, spray drying is the most frequently used, so that about 80%–90% of encapsulates are spray dried (Gharsallaoui, Roudaut, Chambin, Voilley, & Saurel, 2007; Mahdavi et al., 2014). Spray drying is a simple, cost-effective, flexible, and continuous process. It offers the possibility of a rapid transformation of active compounds dissolved in a dispersion of the carrier material into powder (Janiszewska, Jedlińska, & Witrowa-Rajchert, 2014). Moreover, this technique enables the drying of heat-sensitive materials, since the atomization of the liquid mixture in a flow of hot air results in rapid water evaporation that keeps the temperature of the particles relatively low (Samborska et al., 2005; Kalušević et al., 2017).

The advantages of powders in the production of industrial food are easy dosing, long shelf-life (due to low water activity), taking up a small transport and storage space, no need to maintain refrigeration temperatures, product availability throughout the year, easy combining with other dry ingredients (Chauhan & Patil, 2013; Shishir & Chen, 2017; Walkling-Ribeiro, Noci, Cronin, Lyng, & Morgan, 2009).

ANCs can be encapsulated by spray drying directly from plant-related material like fruit/vegetable juice/concentrate/pulp, or the extract prepared of thereof or other sources like herbs and waste materials (e.g., pomace and peel) (Robert & Fredes, 2015). The examples of spray drying experiments performed to encapsulate ANCs from different sources are presented in Table 4.1. The following two sections offer discussions of the spray drying encapsulation of ANCs by different wall materials (Section 4.3.1) and various process conditions (Section 4.3.2).

4.3.1 MATERIAL-BASED APPROACH

The selection of the appropriate carrier material(s) (also known as the coating material, carrier, wall material, encapsulating material, or drying aid) is a crucial stage of the process (Rajabi et al., 2015; Mahdavi et al., 2016). The wall material selected for the production of microcapsules should have desirable emulsifying properties (when working with hydrophobic compounds), low viscosity at high-solids concentrations, low hygroscopicity, and low cost (Santiago et al., 2016; Silva et al., 2013). Different encapsulating agents have various features, including solubility, viscosity, and film-forming properties, which affect the formation rate of a crust on the particle surface (Robert & Fredes, 2015). Moreover, if the raw material is a fruit juice or concentrate, wall material should have the ability to increase glass transition temperature to avoid stickiness problems during drying. Most often, the single encapsulating material does not provide all these properties, and so different drying agents have to be applied to promote the desired properties (Shahidi & Han, 1993). As carriers for microencapsulation of ACNs, the most common materials are maltodextrin and gum Arabic. Sharif, Khoshnoudi-Nia, and Jafari (2020) presented that among 16 studies in the field of spray drying microencapsulation of ANCs carried out during 2016–2019, maltodextrin was the selected wall material (alone or in combination with other materials) in 11 studies. Emulsifying starches, β-cyclodextrin, pullulan, and carbohydrates of added health-promoting properties (e.g., inulin and

TABLE 4.1

Examples of spray drying encapsulation of anthocyanins (ACNs) with different wall materials

Source/Form of Anthocyanins	Wall Material	Drying Temperature	Encapsulation Efficiency (%)	Remarks	Reference
Chokeberry fruit (*Aronia melanocarpa*)/ aqueous extract	Maltodextrin with guar gum, gum Arabic, pectin, β-glucan, inulin	Inlet 140°C	78–92[2]	Stability of ACNs in microcapsules depended on the type of wall material	Pieczykolan and Kurek (2019)
Jussara (*Euterpe edulis* Martius) pulp/extract	Maltodextrins, gum Arabic, maltodextrin+gum Arabic	Inlet 160°C, outlet 97°C	88–98[3]	The blend of maltodextrin and gum Arabic was selected as the optimal wall material	Da Silva Carvalho et al. (2016)
Bordo grape (*Vitis labrusca*) winemaking pomace/extract	Maltodextrin	Inlet 130°C–170°C	[1]	It was possible to obtain pigment powders with functional properties	Souza et al. (2014)
Barberry (*Berberis vulgum Arabicris*) fruits/extract	Maltodextrins, maltodextrin+gum Arabic, maltodextrin+gelatin	Inlet 150°C, outlet 100°C	91–93[2]	Combination of maltodextrin and gum Arabic led to higher efficiency and longest stability	Mahdavi et al. (2016)
Grape (*Vitis labrusca*, cv. BRS Violeta)/juice	Maltodextrin, whey protein concentrate, soy protein isolate	Inlet 140°C	62–99[2] 58–94[3]	Spray dried grape juice powders are a promising food additive for incorporating into functional foods	Moser et al. (2017)
Black carrot/juice	Maltodextrin 20DE, gum Arabic, tapioca starch	Inlet 150°C–225°C, outlet 76°C–112°C	56–87[3]	The inlet temperature of 150 °C was the best for encapsulation of black carrot anthocyanin using both maltodextrin 20 DE and gum Arabic	Murali, Kar, Mohapatra, and Kalia (2015)
Red grape skin variety Prokupac (*Vitis vinifera* L.)/extract	Maltodextrin, gum Arabic, skim milk powder	Inlet 140°C, outlet 65°C	[1]	Grape skin has a significant content of ANCs that make the utilization of such material worthwhile	Kalušević et al. (2017)
Purple corn cob (*Zea mays* L.)/extract	Maltodextrin	Inlet 130°C–170°C	57–89[3]	The amount of carrier agent is critical to enhancing the retention of spray dried ACNs	Lao and Giusti (2017)

(Continued)

TABLE 4.9
(Continued)

Source/Form of Anthocyanins	Wall Material	Drying Temperature	Encapsulation Efficiency (%)	Remarks	Reference
Purple maize (*Zea mays* L. race Cónico) pericarp and aleurone layers fractions/extract	Hydrolyzed normal maize starch, acetylated normal maize starch, hydrolyzed waxy maize starch, acetylated waxy maize starch	Inlet 170°C, outlet 80°C	49–90[3] 77–97[2]	Acetylated starches behave as superior encapsulating agents than hydrolyzed maize starches	García-Tejeda et al. (2015)
Juçara (*Euterpe edulis* M.) pulp	Maltodextrin, gelatin, gum Arabic	Inlet 140°C–190°C, outlet 85°C–126°C	63–82[3]	The spray drying process can be employed as a preservation technique for processing of juçara products where high retention of ACNs is desired	Bicudo et al. (2015)
Pomegranate/juice	Maltodextrin, gum Arabic, modified starch	Inlet 162°C–170°C, outlet 89°C–93°C	35–70[3]	Gum arabic and Capsul™ (1: 1) preserved the ANCs better during the drying process compared with the other mixtures	Santiago et al. (2016)
Jaboticaba (*Myrciaria jaboticaba*) peel/extract	Maltodextrin, maltodextrin + gum Arabic, maltodextrin + Capsul™	Inlet 140°C–180°C, outlet 48°C–89°C	79–100[3]	The use of 30% maltodextrin, the air-drying temperature of 180°C was recommended	Silva et al. (2013)
Iranian borage (*Echinum amoenum*) petal/extract	Maltodextrin, modified maize starch, and blends	Inlet 120°C, outlet 68°C	93–97[2]	The combination of both carriers gave the best protection to ANCs	Mehran, Masoum, and Memarzadeh (2020)

[1] Not determined.
[2] expressed as encapsulation efficiency of wall material.
[3] expressed as retention.

β-glucan) were also applied (Ahmad, Ashraf, Gani, & Gani, 2018; Fernandes, Sousa, Azevedo, Mateus, & de, 2013; Ferreira, Faria, Grosso, & Mercadante, 2009; Pieczykolan & Kurek, 2019). Due to wide availability and good properties for encapsulation, such as emulsification, solubility, film-forming, and water-binding capacity, whey and soy proteins are also considered to have strong potential as encapsulation agents (Moser et al., 2017). Frequently, mixtures of the mentioned materials are used in different proportions, to achieve the proper physical properties of microcapsules, the high efficiency of encapsulation, and to promote physicochemical stability during storage.

Carriers containing certain amounts of proteins, providing film-forming properties, could give better encapsulation and retention of ACNs. It relates to the migration of proteins to the air-water interface during atomization and the formation of the protein-rich film there. Such film is converted into a glassy skin, growing in thickness as the drying progresses, which leads to high microencapsulation efficiency (Bicudo et al., 2015). Among wall materials providing such properties, the most important example is gum Arabic. Maltodextrins do not have such surface-active properties, but are commonly applied as a drying aid due to the ability to increase glass transition temperature helping avoid stickiness problems. Thus, the typical approach is to apply the mixture of both types of wall materials, providing high T_g (glass transition temperature) and film-forming properties, to achieve smooth progress of the drying process and high encapsulation efficiency. The blend of maltodextrin DE 10 and gum Arabic was the optimal wall material for jussara (*Euterpe edulis*) pulp extract microencapsulation, providing the highest ANCs retention, low hygroscopicity, and high T_g (Da Silva Carvalho et al., 2016). Gum Arabic, despite the suitable properties that have facilitated its extensive use as wall material, is an expensive ingredient with availability and costs subjected to fluctuation, which has motivated researchers for alternative encapsulation matrices (Moser et al., 2017).

Tonon, Brabet, and Hubinger (2010) concluded that the degradation of ACNs during spray drying occurred on the surface of the particles, and the entrapped material was protected against the transfer of oxygen through the density of the matrix and the distance from the destructive factor. Thus, it is extremely important to select the proper wall material to create such capsules. Moreover, it has also been reported that intermolecular connections between wall materials and bioactive compounds result in stable microparticles (Fredes, Osorio, Parada, & Robert, 2018; Pieczykolan & Kurek, 2019). Silva et al. (2016) established, by the means of confocal microscopy, that ANCs were homogeneously distributed throughout the structure of capsules obtained with the application of maltodextrin, gum Arabic, and the blend of both materials.

The amount of carrier material is also of vital significance. The ratio of core to the wall is usually 1 : 4, but this can be optimized for each ingredient (Mahdavi et al., 2014). Increasing carrier agent ratio concerning the amount of the core material helps to increase the ACNs retention and drying yield due to the capability to change the hygroscopic and thermoplastic character of the powders (Janiszewska, 2014; Lao & Giusti, 2017). Similar observations were presented by Moser et al. (2017), when spray drying grape juice with maltodextrin + whey protein concentrate. Yousefi, Emam-Djomeh, Mousavi, Kobarfard, and Zbicinski (2015) explained that an increased amount of maltodextrin improved the retention of ANCs from black raspberry during spray drying because of the influence on the particle size. An increased amount of maltodextrin resulted in bigger particles and a smaller area-to-volume ratio, which reduced the oxidation rate. An additional reason for higher heat resistance could be the increase in T_g of the produced particles. However, the opposite phenomenon (decreased retention at higher carrier addition) was also observed. According to Bicudo et al. (2015), a lower proportion of carrier agents (maltodextrin, gelatin, gum Arabic) was useful to obtain higher retention of ACNs during spray drying of juçara pulp. Higher carrier proportion also decreased the retention of bioactive compounds during the spray drying encapsulation of ANCs of cactus pear (Saenz, Tapia, Cha'vez, & Robert, 2009) and watermelon (Quek, Chok, & Swedlund, 2007). According to Souza et al. (2014), Jafari, Ghalenoei, and Dehnad (2017), and Sánchez-Madrigal et al. (2019), the addition of more carriers causes the dilution effect of the dried extract.

The current trend is to apply the carrier materials that can provide additional health benefits, apart from acting as encapsulation agents. An example of such a carrier is inulin of different origin or β-glucan from barley. Sánchez-Madrigal et al. (2019) presented the encapsulation of ANCs from blue corn extract using agave inulin-type fructans that provided prebiotic effect and dietary fiber action. Concentration of 6% (w/v) carrier and inlet air temperature of 150°C favored the best value of ANCs retention (29%) with adequate physical properties for maximum stability during storage. β-glucan has a unique inbuilt honeycomb structure and health-benefiting properties like the lowering of serum cholesterol level, the reduction of glycemic response, enhancing the immune system, and the growth of beneficial microflora in the gut (Ahmad et al., 2018). It was applied recently for the encapsulation of ACNs from saffron (Ahmad et al., 2018) and chokeberry (Pieczykolan & Kurek, 2019). It was concluded that β-glucan had the potential to encapsulate saffron bioactives (although the retention of ANCs was lower than that with the application of β-cyclodextrin) and improved its stability during passage through simulated gastric digestive conditions (Ahmad et al., 2018). According to Pieczykolan and Kurek (2019), the mixture of maltodextrin and β-glucan was characterized by the highest efficiency of encapsulation among tested variants (mixtures of maltodextrin with gum Arabic, inulin, pectin, and gum guar), and also the stability during storage was lower than most of them (only maltodextrin + gum guar variant was more efficient to prevent storage degradation).

In the case of microencapsulating ACNs directly from juices (without purification), the most typical ratio of carrier solids to core material solids is 1:1. Such an approach, applied to avoid powder stickiness on the drying chamber wall, was used by Janiszewska et al. (2014) for beetroot juice, Santiago et al. (2016) for pomegranate juice, Villacrez, Carriazo, and Osorio (2014) for Andes berry juice, and Moser et al. (2017) for grape juice. Other amounts of the carrier were also tested; e.g., 20% (w/w) for tamarillo juice (Ramakrishnan, Adzahan, Yusof, & Muhammad, 2018), 6% for açai pulp (Tonon et al., 2010), the proportion of 1:2 (juice to carrier) for black carrot juice (Murali et al., 2015), proportion of 2:3 (juice to carrier solids) for chokeberry juice (Bednarska & Janiszewska-Turak, 2019; Janiszewska-Turak, Sek, & Witrowa-Rajchert, 2019), and 25%–45% for pomegranate juice (Jafari et al., 2017).

4.3.2 Process-based Approach

Since ACNs are thermosensitive compounds and can undergo oxidation and/or degradation reactions induced by heat, drying conditions should be optimized to achieve the maximum possible encapsulation efficiency and retention (Robert & Fredes, 2015). The inlet and outlet air temperature are the most important parameters to be optimized. Other factors, which can also affect the resulting outlet air temperature, should be also adjusted; i.e., atomization parameters and feed rate.

4.3.2.1 Drying Temperature

Due to the thermolabile nature of ANCs, the most important parameter affecting the retention after spray drying microencapsulation is the drying air temperature. Thus, the optimization of drying temperature is the most common experimental approach. Typical inlet air temperature applied in laboratory or pilot plant spray driers ranges from 150°C to 200°C, while outlet air temperature varies from 70°C to 90°C. This level of temperature combined with very short contact time enables to obtain low degradation and high retention of biologically active compounds, including ANCs (Nath & Satpathy, 1998; Yingngam et al., 2018). However, drying temperature should be optimized for each material individually.

Bicudo et al. (2015) microencapsulated juçara pulp by spray drying at inlet air temperature 140°C–190°C. ACNs retention was affected by temperature due to its high sensitivity. Significantly high retention (>83%) was observed at the treatment conditions of 165°C and 5% of the carrier. On the contrary, Souza et al. (2014) did not notice any significant influence of drying temperature (inlet 130°C, 150°C, and 170°C) on the retention of ANCs of Bordo grape winemaking pomace extract microencapsulated in maltodextrin. Additionally, Lao and Giusti (2017) presented that during the spray drying of purple corn (*Zea mays* L.) cob extract with 5% of maltodextrin, the drying air temperature (inlet 130°C, 150°C, and 170°C ; outlet 89°C, 105°C, and 119°C) did not affect the retention of ANCs. It was in the range of 81% to 89%, and the difference between values obtained at different temperatures was not significant. The retention was higher, and heat change sensitivity lower than what was observed by Murali et al. (2015) for black carrot juice spray drying with maltodextrin at the inlet air temperature 150°C–225°C (retention from 60% to 87%). The researchers (Lao & Giusti, 2017) explained that these results were due to different chemical compositions of ANCs. In other words, ANCs with higher proportions of acylated pigments (e.g., purple corn cob) are more heat resistant than non-acylated ANCs.

According to Jafari et al. (2017), apart from the obvious thermal degradation, the underlying reason for higher ANCs retention after drying at a lower temperature is higher moisture content, which causes a greater degree of aggregation resulting from the natural stickiness of the product, leading to lower oxygen exposure and lower degradation.

Recently, the spray drying process was performed at very low inlet and outlet air temperatures (75°C and 50°C, respectively) (Jedlińska et al., 2019; Samborska et al., 2019). The reduction of drying temperature was possible due to the application of dehumidified air as a drying medium. However, this method has been applied only for the spray drying of honey solutions so far. Since the retention of polyphenolics and antioxidant activity of honey was significantly improved by this novel approach (Samborska et al., 2019), it can be assumed that the application for ANCs microencapsulation would also provide satisfactory results.

4.3.2.2 Atomizing Parameters

The parameters of atomization can affect the final physicochemical properties of microcapsules, because they influence the size of the droplets, the evaporation rate, and outlet air temperature. Two main types of atomizers that have been used for spray drying encapsulation of ANCs include pressure nozzles or rotating discs, depending on the size and geometry of the drying chamber. 0.7, 1.0, 1.5, or 2.0 mm diameter nozzles have been applied (Gagneten et al., 2019; Yingngam et al., 2018; Darniadi, Ifie, Ho, & Murray, 2019; Souza et al., 2014; Da Silva Carvalho et al. 2016, Bicudo et al., 2015; Santiago et al., 2016), while the speed of the rotating disc varied from 15,000 rpm (Ramakrishnan et al., 2018) to 39,000 rpm (Janiszewska, 2014; Janiszewska-Turak et al., 2019).

Villacrez et al. (2014) investigated the influence of nozzle diameter on the characteristics of solids, ANCs content, moisture, and water activity after the microencapsulation of Andes berry extract. The variation of ANCs content in powders with nozzle diameter was <5%. The moisture content in microcapsules was higher when a 1 mm diameter nozzle was used.

4.4 PROPERTIES OF ENCAPSULATED ACNS

4.4.1 Retention and Encapsulation Efficiency

The most important concern during ACNs encapsulation is the retention of thereof after spray drying, as well as the location of the compounds in the microcapsules, which can affect the subsequent storage stability (surface ACNs being more vulnerable for deterioration). Spray drying technique sometimes is considered to be an immobilization method, because some bioactive compounds may be exposed on the surface of the capsules (Robert & Fredes, 2015). Thus, the efficiency of encapsulation can be described by two methods:

1) as the parameter called 'retention' or 'encapsulation productivity', calculated by the comparison between the ACNs concentration in the microcapsules and the initial ACNs concentration in the raw material (García-Tejeda et al., 2015; Kalušević et al., 2017; Rajabi et al., 2015; Turan et al., 2015; Ahmad et al., 2018);

2) as 'real' encapsulation efficiency (EE) of the wall material, taking into account the location of ACNs in the microcapsules, calculated as the ratio between internal ACNs content and total ACNs content in the microcapsules (Idham, Muhamad, & Sarmidi, 2012; Mahdavi et al., 2016; Stoll et al., 2016; Mehran et al., 2020). Low efficiency may result in low stability because of no protection of capsules against unfavorable storage conditions (Mahdavi et al., 2016). Currently, the main emphasis of the microencapsulation of food components has concentrated on improving the EE and extending the shelf-life of the products (Mahdavi et al., 2014).

Successful encapsulation of ANCs should result in an encapsulated powder with minimum surface pigment content on the powder particles and maximum retention and encapsulation efficiency of the core material within the particles (Mahdavi et al., 2014). Retention of ACNs of different origin after spray drying in the research works carried out during the last decade varied from 6%

(Lacerda et al., 2016) to 100% (Weber, Boch, & Schieber, 2017), and even went as high as 113% (Pereira et al., 2020). The retention of 113% was explained by the fact that ANCs are present in the cell vacuoles and, during drying, cellular disruption caused by the high shear and heat may have promoted a greater release of these compounds.

According to Lao and Giusti (2017), a certain amount of carrier agent is critical to enhancing the retention of spray dried ANCs. The retention of pigments in purple corncob after spray drying mainly depended on the amount of maltodextrin applied to the feed pigment solution. ACNs mixed with 2% maltodextrin had the lowest pigment recovery (57.3 ± 8.9%), which was significantly lower than all other groups with a maltodextrin amount of at least 5%. Idham et al. (2012) reported that hibiscus ANCs yielded the best results, with a total retention rate of 99.87% when the mixture of maltodextrin and gum Arabic was applied.

The type of wall material also affected such retention. As mentioned before, carriers containing certain amounts of proteins, providing film-forming properties, could give better retention of ACNs. It was confirmed by Bicudo et al. (2015) that better retention of juçara pulp ANCs with the addition of gelatin could be obtained when gum Arabic was used as the coating material, compared with when maltodextrin was used. Santiago et al. (2016) also showed that a higher (94%) retention of pomegranate ANCs with the addition of gum Arabic than maltodextrin (56%) was achieved. Silva et al. (2013) also reported higher retention of jaboticaba peel extract ANCs with the addition of gum Arabic than maltodextrin.

100% retention of ANCs from blackberry extract was obtained by Weber et al. (2017) under the following conditions: maltodextrin to extract ratio of 5:1 and inlet/outlet air temperature of 150/90°C. As was mentioned before, Pereira et al. (2020) noticed even the increase of ANCs content after spray drying of juçara pulp with no carrier addition. The basic drying parameters affecting the retention of ANCs are the drying temperature, which is due to the high sensitivity for thermal degradation. The examples of retention at different temperatures were presented in Section 4.3.2.2.

The EE of wall materials in the spray drying research works on ANCs carried out during the last decade varied depending on the type of wall materials; from 78% to 83% for tamarillo juice with maltodextrin and gum Arabic (Ramakrishnan et al., 2018), from 78% to 92% for chokeberry fruit extract with the blends of maltodextrin with gum guar, gum Arabic, pectin, β-glucan, and inulin (Pieczykolan & Kurek, 2019), and from 89% to 97% for black glutinous rice extract (Laokuldilok & Kanha, 2017).

EE is strongly dependent on wall material composition and properties. As was mentioned before, the single carrier material does not provide all required properties; thus, the blends are usually applied. Such mechanisms were presented by Mahdavi et al. (2016) during spray drying microencapsulation of barberry extract. Among the three combinations of wall materials (i.e., maltodextrin, maltodextrin + gum Arabic, and maltodextrin + gelatin), maltodextrin + gum Arabic gave the highest EE. It was explained by the fact that maltodextrin and gelatin do not have emulsifying and film-forming properties. On the contrary, gum Arabic, due to its highly branched structure and the presence of a small amount of protein, acts as an excellent film-forming agent better trapping ACNs molecules. Moser et al. (2017) tested the blends of maltodextrin with two types of protein isolates (whey and soy) as encapsulation agents for grape juice ANCs. The samples obtained with whey protein presented EE between 62% and 84%, demonstrating the fair ability of these carrier formulations. In the case of the samples with soy protein, EE was higher (from 97% to 99%). Additionally, EE increased significantly at higher carrier concentration, which was not observed for whey protein. This was explained by the fact that higher amounts of carrier resulted in faster polymer precipitation on the dispersed phase surface, preventing core diffusion across the phase boundary. Moreover, increased viscosity of the solution with higher carrier concentration also delays core diffusion within polymer droplets.

According to Tonon et al. (2010), the solubility of wall material also affects EE. The poorly soluble encapsulating agents present lower values of EE than the highly soluble ones. This fact was verified in the work of García-Tejeda et al. (2015), where the derivatives of acetylated starches, characterized by a higher water solubility index than the hydrolyzed starches, also gave higher total ANCs content and EE after spray drying of purple maize (*Zea mays* L. race Cónico) pericarp and fractions extract of aleurone layers.

4.4.2 COLOR PARAMETERS

Microencapsulated ANCs are produced to be applied as natural food pigments, offering also additional health-promoting properties, contrary to "artificial" pigments. They are of great interest to the food industry since they give rise to a wide range of colors (Mahdavi et al., 2014). It is important to characterize the color parameters after microencapsulation and to provide after drying the material possibly unchanged in terms of color as much as possible. Thus, two ways to express the color parameters are popular. These include the measurement of color parameters of the obtained powder itself, as well as the comparison to the material before drying, involving powder reconstitution. Considering that most applications of powdered pigments involve solubilization in water, typically, before color parameters are measured, samples should be reconstituted to the same solids concentration as the in-feed solution. Pigment compounds are hidden inside the capsules and the measurement without reconstitution can give misleading conclusions (Moser et al., 2017; Silva et al., 2013). Moreover, the successful encapsulation of ANCs should result in an encapsulated powder with minimum surface pigment content (Mahdavi et al., 2014); thus, the measurement of the external color of the spray dried powder can be unrepresentative.

On the other hand, Pieczykolan and Kurek (2019) concluded that the lowest value of a* associated with red color may indicate the best effect of encapsulation. Villacrez et al. (2014), after spray drying of Andes berry extract with different wall materials, analyzed the color parameters of powders and reconstituted solutions, indicating that chroma (C*) and a* values of solutions were higher than of the powders. It indicated a more intense red color after the release of the encapsulated ANCs. However, very often the color parameters of the powders are also measured without reconstitution. In this case, it is possible to evaluate the effect of drying on C* parameter, since it is related to the red color and can be associated with the ANCs content (Da Silva Carvalho et al., 2016; Santiago et al., 2016). The lightness (L*) can also be compared. The powders are usually lighter than raw material before drying due to the presence of white encapsulation agents (Pereira et al., 2020; Santiago et al., 2016). Moser et al. (2017) reported that increasing carrier concentration and ratio resulted in brighter powders, which was called the dilution effect, resulting in color loss. Moreover, an increasing amount of carrier decreased C*, indicating grayish chromaticity, whereas higher chroma values at low carrier content and ratio demonstrated more saturated or vivid color. The examples of color parameters of raw materials and microencapsulated ANCs are presented in Table 4.2.

The overall color difference (i.e., between feed solution and reconstituted powders) is described as $\triangle$E value. The difference of 0–1.5 is considered small (the sample is almost identical to the original by visual observation), and the range of 1.5–5 informs that the change of color can be distinguished, while the evident difference results in the values greater than 5. The type of wall material affects the color degradation. The use of gum Arabic/maltodextrin and modified starch/maltodextrin carriers for spray drying microencapsulation of jaboticaba peels extract resulted in high $\triangle$E (in the range from 9 to 14), while the use of maltodextrin reduced the $\triangle$E to 3–4 (Silva et al., 2013). The blends of soy protein isolate with maltodextrin had a lower influence on grape juice coloration ($\triangle$E = 1–2) after spray drying than a mixture of whey protein concentrate with maltodextrin ($\triangle$E = 9–15) (Moser et al., 2017). Da Silva Carvalho et al. (2016) presented that C* values obtained for microparticles were higher than for those obtained for the extracts of jussra pulp, due to color saturation after drying that is a desirable feature.

Due to the thermolabile nature of ANCs, the drying temperature can also affect color parameters. Laokuldilok and Kanha (2017) microencapsulated black glutinous rice ANCs using maltodextrin produced from broken rice fraction as wall material at the inlet air temperature of 140°C–180°C. Increasing the temperature, as well as DE of maltodextrin, caused the decrease of L* and C* and the

TABLE 4.2

The color parameters of anthocyanins (ANCs) before and after spray drying microencapsulation

Source of ACNs	Form of sample	L* Lightness (0-black, 100-white)	a* Variation of green (−) /red (+) color intensity	b* Variation of blue (−)/yellow (+) color intensity	C* Chroma, color saturation	h* Hue angle Chromatic	Reference
Pomegranate juice	Powder	79.2	11.8	2.1	12.0	10.1	Santiago et al. (2016)
Andes berry (*Rubus glaucus* Benth.)	Fruit	31	56	11	57	11	Villacrez et al. (2014)
	Powder	71–77	29–32	−0.51-1.95	29–32	0.4–3.58	
	Solutions (50 mg solids / mL water)	66–76	21–43	13–22	24–48	27–36	
Jaçara (*Euterpe edulis* Martius) pulp	Pulp	11.5	7.8	−9.8	12.6	308.5	Pereira et al. (2020)
	Powder	14.7	8.0	−4.0	9.0	346.9	
Jussara (*Euterpe edulis* Martius) pulp extract	Extract	1.5	-	-	4.2	18.3	Da Silva Carvalho et al. (2016)
	Powder	36–39	-	-	31–34	2.9–4.5	
Saffron extract	Powder	40–61	13–27	35–81	-	-	Ahmad et al. (2018)
Chokeberry fruit (*Aronia melanocarpa*) extract	Powder	71–74	9–12	4–5			Pieczykolan and Kurek (2019)
Black glutinous rice extract	Powder	37–44	-	-	21–34	12–15	Laokuldilok and Kanha (2017)
Grape	Juice	1.6	3.3	−0.2	-	-	Moser et al. (2017)
	Powder	1–13	2–8	−8 to −1	13–26	305–335	
	Reconstituted	51–57	11–16	−14 to −10	-	-	

increase of h*, so the color was darker and less saturated. The darkening was explained by accelerated non-enzymatic browning at increased temperature and increased amount of sugar.

4.4.3 Physical Properties of Powdered Microcapsules

Apart from the retention and content of ACNs after drying, the physical properties of microcapsules are also important, because they can affect the storage, transportation, and handling. As presented above, the morphology and particle size of particles can also influence the retention, encapsulation efficiency, and storage stability. Typically, the physical properties of powders determined after spray drying include moisture content, bulk density, particle size and morphology, hygroscopicity, and glass transition temperature (T_g).

The moisture content of spray dried powders usually varies from 1% to 5%, although sometimes also higher values are noted. This level of water content is essential for chemical and physical stability during storage (Ramakrishnan et al., 2018). It is the parameter that can be affected by the type and the amount of the applied wall material, as well as by drying parameters, mainly temperature (Da Silva Carvalho et al., 2016). According to Bicudo et al. (2015), the high molecular weight of maltodextrin could be the reason for difficult diffusion of water molecules during spray drying of juçara pulp, resulting in the higher moisture content of the powder (10.17%) as compared with the application of gum Arabic (7.23%) and gelatin (4.94%). However, these findings were not confirmed by Da Silva Carvalho et al. (2016), who showed lower water content of the powder obtained with maltodextrin DE 10 (2.89%) than with gum Arabic (3.68%). Also, Pieczykolan and Kurek (2019) presented that the highest moisture content of microencapsulated chokeberry extract was when gum Arabic was used as wall material. It could be caused by the film-forming activity of gum Arabic that can hinder water evaporation from the droplets (Lee, Taip, & Abdullah, 2018).

Bulk density gives information about the volume needed for the storage of the specific weight of the powder, so it is an important parameter in terms of packaging, storage, and handling organization. Higher values of bulk density are required to reduce the cost of packaging and storage (Pereira et al., 2020). On the other hand, low bulk density, as in agglomerated powders, influences other properties of the particles; e.g., flowability and instant dissolving characteristics (Bicudo et al., 2015). This parameter is affected by some properties of the particles, such as moisture content, particle size, and particle size distribution. Usually, powders of higher moisture content, smaller particle size, and higher particle polydispersity index are characterized by higher bulk density (Chegini & Ghobadian, 2005). The bulk density of microencapsulated ANCs varied from 0.20 to 0.95 g/cm^3 (Table 4.3). Bicudo et al. (2015) presented the above-mentioned relationship of increased bulk density of powders with smaller particles in the case of microencapsulated juçara pulp with the addition of maltodextrin, gum Arabic, and gelatin. Different particle sizes were obtained with different wall materials; thus, the type of carrier indirectly also affects the bulk density. The powders obtained after spray drying of blackcurrant and elderberry extracts with maltodextrin DE 12 had a bulk density of 0.38 g/cm^3 (Gagneten et al., 2019).

Particle size and morphology of spray dried powders affect other physical properties such as flowability and bulk density, as well as influencing the retention, encapsulation efficiency, and storage stability of ANCs (Mahdavi et al., 2014). In the case of small particles (<50 μm), the interparticle adhesive forces are significant, and the roughness on the surface can obstruct the ability of particles to approach each other (Gagneten et al., 2019). According to Gagneten et al. (2019), rough particles may show better flowability than smooth ones. On the contrary, according to Tonon et al. (2010), particles with rough surfaces can have problems in their flow properties, and, additionally, they may be more susceptible to degradation reactions such as oxidation due to large contact areas. This is in accordance with Da Silva Carvalho et al. (2016), who observed that smooth morphology of particle surfaces was favorable for greater ANCs retention, which may be associated with better accommodation of the pigments within the particle. The particles produced of the material of high content of sugars typically have smooth surfaces. It was confirmed by Da Silva Carvalho et al. (2016), who reported that higher hydrolysis of maltodextrin applied as a wall material for jussara pulp extract microencapsulation resulted in more smooth shaped particles. Silva et al. (2013) presented a similar relationship, that more intact and regular particles obtained with maltodextrin and the blend of maltodextrin with gum Arabic resulted in more efficient microencapsulation of jaboticaba peel extract than the application of the blend of maltodextrin with modified starch, which produces more irregular surfaces of angular shapes. Additionally, the same conclusion was presented by Murali et al. (2015), stating that maltodextrin DE 20 as the carrier material, producing smooth particles with hardly any surface cracks in the wall systems, proved to be better in retaining maximum ANCs and antioxidant activity (compared with gum Arabic and tapioca starch). The mean particle size of ANCs-loaded capsules obtained by spray drying is presented in Table 4.3 (in the range of 6–124 μm).

TABLE 4.3

The physicochemical properties of anthocyanins (ACNs)-loaded microcapsules obtained by spray drying

Source of ACNs	Moisture content (%)	Bulk density (g/mL)	Mean particle size (µm)	ACNs content in microparticles	T_g or hygroscopicity	Reference
Chokeberry fruit (*Aronia melanocarpa*) extract	1.29–2.73	0.78–0.95	16–53	Total 1940–3055 mg cyanidin-3-glucoside/100 g	T_g 66°C–69°C	Pieczykolan and Kurek (2019)
Jaçara (*Euterpe edulis* Martius) pulp	1.54	0.42	124.5	Total 7.079 mg/100 g/db	Hygroscopicity 11.6%	Pereira et al. (2020)
Juçara (*Euterpe edulis* M.) pulp	4.94–10.17	0.20–0.37	up to 80	not specified, the only retention shown	Hygroscopicity 10%–14%	Bicudo et al. (2015)
Blackcurrant (*Ribes nigrum*) extract	1.44	0.38	9.1	approx. 90 mg cyanidin-3-glucoside/100 g/db	T_g 50°C	Gagneten et al. (2019)
Raspberry (*Rubus idaeus*) extract	1.40	0.42	8.48	approx. 40 mg cyanidin-3-glucoside/100 g/db	T_g 50°C	Gagneten et al. (2019)
Elderberry (*Sambucus nigra*) extract	0.92	0.38	6.06	approx. 230 mg cyanidin-3-glucoside/100 g/db	T_g 43°C	Gagneten et al. (2019)
Jussara (*Euterpe edulis* M.) extract	2.28–4.18	nd	7.0–11.5	127.8–140.7 mg of cyanidin-3-glucoside/100 g	T_g 37.7–83.4	Da Silva Carvalho et al. (2016)

Laokuldilok and Kanha (2015) presented the effect of process conditions on powder properties of black glutinous rice (*Oryza sativa* L.) bran ANCs. Increasing the inlet air temperature caused the enlargement of particles and the number of smooth surface particles. Gagneten et al. (2019) concluded that the type of fruit extract influences the particle size distribution, the span value in the case of raspberry and blackcurrant powders was higher than in the case of elderberry powder. Hygroscopicity (H) is an important feature of powdered ANCs. Powders with low H are desirable in terms of preservation as they absorb less humidity from the environment. Absorbed humidity may cause subsequent glass transition, softening and the contact of the pigments with atmospheric air and free radicals, thereby causing it to decay (Silva et al., 2013). H depends on the moisture content in powder (so on the drying temperature), as well as the type and amount of wall material. However, as presented by Bicudo et al. (2015) and Ahmed, Akter, and Eun (2010), the general rule of greater capacity for water sorption connected with a lower moisture content of powders cannot be generalized for all spray dried materials. The influence of the type of carrier is sometimes more important, because carriers differ in their chemical structures, and the interactions of the carrier with environmental humidity also vary (Silva et al., 2013). Silva et al. (2013), investigating the properties of jaboticaba peels extract powdered with the addition of maltodextrin, gum Arabic, and the modified starch observed the significant influence of carrier type on the hygroscopicity of the powders. The mixture of maltodextrin and gum Arabic was the matrix that generated the most hygroscopic powders (H approximately 18%). On the contrary, the blend of maltodextrin with modified starch produced powders with the lowest H (about 13%), which could result from the incorporation of the lipophilic component (i.e., octenyl succinate) in its structure. Bicudo et al. (2015) presented similar levels of H (10%–14%) for microencapsulated juçara (*Euterpe edulis* M.) pulp, and the type of wall material affected it significantly. The sample obtained with gelatin had the lowest H, although the moisture content of this sample was also the lowest. Pulp encapsulated with gum Arabic was the most hygroscopic. Powdered jussara pulp extract was the most hygroscopic when maltodextrin DE 30 was applied as the carrier, which could be attributed to the presence of a greater amount of shorter chains and hydrophilic groups as compared with other carriers (gum Arabic and maltodextrin of lower DE) (Da Silva Carvalho et al., 2016).

Glass transition temperature (T_g) of the powders obtained after spray drying is of vital significance in terms of the progress of the spray drying process, as well as in terms of storage stability. Generally, higher T_g values are desirable, because of the influence on the reduction of vulnerability to stickiness during the drying process and the enhancement of storage stability. The value of T_g is the resultant of the T_g values of each compound present in the sample, with the dominant influence of water (strong plasticizer reducing T_g), low molecular sugars (low T_g), and high molecular carriers (high T_g). Thus, the addition of carriers offering high T_g enhances the stability of powders, due to the increase of T_g and decrease of stickiness (Da Silva Carvalho et al., 2016). On the other hand, according to Villacrez et al. (2014), the presence of low-molecular sugars in wall materials, which can act as plasticizers, can prevent shrinkage of the surface during drying.

T_g above the storage temperature increases the storage stability, because the high viscosity of the glassy amorphous matrix slows down the rate of reactions leading to the deterioration of quality. However, the samples should be protected by gum Arabic for moisture absorption to avoid the decrease of T_g down to the storage temperature. T_g of raspberry, blackcurrant, and elderberry extracts spray dried with maltodextrin DE 12 was 50°C, 50°C, and 43°C, respectively. This suggests that according to the authors (Gagneten et al., 2019) the powders could be stored at room temperature without the risk of physical and chemical deterioration. Da Silva Carvalho et al. (2016) presented T_g values of jussara pulp extract microencapsulated with maltodextrin DE 10, 20, and 30, gum Arabic, and the blends of maltodextrin DE 10 with gum Arabic in the range from 37.7°C to 83.4°C. The sample of the highest water and low molecular weight sugars content (obtained with maltodextrin DE 30) had the lowest T_g. The application of mixtures of maltodextrin with gum Arabic, inulin, pectin, gum guar, or β-glucan for the encapsulation of ANCs from chokeberry resulted in the powders characterized by T_g in the range from 66°C to 69°C (Pieczykolan & Kurek, 2019).

4.5 RELEASE BEHAVIOR AND BIOACCESSIBILITY OF MICROENCAPSULATED ACNS

Since the powders produced after spray drying would finally be incorporated in a food system, the releasing and digestion processes of the matrix containing bioactive ingredients should be also investigated. The entrapped core material should pass through the gastrointestinal tract (GIT) and reach the target as intact as possible (Rezvankhah, Emam-Djomeh, & Askari, 2020). However, the variation of pH and temperature, dissolution of the wall material, diffusion, and selective permeability are the factors that can affect the release and bioaccessibility of the core biomaterial from microstructure (Mehran et al., 2020). Bioaccessibility has been defined as the amount of compound that is released from the matrix after digestion; it is usually investigated *in vitro* under simulated gastrointestinal digestion conditions (Vergara et al., 2020). In the case of non-encapsulated biological material, free from any matrix materials, the bioaccessibility theoretically is 100%, but practically it is significantly decreased by the degradation of non-protected molecules/compounds.

According to Fredes et al. (2018), ANCs are unstable in the GIT; the shift from the acidic pH of the stomach to the almost neutral in the duodenum may be responsible for their degradation. On the contrary, microencapsulated ANCs are more protected in the GIT (but the proper wall material should be selected to provide the conditions for the target delivery). It was proved by improved bioaccessibility of ANCs from maqui juice in the simulated GIT after the microencapsulation by spray drying with the addition of inulin and sodium caseinate. Bioaccessibility of ANCs from non-encapsulated juice was 31.1%, while in the case of microcapsules it was 43.0% and 44.1%, when inulin and sodium caseinate were applied as wall materials, respectively. The type of wall material can affect the release behavior of microencapsulated ANCs, which was reported by Mehran et al. (2020). They investigated the release behavior of Persian borage ANCs encapsulated with the blend of maltodextrin and modified maize starch in the simulated gastric and intestinal conditions. The application of acetylated starch was beneficial, because it was not dissociated under an acidic medium. ANCs were partially protected in gastric fluids conditions, the release from microstructure after 30 minutes was 45.2%, and it remained at this level until 120 min (at the same time, the release from not encapsulated extract was 93.2%). In contrast, in the conditions of the intestinal fluid (increased pH) the acetyl groups started to dissociate, which resulted in the rupture of microcapsules and ANCs release. Thus, the combination of maltodextrin with modified starch was suggested for target delivery in the intestine. Similar results of possible controlled release of encapsulated ANCs were presented by Ahmad et al. (2018).

4.6 STABILITY OF MICROENCAPSULATED ACNS

The prolonging of storage stability is the main aim of microencapsulation (Laokuldilok & Kanha, 2017). The works on the storage stability of microencapsulated ACNs typically reveal first-order kinetics behavior, showing that the degradation remains linear with time (Osorio et al., 2010; Souza et al., 2014). Among the parameters that affect the degradation during storage, the storage conditions (e.g., temperature and humidity) and the type and amount of wall material (affecting particle morphology and encapsulation efficiency) are the most important.

The storage temperature plays an important role in the stability of ANCs, which is connected with the influence on reaction rate, glass transition temperature, and possible glassy-rubbery transition (Robert & Fredes, 2015). The change of physical state of microcapsules from amorphous glass to rubbery state increases the molecular mobility of the matrix and the reactants are accelerated, which results in an increased rate of physicochemical changes, such as sticking, collapse, caking, agglomeration, crystallization, and oxidation (Bhandari & Howes, 1999).

Mirhojati, Sharayei, and Ghavidel (2017) have shown that the storage of capsules at room temperature significantly shortens the half-life of ANCs, as compared with storage at refrigeration temperatures. However, Villacrez et al. (2014) presented that ANCs content in Andes berry microcapsules obtained with maltodextrin DE 20 and Hi-Cap™ 100 was not changed for three months of storage at 18°C (60% relative humidity, RH), despite the destruction of the capsules. Similarly, the retention of monomeric ANCs after the storage of powdered pomegranate juice at 25°C for 90 days was at a satisfactory 90% level (Santiago et al., 2016). Yingngam et al. (2018) presented that the remaining total ACNs content in microencapsulated mulberry fruit extract decreased slightly during storage at 4°C and 25°C for up to 30 days, while the greater loss was observed at 45°C. Gaona, Fanzone, and Sari (2019) observed that after 90 days of storage at 38°C, the loss of ANCs from Ancelotta red wine was about 20% (as compared with the initial concentration after spray drying).

For ANCs of *Bactris guineensis* fruit encapsulated by spray drying with maltodextrin, the most suitable conditions for storage were below 37°C and < 76% RH, respectively (Osorio et al., 2010). The influence of the type of source material (juice/pulp or extract), the type and amount of wall material on the stability of microencapsulated ANCs were also presented. According to Santiago et al. (2016) and Robert et al. (2010), the storage stability of ANCs is higher when the juice, not the extract, is spray dried. It is the result of possible antioxidative or encapsulation properties of some juice components (such as vitamin C). Also, Weber et al. (2017) suggest that the presence of pigments (ferulic acid and rutin) greatly enhances the shelf-life stability of spray dried blackberry ACNs, due to possible antioxidative activity, as well as preventing ANCs from being hydrated by the residual or ambient moisture.

According to Mehran et al. (2020), modified acetylated maize starch characterized by thermoprotective nature provides better protection for ANCs during the storage of microcapsules obtained by spray drying. It was proved in the case of microencapsulated Persian borage extract. Moreover, it was reported that the interaction between flavylium cation of ANCs and the acetyl group of modified maize starch prevented the transformation to unstable forms.

The addition of gelatin for the production of microencapsulated ANCs from the juçara pulp caused a 1.6-fold increase in the degradation rate, when compared with gum Arabic. It was explained by the presence of a small amount of protein in the carbohydrate chain, which could better trap the encapsulated molecules owing to the creation of an inflexible barrier (Bicudo et al., 2015). Souza et al. (2014), examining the properties of spray dried extract from Bordo grapes winemaking pomace, observed that ANCs in the samples with more carriers (30%) were more stable during storage. The morphology of particles also affects ANCs stability during storage. Particles of smaller porosity are more effective in protecting against oxidative degradation, which is the major cause of the degradation during storage (Laokuldilok & Kanha, 2017).

4.7 APPLICATION OF ENCAPSULATED ACNS AS FOOD INGREDIENTS

Research shows the possibility of using ANCs-rich extracts in both liquid and powdered forms for various types of food products, including beverages, dairy products (e.g., yogurt, fermented milk, and milkshakes), and sweets (e.g., pancakes, jellies, and chocolate) (Aguilera et al., 2016; Chung, Rojanasasithara, Mutlilangi, & McClements, 2016; Cortez et al., 2017; de Mejia et al., 2015; Gültekin-Özgüven, Karadağ, Dumanc, Özkal, & Özçelik, 2016; Kitts & Tomiuk, 2013; Mahdavi et al., 2016; Pineda-Vadillo et al., 2017; Robert & Fredes, 2015; Shin, Kang, Han, & Park, 2015).

Karaaslan, Ozden, Vardin, and Turkoglu (2011) added ethanol extracts before starting the yogurt production process and thus observed significant ANCs degradation during storage. Sun-Waterhouse, Zhou, and Wadhwa (2013) and Ścibisz, Ziarno, Mitek, and Zaręba (2012) also observed the effect of fermentation and some probiotic cultures on the stability of ANCs from extracts of blackcurrant and cranberry fruit in yogurt. Encapsulating ANCs in the polymer matrix seems to be a good solution. Individual selection of matrix material for the type of product and source of ANCs is an important direction of research. Aguilera et al. (2016) found that the 2% addition of β-cyclodextrin to a sports drink, which was colored with an extract of black bean coatings, reduced ANCs degradation and color change during the storage. Chung et al. (2016) studied the addition of gum Arabic to commercial beverages colored with purple carrot ANCs. It was observed that the addition of gum Arabic improved the stability of ANCs, and the highest color stability was observed with the addition of gum Arabic in the amount of 1.5%. Mahdavi et al. (2016) microencapsulated *Berberis vulgaris* extract with three combinations of wall materials: (1) a combination of acacia and maltodextrin; (2) a combination of maltodextrin and gelatin; and (3) a combination of acacia and maltodextrin. During storage, it was found that all variants of microencapsulated extracts reduced ANCs degradation compared with non-encapsulated. Powdered jelly colored with microencapsulated pigments obtained higher scores of sensory analysis, compared with the commercial version dyed with synthetic colorants.

Gültekin-Özgüven et al. (2016) enriched chocolate with microencapsulated black mulberry ANCs extract, in order to create ANCs-rich chocolate. Chocolate was fortified with microencapsulated ANCs up to 76.8%. Lima et al. (2019) incorporated spray dried microcapsules of ANCs-rich extracts from *Euterpe edulis* M. into dairy beverages. Acidity, pH, and ANCs content in beverages did not change during the 28 days of storage. There was a small, invisible to the naked eye, overall change in the color of the beverages during the storage. Vanegas-Espinoza et al. (2019) applied spray dried microcapsules of ANCs from roselle (*Hibiscus sabdariffa*) as a component of the goldfish diet. The use of ANCs microcapsules showed a significant and positive effect on various growth parameters of fantail goldfish (*Carassius auratus*) and its skin pigmentation. This suggests the potential use of microcapsules of ANCs as a source of nutrients for ornamental fish. Moura et al. (2019) applied microparticles of ANCs from hibiscus (obtained by ionic gelation method by dripping-extrusion and atomization) to jelly candies. The retention of bioactive compounds in sweets was 73% and they gained sensory acceptance among 70% of respondents. Stoll et al. (2016) studied freeze-dried grape ANCs extracts with the addition of maltodextrin and gum Arabic (in various proportions) as an addition to biodegradable films. Thus, it was found that films with the addition of encapsulated ANCs with maltodextrin were characterized by better mechanical properties and better protection against oil oxidation compared with those based on gum Arabic. Moreover, due to the sensitivity of ANCs to pH, which is manifested by a color change, encapsulated ANCs are used as a natural dye for smart food packaging systems (Singh, Gaikwad, & Lee, 2018).

4.8 CONCLUDING REMARKS

Spray drying is an efficient method that can be applied for the production of powdered pigments from extracts or raw materials containing ANCs. For the successful process which yields high retention and stability of ACNs, the selection of appropriate wall material and drying parameters are of vital significance. High-quality powdered pigments can be incorporated in food products, replacing artificial colorants, and providing value-added health benefits for the consumers. The current trend in ANCs microencapsulation is the application of novel wall materials, which can provide even more additional health benefits, apart from acting as encapsulation agents.

REFERENCES

Aguilera, Y., Mojica, L., Rebollo-Hernanz, M., Berhow, M., de Mejia, E.G., Martin-Cabrejas, M.A. 2016. Black bean coats: New source of anthocyanins stabilized by β-cyclodextrin copigmentation in a sport beverage. *Food Chemistry* 212: 561–570.

Ahmad, M., Ashraf, B., Gani, A., Gani, A. (2018). Microencapsulation of saffron anthocyanins using β glucan and β cyclodextrin: Microcapsule characterization, release behaviour & antioxidant potential during in-vitro digestion. *International Journal of Biological Macromolecules* 109: 435–442.

Ahmed, M., Akter, M.S., Eun, J.B. 2010. Impact of alpha-amylase and maltodextrin on physicochemical, functional and antioxidant capacity of spray dried purple sweet potato flour. *Journal of the Science of Food and Agriculture* 90(3): 494–502.

Ataie-Jafari, A.H.S., Karimi, F, Pajouhi, M. 2008. Effects of sour cherry juice on blood glucose and some cardiovascular risk factors improvements in diabetic women: A pilot study. *Nutrition & Food Science* 38(4): 355–360.

Bednarska, M.A., Janiszewska-Turak, E. 2019. The influence of spray drying parameters and carrier material on the physico-chemical properties and quality of chokeberry juice powder. *Journal of Food Science and Technology* 57(2): 564–577.

Bhandari, B.R., Howes, T. 1999. Implication of glass transition for the drying and stability of dried foods. *Journal of Food Engineering* 40: 71–79.

Bicudo, M.O.P., Jó, J., Oliveira, G.A., Chaimsohn, F.P., Sierakowski, M.R., Freitas, R.A., Ribani, R.H. 2015. Microencapsulation of juçara (*Euterpe edulis* M.) pulp by spray drying using different carriers and drying temperatures. *Drying Technology* 33:153–161.

Boldt, J.K., Meyer, M.H., Erwin, J.E. 2014. *Foliar anthocyanins: A horticultural review*. Volume 42, Published 2014 by John Wiley & Sons, Inc, pp. 209–251.

Carochom, M., Barreiro, M.F., Morales, P., Ferreira, I.C.F.R. 2014. Adding molecules to food, pros and cons: A review on synthetic and natural food additives. *Comprehensive Reviews in Food Science and Food Safety* 13: 377–399.

Castañeda, A., Hernandez, L.P., Páez, E., Galán-Vidal, C.A. 2009. Chemical studies of anthocyanins: A review. *Food Chemistry* 113(4):859–871.

Cavalcanti, R.N., Santos, D.T., Meireles, M.A.A. 2011. Non-thermal stabilization of anthocyanins in model and food system – An overview. *Food Research International* 44: 499–509.

Cazzonelli, C.I. 2011. Carotenoids in nature: Insights from plants and beyond. *Functional Plant Biology* 38: 833–847.

Chauhan, A. K., Patil, V. 2013. Effect of packaging material on storage ability of mango milk powder and the quality of reconstituted mango milk drink. *Powder Technology* 239: 86–93.

Chegini, G.R., Ghobadian, B. 2005. Effect of spray drying conditions on physical properties of orange juice powder. *Drying Technology* 23: 3, 657–668.

Chung, C., Rojanasasithara, T., Mutlilangi, W., McClements, D.J. 2016. Stabilization of natural colors and nutraceuticals: inhibition of anthocyanin degradation in model beverages using polyphenols. *Food Chemistry* 212: 596–603.

Clifford, M.N. 2000. Anthocyanins – Nature, occurrence and dietary burden. *Journal of the Science of Food and Agriculture* 80: 1063–1072.

Cooperstone, J.L., Schwartz, S.J. 2016. Recent insights into health benefits of carotenoids. In: Carle R., Schweiggert R.M. (eds) *Handbook on natural pigments in food and beverages: industrial applications for improving food color.* Woodhead Publishing. pp 473–497.

Cortez, R., Luna-Vital, D.A., Margulis, D., Gonzalez de Mejia, E. 2017. Natural pigments: Stabilization methods of anthocyanins for food applications. *Comprehensive Reviews in Food Science and Food Safety* 16: 180–198.

Da Silva Carvalho, A.G., Machado, M.T.C., Silva, V.M., Sartoratto, A., Rodrigues, R.A. F., Hubinger, M.D. 2016. Physical properties and morphology of spray dried microparticles containing anthocyanins of jussara (*Euterpe edulis Martius*) extract. *Powder Technology* 294: 421–428.

Daneshzad, E. D., Shab-Bidar, S., Mohammadpour, Z., Djafarian, K. 2019. Effect of anthocyanin supplementation on cardio-metabolic biomarkers: A systematic review and meta-analysis of randomized controlled trials. *Clinical Nutrition* 38: 1153–1165.

Darniadi, S., Ifie, I. Ho, P., Murray, B.S. 2019. Evaluation of total monomeric anthocyanin, total phenolic content and individual anthocyanins of foam-mat dried and spray dried blueberry powder. *Journal of Food Measurement and Characterization* 13(2): 1599–1606.

Delgado-Vargas, F., Paredes-Lopez, O. 2003. *Natural colorants for food and nutraceutical uses.* CRC Press: Boca Raton, FL.

Delgado-Vargum Arabics, F., Jiménez, A.R., Paredes-López, O. 2000. Natural pigments: Carotenoids, anthocyanins, and betalains – Characteristics, biosynthesis, processing, and stability. *Critical Reviews in Food Science and Nutrition* 40: 173–289.

Duthie, S.J., Jenkinson, A.M., Crozier, A., Mullen, W., Pirie, L., Kyle, J. 2006. The effects of cranberry juice consumption on antioxidant status and biomarkers relating to heart disease and cancer in healthy human volunteers. *European Journal of Nutrition* 45(2): 113–122.

de Mejia, E.G., Dia, V.P., West, L., West, M., Singh, V., Wang, Z., Allen, C. 2015. Temperature dependency of shelf and thermal stabilities of anthocyanins from corn distillers' dried grains with solubles in different ethanol extracts and a commercially available beverage. *Journal of Agricultural and Food Chemistry* 63: 10032–10041.

Feng, R., Ni, H., Wang, S.Y., et al., 2007. Cyanidin-3-rutinoside, a natural polyphenol antioxidant, selectively kills leukemic cells by induction of oxidative stress. *The Journal of Biological Chemistry* 282(18): 13468–13476.

Fernandes, A., Sousa, A., Azevedo, J., Mateus, N., de Freitas, V. 2013. Effect of cyclodextrins on the thermodynamic and kinetic properties ofcyanidin-3-O-glucoside *Food Research International* 51: 748–755.

Fernandes, I., Marques, C., Évora, A., Faria, A., Calhau, C., Mateus, N., de Freitas, V. 2019. Anthocyanins: Nutrition and health. In: Series Editors: Mérillon J.-M., Ramawat K.G. *Bioactive molecules in food.* Springer: Berlin. pp. 1097–1135.

Ferreira, D.S., Faria, A.F., Grosso, C.R.F., Mercadante, A.Z. 2009. Encapsulation of blackberry anthocyanins by thermal gelation of curdlan, *Journal of the Brazilian Chemical Society* 20: 1908–1915.

Fredes, C, Osorio, M.J., Parada, J., Robert, P 2018. Stability and bioaccessibility of anthocyanins from maqui (*Aristotelia chilensis* [Mol.] Stuntz) juice microparticles. *LWT Food Science and Technology* 91: 549–556.

Fukumoto, L., Mazza, G. 2000. Assessing antioxidant and prooxidant activities of phenolic compounds. *Journal of Agricultural and Food Chemistry* 48(8): 3597–3604.

Gagneten, M., Corfield, R., Mattson, M.G., Sozzi, A., Leiva, G., Salvatori, D., Schebor, C. 2019. Spray-dried powders from berries extracts obtained upon several processing steps to improve the bioactive components content. *Powder Technology* 342: 1008–1015.

Gaona, I.J., Fanzone, M., Sari, S. 2019. Spray-dried Ancellotta red wine: Natural colorant with potential for food applications. *European Food Research and Technology* 245: 2621.

Gharsallaoui, A., Roudaut, G., Chambin, O., Voilley, A., Saurel, R. (2007). Applications of spray drying in microencapsulation of food ingredients: An overview. *Food Research International* 40(9): 1107–1121.

Ghosh, D., Konishi, T. 2007. Anthocyanins and anthocyanin-rich extracts: Role in diabetes and eye function. *Asia Pacific Journal of Clinical Nutrition* 16: 200–208.

Gomes, J.W.P., Rigolon, T.C.B., da Silveira Souza, M.S., Alvarez-Leite, J.I., Lucia, C.M.D., Martino, H.S.D., Rosa, D.O.B. 2019. Antiobesity effects of anthocyanins on mitochondrial biogenesis, inflammation, and oxidative stress: A systematic review. *Nutrition* 66: 192–202.

Gould, K.S., Lee, D.W. 2002: *Anthocyanins in leaves*. Academic Press: New York, 119, p. 4.

Gültekin-Özgüven, M., Karadağ, A., Dumanc, S., Özkal, B., Özçelik, B. 2016. Fortification of dark chocolate with spray dried black mulberry (*Morus nigra*) waste extract encapsulated in chitosan-coated liposomes and bioaccessability studies. *Food Chemistry* 201: 205–212.

Hou, D.X., Ose, T., Lin, S. 2003. Anthocyanidins induce apoptosis in human promyelocytic leukemia cells: Structure–activity relationship and mechanisms involved. *International Journal of Oncology* 23(3): 705–712.

Idham, Z., Muhamad, I.I., Sarmidi, M.R. 2012. Degradation kinetics and color stability of spray dried encapsulated anthocyanins from *Hibiscus sabdariffa* L. *Journal of Food Process Engineering* 35: 522–542.

Jafari, S.M., Ghalenoei, M.G., Dehnad, D. (2017). Influence of spray drying on water solubility index, apparent density, and anthocyanin content of pomegranate juice powder. *Powder Technology* 311: 59–65.

Janiszewska, E. 2014. Microencapsulated beetroot juice as a potential source of betalain. *Powder Technology* 264: 190–196.

Janiszewska, E., Jedlińska, A., Witrowa-Rajchert, D., 2014. Effect of homogenization parameters on selected physical properties of lemon aroma powder. *Food and Bioproducts Processing* 94: 554,405–413.

Janiszewska-Turak, E, Sek, A., Witrowa-Rajchert, D. 2019. Influence of the carrier material on the stability of chokeberry juice microcapsules. *International Agrophysics* 33: 517–525.

Jedlińska, A., Samborska, K., Wieczorek, A., Wiktor, A., Ostrowska-Ligęza, E., Jamróz, W., Skwarczyńska-Maj, K., Kiełczewski, D., Błażowski, Ł., Tułodziecki, M., Witrowa-Rajchert, D. 2019. The application of dehumidified air in rapeseed and honeydew honey spray drying – process performance and powders properties considerations. *Journal of Food Engineering* 245: 80–87.

Johnson, E.J. 2002. The role of carotenoids in human health. *Nutrition in Clinical Care* 5: 56–65.

Kähkönen, M.M., Heinonen, M. 2003. Antioxidant Activity of Anthocyanins and Their Aglycons. *Journal of Agricultural and Food Chemistry* 51(3): 628–633.

Kalušević, A.M., Lević, S.M., Ćalija, B.R., Milić, J.R., Pavlović, V.B., Bugarski, B.M., Nedović, V.A. 2017. Effects of different carrier material on physicochemical properties of microencapsulated grape skin extract. *Journal of Food Science and Technology* 54(11): 3411–3420.

Karaaslan, M., Ozden, M., Vardin, H., Turkoglu, H. 2011. Phenolic fortification of yogurt using grape and callus extracts. *LWT- Food Science and Technology* 44: 1065–1072.

Kitts, D.D., Tomiuk, S. 2013. Studies on mitigating lipid oxidation reactions in a value-added dairy product using a standardized cranberry extract. *Agriculture* 3: 236–252.

Konczak, I., Zhang, W. 2004. Anthocyanins-more than nature´ s colours. *Journal of Biomedicine and Biotechnology* 5: 239–240.

Lacerda, E.C.Q., Calado, V.M. de A., Monteiro, M., Finotelli, P.V., Torres, A.G., Perrone, D. 2016. Starch, inulin and maltodextrin as encapsulating agents affect the quality and stability of jussara pulp microparticles. *Carbohydrate Polymers* 151: 500–510.

Landi, M., Tattini, M., Gould, K. 2015. Multiple functional roles of anthocyanins in plant-environment interactions. *Environmental and Experimental Botany*, 119, 4–17.

Lao, F., Giusti, M.M. 2017. The effect of pigment matrix, temperature and amount of carrier on the yield and final color properties of spray dried purple corn (*Zea mays* L.) cob anthocyanin powders. *Food Chemistry* 227: 376–382.

Laokuldilok, T., Kanha, N. 2015. Effects of processing conditions on powder properties of black glutinous rice (*Oryza sativa* L.) bran anthocyanins produces by spray drying and freeze drying. *LWT- Food Science and Technology*, 64, 405–411.

Laokuldilok, T., Kanha, N. 2017. Microencapsulation of black glutinous rice anthocyanins using maltodextrins produced from broken rice fraction as wall material by spray drying and freeze drying. *Journal of Food Processing & Preservation* 41(1): e12877.

Lee, J.K.M., Taip, F.S., Abdullah, Z. 2018. Effectiveness of additives in spray drying performance: A review. *Food Research* 2(6): 486–499.

Li, X., Xu, J., Tang, X., Liu, Y., Yu, X., Wang, Z., Liu, W. 2016. Anthocyanins inhibit trastuzumab resistant breast cancer in vitro and in vivo. *Molecular Medicine Reports* 13: 4007–4013.

Lila, M.A. 2004. Anthocyanins and human health: An in vitro investigative approach. *Journal of Biomedicine and Biotechnology* 5: 306–313.

Lucioli, S. 2012. Anthocyanins: Mechanism of action and therapeutic efficacy. In: Capasso A. (ed)*Medicinal plants as antioxidant agents: Understanding their mechanism of action and therapeutic efficacy.* Research Signpost: India, pp. 27–57. ISBN:97881-308-0509-2.

Mahdavi, S.A., Jafari, S.M., Assadpour, E., Ghorbani, M. 2016. Storage stability of encapsulated barberry's anthocyanin and its application in jelly formulation. *Journal of Food Engineering* 181: 59–66.

Mahdavi, S.A., Jafari, S.M., Ghorbani, M., Assadpoor, E. 2014. Spray-drying microencapsulation of anthocyanins by natural biopolymers: A review. *Drying Technology* 32(5): 509–518.

Mehran, M., Masoum, S., Memarzadeh, M. 2020. Improvement of thermal stability and antioxidant activity of anthocyanins of *Echium amoenum* petal using maltodextrin/modified starch combination as wall material. *International Journal of Biological Macromolecules* 148: 768–776.

Miguel, M.G. 2011. Anthocyanins: Antioxidant and/or anti-inflammatory activities. *Journal of Applied Pharmaceutical Science* 01(06): 7–15.

Mirhojati, H., Sharayei, P., Ghavidel, R.A. 2017. Microencapsulation of anthocyanin pigments obtained from seedless barberry (*Berberis vulgaris* L.) fruit using freeze drying. *Iranian Food Science and Technology Research Journal* 13(3): 14–27.

Moser, P., Nicoletti, Telis V.R., de Andrade Neves, N., García Romero, E., Gómez-Alonso, S., Hermosín-Gutiérrez, I. 2017. Storage stability of phenolic compounds in powdered BRS Violeta grape juice microencapsulated with protein and maltodextrin blends. *Food Chemistry* 2: 4308–4318.

Murali, S., Kar, A., Mohapatra, D., Kalia, P. 2015. Encapsulation of black carrot juice using spray and freeze drying. *Food Science and Technology International* 21: 604–612.

Nath, S., Satpathy, G.R. 1998. A systematic approach for investigation of spray drying processes. *Drying Technology* 16(6): 1173–1193.

Nedović, V., Kalušević, A., Manojlović, V., Petrović, T., Bugarski, B. (2013). Encapsulation systems in the food industry. In: Yanniotis S., Taoukis P., Stoforos N.G., Karathanos V.T. (eds) *Advances in food process engineering research and applications*, 1st edn. Food Engineering Series. Springer: New York, US, pp. 229–253.

Osorio, C., Acevedo, B., Hillebrand, S., Carriazo, J., Winterhalter, P., Morales, A.L. 2010. Microencapsulation by spray drying of anthocyanin pigments from corozo (*Bactris guineensis*) fruit. *Journal of Agricultural and Food Chemistry* 58: 6977–6985.

Pascual-Teresa, S., Sanchez-Ballesta, M.T. 2008. Anthocyanins: From plant to health. *Phytochemistry Reviews* 7: 281–299.

Phan, K., Van Den Broeck, E., Van Speybroeck, V., De Clerck, K., Raes, K., & De Meester, S. (2020). The potential of anthocyanins from blueberries as a natural dye for cotton: A combined experimental and theoretical study. *Dyes and Pigments*, 176, 108180.

Piątkowska, E., Kopeć, A., Leszczyńska, T. 2011. Anthocyanins – their profile, occurrence, and impact on human organism (in Polish). *Żywność. Nauka. Techchnologia. Jakość* 4(77): 24–25.

Pieczykolan, E., Kurek, M.A. 2019. Use of guar gum, gum arabic, pectin, beta-glucan and inulin for microencapsulation of anthocyanins from chokeberry. *International Journal of Biological Macromolecules* 129: 665–671.

Pineda-Vadillo, C., Nau, F., Guerin-Dubiard, C., Jardin, J., Lechevalier, V., Sanz-Buenhombre, M., Guadarrama, A., Toth, T., Csavajda, E., Hajnalka, H., Karakaya, S., Sibakov, J., Capozzi, F., Bordoni, A., Dupont, D. 2017. The food matrix affects the anthocyanin profile of fortified egg and dairy matrices during processing and in vitro digestion. *Food Chemistry* 214: 486–496.

Quek, S.Y., Chok, N.K., Swedlund, P. 2007. The physicochemical properties of spray dried watermelon powder. *Chemical Engineering and Processing* 46(5): 386–392.

Ramakrishnan, Y., Adzahan, N.M., Yusof, Y.A., Muhammad, K. 2018. Effect of wall materials on the spray drying efficiency, powder properties and stability of bioactive compounds in tamarillo juice microencapsulation. *Powder Technology* 328: 406–414.

Rendeiro, C., Vauzour, D., Rattray, M., Waffo-Téguo, P., Mérillon, J.M., Butler, L.T., Williams, C.M., Spencer, J.P.M. 2013. Dietary levels of pure flavonoids improve spatial memory performance and increase hippocampal brain-derived neurotrophic factor. *PLoS One* 8: e63535.

Rezvankhah, A., Emam-Djomeh, Z., Askari, G. 2020. Encapsulation and delivery of bioactive compounds using spray and freeze-drying techniques: A review. *Drying Technology* 38(1–2): 235–258.

Robert, P., Fredes, C. 2015. The encapsulation of anthocyanins from berry-type fruits. Trends in foods. *Molecules* 20: 5875–5888.

Robert, P., Gorena, T., Romero, N., Sepulveda, E., Chavez, J., Saenz, C. (2010). Encapsulation of polyphenols and anthocyanins from pomegranate (*Punica granatum*) by spray drying. *International Journal of Food Science and Technology* 45(7): 1386–1394.

Rodriguez, E.B., Vidallon, M.L.P., Mendoza, D.J.R., Reyes, C.T. 2016. Health-promoting bioactivities of betalains from red dragon fruit (*Hylocereus polyrhizus* (Weber) Britton and Rose) peels as affected by carbohydrate encapsulation. *Journal of the Science of Food and Agriculture* 96(14): 4679–4689.

Roobha, J.J., Marappan, S., Aravindhan, K.M., Devi, P.S. 2011. The effect of light, temperature, pH on stability of anthocyanin pigments in *Musa acuminata* bract. *Research in Plant Biology* 4: 234–245.

Saenz, C., Tapia, S., Cha'vez, J., Robert, P. 2009. Microencapsulation by spray drying of bioactive compounds from cactus pear (*Opuntia ficusindica*). *Food Chemistry* 114(2): 616–622.

Samborska, K., Jedlińska, A., Wiktor, A., Derewiaka, D., Wołosiak, R., Matwijczuk, A., Jamróz, W., Skwarczyńska-Maj, K., Kiełczewski, D., Błażowski, Ł., Tułodziecki, M., Witrowa-Rajchert, D. 2019. The effect of low temperature spray drying with dehumidified air on phenolic compounds, antioxidant activity and aroma compounds of rapeseed honey powders. *Food and Bioprocess Technology* 2: 919–932.

Samborska, K., Wiktor, A., Jedlińska, A., Matwijczuk, A., Jamróz, W., Skwarczyńska-Maj, K., Kiełczewski, D., Tułodziecki, M., Błażowski, Ł., Witrowa-Rajchert, D. 2019. Development and characterization of physical properties of honey-rich powder. *Food and Bioproducts Processing* 115: 78–86.

Sánchez-Madrigal, M.Á., Quintero-Ramos, A., Amaya-Guerra, C.A., Meléndez-Pizarro, C.O., Castillo-Hernández, S.L., Aguilera-González, C.J. 2019. Effect of agave fructans as carrier on the encapsulation of blue corn anthocyanins by spray drying. *Foods* 8: 268.

Santiago, M. C. P. A., Nogueira, R. I., Paim, D. R. S. F., Gouvêa, A. C. M. S., Godoy, R. L. O., Peixoto, F. M., Pacheco, S., Freitas, S. P. 2016. Effects of encapsulating agents on anthocyanin retention in pomegranate powder obtained by the spray drying process. *LWT- Food Science and Technology* 73: 551–556.

Ścibisz, I., Ziarno, M., Mitek, M., Zaręba, D. 2012. Effect of probiotic cultures on the stability of anthocyanins in blueberry yoghurts. *LWT - Journal of Food Science and Technology*. 49: 208–212.

Shahidi, F., Han, X.Q. 1993. Encapsulation of food ingredients. *Critical Reviews in Food Science and Nutrition* 33(6): 501–547.

Sharif, N., Khoshnoudi-Nia, S., Jafari, S.M. (2020). Nano/microencapsulation of anthocyanins; a systematic review and meta-analysis. *Food Research International* 132: 109077.

Shin, B.K., Kang, S., Han, J.I., Park, S. 2015. Quality and sensory characteristics of fermented milk adding black carrot extracts fermented with *Aspergillus oryzae*. *Journal of The Korean Society of Food Culture* 30: 370–376.

Shishir, M.R.I., Chen, W., 2017. Trends of spray drying: A critical review on drying of fruit and vegetable juices. *Trends in Food Science and Technology* 65: 49–67.

Shukitt-Hale, B., Carey, A.N., Jenkins, D., Rabin, B.M.. Joseph, J.A. 2007. Beneficial effects of fruit extracts on neuronal function and behavior in a rodent model of accelerated aging. *Neurobiology of Aging* 28: 1187–1194.

Silva, I., Stringheta, C., Teofilo, F., Nolasco de Oliveira, I. 2013. Parameter optimization for spray drying microencapsulation of jaboticaba (*Myrciaria jaboticaba*) peel extracts using simultaneous analysis of responses. *Journal of Food Engineering* 117(4): 538–544.

Silva, V.O., Freitas, A.A., Maçanitab, L.A., Quinaa, F.H. 2016. Chemistry and photochemistry of natural plant pigments: The anthocyanins. *Journal of Physical Organic Chemistry* 29: 594–599.

Singh, S., Gaikwad, K.K., Lee, Y.S. (2018). Anthocyanin-A natural dye for smart food packaging systems. *Korean Journal of Packaging Science & Technology*, 24(3), 167–180.

Sinopoli, A., Calogero, G., Bartolotta, A. 2019. Computational aspects of anthocyanidins and anthocyanins: A review. *Food Chemistry* 297: 124898.

Smeriglio, A., Barreca, D., Bellocco, E., Trombetta, D. 2016. Chemistry, pharmacology and health benefits of anthocyanins. *Phytotherapy Research* 30: 1265–1286.

Souza, V. B.., Fujita, A., Thomazini, M., da Silva, E.R., Lucon, J.F., Jr., Genovese, M.I., Favaro-Trindade, C.S. 2014. Functional properties and stability of spray dried pigments from Bordo grape (Vitis labrusca) winemaking pomace. *Food Chemistry* 164: 380–386.

Sun-Waterhouse, D., Zhou, J., Wadhwa, S.S. 2013. Drinking yoghurts with berry polyphenols added before and after fermentation. *Food Control* 32: 450–460.

Tachibana, N., Kimura, Y., Ohno, T. 2014. Examination of molecular mechanism for the enhanced thermal stability of anthocyanins by metal cations and polysaccharides. *Food Chemistry* 143: 452–458.

Tonon, R.V., Brabet, C., Hubinger, M.D., 2010. Anthocyanin stability and antioxidant activity of spray dried açai (*Euterpe oleracea* Mart.) juice produced with different carrier agents. *Food Research International* 43: 907–914.

Tsuda, T. 2012. Review Anthocyanins as Functional Food Factors – Chemistry, Nutrition and Health Promotion. *Journal of Food Science and Technology* 18(3): 315–324.

Vergara, C., Pino, M.T., Zamora, O., Parada, J., Pérez, R., Uribe, M., Kalazich, J. 2020. Microencapsulation of anthocyanin extracted from purple flesh cultivated potatoes by spray drying and its effects on in vitro gastrointestinal digestion. *Molecules* 25(3): 722.

Villacrez, J.L., Carriazo, J.G., Osorio, C. 2014. Microencapsulation of Andes Berry (*Rubus glaucus* Benth.) aqueous extract by spray drying. *Food and Bioprocess Technology* 7: 1445–1456.

Walkling-Ribeiro, M., Noci, F., Cronin, D. A., Lyng, J.G., Morgan, D.J. 2009. Shelf life and sensory evaluation of orange juice after exposure to thermo-sonication and pulsed electric fields. *Food and Bioproducts Processing* 87: 102–107.

Wang, H., Cao, G., Prior, R.L. 1997. Oxygen radical absorbing capacity of anthocyanins. *Journal of Agricultural and Food Chemistry* 45(2): 304–309.

Wang, L.S., Stoner, G.D. 2008. Anthocyanins and their role in cancer prevention. *Cancer Letters* 269(2): 281–290.

Weber, F., Boch, K., Schieber, A. 2017. Influence of copigmentation on the stability of spray dried anthocyanins from blackberry. *LWT - Food Science and Technology* 75: 72–77.

Wu, X., Beecher, G.R., Holden, J.M., Haytowitz, D.B., Gebhardt, S.E., Prior, R.L. 2006. Concentrations of anthocyanins in common foods in the United States and estimation of normal consumption. *Journal of Agricultural and Food Chemistry* 54(11): 4069–4075.

Xie, L., Su, H., Sun, C., Zheng, X., Chen, W. (2018). Recent advances in understanding the anti-obesity activity of anthocyanins and their biosynthesis in microorganisms. *Trends in Food Science & Technology*, 72, 13–24.

Yang, M., Koo, S.I., Song, W.O., Chun, O.K. 2011. Food matrix affecting anthocyanin bioavailability: Review. *Current Medicinal Chemistry* 18(2): 291–300.

Yingngam, B., Tantiraksaroj, K., Taweetao, T., Rungseevijitprapa, W., Supaka, N., Brantner, A. H. 2018. Modeling and stability study of the anthocyanin-rich maoberry fruit extract in the fast-dissolving spray dried microparticles. *Powder Technology* 325: 261–270.

Yousefi, S., Emam-Djomeh, Z., Mousavi, M., Kobarfard, F., Zbicinski, I. (2015). Developing spray dried powders containing anthocyanins of black raspberry juice encapsulated based on fenugreek gum. *Advanced Powder Technology* 26(2): 462–469.

Zamora-Ros, R., Knaze, V., Luj'an-Barroso, L., 2011. Estimated dietary intakes of flavonols, flavanones and flavones in the European Prospective Investigation into Cancer and Nutrition (EPIC) 24-hour dietary recall cohort. *British Journal of Nutrition* 106(12): 1915–1925.

Zamora-Ros, R., Knaze, V., Luján-Barroso, L. 2011. Estimation of the intake of anthocyanidins and their food sources in the European Prospective Investigation into Cancer and Nutrition (EPIC) study. *British Journal of Nutrition* 106(7): 1090–1099.

5 Spray Drying Encapsulation of Carotenoids

Emilia Janiszewska-Turak, Dorota Nowak, and
Dorota Witrowa-Rajchert
University of Life Sciences (SGGW), Poland

CONTENTS

5.1   INTRODUCTION

Carotenoids are yellow-orange-red colorants that are synthesized by plants and microorganisms, but not by animals and humans. These compounds can be found in fruits and vegetables, some fungi, and algae (Arunkumar, Calvo, Conrady, & Bernstein, 2018). In plants, the composition of carotenoids is complex, which results in the specific characteristic of a given species or variety. Furthermore, carotenoids as antioxidants can reduce the risk of chronic diseases such as cardiovascular diseases, diabetes, and age-related macular disease. They also support the proper functioning of the immune system (Vila, Chaud, & Balcão, 2015). In addition, due to the unsaturated structure, carotenoids are susceptible to degradation during food processing (e.g., blanching and drying) and storage (Aman, Schieber, & Carle, 2005; Hiranvarachat, Suvarnakuta, & Devahastin, 2008).

The use of natural active substances such as carotenoids in industrial applications can cause problems with their changes observed after exposure to heating, pH, or light (Janiszewska-Turak et al., 2017). In this regard, microencapsulation is one of the methods that can be used for the protection of such natural compounds. The most commonly used microencapsulation technique for bioactive compounds is spray drying. This method allows obtaining a product with the chosen physical properties by applying variable drying parameters such as inlet air temperature, feed flux, and humidity, as well as by changing the type of carriers or pre-treatment of the raw material (Assadpour & Jafari, 2019; Drosou, Krokida, & Biliaderis, 2017). The objective of this chapter is to describe carotenoids in terms of their chemical structure and biological activity, as well as the spray drying process as a feasible and efficient encapsulation technique for the protection and delivery of various carotenoids. The drying conditions and carrier materials are also discussed.

5.2 CAROTENOIDS

5.2.1 STRUCTURE

Carotenoids are lipophilic (only a few are hydrophilic, e.g., crocin in saffron) pigments that are synthesized by plants and microorganisms, but not by animals or in the human body. In plants, the composition of carotenoids is complex, which results in the specific characteristic of a given species or variety. Carotenoids are a group of compounds that consist of over 750 different structures identified in nature, but only about 40 carotenoids are found in major human foods. These food products are a source of carotenoids, which must be supplied to the human body as exogenous nutrients (Arunkumar et al., 2018). Carotenoids are a family of compounds of over 600 plant pigments that contribute to the visible colors (Krinsky & Johnson, 2005). The structure of carotenoids is based on the number of carbons in their backbone, and C_{40} carotenoids are the most commonly found carotenoids in nature (Bell, McEvoy, Tocher, & Sargent, 2000; Fernandes, do, Jacob-Lopes, De, & Zepka, 2018). The basic cyclic structure of these bioactive compounds can be modified by hydrogenation, dehydrogenation, cyclization, and oxidation. Fernandes et al. (2018) reported that based on their chemical composition, carotenoids can be divided into two groups:

1. *Carotenes*: oxygen-free carotenoids, which are non-polar compounds, and only carbon and hydrogen are found in the molecular structure. This group includes β-carotene, α-carotene, lycopene, and neurosporene. This group of carotenoids needs dietary fat to be absorbed in the gastrointestinal tract.
2. *Xanthophylls*: the molecule is made up of carbon, hydrogen atoms, and oxygen. Lutein, zeaxanthin, astaxanthin, and β-cryptoxanthin belong to this group, because they contain at least one oxygen atom. This group is more polar than carotenes.

The various carotenoid types are the result of the transformation in the process of biosynthesis. For example, β-carotene is formed as a result of cyclization of lycopene, and *vice versa*. Another example is β-carotene that can be formed after xanthophylls hydroxylation (Chábera, Fuciman, Hříbek, & Polívka, 2009). More detailed stages of carotenoid biosynthesis are shown in Figure 5.1.

The structure responsible for light absorption, in particular ultraviolet (UV), is the chromophore group, which in carotenoids is characterized by conjugated double bonds. Carotenoids may occur as *cis* or *trans* isomers (E/Z isomers). Therefore, we consume both *cis* and *trans* isomers. It is not known if this affects the functioning of the human body in any way. However, it is obvious that the isomerization process significantly affects the color of carotenoids. Brightening or darkening of the color (during isomerization) may occur. Some studies showed that *trans*-β-carotene is more favorable for intestinal absorption than 9-*cis*-β-carotene, but *cis*-lycopene is more bioavailable than *trans*-lycopene (Failla, Chitchumroonchokchai, & Ishida, 2008). In nature, the *trans* form is more common.

5.2.2 CHEMICAL STABILITY

Due to their highly unsaturated structure, carotenoids are prone to degradation during food processing (e.g., canning, blanching, and drying) and storage, as a result of exposure to high temperatures,

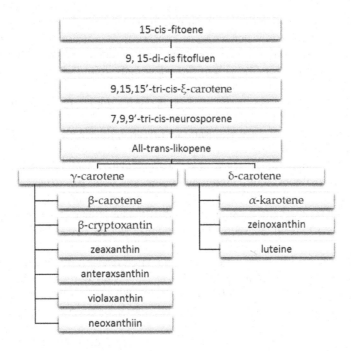

FIGURE 5.1 Schema of carotenoid biosynthesis, based on Lee et al. (2004).

light, ions of metal, pro-oxidant molecules, the physical state of carotenoids (crystalline structures), and the type of fat as solvent (saturated and unsaturated fat content) (Aman et al., 2005;
Hiranvarachat et al., 2008).

The number of *cis* isomers increases with the temperature and time of the process (Hiranvarachat
et al., 2008). *Trans* and *cis* isomers may have different biological properties; e.g., all-*trans*
β-carotene is a better antioxidant than 13-*cis*-β-carotene (Böhm, Puspitasari-Nienaber, Ferruzzi,
& Schwartz, 2002). At high temperatures, carotenoids are degraded; e.g., the degradation rates of
trans-β-carotene and *trans*-lutein increases with temperature. Degradation of *trans*-lutein occurs
at a slower rate than *trans*-β-carotene, due to higher thermal resistance. The degradation products
are isomers such as 13-*cis*- and 9-*cis*-b-carotene, 13-*cis*-, 9-*cis*-, 13′-*cis*-, and 9′-*cis*-lutein (Achir,
Randrianatoandro, Bohuon, Laffargue, & Avallone, 2010; Knockaert et al., 2012). Usually, this process is carried out at temperatures >100°C. Achir et al. (2010) conducted tests at a temperature
range of 120°C to 180°C. According to this study, at 120°C, a slight degradation of carotenoids
in oils (about 5%) was noticed after 20 min thermal processing, while the type of fat significantly
affected the kinetics of degradation. With the increase in temperature, the initial degradation rate of
trans-β-carotene and *trans*-lutein increased faster in the fat environment containing 93% saturated
acids compared with the reaction occurring in fat-containing 61% unsaturated acids. Pérez-Gálvez
and Mínguez-Mosquera (2004) found no significant differences in the degradation of carotenoids
contained in peppers, in high oleic and high linolenic oils when heated at 80°C. Pérez-Gálvez and
Mínguez-Mosquera (2004) assumed that in normal temperatures and time range used in food processing, without the addition of unsaturated fats, the degradation of carotenoids is limited. Achir et
al. (2010) had stated that heat treatment of β-carotene at 120°C, using fresh frying oil, with a low
peroxide value and a high vitamin E content as a solvent, could limit carotenoids oxidation. So, the
degradation of carotenoids should not occur during the spray drying process, because of a short time
and usually not a higher temperature than 80°C in the drying chamber.

Light is another factor, which can cause the degradation of carotenoids to a great extent. Light
can also favor β-carotene isomerization, but to a lower degree in comparison with the temperature
effect (Böhm et al., 2002). Transformation of carotenoids under the influence of light is not obvious

although pure β-carotene degrades after exposure to light. Aman et al. (2005), studying the content of lutein and β-carotene in green vegetable leaves, observed an increase in the content of lutein and β-carotene after exposure to light. It can therefore be concluded that the natural environment in which carotenoids occur protects them from degradation. To limit the degradation of carotenoids under the influence of light, it is necessary to protect the product by packaging in light barrier materials. Barrier films can act as barrier packaging, for example, the active methylcellulose-tea catechins films (Yu, Tsai, Lin, Lin, & Mi, 2015).

Carotenoids, due to the presence of unsaturated hydrocarbon bonds, are chemically unstable and easily degraded under the influence of various factors. Most often, degradation occurs under the influence of heat and in the presence of oxygen. Oxidation can also be caused by the presence of ions, metals, enzymes, light, unsaturated lipids, and other pro-oxidative molecules (Boon, McClements, Weiss, & Decker, 2010). The higher degree of oil unsaturation, the faster the degradation occurs (Achir et al., 2010; Boon et al., 2010). The oxidation mechanism of β-carotene depends on the factors that initiate oxidation. In the presence of free radicals, oxidation is a free radical chain reaction. The degradation of carotenoids causes a loss of the characteristic color, which allows you to easily determine the kinetics of the oxidation reaction by measuring the color of the product (Koca, Burdurlu, & Karadeniz, 2007; Qian, Decker, Xiao, & McClements, 2012), carotenes quantification (Dhuique-Mayer et al., 2007) or on the formation of *cis* isomers (Achir et al., 2010). Determination of kinetics is very important to maintain the declared level of carotenoids in products after the shelf life.

Qian et al. (2012) investigated the effect of antioxidants on the degradation of β-carotene encapsulated in nanoemulsions. Vitamin E, acetate, or coenzyme Q10 were used as fat-soluble antioxidants, as well as ethylenediaminetetraacetic acid (EDTA) and ascorbic acid (soluble in water). It was found that the rate of β-carotene degradation decreased in each variant. The obtained results showed that the creation of such a system allowed to slow down the degradation of carotenoids during storage.

5.2.3 CAROTENOIDS IN PLANTS

Table 5.1 presents information about the natural sources of carotenoids. Carotenoids are accumulated in the green parts of plants, as well as in flowers, fruits, seeds, and roots. Carotenoids are responsible for the yellow, orange, and red colors in many fruits and flowers without chlorophyll. In the plants with high contents of chlorophylls such as green leafy vegetables, the color of carotenoids is masked despite their considerable concentration in these plants (Janiszewska-Turak, 2017). Vegetables rich in carotenoids include kale, spinach (lutein), red pepper (capsantine), and carrot (β-carotene). Fruits containing carotenoids include grapefruit, nectarines, and apricots. These compounds can also be found in large quantities in buriti (*Mauritia vinifera* Mart.), tucuma (*Acrocomia mokayayba* Barb.

TABLE 5.1
Natural dyes containing carotenoids

Name of Colorant	Color	Source	Main Carotenoids
Annatto	Orange-yellow	*Bixa orellana*	Bixin and norbixin
Unrefined palm oil		Palm	α- and β-carotene
Oleoresin	Red	Paprika	Xanthophylls capsanthin and capsorubin
Lycopene	Red-orange	Tomato	Lycopene
Crocin	Yellow	Saffron (*Crocus sativus)* flower pistil moles	Crocin

Rodr.), bocaiuva, acerola, mango, nuts, camu-camu (*Myrciaria dubia*), carrot noodles, rose hip fruits, and palm oil.

β-carotene is the most widespread carotenoid in foods. β-carotene is of particular interest because of its vitamin A dimer structure, potentially providing 100% activity (Mora, Iwata, & von Andrian, 2008). The main sources of β-carotene are the leaves of green vegetables, oranges, and yellow vegetables and fruits. Two of the best sources of β-carotene are carrots and pumpkins. Other important sources of these carotenoids are Brussels sprouts, broccoli, peas, and brassica vegetables in general (Table 5.1). Lycopene is present in the largest amounts in tomatoes and tomato products (e.g., juices, soups, sauces, and ketchup), including their processing waste and peels. Other raw materials rich in lycopene are apricots, watermelons, papaya, cherry, guava, guava products, orange, and grapefruit. The source of lutein and zeaxanthin are mainly green leaves of vegetables, especially, spinach, and cabbage (de, Ortiz, de, & de, 2018).

Lutein and zeaxanthin are present in green and dark green leafy vegetables, like broccoli, Brussels sprouts, spinach, and parsley (Rowles III & Erdman Jr., 2020). An important source of lutein is Marigold petal extract (Carpentier, Knaus, & Suh, 2009). The microalgae *Phaffia rhodozyma*, *Chlorella vulgaris*, and *Haematococcus pluvialis* synthesize large amounts of astaxanthin (Ahmed, Li, Fanning, Netzel, & Schenk, 2015; Hynstova et al., 2018).

Grains may also be a source of carotenoids. The carotenoid profile in grains consists of xanthophylls, lutein (dominant carotenoid), zeaxanthin, and β-cryptoxanthin, as well as small amounts of α- and β-carotene. The highest content of carotenoids is found in the embryo, but quantitatively, the highest content in relation to whole grains exists in the endosperm (Siebenhandl et al., 2007). The cereal with the highest content of carotenoids is corn (*Zea mays* L.) with a yellow genotype (it is up to 63 μg/g) (Panfili, Fratianni, & Irano, 2004), but commonly, grains are not a significant source of carotenoids in food (Panfili et al., 2004).

Carotenoids in the juices pressed from fruits, vegetables, or their homogenized preparations have a higher absorption due to the lack of membranes and cell walls, which hinders their absorption in the digestive tract. Carrot juice may be an important source of carotenoids in the diet. Akhtar and Bryan (2008) studied the content of carotenoids in high-quality juices manufactured for children. International Unit (IU) values of β-carotene in one portion of 240 mL carrot juice were determined at 20.060 and 25.612 for α- and β-carotene, respectively. The use of conversion factors would provide 5940 retinol equivalent (RE) (1672 RE from α-carotene and 4268 from β-carotene), which is 6.6 times more or 660% of the recommended daily intake of 800 RE.

Carotenoids can also be obtained from *Chlorella*, unicellular green algae, which can be cultivated to provide a greater variety of carotenoids (e.g., lutein, zeaxanthin, α-carotene, and β-carotene) than can be obtained from marigold petal extract (Jung, Ok, Park, Kim, & Kwon, 2016; Lee & Kim, 2009). Carotenoids can be classified based on color (de Carvalho et al., 2018; Fernandes et al., 2018). Natural dyes based on carotenoids derived from plants are described in Table 5.2.

5.2.4 BIOACTIVITY AND HEALTH BENEFITS OF CAROTENOIDS

The most important carotenoid to human health is β-carotene, although lycopene, lutein, zeaxanthin, α-carotene, and β-cryptoxanthin are also of great importance. Only 50 from 750 types of carotenoids can be converted into retinol (vitamin A) in animals. Animal organisms can transform some carotenoids into vitamin A. Vitamin A activity is shown only by the compounds containing their molecule, a fragment with the same structure as retinol (i.e., β-ionone). β-carotene has the highest carotenoid activity, because it contains two β-ion rings. Therefore, theoretically, two molecules of retinol (vitamin A) can be obtained from one β-carotene molecule (Skibsted, 2012). However, the efficiency of bioconversion of β-carotene to vitamin A depends on the source of the provitamin. The bioavailability of β-carotene from green leaves of vegetables such as spinach is low. Tang et al. (2005) reported that β-carotene from spinach leaves undergoes an average conversion of 27: 1, so it takes 27 molecules of provitamin to create one molecule of retinol. For carrots, this bioconversion ratio is 15: 1.

TABLE 5.2

Dominant carotenoids in various plants

Type of Tissue	Name	Latin Name	Dominant Carotenoid(s)	References
Fruits	Papaya	*Carica papaya* L.	lycopene, β-cryptoxanthin, and β-carotene	Rivera-Pastrana, Yahia, and Gonzalez-Aguilar (2010)
	Cashew apple	*Anacardium occidentale* L.	Auroxanthin and β-cryptoxanthin	de Abreu et al. (2013)
	Red banana		β-carotene and α-carotene	Lokesh, Divya, Puthusseri, Manjunatha, and Neelwarne (2014)
	Grapefruit	*Citrus paradisi Macf*	Lycopene and β-carotene	Achir et al. (2016)
	Blood orange	*Citrus sinensis* L. Osbeck	*Cis*-violaxanthin Lutein, Zeaxanthin, *cis*-antheraxanthin, and β-cryptoxanthin	Achir et al. (2016)
	Corozo	*Aiphanes aculeate*	β-cryptoxanthin (25.1%) β-carotene (10.9%), and lycopene (12.7%) esters of zeaxanthin (15.1%)	Murillo et al. (2013)
	Sastra	*Garcinia intermedia*	Free zeaxanthin, β-carotene, lutein, α-carotene, and esters of lutein (17%), zeaxanthin (29.8%), and β-cryptoxanthin (10.6%)	Murillo et al. (2013)
	Sapote	*Quararibea cordata*	di-esters of zeaxanthin (42.9%),	Murillo et al. (2013)
	Frutita	*Allophylus psilospermus*	Apo-carotenoid and b-citraurin (29.8%),	Murillo et al. (2013)
Vegetables	Tomato		Lycopene	Jacques, Lyass, Massaro, Vasan, and D'Agostino Sr. (2013) Sidhu, Singh, and Kaur (2019)
	Mayan Gold potato		Violaxanthin and lutein	Morris, Ducreux, Fraser, Millam, and Taylor (2006)
	Carrot	*(Daucus carota* L.)	β-carotene, α-carotene, and lutein	Haas et al. (2019) Janiszewska-Turak and Witrowa-Rajchert (2020) Jayaraj, Devlin, and Punja (2008)
	Lamb's quarters	*Chenopodium album*	Violaxanthin and lutein	Raju, Varakumar, Lakshminarayana, Krishnakantha, and Baskaran (2007)
	Jio	*Commelina benghalensis* L.	Violaxanthin, lutein, and β-Carotene	Raju et al. (2007)
	Indian spinach	*Basella alba* L.	β-carotene, neoxanthin, and lutein	Raju et al. (2007)

Type of Tissue	Name	Latin Name	Dominant Carotenoid(s)	References
Flower	Marigold	*Tagetes erecta* L.	All-*trans* and *cis* isomers of lutein, and lutein esters and all-*trans* and *cis* isomers of zeaxanthin	Jung et al. (2016)
Seeds	Golden Rice		β-carotene, α-carotene, β-cryptoxanthin, zeaxanthin, and lutein	Paine et al. (2005)
	Cereals		Lutein, zeaxanthin, and β-cryptoxanthin	Siebenhandl et al. (2007)

The biotransformation of carotenoids to vitamin A depends on the type of food in which the carotenoids are found, the type of carotenoid, and the body's requirements. When the demand is low, biotransformation is also lower than when there is a retinol deficiency in the body. Technological processes to which vegetables and fruits are subjected also determine the bioavailability of this vitamin. For example, it increases after heat treatment (cooking vegetables) and when they occur in an oil environment in which carotenoids can dissolve. The bioavailability of lycopene in tomatoes is greater from processed tomato products such as juice and sauce than from fresh tomatoes, because it is released from cell wall protein–carotenoid complexes during processing, homogenizing, and heating (van Het, West, Weststrate, & Hautvast, 2000).

Studies show that carotenoids have biological activity and protective function in preventing several diseases (Perera & Yen, 2007), as shown in Figure 5.2. Several processes including digestion, intestinal absorption, tissue distribution, tissue utilization, and excretion are involved in the bioavailability of bioactive compounds such as carotenoids (Holst & Williamson, 2004). Before carotenoids are absorbed, they must be released from the food matrix, incorporated into lipid droplets/micelles, and then these formed micelles must be taken into the enterocytes. In the body, carotenoids are transported through lipoproteins. Low-density lipoproteins (LDL) transport carotenes, while xanthophylls are transported by both LDL and high-density lipoproteins (HDL) because they are more polar than carotenes. This results in a more efficient and faster distribution of xanthophylls throughout the body (Jung et al., 2016).

There are several publications in which antioxidant properties of carotenoids (van den Berg et al., 2000), pro-oxidizing carotenoids, and, in some cases, both properties together (Young & Lowe, 2001), are discussed. There have been many studies about the roles of carotenoids (especially, lycopene and

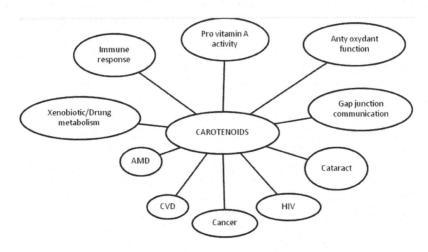

FIGURE 5.2 The bioactivity of carotenoids.

its protection) against DNA damage. However, there is no consensus on the effect of carotenoids on this phenomenon, possibly due to different biomarkers used.

One point of view emphasizes the importance of carotenoid concentration with a switch from anti- to pro-oxidation that is observed in several systems both *in vivo* and *in vitro*; e.g., in the homogeneous lipid solutions, membrane models, and intact cells (Eghbaliferiz & Iranshahi, 2016; Ribeiro, Freitas, Silva, Carvalho, & Fernandes, 2018). The presence of products that result from carotenoid oxidation/reduction was found depending on several factors, including the concentration of carotenoids, type of carotenoid, oxygen pressure, and type of reaction environment (Ribeiro et al., 2018).

Another point of view is related to the reactions of carotenoids with free radicals. They are more complex than those with single oxygen and depend mainly on the nature of the radicals more than on carotenoids. Free radicals have an unpaired electron, and, like all reactions, an unpaired electron is transferred to the carotenoid. As a result, the original radical is extinguished, but a new one is formed, which may have completely different properties than the original form. Whether a carotenoid is considered as an anti-oxidant (beneficial) or a pro-oxidant (damaging) depends on the combination of the properties of the radicals involved in the reaction (in the whole process and in the specific environment of the carotenoids). Therefore, it is not easy to predict and refer to carotenoids as antioxidants, as 'due to the quenching of free radicals' is not always appropriate (Böhm, Edge, & Truscott, 2012; Stahl, Ale-Agha, & Polidori, 2002). As a result of the reaction of the carotenoid with free radical cations, carotenoid radical cations are formed. These cations themselves are oxidizing, which is why they can oxidize important amino acids such as cysteine, tyrosine, tryptophan, and fats. However, regeneration of carotenoid radical cation with ascorbic acid is possible (Böhm et al., 2012). With a large number of free radicals and vitamin C deficiency, carotenoid radical cation may have harmful effects on human health (Edge & Truscott, 2018).

Several studies have focused on the effect of carotenoids on the protection of low-density lipoproteins, and it is known that the addition of carotenoids to blood plasma or isolated from LDL may protect lipoproteins against their oxidation (Dwyer et al., 2001). Many carotenoids have protective properties against lipoproteins; e.g., canthaxanthin and zeaxanthin are just as good antioxidants as lycopene, α-carotene, β-cryptoxanthin, and lutein. However, β-carotene has the most effective antioxidant properties against LDL (Panasenko, Sharov, Briviba, & Sies, 2000). Lycopene, in combination with other antioxidants, shows strong synergism in inhibiting the oxidation of low-density lipoproteins when it has been added to the diet and when it is an isolated LDL. Oleoresin from tomatoes caused very strong inhibition of LDL oxidation by copper ions (Fuhrman, Volkova, Rosenblat, & Aviram, 2000). A high daily dietary intake of tomato juice and ketchup significantly reduced LDL cholesterol levels in healthy normocholesterolaemic adults. In addition, there was an increase in the ability of LDL particles to resist the copper-induced formation of oxidized phospholipids. This may be an example of a synergistic action, because tomatoes are rich in other carotenoids, vitamins, and flavonoids, and these may also have accounted for their cholesterol-lowering effect (Silaste, Alfthan, Aro, Antero Kesäniemi, & Hörkkö, 2007). Moreover, a strong synergism may be observed when lycopene was combined with vitamin E, flavonoids, phenolic compounds, rosemary, carnosic acid from rosemary, or garlic extract containing a mixture of these antioxidants. It has been stated that β-carotene and other carotenoids have protective properties against DNA contained in lymphocytes (Fabiani, De, Rosignoli, & Morozzi, 2001). β-carotene or astaxanthin effectively blocks active forms of oxygen generated as a result of various enzymes in the cancer process, thus lowering the risk of this chronic disease (Kozuki, Miura, & Yagasaki, 2000).

Some soluble dietary fibers can disrupt micellization, and thus reduce the bioavailability of carotenoids. There are two mechanisms by which carotenoids are transported to erythrocytes: through diffusion from the micelle core or by active absorption through receptor-mediated transport in the duodenum (Milani, Basirnejad, Shahbazi, & Bolhassani, 2017). In the active transport, cholesterol, and other lipophilic substances that are taken up selectively by the protein-membrane, β-carotene is transported more efficiently than lycopene.

Carotenoids present in food and dietary supplements differ in absorption kinetics and plasma transport. In addition, factors such as physical form and technological processes may affect the bioavailability of carotenoids (Khalil et al., 2012). Surprisingly, carotenoids derived from microorganisms are characterized by high bioactivity. Jung et al. (2016) compared the bioavailability of lutein, zeaxanthin, α-carotene, and β-carotene from regular chlorella powder, lutein-enriched chlorella powder, and calendula extract in a randomized clinical trial. They stated that chlorella powder was a source of lutein, equivalent to that found in marigold petals. In addition, chlorella powder contains more diverse carotenoids and it is easy to produce at a large scale.

Lutein and zeaxanthin are carotenoids that affect the proper functioning of vision. They and their metabolites are the only carotenoids found in the retina and the lens of the eye (Bernstein et al., 2001; Li et al., 2018). They protect the retina and the lens against the damaging oxidation processes that are initiated by light. As a result of the absorption of a significant proportion of light, these carotenoids prevent photo-oxidation of eye tissues. Lutein and zeaxanthin are found in the macula attached to the retina of the eye and are responsible for its yellow color. Marigold petal extract has long been utilized as a major source of lutein, particularly to reduce the risk of age-related macular degeneration (Carpentier et al., 2009). The risk of cataracts is also reduced by other carotenoids and antioxidants such as β-carotene, vitamin C, E, and zinc (Chylack et al., 2002).

According to statistics, coronary heart disease is the most common cause of death in developed countries, both in men and women. The etiology of the disease is complicated, but elevated LDL is one of the most important causes. Studies show that there is a favorable relationship between the concentration of carotenoids in blood serum and the occurrence of coronary artery disease. The current study of the relationship between lycopene intake and CVD (cardiovascular disease) incidents confirms earlier reports that lycopene is associated with CVD risk (Dwyer et al., 2001). Currently, studies on the health benefits of lycopene are conducted in such a way that patients are given whole tomatoes or their preparations instead of supplements containing isolated lycopene. It is believed that the action of lycopene being in its natural environment is more effective (Jacques et al., 2013).

Due to their oxidizing properties, carotenoids may reduce the risk of Type 2 diabetes. Studies show that a diet high in carotenoids reduces the risk of Type 2 diabetes, but this relationship is unclear. Perhaps this is due to the fact that a diet rich in carotenoids is also a diet rich in vegetables, which is generally beneficial to health and reduces the consumption of reducing sugars (Sluijs et al., 2015).

With regard to cancer, it is also suggested that carotenoids alone, or in combinations with other phytonutrients, at concentrations that can be achieved in the body, can result in significant beneficial health effects, most probably elicited by the synergistic modulation of various transcription systems (Rowles III & Erdman Jr., 2020). Based on the research so far, it is still not possible to determine the effectiveness of carotenoids as a result of their nutritional dose. It cannot be clearly stated how much β-carotene and other carotenoids should be supplied to the body as a daily dose. Nevertheless, it is known that the supplementation of β-carotene or its consumption in large amounts with food reduces the risk of lung cancer in smokers and people working or having contact with carcinogenic asbestos.

Consumption of large amounts of carotenoids is safe for the body. After clinical trials for toxicity, it was observed that the safety level for lycopene intake was 75 mg per day. The observed safety level in this risk assessment for lutein was 20 mg per day (Shao & Hathcock, 2006). Based on "21: Food and Drugs ", 2020), β-carotene is generally recognized as safe (GRAS).

5.3 SPRAY DRYING ENCAPSULATION OF CAROTENOIDS

Carotenoids are very unstable under environmental conditions so their storage can cause problems such as carotenoids degradation (e.g., via their oxidation). Therefore, it is important to protect these bioactive compounds using methods such as encapsulation. Several encapsulation protocols have been suggested for the protection of carotenoids. However, the spray drying method is one of the

FIGURE 5.3 Morphology of carrot juice particles from: a) Scanning electron microscope at a magnification of 5000x, and b) Confocal microscope at a magnification of 100x. The micrographs were taken by authors and have not been published elsewhere.

most popular ones, not only because it is economical and flexible, but also because of the properties of the spray dried particles (Drosou et al., 2017; Janiszewska-Turak, 2017; Ray, Raychaudhuri, & Chakraborty, 2016). By using spray drying as an encapsulation process, obtained powders reach desirable stabilization and can be released in a controlled manner. Particles obtained by the spray drying method are usually matrix-based inside the wall of the capsule and space inside (Figure 5.3). Matrix-based means that the carotenoids are entrapped in a continuous net of a carrier material (Janiszewska-Turak, 2017; Shishir & Chen, 2017).

Because of the simplicity of the spray drying encapsulation process and the stability of the final product, there has been a huge increase in the number of publications on the spray drying encapsulation of carotenoids over the last 20 years (literature search of the ScienceDirect database in May 2020) (Figure 5.4). As can be seen in this figure (i.e., Figure 5.4), over the course of the past decade (i.e., 2009–2019), the published literature on the "spray drying encapsulation of carotenoids" raised from 34 to 211, demonstrating a dramatic increase in the research work in this field.

5.3.1 INFLUENCE OF THE TYPE OF CAROTENOIDS ON THE ENCAPSULATION PROCESS

The specific functional properties of carotenoids are connected to their chemical structure, which is linked to their conjugated double bonds. Unfortunately, these double bonds make carotenoids unstable to oxygen (Haas et al., 2019; Janiszewska-Turak, 2017). For this reason, a growing interest in the protection of these compounds is noticed, regardless of whether single selected carotenoids or carotenoid extracts derived from fruit or vegetable juices are used (Haas et al., 2019).

Spray drying has been used to encapsulate carotenoids such as astaxanthin (Ahmed et al., 2015; Montero, Calvo, Gómez-Guillén, & Gómez-Estaca, 2016), β-carotene (Chuyen, Roach, Golding, Parks, & Nguyen, 2019; Fang, Zhao, Liu, Liang, & Yang, 2019; Haas et al., 2019; Lim, Burdikova, Sheehan, & Roos, 2016; Lim & Roos, 2016), lycopene (Souza et al., 2018), lutein (Álvarez-Henao et al., 2018; Lim et al., 2016; Lim & Roos, 2016; Syamila, Gedi, Briars, Ayed, & Gray, 2019), and norbixin (Tupuna et al., 2018). Moreover, spray drying of vegetable or fruit juices, in which carotenoids occur individually or in combination, has received considerable interest (Table 5.2). Complex carotenoids from juices/extracts or pulps have been tested for spray dried celery juice (Khalilian Movahhed & Mohebbi, 2016), banana passionfruit pulp (Troya, Tupuna-Yerovi, & Ruales, 2018), spinach juice (Çalışkan Koç & Nur Dirim, 2017; Syamila et al., 2019), tamarillo and exotic fruits pulps (García,

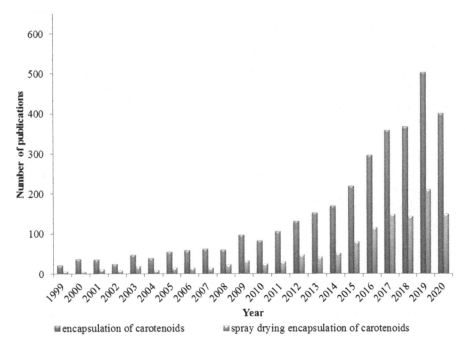

FIGURE 5.4 The number of the literature published from 1999 to 2019, based on the search of phrases including spray drying encapsulation of carotenoids and encapsulation of carotenoids. Sourced from https://www.sciencedirect.com.

Giuffrida, Dugo, Mondello, and Osorio (2018), tamarillo juice (Ramakrishnan, Adzahan, Yusof, & Muhammad, 2018), tomato concentrate (Souza et al., 2018), tomato juice (Sidhu et al., 2019), and carrot juice (Janiszewska-Turak et al., 2017; Janiszewska-Turak and Witrowa-Rajchert, 2021).

Analysis of the literature on the microencapsulation of carotenoids showed that from the beginning the course of obtaining powder process was different. Some authors used single extracted carotenoids (Fang et al., 2019; Ordonez-Santos, Martinez-Giron, & Villamizar-Vargas, 2018), while others used juices obtained from vegetables or fruits rich in carotenoids (Lim et al., 2016; Syamila et al., 2019), or oils rich in carotenoids (Chuyen et al., 2019). However, the later selection of drying parameters and the type of carrier material should be dictated by this first step.

There are also two ways for the feed preparation of carotenoids before spray drying. In the first method (with extract), in the beginning, it is necessary to create an aqueous solution with carrier material and then add the extracted colorant (Fang et al., 2019; Ordonez-Santos et al., 2018; Tupuna et al., 2018). In the second method, in the case of concentrate or puree in which colorant is dried, the carrier material is added directly into the juice (García et al., 2018; Haas et al., 2019; Janiszewska-Turak & Witrowa-Rajchert, 2020; Lim et al., 2016; Ramakrishnan et al., 2018; Sidhu et al., 2019; Syamila et al., 2019; Troya et al., 2018). However, as is shown in the literature (Table 5.3), due to the large variety of carotenoids, it is not possible to determine one 'correct' protocol for the preparation, carotenoid concentration, or type of carrier material prepared for the spray drying process.

5.3.2 INFLUENCE OF THE CARRIER TYPE ON THE ENCAPSULATION PROCESS

One approach of selecting the type and amount of the carrier for spray drying encapsulation of carotenoids is connected to a 'trial and error test' that assesses the effectiveness of the process based on the characteristics of the final product, such as encapsulation yield, powder stability under various

TABLE 5.3

Selected parameters reported for spray drying encapsulation of carotenoids

Sample	Spray Dryer Type	Encapsulation Factors	Effects on Physicochemical Properties of Powders	References
Astaxanthin from shrimps	Model B-290 (Buchi)	IT = 150°C, FFR = 6 mL/min, OT = 95°C MD, GA and its mixtures	Change carrier MD → MD:GA → GA caused L*, hue (↓), d (↕), astaxanthin bioavailability (↑)	Montero et al. (2016)
Banana passionfruit pulp	Model B-290 (Buchi)	IT = 120°C and 150°C Core:Carrier 1:1 or 3:2 MD:GA = 1:9 or 4:6 FFR = 10 mL/min Air pressure 5 bar, nozzle diameter 0.7 mm	Increase in IT → encapsulation efficiency and d50 (↑), moisture content (↓) Increase in core: carrier → encapsulation efficiency, d50 (↕), moisture content (↑) Increase in GA → encapsulation efficiency and moisture content (↕), d50 (↓)	Troya et al. (2018)
Carotenoids from carrot concentrate	Pilot-scale spray dryer, Lab S1 (Anhydro), Semi-industrial spray dryer (Niro)	IT = 195°C OT = 80°C FFR = 30–42 mL/min Suspension type concentrate (SCC) and emulsion type concentrate (ECC) 25% and 45% of dry matter in solutions Nozzle pressure in pilot drier 1 or 2 bar	For SCC 25/2 → 45/1 – semi SD - encapsulation efficiency, carotenoids retention, d4,3 (↑) For ECC 25/2 → 45/1 – semi SD - encapsulation efficiency, carotenoids retention, d4,3 (↑) SD powders with a low surface to volume ratio show high carotenoids retention during storage	Haas et al. (2019)
Carrot juice	Semi-industrial spray dryer, Lab S1 (Anhydro)	IT = 160°C OT = 90°C Disc speed = 39,000 rpm FFR = 0.4 10–6 m³/s MD, GA, WPI and MD: AG = 1:1 or 1:2 or 3:1	Change carrier MD → GA → WPI → water activity (↓), carotenoids content (↕)	Janiszewska-Turak et al. (2017)
Carrot juice	Semi-industrial spray dryer, Lab S1 (Anhydro)	Juice with and without blanching IT = 160°C OT = 90°C Disc speed = 39,000 rpm or 28,000 rpm FFR = 0.4 10–6 m³/s MD, GA, WPI and MD:GA = 1:1 or 1:2	Blanching caused (↓) in carotenoid content and (↑) in densities and diameters Change carrier MD → GA → WPI caused (↑) in diameter d50 and carotenoid content, and (↓) in density values. (↓) in disc speed → MC (↑), carotenoids content (↕)	Janiszewska-Turak and Witrowa-Rajchert (2021)
Gac Peel	Model B-290 (Buchi)	Oil/wall material 10, 20, and 30 g/100 g IT = 170°C, 160°C, and 180°C FFR = 180, 270, and 360 mL/h WPC:GA = 7:3 Air rate = 35 m³/h	Increase in O/W → β-carotene, antioxidant capacity (↑) Increase in IT → β-carotene, antioxidant capacity (↓) Increase in FFR → β-carotene, antioxidant capacity (↓)	Chuyen et al. (2019)

Sample	Spray Dryer Type	Encapsulation Factors	Effects on Physicochemical Properties of Powders	References
Goldenberry (*Physolis peruviana* L.)	Model B-290 (Buchi)	IT = 140°C OT = 70°C FFR = 10 mL/min Air flow = 473 L/h Nozzle 0.5 mm Carrier agents: GA, MD, MS, inulin, cellobiose, alginate carrier addition 20% (w/w)	Change in carrier type from the single carrier into mixture → yield ($\updownarrow$), total solids ($\updownarrow$), encapsulation efficiency ($\updownarrow$) – the exception was alginate (min) and cellobiose (max)	Etzbach et al. (2020)
Lutein	Lab scale spray dryer, LabPlant SD-05	IT = 185°C FFR = 4 mL/min MD: GA: MS in different proportions	The highest lutein concentration was observed in 100% AG and MD: GA: MS = 33.3: 33.3: 33.3; Better protection during storage was observed for the mixture of carrier	Álvarez-Henao et al. (2018)
Norbixin	Model B-290 (Buchi)	Carriers MD: GA proportion = 100: 0, 85: 15, 65: 35, 50: 50, 35: 65, 15: 85, and 0: 100 Norbixin:carrier = 1: 1000 IT = 150°C OT = 75°C FFR = 0.39 L/h	Change in carriers proportion caused → ($\downarrow$) in water solubility for powder, based on proportion 35:65 but for other powders ($\uparrow$) was seen, moisture ($\updownarrow$), size ($\uparrow$), microencapsulation efficiency ($\updownarrow$)	Troya et al. (2018)
Saffron extracts	Model B-90 (Buchi)	IT = 100°C FFR = 100 L/min Nozzle 4 and 7 µm Carrier: MD Core:carrier = 1: 5; 1: 10; 1:20 w/w	Change nozzle diameter 4 → 7 µm caused: product yield ($\uparrow$), encapsulation efficiency ($\updownarrow$), Core: carrier ratio 1: 5 → 1: 10 → 1: 20 → product yield ($\updownarrow$), encapsulation efficiency ($\uparrow$)	Kyriakoudi and Tsimidou (2018)
Spinach juice	Pilot-scale spray dryer mobile Minor, (Niro)	IT = 160°C, 180°C, and 200°C (without carrier agent) Carrier MD, WPI, GA for selected IT Speed of the pump – connected to the IT and carriers (equations in the article)	Increase in IT → Moisture content and water activity ($\updownarrow$), total color ($\downarrow$), total carotenoid, and total chlorophyll ($\downarrow$) MD → WPI → GA → moisture content and water activity ($\updownarrow$), total color ($\downarrow$), total carotenoid, and total chlorophyll ($\updownarrow$)	Çalışkan Koç and Nur Dirim (2017)
Spinach juice	Armfield FT80 Tail form spray dryer (Ringwood, UK)	Juice solid content 6,4 °Brix Storage at 4°C, 20°C, and 40°C with and without light and with or without vacuum package IT = 132°C OT = 65°C FFR = 420 mL/h Air pressure 1.5 bar No carrier	during storage independently form temperature, light, and package → ($\downarrow$) in β-carotene, lutein, and alfa-tocopherol values	Syamila et al. (2019)

(*Continued*)

TABLE 5.3
(*Continued*)

Sample	Spray Dryer Type	Encapsulation Factors	Effects on Physicochemical Properties of Powders	References
Spirulina platensis	Model B-290 (Buchi)	IT = 170°C OT = 95°C FFR = 6 mL/min Nozzle Atomization volume = 150 mL Core:carrier = 1: 1 MD or MD with citric acid	Addition of citric acid → DPPH activity (↓), β-carotene content (↓)	da Silva et al. (2019)
Tamarillo and Mango and Banana pulp Carotenoids	LabPlant SD-06	IT = 130°C and 180°C Air pressure 4 bar Nozzle diameter 2 mm FFR = 486 mL/h MD 2×4 factorial design Yellow tamrillo (YT): MD: Mango pulp (MP): banana pulp (BP) YT: MD: MP: BP = 1: 1:0.5: 0; 1: 1: 0.5: 0.1; 1: 1: 1: 0; 1: 1: 1: 0.1	Changes in proportion and IT → pH (↕), a_w (↕), a*(↕), b* (↕) values Different proportions and IT changed the carotenoids content; utilization for cookies, Indian bread, production as a source of carotenoids	García et al. (2018)
Tamarillo juice	Pilot spray dryer (Niro)	Air flow rate = 900 m³/h IT = 150°C OT = 70°C Disc speed 15,000 rpm Carriers MD, GA, OSA1, OSA2, RMD 20% w/w of juice and carrier	Change from MD → GA → Moisture, water content (↕), Tg (↑), Change from GA → OSA1 → OSA2 → RMD caused (↓) in all properties of powders	Ramakrishnan et al. (2018)
Tomato concentrate	Model B-190 (Buchi)	Pressure 350 kPa, FFR = 34 mL/min IT = 160°C OT = 80°C Carrier:core 3:1 MD, WPI, MS in different proportions (simple centroid design)	The highest carotenoids concentration was stated in powders based on MD and MS, also the smallest degradation of lycopene was observed	Souza et al. (2018)
Tomato juice	Laboratory mini spray dryer (Make: JISL, model LSD-48)	IT = 140°C, 150°C; 160°C MD: juice 30: 70; 40: 60; 50: 50	Increase in IT → moisture content (↓), bulk density, L*, a*/b* (↕), powder recovery (↕); Increase in MD concentration → moisture content, L* (↓); bulk density (↕); powder recovery, a*/b* (↑)	Sidhu et al. (2019)
β-carotene and lutein	Niro 25 spray dryer (Niro)	IT = 185°C OT = 85°C disc speed 18,000 rpm, Trehalose:WPI = 1:1	Degradation of carotenes during storage followed by first-order kinetics. Time and temperature influenced carotenes degradation	Przybysz, Szterk, Symoniuk, Gąszczyk, and Dłużewska (2018)

Sample	Spray Dryer Type	Encapsulation Factors	Effects on Physicochemical Properties of Powders	References
β-carotene	Lab-Plant SD-05, (Huddersfiel, England)	Air flow rate 47 m³/h. Pressure 1.7 bar FFR = 73 m³/h IT = 110°C–200°C. AG 5%–35% experimental design	Air temperature and GA content influenced the drying yield, encapsulation efficiency, and load capacity. Increase in IT → antioxidant activity of β -carotene ($\downarrow$); IT 173 °C and the GA concentration of 11.9% → β -carotene content, encapsulation efficiency, drying yield ($\uparrow$)	Corrêa-Filho, Lourenço, Moldão-Martins, and Alves (2019)
β-carotene	Mini spray dryer YC-015	IT = 185°C Nozzle 0.7 mm FFR = 1100 mL/h Storage temperatures 4L, 37, 60°C OSA and MD = 1:1; 1:2; 1:3 Chitosan as next layer Carrier addition 15, 20 or 25%	Increase MD → d50 ($\uparrow$); zeta-potential for emulsions ($\downarrow$); For SD was chosen 20% of carriers and OSA: MD = 1:2; Good stability of powders in comparison of non-dried β-carotene Increase in storage temperature → β-carotene retention ($\downarrow$)	Fang et al. (2019)
β-carotene and lutein	Niro 25 spray dryer (Niro)	IT = 185°C OT = 85°C Disc speed 24,000 rpm Trehalose:WPI = 2:1	Carotenoid loss followed first-order kinetics Storage temperature increase → carotenoids ($\uparrow$)	Lim et al. (2016)
β-carotene and lutein	Niro 25 spray dryer (Niro)	IT = 185°C OT = 85°C disc speed 18,000 rpm, Trehalose:WPI = 1:1	The carotenoid loss followed first-order kinetics Used of multilayer of carrier materials delayed the degradation rate of carotenoids	Lim and Roos (2016)
β-carotene from carrot-celery juice	Mini-spray dryer (Soroush Sanat)	IT = 120°C, 145°C, 170°C FFR = 36, 44.5, 53 mL/min Air pressure 3 bar Maltodextrin at different concentrations, two-flow nozzle	Increase in IT → MC, a_w, L* a*, b*, β -carotene ($\downarrow$); d ($\uparrow$) Increase in MD → MC, a_w, a*, b* β -carotene ($\downarrow$); d, L* ($\uparrow$) Increase in FFR → MC, a_w, d, a*, b*, β -carotene ($\uparrow$)	Khalilian Movahhed and Mohebbi (2016)

$\uparrow$: increase, $\downarrow$: decrease, $\updownarrow$: no clear trend, SD: spray drier, IT: inlet temperature, OT: outlet temperature, FFR: feed flux rate, MD: maltodextrin, MS: modified starch, WPI: whey protein isolate, GA: gum Arabic, OSA: octenylsuccinate-modified starch, MC: moisture content, L* a* b*: color parameters, a_w: water activity, and d: diameter.

storage conditions, and morphology. It is a kind of response decision, in which the choice of the carrier material is verified based on the obtained product properties and the selection process continues until the best possible product properties are obtained. This approach, and the high sensitivity of the process itself to changes, requires a large number of experiments to obtain the suitable material as the carrier, which further hinders the transfer of the obtained results from one type of active material to another (Anandharamakrishnan & Ishwarya, 2015). For these reasons, many different mixtures of carotenoids as a core material and carrier, as well as different proportions between the used carriers, are used in the tests.

Carrier material(s) can influence the final physical properties of the spray dried powder, meaning that their properties need to be understood before they are employed for the encapsulation of carotenoids. The most important aspect of encapsulation is to protect the core active ingredients (i.e., carotenoids) from environmental factors such as temperature, light, pH, and humidity. After choosing the type of carrier, its concentration should be optimized. The ratio of core to the carrier material, as well as the whole amount of carrier into the final solution, should also be decided (Anandharamakrishnan & Ishwarya, 2015).

5.3.2.1 Carrier Materials

There is a great variety of carrier types choices for the encapsulation of carotenoids using spray drying. In the literature for carotenoid microencapsulation by spray drying, different proportions of carriers in their mixtures are tested (Table 5.3). The most commonly used carriers include maltodextrins (MD) with a different dextrose equivalent (DE), gum Arabic (GA), sucrose, starch, modified starch (e.g., octenylsuccinate-modified starch (OSA)), whey protein isolate (WPI), and whey protein concentrate (WPC). Often a mixture of several carriers is used, due to the lack of an ideal carrier, as well as the need to adjust carriers to ensure the highest possible protection of closed carotenoids. As a relevant parameter for encapsulation process descriptions usually encapsulation efficiency, yield, and some physical properties of the powders, such as diameter, morphology, moisture content, water activity are used (Chranioti, Nikoloudaki, & Tzia, 2015; Janiszewska-Turak, 2017; Shishir & Chen, 2017).

5.3.2.1.1 Maltodextrin

Maltodextrin, as the cheapest material, has been used for the encapsulation of different sources of carotenoids; carotenoids from tamarillo and tropical fruit (García et al., 2018), carotenoids from tamarillo juice (Ramakrishnan et al., 2018), β-carotene from carrot-celery juice (Khalilian Movahhed & Mohebbi, 2016), carotenoids from spinach juice (Çalışkan Koç & Nur Dirim, 2017), and carotenoids from paprika oleoresins (Díaz, Lugo, Pascual-Pineda, & Jiménez-Fernández, 2019).

Sidhu et al. (2019) spray dried tomato juice with MD as a carrier material. They investigated the ratio of juice to carrier material (i.e., MD), in the ratios of 1: 1, 3: 2, and 2.3: 1 (Table 5.3). An increase in carrier (MD) concentration caused an increase in color coefficients of a* and b*and powder yield, as well as a decrease in moisture content and lightness of the powder. This could be related to the increase in the total solids by the addition of the drying carriers that can reduce the amount of water evaporation. Furthermore, MD itself can hurdle the sugars that are highly hygroscopic (Shrestha, Ua-Arak, Adhikari, Howes, & Bhandari, 2007; Sidhu et al., 2019).

Other carrier materials or different drying conditions were also tested for increasing encapsulation efficiency and protection of active ingredients. Díaz et al. (2019) stated that particles from paprika oleoresin with MD were spherical and smooth. A similar observation was made by García et al. (2018) in the case of tropical fruits mixed with tamarillo and Janiszewska-Turak et al. (2017) for carrot juice particles. However, other researchers obtained spherical particles but with a lot of wrinkles, sometimes also destroyed like that observed for tamarillo juice (Ramakrishnan et al., 2018).

5.3.2.1.2 Gum Arabic (GA)

As a single carrier material, GA has also been tested for spray drying encapsulation of carotenoids (Table 5.2). For example, it was found that particles of tamarillo juice with GA and OSA starch had the highest encapsulation efficiency in comparison with MD and resistant maltodextrin. Additionally, the change in carrier material caused differences in moisture content, water activity, drying yield, and morphology. The authors mentioned that it could be related to the molecular weight of used carrier material, from which AG and OSA starch was the highest (Ramakrishnan et al., 2018). Corrêa-Filho et al. (2019) stated that the spray dried particles of β-carotene in GA were a mixture of smooth and wrinkled particles, with a diameter smaller than 10 μm.

5.3.2.1.3   Mixtures of Carriers

The most popular mixtures that have been used for the spray drying encapsulation of carotenoids are MD with AG for the encapsulation of norbixin (Tupuna et al., 2018) or carotenoids from banana (Troya et al., 2018), for example. Other used mixtures include trehalose with WPI for Gac oil encapsulation (Lim et al., 2016), MD with modified starch like OSA for β-carotene (Fang et al., 2019), WPC mixed with GA for Gac peel oil (Chuyen et al., 2019), or in the combination of three or more carriers as a wall matrix for some other carotenoids (Álvarez-Henao et al., 2018; Souza et al., 2018).

The protection of norbixin with a mixture of MD and GA in an experiment by Tupuna et al. (2018) resulted in the conclusion that the single carrier (GA) provided the highest encapsulation efficiency, and independently form carrier ratios, the particles presented similar diameter size, almost smooth surface of particles, and spherical shape. However, moisture content could significantly influence their behavior during storage. Different results were obtained for banana pulp dried with MD: GA mixture in the ratios of 1: 9 or 4: 6 (Troya et al., 2018). The authors stated that an increase in GA content resulted in a decrease in diameter size, but they did not observe a correlation between that and encapsulation efficiency.

Due to GA specific taste and high price, researches have been trying to replace this carrier material with another one. This replacement could be provided by the addition of the third carrier material. For instance, this kind of research was made for lutein by Álvarez-Henao et al. (2018) and lycopene by Souza et al. (2018). As was concluded for lutein, the best formulation was MD: GA: MS in the ratio of 33.3: 33.3: 33.3; however, the highest encapsulation efficiency was not observed in this case (Álvarez-Henao et al., 2018). Moreover, the authors researched the morphology of the manufactured particles and mentioned that the difference was only seen in the size. For lycopene, for which the mixtures of MD, MS, and WPI were used, it was declared that the mixture of MD: MS was the most suitable, in the case of retaining the carotenoids content (Souza et al., 2018).

Different carrier materials affected the size and morphology of carotenoids microcapsules. To visualize the outer morphology of one type of carotenoid source, pictures of spray dried carrot juice with selected carrier materials are presented in Figure 5.5.

5.3.2.1.4   Other Solutions

Nowadays, manufacturers are interested in decreasing or even removing the carrier agent in the spray drying process. Syamila et al. (2019) tested the possibility of obtaining spray dried spinach juice without the addition of any carrier agent. They stated that this was possible and that the manufactured powder could be used as a functional food ingredient. However, the content of carotenoids from spinach (lycopene, β-carotene, and lutein) was decreased during storage time. The protection solution could be reducing the oxygen level by vacuum packaging.

5.3.2.2   Carrier Addition and Type of Carotenoids Solution

After the selection of carrier material, there is a need to decide what ratio of the carrier: core material is needed. The ratio of core: carrier could be 1: 1 (Haas et al., 2019; Janiszewska-Turak et al., 2017; Janiszewska-Turak & Witrowa-Rajchert, 2021; Sidhu et al., 2019; Troya et al., 2018). However, for the encapsulation process of carotenoids by the spray drying method, other ratios have also been tested. Troya et al. (2018) tested smaller additions of carrier agent (MD: GA) into banana passion-fruit pulp and they found that a higher proportion of core to the carrier (3: 2) led to a lower encapsulation efficiency and lower process yield than with the ratio of 1: 1. A similar conclusion was made by Sidhu et al. (2019) for spray drying of tomato juice with MD as a carrier material.

The type of the created solution is also a crucial factor for the spray drying encapsulation of carotenoids. Haas et al. (2019) tested two types of solutions; suspension-type (SCC) and emulsion-type (ECC) carrot concentrate. They mixed the carrier (MD) and carrot concentrate on these solutions in the ratio of 1: 1. The difference between them was that SCC was obtained from orange carrot concentrate in which carotenoids are mainly present as crystalline carotenoids, while ECC

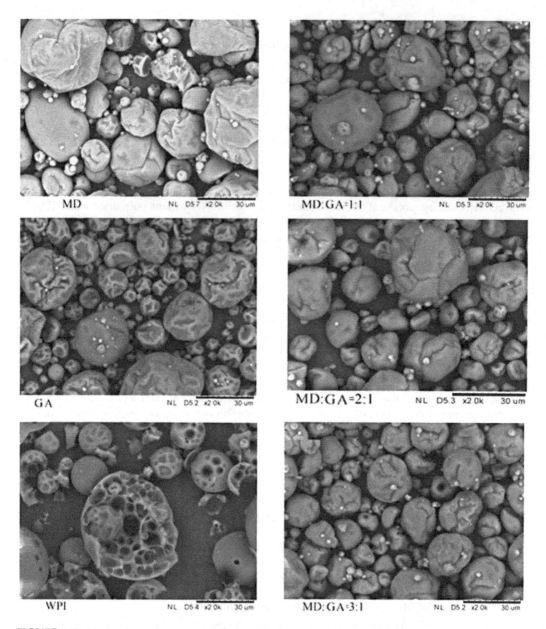

FIGURE 5.5 Morphology of encapsulated carrot juice particles with maltodextrin (MD), gum Arabic (GA), whey protein isolate (WPI), and a mixture of maltodextrin and GA in different proportions (MD: GA of 1: 1, 2: 1, or 3: 1). The micrographs were taken by the authors and have not been published elsewhere.

was obtained from the yellow carrot concentrate and was dissolved in an emulsified sunflower oil. Spray drying of these mixtures resulted in a different morphology of particles and influenced the encapsulation efficiency. Chuyen et al. (2019) spray dried a mixture of Gac peel oil and WPC mixed with GA in different proportions and concentrations and stated that only oil core: carrier material ratio could significantly influence the carotenoids retention.

After the selection of core to carrier addition, a final concentration of the dry substance in the drying solution is being considered (da Silva et al., 2019; Janiszewska-Turak et al., 2017;

Shishir & Chen, 2017). Analyzing the results of the research, it can be stated that researchers used many variations of carrier addition into the carotenoids solutions ratio. It started with 10% for lutein (Álvarez-Henao et al., 2018) and 15% for β-carotene (Álvarez-Henao et al., 2018; Corrêa-Filho et al., 2019; Fang et al., 2019), and can end with 40% for norbixin (Tupuna et al., 2018) and 45% for carotenoids from carrot concentrate (Haas et al., 2019), for example. The main conclusion from increasing carrier material is obtaining a bigger yield after the spray drying process.

5.3.3 Influence of Drying Parameters on the Encapsulation Process

The spray drying method is the simplest method for encapsulation of colorants such as carotenoids. However, a few parameters should be analyzed for the selection of the highest encapsulation efficiency or the most desirable physical properties (i.e., diameter, density, moisture content). Parameters of the drying process which can be changed are inlet air temperature, outlet air temperature, disc speed or pressure of the nozzle, feed flux rate (FFR), and air humidity (Corrêa-Filho et al., 2019; Drosou et al., 2017; Keshani, Daud, Nourouzi, Namvar, & Ghasemi, 2015).

5.3.3.1 Air Temperature

In the process of encapsulation by spray drying, an inlet air temperature must assure the protection of the active ingredient, during water evaporation in the drying chamber. The temperature inside the chamber is a combination of impacts inlet air temperature and amount of the evaporated water; i.e., feed flux rate. The higher inlet air temperatures can affect the physicochemical properties of the spray dried powder such as moisture content, water activity, density, the diameter of particles, and the bioactive content (Drosou et al., 2017; Shishir & Chen, 2017).

Usually, the use of a high inlet air temperature is chosen for the rapid formation of a semipermeable membrane on the surface of the droplet, which is linked to the water evaporation. When the high temperature is used, faster water evaporation is achieved, and that can accelerate the formation of the powder, allowing a proper shrinkage of the outer membrane, and, in turn, resulting in larger particles. On the other hand, a temperature that is too high can damage the particle, as a result of heat accumulation or excessive bubble growth, which causes the particle surface to tear or causes the particles to become sticky and clump (Drosou et al., 2017; Janiszewska-Turak et al., 2017). In addition, with a higher drying temperature, the drying rate is faster, which can reduce moisture content, water activity, and degradation of sensitive bioactives such as carotenoids. Inversely, when the inlet temperature is too low, the recovered powders may not be fully dried and may stick to the walls of the dryer, meaning that the efficiency of the process will decrease (Mahdavi, Jafari, Ghorbani, & Assadpoor, 2014; Rajabi, Ghorbani, Jafari, Sadeghi Mahoonak, & Rajabzadeh, 2015). The inlet air temperature during the spray drying of carotenoids has been reported to be in a range of 120°C to 200°C (Table 5.3). When the temperature was higher, the moisture content in encapsulated tomato juice with MD as a carrier (Sidhu et al., 2019), spinach juice dried without carrier agent (Çalışkan Koç & Nur Dirim, 2017), β-carotene from carrot-celery juice with MD (Khalilian Movahhed & Mohebbi, 2016), and banana passionfruit pulp with MD: GA as the carriers (Troya et al., 2018) decreased.

Another important parameter is encapsulation efficiency that can estimate the carotenoids content of the spray dried materials. For the carotenoids in banana passionfruit pulp encapsulated using the spray with MD and GA, an increase in the encapsulation efficiency and size of the particles with increasing the temperature from 120°C to 150°C was found (Troya et al., 2018). A similar observation was made for lycopene encapsulated with MD as the wall material, where the encapsulation efficiency increased with increasing the inlet air temperature from 110°C to 150°C (Goula & Adamopoulos, 2012). On the other hand, in the case of Gac peel oil encapsulated with WPC:GA as the carriers (temperatures of 160°C, 170°C, or 180°C), a decrease in the encapsulation efficiency as well as the antioxidant activity due to the increase in the inlet air temperatures was stated (Chuyen et al., 2019). In another report, a decrease in the encapsulation efficiency as a result of the increase in the inlet air temperature (160°C, 180°C, and 200°C) was observed for dried spinach juice (Çalışkan Koç & Nur Dirim, 2017). These authors also observed a decrease in the total color parameters, but no clear trend for moisture content was found. Sidhu et al. (2019) also reported no influence of the inlet temperature on moisture content, particle diameter, and color coefficients for the spray dried powders of tomato juice with MD. In general, it can be noticed that the basic assumptions about the spray drying process work in most cases, but that they still, to a great extent, depends on the source of the carotenoids and the type of the carrier(s).

It is also known that the temperature of the outlet air depends on the amount of water evaporated during the spray drying process. Evaporation speed is influenced by the inlet air temperature, the feed flux rate of the solution to the disk, and the speed of the disc/nozzle pressure. As mentioned earlier in this section, higher inlet air temperatures promote faster evaporation and shell formation. For this reason, it is important to consider the material degradation temperature (outlet temperature) to reduce the loss of carotenoids (Drosou et al., 2017; Shishir & Chen, 2017). Moreover, outlet temperature is mentioned to be a factor by which the spray drying process can be controlled.

5.3.3.2 Speed of the Disc/Nozzle Pressure

Spraying is the main goal of the spray drying process and it is important to select the disk speed or suitable pressure for the nozzle atomization. The basic function of the atomizer is to increase the mass and heat exchange surface, owing to which the water evaporates faster. Variations of these parameters linked with the feed flux rate of the solution can result in better physical properties of the formed particles and can then influence the activity of the bioactive compounds (e.g., carotenoids), which are enclosed inside the particles (Janiszewska-Turak, 2017; Shishir & Chen, 2017).

By changing the speed of the atomizer disk, different physical properties of the powders such as particle size (the higher the disk speed, the smaller the particles) and moisture content (the higher the disk speed, the lower the moisture content) can be obtained (Shishir & Chen, 2017). These statements were confirmed by Janiszewska-Turak and Witrowa-Rajchert (2021) during spray drying of carrot juice with the disc rotating at 28,000 to 36,000 rpm. They showed that an increase in disc speed caused a significant decrease in water content and particle diameter value, but no statistically significant effect on carotenoids content.

When using the nozzle, the atomization pressure should be optimized, which has a great impact on the physical properties of the encapsulated spray dried product. After an increase in the atomization pressure, the obtained microcapsules have smaller particles as a result of increasing the exchange surface, which improves the drying process. Besides, a higher atomizer pressure can increase hygroscopicity, process efficiency, the solids content, and bulk density of the final powder microcapsules, which is synonymous with a reduction in moisture content (Janiszewska-Turak, 2017; Shishir & Chen, 2017).

5.3.3.3 Feed Flow Rate

The feed flow rate (FFR) is usually related to the atomizer speed; the higher the disc speed/nozzle pressure, the higher the FFR value. It is also common to use the FFR as a factor for stabilization of outlet temperature, which can increase with the time of drying. A higher FFR speed, at the same disc speed, reduces the heat transfer rate, making drying difficult, because of the slower rate of water evaporation from the created droplets. As a result, particles with higher moisture content and very often with a large diameter are obtained (Mahdavi et al., 2014; Rajabi et al., 2015). Chuyen et al. (2019) tested the influence of FFR on the antioxidant capacity and β-carotene content. They increased FFR from 180 to 270 and 360 mL/h and observed a decrease in the tested values. Khalilian

Movahhed and Mohebbi (2016) spray dried carrot–celery juice and analyzed the effect of FFR on the selected physical and chemical properties of powders. They stated that an increase in FRR from 36 to 53 mL/min increased the moisture content, water activity, particle size, color coefficients a* and b*, and β-carotene content.

5.3.3.4 Air Humidity

Another possible way to obtain spray dried carotenoids with the retained bioactivity is lowering the drying temperature by using dehumidified air as the drying agent. Lower drying temperature, which is linked to the lower air humidity, will also be associated with the higher and faster water removal during drying, because of the higher possibility of taking it from the liquid atomized solution (Goula & Adamopoulos, 2005; Jedlińska et al., 2019; Shishir & Chen, 2017). To date, only a few studies have been published in the field of spray dried fruit/vegetable juices (Goula, Adamopoulos, & Kazakis, 2004) and pulp (Goula & Adamopoulos, 2005) using dehumidified air. Drying with dehumidified air can improve the physical properties of the final products. Goula and Adamopoulos (2005), who spray dried tomato pulp with dehumidified air and inlet temperatures of 110°C–140°C, stated that obtained powders had lower moisture content and higher bulk density, associated with the faster shell formation of the powder particles. There is still a need for further research related to spray drying using air with reduced moisture content, especially for sensitive substances such as colorants.

5.4 APPLICATION OF THE SPRAY DRIED ENCAPSULATED CAROTENOIDS IN FOOD PRODUCTS

It appears that the spray dried encapsulated carotenoids are often used as food colorants (Choudhari, Bajaj, Singhal, & Karwe, 2012; Rocha, Fávaro-Trindade, & Grosso, 2012), mainly because of the stability of carotenoids in powder form. Possible applications of these particles are presented in Table 5.4. Rocha et al. (2012) used spray dried (modified starch as wall material) encapsulated lycopene in a cake formulation. They found that the uniform color of the final product was achievable. Furthermore, Choudhari et al. (2012) showed that the spray dried encapsulated lycopene was stable, and also during the extrusion of rice flour, it resulted in better color retention. Kyriakoudi and Tsimidou (2018) encapsulated saffron in MD and tested its stability under thermal and gastrointestinal conditions. In conclusion, they mentioned that saffron in powder form is suitable for food, pharmaceutical, and other saffron applications.

Encapsulated carotenoids (and other active ingredients) from *Spirulina platensis* was used as a colorant and stabilizer in yogurt production (da Silva et al., 2019). The authors stated that the product of the spray drying encapsulation process of spirulina extract with MD was stable and did not show any toxicity in yogurt. They mentioned that yogurt with spirulina microcapsules have better functional properties, as well as a more stable color. However, they observed that the products did not present a pleasant taste, meaning that further studies are needed for the improvement of its sensorial profile.

In most cases, the authors only mentioned that the obtained carotenoid microcapsules could be used as a natural colorant. The proposition of using microencapsulated carotenoid as a colorant was mentioned by No and Shin (2019) for red paprika powder, encapsulated in modified starch OSA. García et al. (2018) mentioned about possibilities of the application of lycopene from fruit by-products and yellow tamarillo pulp, in dough biscuits or unleavened Indian flatbread. Khalilian Movahhed and Mohebbi (2016) also stated that tamarillo encapsulated powder could be "used as instant beverages, ingredients for bakery or extruded cereal products, snacks, ice cream, yogurt, salad, as well as pharmaceutical tablets and diet drinks." Therefore, further research is yet needed to confirm the efficacy of such microencapsulated colorants.

TABLE 5.4

Practical applications of encapsulated carotenoids in food products

Encapsulated Carotenoids	Application	References
Lycopene powder	As a colorant in a cake formulation, for the uniformity of the color of the final product	Rocha et al. (2012)
Lycopene powder	As a colorant in extruded rice formation	Choudhari et al. (2012)
Red paprika powder encapsulated in OSA modified starch	As a natural colorant	No and Shin (2019)
Saffron powder encapsulated with MD as a carrier	In food and pharmaceutical products as a colorant	Kyriakoudi and Tsimidou (2018)
Spirulina platensis with maltodextrin	In yogurt as a formulation additive and colorant	da Silva et al. (2019)
Carotenoid-rich microencapsulates from tropical fruit by-products and yellow tamarillo	Mentioned only about possibilities of application, as was with nonencapsulated powder, in biscuit dough or unleavened Indian flatbread	García et al. (2018)
Carrot–celery juice powder	Mentioned only in practical application that this powder can be "used as instant beverages, ingredients for bakery or extruded cereal products, snacks, ice cream, yogurt, salad, as well as pharmaceutical tablets and diet drinks"	Khalilian Movahhed and Mohebbi (2016)

5.5 CONCLUDING REMARKS

Foods containing carotenoids are a very important part of the diet, because carotenoids play an important role not only in the color of human's food but also in its functionality/bioactivity. They include part of the defense system against free radicals, the excess of which accelerates the aging process, and the development of several diseases and disorders such as cancer, diabetes, and cardiovascular disease. Carotenoids are also necessary for some vital functions of the body such as eye health, where they are precursors of vitamin A. The most important carotenoids are β-carotene, as well as β-cryptoxanthin and α-carotene, which can be converted into retinol. Carotenoids are used as natural food colorants; however, the problem is that they can easily degrade to their inactive forms under various environmental conditions. Carotenoids are very sensitive to processing and storage conditions and may lose their nutritional value before they are consumed. Therefore, these bioactive compounds must be protected/encapsulated, in order to preserve their bioactivity as much as possible.

The microcapsules formed during the spray drying process protect carotenoids, especially from the access of oxygen, which promotes their stability. However, there are several factors associated with the spray drying encapsulation process, which can influence carotenoid contents (e.g., carrier type, mixing method, inlet and outlet air temperatures, humidity, disc speed, nozzle pressure, and feed flow rate). Therefore, it is necessary to carefully assess the conditions and select the most appropriate carriers and the drying conditions, to obtain a stable product with the highest content of carotenoids. Based on the information available so far, it can be concluded that the most favorable factor for improving the stability of carotenoids is the use of carrier mixture and possibly low process temperature, which can result in the formation of particles that limit the contact of carotenoids with oxygen.

REFERENCES

21: Food and Drugs §Subpart, A. PART 184—INDIRECT FOOD SUB-STANCES AFFIRMED AS GENERALLY RECOGNIZED AS SAFE. (2020). 06.2020. https://www.ecfr.gov/cgi-bin/text-idx?SID=7ffc7 beea43d4e1e0bd451c0d2815e77&mc=true&node=se21.3.184_11245&rgn=div8

Achir, N., Dhuique-Mayer, C., Hadjal, T., Madani, K., Pain, J.-P., & Dornier, M. (2016). Pasteurization of citrus juices with ohmic heating to preserve the carotenoid profile. *Innovative Food Science & Emerging Technologies*, 33, 397–404. https://doi.org/10.1016/j.ifset.2015.11.002

Achir, N., Randrianatoandro, V. A., Bohuon, P., Laffargue, A., & Avallone, S. (2010). Kinetic study of β-carotene and lutein degradation in oils during heat treatment. *European Journal of Lipid Science and Technology*, 112(3), 349–361. https://doi.org/10.1002/ejlt.200900165

Ahmed, F., Li, Y., Fanning, K., Netzel, M., & Schenk, P. M. (2015). Effect of drying, storage temperature and air exposure on astaxanthin stability from *Haematococcus pluvialis*. *Food Research International*, 74, 231–236. https://doi.org/10.1016/j.foodres.2015.05.021

Akhtar, H. M., & Bryan, M. (2008). Extraction and quantification of major carotenoids in processed foods and supplements by liquid chromatography. *Food Chemistry*, 111(1), 255–261. https://doi.org/10.1016/j.foodchem.2008.03.071

Álvarez-Henao, M. V., Saavedra, N., Medina, S., Jiménez Cartagena, C., Alzate, L. M., & Londoño-Londoño, J. (2018). Microencapsulation of lutein by spray-drying: Characterization and stability analyses to promote its use as a functional ingredient. *Food Chemistry*, 256, 181–187. https://doi.org/10.1016/j.foodchem.2018.02.059

Aman, R., Schieber, A., & Carle, R. (2005). Effects of heating and illumination on trans–cis isomerization and degradation of β-Carotene and Lutein in isolated spinach chloroplasts. *Journal of Agricultural and Food Chemistry*, 53(24), 9512–9518. https://doi.org/10.1021/jf050926w

Anandharamakrishnan, C., & Ishwarya, S. P. (2015). Selection of wall material for encapsulation by spray drying. *Spray Drying Techniques for Food Ingredient Encapsulation*, 77–100. https://doi.org/10.1002/9781118863985.ch4 (Wiley Online Books).

Arunkumar, R., Calvo, C. M., Conrady, C. D., & Bernstein, P. S. (2018). What do we know about the macular pigment in AMD: The past, the present, and the future. *Eye (London, England)*, 32(5), 992–1004. https://doi.org/10.1038/s41433-018-0044-0

Assadpour, E., & Jafari, S. M. (2019). Advances in spray-drying encapsulation of food bioactive ingredients: From microcapsules to nanocapsules. *Annual Review of Food Science and Technology*, 10, 103–131. https://doi.org/10.1146/annurev-food-032818-121641

Bell, J. G., McEvoy, J., Tocher, D. R., & Sargent, J. R. (2000). Depletion of α-Tocopherol and Astaxanthin in atlantic salmon (*Salmo salar*) affects autoxidative defense and fatty acid metabolism. *The Journal of Nutrition*, 130(7), 1800–1808. https://doi.org/10.1093/jn/130.7.1800

Bernstein, P. S., Khachik, F., Carvalho, L. S., Muir, G. J., Zhao, D.-Y., & Katz, N. B. (2001). Identification and quantitation of carotenoids and their metabolites in the tissues of the human eye. *Experimental Eye Research*, 72(3), 215–223. https://doi.org/10.1006/exer.2000.0954

Böhm, F., Edge, R., & Truscott, G. (2012). Interactions of dietary carotenoids with activated (singlet) oxygen and free radicals: Potential effects for human health. *Molecular Nutrition & Food Research*, 56(2), 205–216. https://doi.org/10.1002/mnfr.201100222

Böhm, V., Puspitasari-Nienaber, N. L., Ferruzzi, M. G., & Schwartz, S. J. (2002). Trolox Equivalent antioxidant capacity of different geometrical isomers of α-Carotene, β-Carotene, lycopene, and zeaxanthin. *Journal of Agricultural and Food Chemistry*, 50(1), 221–226. https://doi.org/10.1021/jf010888q

Boon, C. S., McClements, D. J., Weiss, J., & Decker, E. A. (2010). Factors influencing the chemical stability of carotenoids in foods. *Critical Reviews in Food Science and Nutrition*, 50(6), 515–532. https://doi.org/10.1080/10408390802565889

Çalışkan Koç, G., & Nur Dirim, S. (2017). Spray drying of spinach juice: Characterization, chemical composition, and storage. *Journal of Food Science*, 82(12), 2873–2884. https://doi.org/10.1111/1750-3841.13970

Carpentier, S., Knaus, M., & Suh, M. (2009). Associations between lutein, zeaxanthin, and age-related macular degeneration: An overview. *Critical Reviews in Food Science and Nutrition*, 49(4), 313–326. https://doi.org/10.1080/10408390802066979

Chábera, P., Fuciman, M., Hříbek, P., & Polívka, T. (2009). Effect of carotenoid structure on excited-state dynamics of carbonyl carotenoids. *Physical Chemistry Chemical Physics*, 11(39), 8795–8803. https://doi.org/10.1039/B909924G

Choudhari, Sheetal, Bajaj, I., Singhal, R., & Karwe, M. (2012). Microencapsulated lycopene for pre-extrusion coloring of foods. *Journal of Food Process Engineering*, *35*(1), 91–103. https://doi.org/10.1111/j.1745-4530.2010.00562.x

Chranioti, C., Nikoloudaki, A., & Tzia, C. (2015). Saffron and beetroot extracts encapsulated in maltodextrin, gum Arabic, modified starch and chitosan: Incorporation in a chewing gum system. *Carbohydrate Polymers*, *127*, 252–263. https://doi.org/10.1016/j.carbpol.2015.03.049

Chuyen, H. V., Roach, P. D., Golding, J. B., Parks, S. E., & Nguyen, M. H. (2019). Encapsulation of carotenoid-rich oil from Gac peel: Optimisation of the encapsulating process using a spray drier and the storage stability of encapsulated powder. *Powder Technology*, *344*, 373–379. https://doi.org/10.1016/j.powtec.2018.12.012

Chylack, L. T., Brown, N. P., Bron, A., Hurst, M., Köpcke, W., Thien, U., & Schalch, W. (2002). The roche European American cataract trial (REACT): A randomized clinical trial to investigate the efficacy of an oral antioxidant micronutrient mixture to slow progression of age-related cataract. *Ophthalmic Epidemiology*, *9*(1), 49–80. https://doi.org/10.1076/opep.9.1.49.1717

Corrêa-Filho, L. C., Lourenço, M. M., Moldão-Martins, M., & Alves, V. D. (2019). Microencapsulation of β-Carotene by spray drying: Effect of wall material concentration and drying inlet temperature [Research Article]. *International Journal of Food Science*, *2019*. https://doi.org/10.1155/2019/8914852

Dhuique-Mayer, C., Tbatou, M., Carail, M., Caris-Veyrat, C., Dornier, M., & Amiot, M. J. (2007). Thermal degradation of antioxidant micronutrients in citrus juice: Kinetics and newly formed compounds. *Journal of Agricultural and Food Chemistry*, *55*(10), 4209–4216. https://doi.org/10.1021/jf0700529

Díaz, D. I., Lugo, E., Pascual-Pineda, L. A., & Jiménez-Fernández, M. (2019). Encapsulation of carotenoid-rich paprika oleoresin through traditional and nano spray drying. *Italian Journal of Food Science*, *31*(1), 125–138. https://doi.org/10.14674/ijfs-1253

Drosou, C. G., Krokida, M. K., & Biliaderis, C. G. (2017). Encapsulation of bioactive compounds through electrospinning/electrospraying and spray drying: A comparative assessment of food-related applications. *Drying technology*, *35*(2), 139–162. https://doi.org/10.1080/07373937.2016.1162797

Dwyer, J. H., Navab, M., Dwyer, K. M., Hassan, K., Sun, P., Shircore, A., Hama-Levy, S., Hough, G., Wang, X., Drake, T., Merz, C. N. B., & Fogelman, A. M. (2001). Oxygenated carotenoid lutein and progression of early atherosclerosis. *Circulation*, *103*(24), 2922–2927. https://doi.org/doi:10.1161/01.CIR.103.24.2922

Edge, R., & Truscott, T. G. (2018). Singlet oxygen and free radical reactions of retinoids and carotenoids - A review. *Antioxidants (Basel, Switzerland)*, *7*(1), 5. https://doi.org/10.3390/antiox7010005

Eghbaliferiz, S., & Iranshahi, M. (2016). Prooxidant activity of polyphenols, flavonoids, anthocyanins and carotenoids: Updated review of mechanisms and catalyzing metals. *Phytotherapy Research*, *30*(9), 1379–1391. https://doi.org/10.1002/ptr.5643

Etzbach, L., Meinert, M., Faber, T., Klein, C., Schieber, A., & Weber, F. (2020). Effects of carrier agents on powder properties, stability of carotenoids, and encapsulation efficiency of goldenberry (*Physalis peruviana* L.) powder produced by co-current spray drying. *Current Research in Food Science*, *3*, 73–81. https://doi.org/10.1016/j.crfs.2020.03.002

de Abreu, F. P., Dornier, M., Dionisio, A. P., Carail, M., Caris-Veyrat, C., & Dhuique-Mayer, C. (2013). Cashew apple (*Anacardium occidentale* L.) extract from by-product of juice processing: A focus on carotenoids. *Food Chemistry*, *138*(1), 25–31. https://doi.org/10.1016/j.foodchem.2012.10.028

Fabiani, R., De Bartolomeo, A., Rosignoli, P., & Morozzi, G. (2001). Antioxidants prevent the lymphocyte DNA damage induced by PMA-stimulated monocytes. *Nutrition and Cancer*, *39*(2), 284–291. https://doi.org/10.1207/S15327914nc392_19

Failla, M. L., Chitchumroonchokchai, C., & Ishida, B. K. (2008). In vitro micellarization and intestinal cell uptake of cis isomers of lycopene exceed those of all-trans lycopene. *The Journal of Nutrition*, *138*(3), 482–486. https://doi.org/10.1093/jn/138.3.482

Fang, S., Zhao, X., Liu, Y., Liang, X., & Yang, Y. (2019). Fabricating multilayer emulsions by using OSA starch and chitosan suitable for spray drying: Application in the encapsulation of β-carotene. *Food Hydrocolloids*, *93*, 102–110. https://doi.org/10.1016/j.foodhyd.2019.02.024

Fernandes, A. S., do Nascimento, T. C., Jacob-Lopes, E., De Rosso, V. V., & Zepka, L. Q. (2018). Carotenoids: A brief overview on its structure, biosynthesis, synthesis, and applications. In Zepka Leila Queiroz, J.-L. Eduardo, & D. R. V. Vera (Eds.), *Progress in Carotenoid Research* (pp. 1). IntechOpen. https://doi.org/10.5772/intechopen.79542

Fuhrman, B., Volkova, N., Rosenblat, M., & Aviram, M. (2000). Lycopene synergistically Inhibits LDL oxidation in combination with vitamin E, glabridin, rosmarinic acid, carnosic acid, or garlic. *Antioxidants & Redox Signaling*, *2*(3), 491–506. https://doi.org/10.1089/15230860050192279

García, J. M., Giuffrida, D., Dugo, P., Mondello, L., & Osorio, C. (2018). Development and characterisation of carotenoid-rich microencapsulates from tropical fruit by-products and yellow tamarillo (*Solanum betaceum* Cav.). *Powder Technology*, *339*, 702–709. https://doi.org/10.1016/j.powtec.2018.08.061

Goula, A. M., & Adamopoulos, K. G. (2005). Spray drying of tomato pulp in dehumidified air: II. The effect on powder properties. *Journal of Food Engineering*, *66*(1), 35–42. https://doi.org/10.1016/j.jfoodeng.2004.02.031

Goula, A. M., & Adamopoulos, K. G. (2012). A New technique for spray dried encapsulation of lycopene. *Drying Technology*, *30*(6), 641–652. https://doi.org/10.1080/07373937.2012.655871

Goula, A. M., Adamopoulos, K. G., & Kazakis, N. A. (2004). Influence of spray drying conditions on tomato powder properties. *Drying Technology*, *22*(5), 1129–1151. https://doi.org/10.1081/DRT-120038584

van den Berg, H., Faulks, R., Granado, H. F., Hirschberg, J., Olmedilla, B., Sandmann, G., Southon, S., & Stahl, W. (2000). The potential for the improvement of carotenoid levels in foods and the likely systemic effects. *Journal of the Science of Food and Agriculture*, *80*(7), 880–912. https://doi.org/10.1002/(SICI)1097-0010(20000515)80:7<880::AID-JSFA646>3.0.CO;2-1

Haas, K., Obernberger, J., Zehetner, E., Kiesslich, A., Volkert, M., & Jaeger, H. (2019). Impact of powder particle structure on the oxidation stability and color of encapsulated crystalline and emulsified carotenoids in carrot concentrate powders. *Journal of Food Engineering*, *263*, 398–408. https://doi.org/10.1016/j.jfoodeng.2019.07.025

van Het Hof, K. H., West, C. E., Weststrate, J. A., & Hautvast, J. G. A. J. (2000). Dietary factors that affect the bioavailability of carotenoids. *The Journal of Nutrition*, *130*(3), 503–506. https://doi.org/10.1093/jn/130.3.503

Hiranvarachat, B., Suvarnakuta, P., & Devahastin, S. (2008). Isomerisation kinetics and antioxidant activities of β-carotene in carrots undergoing different drying techniques and conditions. *Food Chemistry*, *107*(4), 1538–1546. https://doi.org/10.1016/j.foodchem.2007.10.026

Holst, B., & Williamson, G. (2004). A critical review of the bioavailability of glucosinolates and related compounds. *Natural Product Reports*, *21*(3), 425–447. https://doi.org/10.1039/B204039P

Hynstova, V., Sterbova, D., Klejdus, B., Hedbavny, J., Huska, D., & Adam, V. (2018). Separation, identification and quantification of carotenoids and chlorophylls in dietary supplements containing *Chlorella vulgaris* and Spirulina platensis using high performance thin layer chromatography. *Journal of Pharmaceutical and Biomedical Analysis*, *148*, 108–118. https://doi.org/10.1016/j.jpba.2017.09.018

Jacques, P. F., Lyass, A., Massaro, J. M., Vasan, R. S., & D'Agostino, R. B., Sr. (2013). Relationship of lycopene intake and consumption of tomato products to incident CVD. *The British journal of Nutrition*, *110*(3), 545–551. https://doi.org/10.1017/S0007114512005417

Janiszewska-Turak, E. (2017). Carotenoids microencapsulation by spray drying method and supercritical micronization. *Food Research International*, *99*, 891–901. https://doi.org/10.1016/j.foodres.2017.02.001

Janiszewska-Turak, E., Dellarosa, N., Tylewicz, U., Laghi, L., Romani, S., Dalla Rosa, M., & Witrowa-Rajchert, D. (2017). The influence of carrier material on some physical and structural properties of carrot juice microcapsules. *Food Chemistry*, *236*, 134–141. https://doi.org/10.1016/j.foodchem.2017.03.134

Janiszewska-Turak, E., & Witrowa-Rajchert, D. (2020). The influence of carrot pretreatment, type of carrier and disc speed on the physical and chemical properties of spray dried carrot juice microcapsules. *Drying Technology*, 1–11. https://doi.org/10.1080/07373937.2019.1705850

Jayaraj, J., Devlin, R., & Punja, Z. (2008). Metabolic engineering of novel ketocarotenoid production in carrot plants. *Transgenic Research*, *17*(4), 489–501. https://doi.org/10.1007/s11248-007-9120-0

Jedlińska, A., Samborska, K., Wieczorek, A., Wiktor, A., Ostrowska-Ligęza, E., Jamróz, W., Skwarczyńska-Maj, K., Kiełczewski, D., Błażowski, Ł., Tułodziecki, M., & Witrowa-Rajchert, D. (2019). The application of dehumidified air in rapeseed and honeydew honey spray drying - Process performance and powders properties considerations. *Journal of Food Engineering*, *245*, 80–87. https://doi.org/10.1016/j.jfoodeng.2018.10.017

Jung, H. Y., Ok, H. M., Park, M. Y., Kim, J. Y., & Kwon, O. (2016). Bioavailability of carotenoids from chlorella powder in healthy subjects: A comparison with marigold petal extract. *Journal of Functional Foods*, *21*, 27–35. https://doi.org/10.1016/j.jff.2015.11.036

Keshani, S., Daud, W. R. W., Nourouzi, M. M., Namvar, F., & Ghasemi, M. (2015). Spray drying: An overview on wall deposition, process and modeling. *Journal of Food Engineering*, *146*, 152–162. https://doi.org/10.1016/j.jfoodeng.2014.09.004

Khalil, M., Raila, J., Ali, M., Islam, K. M. S., Schenk, R., Krause, J.-P., Schweigert, F. J., & Rawel, H. (2012). Stability and bioavailability of lutein ester supplements from Tagetes flower prepared under food processing conditions. *Journal of Functional Foods*, *4*(3), 602–610. https://doi.org/10.1016/j.jff.2012.03.006

Khalilian Movahhed, M., & Mohebbi, M. (2016). Spray drying and process optimization of carrot–celery juice. *Journal of Food Processing and Preservation*, *40*(2), 212–225.

Knockaert, G., Pulissery, S. K., Lemmens, L., Van Buggenhout, S., Hendrickx, M., & Van Loey, A. (2012). Carrot β-carotene degradation and isomerization kinetics during thermal processing in the presence of oil. *Journal of Agricultural and Food Chemistry*, *60*(41), 10312–10319. https://doi.org/10.1021/jf3025776

Koca, N., Burdurlu, H. S., & Karadeniz, F. (2007). Kinetics of colour changes in dehydrated carrots. *Journal of Food Engineering*, *78*(2), 449–455. https://doi.org/10.1016/j.jfoodeng.2005.10.014

Kozuki, Y., Miura, Y., & Yagasaki, K. (2000). Inhibitory effects of carotenoids on the invasion of rat ascites hepatoma cells in culture. *Cancer Letters*, *151*(1), 111–115. https://doi.org/10.1016/S0304-3835(99)00418-8

Krinsky, N. I., & Johnson, E. J. (2005). Carotenoid actions and their relation to health and disease. *Molecular Aspects of Medicine*, *26*(6), 459–516. https://doi.org/10.1016/j.mam.2005.10.001

Kyriakoudi, A., & Tsimidou, M. Z. (2018). Properties of encapsulated saffron extracts in maltodextrin using the Büchi B-90 nano spray-dryer. *Food Chemistry*, *266*, 458–465. https://doi.org/10.1016/j.foodchem.2018.06.038

de Carvalho, L. M. J., Ortiz, G. M. D., de Carvalho, J. L. V., & de Oliveira, A. R. G. (2018). Carotenoids in raw plant materials. In Zepka Leila Queiroz, J.-L. Eduardo, & D. R. V. Vera (Eds.), *Progress in Carotenoid Research*. IntechOpen. https://doi.org/10.5772/intechopen.78677.

Lee, E. H., Faulhaber, D., Hanson, K. M., Ding, W., Peters, S., Kodali, S., & Granstein, R. D. (2004). Dietary lutein reduces ultraviolet radiation-induced inflammation and immunosuppression. *Journal of Investigative Dermatology*, *122*(2), 510–517. https://doi.org/10.1046/j.0022-202X.2004.22227.x

Lee, H. S., & Kim, M. K. (2009). Effect of *Chlorella vulgaris* on glucose metabolism in wistar rats fed high fat diet. *Journal of Medicinal Food*, *12*(5), 1029–1037. https://doi.org/10.1089/jmf.2008.1269

Li, B., Rognon, G. T., Mattinson, T., Vachali, P. P., Gorusupudi, A., Chang, F.-Y., Ranganathan, A., Nelson, K., George, E. W., Frederick, J. M., & Bernstein, P. S. (2018). Supplementation with macular carotenoids improves visual performance of transgenic mice. *Archives of Biochemistry and Biophysics*, *649*, 22–28. https://doi.org/10.1016/j.abb.2018.05.003

Lim, A. S. L., Burdikova, Z., Sheehan, J. J., & Roos, Y. H. (2016). Carotenoid stability in high total solid spray dried emulsions with gum Arabic layered interface and trehalose–WPI composites as wall materials. *Innovative Food Science & Emerging Technologies*, *34*, 310–319. https://doi.org/10.1016/j.ifset.2016.03.001

Lim, A. S. L., & Roos, Y. H. (2016). Spray drying of high hydrophilic solids emulsions with layered interface and trehalose-maltodextrin as glass formers for carotenoids stabilization. *Journal of Food Engineering*, *171*, 174–184. https://doi.org/10.1016/j.jfoodeng.2015.10.026

Lokesh, V., Divya, P., Puthusseri, B., Manjunatha, G., & Neelwarne, B. (2014). Profiles of carotenoids during post-climacteric ripening of some important cultivars of banana and development of a dry product from a high carotenoid yielding variety. *LWT – Food Science and Technology*, *55*(1), 59–66. https://doi.org/10.1016/j.lwt.2013.09.005

Mahdavi, S. A., Jafari, S. M., Ghorbani, M., & Assadpoor, E. (2014). Spray-drying microencapsulation of anthocyanins by natural biopolymers: A review. *Drying technology*, *32*(5), 509–518. https://doi.org/10.1080/07373937.2013.839562

Milani, A., Basirnejad, M., Shahbazi, S., & Bolhassani, A. (2017). Carotenoids: biochemistry, pharmacology and treatment. *British Journal of Pharmacology*, *174*(11), 1290–1324. https://doi.org/10.1111/bph.13625

Montero, P., Calvo, M. M., Gómez-Guillén, M. C., & Gómez-Estaca, J. (2016). Microcapsules containing astaxanthin from shrimp waste as potential food coloring and functional ingredient: Characterization, stability, and bioaccessibility. *LWT – Food Science and Technology*, *70*, 229–236. https://doi.org/10.1016/j.lwt.2016.02.040

Mora, J. R., Iwata, M., & von Andrian, U. H. (2008). Vitamin effects on the immune system: Vitamins A and D take centre stage. *Nature Reviews Immunology*, *8*(9), 685–698. https://doi.org/10.1038/nri2378

Morris, W. L., Ducreux, L. J., Fraser, P. D., Millam, S., & Taylor, M. A. (2006). Engineering ketocarotenoid biosynthesis in potato tubers. *Metabolic Engineering*, *8*(3), 253–263. https://doi.org/10.1016/j.ymben.2006.01.001

Murillo, E., Giuffrida, D., Menchaca, D., Dugo, P., Torre, G., Melendez-Martinez, A. J., & Mondello, L. (2013). Native carotenoids composition of some tropical fruits. *Food Chemistry*, *140*(4), 825–836. https://doi.org/10.1016/j.foodchem.2012.11.014

No, J., & Shin, M. (2019). Preparation and characteristics of octenyl succinic anhydride-modified partial waxy rice starches and encapsulated paprika pigment powder. *Food Chemistry*, *295*, 466–474. https://doi.org/10.1016/j.foodchem.2019.05.064

Ordonez-Santos, L. E., Martinez-Giron, J., & Villamizar-Vargas, R. H. (2018). Encapsulation of β-carotene extracted from peach palm residues: A stability study using two spray dried processes. *DYNA*, *85*(206), 128–134. https://doi.org/10.15446/dyna.v85n206.68089.

Paine, J. A., Shipton, C. A., Chaggar, S., Howells, R. M., Kennedy, M. J., Vernon, G., Wright, S. Y., Hinchliffe, E., Adams, J. L., Silverstone, A. L., & Drake, R. (2005). Improving the nutritional value of Golden Rice through increased pro-vitamin A content. *Nature Biotechnology*, *23*(4), 482–487. https://doi.org/10.1038/nbt1082

Panasenko, O. M., Sharov, V. S., Briviba, K., & Sies, H. (2000). Interaction of peroxynitrite with carotenoids in human low density lipoproteins. *Archives of Biochemistry and Biophysics*, *373*(1), 302–305. https://doi.org/10.1006/abbi.1999.1424

Panfili, G., Fratianni, A., & Irano, M. (2004). Improved normal-phase high-performance liquid chromatography procedure for the determination of carotenoids in cereals. *Journal of Agricultural and Food Chemistry*, *52*(21), 6373–6377. https://doi.org/10.1021/jf0402025

Perera, C. O., & Yen, G. M. (2007). Functional properties of carotenoids in human health. *International Journal of Food Properties*, *10*(2), 201–230. https://doi.org/10.1080/10942910601045271

Pérez-Gálvez, A., & Mínguez-Mosquera, M. I. (2004). Degradation, under non-oxygen-mediated autooxidation, of carotenoid profile present in Paprika Oleoresins with lipid substrates of different fatty acid composition. *Journal of Agricultural and Food Chemistry*, *52*(3), 632–637. https://doi.org/10.1021/jf0351063

Przybysz, M. A., Szterk, A., Symoniuk, E., Gąszczyk, M., & Dłużewska, E. (2018). α- and β-carotene stability during storage of microspheres obtained from spray dried microencapsulation technology. *Polish Journal of Food and Nutrition Sciences*, *68*, 45–55. https://doi.org/10.1515/pjfns-2017-0006

Qian, C., Decker, E. A., Xiao, H., & McClements, D. J. (2012). Inhibition of β-carotene degradation in oil-in-water nanoemulsions: Influence of oil-soluble and water-soluble antioxidants. *Food Chemistry*, *135*(3), 1036–1043. https://doi.org/10.1016/j.foodchem.2012.05.085

Rajabi, H., Ghorbani, M., Jafari, S. M., Sadeghi Mahoonak, A., & Rajabzadeh, G. (2015). Retention of saffron bioactive components by spray drying encapsulation using maltodextrin, gum Arabic and gelatin as wall materials. *Food Hydrocolloids*, *51*, 327–337. https://doi.org/10.1016/j.foodhyd.2015.05.033

Raju, M., Varakumar, S., Lakshminarayana, R., Krishnakantha, T. P., & Baskaran, V. (2007). Carotenoid composition and vitamin A activity of medicinally important green leafy vegetables. *Food Chemistry*, *101*(4), 1598–1605. https://doi.org/10.1016/j.foodchem.2006.04.015

Ramakrishnan, Y., Adzahan, N. M., Yusof, Y. A., & Muhammad, K. (2018). Effect of wall materials on the spray drying efficiency, powder properties and stability of bioactive compounds in tamarillo juice microencapsulation. *Powder Technology*, *328*, 406–414. https://doi.org/10.1016/j.powtec.2017.12.018

Ray, S., Raychaudhuri, U., & Chakraborty, R. (2016). An overview of encapsulation of active compounds used in food products by drying technology. *Food Bioscience*, *13*, 76–83. https://doi.org/10.1016/j.fbio.2015.12.009

Ribeiro, D., Freitas, M., Silva, A. M. S., Carvalho, F., & Fernandes, E. (2018). Antioxidant and pro-oxidant activities of carotenoids and their oxidation products. *Food and Chemical Toxicology*, *120*, 681–699. https://doi.org/10.1016/j.fct.2018.07.060

Rivera-Pastrana, D. M., Yahia, E. M., & Gonzalez-Aguilar, G. A. (2010). Phenolic and carotenoid profiles of papaya fruit (*Carica papaya* L.) and their contents under low temperature storage. *Journal of the Science of Food and Agriculture*, *90*(14), 2358–2365. https://doi.org/10.1002/jsfa.4092

Rocha, G. A., Fávaro-Trindade, C. S., & Grosso, C. R. F. (2012). Microencapsulation of lycopene by spray drying: Characterization, stability and application of microcapsules. *Food and Bioproducts Processing*, *90*(1), 37–42. https://doi.org/10.1016/j.fbp.2011.01.001

Rowles III, J. L., & Erdman Jr., J. W. (2020). Carotenoids and their role in cancer prevention. *Biochimica et Biophysica Acta (BBA)-Molecular and Cell Biology of Lipids*, 1865, 158613.

da Silva, S. C., Fernandes, I. P., Barros, L., Fernandes, Â., José Alves, M., Calhelha, R. C., Pereira, C., Barreira, J. C. M., Manrique, Y., Colla, E., Ferreira, I. C. F. R., & Filomena Barreiro, M. (2019). Spray dried Spirulina platensis as an effective ingredient to improve yogurt formulations: Testing different encapsulating solutions. *Journal of Functional Foods*, *60*, 103427. https://doi.org/10.1016/j.jff.2019.103427

Shao, A., & Hathcock, J. N. (2006). Risk assessment for the carotenoids lutein and lycopene. *Regulatory Toxicology and Pharmacology*, *45*(3), 289–298.

Shishir, M. R. I., & Chen, W. (2017). Trends of spray drying: A critical review on drying of fruit and vegetable juices. *Trends in Food Science & Technology*, *65*, 49–67. https://doi.org/10.1016/j.tifs.2017.05.006

Shrestha, A. K., Ua-Arak, T., Adhikari, B. P., Howes, T., & Bhandari, B. R. (2007). Glass transition behavior of spray dried orange juice powder measured by differential scanning calorimetry (DSC) and thermal mechanical compression test (TMCT) [research-article]. *International Journal of Food Properties, 10*(3), 661–673. https://doi.org/10.1080/10942910601109218

Sidhu, G. K., Singh, M., & Kaur, P. (2019). Effect of operational parameters on physicochemical quality and recovery of spray dried tomato powder. *Journal of Food Processing and Preservation, 43*(10), e14120. https://doi.org/10.1111/jfpp.14120

Siebenhandl, S., Grausgruber, H., Pellegrini, N., Del Rio, D., Fogliano, V., Pernice, R., & Berghofer, E. (2007). Phytochemical profile of main antioxidants in different fractions of purple and blue wheat, and black barley. *Journal of Agricultural and Food Chemistry, 55*(21), 8541–8547. https://doi.org/10.1021/jf072021j

Silaste, M.-L., Alfthan, G., Aro, A., Antero Kesäniemi, Y., & Hörkkö, S. (2007). Tomato juice decreases LDL cholesterol levels and increases LDL resistance to oxidation. *British Journal of Nutrition, 98*(6), 1251–1258. https://doi.org/10.1017/S0007114507787445

Skibsted, L. H. (2012). Carotenoids in antioxidant networks. colorants or radical scavengers. *Journal of Agricultural and Food Chemistry, 60*(10), 2409–2417. https://doi.org/10.1021/jf2051416

Sluijs, I., Cadier, E., Beulens, J. W. J., van der, A. D. L., Spijkerman, A. M. W., & Van Der Schouw, Y. T. (2015). Dietary intake of carotenoids and risk of type 2 diabetes. *Nutrition, Metabolism and Cardiovascular Diseases, 25*(4), 376–381. https://doi.org/10.1016/j.numecd.2014.12.008

Souza, A. L. R., Hidalgo-Chávez, D. W., Pontes, S. M., Gomes, F. S., Cabral, L. M. C., & Tonon, R. V. (2018). Microencapsulation by spray drying of a lycopene-rich tomato concentrate: Characterization and stability. *LWT - Food Science and Technology, 91*, 286–292. https://doi.org/10.1016/j.lwt.2018.01.053

Stahl, W., Ale-Agha, N., & Polidori, M. C. (2002). Non-Antioxidant properties of carotenoids. *Biological Chemistry, 383*(3–4), 553. https://doi.org/10.1515/BC.2002.056

Syamila, M., Gedi, M. A., Briars, R., Ayed, C., & Gray, D. A. (2019). Effect of temperature, oxygen and light on the degradation of β-carotene, lutein and α-tocopherol in spray dried spinach juice powder during storage. *Food Chemistry, 284*, 188–197. https://doi.org/10.1016/j.foodchem.2019.01.055

Tang, G., Ferreira, A. L. A., Grusak, M. A., Qin, J., Dolnikowski, G. G., Russell, R. M., & Krinsky, N. I. (2005). Bioavailability of synthetic and biosynthetic deuterated lycopene in humans. *The Journal of Nutritional Biochemistry, 16*(4), 229–235. https://doi.org/10.1016/j.jnutbio.2004.11.007

Troya, D., Tupuna-Yerovi, D. S., & Ruales, J. (2018). Effects of wall materials and operating parameters on physicochemical properties, process efficiency, and total carotenoid content of microencapsulated banana passionfruit pulp (*Passiflora tripartita var. mollissima*) by Spray-drying. *Food and Bioprocess Technology, 11*(10), 1828–1839. https://doi.org/10.1007/s11947-018-2143-0

Tupuna, D. S., Paese, K., Guterres, S. S., Jablonski, A., Flôres, S. H., & Rios, A. d. O. (2018). Encapsulation efficiency and thermal stability of norbixin microencapsulated by spray-drying using different combinations of wall materials. *Industrial Crops and Products, 111*, 846–855. https://doi.org/10.1016/j.indcrop.2017.12.001

Vila, M. M. D. C., Chaud, M. V., & Balcão, V. M. (2015). Chapter 19 - Microencapsulation of natural anti-oxidant pigments A2 – Sagis. M.C. Leonard In *Microencapsulation and Microspheres for Food Applications* (pp. 369–389). Academic Press. https://doi.org/10.1016/B978-0-12-800,350-3.00024-8

Young, A. J., & Lowe, G. M. (2001). Antioxidant and prooxidant properties of carotenoids. *Archives of Biochemistry and Biophysics, 385*(1), 20–27. https://doi.org/10.1006/abbi.2000.2149

Yu, S.-H., Tsai, M.-L., Lin, B.-X., Lin, C.-W., & Mi, F.-L. (2015). Tea catechins-cross-linked methylcellulose active films for inhibition of light irradiation and lipid peroxidation induced β-carotene degradation. *Food Hydrocolloids, 44*, 491–505. https://doi.org/10.1016/j.foodhyd.2014.10.022

6 Spray Drying Encapsulation of Vitamins

Zeynep Saliha Gunes and Merve Tomas
Istanbul Sabahattin Zaim University, Turkey

Asli Can Karaca, Beraat Ozcelik, and Esra Capanoglu
Istanbul Technical University, Turkey

CONTENTS

6.1 INTRODUCTION

The use of bioactive compoundds for the fortification purposes in the food and drug industry has been popular nowadays. The crucial aspects of bioactive ingredients, including antimicrobial, anti-fungal, anticancer, and immune system enhancer properties are essential for biomedical purposes. Vitamins are essential nutritional compounds which are included in biochemical reactions in the human body. They have to be taken to the body through the diet because they cannot be synthesized in the body. Deficiency of vitamins can cause health problems and deficiency diseases such as anemia, scurvy, nyctalopia, and dermatitis (Murugesan & Orsat, 2011; Pocobelli, Peters, Kristal, & White, 2009). Like most bioactives, vitamins are sensitive against light, temperature, and oxygen; hence, they can be partially or completely denatured during cooking, processing, and storage. Some vitamins are more sensitive to these environmental conditions than others. For example, high-temperature processing affects heat-sensitive folate, vitamin B6, and vitamin C rather than vitamin A, vitamin B1, B2, and B3. Vitamins can also be damaged because of inconvenient storage conditions, in addition to processing steps, including thawing, chopping, and washing. Therefore, necessary precautions should be taken for protecting vitamins during food processing (Lawson, Hunt, & Glew, 1983; Murugesan & Orsat, 2011; Williams, 1996).

The encapsulation process is defined as entrapping bioactive ingredients into coating materials in nanometer, micrometer, and millimeter-scale and this technology has been properly used to get liquid or solid materials for an influential barrier toward environmental conditions such as temperature, light, oxygen, and free radicals (Burgain, Gaiani, Linder, & Scher, 2011; Desai & Park, 2005;

Lakkis, 2007; Ray, Raychaudhuri, & Chakraborty, 2016). Encapsulation can also be described as a coating process of core or bioactive material within a thin continuous wall (Desai & Park, 2006). Encapsulation allows the bioactive material to be protected against environmental conditions with a wall or matrix material. This technique is effectively used for protecting sensitive core materials, controlled the release of bioactive compounds, and masking undesirable odors. Spray drying is a method that turns the liquid suspensions into solid micro-particles using hot air. Spray drying is widely used in the food industry to produce encapsulated bioactive materials in powder form, because of easy set-up, cost-effectiveness, continuousness, one-step, and compliable processing.

The spray drying method has a basic operation principle that is related to atomization through a hot drying chamber of the solution including the bioactive compound. Solid microparticles are obtained by vaporizing the solvent quickly with a stream of heated air. Spray drying has an advantage over the other encapsulation methods because microparticles are generally produced organic solvent-free (Esposito, Cervellati, Menegatti, Nastruzzi, & Cortesi, 2002; Masters, 1991; Palmieri, Bonacucina, Martino, & Martelli, 2001). Spray drying can be properly applied for water-soluble and insoluble compounds, heat-sensitive, and heat-resistant bioactive complexes, or both hydrophilic and hydrophobic biopolymers (Esposito et al., 2002).

Spray drying process conditions have been shown to affect the characteristic features of the obtained powder. The inlet and outlet temperature, feed flow rate, atomization pressure, and type of atomizers (disk or nozzle) are among the factors manipulated during the drying process. Moisture content and water activity are also important characteristics for spray dried powders to provide information on the efficiency of the drying process (Rezvankhah, Emam-Djomeh, & Askari, 2019).

In the case of lipophilic bioactive ingredients, preparation of the emulsion system before spray drying is essential to get high-quality powder. The correct selection of coating material with the optimum amount of protein, which has good emulsification properties, provides well-preserved microparticles (Rezvankhah et al., 2019). The nutritive value of both hydrophilic and hydrophobic vitamins is remarkably preserved by the spray drying method and choice of wall materials (Hartman, Akeson, & Stahmann, 1967; Ray et al., 2016). Milk proteins, milk serum, and plant-based proteins are among the most commonly used protein-based coating materials used as encapsulating agents, while polysaccharides and lipid-based wall materials such as maltodextrin, starches, gum Arabic, beeswax, and stearic acids are also applied for the spray drying encapsulation of vitamins (Murugesan & Orsat, 2011; Ray et al., 2016; Wilson & Shah, 2007; Zuidam & Heinrich, 2009). Water-soluble vitamins can be effectively coated by ethyl cellulose as a wall material due to its hydrophobicity and the permeability of vitamins is decreased by thicker capsule material (Wilson & Shah, 2007). The summary of processing parameters applied for the encapsulation of water- and fat-soluble vitamins by the spray drying method is presented in Tables 6.1 and 6.2, respectively.

TABLE 6.1

A Summary of Process Parameters Applied for the Encapsulation of Water-Soluble Vitamins by Spray Drying

Core Material	Wall Material(S)	Feed Flow Rate	T_{inlet} (°C)	T_{out} (°C)	Encapsulation Efficiency (EE, %) or Production Yield (PY, %)	Particle Size	References
Vitamin B12	Cyanobacterial polymer or combination of the polymer with gum Arabic	4 ml/ min	120	65	PY: 4–18.8	8 μm	Estevinho et al. (2019)

Core Material	Wall Material(S)	Feed Flow Rate	T_{inlet} (°C)	T_{out} (°C)	Encapsulation Efficiency (EE, %) or Production Yield (PY, %)	Particle Size	References
Vitamin B12	Gum Arabic, cashew nut gum, sodium alginate, carboxymethyl cellulose, and Eudragit® RS100	130 L/min	120	50–60	NA	0.2–5.5 μm	Oliveira, Guimarães, Cerize, Tunussi, and Poço (2013)
Vitamin B12	Chitosan, modified chitosan, and sodium alginate	4 mL/min	120	60	PY: 47	3 μm	Estevinho and Rocha (2016)
Thiamine (vitamin B1) and pyridoxine (Vitamin B6)	Ferulic acid-grafted chitosan	10 mL/min	140	77	PY: 63.58–65.12 EE: 91–83	4.5–4.8 μm	Chatterjee et al. (2016)
Folic acid (Vitamin B9)	Guar gum, whey protein, and resistant starch	140 L/h	90	45	EE: 52.5–83.9	0.2–4.5 μm	Pérez-Masiá et al. (2015)
L-ascorbic acid	Sodium alginate and gum Arabic	2–7 mL/min	140	86	EE: > 82 (gum Arabic) EE: >90 (sodium alginate) PY: 61.7–83.2 (gum Arabic) PY: 37.8–75.7 (sodium alginate)	5.13–14.09 μm (sodium alginate) 2.88–9.73 μm (gum Arabic)	Barra et al. (2019)
Ascorbic acid	Gum Arabic	8 mL/min	150	75	EE: 100.8	9.3 μm	Alvim, Stein, Koury, Dantas, and Carla Léa De Camargo Vianna (2016)
Vitamin C	Chitosan, modified chitosan, and sodium alginate	4 mL/min	120	60	PY: 45	3 μm	Estevinho and Rocha (2016)
L-ascorbic acid	Sodium alginate	20%	110–125	NA	EE: 93.48% PY: 30%	5–100 μm	Marcela, Lucía, Esther, and Elena (2016)
Ascorbic acid	Non-modified and modified soy protein isolate	0.47 m³/h	124	74	EE: 57%–92% PY:92%	4.8–9.3 μm	Nesterenko, Alric, Silvestre, and Durrieu (2014)
Ascorbic acid	Alginic acid and low methoxyl pectin mixed with rice starch	7–14 mL/min	80–120	70–95	NA	10–83 μm	Nizori, Bui, and Small (2012)

(Continued)

TABLE 6.1
(Continued)

Core Material	Wall Material(S)	Feed Flow Rate	T_{inlet} (°C)	T_{out} (°C)	Encapsulation Efficiency (EE, %) or Production Yield (PY, %)	Particle Size	References
Ascorbic acid	Pea protein concentrate carboxymethylcellulose, and maltodextrin	1 l/h	184	93	PY: >84	< 8 μm	Pierucci, Andrade, Baptista, Volpato, and Rocha-Leão (2006)
Ascorbic acid	Modified starch	20 ml/min	190	90	EE: 98–100 PY: 10–20	4–8 μm	Finotelli and Rocha-Leão (2005)
Vitamin C	Malt dextrin and gum Arabic	25–30 mL/min	100–160	85	PY: 24–26	NA	Dib Taxi, De, Santos, and Grosso (2003)
Vitamin C	Methacrylate copolymers	600 l/h	105	80	EE: 98–100	4.73–19.43 μm	Esposito et al. (2002)
Ascorbic acid	β-cyclodextrin and starch	10–20 mL/min	200–300	70–95	PY: 22–45	90–280 μm	Uddin, Hawlader, and Zhu (2001)
Ascorbic acid	Rice starch and gum Arabic	7–15 mL/min	120–150	88–92	EE: 98.8–99.7	<27 μm (for gum Arabic) 55–57 μm (for starch)	Trindade and Grosso (2000)

NA: not available.

TABLE 6.2

A Summary of Process Parameters Applied for the Encapsulation of Fat-Soluble Vitamins by Spray Drying

Core Material	Wall Material(s)	Feed Flow Rate	T_{inlet} (°C)	T_{out} (°C)	Encapsulation Efficiency (EE, %) or Production Yield (PY, %)	Particle Size	References
Vitamin A	Maltodextrin	2 mL/min	120	74	PY: 81–83 EE: 100–23	2–4 μm	Mujica-Álvarez et al. (2020)
Vitamin A palmitate	Modified starches (OSA) and maltodextrin	1–5 ml/min	110–130	55–60	EE: 59–64 PY: 3	1–12 μm	Gangurde and Amin (2017)
β-carotene	Tapioca starch, acid modified tapioca starch, and maltodextrin	NA	170	95	EE: 46.7–82.2	75–250 μm	Loksuwan (2007)
Vitamin A	HI-CAP® 100 (starch octenylsuccinate)	1000 mL/min	182	82	EE: 96.4	66.6–153.5 μm	Xie, Zhou, Liang, He, and Han (2010)

Core Material	Wall Material(s)	Feed Flow Rate	T_{inlet} (°C)	T_{out} (°C)	Encapsulation Efficiency (EE, %) or Production Yield (PY, %)	Particle Size	References
β-carotene	Maltodextrin	NA	170	95	NA	NA	Desobry, Netto, and Labuza (1997)
Vitamin D3	Yeast cells	6.08 mL/min	130	75–77	EE: 31–76	4.47–6.92 µm	Dadkhodazade et al. (2018)
Vitamin D3	Gum Arabic, milk protein isolate, maltodextrin, and modified starch	550 L/h	160–190	90	PY: 16.5–73.9	120 nm	Jafari, Masoudi, and Bahrami (2019)
Vitamin D2	Chitosan and ethylcellulose	5 mL/min	168	NA	EE: >95	2–20 µm	Shi and Tan (2002)
Vitamin E	Maltodextrin	2 mL/min	120	74	PY: 79–84 EE: 48–29	2–4 µm	Mujica-Álvarez et al. (2020)
Vitamin E acetate	Gum Arabic, whey protein, polyvinyl alcohol, modified starch, and maltodextrin	100 L/min	100	38–60	PY: 70–90	0.4–1.1 µm	Li, Anton, Arpagaus, Belleteix, and Vandamme (2010)
Vitamin E	Maltodextrin and sodium caseinate	10 mL/min	110	90	EE: 59.9–71.5	13–29 µm	Selamat, Mohamad, Muhamad, Khairuddin, and Md Lazim (2018)
Vitamin E	Octenyl Succinic Anhydride (OSA) modified starch	10 ml/min	150	85	NA	208–235 nm	Hategekimana, Masamba, Ma, and Zhong (2015)
α-tocopherol	Plant-based protein, maltodextrin-carboxymethylcellulose mixture	1 L/h	180	90	NA	< 7 µm	Pierucci, Andrade, Farina, Pedrosa, and Rocha-Leão (2007)
α-tocopherol	Non-modified and modified soy protein isolate	0.47 m³/h	124	74	EE: 80–95 PY: 80	4.8–9.3 µm	Nesterenko et al. (2014)
α-tocopherol	Gum Arabic and maltodextrin	1.14 L/h	190	100	NA	239–1342 nm	Quintanilla-Carvajal et al. (2014)

NA: not available.

In this chapter, various aspects of spray drying encapsulation of both water-soluble and fat-soluble vitamins are discussed. To begin with, the role of vitamins in the human diet is also briefly explained.

6.2   THE ROLE OF VITAMINS IN THE HUMAN DIET

A functional food is prescribed as a food that has beneficial effects on human health by enhancing the state of well-being and decreasing the risk of illnesses. For instance, breakfast cereals fortified by folate to decrease the risk of neutral tube defects in the fetus is one of the examples of functional foods (Wilson & Shah, 2007).

Vitamins that can degrade the reactive free radicals are emerging because of the consequence of biochemical reactions caused by the free radicals in the body. Injuring proteins, damaging DNA, and affecting the membrane structure and function are among the harmful effects of free radicals in the body (Byers & Perry, 1992; Sun, 1990). Vitamin C is a water-soluble vitamin that can be obtained from fruits and vegetables such as orange, lemon, apple, capsicum, strawberry, grapefruit, mango, and kiwifruit (Reavley, 2000; Wilson & Shah, 2007). The curation and inhibition of cancer have been reported to be achieved by different mechanisms involving vitamin C; e.g., improvement of the immune system by supported lymphocytosis, inducing collagen formation, improvement of the efficiency of certain chemotherapy drugs, reduction of carcinogenic compounds, and lowering the hazard of chemotherapeutic agents (Esposito et al., 2002). Vitamin C, in particular, protects the human body against cancers by enhancing the formation of collagen, suppressing the formation of nitroso (N) compounds in the digestive system, and improving the immune system. The deficiency of vitamin C is attributed to stomach cancer and pancreatic cancer. It was stated that colon and stomach cancer is associated with a low intake of fruits and vegetables (Henson, Block, & Levine, 1991).

Carotenoids are the precursors of the fat-soluble vitamin A found in fruits and plants. Carotenoids can be enzymatically converted into retinol (vitamin A) in the body. One of the crucial properties of vitamin A is preventing cancer by effective scavenging of free radicals and providing the healthy cell division (Moon, 1989). Vitamin A could inhibit mammary tumorigenesis in the lab mouse when fed with high levels of dietary polyunsaturated fatty acids (PUFA) (Ip, 1982; Stoll, 1998).

Vitamin E is a lipophilic compound that has several beneficial effects in the human body in terms of physiologic functions such as antioxidant properties, which take place in the body through some mechanisms. These mechanisms are suppressing the formation of N-nitroso (N) compounds in the digestive system, likewise ascorbic acid, preserving selenium molecules, and protecting the lipid membrane of the cell (Bjørneboe, Bjørneboe, & Drevon, 1990). The retinoid and vitamin E supplemented diet improves the effectiveness of nourishment (Stoll, 1998).

Vitamin B groups cannot be synthesized in the human body; thus, the intake of vitamin B is provided through the diet. Intake of these groups of vitamins, especially vitamin B12, is crucial for the normal operation of the nervous system, brain health, cell metabolism, and blood cell development (Estevinho, Carlan, Blaga, & Rocha, 2016; Estevinho & Rocha, 2016). Therefore, the deficiency of vitamin B12 can cause some diseases such as anemia, weakness, and fatigue. Consuming dietary supplements and vitamin B12-enriched food can help cope with the deficiency problem of this vitamin (Estevinho et al., 2016).

Vitamin B complex is a water-soluble component and plays an important role in human and animal metabolisms and it is very essential for health and human nutrition. Participating in an enzyme-catalyzed reaction as a cofactor is also an important function of the vitamin B complex. Therefore, the intake of vitamin B through the diet is necessary for amino acid and carbohydrate metabolism (Roje, 2007). The deficiency of vitamin B1 leads to beriberi and vitamin B2 deficiency is related to anemia, cancer, nervous system disease, and cardiovascular disease (Lonsdale, 2006; Powers, 2003; Roje, 2007).

6.3   ENCAPSULATION OF WATER-SOLUBLE (HYDROPHILIC) VITAMINS BY SPRAY DRYING

6.3.1   Vitamin B Group

Vitamin B12 is also known as cobalamin due to having cobalt in its molecular structure. This vitamin has a large and chemically complex structure among all the vitamin groups. Encapsulation is one of the best strategies to increase vitamin stability and durability by reducing environmental factors and allowing the controlled release of bioactive core material (Estevinho et al., 2016). A study was conducted to encapsulate vitamin B12 with different wall materials. The spray drying yield of vitamin B12 varied between 41.8% and 55.6%, and the mean size of microcapsules was determined as about 3 μm for sodium alginate (lowest value), chitosan (highest value), and modified chitosan (Estevinho et al., 2016). More recently, Estevinho et al. (2019) prepared vitamin B12 microparticles using a cyanobacterial extracellular polymer or a combination of this polymer with gum Arabic as the encapsulating agent by the spray drying process. The mean size of microparticles was approximately 8 μm for the samples with only cyanobacterial polymer, whereas the microparticles prepared with the combination of gum Arabic and the cyanobacterial polymer showed a smaller size. Moreover, they also reported that the combination of the cyanobacterial polymer with gum Arabic reduced the release time from hours to minutes, and also improved the product yield of the spray dryer, allowing more bioavailable vitamin for the digestive system.

In another study, Chatterjee et al. (2016) investigated thiamine (vitamin B1) and pyridoxine (vitamin B6) loaded ferulic acid-grafted chitosan microspheres. They showed that the encapsulation efficiency (EE) was 91% for thiamine while it was 83% for pyridoxine. The lower EE for pyridoxine could be related to high inlet and outlet temperatures during spray drying. Pyridoxine is also thermally more sensitive than thiamine.

Folic acid (vitamin B9) was encapsulated by the whey protein concentrate matrix and commercial resistant starch using nano spray drying (Pérez-Masiá et al., 2015). Whey protein (83.9%) displayed a greater EE than resistant starch (52.5%), due to the interactions between folic acid and protein that facilitated the incorporation of the bioactive material within the capsules. It was also observed that whey protein concentrate protected the folic acid better against degradation during storage in an aqueous solution, as well as dry storage. The whey protein capsules were able to keep the stability of folic acid almost at 100% in dark conditions, while 40% of non-encapsulated folic acid was degraded. On the other hand, these researchers found that resistant starch led to a great loss of folic acid (higher than 90%), because of the high solubility of resistant starch in water, which caused a very quick release of folic acid, thus, it was no longer protected (Pérez-Masiá et al., 2015).

6.3.2 Vitamin C (Ascorbic Acid)

Vitamin C is known as an essential antioxidant and can decrease the risk of cancer by scavenging reactive oxygen species or free radicals that can damage organelles and DNA. Vitamin C has a vital role as a cofactor that is also important for the healing of wounds, decreasing LDL cholesterol, synthesizing collagen protein, and functioning of immune and nervous systems. Vitamin C deficiency is related to degenerative diseases such as cardiovascular diseases, cancer, cataracts, and scurvy (Reavley, 2000; Jeserich, 1999; Esposito et al., 2002; Wilson & Shah, 2007; Marcela et al., 2016; Devaki & Raveendran, 2017; Estevinho & Rocha, 2016; Saha & Roy, 2017; Barra et al., 2019). The powder form of ascorbic acid is stable, but its stability decreases in aqueous food matrices, due to the effects of environmental factors, including light, temperature, metal ions, acidity, and oxygen (Uddin et al., 2001; Wilson & Shah, 2007). Ascorbic acid is especially vulnerable against heat and light that degrades ascorbic acid to form diketogulonic acid and dehydroascorbic acid by photo-oxidation (Barra et al., 2019).

Food fortification is very popular among the producers for satisfying consumer demands and vitamin C fortification is applied to different types of products. Vitamin C stabilization is one of the crucial points for new product development and this is effectively achieved by encapsulation (Wilson & Shah, 2007). In a study by Uddin et al., (2001), ascorbic acid was encapsulated using the spray drying method using four different wall materials, including β-cyclodextrin, starch, gel, and ethylcellulose. The loss of the ascorbic acid was determined as 20% during drying and the ratio of encapsulated ascorbic acid was found to be less than 50%. Despite the lower recovery values, it was stated that no color change occurred during 30 days of storage at 38°C and 84% relative humidity (RH) (Uddin et al., 2001). In another study (Trindade & Grosso, 2000), ascorbic acid was spray dried in the presence of rice starch and gum Arabic as the wall materials. The loss of ascorbic acid was determined as only 1% and no significant loss for this vitamin was observed at room temperature and 60%–65% RH. Although microencapsulation and coating using gum Arabic did not protect all ascorbic acid, the retention rate maintained steady at 84% for three weeks. Therefore, gum Arabic was determined as the best coating material in spray drying in terms of the stability of ascorbic acid (Trindade & Grosso, 2000).

Camu-camu fruit juice with a high amount of vitamin C content was encapsulated by spray drying and the optimal conditions were determined for coating material (15%) and inlet temperature (150°C). Also, maltodextrin and gum Arabic provided approximately identical yield results for this study (Dib Taxi et al., 2003). The spray drying method was applied to encapsulate linoleic acid by gum Arabic and maltodextrin as wall materials. This study aimed to decrease the oxidation level of linoleic acid using acyl L-ascorbate, which is a lipophilic derivative of vitamin C (L-ascorbic acid), and determining the anti-oxidative activity of acyl L-ascorbate against the encapsulated linoleic acid. The supplementation of acyl L-ascorbate to linoleic acid led to the effective formation of oil droplets with smaller particle sizes and decreased the oxidation of linoleic acid (Watanabe, Fang, Minemoto, Adachi, & Matsuno, 2002). Another study was conducted to determine the potential of sodium alginate and gum Arabic as wall materials for the spray drying encapsulation of L-ascorbic acid (Barra et al., 2019). The encapsulation efficiencies of L-ascorbic acid microparticles were found to be above 82% and 90% for gum Arabic and sodium alginate, respectively. The particle dimensions of encapsulated L-ascorbic acid particles were in the range of 5.13–14.09 µm for sodium alginate-based and 2.88–9.73 µm for gum Arabic-based particles. The obtained different sizes of particles were reported to be a typical feature of the spray drying method and L-ascorbic acid microparticles demonstrated a spherical geometry with regular shapes (Barra et al., 2019). The encapsulation yields of L-ascorbic acid microparticles ranged from 37.8% to 75.7% for sodium alginate-based capsules and from 61.7% to 83.2% for gum Arabic-based capsules. It was observed that the total dispersed solid was correlated with the encapsulation yield, which was decreased as the concentration of sodium alginate increased. However, there was no correlation found between the total dispersed solids and the yield of encapsulation for the gum Arabic-based microcapsules, because of the low viscosities of the gum Arabic solutions (Barra et al., 2019). In addition, the moisture content of L-ascorbic acid microparticles had a significant role in the protection of capsules, storage conditions, and the compatibility of spray dried powder. The microparticles encapsulated by sodium alginate and gum Arabic had moisture content in the range of 2.1%–5.5% and 1.4%–5.4%, respectively (Barra et al., 2019).

Desai and Park (2006) encapsulated vitamin C with tripolyphosphate cross-linked chitosan as a wall material using the spray drying method. Chitosan concentration, amount of tripolyphosphate, inlet temperature, and flow rate of both air and liquid were stated as important processing parameters for microcapsules. The yield and dimension of the encapsulated vitamin C microcapsules ranged from 54.5% to 67.5% and 3.9 to 7.3 mm, respectively. The EE of the tripolyphosphate-chitosan system diminished as the volume of 1% (w/v) tripolyphosphate solution increased (Desai & Park, 2006). The product yield of vitamin C encapsulated by different wall materials changed between 43.6% and 45.4% and a mean diameter of microcapsules was found as 3 µm for sodium alginate, modified chitosan, and chitosan as different wall biopolymers (Estevinho et al., 2016).

A study (Alvim et al., 2016) was conducted to evaluate the incorporation of encapsulated ascorbic acid into a baked product (biscuit) using two different techniques. The EE was calculated as about 100% with the microparticles having a mean diameter of 9.3 µm for spray dried particles. The ascorbic acid loss of biscuits was found as 28%, 11%, and 15% for free ascorbic acid particles, spray dried particles, and spray chilled particles after baking. More than 85% of the active material was protected when spray drying was applied. Lastly, it was stated that microencapsulation by spray drying method and using spray chilling accessory prevented the dark spot formation, which was observed for free ascorbic acid biscuit samples (Alvim et al., 2016).

The degradation of ascorbic acid is reported to begin with the harvest of the produce and proceed during storage (Erdman & Klein, 1982; Murugesan & Orsat, 2011). A study was conducted to encapsulate vitamin C by methacrylate copolymers and the effectiveness of the encapsulation process was found to be between 98% and 100%. Encapsulated vitamin C particles showed good particle size distribution and morphology (Esposito et al., 2002).

Different methods were applied to investigate the encapsulation of ascorbic acid and the loss of ascorbic acid was determined to be minimum in the case of spray drying method (2%), comparing with solvent evaporation, melt dispersion, and phase separation methods. It was also stated that using β-cyclodextrin and starch as coating materials retarded the deterioration of encapsulated ascorbic acid particles (Uddin et al., 2001).

The encapsulated cactus pear juice was investigated for retention of vitamin C. Maltodextrin was used as a wall material in spray drying and maltodextrin with dextrose equivalent (DE) 10 supplied more retention compared to maltodextrin with DE 20 at 205°C and 0.1 MPa (Rodríguez-Hernández, González-García, Grajales-Lagunes, Ruiz-Cabrera, & Abud-Archila, 2005). In another study (Pierucci et al., 2006), plant-based protein and a mixture of maltodextrin and CMC (carboxymethylcellulose) were used as encapsulating wall materials for ascorbic acid. The microparticles obtained by the mix of maltodextrin and CMC had regular and smoother morphology compared to encapsulated particles attained by pea protein; however, the retention of ascorbic acid was found to be higher in the particles encapsulated using pea protein as the wall material (Pierucci et al., 2006). In other studies (de, Maia, de, & de, 2009; Murugesan & Orsat, 2011), the retention of ascorbic acid in the cashew apple juice (rich in vitamin C) and with tree gum of cashew apple and maltodextrin DE 10 as the wall materials was investigated. The inlet temperature of the spray dryer was set at 185°C. Ascorbic acid was found to be preserved at a 95% level using this method (de Oliveria et al., 2009; Murugesan & Orsat, 2011).

Oxidative stability of arachidonic acid during storage was investigated. For this purpose, ascorbic acid blended with arachidonic acid was encapsulated with soybean polysaccharides, maltodextrin, and gum Arabic using a spray drying method. Microencapsulated particles obtained by soybean polysaccharide and gum Arabic had better oxidative stability (Watanabe et al., 2002). Ascorbic acid was encapsulated using modified starch, showing superb stability and emulsion activity due to a lipophilic side obtained by an octenyl succinate chain. Encapsulation efficiency of ascorbic acid was calculated as 100% and after 1 month of storage, the retention of ascorbic acid was maintained at 100%, which decreased to 85% after 45 days of storage. Storage stability of ascorbic acid microparticles obtained by spray drying was considerably affected by the storage temperature and payload of the core material. A higher payload of the core at the beginning increased the storage stability but high-temperature storage (around 45°C) resulted in more reduction instability compared to room temperature storage. This confirmed that ascorbic acid was susceptible to higher temperatures. The higher payload of ascorbic acid at the beginning (20%) demonstrated better storage stability, because a higher amount of bioactive material retarded the penetrating of light and oxygen (Anandharamakrishnan & Ishwarya, 2015; Finotelli & Rocha-Leão, 2005).

6.4   ENCAPSULATION OF FAT-SOLUBLE (OR LIPOPHILIC) VITAMINS BY SPRAY DRYING

6.4.1   Vitamin A

The deficiency of vitamin A is the most prevalent vitamin deficiency all over the world. Fruit and vegetables contain β-carotene that supplies 80% of vitamin A as an important precursor, whereas α-carotene provides only 52.2% of vitamin A as an alternative precursor. β-carotene comes into prominence because of its anticancer activity and its ability to scavenge free radicals (Desobry et al., 1997). Vitamin A is fat-soluble and very sensitive in aqueous media because of its insolubility in such a medium. This vitamin is also very susceptible to oxygen, which accelerates its degradation in the aqueous medium. Therefore, solubility of this vitamin in an aqueous environment can be improved by applying encapsulation techniques for decelerating vitamin A degradation (Anandharamakrishnan & Ishwarya, 2015; Gangurde & Amin, 2017).

Eggs, dairy products, and liver are naturally rich sources of vitamin A. Tomatoes, carrots, squash, and cabbage contain vitamin A in the form of precursors such as lutein, lycopene, and β-carotene; the human body can transform these compounds to vitamin A (retinol). Vitamin A is widely used in beverages, foods, and dietary supplements, due to its strong antioxidant and coloring properties (Gerritsen & Crum, 2002; Potter & Hotchkiss, 1995). Intake of vitamin A and its precursors is important for the defense mechanism of the body, child development in terms of ophthalmic health, good vision, and healthy growth of bones and cells (Reavley, 2000; Gangurde & Amin, 2017).

A study was conducted to determine the effectiveness of three different drying methods for the encapsulation of β-carotene (precursor of vitamin A) (Murugesan & Orsat, 2011). Maltodextrin was used as the wall material for encapsulating pure β-carotene by applying drum drying, freeze drying, and spray drying. Degradation of β-carotene was measured concerning the changes in RH and temperature. It was found that RH did not affect the encapsulated microparticles. The degradation rate for β-carotene was found to be 8%, 11%, and 14% for freeze drying, spray drying, and drum drying, respectively (Murugesan & Orsat, 2011).

Gangurde and Amin (2017) encapsulated vitamin A palmitate, which is the ester of palmitic acid and retinol and available in dry and dispersed form, using a blend of modified starches and maltodextrin. The combination of wall materials was preferred for enhancing the EE. Tween® 80 and Cremophor® RH 40 were used as surfactants for the preparation of emulsions prior to spray drying. The average particle size varied from 5.04 to 7.71 μm. Encapsulated efficiency of vitamin A palmitate ranged from 59% to 64% and it was mentioned that the EE of microparticles increased with increasing maltodextrin concentration. In addition, the solubility of vitamin A palmitate microparticles decreased by decreasing maltodextrin concentration in this study (Gangurde & Amin, 2017). One of the main purposes of the study was investigating the compatibility of vitamin A palmitate microparticles to the cream formulation for topical application. The drug content and concentration of vitamin A palmitate was found to be in the range of 28%–30% and 94%–99%, respectively. It was concluded that encapsulation by spray drying could lead to good reproducibility of drug content and the degradation of drug content was very low during cream preparation in the topical form. Lastly, pH of the formulation was also controlled for compliance with the skin, and pH of vitamin A palmitate formulation changed from 7.1 to 7.7 (Gangurde & Amin, 2017).

Loksuwan (2007) evaluated the most suitable wall material for encapsulating β-carotene. Native tapioca starch, acid-modified tapioca starch, and maltodextrin were used as coating materials. Particle size was found to be in the range of 75–150 μm (56%), 106–250 μm (70%), and 106–250 μm (80%), for modified tapioca starch (MTS), native tapioca starch, and maltodextrin, respectively. The particle size of the capsules obtained with maltodextrin was reported to be larger than that of the capsules obtained with MTS and this was also expressed by caking and agglomeration of powder. According to SEM of spray dried particles, microparticles obtained with MTS demonstrated spherical shape with a wide dented surface which was expressed as rupturing starch granule during the modification process and shrinkage of microcapsules throughout the drying process. However, microcapsules obtained by tapioca native starch and maltodextrin had smooth spherical shapes with few dented surfaces. Coldwater solubility was also investigated for the spray dried microparticles by different wall materials. Cold solubility of MTS was found to be higher than that of tapioca native starch and lower than that of maltodextrin, which was entirely soluble. The reasons for this difference were stated as the acid treatment for MTS obtained with the change of granular structure (Loksuwan, 2007). The total carotene contents after drying were calculated as 82.2%, 68.4%, and 46.7% for MTS, native tapioca starch, and maltodextrin, respectively. It was indicated that MTS had the highest EE compared with other wall materials because of the composition of MTS after steam pressure treatment. The study concluded that MTS was more effective for protecting β-carotene compared to the native tapioca starch and MTS had a high potential for coating β-carotene as a wall material (Loksuwan, 2007).

More recently, Mujica-Álvarez et al. (2020) evaluated the effect of some encapsulating agents (sodium caseinate-SC, Capsul-CAP®) in combination with maltodextrin (MD) as wall materials and Tween® 80 (TW) as an emulsifier on the physicochemical properties of emulsions and powders. They showed that the EE for vitamin A using MD: CAP: TW was 100%, whereas the use of MD: SC: TW resulted in a value of 23%. This lower EE value for sodium caseinate could be explained by the fact that proteins could be partially denatured during ultrasonication.

6.4.2 Vitamin D

Vitamin D is one of the fat-soluble vitamins and a steroid hormone that has long been known for its role in calcium homeostasis, and in the mineralization of bone. Vitamin D can be taken into the body with the diet in the form of ergocalciferol (Vitamin D2) found in plants and in the form of chole-calciferol (Vitamin D3), which is found in animal tissues and can be synthesized in the body by the skin when exposed to sunlight. Fish, liver, and egg yolk are well-known natural sources of vitamin D (Maurya, Bashir, & Aggarwal, 2020; Ongen, Kabaroglu, & Parıldar, 2008). Dietary vitamin D2 and vitamin D3 are combined with chylomicrons transported by venous circulation to participate in the lymphatic system. Endogenous and also dietary vitamin D2 and vitamin D3 are stored in fat cells in the body and released into the circulation when necessary (Holick, 2007; Ongen et al., 2008).

Vitamin D deficiency is defined with the 25-hydroxyvitamin D (25(OH)D) level, which is the bio-activated form of vitamin D2 and vitamin D3, being less than 20 ng/ml. The deficiency of this vitamin mainly leads to rheumatism, which causes bone demineralization in children and osteomalacia in adults. It is estimated that 1 billion people in the world suffer from vitamin D deficiency (Holick, 2006; Malabanan, Veronikis, & Holick, 1998; Ongen et al., 2008). Diabetes, rickets, and calcium-phosphorus and parathyroid imbalance are the other diseases observed because of the deficiency of vitamin D (Maurya et al., 2020). Vitamin D deficiency emerges due to several factors including geographical location, the limited natural source of vitamin D, and the low bioavailability of this vitamin (Borel, Caillaud, & Cano, 2013; Maurya et al., 2020).

Vitamin D is susceptible to light, moisture, heat, and oxygen when the food is exposed to processing and storage. The Vitamin D content of products is significantly reduced during thermal treatments such as oven drying, boiling, smoking, roasting, steaming, etc. The reduced vitamin D contents via these processes have been observed to finally influence the bioavailability of vitamin D in the body (Jakobsen & Knuthsen, 2014; Maurya et al., 2020).

Encapsulation technology is one of the effective methods which protects vitamin D and other bioactive materials against degradation during the above-mentioned processes and improves the bioavailability of vitamin D. Providing a wide range of wall materials by spray drying as an encapsulation method is the advantage of this technique. Despite its advantages for the encapsulation of vitamin D, the number of studies using spray drying is limited because of obtaining porous microcapsules of vitamin D in this method, which causes degradation (Jafari et al., 2019; Maurya et al., 2020; Moeller, Martin, Schrader, Hoffmann, & Lorenzen, 2018; Shi & Tan, 2002).

In a recent study by Jafari et al. (2019), fortification of yogurt with encapsulated vitamin D was investigated, and nano-liposomal encapsulation with spray drying was applied. The inlet temperature, milk protein concentrate, modified starch content, gum Arabic, and maltodextrin amounts were the independent variables of the Taguchi method (Jafari et al., 2019). It was stated that the optimum conditions for the most suitable moisture content of the spray dried powder was 190°C inlet temperature, 3% milk protein concentrate, modified starch, or gum Arabic, and 25% maltodextrin. The highest solubility was obtained under the following conditions; 170°C inlet temperature, 0% milk protein concentrate-gum Arabic, 3% modified starch, and 20% maltodextrin. In order to provide yogurt powder fortified with acceptable levels of vitamin D the conditions should be as follows; 180°C inlet temperature, 0% milk protein concentrate, 3% modified starch, 1% gum Arabic, and 15% maltodextrin (Jafari et al., 2019).

6.4.3 Vitamin E

Vitamin E has an important role in human health and well-being. Vitamin E is generally applied for food fortification, cosmetic, and drug formulations, but it is very sensitive to oxygen and heat (Lazim & Muhamad, 2017). Free radicals and oxygen easily cause the degradation of this vitamin (Anandharamakrishnan & Ishwarya, 2015; Lazim & Muhamad, 2017). Vitamin E is hydrophobic and the utilization of this vitamin is difficult in aqueous food systems, which causes lower absorbance and poor bioavailability. Vitamin E mainly composes tocopherol and tocotrienol bioactive compounds and this vitamin has a cholesterol-lowering effect, protects atherosclerosis, and high antioxidant activity to scavenge free radicals for preventing cardiac disease and cancers (Selamat et al., 2018; Yoo, Song, Chang, & Lee, 2006). Vitamin E can be highly preserved by encapsulation methods against the adverse effects of environmental factors (Lazim & Muhamad, 2017).

In order to attain favorable chemical or physical stability and bioavailability, microemulsions, nanoemulsions, or conventional emulsions were prepared for Vitamin E fat-in-water suspensions concerning droplet size, including d < 100 nm, d < 200 nm, and d > 200 nm, respectively (Hategekimana et al., 2015; McClements, 2012). On the other hand, choosing the wall material is crucial for the EE and stability of the nanocapsules. Solubility, diffusibility, molecular weight, and melting and glass transition temperatures are the important physicochemical properties for selecting the appropriate wall material(s) (Hategekimana et al., 2015).

Spray drying is a convenient method for the encapsulation of vitamins because of the availability of the equipment and cost-effectiveness, compared with several other encapsulation techniques. In a study performed by Hategekimana et al. (2015), obtaining stable vitamin E nanocapsules and determining physicochemical properties of Octenyl Succinic Anhydride (OSA) modified starches as a wall material (because of their good film-forming, emulsifying, and oxygen-barrier properties) was aimed and spray drying was used as the encapsulation technique. The results of this study demonstrated that the modified starch types, which have low interfacial tension, low molecular weight, and high degree of substitution, were able to generate vitamin E nanocapsules with a retention value of about 50% of its initial amount at 4°C–35°C after 60 days. The authors stated that stable vitamin E nanoparticles were produced and could be effectively applied for drug formulations and beverages (Hategekimana et al., 2015).

A study was carried out to encapsulate a mixture of natural tocopherols obtained by palm and tocotrienol by spray drying method. The wall material was selected as a blend of maltodextrin and sodium caseinate. It was indicated that EE of vitamin E extracted from palm oil decreased from 71.5% to 59.9% as the core/wall ratio raised from 0.6 to 1.0. The hydrophobic nature of the core material was assumed to be the reason for the decrease in EE due to the formation of poor emulsion capacity and stability prior to atomization and low oil retention of maltodextrin throughout the drying treatment. A leaner thickness of wall material between each encapsulated microparticle could be the reason for the decrease in EE (Anandharamakrishnan & Ishwarya, 2015; Selamat et al., 2018).

The development of wall materials by modification of the lipophilic nature of vitamins contribute to coping with the challenge of incompatibility of the bioactive core material with the wall material. Proteins were stated as proper compounds for encapsulation due to their amphiphilic structure. Plant proteins could be appropriate materials, particularly proteins extracted from soybeans, as a wall material of microcapsules due to having good oil/water absorption properties and emulsification ability (Anandharamakrishnan & Ishwarya, 2015).

In another study (Nesterenko et al., 2014), lipophilic α-tocopherol was encapsulated by a modified soy protein isolate. The modification was achieved by adding fatty acid chains to soy protein isolate, which enhanced the affinity of α-tocopherol through the wall matrix. According to the results of the study, the EE of α-tocopherol raised from 79.7% to 94.8% if modified soy protein was used instead of natural soy protein (Nesterenko et al., 2014).

On the other hand, Quintanilla-Carvajal et al. (2014) investigated the effect of microfluidizer on encapsulated α-tocopherol nanoparticles obtained by the spray drying method. It was indicated that the EE was affected by the preparation of the emulsion. The average diameter of nanoparticles decreased from 1341.8 to 239.0 nm after the first cycle of microfluidization and increased to 412.0 nm with the second cycle of microfluidization. It was also reported that the encapsulated nanoparticle dimensions were decreased when the number of microfluidizer cycle were increased in the case of lower levels of α-tocopherol (Quintanilla-Carvajal et al., 2014). The size of the emulsified droplets is important for controlled release of vulnerable compounds. As the particle size of the emulsified droplets is decreased, the retention of vitamins is enhanced, decreasing the oxidation of coated materials and preventing the adverse impacts on flow properties of the obtained powder. As the feed concentration was increased, the retention of α-tocopherol was reduced and the reason was considered as over-processing and recoalescence. It was inferred from the study that the loading of α-tocopherol and cycle number of microfluidizer could significantly influence the extent of non-coated α-tocopherol on the facet of the particles (Quintanilla-Carvajal et al., 2014).

In another study, vitamin E acetate (formulated in a nanoemulsion) was encapsulated by different wall materials (gum Arabic, whey protein, polyvinyl alcohol, modified starch, and maltodextrin) using nano-spray drying (Li et al., 2010). Their results showed that optimum conditions were inlet and outlet temperatures of 100°C and 40°C–60°C, respectively, with a 1% aqueous wall material solution and a weight ratio of 1: 4 for nanoemulsion: wall material (w/w). Under these conditions, high EE, approximately 100%, was observed and the spray drying process did not induce the degradation of the encapsulated molecule and preserved the integrity of the nanoemulsion system.

6.5 CHARACTERISTICS OF SPRAY DRIED ENCAPSULATED VITAMINS

Characterization of encapsulated vitamins is generally performed using techniques such as scanning electron microscopy (SEM), Fourier transform infrared spectroscopy (FTIR), X-ray diffraction (XRD), transmission electron microscopy (TEM), differential scanning calorimetry (DSC), and thermogravimetric analysis (TGA). The morphology of the encapsulated structures is examined using SEM. For example, vitamin B12 was encapsulated with three different biopolymers (chitosan, modified chitosan, and sodium alginate) (Estevinho et al., 2016). According to the results, spherical microparticles were formed with modified chitosan or sodium alginate and presented a very smooth surface, whereas the particles formed with chitosan presented a very rough surface. In another study, Dadkhodazade et al. (2018) showed that the XRD characteristic peaks of vitamin D were not detected. Therefore, these researchers suggested that the cholecalciferol was not in a crystalline state when incorporated into yeast microcapsules. Pérez-Masiá et al. (2015) observed that the spray drying technique caused a greater displacement of the spectral bands due to the stronger molecular bonding of the whey protein concentrate chains using FTIR. The capsules containing folic acid presented narrower bands than those without the bioactive material, suggesting that the incorporation of folic acid caused a greater molecular order. Moreover, the incorporation of folic acid produced a greater displacement of the amide II band when compared to those without this bioactive compound (Pérez-Masiá et al., 2015).

The shape, size, surface structure, and integrality of microcapsules are important parameters to interpret any breakage and damage. The spherical straight shapes of microparticles with convenient particle size supply controlled the release of bioactive material, better preservation, and lower permeability of moisture and gas (Nizori et al., 2012). The encapsulated ascorbic acid microparticles had small uniform size distribution in the range of 5–100 μm and the particle dimension tended to increase by a higher rate of feed flow and lower inlet temperature (Nizori et al., 2012).

The inlet temperature of the process should be properly adjusted, because higher temperatures can damage the vulnerable bioactive compounds whereas the lower inlet temperatures can result in no evaporation of the prepared emulsion. Besides, a higher mass ratio, lower inlet temperature, and lower suction power lead to lower production yield because of the low evaporation rate, which brings on the formation of flocculated microparticles with higher water content and poor mobility (Marcela et al., 2016). The encapsulated L-ascorbic acid microparticles had a more spherical shape when suction power and the inlet temperature of the process were reduced. The mass ratio (2: 3.5, w/w) of L-ascorbic acid to alginate as a wall material plays an important role in the morphology of microcapsules (Marcela et al., 2016).

Vitamin C and vitamin B12-loaded microparticles made up of sodium alginate and modified chitosan demonstrated a straight surface, whereas the microparticles formed with chitosan showed a rough surface. The mean size of the capsulated particles was monitored approximately as 3 µm for the whole wall materials used in the study (Estevinho & Rocha, 2016). Besides, SEM micrographs of fortified yogurt powder with vitamin D demonstrated spherical-shaped particles with even distribution when maltodextrin was loaded to the formulations; on the other hand, adding modified starch, milk protein concentrates, and gum Arabic resulted in a dented surface of the powder particles with inequality in the size and shape (Jafari et al., 2019). In another study, SEM micrographs of encapsulated vitamin D2 particles with chitosan and ethylcellulose showed that chitosan particles resulted in a smooth surface, however, the surface of microcapsules coated with ethylcellulose was not smooth because of the heterogeneous accumulation of ethylcellulose on the facet of the bioactive material (vitamin D2) (Shi & Tan, 2002).

6.6   RELEASE AND BIOAVAILABILITY OF ENCAPSULATED VITAMINS

The encapsulated particles obtained by the spray drying technique are ultimately used for food applications and the pharmaceutical industry. However, the release mechanism of vitamins from these particles should be taken into consideration. Thus, the selection of the wall material and preparation of the matrix in correct proportion before drying is the crucial process for providing the desired delivery of the core material (McClements, 2017).

Stomach and small intestine conditions affect the release mechanism of coated vitamins and these mechanisms can be lined up as diffusion, osmosis, bead disintegration, biodegradation, and bead swelling. The molecular interference, enlargement of pore diameter, and disintegration of molecular connection finally cause the releasing of bioactive material entrapped into convenient wall material. It has been stated that the release of bioactive material is faster for small active core materials than the larger ones, because large core materials cannot easily disintegrate within the strong wall materials. *In vitro* tests applied for the controlled release of essential active material such as vitamins are very practical to prevent time and money consumption (Rezvankhah et al., 2019).

The encapsulated bioactive material for the nutraceutical delivery can be preserved with a coating material resistant against the acidic pH of the stomach conditions. The encapsulated particles should be stable with a low absorption ratio when arrived at the stomach and the encapsulated bioactive microparticles should maintain the controlled release of the core material at the desired rate when the capsules reach the intestine (Lazim & Muhamad, 2017; Yoo et al., 2006). The desired release mechanisms can be achieved by first-order kinetic, half-order kinetic, and zero-order (constant rate) kinetic that supplies the release of the core material as a pure substance (Estevinho & Rocha, 2016).

Assadpour, Jafari, and Maghsoudlou (2017) investigated the release of folic acid from spray dried powder particles of pectin-whey protein nanocapsules. They reported that folic acid nanocapsules prepared with sorbitan monooleate as the surfactant had the lowest release rate in acidic conditions (pH = 4) and the highest release in the alkaline conditions (pH = 11). On the other hand, Chatterjee et al. (2016) evaluated the *in vitro* release behavior of thiamine and pyridoxine from the encapsulated particles. After 2 h, 24% of thiamine and pyridoxine was released, whilst at 100 h, a maximum of 87% of thiamine and pyridoxine was found to be released from the microspheres. The authors concluded that the slow release of thiamine and pyridoxine in acidic conditions throughout 100 h shows the effectiveness of the encapsulation methodology. In another study, vitamin B12 was encapsulated within gum Arabic, cashew nut gum, sodium alginate, carboxymethyl cellulose, and Eudragit® RS100 through nanospray drying (Oliveira et al., 2013). Moreover, two different pH conditions (pH = 1 and pH = 8) were assessed to simulate the gastric (acid) and the enteric (alkaline) conditions. The results indicated that the Eudragit® RS100 presented the most promising results in terms of particle morphology.

A study was carried out to investigate the controlled release profiles of encapsulated microparticles. Vitamin B12 and vitamin C were encapsulated with chitosan, modified chitosan, and sodium alginate by spray drying. Five kinetic-dependent models were considered and applied for the release of vitamins. It was stated that the zero-order kinetic equation fit the experimental data well and also that the Weibull model and the Korsmeyer-Peppas equations had high correlation coefficients. The swelling of the wall materials was mentioned as the main release mechanism of vitamin microcapsules (Estevinho & Rocha, 2016). In another study, Garcia-Arieta, Torrado-Santiago, Goya, and Torrado (2001) examined the nasal absorption of spray dried powders containing cyanocobalamin (vitamin B12) in rabbits. The bioavailability of cyanocobalamin in rabbits followed the order of spray dried-microcrystalline cellulose (25.1%) > spray dried-crospovidone (14.6%) > spray dried-dextran microspheres (6.9%). It was observed that the spray-dried powders showed higher bioavailability than the nasal reference formulations.

Encapsulation can effectively procure the controlled release of bioactive material at the desired time, desired site, and the right place. The protection and release of core material enhance the efficiency of coated material, expand the application areas of food constituents, and provide optimum dosage. Vitamins and other sensitive volatile compounds become more stable by the process of encapsulation (Desai & Park, 2006; Estevinho & Rocha, 2016; Gouin, 2004). The inlet temperature and feed flow of the process are the important factors affecting the release of the core materials. It was also stated that the increased inlet temperature brought about the increase of retention of ascorbic acid because of lower moisture content causing aggregates to protect powder against oxygen. The lower rate of feed flow and the higher inlet temperature also led to an increase in the production yield of the spray drying process (Nizori et al., 2012).

Enzymes, colors, sweeteners, and preservatives are the food additives that can be encapsulated for the controlled release mechanisms of capsules such as temperature, moisture, and thermal release. The *in vitro* release profile of encapsulated vitamin A palmitate microcapsules was studied and less than 20% of the drug release was observed at the previous 2 h, due to the slow release of the core material from the wall material. The rate of release accelerated after 2 h to 8 h. It was indicated that the release of vitamin A palmitate encapsulated particles changed in the range of 80.2%–83.4% at the end of 24 h. The stability of the encapsulated vitamin A palmitate was also examined and the microcapsules were kept in amber glass bottles. The microcapsules were found to be stable because there was no significant change in the contents of drug and moisture after three months of storage at accelerated conditions (Gangurde & Amin, 2017).

In another study, the release rate of encapsulated vitamin D2 was investigated and the release of ethylcellulose-coated microcapsules was observed to be very limited during the swelling in the gastric juice. The wall material: core ratio was considered to be effective in the release of vitamin D2. It was stated that decreasing the wall material: core ratio could increase the release rate, because the resistance of the core material for diffusion was decreased by reducing the wall material layer. It was inferred that ethylcellulose as a wall material could delay the release of vitamin D2 in artificial gastric juice, and the drug loading of the studied system was calculated to be more than 86% (Shi & Tan, 2002).

6.7 CONCLUDING REMARKS

Vitamins have countless biological functions in the human body; they are essential for the immune system, development of nerves and muscle system, the use of proteins, carbohydrates, and fats in the body, bone, and tooth health. Furthermore, vitamins are antioxidant micronutrients that play several important roles in preserving the body against cancer and other chronic diseases, by preventing the formation of carcinogens and oxidative damage caused by free radicals. Biologically and chemically active compounds such as vitamins are very sensitive against environmental conditions including light, heat, moisture, and oxygen. Thus, vitamins can be easily degraded during food processing (e.g., boiling, steaming, frying, and pressure cooking) and storage. In order to prevent the degradation of vitamins, encapsulation technologies are applied for protecting these substantial bioactive materials. This chapter focuses on spray drying encapsulation of vitamins, which is one of the most commonly-used methods for the encapsulation of bioactive compounds. Spray drying is widely used for its advantages, including continuous manufacturing, easy scale-up, better control of the size and shape of the obtained particles. The atomizer type, feed flow rate, air pressure, inlet, and outlet temperature affect the EE and yield. To provide highly retained and protected particles after drying, vitamins should be homogeneously distributed in the emulsions containing wall materials before drying via hot air. The retention of vitamins in a food matrix and their controlled release during digestion can also be improved using the spray drying encapsulation method. It appears that there are only limited studies available for the spray drying encapsulation of vitamin K and vitamin D, while the number of studies related to the bioavailability and controlled release of vitamins, in general, are restricted. Thus, further research is required to better understand the effect of spray drying parameters on the retention of these vitamins.

REFERENCES

Alvim, I. D., Stein, M. A., Koury, I. P., Dantas, F. B. H., & Carla Léa De Camargo Vianna Cruz. (2016). Comparison between the spray drying and spray chilling microparticles contain ascorbic acid in a baked product application. *LWT - Food Science and Technology*, 65, 689–694. doi: 10.1016/j.lwt.2015.08.049

Anandharamakrishnan, C., & Ishwarya, S. P. (2015). *Spray Drying Techniques for Food Ingredient Encapsulation*. John Wiley & Sons, Ltd., West Sussex, UK, & the Institute of Food Technologists, Chicago, IL. doi: 10.1002/9781118863985

Assadpour, E., Jafari, S. M., & Maghsoudlou, Y. (2017). Evaluation of folic acid release from spray dried powder particles of pectin-whey protein nano-capsules. *International Journal of Biological Macromolecules*, 95, 238–247.

Barra, P., Márquez, K., Gil-Castell, O., Mujica, J., Ribes-Greus, A., & Faccini, M. (2019). Spray-drying performance and thermal stability of l-ascorbic acid microencapsulated with sodium alginate and gum Arabic. *Molecules*, 24, 2872. doi: 10.20944/preprints201906.0197.v1

Bjørneboe, A., Bjørneboe, G.-E. A., & Drevon, C. A. (1990). Absorption, transport and distribution of vitamin E. *The Journal of Nutrition*, 120(3), 233–242. doi: 10.1093/jn/120.3.233

Borel, P., Caillaud, D., & Cano, N. J. (2013). Vitamin D bioavailability: State of the art. *Critical Reviews in Food Science and Nutrition*, 55(9), 1193–1205. doi: 10.1080/10408398.2012.688897

Burgain, J., Gaiani, C., Linder, M., & Scher, J. (2011). Encapsulation of probiotic living cells: From laboratory scale to industrial applications. *Journal of Food Engineering*, 104(4), 467–483. doi: 10.1016/j.jfoodeng.2010.12.031

Byers, T., & Perry, G. (1992). Dietary carotenes, vitamin C, and vitamin E as protective antioxidants in human cancers. *Annual Review of Nutrition*, 12(1), 139–159. doi: 10.1146/annurev.nu.12.070192.001035

Chatterjee, N. S., Anandan, R., Navitha, M., Asha, K. K., Kumar, K. A., Mathew, S., & Ravishankar, C. N. (2016). Development of thiamine and pyridoxine loaded ferulic acid-grafted chitosan microspheres for dietary supplementation. *Journal of Food Science and Technology*, 53(1), 551–560.

Dadkhodazade, E., Mohammadi, A., Shojaee-Aliabadi, S., Mortazavian, A. M., Mirmoghtadaie, L., & Hosseini, S. M. (2018). Yeast cell microcapsules as a novel carrier for cholecalciferol encapsulation: Development, characterization and release properties. *Food Biophysics*, 13(4), 404–411.

Desai, K. G., & Park, H. J. (2006). Effect of manufacturing parameters on the characteristics of vitamin C encapsulated tripolyphosphate-chitosan microspheres prepared by spray drying. *Journal of Microencapsulation*, 23(1), 91–103. doi: 10.1080/02652040500435436

Desai, K. G. H., & Park, H. J. (2005). Recent developments in microencapsulation of food ingredients. *Drying Technology*, 23(7), 1361–1394. doi: 10.1081/drt-200,063,478

Desobry, S. A., Netto, F. M., & Labuza, T. P. (1997). Comparison of spray drying, drum-drying and freeze-drying for β-carotene encapsulation and preservation. *Journal of Food Science*, 62(6), 1158–1162. doi: 10.1111/j.1365-2621.1997.tb12235.x

Devaki, S. J., & Raveendran, R. L. (2017). Vitamin C: sources, functions, sensing and analysis. *Vitamin C.* InTech: London, UK. doi: 10.5772/intechopen.70162

Dib Taxi, C. M. A. D., De Menezes, H. C., Santos, A. B., & Grosso, C. R. F. (2003). Study of the microencapsulation of camu-camu (*Myrciaria dubia*) juice. *Journal of Microencapsulation*, 20(4), 443–448. doi: 10.1080/0265204021000060291

Erdman, J. W., & Klein, B. P. (1982). Harvesting, processing, and cooking influences on vitamin C in foods. In *Ascorbic Acid: Chemistry, Metabolism, and Uses*, P. A. Saib & B. M. Tolbert (Eds), pp. 499–532. Advances in Chemistry; American Chemical Society: Washington, DC.

Esposito, E., Cervellati, F., Menegatti, E., Nastruzzi, C., & Cortesi, R. (2002). Spray dried Eudragit microparticles as encapsulation devices for vitamin C. *International Journal of Pharmaceutics*, 242(1–2), 329–334. doi: 10.1016/s0378-5173(02)00176-x

Estevinho, B. N., Carlan, I., Blaga, A., & Rocha, F. (2016). Soluble vitamins (vitamin B12 and vitamin C) microencapsulated with different biopolymers by a spray drying process. *Powder Technology*, 289, 71–78. doi: 10.1016/j.powtec.2015.11.019

Estevinho, B. N., Mota, R., Leite, J. P., Tamagnini, P., Gales, L., & Rocha, F. (2019). Application of a cyanobacterial extracellular polymeric substance in the microencapsulation of vitamin B12. *Powder Technology*, 343, 644–651.

Estevinho, B. N., & Rocha, F. (2016). Kinetic models applied to soluble vitamins delivery systems prepared by spray drying. *Drying Technology*, 35(10), 1249–1257. doi: 10.1080/07373937.2016.1242015

Finotelli, P.V., & Rocha-Leão, M.H.M. (2005). *Microencapsulation of ascorbic acid in maltodextrin and Capsul using spray-drying.* Paper presented at the *Empromer/2nd Mercosur Congress on Chemical Engineering/4th Mercosur Congress on Process Systems Engineering*, Costa Verde, RJ.

Gangurde, A. B., & Amin, P. D. (2017). Microencapsulation by spray drying of vitamin A palmitate from oil to powder and its application in topical delivery system. *Journal of Encapsulation and Adsorption Sciences*, 07(01), 10–39. doi: 10.4236/jeas.2017.71002

Garcia-Arieta, A., Torrado-Santiago, S., Goya, L., & Torrado, J. J. (2001). Spray-dried powders as nasal absorption enhancers of cyanocobalamin. *Biological and Pharmaceutical Bulletin*, 24(12), 1411–1416.

Gerritsen, J. and Crum, F. 2002. Getting the best from beta-carotene. *International Food Ingredients*, 3, 40–41.

Gouin S. 2004. Microencapsulation: Industrial appraisal existing technologies and trends. *Trends in Food Science and Technology*, 15, 330–347.

Hartman, G. H., Akeson, W. R., & Stahmann, M. A. (1967). Leaf protein concentrate prepared by spray drying. *Journal of Agricultural and Food Chemistry*, 15(1), 74–79. doi: 10.1021/jf60149a012

Hategekimana, J., Masamba, K. G., Ma, J., & Zhong, F. (2015). Encapsulation of vitamin E: Effect of physicochemical properties of wall material on retention and stability. *Carbohydrate Polymers*, 124, 172–179. doi: 10.1016/j.carbpol.2015.01.060

Henson, D. E., Block, G., & Levine, M. (1991). Ascorbic acid: Biologic functions and relation to cancer. *JNCI Journal of the National Cancer Institute*, 83(8), 547–550. doi: 10.1093/jnci/83.8.547

Holick, M. F. (2006). High prevalence of vitamin D inadequacy and implications for health. *Mayo Clinic Proceedings*, 81(3), 353–373. doi: 10.4065/81.3.353

Holick, M. F. (2007). Vitamin D deficiency medical progress. *The New England Journal of Medicine*, Boston, 357(3), 266.

Ip, C. (1982). Dietary vitamin E intake and mammary carcinogenesis in rats. *Carcinogenesis*, 3(12), 1453–1456. doi: 10.1093/carcin/3.12.1453

Jafari, S. M., Masoudi, S., & Bahrami, A. (2019). A Taguchi approach production of spray-dried whey powder enriched with nanoencapsulated vitamin D3. *Drying Technology*, 37(16), 2059–2071. doi: 10.1080/07373937.2018.1552598

Jakobsen, J., & Knuthsen, P. (2014). Stability of vitamin D in foodstuffs during cooking. *Food Chemistry*, 148, 170–175. doi: 10.1016/j.foodchem.2013.10.043

Jeserich, M. (1999). Vitamin C improves endothelial function of epicardial coronary arteries in patients with hypercholesterolaemia or essential hypertension—assessed by cold pressor testing. *European Heart Journal*, 20(22), 1676–1680. doi: 10.1053/euhj.1999.1689

Lakkis, J. M. (2007). Introduction. *Encapsulation and Controlled Release Technologies in Food Systems*, Blackwell Publishing: Oxford, UK. 1–11. doi: 10.1002/9780470277881.ch1

Lawson, J. M., Hunt, C., & Glew, G. (1983). Nutrition in catering. *Nutrition Bulletin*, 8(2), 93–104. doi: 10.1111/j.1467-3010.1983.tb01317.x

Lazim, N. A. M., & Muhamad, I. I. (2017). Encapsulation of vitamin E using maltodextrin/sodium caseinate/selenomethionine and its release study. *Chemical Engineering Transactions*, 56, 1951–1956. doi: 10.3303/CET1756326

Li, X., Anton, N., Arpagaus, C., Belleteix, F., & Vandamme, T. F. (2010). Nanoparticles by spray drying using innovative new technology: The Büchi nano spray dryer B-90. *Journal of Controlled Release*, 147(2), 304–310.

Loksuwan, J. (2007). Characteristics of microencapsulated β-carotene formed by spray drying with modified tapioca starch, native tapioca starch and maltodextrin. *Food Hydrocolloids*, 21(5–6), 928–935. doi: 10.1016/j.foodhyd.2006.10.011

Lonsdale, D. (2006). A review of the biochemistry, metabolism and clinical benefits of thiamin(e) and its derivatives. *Evidence-Based Complementary and Alternative Medicine*, 3(1), 49–59. doi: 10.1093/ecam/nek009

de Oliveria, M. A., Maia, G. A., de Figueiredo, R. W., de Souza, A. C. R., … (2009). Addition of cashew tree gum to maltodextrin-based carriers for spray drying of cashew apple juice. *International Journal of Food Science and Technology*, 44, 641–645. doi: 10.1111/j.1365-2621.2008.01888.x

Malabanan, A., Veronikis, I., & Holick, M. (1998). Redefining vitamin D insufficiency. *The Lancet*, 351(9105), 805–806. doi: 10.1016/s0140-6736(05)78933-9

Marcela, F., Lucía, C., Esther, F., & Elena, M. (2016). Microencapsulation of L-ascorbic acid by spray drying using sodium alginate as wall material. *Journal of Encapsulation and Adsorption Sciences*, 06(01), 1–8. doi: 10.4236/jeas.2016.61001

Masters, K., 1991. *The Spray Drying Handbook*. Longman Scientific and Technical: New York.

Maurya, V. K., Bashir, K., & Aggarwal, M. (2020). Vitamin D microencapsulation and fortification: Trends and technologies. *The Journal of Steroid Biochemistry and Molecular Biology*, 196, 105489. doi: 10.1016/j.jsbmb.2019.105489

McClements, D. J. (2012). Nanoemulsions versus microemulsions: Terminology, differences, and similarities. *Soft Matter*, 8(6), 1719–1729. doi: 10.1039/c2sm06903b

McClements, D. J. (2017). Designing biopolymer microgels to encapsulate, protect and deliver bioactive components: Physicochemical aspects. *Advances in Colloid and Interface Science*, 240, 31–59. doi: 10.1016/j.cis.2016.12.005

Moeller, H., Martin, D., Schrader, K., Hoffmann, W., & Lorenzen, P. C. (2018). Spray- or freeze-drying of casein micelles loaded with Vitamin D2: Studies on storage stability and in vitro digestibility. *LWT*, 97, 87–93. doi: 10.1016/j.lwt.2018.04.003

Moon, R. C. (1989). Comparative aspects of carotenoids and retinoids as chemopreventive agents for cancer. *The Journal of Nutrition*, 119(1), 127–134. doi: 10.1093/jn/119.1.127

Mujica-Álvarez, J., Gil-Castell, O., Barra, P. A., Ribes-Greus, A., Bustos, R., Faccini, M., & Matiacevich, S. (2020). Encapsulation of vitamins A and E as spray-dried additives for the feed industry. *Molecules*, 25(6), 1357.

Murugesan, R., & Orsat, V. (2011). Spray drying for the production of nutraceutical ingredients—A review. *Food and Bioprocess Technology*, 5(1), 3–14. doi: 10.1007/s11947-011-0638-z

Nesterenko, A., Alric, I., Silvestre, F., & Durrieu, V. (2014). Comparative study of encapsulation of vitamins with native and modified soy protein. *Food Hydrocolloids*, 38, 172–179. doi: 10.1016/j.foodhyd.2013.12.011

Nizori, A., Bui, L. T. T., & Small, D. M. (2012). Microencapsulation of ascorbic acid by spray drying: Influence of process conditions. *International Journal of Chemical and Molecular Engineering*, 6(12), 1123–1127.

Oliveira, A. M., Guimarães, K. L., Cerize, N. N., Tunussi, A. S., & Poço, J. G. (2013). Nano spray drying as an innovative technology for encapsulating hydrophilic active pharmaceutical ingredients (API). *Journal of Nanomedicine & Nanotechnology*, 4(6), 1–6.

Ongen, B., Kabaroglu, C., & Parıldar, Z. (2008). Biochemical and laboratory evaluation of vitamin D. *Türk Klinik Biyokimya Dergisi*, 6(1), 23–31.

Palmieri, G. F., Bonacucina, G., Martino, P. D., & Martelli, S. (2001). Spray-drying as a method for microparticulate controlled release systems preparation: Advantages and limits. I. Water-soluble drugs. *Drug Development and Industrial Pharmacy*, 27(3), 195–204. doi: 10.1081/ddc-100,000,237

Pérez-Masiá, R., López-Nicolás, R., Periago, M. J., Ros, G., Lagaron, J. M., & López-Rubio, A. (2015). Encapsulation of folic acid in food hydrocolloids through nanospray drying and electrospraying for nutraceutical applications. *Food Chemistry*, 168, 124–133.

Pierucci, A. P. T. R., Andrade, L. R., Baptista, E. B., Volpato, N. M., & Rocha-Leão, M. H. M. (2006). New microencapsulation system for ascorbic acid using pea protein concentrate as coat protector. *Journal of Microencapsulation*, 23(6), 654–662. doi: 10.1080/02652040600776523

Pierucci, A. P. T. R., Andrade, L. R., Farina, M., Pedrosa, C., & Rocha-Leão, M. H. M. (2007). Comparison of α-tocopherol microparticles produced with different wall materials: Pea protein a new interesting alternative. *Journal of Microencapsulation*, 24(3), 201–213. doi: 10.1080/02652040701281167

Pocobelli, G., Peters, U., Kristal, A. R., & White, E. (2009). Use of supplements of multivitamins, vitamin C, and vitamin E in relation to mortality. *American Journal of Epidemiology*, 170(4), 472–483. doi: 10.1093/aje/kwp167

Potter, N. N., & Hotchkiss, J. H. (1995). *Food Science*. Food Science Text Series. Chapman and Hall: New York. doi: 10.1007/978-1-4615-4985-7

Powers, H. J. (2003). Riboflavin (vitamin B-2) and health. *The American Journal of Clinical Nutrition*, 77(6), 1352–1360. doi: 10.1093/ajcn/77.6.1352

Quintanilla-Carvajal, M. X., Hernández-Sánchez, H., Alamilla-Beltrán, L., Zepeda-Vallejo, G., Jaramillo-Flores, M. E., Perea-Flores, M. D. J., ... Gutiérrez-López, G. F. (2014). Effects of microfluidisation process on the amounts and distribution of encapsulated and non-encapsulated α-tocopherol microcapsules obtained by spray drying. *Food Research International*, 63, 2–8. doi: 10.1016/j.foodres.2014.05.025

Ray, S., Raychaudhuri, U., & Chakraborty, R. (2016). An overview of encapsulation of active compounds used in food products by drying technology. *Food Bioscience*, 13, 76–83. doi: 10.1016/j.fbio.2015.12.009

Reavley, N. (2000). *Vitamin Counter: The Vitamin Content of Common Foods*. Bookman Press: Melbourne.

Rezvankhah, A., Emam-Djomeh, Z., & Askari, G. (2019). Encapsulation and delivery of bioactive compounds using spray and freeze-drying techniques: A review. *Drying Technology*, 38(1–2), 235–258. doi: 10.1080/07373937.2019.1653906

Rodríguez-Hernández, G., González-García, R., Grajales-Lagunes, A., Ruiz-Cabrera, M., & Abud-Archila, M. (2005). Spray-drying of cactus pear juice (*Opuntia streptacantha*): Effect on the physicochemical properties of powder and reconstituted product. *Drying Technology*, 23(4), 955–973. doi: 10.1080/drt-200054251

Roje, S. (2007). Vitamin B biosynthesis in plants. *Phytochemistry*, 68, 1904–1921. doi: 10.1016/j.phytochem.2007.03.038

Saha, S., & Roy, M. N. (2017). Encapsulation of vitamin C into β-Cyclodextrin for advanced and regulatory release. In *Vitamin C*, A. H. Hamza (ed), InTech: Rijeka. doi: 10.5772/intechopen.70035

Selamat, S. N., Mohamad, S. N. H., Muhamad, I. I., Khairuddin, N., & Md Lazim, N. A. (2018). Characterization of spray-dried palm oil vitamin E concentrate. *Arabian Journal for Science and Engineering*, 43(11), 6165–6169. doi: 10.1007/s13369-018-3362-4

Shi, X.-Y., & Tan, T.-W. (2002). Preparation of chitosan/ethylcellulose complex microcapsule and its application in controlled release of vitamin D2. *Biomaterials*, 23(23), 4469–4473. doi: 10.1016/s0142-9612(02)00165-5

Stoll, B. (1998). Breast cancer and the Western diet: Role of fatty acids and antioxidant vitamins. *European Journal of Cancer*, 34(12), 1852–1856. doi: 10.1016/s0959-8049(98)00204-4

Sun, Y. (1990). Free radicals, antioxidant enzymes, and carcinogenesis. *Free Radical Biology and Medicine*, 8(6), 583–599. doi: 10.1016/0891-5849(90)90156-d

Trindade, M. A., & Grosso, C. R. (2000). The stability of ascorbic acid microencapsulated in granules of rice starch and in gum Arabic. *Journal of Microencapsulation*, 17(2), 169–176. doi: 10.1080/026520400288409

Uddin, M. S., Hawlader, M. N., Zhu, H. J. (2001). Microencapsulation of ascorbic acid: Effect of process variables on product characteristics. *Journal of Microencapsulation*, 18(2), 199–209. doi: 10.1080/02652040010000352

Watanabe, Y., Fang, X., Minemoto, Y., Adachi, S., & Matsuno, R. (2002). Suppressive effect of saturated acyll-ascorbate on the oxidation of linoleic acid encapsulated with maltodextrin or gum Arabic by spray drying. *Journal of Agricultural and Food Chemistry*, 50(14), 3984–3987. doi: 10.1021/jf011656u

Williams, P. G. (1996). Vitamin retention in cook/chill and cook/hot-hold hospital foodservices. *Journal of the American Dietetic Association*, 96(5), 490–498. doi: 10.1016/s0002-8223(96)00135-6

Wilson, N., & Shah, N. P. (2007). Microencapsulation of vitamins. *ASEAN Food Journal*, 14(1), 1–14.

Xie, Y.-L., Zhou, H.-M., Liang, X.-H., He, B.-S., & Han, X.-X. (2010). Study on the morphology, particle size and thermal properties of vitamin A microencapsulated by starch octenylsucciniate. *Agricultural Sciences in China*, 9(7), 1058–1064. doi: 10.1016/s1671-2927(09)60190-5

Yoo, S.-H., Song, Y.-B., Chang, P.-S., & Lee, H. G. (2006). Microencapsulation of α-tocopherol using sodium alginate and its controlled release properties. *International Journal of Biological Macromolecules*, 38(1), 25–30. doi: 10.1016/j.ijbiomac.2005.12.013

Zuidam, N. J., & Heinrich, E. (2009). Encapsulation of aroma. *Encapsulation Technologies for Active Food Ingredients and Food Processing*, 127–160. doi: 10.1007/978-1-4419-1008-0_5.

7 Spray Drying Encapsulation of Minerals

Handan Basunal Gulmez, Emrah Eroglu, and Ayhan Topuz
Akdeniz University, Turkey

CONTENTS

7.1 INTRODUCTION

Minerals, which make up approximately 4–6% of the human body, are inorganic compounds that play a vital role in many basic body systems and functions such as skeletal, muscular and tissue, nervous, hormonal, and circulatory systems. While some minerals (e.g., Ca, Cl, Mg, P, K, and Na) are abundant in the body, others (e.g., Cu, Fe, F, Mn, Se, and Zn) are found in lower quantities. Generally speaking, these compounds are classified into macro, micro (trace), and ultratrace minerals. For the body to function fully and properly, humans need to consume certain amounts of 20–25 different mineral substances together with other nutrients (Gupta & Gupta, 2014). Minerals with an amount of more than 100 mg required by our body are grouped as major minerals; those below this level are known as micro or trace minerals. In this context, calcium, magnesium, phosphorus, potassium, sodium, and sulfur are grouped as macrominerals and other minerals needed by our body as regarded as microminerals (Gharibzahedi & Jafari, 2017a; Medeiros & Wildman, 2019). The term 'ultratrace' is also used for the description of the minerals that are found in the diet in extremely small quantities (e.g., arsenic, boron, nickel, silicon, and vanadium). Humans consume these minerals in micrograms every day and they are present in human tissue.

Minerals give structure and hardness to various systems in the human body, while they also take part in critical vital functions directly or indirectly. In the case of a mineral deficiency, the metabolic activities, in which the relevant mineral is involved, do not take place properly, resulting in the occurrence of various disorders/diseases (Table 7.1). A balanced diet with diverse food ingredients is important to avoid mineral deficiency in the body. Although we need fewer microminerals than macrominerals, malnutrition in terms of microminerals is more common (DiSilvestro, 2004; Gupta & Gupta, 2014; Medeiros & Wildman, 2019; Samman et al., 2017).

TABLE 7.1

Health Problems Related to the Common Mineral Deficiencies

Deficient Minerals	Health Problems	References
Calcium	Osteomalacia, osteoporosis	Haas, Canary, Kyle, Meyer, and Schaaf (1963); Nordin (1960)
Magnesium	Fatigue, insomnia, involuntary tremors of muscles, brittleness of hair and nails	Faryadi (2012); Medeiros and Wildman (2019)
Sodium	Hyponatremia, muscle disorder (including the cardiac muscle)	Graudal, Hubeck-Graudal, and Jurgens (2017); Pohl, Wheeler, and Murray (2013)
Potassium	Cardiovascular diseases, labored breathing, glucose intolerance	DiSilvestro (2004); Fairweather-Tait and Cashman (2015)
Phosphorous	Loss of calcium, loss of appetite, muscle pain	Takeda, Ikeda, and Nakahashi (2012)
Iron	Anemia, mental disorders, fatigue, dizziness, intestinal malabsorption	Strain and Cashman (2009); Welch (2002)
Zinc	Dwarfism, slowing of skeletal development, dermatitis, diarrhea, baldness, weakening of the immune system, loss of appetite, neuropsychiatric disorders, delayed healing of wounds	DiSilvestro (2004); Guo et al. (2014); Medeiros and Wildman (2019); Strain and Cashman (2009); Welch (2002)
Selenium	Keshan's disease, muscular weakness, premature aging, nervous system disorders, and mental retardation	Contempré et al. (1992); Strain and Cashman (2009)
Copper	Low birth weight, bone fractures, anemia, premature baby	Medeiros and Wildman (2019); Strain and Cashman (2009)
Iodine	Goiter, thyroid failure, mental dysfunction, stillbirth, congenital abnormalities, cretinism	Samman, Skeaff, Thomson, and Truswell (2017); Strain and Cashman (2009)

7.2   MINERAL DEFICIENCY IN HUMANS

The requirement of minerals in the body is less than other nutrients. Although their requirements are less, some minerals are less than the required amount in the body due to various reasons and this situation negatively affects body functions. In addition, our need for minerals may vary according to lifestyle, health status, age, smoking, and alcohol consumption (Campbell, 2001; Eichholzer, 2003). The average recommended daily intake (RDI) amount of minerals for an adult and healthy person are given in Table 7.2.

Today, millions of people are fed insufficiently in terms of micronutrients; thus, various disorders occur because of this malnutrition. Globally speaking, iron, zinc, iodine, selenium, calcium, magnesium, and copper deficiency are among the mineral substances most observed (Stein, 2010; White & Broadley, 2009). Mineral malnutrition is more common, especially in developing countries. As mentioned above, the mineral requirement of individuals is being affected by lifestyle, nutritional habits (the excessive or inadequate consumption of certain food groups, vegetarian diet and nutritional disorders, medical conditions), and social factors (economic conditions of countries, education levels, current food production policies, health systems, and nutritional habits). Furthermore, the chemical structure and solubility of minerals, nutrients interactions affect the absorption of minerals in the body (Black, 2003; Bogden & Klevay, 2000; Bouis, Boy-Gallego, & Meenakshi, 2012; Farrar et al., 2013; Jamison et al., 2006; Müller & Krawinkel, 2005).

TABLE 7.2

The Recommended Daily Intake (RDI) Values of Some Minerals. Modified from Gupta and Gupta (2014)

Mineral(s)	Recommended Daily Intake (RDI, mg/day)
Calcium	1000
Magnesium	350
Sodium	2400
Potassium	3500
Chloride	3400
Phosphorus	1000
Iron	15
Zinc	15
Selenium	0.035
Copper	2
Manganese	5
Iodine	0.15
Other trace minerals (total)	20–25

7.2.1 NUTRITIONAL HABITS AND MINERAL DEFICIENCY

The nutritional habits of individuals directly affect the mineral composition and their amount in the body. Minerals that are necessary for our body are obtained from foods through diet. Mineral composition of foodstuffs and their bioavailability may vary according to the type of foods, their origins, and processing conditions. For this reason, choices such as excessive or inadequate consumption of certain food groups or not consuming them at all are effective on mineral deficiency (Gupta & Gupta, 2014; Medeiros & Wildman, 2019; Müller & Krawinkel, 2005).

In the presence of excess fat or fiber in the diet, minerals such as calcium can form complex compounds with these components, which has a reducing effect on calcium absorption in the body. The absorption of iron (heme iron) found in animal foods is much more common than iron (non-heme iron) in vegetable sources. For this reason, iron deficiency in humans is more common with meat-free, vegetarian, and vegan-style diets (Medeiros & Wildman, 2019). In parallel with increasing the technology and development levels of the countries, there has also been an increase in the production and consumption of foods that are overly processed and contain a large amount of fat and sugar. These foods contain lower levels of manganese and people who are in a habit of consuming such foods have a higher incidence of manganese deficiency (Freeland-Graves, Mousa, & Kim, 2016).

7.2.2 INTERACTIONS OF MINERALS AND NUTRIENTS

Nowadays, mineral supplementation/fortification (e.g., calcium, iron, magnesium, manganese, and zinc) is attracting increasing attention. However, the direct addition of mineral salts has many negative consequences (Augustin, 2003). Some substances in the food can affect the bioavailability of minerals during digestion. These organic or inorganic substances have either positive or negative effects on the bioavailability of various minerals (Fairweather-Tait, 1996). Therefore, in many cases, mineral deficiency may occur, even in the case of people consuming a mineral-rich diet or direct mineral enrichments.

Phytic acid, tannins, and polyphenols are compounds that are reported to have negative effects on the absorption of minerals. Phosphate groups, which are found in the structure of phytic acid, can be partially ionized negatively at acidic pH values in the stomach and become converted into phytate anion. Positively charged minerals such as calcium, iron, and zinc can react with these functional groups, forming a salt and limiting the bioavailability of the specific mineral (Barrientos & Murthy, 1996; Gupta & Gupta, 2014; Lopez, Leenhardt, Coudray, & Remesy, 2002; Schlemmer, Frølich, Prieto, & Grases, 2009). In addition to phytic acid, some groups of compounds such as tannins, phenolic compounds, and vitamins also have significant effects on mineral bioavailability. Tannins found in herbal foods and phenolic compounds found in tea may limit the bioavailability of iron (Akhter, Mohammad, Orfi, Ahmad, & Rehman, 2005; Fairweather-Tait, 1996). Vitamin C, which forms iron ascorbate with non-heme iron, can increase the bioavailability of iron by making iron digestible in a wide pH range (Lopez et al., 2002). Similarly, vitamin D plays a critical role in the absorption of calcium. As a result of various metabolic activities in the presence of vitamin D in the body, calbindin, a calcium-binding protein, is produced and regulates calcium absorption (Lopez et al., 2002; Medeiros & Wildman, 2019).

As well as the interaction of minerals with other compounds, their interactions with other minerals are also effective on their bioavailability. Minerals that are chemically similar in nutrition and/ or have the same transport system may have either an increasing or reducing effect on each other's absorption (Lopez et al., 2002; SandstroÈm, 2001). Increasing the amount of calcium and polyphosphate in the diet has an adverse effect on iron and zinc absorption, depending on concentration. While the absorption of iron and copper is positively dependent on each other (up to certain concentrations), the presence of iron and zinc has a partially negative effect on zinc and copper absorption, respectively. Similarly, other microminerals have increasing or decreasing effects on each other's absorption (SandstroÈm, 2001; Strain & Cashman, 2009).

As mentioned above, mineral deficiency can be observed in individuals due to various factors, and this can lead to health problems and even, in severe cases, may result in death. It is reported that, especially in the case of developing countries, iron, zinc, and iodine deficiencies are common, and all have negative effects on human health. In addition, calcium and selenium deficiencies are also frequent. In addition to these minerals, magnesium, and copper deficiencies are also regionally reported (Stein, 2010). To overcome these drawbacks of mineral deficiency, different food processing approaches are still under investigation.

7.2.3 Overcoming Mineral Deficiency in Humans

So far, different techniques such as mineral-rich agricultural production (Bouis, 2007; Frossard, Bucher, Mächler, Mozafar, & Hurrell, 2000; Grusak & DellaPenna, 1999; Yang, Alidoust, & Wang, 2020), drug and food supplements (Churio, Pizarro, & Valenzuela, 2018; Diosady, Alberti, & Mannar, 2002), and increasing the bioavailability of minerals (Blanco-Rojo & Vaquero, 2019) have been applied in order to prevent/lessen mineral deficiency. Food fortification has been carried out by the food industry for years as a feasible and cost-effective approach to minimize micronutrient deficiencies (Black, 2003; Gharibzahedi & Jafari, 2017a). However, some changes in the micronutrients, such as stability, solubility, and interactions with other nutrients, are restricting its direct use as a food fortification agent (Nedovic, Kalusevic, Manojlovic, Levic, & Bugarski, 2011).

Encapsulated minerals, which are the preliminary stage of fortification, can be successfully used in mineral fortification in the food industry. Different encapsulation techniques increase the bioavailability of the minerals in nutrition by limiting their interaction with nutrients in foods (Gharibzahedi & Jafari, 2017b; Li, González, & Diosady, 2014). Furthermore, the encapsulation of micronutrients such as minerals preserves the desired taste and aroma of the products to which it is added, by masking undesired characteristics of the micronutrient (if there is any) (Gharibzahedi & Jafari, 2017a; Gouin, 2004). This technique can also play a critical role in the fortification of vegans' and vegetarians' diets. These foods can be enriched with encapsulated minerals such as calcium, iron, zinc, etc.

Numerous encapsulation techniques have been developed to produce fortified and functional foods. In the literature, it is reported that minerals can be encapsulated by physicochemical methods, such as entrapping in liposomes (Jizomoto & Hirano, 1989) and noisome (Gutierrez et al., 2016), esterification with fatty acids (Kwak, Yang, & Ahn, 2003), coacervation (Meng et al., 2019), emulsification (Jiménez-Alvarado, Beristain, Medina-Torres, Román-Guerrero, & Vernon-Carter, 2009), freeze drying (Bagci & Gunasekaran, 2017), fluidized bed drying, extrusion (Li, Yadava, Lo, Diosady, & Wesley, 2011), spray cooling (Wegmüller, Zimmermann, Bühr, Windhab, & Hurrell, 2006), and spray drying (Filiponi, Gaigher, Caetano-Silva, Alvim, & Pacheco, 2019). Of these encapsulation techniques, spray drying is one of the most preferred and common methods for the encapsulation of minerals, due to its advantages such as being flexible, continuous, and economic (Nedovic et al., 2011). For these reasons, the spray drying encapsulation technique is an innovative approach that can be applied in preventing mineral deficiency through enriching foods with encapsulated minerals with increased bioavailability (Fathima, Nallamuthu, & Khanum, 2017).

7.3 ENCAPSULATION OF MINERALS BY SPRAY DRYING

Spray drying is a versatile food processing technique which is suitable for producing functional products in powder form from several liquid nutritious materials. A spray drying unit involves several subunits, such as an air heater, an atomizer, a feeding pump, a cyclone separator, and a collection vessel. Murugesan and Orsat (2012) reviewed that the spray drying process enables the production of a uniform fine powder with low water activity, high solubility, and high stability during storage and transportation. The properties of the final product mainly depend on feed flowrate, the viscosity of the feeding solution, inlet and outlet temperatures, type, and pressure of atomizer (Tonon, Brabet, & Hubinger, 2008). Recently, the use of the spray drying technique in the encapsulation of minerals has become increasingly common, since it is a rapid process and particle size distribution and other characteristics of the final products can be controlled easily (Obón, Castellar, Alacid, & Fernández-López, 2009).

Like many functional components, the encapsulation of essential minerals is becoming increasingly popular and has been studied extensively in recent years. Generally, the spray drying encapsulation of minerals is carried out in two consecutive stages. In the first stage, the hydrated mineral is mixed with a suitable carrier/wall/coating material/agent. In the second stage, the mixture is homogenized by various homogenization techniques to entrap the minerals within the carrier agent. The type and efficiency of the homogenization pretreatment are related to the particle size distribution of the mixtures, and, thereby, the features of the final product. The mixture is then fed into the atomizer of the spray drying unit to evaporate water from the product, in order to dry up the particles and complete the encapsulation process (Abd El-Kader & Abu Hashish, 2020; Murugesan & Orsat, 2012; Risch, 1995).

In the spray drying encapsulation process, several factors can affect the encapsulation efficiency, and product yield; e.g., temperature and relative humidity (RH) of the air, airflow rate, atomization rate, feeding rate, and dry matter content of the mixture (Murugesan & Orsat, 2012). The capsules manufactured by the spray drying process must have the qualities of being able to protect the encapsulated material from the external factors, maintain its properties throughout storage, ensure its release under the desired conditions, and not change the taste and odor properties of the food into which it is incorporated (Gharibzahedi & Jafari, 2017a; Nedovic et al., 2011).

7.3.1 WALL MATERIALS

The wall materials are generally film-forming biological molecules that cover core materials and prevent them from external effects (Murugesan & Orsat, 2012; Shahidi & Han, 1993). In order to use any biological source as wall materials, first, it must be on the GRAS list (Abd El-Kader & Abu Hashish, 2020). Individual carbohydrates, proteins, lipids, or their combinations can be selected as the wall material, by considering the core material to be encapsulated and the desired properties of the end products. In addition, several polymers suitable for use in the food industry can also be used as wall material, either alone or in combination with the former materials (Gharibzahedi & Jafari, 2017a; Murugesan & Orsat, 2012). The wall material can directly affect core material stability and its bioavailability (Fathima et al., 2017; Khan, Iqbal, Khan, & Khalid, 2020; Murugesan & Orsat, 2012; Shahidi & Han, 1993).

A carrier (wall material) to be used in the encapsulation of minerals is expected to inhibit their oxidation reactions, restrict their interactions with other nutrients, and release the minerals at the target part of the digestive tract. In the literature, different wall materials such as cellulose (Oneda & Ré, 2003), whey protein isolate (Azzam, 2009; Cavallini & Rossi, 2009; Jayalalitha, Balasundaram, & Kumar, 2012), oil (Jalili, 2016; Nkhata, 2013), carboxymethyl cellulose (CMC), maltodextrin, and gum Arabic (Bryszewska et al., 2019; Choi, Decker, & McClements, 2009; Diosady et al., 2002; Dueik & Diosady, 2017; Kim, Kim, Jeong, & Kim, 2006; Li, Diosady, & Wesley, 2009; Porrarud & Pranee, 2010; Romita, Cheng, & Diosady, 2011) have been used.

7.4 BIOAVAILABILITY OF THE ENCAPSULATED MINERALS

One of the most important factors determining product quality in the encapsulation process is the improvement in the bioavailability of the encapsulated mineral(s). The balance of minerals digested and excreted in the gastrointestinal tract is expressed by its bioavailability (McGhie & Walton, 2007; Moretti & Zimmermann, 2016). The stability and bioavailability of an encapsulated mineral vary depending on the encapsulated mineral, wall material, wall material to core ratio, and encapsulation efficiency. The digestion of the encapsulated minerals is mainly carried out by dissolving in digestive conditions and releasing the core from capsules and the absorption by the gastrointestinal tract and transfer to tissues (Moretti & Zimmermann, 2016). In general, the bioavailability of minerals is increased by increasing their water solubility once they are encapsulated (Khan et al., 2020). However, the presence of inhibitory substances in the digestion medium and external factors such as the individual's gender, age, and health status are also effective on mineral bioavailability (Teucher, Olivares, & Cori, 2004).

7.5 APPLICATIONS OF THE SPRAY DRIED ENCAPSULATED MINERALS IN FUNCTIONAL FOOD PRODUCTS

In a study on the enrichment of bread by iron sulfate and iron fumarate, these iron salts were encapsulated in heat-resistant modified starch as a carrier. The results showed that the encapsulation process increased the bioavailability of iron (Bryszewska et al., 2019). Fortification of salt with iron is also a very useful approach to eliminate iron deficiency. However, iron and iodine interact with each other in the salt. In order to minimize such an interaction, the encapsulation process by spray drying is a very useful approach. The effect of inlet air temperature (125°C–170°C) and flow rate of feeding (0.36–0.90 L/h) on yield was determined by spray drying ferrous fumarate suspension containing the coating materials HPMC E15 (2% w/v) and Dextrin DE7 (8% w/v). The study demonstrated that ferrous fumarate microcapsules, when incorporated in iodized salt, produce double-fortified salt that is stable for six months at elevated temperature (40°C) and humidity (40%–60%) (Romita et al., 2011).

Singh, Siddiqui, and Diosady (2018) studied the release of reverse enteric spray dried iron sulfate microcapsules with ready-mix polymers (Eudragit E PO, Evonik Nutrition & Care GmbH, Germany) and chitosan under different pH conditions (1, 4, and 7). The experimental conditions of spray drying were set as follows for chitosan; inlet air temperature of 150°C, feed flow rate of 3 mL/min, atomizer gas flow rate of 667 L/h, the pressure 618 kPa, and the aspirator rate at 4.5 Pa. Eudragit EPO solution was dried at an inlet air temperature of 110°C and the other conditions were kept same. It is reported that they obtained microcapsules with minimal release at neutral pH. Microcapsules manufactured using ready-to-mix polymer containing 40% iron sulfate and chitosan encapsulated with 30% iron sulfate were a good candidate to produce ferrous sulfate capsules.

Fortification of food with calcium is another challenge in the food industry. Food is generally fortified with different forms of calcium, such as calcium triphosphate, calcium citrate, and calcium lactate. In the case of direct fortification of calcium in food formulation, especially the foods rich in protein, there are common precipitation and dissociation problems of naturally occurring protein, as stated for soymilk. In a previous calcium fortification study on soymilk, it was reported that the precipitation problem was eliminated by the encapsulation of calcium with lecithin (Schrooyen, van der, & De, 2001). In one of the later studies (Oneda & Ré, 2003), calcium citrate monohydrate and calcium lactate were encapsulated with different coating materials (sodium carboxymethylcellulose, hydroxy methylcellulose, and Eudragit RNE 30D) by using the spray drying process at different operating conditions. The calcium microcapsules were tested for the release characteristics in a model gastrointestinal tract. Additionally, some physical properties of the microcapsules were determined. The results showed that the rate of release from both calcium sources could be controlled by polymer type, formulation conditions (calcium source/polymer ratio), and pH regulations (Oneda & Ré, 2003).

Iodine fortification of salt is another common commercial application in practice and there are some stability problems like rapid releasing of this mineral, due to impurities and the presence of moisture. To overcome this problem, potassium iodide was encapsulated in different proportions (0.5%, 1%, and 2%) by spray drying, where gelatin and modified starch were used as wall materials. It was observed that the lowest iodine loss in the powder obtained was in the microcapsules containing modified starch. The samples had more than 75% of their iodine content at 40°C and 100% RH for 12 months (Diosady et al., 2002).

There have been similar fortification efforts on maize flour and tortillas with ferrous fumarate. The ferrous fumarate accompanied by folic acid was encapsulated in acetylated normal and waxy maize starches by using spray drying. The encapsulation process was successfully performed at different operating conditions. Although the encapsulated ferrous fumarate content was increased (almost threefold fortification), the fresh and stored tortillas were not affected in terms of their physical properties such as moisture content, cutting values, tensile strength, and color (Martínez-Bustos, Cruz Sánchez, Ortega-Martínez, & Aguilar-Palazuelos, 2018).

Dueik and Diosady (2017) studied the formation of reversed enteric-coated iron microparticles with either Eudragit EPO or chitosan as carriers to stabilize iron in the presence of moisture and iodine. In this study, after the determination of the maximum amount of iron for encapsulation, the spray drying conditions were optimized as follows; inlet air temperature of 130°C and liquid flow rate of 2.3 mL/min. The results showed that Eudragit EPO was not suitable for iron encapsulation, but chitosan was an effective iron coating wall material in terms of both functionality and morphology, with loadings up to 25%. It was also determined that iodine in double-fortified salt showed desirable stability for 12 weeks of storage by retaining 90% of iodine. A previous study was also carried out to produce and characterize different types of heme and non-heme iron-maltodextrin microparticles (Churio & Valenzuela, 2018). Different concentrations of bovine erythrocytes as heme iron and ferrous sulfate as non-heme iron were encapsulated using an optimal maltodextrin concentration (40% w/v) by spray drying, where the process conditions were set as follows; inlet air temperature of 130°C–140°C, outlet air temperature of 90°C–100°C, an airflow of 500 L/h, feeding rate of 8 mL/min, and an atomization pressure of 20 psi. According to the results, the iron content of ferrous sulfate microparticles was more than the iron content of the bovine erythrocytes as expected. However, in this study, it was recommended to use the combination of ferrous sulfate and bovine erythrocytes microparticles together, concerning a higher bioavailability (Churio & Valenzuela, 2018).

In another study (Estrada et al., 2018), iron was encapsulated in pea protein and soy protein to test their oxidation difference. The iron incorporations were performed by spray drying maintained at inlet and outlet air temperatures of 180°C and 100°C, respectively. The feeding solution was pumped at 6 mL/min (20%) to the nozzle and the atomized solution was dried at an aspiration rate of 90%. The physicochemical stability of iron-loaded pea protein particles and the iron-fortified soy protein-based fibrous structures were measured in protein oxidation. The incorporation of iron (free or encapsulated) did not affect carbonyl content in the fibrous products, but the process conditions for making such products induced the formation of carbonyls (Estrada et al., 2018).

Bai, Hong, He, Hong, and Tan (2017) studied selenium encapsulation with chitosan as wall material using spray drying technology. The results showed that the encapsulation of selenium with chitosan increased selenium efficiency and decreased selenium toxicity in the body. In addition, the encapsulated selenium was determined to be stable for 28 days. Mulia, Putri, Krisanti, and Handayani (2019) also investigated the iron release from spray dried capsules, prepared with chitosan as wall material and tripolyphosphate (0%, 1%, 2%, and 3%) as the cross-linking agent, in the gastric and intestinal fluid system. Some physicochemical properties of the capsules (e.g., yield, loading capacity, encapsulation efficiency, and particle size) were also determined. The results showed that tripolyphosphate could be used for the controlled release of iron, as the release time of iron increased due to the increase in the ratio of tripolyphosphate.

Sodium chloride was spray dried at different inlet temperatures (130°C,140°C,150°C,160°C and 170°C) and atomization pressure (60, 100, 140, 150, and 220 kPa) using maltodextrin as wall material (Cho, Kim, Chun, & Choi, 2015). Spray drying conditions were as follows: blow power of 0.6 m³/min, feeding rate of 500 mL/h, and outlet temperature of 80°C–90°C. Atomization pressure was more effective than inlet temperature to produce the smallest particle of maltodextrin/NaCl complexes. It was concluded that the microcapsules could increase the salinity density with a small amount of salt (Cho et al., 2015).

7.6   CONCLUDING REMARKS

For the vital functions of the human body to continue as required, individuals need to consume certain amounts of mineral substances along with other nutrients. However, a significant number of people still suffer from mineral deficiencies, due to the low amount of some minerals in foods and/or their low bioavailability. Encapsulation applications, which can be achieved by different techniques (e.g., spray drying, fluid bed drying, liposome entrapping, coacervation, and freeze drying), increase the stability, controlled release, and bioavailability of certain minerals. Spray drying is one of the techniques that is commonly used for the encapsulation of bioactive components such as minerals. There have been several efforts to encapsulate minerals by spray drying to prevent mineral–nutrient interactions and increase their bioavailability. However, several systematic studies are yet required to determine the detailed interactions of minerals with inhibitors, activators, and nutrients, as well as the investigation of their properties such as loading capacity and releasing rate. Moreover, the efficiency of different carrier substances in mineral encapsulation needs to be investigated in terms of physicochemical properties. All these efforts should be supported by both *in vitro* and *in vivo* bioavailability experiments.

REFERENCES

Abd El-Kader, A., & Abu Hashish, H. (2020). Encapsulation techniques of food bioproduct. *Egyptian Journal of Chemistry*, *63*(5), 1881–1909.

Akhter, P., Mohammad, D., Orfi, S., Ahmad, N., & Rehman, K. (2005). Assessment of daily iron intake for the Pakistani population. *Nutrition & Food Science*, *35*(2), 109–117.

Augustin, M. (2003). The role of microencapsulation in the development of functional dairy foods. *Australian Journal of Dairy Technology*, *58*(2), 156.

Azzam, M. (2009). Effect of fortification with iron-whey protein complex on quality of yoghurt. *Egyptian Journal of Dairy Science*, *37*(1), 55–63.

Bagci, P. O., & Gunasekaran, S. (2017). Iron-encapsulated cold-set whey protein isolate gel powder - Part 1: Optimisation of preparation conditions and invitro evaluation. *International Journal of Dairy Technology*, *70*(1), 127–136.

Bai, K., Hong, B., He, J., Hong, Z., & Tan, R. (2017). Preparation and antioxidant properties of selenium nanoparticles-loaded chitosan microspheres. *International Journal of Nanomedicine*, *12*, 4527.

Barrientos, L. G., & Murthy, P. P. (1996). Conformational studies of myo-inositol phosphates. *Carbohydrate Research*, *296*(1–4), 39–54.

Black, R. (2003). *Micronutrient deficiency: an underlying cause of morbidity and mortality* (Vol. 83, pp. 79): Bulletin of the World Health Organization.

Blanco-Rojo, R., & Vaquero, M. P. (2019). Iron bioavailability from food fortification to precision nutrition. A review. *Innovative Food Science & Emerging Technologies*, *51*, 126–138.

Bogden, J. D., & Klevay, L. M. (2000). *Clinical nutrition of the essential trace elements and minerals: the guide for health professionals* (1 ed.), Totowa, NJ: Springer Science & Business Media.

Bouis, H., Boy-Gallego, E., & Meenakshi, J. V. (2012). Micronutrient malnutrition: causes, prevalence, consequences, and interventions In T. W. Bruulsema, P. Heffer, R. M. Welch, I. Cakmak, & K. Moran (Eds.), *Fertilizing Crops to Improve Human Health: A Scientific Review* (1 ed., pp. 97–122), Norcross, GA: International Plant Nutrition Institute, and IFIA.

Bouis, H. E. (2007). The potential of genetically modified food crops to improve human nutrition in developing countries. *The Journal of Development Studies*, *43*(1), 79–96.

Bryszewska, M. A., Tomás-Cobos, L., Gallego, E., Villalba, M., Rivera, D., Taneyo Saa, D. L., & Gianotti, A. (2019). In vitro bioaccessibility and bioavailability of iron from breads fortified with microencapsulated iron. *LWT*, *99*, 431–437.

Campbell, J. (2001). Lifestyle, minerals and health. *Medical Hypotheses*, *57*(5), 521–531.

Cavallini, D. C. U., & Rossi, E. A. (2009). Soy yogurt fortified with iron and calcium: Stability during the storage. *Alimentos e Nutrição Araraquara*, *20*(1), 7–13.

Cho, H.-Y., Kim, B., Chun, J.-Y., & Choi, M.-J. (2015). Effect of spray-drying process on physical properties of sodium chloride/maltodextrin complexes. *Powder Technology*, *277*, 141–146.

Choi, S. J., Decker, E. A., & McClements, D. J. (2009). Impact of iron encapsulation within the interior aqueous phase of water-in-oil-in-water emulsions on lipid oxidation. *Food Chemistry*, *116*(1), 271–276.

Churio, O., Pizarro, F., & Valenzuela, C. (2018). Preparation and characterization of iron-alginate beads with some types of iron used in supplementation and fortification strategies. *Food Hydrocolloids*, *74*, 1–10.

Churio, O., & Valenzuela, C. (2018). Development and characterization of maltodextrin microparticles to encapsulate heme and non-heme iron. *LWT*, *96*, 568–575.

Contempré, B., Duale, N., Dumont, J. E., Ngo, B., Diplock, A., & Vanderpas, J. (1992). Effect of selenium supplementation on thyroid hormone metabolism in an iodine and selenium deficient population. *Clinical Endocrinology*, *36*(6), 579–583.

Diosady, L., Alberti, J., & Mannar, M. V. (2002). Microencapsulation for iodine stability in salt fortified with ferrous fumarate and potassium iodide. *Food Research International*, *35*(7), 635–642.

DiSilvestro, R. A. (2004). *Handbook of minerals as nutritional supplements*. Boca Raton, FL: CRC Press.

Dueik, V., & Diosady, L. L. (2017). Microencapsulation of iron in a reversed enteric coating using spray drying technology for double fortification of salt with iodine and iron. *Journal of Food Process Engineering*, *40*(2), e12376.

Eichholzer, M. (2003). Micronutrient deficiencies in Switzerland: Causes and consequences. *Journal of Food Engineering*, *56*(2), 171–179.

Estrada, P. D., Berton-Carabin, C. C., Schlangen, M., Haagsma, A., Pierucci, A. P. T., & van der Goot, A. J. (2018). Protein oxidation in plant protein-based fibrous products: Effects of encapsulated iron and process conditions. *Journal of Agricultural and Food Chemistry*, *66*(42), 11105–11112.

Fairweather-Tait, S. J. (1996). Bioavailability of dietary minerals. *Biochemical Society Transactions*, *24*(3), 775–780. doi:10.1042/bst0240775

Fairweather-Tait, S. J., & Cashman, K. (2015). Minerals and Trace Elements. In D. M. Bier, J. Mann, D. H. Alpers, H. H. E. Vorster, & M. J. Gibney (Eds.), *Nutrition for the Primary Care Provider* (pp. 45–53), Ettlingen, Germany: Karger.

Farrar, J., Hotez, P. J., Junghanss, T., Kang, G., Lalloo, D., & White, N. J. (2013). *Manson's Tropical Diseases E-Book*: Elsevier Health Sciences.

Faryadi, Q. (2012). The magnificent effect of magnesium to human health: A critical review. *International Journal of Applied*, *2*(3), 118–126.

Fathima, S. J., Nallamuthu, I., & Khanum, F. (2017). 12 - Vitamins and minerals fortification using nano-technology: bioavailability and recommended daily allowances. In A. M. Grumezescu (Ed.), *Nutrient Delivery* (pp. 457–496), London, UK: Academic Press.

Filiponi, M. P., Gaigher, B., Caetano-Silva, M. E., Alvim, I. D., & Pacheco, M. T. B. (2019). Microencapsulation performance of Fe-peptide complexes and stability monitoring. *Food Research International, 125,* 108505.

Freeland-Graves, J. H., Mousa, T. Y., & Kim, S. (2016). International variability in diet and requirements of manganese: Causes and consequences. *Journal of Trace Elements in Medicine and Biology, 38,* 24–32.

Frossard, E., Bucher, M., Mächler, F., Mozafar, A., & Hurrell, R. (2000). Potential for increasing the content and bioavailability of Fe, Zn and Ca in plants for human nutrition. *Journal of the Science of Food and Agriculture, 80*(7), 861–879.

Gharibzahedi, S. M. T., & Jafari, S. M. (2017a). Chapter 9 - Nanoencapsulation of Minerals. In S. M. Jafari (Ed.), *Nanoencapsulation of Food Bioactive Ingredients* (pp. 333–400), Eastbourne, UK: Academic Press.

Gharibzahedi, S. M. T., & Jafari, S. M. (2017b). The importance of minerals in human nutrition: Bioavailability, food fortification, processing effects and nanoencapsulation. *Trends in Food Science & Technology, 62,* 119–132.

Gouin, S. (2004). Microencapsulation: Industrial appraisal of existing technologies and trends. *Trends in Food Science & Technology, 15*(7), 330–347.

Graudal, N. A., Hubeck-Graudal, T., & Jurgens, G. (2017). Effects of low sodium diet versus high sodium diet on blood pressure, renin, aldosterone, catecholamines, cholesterol, and triglyceride. *Cochrane Database of Systematic Reviews,* (4), PMC6478144.

Grusak, M. A., & DellaPenna, D. (1999). Improving the nutrient composition of plants to enhance human nutrition and health. *Annual Review of Plant Biology, 50*(1), 133–161.

Guo, L., Harnedy, P. A., Li, B., Hou, H., Zhang, Z., Zhao, X., & FitzGerald, R. J. (2014). Food protein-derived chelating peptides: Biofunctional ingredients for dietary mineral bioavailability enhancement. *Trends in Food Science & Technology, 37*(2), 92–105.

Gupta, U., & Gupta, S. (2014). Sources and deficiency diseases of mineral nutrients in human health and nutrition: A review. *Pedosphere, 24*(1), 13–38.

Gutierrez, G., Matos, M., Barrero, P., Pando, D., Iglesias, O., & Pazos, C. (2016). Iron-entrapped niosomes and their potential application for yogurt fortification. *LWT- Food Science and Technology, 74,* 550–556. doi:10.1016/j.lwt.2016.08.025

Haas, H. G., Canary, J. J., Kyle, L. H., Meyer, R. J., & Schaaf, M. (1963). Skeletal calcium retention in osteo-porosis and in osteomalacia. *The Journal of Clinical Endocrinology & Metabolism, 23*(7), 605–614.

Jalili, M. (2016). Chemical composition and sensory characteristics of feta cheese fortified with iron and ascorbic acid. *Dairy Science & Technology, 96*(4), 579–589.

Jamison, D. T., Breman, J. G., Measham, A. R., Alleyne, G., Claeson, M., Evans, D. B., Musgrove, P. (2006). *Disease control priorities in developing countries,* Washington, DC: The World Bank.

Jayalalitha, V., Balasundaram, B., & Kumar, C. (2012). Fortifiication of encapsulated iron in probiotic yoghurt. *International Journal of Agriculture: Research and Review, 2*(2), 80–84.

Jiménez-Alvarado, R., Beristain, C. I., Medina-Torres, L., Román-Guerrero, A., & Vernon-Carter, E. J. (2009). Ferrous bisglycinate content and release in W1/O/W2 multiple emulsions stabilized by protein–polysaccharide complexes. *Food Hydrocolloids, 23*(8), 2425–2433.

Jizomoto, H., & Hirano, K. (1989). Encapsulation of drugs by lyophilized empty dipalmitoylphosphatidylcho-line liposomes - Effect of calcium-ion. *Chemical & Pharmaceutical Bulletin, 37*(11), 3066–3069.

Khan, S., Iqbal, R., Khan, R. S., & Khalid, N. (2020). Chapter Twelve - Bioavailability of nanoencapsulated food bioactives. In S. M. Jafari (Ed.), *Release and Bioavailability of Nanoencapsulated Food Ingredients* (Vol. 5, pp. 449–481), E-Book: Academic Press.

Kim, J.-M., Kim, Y.-J., Jeong, J., & Kim, C.-J. (2006). Meat tenderizing effect of injecting encapsulated Ca^{2+} in liposome into rabbit before slaughter. *Bioscience, Biotechnology, and Biochemistry, 70*(10), 2381–2386.

Kwak, H. S., Yang, K. M., & Ahn, J. (2003). Microencapsulated iron for milk fortification. *Journal of Agricultural and Food Chemistry, 51*(26), 7770–7774.

Li, Y., Yadava, D., Lo, K. L., Diosady, L. L., & Wesley, A. S. (2011). Feasibility and optimization study of using cold-forming extrusion process for agglomerating and microencapsulating ferrous fumarate for salt double fortification with iodine and iron. *Journal of Microencapsulation, 28*(7), 639–649.

Li, Y. O., Diosady, L. L., & Wesley, A. S. (2009). Iron in vitro bioavailability and iodine storage stability in double-fortified salt. *Food and Nutrition Bulletin, 30*(4), 327–335.

Li, Y. O., González, V. P. D., & Diosady, L. L. (2014). Microencapsulation of vitamins, minerals, and nutraceuticals for food applications *Microencapsulation in the Food Industry* (pp. 501–522), E-Book: Elsevier.

Lopez, H. W., Leenhardt, F., Coudray, C., & Remesy, C. (2002). Minerals and phytic acid interactions: Is it a real problem for human nutrition? *International Journal of Food Science & Technology*, *37*(7), 727–739.

Martínez-Bustos, F., Cruz Sánchez, A., Ortega-Martínez, A. D. C., & Aguilar-Palazuelos, E. (2018). Study of the functionality of nixtamalized maize flours and tortillas added with microcapsules of ferrous fumarate and folic acid. *Cereal Chemistry*, *95*(5), 699–707. doi:10.1002/cche.10084

McGhie, T. K., & Walton, M. C. (2007). The bioavailability and absorption of anthocyanins: Towards a better understanding. *Molecular Nutrition & Food Research*, *51*(6), 702–713. doi:10.1002/mnfr.200700092

Medeiros, D. M., & Wildman, R. E. (2019). *Advanced human nutrition* (4 ed.), Burlington, MA: Jones & Bartlett Learning.

Meng, F. B., Li, M., Wang, S. J., Liu, X., Gao, W. Y., Ma, Z., Li, J. C. (2019). Encapsulation of potassium persulfate with ABS via coacervation for delaying the viscosity loss of fracturing fluid. *Journal of Applied Polymer Science*, *136*(27), 47734.

Moretti, D., & Zimmermann, M. (2016). Assessing bioavailability and nutritional value of microencapsulated minerals. In J. M. Lakkis (Ed.), *Encapsulation and Controlled Release Technologies in Food Systems*, (pp. 289–308), E-Book: Wiley-Blackwell.

Mulia, K., Putri, T., Krisanti, E. A., & Handayani, N. A. (2019). *Preparation and evaluation of chitosan biopolymers encapsulated iron gluconate using spray drying method.* Paper presented at the *AIP Conference Proceedings*, July 8–12, 2019, Grenoble, France.

Müller, O., & Krawinkel, M. (2005). Malnutrition and health in developing countries. *CMAJ*, *173*(3), 279–286.

Murugesan, R., & Orsat, V. (2012). Spray drying for the production of nutraceutical ingredients—A review. *Food and Bioprocess Technology*, *5*(1), 3–14.

Nedovic, V., Kalusevic, A., Manojlovic, V., Levic, S., & Bugarski, B. (2011). An overview of encapsulation technologies for food applications. *Procedia Food Science*, *1*, 1806–1815.

Nkhata, S. G. (2013). *Iron fortification of yogurt and pasteurized milk.* (M.Sc Thesis), Michigan State University, East Lansing, MI.

Nordin, B. (1960). Osteomalacia, osteoporosis and calcium deficiency. *Clinical Orthopaedics and Related Research*, *17*, 235–258.

Obón, J., Castellar, M., Alacid, M., & Fernández-López, J. (2009). Production of a red–purple food colorant from *Opuntia stricta* fruits by spray drying and its application in food model systems. *Journal of Food Engineering*, *90*(4), 471–479.

Oneda, F., & Ré, M. (2003). The effect of formulation variables on the dissolution and physical properties of spray-dried microspheres containing organic salts. *Powder Technology*, *130*(1–3), 377–384.

Pohl, H. R., Wheeler, J. S., & Murray, H. E. (2013). Sodium and potassium in health and disease *Interrelations between essential metal ions and human diseases* (pp. 29–47), Dordrecht, Holland: Springer.

Porrarud, S., & Pranee, A. (2010). Microencapsulation of Zn-chlorophyll pigment from Pandan leaf by spray drying and its characteristic. *International Food Research Journal*, *17*(4), 1030–1042.

Risch, S. J. (1995). Encapsulation: Overview of uses and techniques *Encapsulation and Controlled Release of Food Ingredients* (Vol. 590, pp. 2–7), Washington, DC: American Chemical Society.

Romita, D., Cheng, Y.-L., & Diosady, L. L. (2011). Microencapsulation of ferrous fumarate for the production of salt double fortified with iron and iodine. *International Journal of Food Engineering*, *7*(3), 5.

Samman, S., Skeaff, S., Thomson, C., & Truswell, S. (2017). Trace elements. In J. Mann & A. S. Truswell (Eds.), *Essentials of Human Nutrition*, Hamspshire, UK: Oxford University Press.

SandstroÈm, B. (2001). Micronutrient interactions: Effects on absorption and bioavailability. *British Journal of Nutrition*, *85*(S2), S181–S185.

Schlemmer, U., Frølich, W., Prieto, R. M., & Grases, F. (2009). Phytate in foods and significance for humans: Food sources, intake, processing, bioavailability, protective role and analysis. *Molecular Nutrition & Food Research*, *53*(S2), S330–S375.

Schrooyen, P. M., van der Meer, R., & De Kruif, C. (2001). Microencapsulation: Its application in nutrition. *Proceedings of the Nutrition Society*, *60*(4), 475–479.

Shahidi, F., & Han, X. Q. (1993). Encapsulation of food ingredients. *Critical Reviews in Food Science and Nutrition*, *33*(6), 501–547.

Singh, A. P., Siddiqui, J., & Diosady, L. L. (2018). Characterizing the pH-dependent release kinetics of food-grade spray drying encapsulated iron microcapsules for food fortification. *Food and Bioprocess Technology*, *11*(2), 435–446.

Stein, A. J. (2010). Global impacts of human mineral malnutrition. *Plant and Soil*, *335*(1–2), 133–154.

Strain, J., & Cashman, K. (2009). Minerals and Trace Elements. In M. J. Gibney, S. A. Lanham-New, A. Cassidy, & H. H. Vorster (Eds.), *Introduction to Human Nutrition* (2 ed.), Loyangi, Singapore: Wiley-Blackwell.

Takeda, E., Ikeda, S., & Nakahashi, O. (2012). Lack of phosphorus intake and nutrition. *Clinical Calcium, 22*(10), 1487–1491.

Teucher, B., Olivares, M. & Cori, H. (2004). Enhancers of iron absorption: Ascorbic acid and other organic acids. *International Journal for Vitamin and Nutrition Research, 74*(6), 403–419.

Tonon, R. V., Brabet, C., & Hubinger, M. D. (2008). Influence of process conditions on the physicochemical properties of açai (*Euterpe oleraceae* Mart.) powder produced by spray drying. *Journal of Food Engineering, 88*(3), 411–418.

Wegmüller, R., Zimmermann, M. B., Bühr, V. G., Windhab, E. J., & Hurrell, R. F. (2006). Development, stability, and sensory testing of microcapsules containing iron, iodine, and vitamin a for use in food fortification. *Journal of Food Science, 71*(2), S181–S187.

Welch, R. M. (2002). The impact of mineral nutrients in food crops on global human health. *Plant and Soil, 247*(1), 83–90.

White, P. J., & Broadley, M. R. (2009). Biofortification of crops with seven mineral elements often lacking in human diets – iron, zinc, copper, calcium, magnesium, selenium and iodine. *New Phytologist, 182*(1), 49–84.

Yang, X. L., Alidoust, D., & Wang, C. Y. (2020). Effects of iron oxide nanoparticles on the mineral composition and growth of soybean (*Glycine max* L.) plants. *Acta Physiologiae Plantarum, 42*, 128.

8 Spray Drying Encapsulation of Essential Oils

Haroldo Cesar Beserra Paula, Regina Celia Monteiro De Paula, and Irisvan Da Silva Ribeiro
Federal University of Ceará, Brazil

Selene Maia De Morais
State University of Ceará, Brazil

CONTENTS

8.1 INTRODUCTION

Over the last few decades, natural compounds have reached the Pantheon of all commercial products, being described as the "most desirable market products" in several fields such as pharmacy, food, cosmetics, and nutraceuticals. Essential oils (EOs) are obtained from plants and are composed of a mixture of substances that are dependent on several factors such as soil, climate, country, planting, and harvesting conditions. Moreover, they are labile under various conditions, including environment, heat, and pH. Although EOs are readily available, their origin and variable composition point to the need for their protection from factors likely to cause degradation, before their application in any of the aforementioned research and market fields. Applications of EOs as natural preservatives, for example, has the potential to be used in food preservation, pharmaceutical, and cosmetic industries. Factors that are responsible for food spoilage and other health-related problems are still demanding solutions, leading to the need for developing sustained preservation and public health relief techniques. From green extraction and synthesis to sustainable applications and utilization, the interest in EOs is increasing day by day, as they are economically effective and resourceful, and are preferred by consumers as flavor and fragrance materials over synthetic chemicals. EO utilization is expanding rapidly in different areas such as alternatives to synthetic plant hormones and growth regulators, as biopesticides, as agents to reduce methane production in ruminant animals and overall anthropogenic greenhouse production, and the development of new drugs and products (Costa et al., 2013; Fernandes, Marques, Borges, & Botrel, 2014; Herman, Ayepa, Shittu, Fometu, & Wang, 2019; Martínez-Ballesta, Gil-Izquierdo, García-Viguera, & Domínguez-Perles, 2018; Peterfalvi et al., 2019; Ziosi et al., 2010). Moreover, in herbal medicine, in the prevention and treatment of diseases through different routes of administration (aromatherapy), two specific fields, EOs dietary supplementation for livestock and fish, and forest bathing, have been explored. Some EOs, particularly eucalyptus and ginger, seem to have immune function-enhancing properties (Peterfalvi et al., 2019). Special attention has been drawn worldwide to *Eucalyptus* plants, mainly due to their action against disease vectors such as *Aedes aegypti* in Dengue and *Lutzomyia longipalpis* in Leishmaniasis. Great attention has also been given to its nanotechnological applications by the food and pharmaceutical industries. Nanoemulsions containing *Eucalyptus globulus* oil have been recognized for their antimicrobial and antibiofilm effects against gram-negative bacteria and the major microorganisms responsible for causing fungal infections worldwide (*Candida albicans*) (Maciel et al., 2010; Salehi et al., 2019). *Origanum vulgare* EO has been used widely in Southern Europe, as well as in the American continent as a spice. It has shown potent bioactivities owing to its major constituents, including carvacrol, thymol, and monoterpenes. Several preclinical studies have confirmed its pharmacological potential as antiproliferative or anticancer, antidiabetic, antihyperlipidemic, anti-obesity, renoprotective, anti-inflammatory, vasoprotective, cardioprotective, antinociceptive, insecticidal, and hepatoprotective properties. Its nanotechnological applications as promising pharmaceutical agents are also claimed to enhance drug solubility, physicochemical stability, and the rate of accumulation (Sharifi-Rad et al., 2020).

Micro- and nanoencapsulation by spray drying are techniques that can achieve these tasks easily, preventing the deterioration of EOs and maintaining their desirable features such as bioactivity, flavor, and stability. Among several variables involved in a spray drying process, this chapter discusses those that are essential for process optimization and product analysis of the encapsulation of EOs. These include wall materials, their components, and composition, essential oil loading, encapsulation efficiency, encapsulation yield, particle size distribution, ζ-potential, morphology, oil release profile, and thermal properties.

8.2 CHEMISTRY, STRUCTURE, AND FUNCTIONALITY OF EOS

Essential oils can be defined as liquid mixtures of fragrant or odorless substances, which are usually lipophilic and highly volatile in character. These substances are mostly mono- and sesquiterpenes, but other types of compounds, such as allyl, isoallyl phenols, and coumarins, may also be present. The volatile components are secondary plant metabolites, having low molar masses (below 300 Da) and can also be present in a functionalized form with groups such as alcohols, ethers, aldehydes and ketones, esters and thiols, arylpropanoids and, to a lesser extent, short-chain alcohols and fatty acid esters (Figure 8.1).

Essential oil compositions vary greatly, sometimes for genetic reasons, but also because of climate, rainfall, or geographic origin. Essential oils from different plant species contain more than 200 constituents, comprising both volatile and non-volatile components (Aziz et al., 2018). Many aromatic oils from plants contain a few major constituents, several minor ones, and a larger number of trace compounds. It is the synergism of the specific combination of hundreds of constituents naturally present in each plant (including trace compounds) that give the EOs their valuable

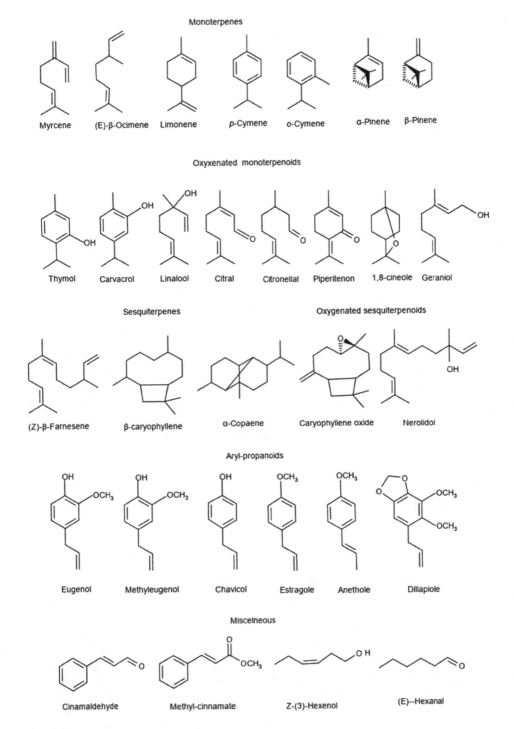

FIGURE 8.1 Essential oil constituents.

therapeutic/healing properties (Vankar, 2004). Essential oils such as Clary sage (*Salvia sclarea*), German chamomile (*Matricaria chamomilla*), lemongrass (*Cymbopogon flexuosus*), lavender (*Lavandula angustifolia*), geranium (*Pelargonium graveolens*), and frankincense (*Boswellia carterii*) are all known for their impressive and significant health benefits. Lavender is possibly the best known of these, thanks to its soothing properties which heal skin irritation, making it an excellent choice for ointments, lotions, and salves. Frankincense, meanwhile, supports healthy cellular function, reduces skin imperfections, and alleviates inflammation, making it a valuable addition to any makeup, lotion, toner, or face wash.

Skin is constantly exposed to environmental oxidative stressors such as ultraviolet radiation, air pollutants, chemical oxidants, and aerobic microorganisms. The skin possesses an innate antioxidant defense; nevertheless, supplying exogenous antioxidants is a valuable approach to preventing or minimizing reactive oxygen species (ROS)-induced photoaging. Cosmetic formulations containing *Ocimum micrantum* EO, a substance rich in eugenol, were shown to be effective in an antioxidant experiment (Ziosi et al., 2010).

8.2.1 PHYSICOCHEMICAL ASPECTS OF EOs

Encapsulation processes have been largely applied by industry in the search for new oral delivery alternatives obtained upon the modification of the solubility properties of bioactive compounds. Many scientific efforts have been made in recent years demonstrating the usefulness of nanoparticles for improving their therapeutic value, while avoiding toxicity, by releasing bioactive compounds specifically to target tissues affected by specific chemical and pathophysiological settings. The contribution of nanoencapsulation is also important to protect bioactive compounds from degradation as a result of gastrointestinal digestion and cellular metabolism, and to allow them to target those tissues affected by biological disturbances (Martínez-Ballesta et al., 2018).

The properties and performances of EOs are dependent on their extraction procedures. There are a variety of methods used for the extraction of EOs; e.g., steam distillation, hydrodistillation, microwave extraction, organic solvent extraction, ultrasound extraction, and supercritical fluid extraction. Each of these methods has certain advantages that determine the biological and physicochemical properties of the extracted oil. Aiming to widespread the use of EOs, their encapsulation is a suitable alternative, since it protects them from evaporation loss and improves their conservation and transport facilitation.

8.3 PROCESS KEY PARAMETERS AND WALL MATERIALS FOR SPRAY DRYING ENCAPSULATION OF EOS

Initially, it is important to be aware of some EOs features that should be taken into consideration when the spray drying encapsulation of EOs is being carried out. These oils possess various chemical compositions and a myriad of substances presenting different molar masses, polarity, and vapor pressures, which are the more relevant parameters regarding their encapsulation by spray drying. Once encapsulated, oil molecules must diffuse through the wall material(s), to be released into the medium. Similarly, during the evaporation, the low molar mass molecules are lost on a greater scale than the high molar mass ones. Therefore, the oil composition before and after spray drying may vary for some systems. For instance, peppermint (*Mentha piperita*) essential oil encapsulated in a maltodextrin matrix, had its most volatile constituents (monoterpenes) lost during emulsion preparation (Adamiec & Kalemba, 2006), while elemi EO, processed under the same conditions, showed no losses, even in the final microencapsulated powder. Rosemary EO encapsulated in different matrices of maltodextrin/modified starch and inulin (Fernandes, Borges, & Botrel, 2014) exhibited lower levels of 1,8-cineol and beta-pinene in the microcapsules than in pure oil, while the other constituents were present in higher relative concentrations, a fact that was attributed to evaporation losses. Carvacrol, a major component of oregano EO, encapsulated in gum Arabic/maltodextrin/modified starch matrix was reported to have a higher concentration in microparticles than in the unprocessed oil (Partheniadis, Vergkizi, Lazari, Reppas, & Nikolakakis, 2019), the reduction being due to the volatility of minor ingredients with lower boiling points than other compounds. The use of a pectin/alginate matrix revealed that employing high inlet temperatures (160 °C–190 °C) decreased carvacrol retention efficiency accordingly (Sun, Cameron, & Bai, 2020). Pink pepper EO encapsulated in a pectin/maltodextrin yielded microcapsules that showed no cracking or pores, which are usually responsible for oil loss (Pereira, Gonçalves Cattelan, & Nicoletti, 2019). It is shown that the molar mass, vapor pressure, and volatility, along with wall material blends, which can prevent particle pores and cracking, are the most important parameters to be taken into consideration regarding the properties of EOs.

One of the most important parameters of encapsulated substances is the wall material, mainly because its physicochemical properties play a paramount role in many variables, including system stability, particle size distribution, ζ-potential, the encapsulation efficiency, morphology, and release profile. A good wall material must protect active principles from external threats such as oxidative agents, heat, moisture, and other deleterious substances. It must also be readily available, at a low or reasonable cost. Since encapsulated substances have broad applications in different areas, including food, pharmaceutical, and medical, wall materials should be also biocompatible and biodegradable. Taking that into consideration, polysaccharides and proteins are substances that meet the most demanding features of appropriated wall materials for the spray drying encapsulation of EOs. Table 8.1 presents the most frequently employed matrix components used for the spray drying encapsulation of EOs.

TABLE 8.1

Process Initial Parameters Used For Spray Drying Encapsulation Of Essential Oils

Matrix	Essential Oil(s)	Feed Flow Rate (L h⁻¹)	Drying Air Flow (m³ h⁻¹)	Solid Concentration (%)	EO: Wall Material Ratio (%)	Spray Drying Temperature (°C) Inlet	Outlet	References
Acacia gum	Citronella	0.36	31	10–30	1:10 to 1:3.3	136–203	-	(Yingngam et al., 2019)
Gum Arabic	Basil	0.7	2.4	30	10–25	180	110	(Garcia, Tonon, & Hubinger, 2012)
	Rosemary	0.5–1.0	2.4	10–30	1: 4	135–195	-	(Fernandes et al., 2013)
	Lavender	0.3	0.084	25, 30 and 35	1: 6, 1: 5, and 1: 4	150	110	(Burhan, Abdel-Hamid, Soliman, & Sammour, 2019)
Gum Arabic and maltodextrin	Indian clove	0.3	35	30	1: 4	180	117	(Teodoro et al., 2019)
	Schinus molle	0.3	0.600	-	1: 4	160	100	(López et al., 2014)
	Lippia sidoides	-	60	30, 40, 50 and 60	1: 3 and 1: 4	140, 150, and 160	-	(Fernandes et al., 2008)
	Lippia sidoides	1.05	60	50	1: 4	160	-	(Fernandes, Oliveira, Sztatisz, & Novák, 2008)

(Continued)

TABLE 8.1
(Continued)

Matrix	Essential Oil(s)	Feed Flow Rate (L h⁻¹)	Drying Air Flow (m³ h⁻¹)	Solid Concentration (%)	EO: Wall Material Ratio (%)	Spray Drying Temperature (°C) Inlet	Outlet	References
Gum Arabic and modified starch	Origanum vulgare L.	1.07	2.088	-	1: 10	180	105	(Hijo et al., 2015)
Gum Arabic, maltodextrin, and modified starch	Origanum vulgare L.	0.3	0.6	30	3 and 6	180	117	(Partheniadis et al., 2019)
Gum Arabic, maltodextrin, and modified starch	Origanum vulgare L.	1.07	2.088	-	1: 10	180	105	(Costa et al., 2013)
Gum Arabic, maltodextrin, and cellulose nanofibrils	Sweet orange	0.9	2.4	-	1: 4	160	-	(Souza et al., 2018)
Gum Arabic, starch, maltodextrin, and inulin	Rosemary	0.9	-	10–30	1: 4	190	-	(Fernandes, Borges, & Botrel, 2014)
Whey protein concentrate, mesquite gum and, gum Arabic	Chia	2.4	-	30–40	1: 2 and 1: 3	135	80	(Rodea-González et al., 2012)
Maltodextrin and pectin	Pink pepper	0,12	35	35	10	140	98	(Pereira et al., 2019)
	Citral	0.5	-	20	2.5–5.0	180	80	(Wang, Oussama Khelissa, Chihib, Dumas, & Gharsallaoui, 2019)
		1.2	-	43	-	180	80	(Talón et al., 2019)
Maltodextrin	Eugenol	0.4	35	-	-	150	80–90	(Shah, Davidson, & Zhong, 2012)
	Elemi Peppermint	-	-	30	10–30	150	80	(Adamiec & Kalemba, 2006)
	Caraway	-	-	30	15	180	90	(Bylaitë, Venskutonis, & Mapdpierienë, 2001)

(Continued)

Matrix	Essential Oil(s)	Feed Flow Rate (L h⁻¹)	Drying Air Flow (m³ h⁻¹)	Solid Concentration (%)	EO: Wall Material Ratio (%)	Spray Drying Temperature (°C)		References
						Inlet	Outlet	
Maltodextrin and modified starch	Rosemary	0.7	2.4	10–30	10–30	190	-	(Fernandes, Marques, et al., 2014)
Alginate and cashew gum	Lippia sidoides	0.3	35	0.2, 0.5 and 1.0	1: 10	170	65	(Oliveira, Paula, & Paula, R. C. M. d., 2014)
Cashew gum	Eucalyptus staigeriana	0.3	35	-	1: 2 and 1: 4	160	70	(Herculano, de, de, Dias, & Pereira, V. de A., 2015)
Cashew gum and inulin	Ginger	0.8	2.1	20	1: 4	170	-	(Fernandes et al., 2016)
Chitosan and cashew gum	Lippia sidoides	0.3	35	-	1: 10 and 2: 10	170	65	(Abreu, Oliveira, Paula, & De, 2012)
Chitosan and angico gum		0.3	35	-	1: 2, 1: 4, 1: 10 and 1: 20	160	70	(Paula, Sombra, Abreu, & De, 2010)
Chitosan and cashew, angico, and Chichá gums	Lippia sidoides	0.3	35	-	-	160	70	(Paula, Oliveira, Carneiro, & De, 2017)
Chitosan	Coriander	0.3	0.6	-	1: 1	120	68–71	(Duman & Kaya, 2016)
Modified starch	Rosmaris officinalis and Zataria multiflora	1.2	0.04	-	0.05: 1–1: 5	120, 140 and 160	81, 85 and 89	(Ahsaei et al., 2020)
Rice starch and maltodextrin	Orange	0.225	-	30	15	180	85	(Márquez-Gómez, Galicia-García, Márquez-Meléndez, Ruiz-Gutiérrez, & Quintero-Ramos, 2018)
Rice starch, inulin, and gelatin/sucrose	Oregano	-	-	5–10	15	120–190	-	(Costa et al., 2012)
Modified starches and hydrolyzed starches	Peppermint	-	-	-	15.25	200	120	(Baranauskiené, Bylaité, Juraté, & Venskutonis, 2007)
Pectin/zein/caseinate	Eugenol	-	7.2	0.05–0.2	1: 2	100	-	(Veneranda et al., 2018)

(Continued)

TABLE 8.1
(*Continued*)

Matrix	Essential Oil(s)	Feed Flow Rate (L h⁻¹)	Drying Air Flow (m³ h⁻¹)	Solid Concentration (%)	EO: Wall Material Ratio (%)	Spray Drying Temperature (°C) Inlet	Outlet	References
Pectin and alginate	Carvacrol	0,42	-	3	1: 3	100–190	-	(Sun et al., 2020)
		0,42	-	-	-	110	-	(Sun, Cameron, & Bai, 2019)
Inulin	Oregano	3.48	-	5, 15 and 25	-	120–190	-	(Costa, Duarte, et al., 2013)
Whey protein and inulin	Ginger	0.8	2.1	20, 25 and 30	1: 4	140, 155 and 170	-	(Fernandes et al., 2017)
Whey protein and inulin	Rosemary	0.9	-	20	25	170	-	(Fernandes et al., 2017)
Mesquite (Prosopis juliflora) gum	Cardamom	-	-	-	1: 3, 1: 4 and 1: 5	200	110	(Beristain, García, & Vernon-Carter, 2001)
Methylcellulose and hydroxypropyl methylcellulose phthalate	Thymol	1.74	31.3	2 and 5	1: 2	50	42–45	(Rassu et al., 2014)
β-cyclodextrin	Lippia sidoides	-	60	50	1: 10, 1.33: 10 and 2: 10	160	-	(Fernandes, Oliveira, Sztatisz, Szilágyi, & Novák, 2009)
Cyclodextrins - whey protein concentrate	Eugenol and carvacrol	0.3	-	11	2: 1	150	105	(Barba, Eguinoa, & Maté, 2015)
Whey protein concentrate Skimmed milk powder	Oregano, citronella, marjoram, and	-	-	-	20	190	90	(Baranauskiene, Venskutonis, Dewettinck, & Verhé, 2006)
Skim milk powder	Holy basil	0.36	35	20	-	130 140 150	74 79 84	(Rodklongtan & Chitprasert, 2017)
Bovine serum albumin, rice bran protein, and carrageenan	Eugenol	0.6	35	1	1: 20	150	-	(Scremin et al., 2018)
Zein and Casein	Eugenol and thymol	-	-	2	1: 4	105	60	(Chen, Zhang, & Zhong, 2015)

8.3.1 Gum Arabic

Gum Arabic is one of the most-employed polysaccharides as wall material for the spray drying encapsulation of EOs, either alone or combined with maltodextrin (a starch derivative polysaccharide). The composition of this gum includes D-galactose, L-arabinose, L-rhamnose, and D-glucuronic acid, attached to linear and branched chains. Gum Arabic has good emulsifying properties, which enables it to form stable emulsions with EOs, in a variety of pHs. Upon drying, gum Arabic forms films that contributed to the protection of labile molecules such as EO components. Gum Arabic was employed for the encapsulation of basil EO (*Ocimum basilicum L.*) at concentrations varying from 22.5% to 27% (w/w) (Garcia et al., 2012). The ratio of wall material to an essential oil (WM: EO) is a very important variable regarding the design of an encapsulation procedure. WM: EO ratios in the range 10%–25% were applied for an emulsion used for the encapsulation of basil EO, which occurred through a high-pressure homogenizer with pressures ranged from 5 to 100 MPa. An emulsion droplet size of 0.40 μm was obtained. The best conditions for basil oil encapsulation to achieve the highest oil retention were oil concentration in the range 10%–14% and homogenization pressures greater than 50 MPa. Particle size distribution is a function of the physicochemical variables of the chosen matrix and the features of the spray drying equipment such as inlet and outlet temperatures, airflow, and feed speed. Particle size distribution was found to be bimodal, with sizes in the range of 0.5–0.9 μm. It was reported the observation of a linear relation between particle mean diameter and emulsion viscosity (Garcia et al., 2012). Scanning electronic microscopy (SEM) images showed microparticles (Figure 8.2a) with spherical shapes, holes, and irregular surfaces.

The essential oil of *Rosmarinus officinalis L.* was also encapsulated using gum Arabic as wall material (Fernandes et al., 2013) in the concentration range of 10%–30%, with a ratio of EO: WM= 1: 4. The optimal WM concentration was reported to be about 19%, which provided minimum oil degradation. Acacia gum, also known as gum Arabic, was employed as WM for citronella (*Cymbopogon nardus L.* Rendle) oil encapsulation (Yingngam et al., 2019). Acacia gum concentration varied from about 10% to 30% in the feed, with an EO: WM from 1: 10 to 1: 3.3. The best conditions were found to be gum concentration of 26% and EO: WM ratio of 1: 7. Powders were suggested to be used in the development of citronella essential oil-based cosmetic textiles.

Once the process of spray drying is finished, the amount of powder present in the equipment collector is compared to that of solid concentration in the feed, resulting in the corresponding process

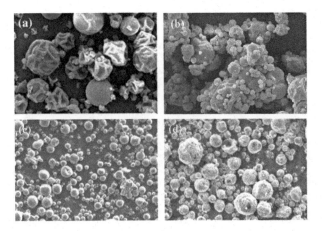

FIGURE 8.2 Scanning electron micrographs of particles of different encapsulated essential oils: Basil essential oil encapsulated with gum Arabic (a, magnification: 5000x), eugenol encapsulated with whey protein (b, magnification: 1500x), carvacrol in pectin/sodium alginate matrix microcapsules with an inlet air temperature of 100°C and 160°C (c and d, respectively, magnification: 2000x).

yield (powder recovery). This depends mostly on the nature of wall material, solid concentration, and WM: EO ratio, as well as operating temperature (Table 8.1). The yield of encapsulation of citronella EO was in the range 20.58%–49.58% and 12% loading (Yingngam et al., 2019); by contrast, for encapsulating rosemary EO, lower yields (17.25%–33.96%) were reported, as shown in Table 8.2 (Fernandes et al., 2013). Taking into consideration that other parameters are roughly similar, the yield drop can be explained by the fact that the optimal inlet temperature for citronella EO

TABLE 8.2

Process Outcome Parameters Used For Spray Drying Encapsulation Of Essential Oils

Matrix	Essential Oil(s)	Yield (%)	Oil Retention (Loading) (%)	Encapsulation Efficiency (%)	Release Profile	Size (µm)	• (mV)	References
Acacia gum	Citronella	21–50	93–97	26–82	-	6–8	-	(Yingngam et al., 2019)
Gum Arabic	Basil	-	57–91	-	-	0.5–0.9	-	(Garcia et al., 2012)
	Rosemary	17–34	7–48	-	-	14	-	(Fernandes et al., 2013)
Acacia gum and maltodextrin	Lavender	42–63	7–15	29–78	-	16–24	-	(Burhan et al., 2019)
	Indian clove	-	-	58–82 44–88	-	16–23 16–24	-	(Teodoro et al., 2019)
	Schinus molle	75–90	71	96–100	-	0.2–40	-	(López et al., 2014)
Gum Arabic and maltodextrin	Lippia sidoides	14–63	43–63	-	-	8–16	-	(Fernandes, Turatti, et al., 2008)
	Lippia sidoides	-	10–16	45–63	-	-	-	(Fernandes, Oliveira, et al., 2008)
Gum Arabic and modified starch	Origanum vulgare L.	-	62	86	-	2, 8 and 16	-	(Hijo et al., 2015)
Gum Arabic, maltodextrin, and modified starch	Origanum vulgare L.	-	75 and 64	98 and 98	Fast	14 and 13	−24 and −23	(Partheniadis et al., 2019)
Gum Arabic, maltodextrin, and modified starch	Origanum vulgare L.	-	84	87	-	-	-	(Costa, Borges, et al., 2013)
Gum Arabic, maltodextrin, and cellulose nanofibrils	Sweet orange	-	62–84	62–71 74–84	Slow	12–37	-	(Souza et al., 2018)

(Continued)

Matrix	Essential Oil(s)	Yield (%)	Oil Retention (Loading) (%)	Encapsulation Efficiency (%)	Release Profile	Size (μm)	• (mV)	References
Gum Arabic, starch, maltodextrin, and Inulin	Rosemary	-	29–60	26–62	-	13	-	(Fernandes, Borges, & Botrel, 2014)
Whey protein concentrate, mesquite gum, and gum Arabic	Chia	-	-	71–81	-	13–28	-	(Rodea-González et al., 2012)
Maltodextrin and pectin	Pink pepper	46 and 46	42 and 49	-	-	4 and 6	-	(Pereira et al., 2019)
	Citral	-	-	62.3–99.6	-	3–15	+25 +12 –10	(Wang et al., 2019)
		-	-	22–98	Fast	0.1–100	–47 to +61	(Talón et al., 2019)
Maltodextrin	Eugenol	68–83	21–44	14–36	-	0.4–5		(Shah et al., 2012)
	Elemi peppermint	29–41 57–71	7.06–18.17	57.2–70.6	-	-	-	(Adamiec & Kalemba, 2006)
	Caraway	-	69–88	68–86	Slow	-	-	(Bylaitė et al., 2001)
Maltodextrin and modified starch	Rosemary	-	9.1–54	-	-	12	-	(Fernandes, Marques, et al., 2014)
Alginate and cashew gum	Lippia sidoides	-	1.9–4.4	21–48	Fast and Slow	0.223–0.399	–30 to –36	(Oliveira et al., 2014)
Cashew gum	Eucalyptus staigeriana	-	5–7	25–27	Fast	0.0297–0.4327	–10 to –24	(Herculano et al., 2015)
Cashew gum and inulin	Ginger	-	-	16–31	-	14–18	-	(Fernandes et al., 2016)
Chitosan and cashew gum	Lippia sidoides	45–60	5–11.0	40–70	Fast	0.335–0.558	+4 to +50	(Abreu et al., 2012)
Chitosan and angico gum		30–61	3–6	16–78	Slow	0.012–0.271	–15 to –21	(Paula, Sombra, et al., 2010)
Chitosan and cashew, angico, and Chichá gums	Lippia sidoides	-	8–16 6–13 4–15	34–60 25–50 15–59	Fast	0.181–0.483 0.019–0.472 0,017–0.429	+32 to +40 +1 to +30 –41 to +26	(Paula et al., 2017)
Chitosan	Coriander	-	-	28.4	Bi-phasic (fast and slow)	0.4–7	-	(Duman & Kaya, 2016)

(Continued)

TABLE 8.2
(*Continued*)

Matrix	Essential Oil(s)	Yield (%)	Oil Retention (Loading) (%)	Encapsulation Efficiency (%)	Release Profile	Size (μm)	• (mV)	References
Modified starch	Rosmaris officinalis and Zataria multiflora	12–61	-	68–88	Slow	8–11	-	(Ahsaei et al., 2020)
Rice starch and maltodextrin	Orange	29–87	-	36.76–98.82	-	30 and 40	-	(Márquez-Gómez et al., 2018)
Rice starch, inulin, and gelatin/sucrose	Oregano	-	-	-	Biphasic (fast and slow)	-	-	(Costa et al., 2012)
Modified starches and hydrolyzed starches	Peppermint	-	6–15	39–97	Slow	12–138	-	(Baranauskiené et al., 2007)
Pectin/zein/caseinate	Eugenol	-	0.3	-	-	0.409–0.786	−47 to −49	(Veneranda et al., 2018)
Pectin and alginate	Carvacrol	-	10–89	43–83	-	3, 4, and 5	-	(Sun et al., 2020)
		-	-	78	Fast	1–3	-	(Sun et al., 2019)
Inulin	Oregano	-	-	-	Fast	6–9	-	(Costa, Duarte, et al., 2013)
Whey protein and inulin	Ginger	-	-	45		12–24	-	(Fernandes, Botrel, et al., 2017)
	Rosemary	-	-	40	-	-	-	(Fernandes, Guimarães, et al., 2017)
Mesquite (prosopis juliflora) gum	Cardamom	-	67–84	84	-	2–3	-	(Beristain et al., 2001)
Methylcellulose and hydroxypropyl methylcellulose phthalate	Thymol	37 and 53	22–26 20–24	65 and 79 58 and 73	Slow	5–6	-	(Rassu et al., 2014)
β-cyclodextrin	Lippia sidoides	-	35–70	70	-	10–12	-	(Fernandes et al., 2009)

(*Continued*)

Matrix	Essential Oil(s)	Yield (%)	Oil Retention (Loading) (%)	Encapsulation Efficiency (%)	Release Profile	Size (µm)	• (mV)	References
Cyclodextrins - whey protein concentrate	Eugenol/ carvacrol	≤ 50	50.4 (engenol) 79.6 (carvacol)	-	Fast	-	-	(Barba et al., 2015)
Whey protein concentrate and skimmed milk powder	Oregano citronella marjoram	-	73–85	54–80	-	6–280	-	(Baranauskiene et al., 2006)
Skim milk powder	Holy basil	-	-	47 and 44 48 and 44 45 and 42	-	9–12	-	(Rodklongtan & Chitprasert, 2017)
Bovine serum albumin, rice bran protein, and carrageenan	Eugenol	-	-	24–79	-	3–5	-	(Scremin et al., 2018)
Zein and casein	Eugenol and thymol	-	5–8	16–26	Fast	0.131– 0.174	-	(Chen et al., 2015)

was 150°C and for rosemary EO it was 171°C. Under these conditions, high temperatures did not fasten the evaporation, which led to a higher yield.

The amount of EOs entrapped in the spray dried particles compared to the EOs present in the feed emulsion, provides the factor known as the encapsulation efficiency (EE) (Costa, Borges, et al., 2013). EE of citronella EO in the gum Arabic matrix was in the range 26.38%–81.86% (Yingngam et al., 2019). Thermal gravimetric analysis (TGA) was used to characterize the thermal properties of the obtained spray dried matrices. Thermograms of free and encapsulated citronella oil are shown in Figure 8.3 a–c. It can be seen that the oil evaporates at room temperature and that free citronella oil exhibited a single-step weight loss, while citronella oil microcapsules presented a three-step weight loss, in the 50°C–200°C temperature range, clear evidence that oil thermal stability was improved (Yingngam et al., 2019).

Acacia gum (25%–35%) was also used as the wall material for the spray drying encapsulation of lavender (*Lavandula angustifolia*), at EO: WM ratios of 1: 6, 1: 5, and 1: 4. The optimal oil protection was achieved for the 25% WM content and EO: WM= 1: 4 (Burhan et al., 2019).

Mesquite (*Prosopis juliflora*) gum is a polysaccharide very similar to gum Arabic and has been employed for the encapsulation of cardamom (*Elettaria cardamomum*) seed EO (Beristain et al., 2001) using different EO: WM ratios. The best performance was reported to be the 1: 4 ratio at 200°C, with 67%–84% microparticle loading and 84% EE (Beristain et al., 2001). Cashew gum is a polysaccharide with properties analogous to gum Arabic, exhibiting low viscosity at high concentrations and film-forming features. *Eucalyptus staigeriana* EO, obtained from leaves of the eucalyptus tree, was encapsulated by cashew gum with EO: WM= 1: 2 and 1: 4, at 160°C (Fernandes et al., 2016).

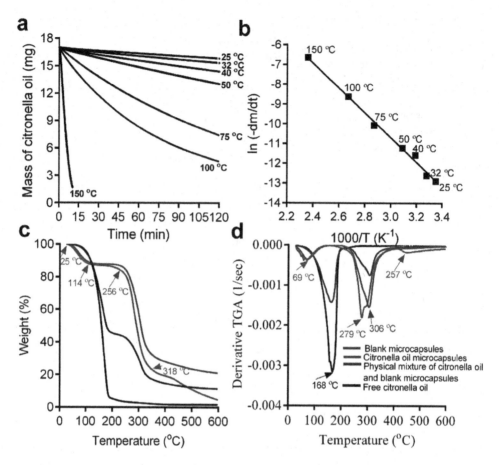

FIGURE 8.3 Thermogravimetric analysis (TGA) thermograms of mass loss of free citronella oil at different temperatures over 120 min (a), an Arrhenius plot showing the evaporation of free citronella oil at various temperatures (b), the citronella oil microcapsule and its ingredients (c), and citronella oil microcapsule and its ingredients (d).

8.3.2 BLENDS OF GUM ARABIC

A suitable wall material must present emulsifying properties along with film-forming capability, features that are not usually exhibited by one single wall material component; hence, blends of wall materials are used in such cases. Teodoro et al. (2019) used maltodextrin, a starch derivative, blended with gum Arabic in different proportions, for the encapsulation of Indian clove (*Caryophyllus aromaticus* L) EO, using a 30% total solids concentration and EO: WM= 1: 4. Process parameters were found to improve with increasing the gum content by up to 75%. The EO of *Schinus molle L* has been proposed as an agent for fighting horn fly, a blood-sucking insect of pastured cattle. The oil was encapsulated with employing a blend of maltodextrin and gum Arabic in the ratios of 4: 1, 3: 2, 2: 3, and 1: 1 and an EO: WM of 1: 4 (López et al., 2014). The findings revealed favorable retention and protection of the encapsulated EO, besides the improvement of its insecticidal activity.

A blend of gum Arabic and maltodextrin containing lavender EO presented EE values of 77.89% (Burhan et al., 2019). DSC (differential scanning calorimetry) thermogram of that matrices revealed an endothermic event at 45°C, attributed to the likely loss of the most volatile oil components. No further losses were reported, which seemed to indicate that the thermal stability was provided by the matrix components.

The essential oil of oregano (*Oreganum vulgare* L.) is mainly used as a food additive and was encapsulated in a maltodextrin and gum Arabic matrix in different ratios, with an EO: WM of 1: 10 (Costa, Borges, et al., 2013). Optimized conditions yielded encapsulation efficiency and oil retention of 93 % and 77.39 %, respectively, for a matrix containing 74.5% gum Arabic and 12.7% maltodextrin. Matrices of gum Arabic and maltodextrin used for Indian clove encapsulation resulted in particle sizes in the range of 16–24 μm, with low PDI (polydispersity index) values, indicating a homogeneous droplet distribution (Teodoro et al., 2019). It was stated that by decreasing the amount of gum Arabic, emulsions with higher PDI values were obtained (Teodoro et al., 2019). This is likely due to the fact that polysaccharides exhibit emulsifying properties that ultimately lead to more homogeneous emulsions.

Blends of maltodextrin and gum Arabic in different proportions containing *Schinus molle* EO were spray dried, yielding microparticles that showed slow liberation profile throughout 366 h, with 71% of EO retention, along with a time-dependent insecticidal activity (López et al., 2014). In another study (Fernandes, Turatti, et al., 2008), a blend of maltodextrin and gum Arabic in the ratios varying from 4: 1 to 4: 0 and EO: WM of 1: 3 to 1: 4 was employed for *Lippia sidoides* EO encapsulation. This resulted in the production of microparticles that were used as an antifungal agent, and the best performance was obtained for a matrix containing gum Arabic as a sole component and an EO: WM ratio of 1:4 (Fernandes, Turatti, et al., 2008).

Schinus molle EO encapsulated in gum Arabic/maltodextrin matrix (1: 1 ratio) showed higher yield (75%–90%) (López et al., 2014) than *Lippia sidoides* EO (1: 0 ratio) (Fernandes, Turatti, et al., 2008) and lavender EO (Burhan et al., 2019). In the case of the latter study, it was claimed that yield could be affected by solid concentration, oil loading, and gum Arabic concentration; a high WM: EO ratio led to low yields. For *Lippia sidoides* EO system, powder recovery was reported to be affected by WM ratio. High feed solution concentration (>30%) of gum Arabic/maltodextrin matrix was reported to lead to the lower yields, due to increasing the solution viscosity, which ultimately resulted in the adhesion of the powder to the collecting chamber (Yingngam et al., 2019). On the encapsulation of eugenol, it was observed that the smaller whey protein: maltodextrin ratio resulted in a higher yield (Shah et al., 2012).

Most blends exhibit better performance than their sole component, even when applied to EOs (Souza et al., 2018). The highest EE for this matrix was observed for the encapsulation of *Schinus molle* EO (96%–100%) (López et al., 2014); by contrast, *Lippia sidoides* EO showed an EE in the range of 45%–63% (Fernandes, Oliveira, et al., 2008). In another study (Fernandes, Oliveira, et al., 2008), the thermal analysis of the microparticles of *Lippia sidoides* EO encapsulated in the gum Arabic/maltodextrin matrix revealed that the thermal stability of the entrapped oil has augmented and that by increasing the amount of gum Arabic in the matrix, the thermal stability of the manufactured microparticles increased accordingly (Fernandes, Oliveira, et al., 2008). In another approach, tablets of a blend of gum Arabic, maltodextrin, and modified starch (octenyl succinate) were prepared and employed for oregano EO encapsulation (Partheniadis et al., 2019). Solid content of 30% and WM: EO of 3 and 6 were reported with satisfactory properties of the tablets, with minimum oil loss, offering a promising alternative to oral delivery.

In a few studies (Costa, Borges, et al., 2013; Hijo et al., 2015; Partheniadis et al., 2019), starch modified with octenyl succinate was incorporated in a matrix of gum Arabic and maltodextrin and employed for the encapsulation of EO of *Oreganum vulgare* L. EE was reported to be in the range of 86.8%–98.3%, where the highest value was obtained for the bigger WM: EO ratio. Interestingly, the preparation of the tablets from the spray dried powder containing EO did not alter tablet properties significantly and did not lead to oil loss (Partheniadis et al., 2019). Hijo et al. (2015) reported, for the same matrix, particle sizes in the range of 2–16 μm, with high polydispersity, cracking, and imperfections (formed likely because of shrinkages during the drying and cooling processes). DSC thermograms revealed high values of Tg, which is an indicative of high thermal stability.

Gum Arabic, maltodextrin, and modified starch matrices used for carvacrol encapsulation were prepared from 3% and 6% w/w emulsions with a ζ-potential of about −24 mV (Figure 8.4)

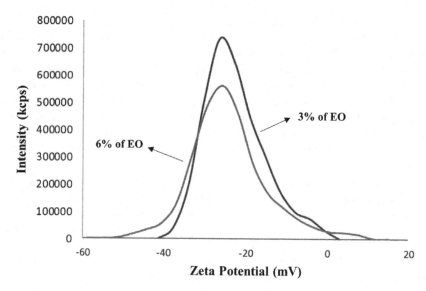

FIGURE 8.4 ζ-potential of carvacrol encapsulated emulsions before and after spray drying (3% EO in blue, 6% EO in brown).

(Partheniadis et al., 2019). It is shown that the greater the concentration, the larger is the ζ-potential distribution. Most particles or emulsions are considered stable when they present a ζ-potential of higher than ±30 mV.

Spray-dried oregano EO encapsulated in maltodextrin/gum Arabic matrix and compressed to tablets showed a fast release of EO for the spray dried powder and slower rates for tablets with cross-linked carboxymethyl cellulose and with starch glycolate (Partheniadis et al., 2019). SEM micrographs revealed that holes augment with increasing EO: WM ratio and more holes and cracks were observed on the microparticles with higher oil uploads (17.5% and 25% w/w). Homogenization pressure was found to vary inversely with crack and holes incidence. Carvacrol EO encapsulated in gum Arabic/maltodextrin and modified starch resulted in particles possessing a morphology similar to that of gum Arabic/Holy basil EO matrices (Figure 8.2a); however, the microparticles were present as agglomerates (Partheniadis et al., 2019).

Gum Arabic, maltodextrin, and cellulose nanofibrils in different ratios were used for the encapsulation of sweet orange EO (Souza et al., 2018), at the EO: WM ratio of 1: 4. It was reported that cellulose nanofibrils enhanced the encapsulation process and its outcome variables. EE was considered satisfactory, ranging from 74.33% to 84.30%. In another study, inulin (a fructan) was added to gum Arabic/maltodextrin blends, aiming at rosemary (*Rosmarinus officinalis* L.) EO encapsulation (Fernandes, Borges, & Botrel, 2014). Feed concentration varied from 10 to 30%, at an EO: WM ratio of 1: 4. Although inulin did not increase oil retention, it was suggested to be a favorable alternative for the use in food products. Inulin is a fructan obtained from chicory, dahlia, and Jerusalem artichoke roots, and its structure is composed of a linear fructose chain with glucose at the end. This polysaccharide is considered a dietary fiber (Fernandes, Borges, & Botrel, 2014).

Chia (*Salvia hispanica* L.) seeds EOs are used in foods and in alternative medicine products. Blends of gum Arabic/mesquite gum (*Prosopis juliflora*) and whey protein (altogether, 30%–40% total solid content), and EO: WM= 1: 2 and 1: 3 were produced, to fabricate a stable system. The diameter of microcapsules was reported to be smaller when the core: wall material ratio was higher (Rodea-González et al., 2012). EE was found to be in the range of 70.70% to 80.70% for both whey protein/mesquite and gum Arabic blends, and it was reported to increase when EO: WM increased from 1: 2 to 1: 3. Spray dried particles presented unimodal size distributions and particle size decreased as the EO: WM ratio increased from 1:2 to 1:3, for all studied matrices (Rodea-González et al., 2012).

8.3.3 MALTODEXTRIN

Maltodextrin (MD) is a product of the starch industry, obtained from the partial hydrolysis of starch, and is denoted by its DE (dextrose equivalent). MD has widely been used as a food ingredient, as well as in the cosmetic and pharmaceutical industries. In recent research (Talón et al., 2019), MD was blended with whey protein and lecithin for the spray drying encapsulation of eugenol (2-methoxy-4-(2-propenyl)phenol) in a 1: 42 (w/w) ratio (whey protein: MD + lecithin), and a solid concentration of 43%. The data revealed that the formulations with whey protein or lecithin alone gave better results as an antibacterial agent against *E. coli*. The mean particle diameter was 15 μm, regardless of the wall material used. The addition of chitosan to the matrices resulted in the broad multimodal particle size distributions (0.1–100 μm). EE was in the range of 22%–98% and ζ-potential varied from –45.7 mV to +61.5 mV, leading to a higher particle size. ζ-potential is a measure of particle stability in a given medium. For whey protein systems, chitosan addition also led to positive ζ-potential; however, with greater polydispersity and aggregation.

The release profile of the spray dried encapsulated EO particles is dependent on the matrix components, composition, and oil retention. Talón et al. (2019) encapsulated eugenol in matrices of MD/whey protein and MD/lecithin and investigated their release kinetics in different food simulants. There was found a fast release rate (< 20 min) for all matrices, where equilibrium was achieved with 84%–100% of oil released. Adding chitosan and oleic acid to the matrices resulted in slower release kinetics. Wall material, pH, or simulant medium polarity were reported not to affect eugenol release (Talón et al., 2019). SEM micrographs (Figure 8.2b) revealed irregular surfaces and multimodal particle sizes distribution. On the other hand, after the addition of acid oleic to the matrix, particle shape became spherical, with few irregularities observed (Talón et al., 2019). As chitosan was added to the matrix, larger particles were obtained, with large agglomerates. The authors reported a linear relationship between emulsion viscosity and the formation of large particles. Talón et al. (2019) also blended MD with whey protein or lecithin and added oleic acid or chitosan for encapsulation of eugenol. TGA thermograms showed eugenol evaporation in the range of 200°C–250°C. Matrices containing lecithin were found to lead to lower thermal stability.

EO of elemi from *Cannarium commune* tree and peppermint were also encapsulated by MD, with a 30% solid concentrations and EO: WM of 10%–30% (Adamiec & Kalemba, 2006). The findings were reported to be promising, although further research was suggested for confirmation. EE was found to be in the range of 57.2%–70.6%, with loading varying from 7.06% to 18.17%. In another study (Bylaitë et al., 2001), caraway (*Carum carvi* L.) EO in a 15% ratio (EO: WM) was encapsulated using MD and dairy proteins, with a feed rate of 30%. Whey protein/MD matrices were reported to be more effective as encapsulating agents. Data showed an excellent figure for oil loading (69%–88%) and EE in the range of 68%–86%, with the highest reported for MD/whey protein matrices.

Pink pepper EO (*Schinus terebinthifolia Raddi*) was encapsulated in a matrix containing 20% MD with a 5% EO: WM ratio. Data revealed a bimodal particle size distribution (Figure 8.5) with a large peak below 100 μm and a smaller peak ranging between 200 and 800 μm. The bigger particles were found to exhibit greater EE (Pereira et al., 2019). It was claimed that the addition of pectin that resulted in the formation of double-layer emulsions, lowered the degree of aggregation between particles as indicated by the low PDI values. The use of pectin for the spray drying encapsulation of pink pepper EO revealed the presence of spherical particles, with slight depressions without cracking or pores (Pereira et al., 2019).

Aldehyde 3,7-dimethyl-2,6-octadienal, with the commercial name 'Citral' and widely used in the food, cosmetic, and beverage industries, was encapsulated in an MD matrix, using a 20% feed concentration and EO: WM ratios of 2.5% and 5% (Wang et al., 2019). The emulsion was stabilized by 0.5% sodium caseinate and 0.8% pectin, and it was suggested that the additional pectin layer provided improvements to MD matrix. EE as high as 99.6% was reported, along with a positive ζ-potential (+25 mV; resulted from the sodium caseinate layer) and a negative ζ-potential (-10 mV;

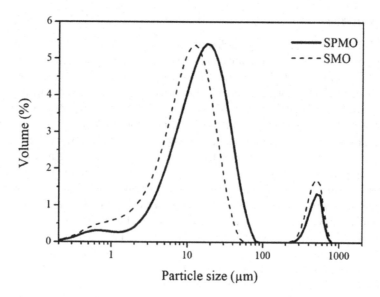

FIGURE 8.5 Particle size distribution of spray dried microcapsules of pink pepper essential oil produced by single-layer (SMO) and double-layer (SPMO) emulsions.

resulted from the pectin layer). After spray drying, no changes were observed in the ζ-potential for pectin matrices; whereas, the matrices covered with sodium caseinate showed lower values of ζ-potential (Wang et al., 2019).

8.3.4 CHITOSAN

Chitosan is a polysaccharide composed of d-glucosamine and N-acetyl-d-glucosamine residues, possessing excellent film-forming properties. Chitosan has been used either alone or with polysaccharides for the spray drying encapsulation of EOs. For example, Abreu et al. (2012) used a blend of chitosan/cashew gum for *Lippia sidoide* EO encapsulation with the following conditions; EO: WM = 1:10 and 2:10, and operating temperature of 170°C. The spray dried nanoparticles presented positive potentials, ranging from +4 mV to +49 mV, the highest figure being obtained for gum: chitosan ratio of 1: 10. The authors (Abreu et al., 2012) reported that chitosan molecules were located on the surface of the manufactured nanoparticles, and could contribute more effectively to nanogel stabilization.

Chitosan was also blended with angico gum and employed for the microencapsulation of the same oil (*Lippia sidoides*), in EO: WM= 1: 2, 1: 4, 1: 10, and 1: 20, under similar operating conditions. Microparticles were found to be effective as a larvicidal agent (Paula, Sombra, et al., 2010). Particles presented negative ζ-potential ranging from -14 to -21 mV, regardless of the chitosan: gum ratio, which seemed to indicate that angico gum carboxylated groups were located in the particle outer shell, and, therefore, chitosan and oil were located in the particle core. Thermal analysis of the nanoparticles of chitosan/angico gum loaded with *Lippia sidoides* EO showed a shift of matrix component degradation temperatures toward lower values, with the increase in the oil proportion in the initial emulsion. In another study (Paula et al., 2017), *Lippia sidoides* was also encapsulated by a blend of chicha gum (a polysaccharide from *Sterculia striata* trees), cashew gum, and angico gum and chitosan, in order to investigate the matrix effect on the encapsulation parameters. These authors (Paula et al., 2017) concluded that both chitosan: gum ratio and polysaccharide properties were important factors during the nanoencapsulation processes. Yields were reported to be in the range of 30%–61% and changes were attributed to the nature of the polysaccharides in the matrices. Cashew gum presented EE in the range of 40%–70% (Abreu et al., 2012), while it was 16%–78% for angico gum (Paula, Sombra, et al., 2010), and 15%–59% for chicha gum (Paula et al., 2017). Loadings were reported to be in the range of 6.7%–15.6%, higher values associated with cashew gum and chicha gum. It should be noticed that by blending chitosan with chicha gum, a wide range of ζ-potential values was obtained (i.e., from −40 to +26 mV), where the negative ζ-potentials were known to be due to a large number of carboxylate groups present in chicha gum (around 40%) (Paula et al., 2017).

A matrix composed of only cashew gum, used for *Lippia sidoides* EO encapsulation, resulted in the fabrication of nanoparticle loading in the range of 5%–7% with an EE of 25%–27% (Herculano et al., 2015). The system was revealed to show good antimicrobial activity against gram-positive bacteria. The ζ-potential varied from −10.45 to −24.50 mV, where the negative figures were claimed to be attributed to the carboxylic acid groups in the carboxylated form (-COO$^-$) of cashew gum (Herculano et al., 2015). Lastly, the EO from seeds of coriander (*Coriandrum sativum* L.), was encapsulated by chitosan at a high ratio of EO: WM= 1:1, under mild operation conditions (inlet temperature of 120°C, outlet temperature of 68°C–71°C). Microparticles were reported to have higher antioxidant activity than the oil and the pure chitosan presented a low EE of 28.4%, a figure that was attributed to the high molar mass of chitosan (Duman & Kaya, 2016). TGA analysis revealed that thermograms showed the oil decomposition in the range 100°C–200°C, with a peak temperature at 115.9°C.

8.3.5 STARCH

Starch, a polysaccharide composed of chains of amylopectin and amylose, was reacted with octenyl succinic anhydride and the derivative was used as wall material for encapsulation of EOs of *Rosmaris officinalis* and *Zataria multiflora* (EO: WM= 0.05: 1 and 1: 5 ratios), under mild operating temperatures (i.e., 120°C–160°C) (Ahsaei et al., 2020). Microparticles were proposed for agricultural applications such as a pesticide against *T. confusum*. Octenyl succinic anhydride (OSA)-starch possesses high emulsification properties, good volatile retention, and can be used at a high feed solids concentration. Drying yield (12.4%–50.9%) was reported to increase linearly with the ratio of modified starch to EO, and microparticles presented mean particle sizes from 8.29 to 11.35 μm, with narrow particle size distribution. EE ranged from 68% to 88%, showing insecticidal activity (mortality) of 35%–46%. SEM showed that the particles exhibited a sphere-like shape, without any obvious cracks or breaks, likely due to the film-forming ability of OSA-starch. The same matrix was used for peppermint (*Mentha piperita* L.) EO encapsulation (Baranauskiené et al., 2007), where the EE reached 97%, with microparticle loading varying from 6% to 15%. Similar figures were obtained for hydrolyzed starch (i.e., dextrin).

Márquez-Gómez et al. (2018) used rice starch, modified starch, maltodextrin, and hydrolyzed protein for the spray drying encapsulation of EOs from orange at 30% solid concentration and 180ºC. It was reported that the best results were obtained for the matrix with rice starch, as the main component (>50%) of the system. EE was in the range of 36.76%–98.82% and SEM showed particles with irregular or collapsed morphology, an indication of the need for improvement of the interactions between the wall materials. In another approach, Costa et al. (2012) used rice starch, inulin, gelatin/sucrose matrices for oregano EO microencapsulation, aiming at the investigation of matrix effect on the process parameters (Costa et al., 2012). In another research (Fernandes, Borges, & Botrel, 2014), starch (20%–35% w/w), gelatin (0%–1% w/w), and inulin (5%–25% w/w) were tested for the spray drying encapsulation of rosemary EO at the EO: WM ratio of 15%. The authors (Fernandes, Borges, & Botrel, 2014) evaluated matrices according to their stability and antimicrobial and antioxidant activities. Starch was blended with MD in a 1: 1 ratio (w/w), a solid concentration of 10%–30%, and EO: WM of 10%–30% at 190°C (Fernandes, Marques, et al., 2014). The results showed that the optimal wall concentration and oil load were 20.9% and 29.4%, respectively. SEM micrographs showed that the particles were spherical with variable sizes, where most of the particles showed dented surfaces, with continuous walls with no fissures or cracks. Spray-dried powders could represent options as ingredients for new food formulations.

8.3.6   Pectin

Pectin is a heteropolysaccharide composed of galacturonic acid residues, and it is usually obtained from citrus fruits. A blend of pectin/alginate matrix, with a pectin: alginate ratio of 1: 1 (w/w), a solid concentration of 3%, and EO: WM of 1: 3 was employed for the encapsulation of carvacrol (5-Isopropyl-2-methylphenol) (Sun et al., 2020). High oil retention (88%) and EE of 82.79% was reported. Microparticles were claimed to exhibit high antioxidant capacity and antimicrobial activity on *E. coli*. The release kinetics in phosphate buffer solution at 37°C showed a fast release, where about 60% of the oil was released within 3 h. SEM data presented particles with smooth surfaces and spherical shapes, provided that the inlet temperature was kept in the range of 100°C to 130°C, because the higher temperatures (160°C and 190°C) resulted in the cracked particles, as presented in Figures 8.2c and d (Sun et al., 2020). The matrix is reported to show excellent performance on oil retention and could be used in processed foods, dairy products, or nutraceutical supplements. A mixture of the polysaccharide pectin, corn protein, and caseinate, in different proportions, was employed for the nanoencapsulation of eugenol, where a solid concentration in the range of 0.05–0.2 and EO: WM of 1: 2 were achieved (Veneranda et al., 2018).

8.3.7   Alginate

Alginate or Alginic acid is a linear copolymer containing D-mannuronate and α-L-guluronate units, covalently linked via covalent bonds in different sequences or blocks. Its main source is brown seaweeds, the species and origin of which influence polysaccharide properties. Oliveira et al. (2014) blended alginate with cashew gum and used it for *Lippia sidoides* EO encapsulation, at a solid concentration of 0.2%, 0.5%, and 1.0% and EO: WM ratio of 1: 10. This process yielded an EE of 55% and loading in the range of 1.9% to 4.4%. The low loading product was proven to be effective as an insecticidal agent. The ζ-potential varied from −30.0 to −36.2 mV, the highest value being obtained for alginate: cashew gum with the ratio of 1: 1 blend. Thermograms of spray dried nanoparticles showed that EO evaporates at temperature < 130°C and lower cashew gum: alginate ratio led to higher thermal stability. The authors (Oliveira et al., 2014) claimed that nanoparticles possessed the potential to be used as a larvicide on fighting *A. aegypti* mosquito.

8.3.8   Cyclodextrins

Cyclodextrin is a cyclic polysaccharide that provides cavities for the molecular inclusion of bioactive substances, enabling several applications in the food, pharmaceutical, and cosmetic industries. In this sense, β-cyclodextrin has been used for *Lippia sidoides* EO encapsulation (Fernandes et al., 2009), at feed concentration of 50%, EO: WM = 1: 10, 1.33: 10 and 2: 10, at 160°C. The results showed that a 1: 10 ratio provided the most stable complex. Oil loading was in the range of 35%–70% and EE of 70% (Fernandes et al., 2009). In another approach, β-cyclodextrin and MD were employed for eugenol and carvacrol complex preparation (Barba et al., 2015), followed by spray drying at 11% feed concentration, at EO: WM= 1:3, and at 150°C. Yields were lower than 50% for eugenol and 79.6% for carvacrol, indicating the need for process improvement.

8.3.9   Other Wall Materials

Several proteins or protein-rich materials have been used as encapsulating agents for spray drying encapsulation of various EOs; e.g., whey protein (Fernandes, Botrel, et al., 2017; Fernandes, Guimarães, et al., 2017), skimmed milk powder (Baranauskiene et al., 2006; Rodklongtan &

Chitprasert, 2017), bovine serum albumin, rice bran protein (Scremin et al., 2018), and zein (Chen et al., 2015). For example, the EO from Holy basil (*Ocimum sanctum* Linn.) was encapsulated using skimmed milk powder (Rodklongtan & Chitprasert, 2017), with 20% feed concentration, at different temperatures (i.e., 130°C, 240°C, and 150°C). Spray dried particles showed EE in the range of 42%–48%, where the higher the temperature, the lower was the EE. Additionally, surface oil was found to increase with temperature (Rodklongtan & Chitprasert, 2017).

Bovine serum albumin and rice bran proteins were blended with the polysaccharide carrageenan in different ratios and employed for eugenol encapsulation, at a feed concentration of 1%, EO: WM= 1: 20, and at 150°C (Scremin et al., 2018). The blend containing all three components yielded the best data, with EE varying from 24% to 79.4%. Zein, a corn protein of the prolamins group, was used as an encapsulated agent of eugenol and thymol along with casein, forming a complex (Chen et al., 2015), which was spray dried at 2% solid concentration and EO: WM= 1: 4. Data revealed loading of 5%–8% and EE varying from 16% to 26%, where the nanoparticles were stable in terms of dispersibility with good hydration degrees, and could be used as antimicrobial preservatives in food products. In another approach, blends of zein with pectin and caseinate (Veneranda et al., 2018) were employed for eugenol encapsulation. The authors reported nanoparticles formation, with sizes of 140 nm. Complex nanoparticles showed ζ-potentials ranging from –47 to –49 mV, indicating a strong stability that was attributed to the added sodium caseinate.

Methylcellulose and hydroxypropyl methylcellulose phthalate matrices were also used for thymol encapsulation (Rassu et al., 2014), and presented a yield of 37% and 53%, respectively, where the yield depended on the polymer used. In another study (Fernandes, Botrel, et al., 2017), ginger EO was encapsulated by a blend of whey protein and inulin that resulted in particle sizes in the range of 12–24 µm. It was reported that the wall materials ratio had a direct effect on the particle size distribution and the corresponding polydispersity. SEM data revealed the presence of spherical shapes with cavities and folds on the surface (Fernandes, Botrel, et al., 2017). Further, the SEM micrographs of particles of cashew gum/inulin matrix used for ginger EO encapsulation showed no cracks or fissures in the particles with cashew gum and cashew gum/inulin at the 3: 1 ratio. On the other hand, using high inulin content in the matrices resulted in cracks on the surface of the particles, possibly due to the low viscosity of inulin solutions (Fernandes et al., 2016). TGA revealed that about 85% of the EO was lost at temperatures lower than 180°C, and the matrices containing greater cashew gum proportion were found to be more stable.

8.4   APPLICATIONS OF SPRAY DRIED ENCAPSULATED EOS IN VARIOUS PRODUCTS

The applications of EOs are diverse as they are widely used in cosmetics and perfumes. They also have medicinal applications, due to their therapeutic properties, as well as agricultural and food uses, because of their antimicrobial and antioxidant effects (Ríos, Stashenko, & Duque, 2017). In the food, pharmaceutical, and cosmetic industries, EOs have been used since antiquity as condiments, spices, antimicrobial, insecticidal, and agents to protect stored products (Mejri, Aydi, Abderrabba, & Mejri, 2018).

8.4.1   EOs USED AS ANTIMICROBIAL AGENT

Terpenoids such as carvacrol and thymol have been reported as the most active monoterpenoids against a large number of microorganisms (Hyldgaard, Mygind, & Meyer, 2012). It has been shown that 25 different bacterial strains were affected by all aromatic terpenoids, except borneol and carvacrol methyl ester. The most active compound was carvacrol followed by thymol, with their highest MIC being 300 and 800 µg/mL, respectively. Authors stated that these aromatic monoterpenes can be used in food preservation or for general antimicrobial use.

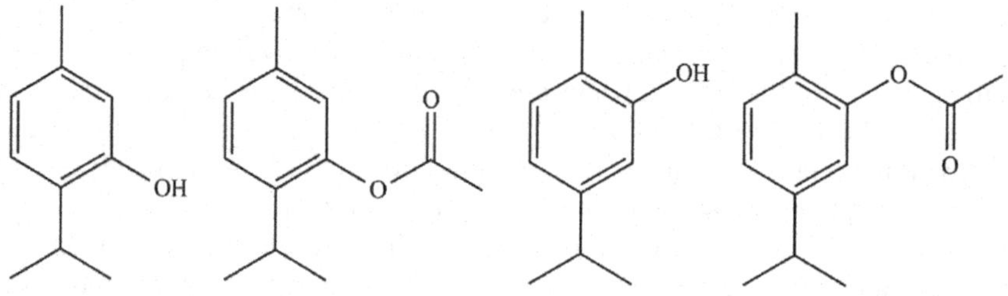

FIGURE 8.6 Thymol and carvacrol acetylated derivatives.

Fontenelle et al. (2007) used the essential oil from *Lippia sidoides* for new antifungal drugs due to its efficacy and low toxicity. The main constituents of *L. sidoides* essential oil were thymol (59.65%), E-caryophyllene (10.60%), and *p*-cymene (9.08%). The acute administration of the essential oil by the oral route to mice was devoid of overt toxicity (Fontenelle et al., 2007).

Owing to the resistance exhibited by microorganisms to antifungal drugs, EOs have been investigated as alternative to those drugs. Five Brazilian *Ocimum* species, *Ocimum americanum, Ocimum basilicum var. purpurascens, O. basilicum var. minimum, Ocimum micranthum* and *Ocimum selloi*, were tested against *Candida* species, yielding promising data, which points to their use to treat diseases such as mycosis. The activity was likely to be due to the presence of eugenol and anethole (Vieira et al., 2014).

The essential oil compounds (Figure 8.6) such as thymol, eugenol, estragole, anethole, and some *O*-methyl-derivatives (methylthymol and methyleugenol) have been tested against *Candida* spp. and *Microsporum canis* and showed *in vitro* antifungal activities (Fontenelle et al., 2011). The EO from *Coriandrum sativum* L. contains linalool was known as the main constituent (58.22%). The oil was considered bioactive, demonstrating good antifungal activity (Soares et al., 2012).

8.4.2 ANTILEISHMANIAL ACTIVITY

The increased incidence of visceral leishmaniasis (VL) in many countries like Brazil is due to a lack of effective disease control actions. In addition, no effective treatment exists for canine VL in response to synthetic drugs. In the search for new antileishmanial agents, EOs from four *Croton* species (*C. argyrophylloides, C. jacobinensis, C. nepetifolius* and *C. sincorensis*) were evaluated *against Leishmania infantum chagasi, L. amazonensis* and *L. braziliensis*. The major constituents of these EOs were spathulenol, β-caryophyllene, β-caryophyllene oxide, 1,8-cineole, and methyl eugenol (Morais et al., 2019).

8.4.3 ANTHELMINTIC ACTIVITY

Phytotherapy can be an alternative for the control of gastrointestinal parasites of small ruminants. The efficacies of *Coriandrum sativum, Alpinia zerumbet, Tagetes minuta* and *Lantana camara* EOs were demonstrated by two *in vitro* assays on *Haemonchus contortus* (Macedo et al., 2013), where the main component of *A. zerumbet* was known to be 1,8-cineol, for *L. camara* was caryophyllene oxide, for *C. sativum* was linalool and the major constituent of *T. minuta was* piperitone. Thymol and carvacrol (Figure 8.6) are monoterpenes and their acetylation form has been found to present low toxicities. Carvacrol and thymol acetate (TA) were assayed on egg, larvae, and adult *Haemonchus contortus* (Andre et al., 2016; André et al., 2017).

The anthelmintic activity of *Eucalyptus citriodora* EO and citronellal was observed on sheep gastrointestinal nematodes. Citronellal was confirmed as the essential oil major constituent (63.9%).

The essential oil and citronellal completely inhibited *Haemonchus contortus* motility at 6 h post-exposure (Araújo-Filho et al., 2019).

8.4.4 LARVICIDAL AGENTS AGAINST AEDES AEGYPTI

The search for new insecticides to control dengue fever (Paula et al., 2010; Paula, De, & Bezerra, 2006) chikungunya, and Zika vectors has gained relevance in recent decades. The larvicidal action of EOs (EOs) from *Thymus vulgaris, Salvia officinalis, Lippia origanoides, Eucalyptus globulus, Cymbopogon nardus, Cymbopogon martinii, Lippia alba, Pelargonium graveolens, Turnera diffusa*, and *Swinglea glutinosa* on *Aedes* (Stegomyia) *aegypti* were evaluated. All EOs achieved larvicidal activity at medium lethal concentration LC_{50} values lower than 115 mg/L. The main compounds of the EOs with the highest larvicidal activity were thymol (42%) and p-cymene (26.4%) (Ríos et al., 2017).

8.4.5 IN FOOD PRODUCTS

Pectin/sodium alginate microcapsules containing carvacrol were used as a controlled-release decontaminator in food and nutraceutical processing industries (Sun et al., 2019) and as antimicrobial and antioxidant agents (Sun et al., 2020).

Talón et al. (2019) reported that the encapsulation of eugenol using maltodextrin and incorporation as an ingredient in foodstuffs enables preservation against oxidative or microbial decay, thus extending the shelf-life of the foodstuffs.

Wang et al. (2019) showed that maltodextrin microcapsules containing encapsulated citral exhibit antibacterial activity which, upon storage at low temperatures, can prolong the shelf-life of perishable food products.

Microencapsulated rosemary essential oil was used as a preservative in cheese, thereby extending the product shelf-life (Fernandes, Guimarães, et al., 2017). Coriander essential oil encapsulated in a chitosan matrix was employed as a natural antioxidant and antimicrobial agent in the food industries (Duman & Kaya, 2016). Herculano et al. (2015) reported that the nanoencapsulation of *Eucalyptus staigeriana* essential oil using cashew gum is a promising alternative to the use of chemical preservatives, being a good candidate for natural food preservation. Zein–casein nanocapsules with co-encapsulated eugenol and thymol were shown to have the potential to be used as antimicrobial preservatives in food products (Chen et al., 2015).

8.5 CONCLUDING REMARKS

Among all the variables involved in the optimization of EO encapsulation using spray drying, the wall material plays a paramount role. Its choice, along with the use of the right operation parameters, ensures oil components protection, high yield, and loading and encapsulation efficiency. Polysaccharide-based matrices have been most cited in the literature, whereby gums are by far largely employed either alone or blended with starch derivatives, for example. The biological health-promoting properties of EOs such as antimicrobial, antileishmanial, and anthelmintic activities may not be affected by the spray drying process, rather being enhanced as in the case of nanoparticulated systems. As a remark, authors may publish their work in a fashion with more standardized data, allowing for a better comparison and referencing to future research.

ACKNOWLEDGEMENT

The authors would like to thank CNPQ, CAPES, and INOMAT, for financial support and scholarships.

REFERENCES

Abreu, F. O. M. S., Oliveira, E. F., Paula, H. C. B., & De Paula, R. C. M. (2012). Chitosan/cashew gum nanogels for essential oil encapsulation. *Carbohydrate Polymers*, *89*(4), 1277–1282. https://doi.org/10.1016/j. carbpol.2012.04.048

Adamiec, J., & Kalemba, D. (2006). Analysis of microencapsulation ability of essential oils during spray drying. *Drying Technology*, *24*(9), 1127–1132. https://doi.org/10.1080/07373930600778288

Ahsaei, S. M., Rodríguez-Rojo, S., Salgado, M., Cocero, M. J., Talebi-Jahromi, K., & Amoabediny, G. (2020). Insecticidal activity of spray dried microencapsulated essential oils of *Rosmarinus officinalis* and *Zataria multiflora* against *Tribolium confusum*. *Crop Protection*, *128*, 104996. https://doi.org/10.1016/j. cropro.2019.104996

André, W. P. P., Cavalcante, G. S., Ribeiro, W. L. C., Dos Santos, J. M. L., Macedo, I. T. F., De Paula, H. C. B., De Morais, S. M., De Melo, J. V., & Bevilaqua, C. M. L. (2017). Anthelmintic effect of thymol and thymol acetate on sheep gastrointestinal nematodes and their toxicity in mice. *Revista Brasileira de Parasitologia Veterinária*, *26*(3), 323–330. https://doi.org/10.1590/S1984-29612017056

Andre, W. P. P., Ribeiro, W. L. C., Cavalcante, G. S., Santos, J. M. L. do., Macedo, I. T. F., Paula, H. C. B. D., de Freitas, R. M., de Morais, S. M., Melo, J. V. d., & Bevilaqua, C. M. L. (2016). Comparative efficacy and toxic effects of carvacryl acetate and carvacrol on sheep gastrointestinal nematodes and mice. *Veterinary Parasitology*, *218*, 52–58. https://doi.org/10.1016/j.vetpar.2016.01.001

Araújo-Filho, J. V., Ribeiro, W. L. C., André, W. P. P., Cavalcante, G. S., Rios, T. T., Schwinden, G. M., Da Rocha, L. O., Macedo, I. T. F., de Morais, S. M., Bevilaqua, C. M. L., & De Oliveira, L. M. B. (2019). Anthelmintic activity of Eucalyptus citriodora essential oil and its major component, citronellal, on sheep gastrointestinal nematodes. *Revista Brasileira de Parasitologia Veterinária*, *28*(4), 644–651. https://doi. org/10.1590/s1984-29612019090

Aziz, Z. A. A., Ahmad, A., Setapar, S. H. M., Karakucuk, A., Azim, M. M., Lokhat, D., Rafatullah, M., Ganash, M., Kamal, M. A., & Ashraf, G. M. (2018). Essential oils: Extraction techniques, pharmaceutical and therapeutic potential - A review. *Current Drug Metabolism*, *19*(13), 1100–1110. https://doi.org/10.2174 /1389200219666180723144850

Baranauskiené, R., Bylaité, E., Juraté, Ž., & Venskutonis, R. P. (2007). Flavor retention of peppermint (*Mentha piperita* L.) essential oil spray dried in modified starches during encapsulation and storage. *Journal of Agricultural and Food Chemistry*, *55*(8), 3027–3036. https://doi.org/10.1021/jf062508c

Baranauskiene, R., Venskutonis, P. R., Dewettinck, K., & Verhé, R. (2006). Properties of oregano (*Origanum vulgare* L.), citronella (*Cymbopogon nardus* G.) and marjoram (*Majorana hortensis* L.) flavors encapsulated into milk protein-based matrices. *Food Research International*, *39*(4), 413–425. https://doi. org/10.1016/j.foodres.2005.09.005

Barba, C., Eguinoa, A., & Maté, J. I. (2015). Preparation and characterization of β-cyclodextrin inclusion complexes as a tool of a controlled antimicrobial release in whey protein edible films. *LWT- Food Science and Technology*, *64*(2), 1362–1369. https://doi.org/10.1016/j.lwt.2015.07.060

Beristain, C. I., García, H. S., & Vernon-Carter, E. J. (2001). Spray-dried encapsulation of cardamom (*Elettaria cardamomum*) essential oil with mesquite (*prosopis juliflora*) gum. *LWT- Food Science and Technology*, *34*(6), 398–401. https://doi.org/10.1006/fstl.2001.0779

Burhan, A. M., Abdel-Hamid, S. M., Soliman, M. E., & Sammour, O. A. (2019). Optimisation of the microencapsulation of lavender oil by spray drying. *Journal of Microencapsulation*, *36*(3), 250–266. https://doi. org/10.1080/02652048.2019.1620355

Bylaitë, E., Venskutonis, P. R., & Mapdpierienë, R. (2001). Properties of caraway (*Carum carvi* L.) essential oil encapsulated into milk protein-based matrices. *European Food Research and Technology*, *212*(6), 661–670. https://doi.org/10.1007/s002170100297

Chen, H., Zhang, Y., & Zhong, Q. (2015). Physical and antimicrobial properties of spray dried zein-casein nanocapsules with co-encapsulated eugenol and thymol. *Journal of Food Engineering*, *144*, 93–102. https://doi.org/10.1016/j.jfoodeng.2014.07.021

Costa, J. M. G. da, Borges, S. V., Hijo, A. A. C. T., Silva, E. K., Marques, G. R., Cirillo, M. Â., & De Azevedo, V. M. (2013). Matrix structure selection in the microparticles of essential oil oregano produced by spray dryer. *Journal of Microencapsulation*, *30*(8), 717–727. https://doi.org/10.3109/02652048.2013.778909

Costa, S. B., Duarte, C., Bourbon, A. I., Pinheiro, A. C., Januário, M. I. N., Vicente, A. A., Beirão-da-Costa, M. L., & Delgadillo, I. (2013). Inulin potential for encapsulation and controlled delivery of Oregano essential oil. *Food Hydrocolloids*, *33*(2), 199–206. https://doi.org/10.1016/j.foodhyd.2013.03.009

Costa, S. B., Duarte, C., Bourbon, A. I., Pinheiro, A. C., Serra, A. T., Martins, M. M., Januário, M. I. N., Vicente, A. A., Delgadillo, I., Duarte, C., & Da Costa, M. L. B. (2012). Effect of the matrix system in the delivery and in vitro bioactivity of microencapsulated Oregano essential oil. *Journal of Food Engineering*, *110*(2), 190–199. https://doi.org/10.1016/j.jfoodeng.2011.05.043

Duman, F., & Kaya, M. (2016). Crayfish chitosan for microencapsulation of coriander (*Coriandrum sativum* L.) essential oil. *International Journal of Biological Macromolecules*, *92*, 125–133. https://doi.org/10.1016/j.ijbiomac.2016.06.068

Fernandes, L. P., Oliveira, W. P., Sztatisz, J., & Novák, C. (2008). Thermal properties and release of *Lippia sidoides* essential oil from gum arabic/maltodextrin microparticles. *Journal of Thermal Analysis and Calorimetry*, *94*(2), 461–467. https://doi.org/10.1007/s10973-008-9170-4

Fernandes, L. P., Oliveira, W. P., Sztatisz, J., Szilágyi, I. M., & Novák, C. (2009). Solid state studies on molecular inclusions of lippia sidoides essential oil obtained by spray drying. *Journal of Thermal Analysis and Calorimetry*, *95*(3), 855–863. https://doi.org/10.1007/s10973-008-9149-1

Fernandes, L. P., Turatti, I. C. C., Lopes, N. P., Ferreira, J. C., Candido, R. C., & Oliveira, W. P. (2008). Volatile retention and antifungal properties of spray dried microparticles of *Lippia sidoides* essential oil. *Drying Technology*, *26*(12), 1534–1542. https://doi.org/10.1080/07373930802464034

Fernandes, R. V. de B., Borges, S. V., Botrel, D. A., Silva, E. K., Costa, J. M. G. da, & Queiroz, F. (2013). Microencapsulation of rosemary essential oil: Characterization of particles. *Drying Technology*, *31*(11), 1245–1254. https://doi.org/10.1080/07373937.2013.785432

Fernandes, R. V. de B., Botrel, D. A., Silva, E. K., Borges, S. V., Oliveira, C. R. de, Yoshida, M. I., Feitosa, J. P. De A., & de Paula, R. C. M. (2016). Cashew gum and inulin: New alternative for ginger essential oil microencapsulation. *Carbohydrate Polymers*, *153*, 133–142. https://doi.org/10.1016/j.carbpol.2016.07.096

Fernandes, R. V. de B., Botrel, D. A., Silva, E. K., Pereira, C. G., Carmo, E. L. do, Dessimoni, A. L. de A., & Borges, S. V. (2017). Microencapsulated ginger oil properties: Influence of operating parameters. *Drying Technology*, *35*(9), 1098–1107. https://doi.org/10.1080/07373937.2016.1231690

Fernandes, R. V. de B., Guimarães, I. C., Ferreira, C. L. R., Botrel, D. A., Borges, S. V., & de Souza, A. U. (2017). Microencapsulated rosemary (*Rosmarinus officinalis*) essential oil as a biopreservative in minas frescal cheese. *Journal of Food Processing and Preservation*, *41*(1), 1–9. https://doi.org/10.1111/jfpp.12759

Fernandes, R. V., Marques, G. R., Borges, S. V., & Botrel, D. A. (2014). Effect of solids content and oil load on the microencapsulation process of rosemary essential oil. *Industrial Crops and Products*, *58*, 173–181. https://doi.org/10.1016/j.indcrop.2014.04.025

Fernandes, R. V. D. B., Borges, S. V., & Botrel, D. A. (2014). Gum arabic/starch/maltodextrin/inulin as wall materials on the microencapsulation of rosemary essential oil. *Carbohydrate Polymers*, *101*(1), 524–532. https://doi.org/10.1016/j.carbpol.2013.09.083

Fontenelle, R. O.S., Morais, S. M., Brito, E. H. S., Kerntopf, M. R., Brilhante, R. S. N., Cordeiro, R. A., Tomé, A. R., Queiroz, M. G. R., Nascimento, N. R. F., Sidrim, J. J. C., & Rocha, M. F. G. (2007). Chemical composition, toxicological aspects and antifungal activity of essential oil from *Lippia sidoides* Cham. *Journal of Antimicrobial Chemotherapy*, *59*(5), 934–940. https://doi.org/10.1093/jac/dkm066

Fontenelle, Raquel O.S., Morais, S. M., Brito, E. H. S., Brilhante, R. S. N., Cordeiro, R. A., Lima, Y. C., Brasil, N. V. G. P. S., Monteiro, A. J., Sidrim, J. J. C., & Rocha, M. F. G. (2011). Alkylphenol activity against candida spp. and microsporum canis: A focus on the antifungal activity of thymol, eugenol and o-methyl derivatives. *Molecules*, *16*(8), 6422–6431. https://doi.org/10.3390/molecules16086422

Garcia, L. C., Tonon, R. V., & Hubinger, M. D. (2012). Effect of homogenization pressure and oil load on the emulsion properties and the oil retention of microencapsulated basil essential oil (*Ocimum basilicum* L.). *Drying Technology*, *30*(13), 1413–1421. https://doi.org/10.1080/07373937.2012.685998

Herculano, E. D., de Paula, H. C. B., de Figueiredo, E. A. T., Dias, F. G. B., & Pereira, V. de A. (2015). Physicochemical and antimicrobial properties of nanoencapsulated Eucalyptus staigeriana essential oil. *LWT- Food Science and Technology*, *61*(2), 484–491. https://doi.org/10.1016/j.lwt.2014.12.001

Herman, R. A., Ayepa, E., Shittu, S., Fometu, S. S., & Wang, J. (2019). Essential oils and their applications – A mini review. *Advances in Nutrition & Food Science*, *4*(4). https://doi.org/10.33140/anfs.04.04.08

Hijo, A. A. C. T., Costa, J. M. G., Silva, E. K., Azevedo, V. M., Yoshida, M. I., & Borges, S. V. (2015). Physical and thermal properties of oregano (*Origanum vulgare* L.) essential oil microparticles. *Journal of Food Process Engineering*, *38*(1), 1–10. https://doi.org/10.1111/jfpe.12120

Hyldgaard, M., Mygind, T., & Meyer, R. L. (2012). Essential oils in food preservation: Mode of action, synergies, and interactions with food matrix components. *Frontiers in Microbiology*, *3*, 1–24. https://doi.org/10.3389/fmicb.2012.00012

López, A., Castro, S., Andina, M. J., Ures, X., Munguía, B., Llabot, J. M., Elder, H., Dellacassa, E., Palma, S., & Domínguez, L. (2014). Insecticidal activity of microencapsulated Schinus molle essential oil. *Industrial Crops and Products*, *53*, 209–216. https://doi.org/10.1016/j.indcrop.2013.12.038

Macedo, I. T. F., de Oliveira, L. M. B., Camurça-Vasconcelos, A. L. F., Ribeiro, W. L. C., dos Santos, J. M. L., de Morais, S. M., de Paula, H. C. B., & Bevilaqua, C. M. L. (2013). In vitro effects of *Coriandrum sativum*, *Tagetes minuta*, *Alpinia zerumbet* and *Lantana camara* essential oils on *Haemonchus contortus*. *Revista Brasileira de Parasitologia Veterinária*, 22(4), 463–469. https://doi.org/10.1590/s1984-29612013000400004

Maciel, M. V., Morais, S. M., Bevilaqua, C. M. L., Silva, R. A., Barros, R. S., Sousa, R. N., Sousa, L. C., Brito, E. S., & Souza-Neto, M. A. (2010). Chemical composition of Eucalyptus spp. essential oils and their insecticidal effects on *Lutzomyia longipalpis*. *Veterinary Parasitology*, 167(1), 1–7. https://doi.org/10.1016/j.vetpar.2009.09.053

Márquez-Gómez, M., Galicia-García, T., Márquez-Meléndez, R., Ruiz-Gutiérrez, M., & Quintero-Ramos, A. (2018). Spray-dried microencapsulation of orange essential oil using modified rice starch as wall material. *Journal of Food Processing and Preservation*, 42(2). https://doi.org/10.1111/jfpp.13428

Martínez-Ballesta, M. C., Gil-Izquierdo, Á., García-Viguera, C., & Domínguez-Perles, R. (2018). Nanoparticles and controlled delivery for bioactive compounds: Outlining challenges for new "smart-foods" for health. *Foods*, 7(5), 72–101. https://doi.org/10.3390/foods7050072

Mejri, J., Aydi, A., Abderrabba, M., & Mejri, M. (2018). Asian Journal of Green Chemistry review article emerging extraction processes of essential oils: A review. *Asian Journal of Green Chemistry*, 2, 246–267. https://doi.org/10.22631/ajgc.2018.119980.1053

Morais, S. M., Cossolosso, D. S., Silva, A. A. S., de Moraes Filho, M. O., Teixeira, M. J., Campello, C. C., Bonilla, O. H., de Paula, V. F., & Vila-Nova, N. S. (2019). Essential oils from Croton species: Chemical composition, in vitro and in silico antileishmanial evaluation, antioxidant and cytotoxicity activities. *Journal of the Brazilian Chemical Society*, 30(11), 2404–2412. https://doi.org/10.21577/0103-5053.20190155

Oliveira, E. F., Paula, H. C. B., & Paula, R. C. M. d. (2014). Alginate/cashew gum nanoparticles for essential oil encapsulation. *Colloids and Surfaces B: Biointerfaces*, 113, 146–151. https://doi.org/10.1016/j.colsurfb.2013.08.038

Partheniadis, I., Vergkizi, S., Lazari, D., Reppas, C., & Nikolakakis, I. (2019). Formulation, characterization and antimicrobial activity of tablets of essential oil prepared by compression of spray dried powder. *Journal of Drug Delivery Science and Technology*, 50, 226–236. https://doi.org/10.1016/j.jddst.2019.01.031

Paula, H. C. B., De Paula, R. C. M., & Bezerra, S. K. F. (2006). Swelling and release kinetics of larvicide-containing chitosan/cashew gum beads. *Journal of Applied Polymer Science*, 102(1), 395–400. https://doi.org/10.1002/app.24009

Paula, H. C. B., Oliveira, E. F., Abreu, F. O. M. S., De Paula, R. C. M., Morais, S. M., & Forte, M. M. C. (2010). Esferas (Beads) de alginato como agente encapsulante de óleo de Croton Zehntneri Pax et Hoffm. *Polimeros*, 20(2), 112–120. https://doi.org/10.1590/S0104-14282010005000019

Paula, H. C. B., Oliveira, E. F., Carneiro, M. J. M., & De Paula, R. C. M. (2017). Matrix effect on the spray drying nanoencapsulation of *Lippia sidoides* essential oil in chitosan-native gum blends. *Planta Medica*, 83(5), 392–397. https://doi.org/10.1055/s-0042-107470

Paula, H. C. B., Sombra, F. M., Abreu, F. O. M. S., & De Paula, R. C. M. (2010). Lippia sidoides essential oil encapsulation by angico gum/chitosan nanoparticles. *Journal of the Brazilian Chemical Society*, 21(12), 2359–2366. https://doi.org/10.1590/S0103-50532010001200025

Pereira, A. R. L., Gonçalves Cattelan, M., & Nicoletti, V. R. (2019). Microencapsulation of pink pepper essential oil: Properties of spray dried pectin/SPI double-layer versus SPI single-layer stabilized emulsions. *Colloids and Surfaces A: Physicochemical and Engineering Aspects*, 581, 123806. https://doi.org/10.1016/j.colsurfa.2019.123806

Peterfalvi, A., Miko, E., Nagy, T., Reger, B., Simon, D., Miseta, A., Czéh, B., & Szereday, L. (2019). Much more than a pleasant scent: A review on essential oils supporting the immune system. *Molecules*, 24(24), 1–16. https://doi.org/10.3390/molecules24244530

Rassu, G., Nieddu, M., Bosi, P., Trevisi, P., Colombo, M., Priori, D., Manconi, P., Giunchedi, P., Gavini, E., & Boatto, G. (2014). Encapsulation and modified-release of thymol from oral microparticles as adjuvant or substitute to current medications. *Phytomedicine*, 21(12), 1627–1632. https://doi.org/10.1016/j.phymed.2014.07.017

Ríos, N., Stashenko, E. E., & Duque, J. E. (2017). Evaluation of the insecticidal activity of essential oils and their mixtures against *Aedes aegypti* (Diptera: Culicidae). *Revista Brasileira de Entomologia*, 61(4), 307–311. https://doi.org/10.1016/j.rbe.2017.08.005

Rodea-González, D. A., Cruz-Olivares, J., Román-Guerrero, A., Rodríguez-Huezo, M. E., Vernon-Carter, E. J., & Pérez-Alonso, C. (2012). Spray-dried encapsulation of chia essential oil (*Salvia hispanica* L.) in whey protein concentrate-polysaccharide matrices. *Journal of Food Engineering*, 111(1), 102–109. https://doi.org/10.1016/j.jfoodeng.2012.01.020

Rodklongtan, A., & Chitprasert, P. (2017). Combined effects of holy basil essential oil and inlet temperature on lipid peroxidation and survival of Lactobacillus reuteri KUB-AC5 during spray drying. *Food Research International*, *100*, 276–283. https://doi.org/10.1016/j.foodres.2017.07.016

Salehi, B., Sharifi-Rad, J., Quispe, C., Llaique, H., Villalobos, M., Smeriglio, A., Trombetta, D., Ezzat, S. M., Salem, M. A., Zayed, A., Salgado Castillo, C. M., Yazdi, S. E., Sen, S., Acharya, K., Sharopov, F., & Martins, N. (2019). Insights into Eucalyptus genus chemical constituents, biological activities and health-promoting effects. *Trends in Food Science and Technology*, *91*, 609–624. https://doi.org/10.1016/j.tifs.2019.08.003

Scremin, F. R., Veiga, R. S., Silva-Buzanello, R. A., Becker-Algeri, T. A., Corso, M. P., Torquato, A. S., Bittencourt, P. R. S., Flores, E. L. M., & Canan, C. (2018). Synthesis and characterization of protein microcapsules for eugenol storage. *Journal of Thermal Analysis and Calorimetry*, *131*(1), 653–660. https://doi.org/10.1007/s10973-017-6302-8

Shah, B., Davidson, P. M., & Zhong, Q. (2012). Encapsulation of eugenol using Maillard-type conjugates to form transparent and heat stable nanoscale dispersions. *LWT- Food Science and Technology*, *49*(1), 139–148. https://doi.org/10.1016/j.lwt.2012.04.029

Sharifi-Rad, M., Berkay Yılmaz, Y., Antika, G., Salehi, B., Tumer, T. B., Kulandaisamy Venil, C., Das, G., Patra, J. K., Karazhan, N., Akram, M., Iqbal, M., Imran, M., Sen, S., Acharya, K., Dey, A., & Sharifi-Rad, J. (2020). Phytochemical constituents, biological activities, and health-promoting effects of the genus *Origanum*. *Phytotherapy Research*, 1–27. https://doi.org/10.1002/ptr.6785

Soares, B. V., Morais, S. M., Dos Santos Fontenelle, R. O., Queiroz, V. A., Vila-Nova, N. S., Pereira, C. M. C., Brito, E. S., Neto, M. A. S., Brito, E. H. S., Cavalcante, C. S. P., Castelo-Branco, D. S. C. M., & Rocha, M. F. G. (2012). Antifungal activity, toxicity and chemical composition of the essential oil of *coriandrum sativum* L. Fruits. *Molecules*, *17*(7), 8439–8448. https://doi.org/10.3390/molecules17078439

Souza, H. J. B. de, Fernandes, R. V. de B., Borges, S. V., Felix, P. H. C., Viana, L. C., Lago, A. M. T., & Botrel, D. A. (2018). Utility of blended polymeric formulations containing cellulose nanofibrils for encapsulation and controlled release of sweet orange essential oil utility of blended polymeric formulations containing cellulose nanofibrils for encapsulation and controlled R. *Food and Bioprocess Technology*, *11*, 1188–1198. https://doi.org/10.1007/s11947-018-2082-9

Sun, X., Cameron, R. G., & Bai, J. (2019). Microencapsulation and antimicrobial activity of carvacrol in a pectin-alginate matrix. *Food Hydrocolloids*, *92*, 69–73. https://doi.org/10.1016/j.foodhyd.2019.01.006

Sun, X., Cameron, R. G., & Bai, J. (2020). Effect of spray-drying temperature on physicochemical, antioxidant and antimicrobial properties of pectin/sodium alginate microencapsulated carvacrol. *Food Hydrocolloids*, *100*, 105420. https://doi.org/10.1016/j.foodhyd.2019.105420

Talón, E., Lampi, A. M., Vargas, M., Chiralt, A., Jouppila, K., & González-Martínez, C. (2019). Encapsulation of eugenol by spray-drying using whey protein isolate or lecithin: Release kinetics, antioxidant and antimicrobial properties. *Food Chemistry*, *295*, 588–598. https://doi.org/10.1016/j.foodchem.2019.05.115

Teodoro, R. A. R., do Carmo, E. L., Borges, S. V., Botrel, D. A., Marques, G. R., Campelo-Felix, P. H., Silva, E. K., & Fernandes, R. V. de B. (2019). Effects of ultrasonication on the characteristics of emulsions and microparticles containing Indian clove essential oil. *Drying Technology*, *37*(9), 1162–1172. https://doi.org/10.1080/07373937.2018.1492611

Vankar, P. S. (2004). Essential oils and fragrances from natural sources. *Resonance*, *9*(4), 30–41. https://doi.org/10.1007/bf02834854

Veneranda, M., Hu, Q., Wang, T., Luo, Y., Castro, K., & Madariaga, J. M. (2018). Formation and characterization of zein-caseinate-pectin complex nanoparticles for encapsulation of eugenol. *LWT - Food Science and Technology*, *89*, 596–603. https://doi.org/10.1016/j.lwt.2017.11.040

Vieira, P. R. N., De Morais, S. M., Bezerra, F. H. Q., Travassos, P. A., Oliveira, Í. R., & Silva, M. G. V. (2014). Chemical composition and antifungal activity of essential oils from *Ocimum* species. *Industrial Crops and Products*, *55*, 267–271. https://doi.org/10.1016/j.indcrop.2014.02.032

Wang, J., Oussama Khelissa, S., Chihib, N. E., Dumas, E., & Gharsallaoui, A. (2019). Effect of drying and interfacial membrane composition on the antimicrobial activity of emulsified citral. *Food Chemistry*, *298*, 1–7. https://doi.org/10.1016/j.foodchem.2019.125079

Yingngam, B., Kacha, W., Rungseevijitprapa, W., Sudta, P., Prasitpuriprecha, C., & Brantner, A. (2019). Response surface optimization of spray dried citronella oil microcapsules with reduced volatility and irritation for cosmetic textile uses. *Powder Technology*, *355*, 372–385. https://doi.org/10.1016/j.powtec.2019.07.065

Ziosi, P., Manfredini, S., Vertuani, S., Ruscetta, V.; Radice, M., Sacchetti, G. & Bruni, R. (2010). Evaluating essential oils in cosmetics: Antioxidant capacity and functionality. *Cosmetics & Toiletries*, *125*, 32–40.

9 Spray Drying Encapsulation of Essential Fatty Acids and Functional Oils

*Mansoureh Geranpour, Elham Assadpour,
and Seid Mahdi Jafari*
Gorgan University of Agricultural Sciences and Natural Resources, Iran

Cordin Arpagaus
Eastern Switzerland University of Applied Sciences, Switzerland

CONTENTS

9.1 INTRODUCTION

Fatty acids (FAs) are simple lipids and the base components of other different kinds of lipids, such as phospholipids, cholesteryl esters, and triglycerides (Lawrence, 2010). Although the human body can produce most of these FAs, it is unable to generate some others, which are designated as 'essential fatty acids' (EFAs). Therefore, humans should obtain these FAs through the consumption of a variety of foods (Rodriguez, Martin, Ruiz, & Clares, 2016). The essential fatty acid deficiency will cause health risks like inflammatory bowel disease, fistulas involving the small bowel, cystic fibrosis, pancreatic insufficiency, carnitine deficiency, and so on (Mogensen, 2017).

According to the definition offered by Biesalski et al. (2009), the non-essential and essential compounds, which are derived from the food components and affect human health, are known as "bioactive compounds." These components are chemicals found in certain foods, including marines, vegetables, nuts, seeds, and oils, and can promote human health (Guaadaoui, Benaicha, Elmajdoub, Bellaoui, & Hamal, 2014). Bioactive FAs differ from simple FAs by a more complex molecular structure consisting of short-, medium-, and long-chain FAs, with regulatory effects on inflammatory processes and other influences such as cell proliferation, energy homeostasis, and metabolic homeostasis (Hernández, Chávez-Tapia, Uribe, & Barbero-Becerra, 2016). Among various FAs, polyunsaturated fatty acids (PUFAs), which are categorized in the long-chain fatty acids groups, dominate in terms of health promotion and disease prevention (Prato et al., 2018). They potentially prevent cardiovascular diseases (Ajith & Jayakumar, 2019; Lavie, Milani, Mehra, & Ventuar, 2009), neurological disorders (Mazza, Pomponi, Janiri, Bria, & Mazza, 2007), and also certain cancers (Gerber, 2012).

On the other hand, the oils that can possess more health benefits than their nutritional value are defined as 'functional oils' (Murakami, Eyng, & Torrent, 2014). It is imperative to conserve these bioactive compounds to profit from their nutritional and pharmaceutical advantages. In this case, encapsulation is a well-established approach to preserve these functional oils by entrapping the internal bioactive substance in an outer layer recognized as wall material or capsule (Katouzian & Jafari, 2016; Nedovic, Kalusevic, Manojlovic, Levic, & Bugarski, 2011). Microencapsulation is a technique that is performed for the protection of the EFAs against deterioration/oxidation, increasing the stability of EFAs, and decreasing the key challenges of food enrichment with ω-3 and ω-6 PUFA (e.g., fishy flavor and oxidation of PUFAs) (; Fakir, Gadhave, & Waghmare, 2015). Spray drying is a common and recognized technique for the microencapsulation of bioactive compounds such as EFAs that leads to high-quality final powder products. This process is applicable for liquid substances, which consist of a complex of wall and core components. During the spray drying process, the wall materials embrace the core material while trying to preserve it. This process consists of several steps, which need to be organized to reach an efficient process.

9.2 ESSENTIAL FATTY ACIDS AND FUNCTIONAL OILS

The functional oils are naturally unstable and can deteriorate easily. These oils oxidize quickly when exposed to oxygen, light, and high temperature (Chew et al., 2019). Moreover, PUFAs are more susceptible and vulnerable to oxidation in comparison to saturated fatty acids (Zhang, Shi, & Jeff, 2007).

EFAs are categorized into two groups: 1) the PUFAs that cannot be synthesized in the body yet are necessary for health, so they must be gained from human's diet; and 2) PUFAs with essential functions that can be synthesized in the human body (Shireman, 2003). The former refers to two types of EFAs, the omega-3 (ω-3) FAs and omega-6 (ω-6) FAs, while the latter mentions those FAs that are not essential because of their prompt function, but because they are the precursor of other FAs such as linoleic acid. Not only these FAs are a type of omega-6 fatty acids, but they are also the precursor of arachidonic acid (20:4n-6), which is essential for the function of whole cells, muscles, immune and nervous systems (Tallima & El, 2018). Generally, ω-3 series of FAs have a double bond of C=C on the third-number carbon away from the methyl end of the FA (in cis-configuration), whereas the ω-6 series of FAs have it on the sixth-number carbon, away from the methyl end of the FA (Kaur, Chagh, & Gupt, 2014). Moreover, there is another group of FAs known as the ω-9 series of FAs that are derived from oleic acid (OA, 18:1). This type of FA is not EFA, yet alongside the EFAs they are classified as functional oils, due to their various properties (Das, 2006). Thus, most of the current and past research concentrated on the ω-3 and ω-6 EFAs because of their proven necessity. Figure 9.1 shows the general formula and structure of the ω-3 and ω-6 EFAs.

The ω-3 PUFAs include three main EFAs and contain docosahexaenoic acid (DHA, 22:6, n-3), eicosapentaenoic acid (EPA, 20:5,n-3), and α-linolenic acid (ALA, 18:3,n-3) (Gammone, Riccioni, Parrinello, & D'Orazio, 2019). Among them, ALA is the principal EFA, as it is the precursor of DHA and EPA (Di, 2009). The primary sources of ALA are nuts, linseed, canola, flax, pumpkin, and other seed oils (Kaur et al., 2014; Gammone et al., 2019). Long-chain EFAs of DHA and EPA can be found in marine fish oils and marine algae (Kaur et al., 2014). It is proved that the ω-3 PUFAs play an essential role in developing the brain, and decreasing the risk of some kinds of cancer and cardiovascular disease (Garg, Wood, Singh, & Moughan, 2006; Eilander, Hundscheid, Osendarp, Transler, & Zock, 2007; Lavie et al., 2009). In addition, the ω-3 PUFAs are converted in the body to the anti-inflammatory mediator that prevents diseases, such as neurodegenerative, Alzheimer's, and diabetes (Lavie et al., 2009; Kralovec, Zhang, Zhang, & Barrow, 2012).

The ω-6 PUFAs include linoleic acid (LA, 18:2, n-6), gamma-linolenic acid (Υ-LA, 18:3,n-6), dihomo-gamma-linolenic acid (DHΥ-LA, 20:3, n-6), and arachidonic acid (ARA, 20:4,n-6) (Das, 2006, Shireman, 2003; Figure 9.1). LA is found plentifully in the diet; its main sources are vegetable oils such as sunflower oil, corn oil, and safflower oil. In addition, it is present in the seeds of most plants (Patterson, Wall, Fitzgerald, Ross, & Stanton, 2012). Υ-LA can be found in small quantities in human milk and meats, whereas it is obtained in the larger quantities from evening primrose, black current, borage, and hemp seed oil (Kaur et al., 2014; Patterson et al., 2012; Horrobin, 1990). ARA can be obtained by consuming meat, eggs, and dairy products. DHΥ-LA is enhanced in the cellular membrane after synthesizing from dietary Υ-LA, which can be converted to mediator compounds, which are anti-inflammatory and anti-proliferative (Kaur et al., 2014; Fan & Chapkin, 1998).

The balance between ω-3 and ω-6 is crucial, especially for the health influences they have on the human body. In the past decades, the ratio of ω-3 FAs to ω-6 FAs was 1: 1 in the human diet

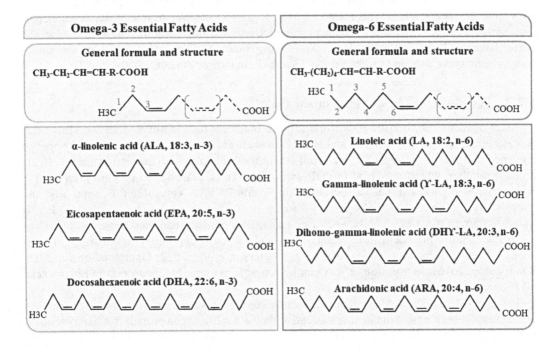

FIGURE 9.1   The general formula and structure of ω-3 and ω-6 Essential Fatty Acids alongside the main fatty acids in each group.

(Simopoulos, 2006), whereas it has changed to approximately 1: 15 in the modern diet. The main reasons for such a dramatic change in the balance are the increase in the desire to consume the land-based seed and vegetable oils with high amounts of LA (ω-6 FAs) and the decrease in the use of marine sources. Although these FAs have potential influences to cure diseases, the high ratio of ω-6 FAs to ω-3 FAs promotes the pathogenesis of diseases such as inflammatory, cardiovascular, brain disorders, and autoimmune diseases (Simopoulos, 2002; Mozaffarian, 2005). It has been reported that the transformation of ALA to DHA and EPA is disrupted when the proportion of ω-6 FAs is higher than the value of ω-3 FAs. The ratio of 1: 1 between LA and ALA caused the uppermost rate of formation of DHA and EPA (Harnack, Andersen, & Somoza, 2009).

9.2.1 MARINE-BASED EFAs AND FUNCTIONAL OILS

The long-chain PUFAs of DHA and EPA are introduced into our diet primarily from marine sources (Islam, Mahmud, Nawas, Fang, & Xia, 2018; Ackman, 2008). Marine fish, including salmon, herring, sardine, smelt, and mackerel, are superb sources of DHA and EPA, and fish oils containing 60% DHA and EPA are good sources of the essential ω-3 FAs (Ackman, 2008). Other marine sources, like sponges, bacteria, fungi, plants, autotrophic macroalgae, and microalgae, are other alternative founts of DHA and EPA that are used to produce ω-3 FAs commercially (Gammone et al., 2019). Furthermore, cold-water fish have a higher proportion of long-chain PUFAs than warm-water fish, which helps them to adapt to the cold environment. Similarly, the marine (wild) fish accumulate more ω-3 PUFAs than the cultivated (farmed)fish, since they usually feed on phytoplankton and zooplankton, which are rich in ω-3 FAs. In contrast, farmed fish consume cereal and vegetables that contain more concentration of ω-6 FAs (Saini & Keum, 2018). The consumption of about 500 mg per day of DHA and EPA from two fish meals per week is recommended for the reduction of cardiovascular disease risk. In general, a standard dose of 1 g per day EPA and DHA is recommended by cardiac societies (Saini & Keum, 2018).

Marine-based oils are extremely sensitive to oxidation during processing and storage, and some indications are reported for their oxidation during gastrointestinal digestion (Tullberg, Vegarud, Undeland, & Scheers, 2017). This phenomenon is because of the high degree of unsaturation, thus limiting the shelf-life of products containing ω-3 PUFAs and impairing the nutritional value and flavor (Islam et al., 2018; Tullberg et al., 2017). The oxidation of FAs causes off-tastes and smells such as bean, grass, fish, and cardboard-like flavors (Beindorff & Zuidam, 2010).

9.2.2 PLANT-BASED EFAs AND FUNCTIONAL OILS

As mentioned before, marine-based oils are the main sources of the ω-3 EFAs, since they contain the highest levels of DHA and EPA. However, growing the fisheries to supply needed fish and other marine animals for industrial and individual purposes causes substantial effects on their number, which may lead to their extinction. Thus, plant-based oils are increasingly being seen as a good alternative source of long-chain PUFAs. Vegetable oils, seed oils, and nut oils have both types of EFAs (i.e., ω-3 and ω-6). Although ALA, which is the ω-3 FA, is found in certain plant-based oils, such as canola, linseed, flaxseed, and nuts, LA, which is a ω-6 FA, is more prevalent in the human diet and can be found easily in the seeds of most plants, except for coconut, cocoa, and palm (Patterson et al., 2012; Gammone et al., 2019; Simopoulos, 2016). In addition, Υ-LA can be found in borage, blackcurrant, and hemp seed oil (Patterson et al., 2012).

To sum up, foods rich in ALA, such as particular vegetables, seeds, and nuts, should be included in the human diet alongside the marine sources, as they are safe and can provide the dietary requirements to maintain the best ratio of ω-6 FAs to ω-3 FAs. Furthermore, those plant-based oils also contain other beneficial and biological substances, such as oil-soluble vitamins, phytosterols, tocopherols, and pigments. The levels of phytosterols, β-sitosterol, campesterol, and stigmasterol in plant-based oils vary from 0.5% to 1.5%, which decreases the cholesterol level and increases the secretion of bile (Gunstone, 2004). All of the isomers of tocopherol (α, β, γ, and δ) are powerful antioxidants with the function of radical scavenging against lipoperoxyl groups. However, most research conducted on vitamin E has focused on αT because it is the dominant form of vitamin E, whose low consumption causes vitamin E deficiency (Jiang, 2014).

9.3 SPRAY DRYING ENCAPSULATION OF EFAS AND FUNCTIONAL OILS

Spray drying encapsulation is one of the most popular techniques and is used frequently for the drying the pre-prepared emulsion containing EFAs (Ray, Raychaudhuri, & Chakraborty, 2016; Tonon, Grosso, & Hubinger, 2011; Watanabe, Fang, Adachi, Fukami, & Matsuno, 2004). Based on the report released by Assadpour and Jafari (2019), the number of publications related to spray drying encapsulation of bioactive food components has recorded a growth of 500% over the past 15 years.

This method is appropriate for drying liquid substances (Costa et al., 2015) that are produced from the initial wall and core materials. The size range of the final dried powders varies between 10 and 120 µm, which can have hollow structures (Subtil et al., 2014; Comunian & Favaro-Trindade, 2016). Sizes below 100 µm are highly applicable for the fortification of the food matrix without unpleasant mouthfeel and sensory problems (Feizollahi, Hadian, & Honarvar, 2018).

The spray drying technique for bioactive encapsulation is a promising process that can produce high-quality powders (Calvo, Hernández, Lozano, & González-Gómez, 2010) with low water activities (Rodriguez et al., 2016), and with more than 90% microencapsulation efficiency (MEE; Chew et al., 2019). The results from the recent research conducted by Shamaei, Seiiedlou, Aghbashlo, Tsotsas, and Kharaghani (2017), Wang, Adhikari, and Barrow (2018), Yu et al. (2017), Chew, Tan, and Nyam (2018), Silva-James, Nogueira, and Freitas (2018) on spray drying microencapsulation of walnut oil, tuna oil, fish oil, refined kenaf seed oil, and combination of pomegranate seed oil and soybean oil, respectively, support this claim.

The spray drying process possesses several benefits and some drawbacks that will be mentioned separately in the following.

The efficiency of the spray drying process to produce a high-quality product is not just limited to the drying stage, which is held in the drying chamber. Choosing the best and most appropriate wall material, formulating the pre-emulsion containing an appropriate ratio of wall and core material, feeding and drying process in spray dryer, and, ultimately, collection and storage of the final product are the four principal stages that are crucial for the spray drying microencapsulation of sensitive EFAs (Corrêa-Filho, Moldão-Martins, & Alves, 2019; Minemoto et al., 2002; Shamaei et al., 2017; Esfahani, Jafari, Jafarpour, & Dehnad, 2019).

Figure 9.2 illustrates the schematic of two general stages, including emulsification and drying alongside the list of flexible parameters of a spray dryer. In addition, the adjustable and fixed parameters during the experiment are indicated. The design of a spray dryer consists of several main components, including the feed pump, the atomizer, drying chamber, cyclone, and product container. In short, the function of this process contains the atomization of the feed emulsion and the simultaneous entry of the hot air into the drying chamber (Dias, Botrel, Fernandes, & Borges, 2017). The atomizer type can be ultrasonic, centrifugal, or a two-fluid nozzle (Assadpour & Jafari, 2019). The direction of entering gas flow can be co-current, counter-current, or mixed current (Islam et al., 2018). The co-current configuration is more appropriate for the production of sensitive EFAs microparticles, with a moderate temperature of 50°C to 80°C. The powders typically will be produced as a result of countering the initial liquid feed with the hot drying air of 150°C to 220°C (Murugesan & Orsat, 2012; Gharsallaoui, Roudaut, Chambin, Voilley, & Saurel, 2007). Thus, the droplets dry in a few seconds and are collected in the powder container (Figure 9.2).

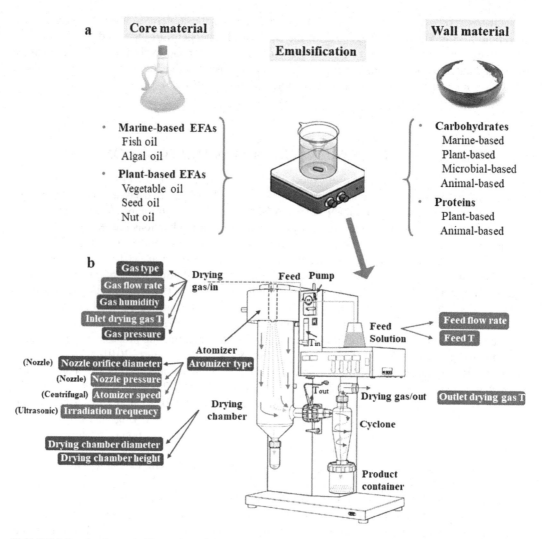

FIGURE 9.2 A schematic illustration of the spray drying procedure: (a) emulsification step, (b) drying step, and the list of flexible process parameters of a spray dryer: the parameters shown in green are adjustable during the experiment, and the parameters shown in blue are fixed during the experiments; T: temperature.

9.3.1 Advantages of Spray Drying Encapsulation of EFAs and Functional Oils

Spray drying encapsulation is a cost-effective, simple, efficient, and flexible technology (Tonon et al., 2011; Ray et al., 2016; Dias et al., 2017; Bakry et al., 2015), which utilizes less electricity than freeze drying and can be easily operated both continuously and on an industrial scale (Islam et al., 2018; Rodriguez et al., 2016). This process is useful for the encapsulation of the hydrophobic and hydrophilic components and produces high-quality powders (Calvo et al., 2010), which are mostly spherical and homogeneous (Rodriguez et al., 2016; Chen, Zhong, Wen, McGillivray, & Quek, 2013). The less moisture extent of micropowders causes their resistance to oxidative and microbiological degradation (Shishir, Taip, Aziz, Talib, & Sarker, 2016).

This method of encapsulation can preserve diverse vegetable and animal oils against oxidation (Kolanowski, Ziolkowski, Weißbrodt, Kunz, & Laufenberg, 2005) and also external active factors such as humidity, light, and temperature (Jafari, 2017). In terms of the processing time, it takes a few seconds; this is appropriate for protecting ingredients like EFAs, which are sensitive to heat

(Kaushik, Dowling, Barrow, & Adhikari, 2015). In addition, this rapid drying process leads to the maintenance of the enclosed core temperature at a far below temperature, in comparison with the high inlet temperature of the spray drying process (typically 150°C to 220°C for aqueous solutions, emulsions, and suspensions), which, in turn, perform a protective effect against lipid degradation (Islam et al., 2018; Barrow, Wang, Adhikari, et al., 2013). Moreover, the capability of reducing the oil at the particle's surface and subsequently increasing the MEE is another characteristic of encapsulation by spray drying (Jafari, Assadpoor, He, & Bhandari, 2008; Frascareli, Silva, Tonon, & Hubinger, 2012). Thus, this technology can retard the oxidation of FAs, enhance their stability and shelf-life, restrict the creation of off-flavors, and manage their release in a food matrix (Encina, Vergara, Giménez, Oyarzún-Ampuero, & Robert, 2016).

A spray drying process can be fully controlled (automatically) and operated by recording and monitoring the variables of the procedure at the same time. It can also be optimized by applying different designs of the experiment (DOE). Moreover, the design and stages of the process can be modified to improve the quality of the final product(s). The layout of the spray dryer can be an open cycle, a closed cycle, or a semi-closed cycle, and the drying can be operated in either a single-stage or a two-stage process, which has a complementary drying stage in a fluidized bed dryer (Anandharamakrishnan & Padma Ishwarya, 2015).

Another superiority of this technology is the flexibility of MEE (Shamaei et al., 2017), which can be modified by changing and adjusting the effective factors, such as wall material, emulsification method, parameters of initial emulsion, and flexible parameters of spray drying. The ability to use a wide range of materials (e.g., from heat-sensitive proteins to stabilizing carbohydrates) as an encapsulant or wall material is an important advantage of this method (Feizollahi et al., 2018). Besides, the initial entering liquid feed may be highly viscous or sticky, which can be dried successfully by taking some measures such as preheating or using surfactants and carriers, respectively (Assadpour & Jafari, 2019).

9.3.2 Challenges for the Spray Drying Encapsulation of EFAs and Functional Oils

Although microencapsulating of EFAs by the spray drying technique brings several advantages, there are some challenges that this method of drying may cause while producing oil containing powders. The spray drying of PUFAs may increase the oxidation due to high temperature and air contact of oil within the drying chamber (Beindorff & Zuidam, 2010; Kolanowski, Jaworska, Weissbrodt, & Kunz, 2007; Kaushik et al., 2015). A high temperature resulted in porous particles in some studies, which, in turn, facilitate the oxidation and decreased the shelf-life of the final product (Rodriguez et al., 2016). Similarly, Martínez et al. (2015) reported a higher oxidation rate and a diminished shelf-life of powders after the spray drying encapsulation of chia oil.

Moreover, the high moisture content of the spray dried microparticles may affect their shelf-life (Baik et al., 2004; Drusch, Serfert, & Schwarz, 2006). The higher the moisture content in the drying chamber, the lower the shelf-life of the powders, because of the lower drying efficiency that retards the crust shaping surrounding the particles. In terms of the effect of particle size on the shelf-life, controversial influences are reported. The delay in crust formation may happen in larger particles in comparison with smaller ones, which leads to more surface oil and consequently a shorter shelf-life (Aghbashlo, Mobli, Madadlou, & Rafiee, 2013b). On the other hand, according to Fang, Shima, and Adachi (2005), larger microcapsules are more resistant to diffusion of oxygen and more stable than smaller ones.

This method of encapsulating tends to vary from process to process and it can be difficult to control the particle size (Islam et al., 2018). The equipment of the process is rather complex, and the process conditions are not uniform (Feizollahi et al., 2018). The wall materials, which are being used for the encapsulation by spray drying, should have acceptable water solubility and suitable viscosity (Bakry et al., 2015; Gonçalves, Estevinho, & Rocha, 2016; Corrêa-Filho et al., 2019).

Moreover, the amount of oil that is usually loaded in microcapsules is limited, varying typically between 20% and 30% and may reach a maximum of about 50% (Barrow et al., 2013).

Despite the mentioned challenges towards the spray drying encapsulation of functional oils, there are some viable strategies for extending the shelf-life and storage stability of the microencapsulated sensitive EFAs. These include:

- Choosing an alternative neutral drying gas like nitrogen;
- Adding natural antioxidants into the primary emulsion;
- Application of dehumidified air as the drying gas;
- Spray drying at lower temperatures;
- Combination of different wall materials to improve their efficiency;
- And, more importantly, optimizing the process parameters to achieve the maximum MEE (Assadpour & Jafari, 2019; Arpagaus, Collenberg, Rütti, Assadpour, & Jafari, 2018).

9.3.3 SPRAY DRYING ENCAPSULATION OF MARINE-BASED OILS

Since marine-based oils are highly susceptible to oxidation, spray drying encapsulation is used as a preservative method to protect these sensitive oils in most of the studies and industrial applications. In recent years, several studies investigated the influences of this method on the stability and shelf-life of marine-based oils Table 9.1. Takeungwongtrakul and Benjakul (2016) reported the ability of the spray drying microencapsulation in producing the shrimp oil powder with good oxidative stability and high MEE. In this study, the potential of combining three wall materials of gelatin, sodium caseinate, and glucose syrup with a ratio of 4: 1: 1 was demonstrated to maintain the stability of PUFAs. Similarly, Nawas et al. (2019) observed the increased oxidative stability of silver carp oil after spray drying microencapsulation. The stability of the final product was measured after two months of storage at room temperature. The combination of all different wall materials (i.e., gum Arabic, maltodextrin, modified starch Hi-cap, and inulin) increased the stability of this marine-based oil. However, it was found that modified starch was more effective than gum Arabic. Likewise, both Botrel et al. (2017) and Vishnu et al. (2017) studied the suitability of new wall materials of cashew gum and vanillic acid-grafted chitosan on the microencapsulation of fish oil and sardine oil, respectively. Both studies reported high oxidative stability of marine-based oils during storage. In fact, the investigations were successful to suggest new potential wall materials for spry drying encapsulation of marine-based oils. These results show that choosing the right components as wall materials add to the efficiency of the process and increase the oxidative stability of the PUFAs. These findings agree with those reported by Calvo, Castaño, Lozano, and González-Gómez (2012). Accordingly, by choosing an appropriate wall material, the capability to protect and deliver encapsulated components will be predetermined. Moreover, wall materials affect the stability and flowability of the initial emulsion and the shelf-life of the final microcapsules (Chang & Nickerson, 2018). On the other hand, another research study, which evaluated the effects of outlet temperature on the oxidative stability of microencapsulated combined fish oil and chia oil with whey protein, found that an outlet temperature of 55 °C showed better powder characteristics (Lavanya et al., 2019). In addition, Taktak et al. (2019) microencapsulated the European eel oil by spray drying and used European eel protein isolate as wall material. The study investigated the influences of the emulsion preparation method (ultrasonication or/and homogenization) in addition to the effects of the core/wall ratio on oxidative stability and physicochemical properties of both emulsions and powders. It was found that oxidative stability after three hours of incubation at 80°C was higher for the microparticles prepared by the combined ultrasonication and homogenization processes. In addition, the European eel protein isolate proved to be an efficient wall material for the encapsulation of European eel oil as it was able to increase its antioxidant activity, oxidative stability, and thermo-stability compared to the bulk oil. Hence, it can be concluded not only that the spray drying encapsulation is highly suitable for increasing the shelf-life of the marine-based oils, but also that

TABLE 9.1

A summary of recent studies on spray drying microencapsulation of functional oils in terms of oxidative stability and encapsulation efficiency

Type of EFA / Functional Oil	Wall Materials	Special Remarks	Reference
Marine-based oils			
Shrimp oil	Whey protein concentrate, sodium caseinate, glucose syrup, and gelatin	- The capability of spray drying encapsulation to produce a final powder with good oxidative stability and high MEE. - The used wall materials had the potential to maintain the stability of shrimp oil PUFAs.	Takeungwongtrakul and Benjakul (2016)
Silver carp oil	Gum Arabic, maltodextrin, inulin, and modified starch Hi-Cap	- The oxidative stability of Silver carp oil micropowders increased with a combination of all these four wall materials. But, modified starch Hi-Cap was found to be more effective than gum Arabic.	Nawas, Azam, Ramadhan, Xu, and Xia (2019)
Fish oil	Gum Arabic, cashew gum, and modified starch	- Cashew gum was a potential wall material for the encapsulation of fish oil. - Cashew gum increased oxidative stability and EE of micropowders compared to gum Arabic.	Botrel et al. (2017)
Sardine oil	Vanillic acid-grafted chitosan (Va-g-Ch)	- The Va-g-Ch improved the EE and oxidative stability of sardine oil micropowders during the storage.	Vishnu et al. (2017)
European eel oil	European eel protein isolate	- European eel protein isolate increased the antioxidant activity, oxidative stability, and thermo-stability of the European eel oil microcapsules compared to its bulk oil. - The microparticles prepared with combined ultrasonication and homogenization processes (as emulsion preparation method) showed higher oxidative stability after three hours of incubation in 80°C, when compared to each method of emulsion preparation.	Taktak et al. (2019)
Plant-based oil			
Sunflower oil	Sucrose (99%), D-lactose monohydrate, sodium caseinate, maltodextrin: DE10, and gelatine (from bovine milk)	- In comparison between the spray drying and freeze drying as methods of encapsulation, the sunflower microparticles prepared by the spray drying method showed higher oxidative stability. - The wall material matrix comprised of D-lactose monohydrate and sodium caseinate was stable and caused small and homogeneous droplets in both methods of drying. - The spray dried powder with maltodextrin, sucrose, and gelatin had low stability.	Holgado, Márquez-Ruiz, Ruiz-Méndez, and Velasco (2019)
Pumpkin seed oil	Gum Arabic, maltodextrin: DE 6, and whey protein concentrate: 80% protein	- Spray drying microencapsulation of pumpkin seed oil was effective to protect EFAs. - The total unsaturated fatty acids in encapsulated pumpkin seed oil did not reduce significantly after the process compared to the bulk oil. - The inlet air temperature of 141.51°C, aspirator rate of 75%, and feed pump rate of 15% reported being optimal conditions for encapsulation of pumpkin seed EFAs by spray drying.	Geranpour, Emam-Djomeh, and Asadi (2019a)

(Continued)

TABLE 9.1
(Continued)

Type of EFA / Functional Oil	Wall Materials	Special Remarks	Reference
Walnut oil	Skim milk powder, maltodextrin, and Tween 80	- The highest microencapsulation efficiency (91.01%) was related to the micropowders produced with skim milk powder + Tween 80 as wall material at 180°C drying air temperature and 3 bar atomization pressure. - Adding Tween 80 to the emulsion formulation increased microencapsulation efficiency. - The micropowders with the wall materials of skim milk powder + Tween 80 and skim milk powder + maltodextrin had a higher bulk density than those with skim milk powder.	Shamaei et al. (2017)
Combined pomegranate seed oil and soybean oil	Gum Arabic and maltodextrin: DE 5	- The high-quality powders with 95% encapsulation efficiency were obtained at a drying temperature of 130°C and an airflow rate of 25 kg/h.	Silva-James et al. (2018)
Refined kenaf seed oil	β-cyclodextrin, gum Arabic, and sodium caseinate	- The encapsulation efficiency of micropowders reported being more than 90% for all three-model combination of wall materials (i.e., β-cyclodextrin & sodium caseinate, β-cyclodextrin & gum Arabic, β-cyclodextrin & sodium caseinate & gum Arabic) - All produced micropowders were spheres without cracks.	Chew et al. (2018)
Mixed marine-based and plant-based oil			
Combined fish oil and chia oil	Whey protein concentrate 80%	- The outlet temperature of 55°C showed better powder characteristics in terms of oxidative stability.	Lavanya, Kathiravan, Moses, and Anandharamakrishnan (2019)

it can improve various process-related parameters such as process conditions, emulsion properties, and emulsification characteristics. These parameters highly affect the stability and efficiency of the produced encapsulated EFAs.

9.3.4 THE SPRAY DRYING ENCAPSULATION OF PLANT-BASED OILS

Spray drying encapsulation provides positive effects on the conservation and the stability of vegetable oils. For instance, Holgado et al. (2019) investigated the effects of encapsulation by spray drying and freeze drying on the oxidative stability of sunflower oil microcapsules and reported that the spray drying method caused the encapsulated and free oil fractions to be more stable against oxidation in comparison with the freeze drying method. D-lactose monohydrate, sodium caseinate, maltodextrin, and gelatin (from bovine milk) were used as wall materials and it was found that the matrix made with D-lactose monohydrate and sodium caseinate and lactose was stable, and caused droplets to be homogeneous and small in both methods of drying. In another study, Geranpour et al. (2019a) revealed the ability of spray drying encapsulation to protect EFAs, as the number of total unsaturated fatty acids in encapsulated pumpkin seed oil did not reduce significantly after the process in comparison with the bulk oil. On the other hand, it is reported that the spray drying technique encapsulation is a promising process that can produce high-quality powders (Calvo et al., 2010) with more than 90% EE (Chew et al., 2019). In Shamaei et al., 2017, Shamaei et al. studied the microencapsulation of walnut oil by spray drying. In this study, different process parameters and wall materials were investigated, and 91.09% EE for microencapsulation of walnut oil by spray drying using skim milk powder and Tween 80 as wall materials was obtained at 180°C drying air temperature and 3 bar atomization pressure. Similarly, Silva-James et al. (2018) investigated the influences of process conditions and wall material composition on physicochemical characteristics of pomegranate seed oil and soybean oil (combined) micropowders and attained an EE of 95% at a drying temperature of 130°C and an airflow rate of 25 kg/h. Chew et al. (2018) also achieved more than 90% EE by spray drying encapsulation of refined kenaf seed oil with all three model combinations of studied wall materials (β-cyclodextrin & sodium caseinate, β-cyclodextrin + gum Arabic, β-cyclodextrin + sodium caseinate, and gum Arabic). These achievements from spray drying microencapsulation of different plant-based oils in recent years demonstrate the remarkable capability of the spray drying method for the manufacture of the micropowders with more than 90% EE. Table 9.1 indicates a summary of conducted research with an investigation of the encapsulation efficiency and oxidative stability of microencapsulated functional oils in recent years. Furthermore, the examples of scanning electron microscopy (SEM) images indicating micropowder structure are shown in Figure 9.3.

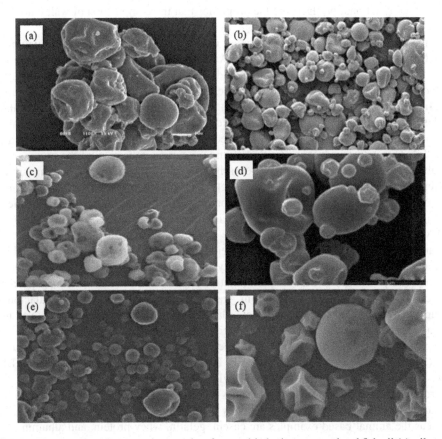

FIGURE 9.3   The scanning electron micrographs of spray dried microencapsulated fish oil (a), silver carp oil (b), sardine oil (c), grapeseed oil (d), Ganoderma Lucidum spores oil (e), and chili seed oil (f), at magnifications × 1100, 1200, 1500, 5000, 1000, and 8000, respectively. Reproduced with permission from Voucher et al. (2019), Nawas et al. (2019), Vishnu et al. (2017), Böger, Georgetti, and Kurozawa (2018), Zhou, Zhou, Ma, and Ge (2019), and Wang et al. (2017).

9.4 WALL MATERIALS FOR THE SPRAY DRYING ENCAPSULATION OF EFAS AND FUNCTIONAL OILS

As is known, the wall material or the capsule is the layer that embraces the core material (bioactive and valuable compounds) with the aim of its protection (Mahdavi, Jafari, Ghorbani, & Assadpoor, 2014; Rajabi, Ghorbani, Jafari, Sadeghi Mahoonak, & Rajabzadeh, 2015). Wall materials are also counted as an emulsifying agent that is essential to disperse core material (oil phase) in water (Islam et al., 2018). The wall material has a direct influence on the stability, protection efficiency, encapsulation/entrapment efficiency of the core material, and, ultimately, the MEE of the process, so it is imperative to have an ideal and compatible choice of this material (Chang & Nickerson, 2018). The chosen wall components should be food-grade, biodegradable, inexpensive, tasteless, soluble in the typical solvents, and nonreactive with the loaded core material. It should also have low viscosity at high content quantity, low hygroscopicity, good emulsifying features, and favorable covering properties such as strength, stability, flexibility, and impermeability are of importance for choosing the ideal wall materials (Desai & Park, 2005; Nedović, Kalušević, Manojlović, Petrović, & Bugarski, 2013; Jafari et al., 2008; Özbek & Ergönül, 2017; Pereira et al., 2018; Veiga, Aparecida Da Silva-Buzanello, Corso, & Canan, 2019). Generally, the ability to encompass a large amount of EFAs, protect against the chemical degradation (oxidation or hydrolysis) during storage, release at a specific site, release at a controlled rate, compatibility with the targeted food matrix without adverse changes of final functional products are the properties that the composition of wall material should own (Chang & Nickerson, 2018; McClements, Decker, & Weiss, 2007).

Three main groups of components are being used as wall material for encapsulation, including carbohydrates, proteins, and lipids/waxes. Among these groups, proteins and carbohydrates are being used in spray drying encapsulation and lipids/waxes are being utilized for other methods of encapsulation like emulsion and liposomes (Ton, Tran, & Le, 2016; Desai & Park, 2005; Anandharamakrishnan & Padma Ishwarya, 2015). Choosing the appropriate wall material depends on the desired properties of the final micropowders and the core material (Bakry et al., 2015). The carbohydrates and proteins are usually mixed to eliminate the shortcomings of each (if used individually). This action improves the emulsifying properties, efficiency of the process, and filmmaking abilities throughout spray drying (Pourashouri et al., 2014; Augustin, Sanguansri, & Bode, 2006; Mendanha et al., 2009). Murali, Kar, Patel, et al. (2017) investigated the encapsulation of rice bran oil with jackfruit seed starch and whey protein isolate as the wall material complex. Using the combined protein and starch resulted in efficient microencapsulation with high MEE and low peroxide value.

The carbohydrate group is proved to have diverse functionalities, excellent water solubility, low viscosity at high concentrations, low cost, diversity, high availability, bland in flavor, and poor emulsifying properties. This group derives from plants, marines, microbes, and animals. Examples of each are cellulose, maltodextrin, starch, gum Arabic, mesquite gum, guar gum, galactomannans, corn syrup, cyclodextrin, and pectin, which are plant-based carbohydrates, while the wall materials such as carrageenan and alginate are marine-based carbohydrates. Other products such as xanthan, gellan, dextran, lactose, and chitosan fall into the category of microbial and animal-based carbohydrates (Sagiri, Anis, & Pal, 2015; Veiga et al., 2019; Anandharamakrishnan & Padma Ishwarya, 2015; Nesterenko, Alric, Silvestre, & Durrieu, 2013; Alvarenga Botrel et al., 2012).

The protein group derives from two subgroups of plant-based and animal-based proteins. Plant-based examples include examples such as barley protein, soy protein, zein, pea protein, and gluten, whereas the animal-based examples are casein, gelatin, whey protein, and albumin. The most important features that lead to the selection of this group as an appropriate wall material are their biocompatibility, availability, biodegradability, functional properties, very good emulsion-forming ability, film-forming properties, molecule chain flexibility, and probable non-enzymatic browning reaction. The plant-based proteins are less allergenic, cheaper, and more available (as compared to animal-based). However, the animal-based materials are restricted by diet limitations and religious reasons (Ton et al., 2016; Li, Zhu, Zhou, & Peng, 2012; Nesterenko et al., 2013; Jafari et al., 2008; Jenkins, Breiteneder, & Mills, 2007).

Some wall materials are very common for the encapsulation of functional oils; starch, malto-dextrin, sugars, whey proteins, gelatin, plant gums, sodium caseinate, and chitosan (Pourashouri et al., 2014; Karthik & Anandharamakrishnan, 2013). According to a research study conducted by Geranpour, Assadpour, and Jafari (2020), gum Arabic, maltodextrin, sodium caseinate, and whey protein isolate were the most used wall materials for the microencapsulation of functional oils between the years 2016 and 2019. A further selection criterion in the choice of wall materials is their suitability for spray drying concerning temperature properties. The glass transition temperature should preferably be higher than the outlet temperature of the drying gas; otherwise, the particles can soften and stick to the wall of the spray drying chamber, which reduces the overall yield.

9.5 EFFECTIVE FACTORS ON THE MICROENCAPSULATION EFFICIENCY OF EFAS AND FUNCTIONAL OILS

The MEE of functional oils is the ratio of the trapped oil to the total amount of oil in microcapsules (after spray drying process), that the total amount of oil in microcapsules comprises the encapsulated oil and the surface oil which is known as non-encapsulated oil (Anandharamakrishnan & Padma Ishwarya, 2015; Gallardo, Guida, Martinez, et al., 2013). The MEE reflects the oil amount trapped in the microcapsules and the grade to which the complex can inhibit oil diffusing throughout the wall (Akhavan Mahdavi, Jafari, Assadpour, & Ghorbani, 2016; Sarabandi, Gharehbeglou, & Jafari, 2020). Thus, to reach the highest MEE, the surface oil should be minimum, and the trapped oil should be maximum. The percentages of the MEE and the oil retention are two important parameters used to assess the encapsulation process, both determining the oil content in powders. However, the oil retention represents the ratio between the total amount of oil in the microcapsules (after the spray drying process) and the total amount of oil in the initial feed emulsion (Encina et al., 2016).

The influential factors on MEE will be categorized into two broad groups of emulsification characteristics and spray drying conditions that will be described in the following. Figure 9.4 depicts the shape of a common laboratory-scale spray dryer and lists the effective factors on the microencapsulation efficiency of EFAs and functional oils.

Emulsification characteristics		Spray drying conditions
- Type of wall materials - The composition of wall materials - Emulsification method - Droplet size - Total solid concentration - Viscosity - Stability		- Inlet drying air temperature - The type of the atomizer - The conditions of the atomizer - Feed flow rate - Drying air flow rate

FIGURE 9.4 A laboratory spray dryer (Mini Spray Dryer B-290 from Büchi Labortechnik AG), alongside a list of effective factors on microencapsulation efficiency of essential fatty acids and functional oils. The spray dryer works with a two-fluid or ultrasonic nozzle and has a water evaporation capacity of around 1 L/h for feasibility studies.

9.5.1 Emulsification Conditions

Generally, an aqueous colloidal system made by the dispersion of one liquid (core material) into another (wall material), which is immiscible, is defined as "emulsion". This composition may also contain emulsifiers to avoid coalescence (McClements et al., 2007; Jacobsen, 2016). Various factors are involved in making the initial emulsion and can affect the MEE (Figure 9.4). This includes the type of wall materials, the factors involved in the emulsification of initial feed such as the emulsification method, and the features of the prepared emulsion such as viscosity, stability, droplet size, and total solid concentration (Jafari et al., 2008; Karthik & Anandharamakrishnan, 2013; Hosseini et al., 2019). These factors will be discussed in the following sections. Figure 9.2(a) shows a simplified schematic of an emulsification step, a list of functional oils, and typical wall materials that are used for spray drying encapsulation. Several researchers have conducted studies recently concerning the emulsification step of microencapsulation spray drying of different-based EFAs. For instance, Albert, Vatai, and Koris (2017), Di and Mauri (2018), Taktak et al. (2019), Shao, Pan, Liu, Teng, and Yua (2018), Takashige et al. (2018), Le et al. (2019), Wang et al. (2017), Fuentes-Ortega et al. (2017), Sanchez-Reinoso and Gutiérrez (2017), microencapsulated various oils extracted from sunflower seed, fish, European eel, algal, krill, sunflower, chili seed, sesame, and *sacha inchi*, respectively, by spray drying, in order to investigate the emulsification-related parameters on the properties of the final micropowders.

9.5.1.1 The Type and Composition of Wall Materials

As previously described, selecting the suitable wall material has the most important and direct influence on the MEE of functional oils. By choosing an appropriate wall material, the capability to protect and deliver the encapsulated components will be predetermined (Calvo et al., 2012). Moreover, it affects the stability and flowability of the initial emulsion, and the shelf-life of the final microcapsules (Chang & Nickerson, 2018). There are two methodologies to choose the suitable wall materials for encapsulating EFAs, which are named the 'reactive method' and 'proactive approach'. The *reactive method* depends on doing numerous experiments and errors, and the suitable wall material is recognized based on the properties of final microcapsules in terms of MEE, stability during storage, yield of process, and structure. Indeed, if these properties of the produced microcapsules with a chosen wall material are satisfactory, that choice could be appropriate for being an encapsulation agent (Anandharamakrishnan & Padma Ishwarya, 2015; Prata, Garcia, Tonon, & Hubinger, 2013). This method demands a high amount of time, energy, material, and effort. However, as it is proven to be an effective approach, many researchers are identifying and evaluating the effects of newly-suggested compounds or various combinations of different components as wall materials on the properties of the final micropowders in the spray drying encapsulation of different-based EFAs, to offer potential wall material composition. Here are some examples of the related research, which have been carried out in recent years: Botrel et al. (2017), Vishnu et al. (2017), and Hoyos-Leyva, Bello-Perez, Agama-Acevedo, Alvarez Ramirez, and Jaramillo-Echeverry (2019) studied the suitability of new wall materials of cashew gum, vanillic acid-grafted chitosan, and taro starch on the microencapsulation of fish oil, sardine oil, and almond oil, respectively; and Nawas et al. (2019) and Chang et al. (2018) investigated the influence of combined wall materials on the physicochemical properties of fish oils. The former used gum Arabic, maltodextrin, modified starch Hi-cap, and inulin, while the latter utilized a mixture of maltodextrin, chitosan, and thiol-modified β-lactoglobulin. Additionally, Böger et al. (2018) and Us-Medina, Julio, Segura-Campos, Ixtaina, and Tomás (2018) evaluated the influence of combined wall materials on the physicochemical properties of seed oils. In this case, Böger et al. (2018) microencapsulated a grape seed oil by the mixture of maltodextrin DE = 10 and gum Arabic, whereas Us-Medina et al. (2018) microencapsulated a chia seed oil through the mixture of maltodextrin with DE of 13%–17%, lactose, sodium caseinate, chia mucilage, and the protein-rich fraction of chia flour. The chia mucilage was reported to be an efficient wall material that could delay the oxidation of chia seed oil and increase MEE. On the other hand, the *proactive approach* acts directly based on the wall material efficiency, which can be estimated by their drying characteristics, drying kinetics, and determining their other characteristics such as solubility, emulsification property, film-forming ability, viscosity, glass transition, and degree of crystallinity before the selection for spray drying encapsulation (Anandharamakrishnan & Padma Ishwarya, 2015; Prata et al., 2013).

9.5.1.2   Emulsification Method

The emulsification technique is one of the effective factors on emulsion stability and accordingly the microencapsulation stability and MEE of EFAs (Serdaroglu, Kerimoğlu, & Kara, 2015; Danviriyakul, McClements, Decker, Nawar, & Chinachoti, 2002). This technique can substantially impact the droplet size distribution of the initial emulsion, and, accordingly, it plays an important role in MEE (Ramakrishnan & Ferrando, 2014).

Firstly, the chosen wall material should be dissolved in distilled water (usually with a stirrer) and kept one night at ambient temperature or under refrigeration for rehydration. This rehydration is useful for the full saturation of the wall polymers and to render their emulsifying features (Bakry et al., 2015). As discussed in Section 9.4, carbohydrates have poor emulsifying properties, while most of the proteins are good emulsifiers. Secondly, the core material (i.e., EFA) is added and dispersed in the mixture of distilled water and wall material. As this system is mainly aqueous-based, it forms an oil-in-water emulsion (McClements, 2005). According to some destabilization mechanisms like coalescence, gravitational separation, and flocculation, this emulsion tends to break down, especially by passing the time (Komaiko & McClements, 2016). In the following, by choosing an appropriate emulsification method, a stable emulsion is being prepared. The emulsification methods can be low speed/energy or high speed/energy, which the latter is more recommended and usually applied before spray drying encapsulation. Isothermal methods including emulsion phase inversion and spontaneous emulsification and thermal method of phase inversion temperature are examples of low speed/energy emulsification, while high-pressure homogenization, ultrasound, and high-speed homogenization are examples of high speed/energy methods of emulsification (Komaiko & McClements, 2016; Anandharamakrishnan & Padma Ishwarya, 2015). Recently, Di and Mauri (2018) and Taktak et al. (2019) investigated the influences of the emulsion preparation method on the physicochemical properties of the microencapsulated marine powders. Di and Mauri (2018) implemented the ultra-turrax and ultrasound for the emulsification of fish oil in soybean protein, while Taktak et al. (2019) utilized the ultrasonication and homogenization for the emulsification of European eel oil in European eel protein isolate.

9.5.1.3   Droplet Size

The droplet size of the prepared emulsion is one of the factors that affects MEE (Hosseini et al., 2019). This parameter depends strongly on the emulsification method (Ramakrishnan & Ferrando, 2014). Moreover, the size and size distribution of the final micropowders depends on the droplet size of the initial emulsion (Freitas, Merkle, & Gander, 2005). The "nano-" prefix is used for the droplets with a size ranging from 0 to 1000 nm, but if the droplet size of an emulsion ranges between 1 and 100 nm, that emulsion would be a real nanoemulsion with superior stability before atomizing into the spray dryer (Anandharamakrishnan & Padma Ishwarya, 2015). On the other hand, the sprayed droplet size using conventional atomizers (Section 9.5.2.2.) usually ranging between 10 and 100 μm in a lab-scale spray dryer. Thus, if 100 nm is considered as an ideal emulsion droplet size and 100-micron accounts for the ideal spray droplet size, the optimal ratio of emulsion droplet size/ sprayed droplet size would be 1: 1000.

It is reported that the smaller droplet size distribution in the emulsion leads to a higher MEE in the final product (Soottitantawat, Yoshii, Furuta, Ohkawara, & Linko, 2003; Ramakrishnan et al., 2012; Ramakrishnan & Ferrando, 2014). A larger emulsion droplet size leads to core loss during atomizing the initial feed into the spray drying chamber. Indeed, large emulsion droplets experience shear and breakdown during atomization, which enhances the oil at the surface of droplets. This phenomenon results in less core oil retention and accordingly less MEE (Soottitantawat et al., 2003; Danviriyakul et al., 2002). Therefore, smaller droplets are less likely to lose their core material. Furthermore, the drying speed and the formation of a semi-permeable coating is higher in smaller atomized droplets, which, in turn, increases the MEE due to retarding the oil penetration on the surface (Jafari, He, & Bhandari, 2007).

9.5.1.4 Total Solids Concentration

Increasing the wall material or in other words, increasing the total solid concentration in the initial emulsion leads to enhanced MEE (Soottitantawat et al., 2005; Bakry et al., 2015). Rubilar et al. (2012) reported about a 36% increase in MEE of flaxseed oil by a 5% increase in the total wall material from 25% to 30% (wall material: maltodextrin and gum Arabic). By increasing the total solid concentration in an emulsion system, the initial semi-permeable crust forms more quickly and inhibits the oil leaching to the coated surface, which means better protection, more oil retention, and ultimately more MEE (Jafari et al., 2008; Anandharamakrishnan & Padma Ishwarya, 2015). According to Reineccius (1988), the ratio of 4 walls to 1 core (4: 1 w/w) provides suitable protection of the bioactive core material. Similarly, Toure, Xiaoming, and Zhijian (2007), Soottitantawat et al. (2005), and Madene, Jacquot, Scher, and Desobry (2006) reported this ratio as an optimal ratio for the microencapsulation of essential oils and flavors, especially with carbohydrate-based wall materials like maltodextrin and gum Arabic. For food applications, in general, the rate of the wall to core material is accepted to be in the range of 1: 1 to 4: 1 (w/w) (Roccia, Mart'ınez, Llabot, & Ribotta, 2014). Di and Mauri (2018) also used a ratio of 4: 1 (wall: core) for the spray drying microencapsulation of fish oil using soybean protein isolate and reported efficient protection of fish oil via this formulation and high MEE (88%) for the final micropowders. Moreover, Sanchez-Reinoso and Gutiérrez (2017) reported that changes in the ratio of wall to core material could influence the physicochemical properties and peroxide value of microencapsulated *sacha inchi* oil (a kind of vegetable oil), in addition to MEE. A lower MEE was found in the case of the powders with high core loading, where increasing the loaded oil caused lower flowability of the powders and a more regular shape. In spray drying, a higher solid concentration leads to larger dried particles, simply due to the mass balance.

9.5.1.5 Viscosity

Emulsion viscosity is affected by several factors such as the type of wall material, droplet size, total solid content, and even the type of used emulsifier. Emulsion viscosity is directly related to total solid concentration and droplet size of emulsion and increases with the enhancement of these parameters. Increasing the viscosity of the initial emulsion has positive effects on the MEE. This increment should be such to reduce the flow movement inside the emulsion droplets (Drusch, 2007; Bakry et al., 2015), which leads to the rapid formation of the semi-permeable coating on the surface of the droplets, better core protection, and, accordingly, a higher MEE. With a very large increase in emulsion viscosity, the spray drying process would not have an efficient drying rate because of the larger particles. Although the atomization process supplies energy for the breakdown of the droplets into smaller ones, viscous emulsions usually confront this energy and prevent the formation of small particles (Anandharamakrishnan & Padma Ishwarya, 2015). As explained in Section 9.5.1.3, large emulsion droplets result in micropowders with low MEE. Therefore, it is indispensable to ensure having a stable emulsion with an appropriate and ideal droplet size (Section 9.5.1.3), total solid concentration, and sufficient viscosity before its atomization into the drying chamber.

9.5.1.6 Stability

The more the initial emulsion is stable, the greater the obtained MEE will be for microencapsulated oils (Danviriyakul et al., 2002). The emulsion stability depends on viscosity, particle size, and total solid concentration of the emulsion, which not only affect the emulsion stability and MEE but also influence each other. According to Faldt and Bergenstahl (1995), the stability of the emulsion before spraying in the drying chamber is critical. The emulsion droplet size and emulsion viscosity are effective in inhibiting the occurring sedimentation in the initial emulsion. A viscous emulsion with small droplets accounts for a stable emulsion, in which the sedimentation of the particles and coalescence of droplets do not happen (Williams & Phillips, 2003). The explaining mechanisms stating how the mentioned factors (e.g., particle size, total solid concentration, and viscosity) influence on MEE are elucidated in previous Sections 9.5.1.3, 9.5.1.4, and 9.5.1.5, respectively.

9.5.2 SPRAY DRYING CONDITIONS

One of the favorable aspects of the spray drying conditions is the fact that various parameters can be adjusted in the process in order to maximize the MEE and subsequently increase the stability and shelf-life of EFAs. The most important spray drying parameters include:

- Inlet drying air temperature;
- Atomizer type (i.e., one-/two-fluid nozzle, centrifugal, or ultrasonic);
- Atomizer conditions based on its type (e.g., orifice diameter and pressure for pressure/nozzle atomizers, rotational speed for centrifugal atomizers, frequency of irradiation for ultrasonic atomizers);
- Feed flow rate;
- Drying air flow rate or gas flow rate.

These parameters can directly affect the physicochemical properties of the dried micropowders such as MEE (Figure 9.4). Accordingly, several studies have been conducted to specify the optimal point for spray drying conditions, in order to maximize the MEE in the spray drying of EFAs. For instance, optimizing the process conditions in the microencapsulating of pumpkin seed oil (Geranpour et al., 2019a), soybean oil (Bai et al., 2019), sesame oil (Fuentes-Ortega et al., 2017), rice bran oil (Murali et al., 2017), flaxseed oil (Bhushan, Mani, Kar, & Datta, 2017), *Brucea javanica* oil (Hu et al., 2016), fish oil (Encina et al., 2018), and algal oil (Shao et al., 2018) are among the case studies about the microencapsulation of EFAs in recent years.

9.5.2.1 Inlet Drying Air Temperature

A high adequate inlet air temperature in a spray dryer leads to a swift development of semi-permeable crust on the external layer of the droplet, inhibiting further leaching of oil to the surface of microcapsules, and, subsequently, better retention of oil and higher MEE (Aghbashlo, Mobli, Madadlou, & Rafiee, 2013a; Rezvankhah, Emam-Djomeh, & Askari, 2019). Murali et al. (2016), Shamaei et al. (2017), and Bhushan et al. (2017) observed an enhanced MEE by rising the inlet air temperature, averagely from 140 to 180°C in their studies on the microencapsulation of various functional oils (rice bran oil, walnut oil, and flaxseed oil, respectively). Kalkan, Vanga, Murugesan, Orsat, and Raghavan (2017) increased the above-mentioned inlet air temperature to 220°C and observed the same results for MEE in the spray drying microencapsulation of hazelnut oil. Moreover, Huang et al. (2014) reported the positive influence of increasing the inlet temperature of the spray dryer in lower ranges (from 110°C to 120°C) on enhancing the MEE of tilapia oil encapsulation by the spray dryer. On the contrary, Frascareli et al. (2012) observed the opposite effect of enhancing inlet air temperature on MEE of coffee oil micropowders between the range of 150°C to 190°C. Gharsallaoui et al. (2007) stated that the formation of probable cracks on the surface of the microcapsules (manufactured using spray drying) is the result of excessive evaporation in the drying chamber. This, in turn, leads to greater degradation and further release of bioactive compounds, and therefore, the reduction in MEE. Furthermore, a too high inlet air temperature may decrease MEE because of the possible surface defects and additional bubble extension, a phenomenon known as "ballooning" (Goula & Adamopoulos, 2012).

Therefore, it can be concluded that although in most of the studies the positive effect of increasing inlet air temperature on MEE is reported, the adjustment of this process parameter is very important to obtain high-quality micropowders. In this case, using the results of similar previous research and implementing some pre-tests are considered as influential strategies.

9.5.2.2   The Type and Conditions of the Atomizer

The atomizer type and conditions have a direct influence on the droplet size of the microparticles at the primary stages of drying in the drying chamber. The pressure/nozzle and rotary/centrifugal atomizers are the most common atomizers used in the food industry for encapsulation purposes and are known as 'conventional atomizers' (Luz, Pires, & Serra, 2007). The two-fluid nozzle atomizer is able to produce droplets with a size ranging from 30 to 150 µm, while the latter can produce particles with a size between 30 and 120 µm (Anandharamakrishnan & Padma Ishwarya, 2015).

Increasing the atomizer speed and nozzle pressure decreases the size of the droplets entering the drying chamber and results in smaller particles (Huang & Mujumdar, 2008; Goula & Adamopoulos, 2005). Moreover, decreasing the diameter of the nozzle orifice, naturally, decreases the droplet size and, accordingly, particle size. With regard to the ultrasonic atomizer, which is a viable alternative for the mentioned conventional atomizers, it is demonstrated that it can produce more uniform and spherical nanodroplets with narrow droplet size distributions compared to them (Luz et al., 2007; Tatar Turan, Cengiz, Sandıkçı, Dervisoglu, & Kahyaoglu, 2016). The higher the frequency of irradiation in the ultrasonic atomizer, the smaller the sprayed droplet (Gogate, 2015). According to Jafari et al. (2007), the smaller the atomized droplet, the higher the drying speed and the smaller the drying time. Therefore, the amount of surface oil is minimizing due to the fast development of a semi-permeable crust, which means the subsequent enhancement of MEE. In addition, the surface changes, such as depression and shrinkage in smaller particles, are less common (compared to the large ones), because of the higher and faster mass and feed transfers inside the drying chamber, which, in turn, increases the MEE.

Tatar Turan et al. (2016) reported a higher range of obtained MEE with an ultrasonic atomizer (81%–98%), in comparison with the conventional nozzle atomizer (about 70%–91%), for blueberry micropowders. However, Legako and Dunford (2010) observed a reduced MEE of the encapsulated fish oil with ultrasonic atomizer compared to the pressure nozzle. This was attributed to the nozzle clogging during the process and some modifications in the design of the ultrasonic nozzle – the operating parameters are suggested to solve the clogging problem. Therefore, it can be concluded that high MEE can normally be achieved by decreasing the nozzle orifice diameter, increasing the atomization pressure, increasing the rotational speed (for centrifugal atomization), and increasing the frequency of irradiation (for ultrasonic atomization).

9.5.2.3   Feed Flow Rate

Regarding the effect of the feed flow rate on the MEE, various observations have been reported. Raising the feed flow rate forms larger droplets inside the drying chamber (Chegini & Ghobadian, 2005). For larger droplets, the crust layer formation retards, leaching more oil on the capsule's surface and decreasing the MEE. Similar results were reported by Aghbashlo et al. (2013b), who microencapsulated fish oil by spray drying. They investigated the influence of operational variables including inlet drying air temperature, drying air mass flow rates, liquid mass flow rates, and spraying air volume flow rates on microcapsule properties. The results showed that the encapsulation efficiency of fish oil microparticles decreased significantly ($P < 0.05$) by increasing the feed mass flow rate (peristaltic pump rate). They argued that the easier diffusion of oils toward the surface of droplets, due to retarded crust formation in large droplet size, was responsible for decreasing the MEE. On the other hand, some studies reported an increase of MEE by increasing the feed flow rate, which was attributed to the formation of small particles in the effect of particle collision during the process within the drying chamber (Seddighi Pashaki, Emam-Djomeh, & Askari, 2016; Geranpour, Emam-Djomeh, & Asadi, 2019b).

9.5.2.4 Drying Air Flow Rate

Different results have been reported on the influences of the drying air flow rate on MEE. Geranpour et al. (2019b) observed that a higher drying air flow rate, also known as the aspirator rate, caused an increment of MEE. It was speculated that the formation of crust surrounding the droplets was faster because of the high amount of energy entering the drying chamber at a high airflow rate. On the contrary, Huang et al. (2014) and Aghbashlo et al. (2013b) unexpectedly observed that the MEE decreased by increasing the drying air flow rate. Aghbashlo et al. (2013b) stated that the change in the enduring time of particles in the chamber of a spray dryer and the hard separation of the components within the atomized droplets were the probable reasons for decreasing MME. It can be concluded that the drying gas flow rate is an optimizing parameter to achieve the highest MEE.

9.6 THE OXIDATIVE STABILITY OF SPRAY DRIED ENCAPSULATED EFAS AND FUNCTIONAL OILS

It has been proven that spray drying microencapsulation of EFAs has a significant impact upon lengthening their oxidative stability, due to the creation of a physical barrier of wall material that protects oils against oxygen incursion (Islam et al., 2018). Nevertheless, some studies have revealed lower oxidative stability for the spray dried microencapsulated oils, which could be related to the high surface area of the microencapsulated oil compared to the bulk oil (Kolanowski et al., 2005; Encina et al., 2016). Therefore, oxidative stability is a challenge that should be considered as a primary goal of the microencapsulation of functional oils by spray drying. Generally, the oxidative stability of oils and fats is evaluated by measuring the formed hydroperoxides and their decomposition compounds as oxidation parameters by different analytical methods, including chemical methods, instrumental methods, and sensory analysis (Shahidi & Zhong, 2020). Peroxide value (PV), anisidine value (pAV), thiobarbituric acid test (TBA), and Kreis test are chemical methods, while spectrometric methods like UV absorption and infrared (IR) spectroscopy, and the chromatographic methods such as high-performance liquid chromatography (HPLC), gas chromatography–mass spectrometry (GC-Mas), and gas chromatography, alongside other oxygen measurements like Warburg's manometer and dissolved oxygen meter, are some examples of the instrumental methods (Shahidi & Zhong, 2020; Nanditha & Prabhasankar, 2009).

Many researchers have implemented different viable strategies such as those mentioned in Section 9.3.2 to prevent the oxidation of spray dried microencapsulated EFAs and increase their oxidative stability. For example, Voucher et al. (2019) investigated the spray drying microencapsulation of fish oil by the combination of different wall materials (gum Arabic, maltodextrin, and casein-pectin) and found that using casein-pectin combination was more effective than gum Arabic alone in lessening the oxidation of fish oil. These researchers observed that the highest rate of peroxide value (17.4 mmol/kg oil) was obtained after 28 days under harsh conditions of storage (40°C/75% RH). Similarly, Botrel et al. (2017) used gum Arabic, cashew gum, and modified starch for the microencapsulation of fish oil by spray drying and reported that cashew gum could effectively delay the oxidation of core oil compared to gum Arabic. Thus, it can be used as a potential wall material to preserve bioactive oils. Martínez et al. (2015), Takashige et al. (2018), and Binsi et al. (2017) could decrease the challenge of oxidation in spray dried microcapsules by adding natural antioxidants and stabilizers into the primary emulsion.

Martínez et al. (2015) and Takashige et al. (2018) added rosemary extract as an antioxidant into the initial emulsion of chia oil and krill oil before the spray drying process, respectively. While, Binsi et al. (2017) used sage extract as a stabilizer in the fish oil emulsion, which led to the enhancement of the MEE, the reduction of surface oil content, and, ultimately, the enhancement of oxidative stability. Wang et al. (2018) used an innovative strategy to create double-shell and multi-core microcapsules of tuna oil with whey protein isolate, agar gum, and gellan gum as the wall materials. First, a complex coacervation process was implemented; then the obtained microcapsules were spray dried. The final spray dried tuna oil microcapsules possessed significant oxidative stability with an MEE of 95.8%. Moreover, several research examples of optimizing the process parameters to achieve the maximum MEE as a viable strategy to decrease the challenge of oxidation are presented in Section 9.5.2.

Furthermore, it is important to control the storage conditions of spray dried micropowders for reaching the optimum results for the encapsulation of functional oils; i.e., the increment of functional oil stability and shelf-life by minimizing the oxidation of the core functional oils. In this case, adjusting the light, temperature, oxygen, relative humidity, and packaging systems are among the measures that should be considered. Measuring the PV, pAV, and TBA of the encapsulated oil powder are typically used to determine the oxidative stability after a certain time or under special conditions to show the efficiency and effectiveness of the spray drying microencapsulation process (Jeyakumari, Zynudheen, Parvathy, & Binsi, 2018). The PV determines the primary oxidative compounds, while pAV and TBA indicate the secondary oxidation products. In addition, the TOTOX value, which is calculated by the formulation of 2*PV + pAV, assesses the total oxidation products (Shahidi & Zhong, 2020). The maximum degradation limits for marine oils and vegetable oils are defined by different food regulation standards. For example, the Global Organization for EPA and DHA Omega-3 s (GOED, 2015), the Canada Natural Health Product (NHP), and the Codex Alimentarius Commission defined the PV/pAV limit of 5 mEq/kg and 20, respectively as the maximum degradation limits for marine and vegetable oils (CAC, 1999, 2017). The Australian government guidelines, EPA/DHA-Rich Schizochytrium Algal Oil, defined the PV limit of 5 mEq/kg and pAV limit of 20 for algal oils as the maximum degradation limits. The CODEX/FAO defined the PV limit of 10 mEq/kg and 15 mEq/kg as the maximum degradation limits for Refined Vegetable Oils and Cold pressed/virgin vegetable oils, respectively (CAC, 1999, 2017). It has been reported that the oxidation of fats and oils is accelerated under extreme conditions such as high storage temperatures and higher air access during the storage time (Frankel, Satu'e-Gracia, Meyer, & German, 2002; Kolanowski et al., 2005). Thus, accelerated storage conditions are usually applied to predict the oxidative stability of spray dried microencapsulated functional oils, and the associated test is called 'accelerated shelf-life testing' (ASLT), although it has also been called 'accelerated oxidative stability testing', 'accelerated storage testing', or 'accelerated oxidation testing' in different studies related to oil microencapsulation (Ton et al., 2016; Li et al., 2019). These experiments apply storage conditions severe than normal ones in a short period, which helps to save time and estimate the peroxidation in a relatively short time (Ganje et al., 2016). In general, the time needed to estimate the shelf-life of food products is reasonably longer than 10–12 months, whereas that of research studies is often shorter than six months. Accelerating factors of ALST, which are used for determining the shelf-life of spray dried microparticles, can include a variety of environmental factors like temperature, light, oxygen, and relative humidity. However, the temperature is an accelerating factor that is being used frequently for ASLT (Calligaris, Manzocco, Anese, & Nicoli, 2019).

In the following, a look at recent research that has measured the oxidative stability of spray dried microencapsulated oils by using temperature as an accelerating factor is given. Zhou et al. (2019) measured the oxidative stability of *ganoderma lucidum* spore oil by setting the temperature at 65 and 45 °C for seven days. By comparing the results, they found that spray dried microcapsules with GA/MD-MCs as wall material provided considerable oxidative protection to this encapsulated vegetable oil over the oxidation process. Similarly, Mohammed, Meor Hussin, Tan, Abdul Manap, and Alhelli (2017) set the temperature of 65°C for 24 days on bulk and microencapsulated *Nigella sativa* oil and reported the superiority of the spray drying microencapsulation process to increase the oxidative stability of oils. It is reported that the reason for the priority of the temperature to be selected in most research as an accelerating factor is the fact that there is a recognized scientific model (Arrhenius), which can easily describe the dependence of temperature and the quality loss as a result of oxidation (Manzocco, Panozzo, & Calligaris, 2011; Mizrahi, 2000; Shafiei et al., 2020).

One standard method which is used extensively for implementing the ASLT to determine the oxidative stability of fats and oils is Rancimat (Hidalgo, Leon, & Zamora, 2006; Mancebo-Campos, Salvador, & Fregapane, 2007; Wan, 1995), which can also be used to measure the oxidation of the spray dried microparticles. It assesses the oxidation process by subjecting the oil to the standardized oxidation conditions and report the result in the time of oxidation (Comunian & Favaro-Trindade, 2016). Besides, while this method is an expensive analysis, it can analyze several samples at the same time (Kaushik et al., 2015). According to Velasco, Dobarganes, and Márquez-Ruiz (2000); Velasco, Dobarganes, Holgado, and Márquez-Ruiz (2009), who applied Rancimat tests directly to the dried microencapsulated oils, this method is suitable and useful to study the oxidative stability of microencapsulated oils as the results were similar to the results obtained from bulk oil and storage under ambient conditions.

9.7 THE FORTIFICATION OF FOOD PRODUCTS WITH SPRAY DRIED FUNCTIONAL OILS

While there are some challenges, such as the degradation/oxidation of functional oils and bioactive components under food processing (and storage) conditions and their reaction with other food elements, the microencapsulation process by spray drying can solve such application limits in food fortification. Spray dryers can produce dried microparticles in various size range, which is highly applicable to the fortification of a food matrix without unpleasant mouthfeel and sensory problems (Feizollahi et al., 2018). It is reported that solid foods are more appropriate to be incorporated with EFAs powders than liquid foods. The reason may be the fact that the wall materials used for spray drying encapsulation are water-soluble and adding microparticles to liquid foods may cause core release, which leads to subsequent deterioration/oxidation (Encina et al., 2016). According to Hinriksdottir, Jonsdottir, Sveinsdottir, et al. (2015) and Mu and Mullertz (2015), the bioavailability of PUFAs in microparticles is highly similar to that of the foods enriched with liquid oils. Moreover, people tend to use functional foods more nowadays, as their awareness of the health benefits of such products has increased and continues to grow. Therefore, the production of microencapsulated EFAs by spray drying has been scaled up in the food industry, in addition to its utilization in target food products commercially. Several manufactures around the world produce ω-3 microcapsules by spray drying commercially with different brand names such as Dry n-3 (BASF, Germany), Supercoat omega-3 (The Wright Group, USA), and Omega-3 powder (Arjuna, India), which are used as food additives in the food industry. In addition, several food producers around the world contain microcapsules of EFAs. However, most manufactures usually do not reveal their information about the method of microencapsulation of the ω-3 for fortifying the food products. Infant foods, pet foods, animal feeds, milk and dairy products, seafood, pasta, bread, cereal bars, bakery products, eggs, meat, and meat products are typical examples of fortified foods with EFAs like DHA and EPA that are available commercially (Kaur, Basu, & Shivhare, 2015).

Fortification of foods with microencapsulated powder improves the oxidative stability of the final product. Aquilani et al. (2018) and Jiménez-Martín, Pérez-Palacios, Carrascal, and Rojas (2016) fortified pork burgers and chicken nuggets, respectively, with spray dried microencapsulated fish oil and both reported improved oxidative stability of the final product. In addition, this could improve the physicochemical properties of food products. Takeungwongtrakul and Benjakul (2017) reported a reduced dough shrinkage in sheeting of biscuit-making after the fortification of biscuits with spray dried microencapsulated shrimp oil. In another study, Abedi, Rismanchi, Shahdoostkhany, Mohammadi, and Hosseini (2016) found that the fortification of a stirred-type yogurt with spray dried microencapsulated *Nigella sativa* seeds oil did not change the pH and acidity ranges of yogurt during four weeks of storage. Moreover, they observed satisfactory stability of microencapsulated EFAs in yogurt over four weeks, which is beyond the product shelf-life (i.e., two weeks).

9.8 CONCLUDING REMARKS

The omega-3 and omega-6 FAs are essential for health and must be obtained from the human diet. These essential fatty acids are extremely sensitive to oxidation during processing and storage, which causes off-tastes and smells. Microencapsulation by spray dryer is an efficient and successful technique to protect the EFAs against deterioration/oxidation and produce high-quality micropowders. It can increase the stability of EFAs and decline the challenges of food enrichment with ω-3 and ω-6 EFAs. The efficiency of spray drying microencapsulation of functional oils depends on several factors such as choosing appropriate wall materials, optimal emulsion formulation, optimal drying stage considering all influential spray drying parameters, and ultimately, collection and storage of the final product.

A wide range of wall materials from heat-sensitive proteins to stabilizing carbohydrates are used for spray drying encapsulation of EFAs and functional oils, with the option of mixing different components of these two groups for improving their functionality and efficiency as an ideal wall material toward the efficient protection of the encapsulated EFAs. Generally, the selected wall material should be food-grade, inexpensive, tasteless with acceptable water solubility, suitable viscosity, and compatibility with the targeted food matrix. Choosing the most suitable type of wall materials with the reactive/proactive approaches, optimal emulsion formulation conditions including the emulsification method, the optimal ratio of wall to core materials (in the range of 1: 1 to 4: 1 (w/w) for food applications), and optimal spray drying parameters are necessary in order to achieve efficient protection of EFAs and functional oils via spray drying process. Adopting these components is also highly effective in improving microencapsulation efficiency and subsequent high oxidative stability of the microencapsulated EFAs.

The incorporation of the spray dried functional oil microcapsules in the food matrix is highly applicable according to the appropriate size of microparticles that inhibits the unpleasant mouthfeel and sensory problems. Nowadays, fortification of food products with functional oil microcapsules is being implemented widely around the world, not only because of their health-promoting effects but also since these products improve the oxidative stability and physicochemical properties of the final food products. Accordingly, infant foods, pet foods, animal feeds, milk and dairy products, seafood, pasta, bread, cereal bars, bakery products, eggs, meat, and meat products containing spray dried encapsulated EFAs and functional oils have become commercially available.

REFERENCES

Abedi, A.-S., Rismanchi, M., Shahdoostkhany, M., Mohammadi, A., Hosseini, H. (2016). Microencapsulation of *Nigella sativa* seeds oil containing thymoquinone by spray-drying for functional yogurt production. *International Journal of Food Science & Technology*, 51(10), 2280–2289. doi:10.1111/ijfs.13208

Ackman, R. G. (2008). Fatty acids in fish and shellfish. In: Chow, C. K., ed., *Fatty Acids in Foods and Their Health Implications*, London, UK: CRC Press, pp. 155–185.

Aghbashlo, M., Mobli, H., Madadlou, A., Rafiee, S. (2013a). Influence of wall material and inlet drying air temperature on the microencapsulation of fish oil by spray drying. *Food and Bioprocess Technology*, 6, 1561–1569.

Aghbashlo, M., Mobli, H., Madadlou, A., Rafiee, S. (2013b). Fish oil microencapsulation as influenced by spray dryer operational variables. *International Journal of Food Science & Technology*, 48(8), 1707–1713. doi:10.1111/ijfs.12141

Ajith, T. A., Jayakumar, T. G. (2019). Omega-3 fatty acids in coronary heart disease: Recent updates and future perspectives. *Clinical and Experimental Pharmacology & Physiology*, 46, 11–18.

Akhavan Mahdavi, S., Jafari, S. M., Assadpour, E., Ghorbani, M. (2016) Storage stability of encapsulated barberry's anthocyanin and its application in jelly formulation. *Journal of Food Engineering*, 181, 59–66.

Albert, K., Vatai, G., Koris, A. (2017). Microencapsulation of vegetable oil: Alternative approaches using membrane technology and spray drying. *Hungarian Journal of Industry and Chemistry*, 45(2), 29–33.

Alvarenga Botrel, D., Vilela Borges, S., Fernandes, R.V.B., Dantas Viana, A., Costa, J. M. G. Marques, G. R. (2012). Evaluation of spray drying conditions on properties of microencapsulated oregano essential oil microencapsulated oregano essential oil. *International Journal of Food Science & Technology*, 47(11), 2289–2296.

Anandharamakrishnan, C., Padma Ishwarya, S. (2015). *Spray Drying Techniques for Food Ingredient Encapsulation*. IFT press, Wiley Blackwell.

Aquilani, C., Pérez-Palacios, T., Jiménez Martín, E., Antequera, T., Bozzi, R., Pugliese, C. (2018). Cinta senese burgers with omega-3 fatty acids: Effect of storage and type of enrichment on quality characteristics. *Archivos de Zootecnia*, 67, 217–220.

Arpagaus, C., Collenberg, A., Rütti, D., Assadpour, E., Jafari, S. M. (2018) Nano spray drying for encapsulation of pharmaceuticals *International Journal of Pharmaceutics*, 546, 194–214.

Assadpour, E., Jafari, S.M. (2019). Advances in spray-drying encapsulation of food bioactive ingredients: From microcapsules to nanocapsules. *Annual Review of Food Science and Technology*, 10, 103–131.

Augustin, M.A., Sanguansri, L., Bode, O., 2006. Maillard reaction products as encapsulants for fish oil powders. *Journal of Food Science*, 71, 25–32.

Bai, X., Li, C., Yu, L., Jiang, Y., Wang, M., Lang, S., Liu, D. (2019). Development and characterization of soybean oil microcapsules employing kafirin and sodium caseinate as wall materials. *LWT - Food Science and Technology*, 111, 235–241.

Baik, M. Y., Suhendro, E. L., Nawar, W. W., McClements, D. J., Decker, E. A., Chinachoti, P. (2004). Effects of antioxidants and humidity on the oxidative stability of microencapsulated fish oil. *Journal of the American Oil Chemists' Society*, 81(4), 355–360.

Bakry, A. M., Abbas, S., Ali, B., Majeed, H., Abouelwafa, M. Y., Mousa, A., Liang, L. (2015). Microencapsulation of oils: A comprehensive review of benefits, techniques, and applications. *Comprehensive Reviews in Food Science and Food Safety*, 15(1), 143–182.

Barrow, C., Wang, B., Adhikari, B., et al. (2013). Spray drying and encapsulation of omega-3 oils. In: Jacobsen, C., Nielsen, N. S., Frisenfeldt, H. A., Sørensen, M. A. -D., Eds. *Food Enrichment with Omega-3 Fatty Acids*. Cambridge: Woodhead Publishing Limited.

Beindorff, C. M., Zuidam, N. J. (2010). Microencapsulation of fish oil. In Zuidam N., Nedovic V. (eds) *Encapsulation technologies for active food ingredients and food processing*, Springer, New York, NY. https://doi.org/10.1007/978-1-4419-1008-0_6

Bhushan, B., Mani, I., Kar, A., Datta, A. (2017). Optimization of jackfruit seed starch-soya protein isolate ratio and process variables for flaxseed oil encapsulation. *Indian Journal of Agricultural Sciences*, 87(12),1657–1663.

Biesalski, H. K. et al. (2009). Bioactive compounds: Definition and assessment of activity. *Nutrition*, 25(11–12), 1202–1205.

Binsi, P. K., Nayak, N., Sarkar, P. C., Jeyakumari, A., Muhamed Ashraf, P., et al. (2017). Structural and oxidative stabilization of spray dried fish oil microencapsulates with gum Arabic and sage polyphenols: Characterization and release kinetics. *Food Chemistry*, 219, 158–168.

Böger, B. R., Georgetti, S. R., Kurozawa, L. E. (2018). Microencapsulation of grape seed oil by spray drying. *Food Science and Technology*, 38(2), 263–270.

Botrel, D. A., Borges, S. V., de Barros Fernandes, R. V., Antoniassi, R., de Faria-Machado, A. F., et al. (2017). Application of cashew tree gum on the production and stability of spray-dried fish oil. *Food Chemistry*, 221, 1522–1529.

Calligaris, S., Manzocco, L., Anese, M., Nicoli, M. C. (2019). Accelerated shelf-life testing. In: Charis M. Galanakis, ed., *Food Quality and Shelf Life*, Academic Press. Chapter 12, 359–392. doi: 10.1016/B978-0-12-817190-5.00012-4.

Calvo, P., Castaño, Á. L., Lozano, M., González-Gómez, D. (2012). Influence of the microencapsulation on the quality parameters and shelf-life of extra-virgin olive oil encapsulated in the presence of BHT and different capsule wall components. *Food Research International*, 45(1), 256–261.

Calvo, P., Hernández, T., Lozano, M., González-Gómez, D. (2010). Microencapsulation of extra-virgin olive oil by spray-drying: Influence of wall material and olive quality. *European Journal of Lipid Science and Technology*, 112, 852–858.

Chang, C., Nickerson, M. T. (2018). Encapsulation of omega 3-6-9 fatty acids-rich oils using protein-based emulsions with spray drying. *Journal of Food Science and Technology*, 55(8), 2850–2861.

Chang, H. W., Tan, T. B., Tan, P. Y., Abas, F., Lai, O. M., et al. (2018). Microencapsulation of fish oil using thiol-modified β-lactoglobulin fibrils/chitosan complex: A study on the storage stability and in vitro release. *Food Hydrocolloids*, 80, 186–194.

Chegini, G. R., Ghobadian, B. (2005). Effect of spray-drying conditions on physical properties of orange juice powder. *Drying Technology*, 23(3), 657–668. doi:10.1081/drt-20,005,416

Chen, Q., Zhong, F., Wen, J., McGillivray, D., Quek, S. Y. (2013). Properties and stability of spray-dried and freeze-dried microcapsules co-encapsulated with fish oil, phytosterol esters, and limonene. *Drying Technology*, 31(6), 707–716.

Chew, S., Tan, C., Pui, L., Chong, P., Gunasekaran, B., Lin, N. (2019). Encapsulation Technologies: A Tool for Functional Foods Development. *International Journal of Innovative Technology and Exploring Engineering*, 8(5s), 154–160.

Chew, S. C., Tan, C. P., Nyam, K. L. (2018). Microencapsulation of refined kenaf (*Hibiscus cannabinus* L.) seed oil by spray drying using β-cyclodextrin/gum arabic/sodium caseinate. *Journal of Food Engineering*, 237, 78–85.

Codex Alimentarius Commission, Food and Agriculture Organization of the United Nations, STANDARD FOR NAMED VEGETABLE OILS CXS 210–1999, Adopted in 1999. Revised in 2001, 2003, 2009, 2017, 2019. Amended in 2005, 2011, 2013, 2015, 2019.

Codex Alimentarius Commission, Food and Agriculture Organization of the United Nations, STANDARD FOR FISH OILS CXS 329–2017.

Comunian, T. A., Favaro-Trindade, C. S. (2016). Microencapsulation using biopolymers as an alternative to produce food enhanced with phytosterols and omega-3 fatty acids: A review. *Food Hydrocolloids*, 61, 442–457.

Corrêa-Filho, L., Moldão-Martins, M., Alves, V. (2019). Advances in the application of microcapsules as carriers of functional compounds for food products. *Applied Sciences*, 9(3), 571. doi:10.3390/app9030571.

Costa, S. S., Machado, B. A. S., Martin, A. R., Bagnara, F., Ragadalli, S. A., Alves, A. R. C. (2015). Drying by spray drying in the food industry: Micro-encapsulation, process parameters and main carriers used. *African Journal of Food Science*, 9(9), 462–470.

Danviriyakul, S., McClements, D.J., Decker, E., Nawar, W.W., Chinachoti, P. (2002). Physical stability of spray dried milk fat emulsion as affected by emulsifiers and processing conditions. *Journal of Food Science*, 67(6), 2183–2189.

Das, U. N. (2006). Essential fatty acids – A review. *Current Pharmaceutical Biotechnology*, 7, 467–482.

Desai, K. G. H.; Park, H. J. (2005). Recent developments in microencapsulation of food ingredients, *Drying Technology*, 23(7), 1361–1394.

Di Giorgio, L., Mauri, P. R. S. A. N. (2018). Encapsulation of fish oil in soybean protein particles by emulsification and spray drying. *Food Hydrocolloids*. doi: 10.1016/j.foodhyd.2018.09.024

Di Pasquale, M. G. (2009). The essentials of essential fatty acids. *Journal of Dietary Supplements*, 6(2), 143–161.

Dias, D. R., Botrel, D. A., Fernandes, R. V. D. B., Borges, S. V. (2017). Encapsulation as a tool for bioprocessing of functional foods. *Current Opinion in Food Science*, 13, 31–37.

Drusch, S. (2007). Sugar beet pectin: A novel emulsifying wall component for microencapsulation of lipophilic food ingredients by spray-drying. *Food Hydrocolloids*, 21, 1223–1228.

Drusch, S., Serfert, Y., Schwarz, K. (2006). Microencapsulation of fish oil with n-octenyl succinate derivatised starch: Flow properties and oxidative stability. *European Journal of Lipid Science and Technology*, 108, 501–512.

Eilander, A., Hundscheid, D. C., Osendarp, S. J., Transler, C., Zock, P. L. (2007) Effects of n-3 long chain polyunsaturated fatty acid supplementation on visual and cognitive development throughout childhood: A review of human studies. *Prostaglandins, Leukotrienes, and Essential Fatty Acids*, 76, 189–203.

Encina, C., Márquez-Ruiz, G., Holgado, F., Giménez, B., Vergara, C., Roberta, P. (2018). Effect of spray-drying with organic solvents on the encapsulation, release and stability of fish oil. *Food Chemistry*, 263, 283–291.

Encina, C., Vergara, C., Giménez, B., Oyarzún-Ampuero, F., Robert, P. (2016). Conventional spray-drying and future trends for the microencapsulation of fish oil. *Trends in Food Science & Technology*, 56, 46–60.

Esfahani, R., Jafari, S. M., Jafarpour, A., Dehnad, D. (2019). Loading of fish oil into nanocarriers prepared through gelatin-gum Arabic complexation. *Food Hydrocolloids*, 90, 291–298.

Fakir, A., Gadhave, A., Waghmare, J. (2015). A review on a microencapsulation of fish oil to improve oxidative stability. *Asian Journal of Science and Technology*, 6(3), 1197–1204.

Faldt, P., Bergenstahl, B. (1995). Fat encapsulation in spray-dried food powders. *Journal of the American Oil Chemist's Society* 72(2), 171–176.

Fan, Y. Y., Chapkin, R. S. (1998) Importance of dietary gamma-linolenic acid in human health and nutrition. *The Journal of Nutrition*, 128, 1411–1414.

Fang, X., Shima, M., Adachi, S. (2005). Effects of drying conditions on the oxidation of lineoleic acid encapsulated with gum arabic by spray-drying. *Food Science and Technology Research*, 11(4), 380–384.

Feizollahi, E., Hadian, Z., Honarvar, Z. (2018). Food fortification with omega-3 fatty acids; microencapsulation as an addition method. *Current Nutrition & Food Science*, 14(2), 90–103.

Frankel, E. N., Satu'e-Gracia, T., Meyer, A. S., German, J. B. (2002). Oxidative stability of fish and algae oils containing long-chain polyunsaturated fatty acids in bulk and in oil-in-water emulsions. *Journal of Agricultural and Food Chemistry*, 50, 2094–2099.

Frascareli, E. C., Silva, V. M., Tonon, R. V., Hubinger, M. D. (2012). Effect of process conditions on the microencapsulation of coffee oil by spray drying. *Food and Bioproducts Processing*, 90(3), 413–424.

Freitas, S., Merkle, H. P., Gander, B. (2005). Microencapsulation by solvent extraction/evaporation: Reviewing the state of the art of microsphere preparation process technology. *Journal of Controlled Release*, 102, 313–332.

Fuentes-Ortega, T., Martínez-Vargas, S. L., Cortés-Camargo, S., Guadarrama-Lezama, A. Y., Gallardo-Rivera, R., Baeza-Jimenéz, R., Pérez-Alonso, C. (2017). Effects of the process variables of microencapsulation sesame oil (*Sesamum indica* L.) by spray drying, *Revista Mexicana de Ingeniería Química*, 16(2), 477–490.

Gallardo, G., Guida, L., Martinez, V., et al. (2013). Microencapsulation of linseed oil by spray drying for functional food application. *Food Research International*, 52, 473–482.

Gammone, M. A., Riccioni, G., Parrinello, G., D'Orazio, N. (2019). Omega-3 polyunsaturated fatty acids: Benefits and endpoints in sport, review, *Nutrients*, 11, 46.

Ganje, M., Jafari, S. M., Dusti, A., Dehnad, D., Amanjani, M., Ghanbari, V. (2016). Modeling quality changes in tomato paste containing microencapsulated olive leaf extract by accelerated shelf-life testing, *Food and Bioproducts Processing*, 97, 12–19.

Garg, M. L., Wood, L. G., Singh, H., Moughan, P. J. (2006) Means of delivering recommended levels of long chain n-3 polyunsaturated fatty acids in human diets. *Journal of Food Science*, 71(5), 66–71.

Geranpour, M., Assadpour, E., Jafari, S. M. (2020). Recent advances in spray drying encapsulation of essential fatty acids and functional oils. *Journal of Trends in Food Science and Technology*, 102, 71–90. doi: 10.1016/j.tifs.2020.05.028.

Geranpour, M., Emam-Djomeh, Z., Asadi, G. (2019a). *Microencapsulation of pumpkin seed oil by spraydryer under various process conditions and determination of the optimal point by RSM. IDS'2018 - 21st International Drying Symposium*, València, Spain. doi: 10.4995/ids2018.2018.7332

Geranpour, M., Emam-Djomeh, Z., Asadi, G. (2019b). Investigating the effects of spray drying conditions on the microencapsulation efficiency of pumpkin seed oil. *Journal of Food Processing and Preservation*, e13947. doi:10.1111/jfpp.13947.

Gerber, M. (2012), Omega-3 Fatty Acids and Cancers: A Systematic Update Review of Epidemiological Studies. *British Journal of Nutrition*, 107, S228–S239.

Gharsallaoui, A., Roudaut, G., Chambin, O., Voilley, A., Saurel, R. (2007). Applications of spraydrying in microencapsulation of food ingredients: An overview. *Food Research International*, 40, 1107–1121.

Global Organization for EPA and DHA Omega-3. GOED Voluntary Monograph, Version 5 (2015). http://www.goedomega3.com/index.php/files/download/350 (Date of access: 10/03/2017)

Gogate, P. R. (2015). The use of ultrasonic atomization for encapsulation and other processes in food and pharmaceutical manufacturing. In: Gallego-Juárez, J. A., Graff, K. F., eds., *Power Ultrasonics: Applications of High-Intensity Ultrasound*. Chapter 30, 911–935. doi: 10.1016/B978-1-78242-028-6.00030-2

Gonçalves, A., Estevinho, B. N., Rocha, F. (2016). Microencapsulation of vitamin A: A review. *Trends in Food Science and Technology*, 51, 76–87.

Goula, A. M., Adamopoulos, K. G. (2005). Spray drying of tomato pulp in dehumidified air: II. The effect on powder properties. *Food Engineering*, 66, 35–42.

Goula, A. M., Adamopoulos, K. G. (2012). A new technique for spray-dried encapsulation of lycopene. *Drying Technology*, 30, 641–652. doi: 10.1080/07373937.2012.655871.

Guaadaoui, A., Benaicha, S., Elmajdoub, N., Bellaoui, M., Hamal, A. (2014). What is a bioactive compound? A combined definition for a preliminary consensus. *International Journal of Nutrition and Food Sciences*, 3(3), 174–179.

Gunstone, F., (2004), *The Chemistry of Oils and Fats*. Wiley-Blackwell, 140–210.

Harnack, K., Andersen, G., Somoza, V. (2009). Quantitation of alpha-linolenic acid elongation to eicosapentaenoic and docosahexaenoic acid as affected by the ratio of n6/n3 fatty acids. *Nutrition and Metabolism*, 6(1), 8, doi:10.1186/1743-7075-6-8.

Hernández, E. J., Chávez-Tapia, N. C., Uribe, M., Barbero-Becerra, V. J. (2016). Role of bioactive fatty acids in nonalcoholic fatty liver disease. *Nutrition Journal*, 15, 72.

Hidalgo, F. J., Leon, M. M., Zamora, R. (2006). Antioxidative activity of amino phospholipids and phospholipid/amino acid mixtures in edible oils as determined by the Rancimat method. *Journal of Agricultural and Food Chemistry*, 54, 5461–5467.

Hinriksdottir, H., Jonsdottir, V., Sveinsdottir, K., et al. (2015). Bioavailability of long-chain n-3 fatty acids from enriched meals and from microencapsulated powder. *European Journal of Clinical Nutrition*, 69(3), 344–348.

Holgado, F., Márquez-Ruiz, G., Ruiz-Méndez, M. V., Velasco, J. (2019). Effects of the drying method on the oxidative stability of the free and encapsulated fractions of microencapsulated sunflower oil. *International Journal of Food Science & Technology*. doi:10.1111/ijfs.14162

Horrobin, D. F. (1990) Gamma-linolenic acid. *Reviews in Contemporary Pharmacotherapy*, 1, 1–41.

Hosseini, H., Ghorbani, M., Jafari, S. M., Sadeghi Mahoonak, A. (2019). Encapsulation of EPA and DHA concentrate from Kilka fish oil by milk proteins and evaluation of its oxidative stability. *Journal of Food Science and Technology*. doi: 10.1007/s13197-018-3455-9

Hoyos-Leyva, J. D., Bello-Perez, L. -A., Agama-Acevedo, J. E., Alvarez Ramirez, J., Jaramillo-Echeverry, L. M. (2019). Characterization of spray drying microencapsulation of almond oil into taro starch spherical aggregates. *LWT - Food Science and Technology*, 101, 526–533. doi: 10.1016/j.lwt.2018.11.079.

Hu, L., Zhang, J., Hu, Q., Gao, N., Wang, S., Sun, Y., Yang, X. (2016). Microencapsulation of *brucea javanica* oil: Characterization, stability and optimization of spray drying conditions. *Journal of Drug Delivery Science and Technology*, 36, 46–54. doi:10.1016/j.jddst.2016.09.008

Huang, H., Hao, S., Li, L., Yang, X., Cen, J., Lin, W., Wei, Y. (2014). Influence of emulsion composition and spray-drying conditions on microencapsulation of tilapia oil. *Journal of Food Science and Technology*, 51(9), 2148–2154.

Huang, L. X., Mujumdar, A. S. (2008). The effect of rotary disk atomizer RPM on particle size distribution in a semi-industrial spray dryer. *Drying Technology*, 26, 1319–1325.

Islam, M., Mahmud, N., Nawas, T., Fang, Y., Xia, W. (2018). Health benefits and spray drying microencapsulation process of fish oil (Omega-3). *American Journal of Food Science and Nutrition Research*, 5(2), 29–42.

Jacobsen, C. (2016). Oxidative Stability and Shelf-life of Food Emulsions. *Oxidative Stability and Shelf-life of Foods Containing Oils and Fats*, 287–312. doi:10.1016/b978-1-63,067-056-6.00008-2

Jafari, S. M. (2017). An overview of nanoencapsulation techniques and their classification. In: Jafari, S. M., ed., *Nanoencapsulation Technologies for the Food and Nutraceutical Industries*, Cambridge, MA: Academic, pp. 1–34.

Jafari, S. M., Assadpoor, E., He, Y., Bhandari, B. (2008). Encapsulation efficiency of food flavours and oils during spray drying. *Drying Technology*, 26(7), 816–835.

Jafari, S. M., He, Y., Bhandari, B. (2007). Role of powder particle size on the encapsulation efficiency of oils during spray drying. *Drying Technology*, 25(6), 1081–1089.

Jenkins, J. A., Breiteneder, H., Mills, E. N. C. (2007). Evolutionary distance from human homologs reflects allergenicity of animal food proteins. *The Journal of Allergy and Clinical Immunology*, 120(6), 1399–1405.

Jeyakumari, A., Zynudheen, A. A., Parvathy, U., Binsi, P. K. (2018). Impact of chitosan and oregano extract on the physicochemical properties of microencapsulated fish oil stored at different temperature. *International Journal of Food Properties*, 21(1), 942–955. doi:10.1080/10942912.2018.1466319

Jiang, Q. (2014). Natural forms of vitamin E: Metabolism, antioxidant, and anti-inflammatory activities and their role in disease prevention and therapy. *Free Radical Biology and Medicine*, 72, 76–90.

Jiménez-Martín, E., Pérez-Palacios, T., Carrascal, J. R., Rojas, T. A. (2016). Enrichment of chicken nuggets with microencapsulated omega-3 fish oil: Effect of frozen storage time on oxidative stability and sensory quality. *Food and Bioprocess Technology*, 9(2), 285–297.

Kalkan, F., Vanga, S. K., Murugesan, R., Orsat, V., Raghavan, V. (2017). Microencapsulation of hazelnut oil through spray drying. *Drying Technology*, 35(5), 527–533.

Karthik, P., Anandharamakrishnan, C. (2013) Microencapsulation of docosahexaenoic acid by spray-freeze-drying method and comparison of its stability with spray-drying and freeze-drying methods. *Food and Bioprocess Technology*, 6(10), 2780–2279.

Katouzian, I., Jafari, S. M. (2016). Nano-encapsulation as a promising approach for targeted delivery and controlled release of vitamins. *Trends in Food Science and Technology*, 53, 34–48.

Kaur, M., Basu, S., Shivhare, U. S., (2015). Omega-3 fatty acids: Nutritional aspects, sources and encapsulation strategies for food fortification. *Direct Research Journal of Health and Pharmacology*, 3(1), 12–31.

Kaur, N., Chagh, V., Gupt, A. K. (2014). Essential fatty acids as functional components of foods – A review. *Journal of Food Science and Technology*, 51(10), 2289–2303.

Kaushik, P., Dowling, K., Barrow, C. J., Adhikari, B. (2015). Microencapsulation of omega-3 fatty acids: A review of microencapsulation and characterization methods. *Journal of Functional Foods*, 19, 868–881.

Kolanowski, W., Jaworska, D., Weissbrodt, J., Kunz, B. (2007). Sensory assessment of microencapsulated fish oil powder. *Journal of the American Oil Chemists' Society*, 84(1), 37–45.

Kolanowski, W., Ziolkowski, M., Weißbrodt, J., Kunz, B., Laufenberg, G. (2005). Microencapsulation of fish oil by spray drying—Impact on oxidative stability. Part 1. *European Food Research and Technology*, 222(3–4), 336–342.

Komaiko, J. S., McClements, D. J. (2016). Formation of food-grade nanoemulsions using low-energy preparation methods: A review of available methods. *Comprehensive Reviews in Food Science and Food Safety*, 15(2), 331–352.

Kralovec, J. A., Zhang, S., Zhang, W., Barrow, C. J. (2012). A review of the progress in enzymatic concentration and microencapsulation of omega-3 rich oil from fish and microbial sources. *Food Chemistry*, 131, 639–644.

Lavanya, M. N., Kathiravan, T., Moses, J. A., Anandharamakrishnan, C. (2019). Influence of spray-drying conditions on microencapsulation of fish oil and chia oil, *Drying Technology*. doi: 10.1080/07373937.2018.1553181

Lavie, C. J., Milani, R. V., Mehra, M. R., Ventuar, H. O. (2009). Omega-3 polyunsaturated fatty acids and cardiovascular diseases. *Journal of the American College of Cardiology*, 54(7), 585–594.

Lawrence, G. D. (2010). *The Fats of Life: Essential Fatty Acids in Health and Disease*. Rutgers University Press.

Le Priol, L., Dagmey, A., Morandat, S., Saleh, K., El Kirat, K., Nesterenko, A. (2019). Comparative study of plant protein extracts as wall materials for the improvement of the oxidative stability of sunflower oil by microencapsulation. *Food Hydrocolloids*, 95, 105–115.

Legako, J., Dunford, N. T. (2010). Effect of spray nozzle design on fish oil-whey protein 732 microcapsule properties. *Journal of Food Science*, 75, 394–400.

Li, D., Xie, H., Liu, Z., Li, A., Li, J., Liu, B., ... Zhou, D. (2019). Shelf-life prediction and changes in lipid profiles of dried shrimp (*Penaeus vannamei*) during accelerated storage. *Food Chemistry*, 297, 124951.

Li, H., Zhu, K., Zhou, H., Peng, W. (2012). Effects of high hydrostatic pressure treatment on allergenicity and structural properties of soybean protein isolate for infant formula. *Food Chemistry*, 132(2), 808–814.

Luz, P. P., Pires, A. M., Serra, O. A. (2007). A low cost ultrasonic spray dryer to produce spherical microparticles from polymeric matrices. *Quimica Nova*, 30, 1744–1746.

Madene, A., Jacquot, M., Scher, J., Desobry, S. 2006. Flavour encapsulation and controlled release – a review. *International Journal of Food Science and Technology*, 41(1), 1–21.

Mahdavi, S. A., Jafari, S. M., Ghorbani, M., Assadpoor, E. (2014). Spray-drying microencapsulation of anthocyanins by natural biopolymers: A review. *Drying Technology*, 32, 509–518.

Mancebo-Campos, V., Salvador, M. D., Fregapane, G. (2007). Comparative study of virgin olive oil behavior under Rancimat accelerated oxidation conditions and long-term room temperature storage. *Journal of Agricultural and Food Chemistry*, 55, 8231–8236.

Manzocco, L., Panozzo, A., Calligaris, S. (2011). Accelerated Shelf-life Testing (ASLT) of oils by light and temperature exploitation. *Journal of the American Oil Chemists' Society*, 89(4), 577–583.

Martínez, M. L., Curti, M. I., Roccia, P., Llabot, J. M., Penci, M. C., et al. (2015). Oxidative stability of walnut (*Juglans regia* L.) and chia (*Salvia hispanica* L.) oils microencapsulated by spray drying. *Powder Technology*, 270, 271–277.

Mazza, M., Pomponi, M., Janiri, L., Bria, P., Mazza, S. (2007). Omega-3 fatty acids and antioxidants in neurological and psychiatric diseases: An overview. *Progress in Neuro-Psychopharmacology & Biological Psychiatry*, 31, 12–26.

McClements, D. J. (2005). *Food Emulsions; Principles, Practice, and Techniques*, 2nd Edition; CRC Press.

McClements, D. J., Decker, E. A., Weiss, J. (2007). Emulsion-based delivery systems for lipophilic bioactive components. *Journal of Food Science*, 72, 109–124.

Mendanha, D. V., Ortiz, S. E. M., Favaro-Trindade, C. S., Mauri, A., Monterrey-Quintero, E. S., Thomazini, M., 2009. Microencapsulation of casein hydrolysate by complex coacervation with SPI/pectin. *Food Research International*, 42, 1099–1104.

Minemoto, Y., Fang, X., Hakamata, K., Watanabe, Y., Adachi, S., Kometani, T., Matsuno, R. (2002). Oxidation of linoleic acid encapsulated with soluble soybean polysaccharide by spray-drying. *Bioscience, Biotechnology, and Biochemistry*, 66(9), 1829–1834.

Mizrahi, S. (2000). Accelerated shelf-life testing of foods In: Kilcast, D., Subramaniam, P., eds., *The Stability and Shelf-Life of Food*, A volume in Woodhead Publishing Series in Food Science, Technology and Nutrition. Chapter 5, 107–128. doi:10.1533/9781855736580.1.107.

Mogensen, K. M. (2017), Essential fatty acid deficiency. *Practical Gastroenterology*, Nutrition Issues in Gastroenterology. Series 164, 37–44.

Mohammed, N. K., Meor Hussin, A. S., Tan, C. P., Abdul Manap, M. Y., Alhelli, A. M. (2017). Quality changes of microencapsulated *Nigella sativa* oil upon accelerated storage. *International Journal of Food Properties*, 20(sup3), S2395–S2408.

Mozaffarian, D. (2005). Does alpha-linolenic acid intake reduce the risk of coronary heart disease? A review of the evidence. *Alternative Therapies in Health and Medicine*, 11, 24–30.

Mu, H., Mullertz, A. (2015). Marine lipids and the bioavailability of omega-3 fatty acids. *Current Nutrition & Food Science*, 11(3), 177–187.

Murakami, A. E., Eyng, C., Torrent, J. (2014). Effects of functional oils on coccidiosis and apparent metaboliz-able energy in broiler chickens. *Asian-Australasian Journal of Animal Sciences*, 27(7), 981–989.

Murali, S., Kar, A., Patel, A., et al. (2017). Optimization of rice bran oil encapsulation using jackfruit seed starch – Whey protein isolate blend as wall material and its characterization. *International Journal of Food Engineering*, 13(4), 25–35.

Murali, S., Kar, A., Patel, A. S., Kumar, J., Mohapatra, D., Dash, S. K. (2016). Encapsulation of rice bran oil in tapioca starch-soya protein isolate complex using spray drying. *Indian Journal of Agricultural Sciences*, 86(8), 984–991.

Murugesan, R., Orsat, V. (2012). Spray drying for the production of nutraceutical ingredients- A review. *Food and Bioprocess Technology*, 5, 3–14.

Nanditha, B., Prabhasankar, P. (2009). Antioxidants in bakery products: A review, *Critical Reviews in Food Science and Nutrition*, 49(1), 1–27.

Nawas, T., Azam, M. S., Ramadhan, A. H., Xu, Y., Xia, W. (2019). Impact of wall material on the physiochemical properties and oxidative stability of microencapsulated spray dried silver carp oil. *Journal of Aquatic Food Product Technology*, 28(1), 49–63. doi:10.1080/10498850.2018.1560380

Nedovic, V., Kalusevic, A., Manojlovic, V., Levic, S., Bugarski, B. (2011). An overview of encapsulation tech-nologies for food applications. *Procedia Food Science*, 1(1), 1806–1815.

Nedović, V., Kalušević, A., Manojlović, V., Petrović, T., Bugarski, B. (2013). Encapsulation systems in the food industry. In: Yanniotis, S., Taoukis, P., Stoforos, N. G., Karathanos, V. T., Eds., *Advances in Food Process Engineering Research and Applications*, New York, NY: Springer Science+Business Media, pp. 229–253.

Nesterenko, A., Alric, I., Silvestre, F., Durrieu, V. (2013). Vegetable proteins in microencapsulation: A review of recent interventions and their effectiveness. *Industrial Crops and Products*, 42(1), 469–479.

Özbek, Z. A., Ergönül, P. G., (2017). A Review on Encapsulation of Oils. *CBU Journal of Science*, 13(2), 293–309. doi: 10.18466/cbayarfbe.313358

Patterson, E., Wall, R., Fitzgerald, G. F., Ross, R. P., Stanton, C. (2012). Health implications of high dietary omega-6 polyunsaturated fatty acids. *Journal of Nutrition and Metabolism*. doi: 10.1155/2012/539426

Pereira, K.C., Ferreira, D. C. M., Alvarenga, G. F., Pereira, M. S. S., Barcelos, M. C. S., Costa, J. M. G. (2018). Microencapsulation and release controlled by the diffusion of food ingredients produced by spray drying: A review. *Brazilian Journal of Food Technology*, 211, e2017083. doi: 10.1590/1981-6723.08317

Pourashouri, P., Shabanpour, B., Razavi, S. H., Jafari, S. M., Shabani, A., Aubourg, S. P. (2014). Impact of wall materials on physicochemical properties of microencapsulated fish oil by spray drying. *Food and Bioprocess Technology*, 7(8), 2354–2365.

Prata, A. S., Garcia, L., Tonon, R. V., Hubinger, M. D. (2013). Wall material selection for encapsulation by spray drying. *Journal of Colloid Science and Biotechnology*, 2, 1–7.

Prato, E., Biandolino, F., Parlapiano, I., Papa, L., Kelly, M., & Fanelli, G. (2018), Bioactive fatty acids of three commercial scallop species. *International Journal of Food Properties*, 21(1), 519–532.

Rajabi, H., Ghorbani, M., Jafari, S. M., Sadeghi Mahoonak, A., Rajabzadeh, G. (2015) Retention of saffron bioactive components by spray drying encapsulation using maltodextrin, gum Arabic and gelatin as wall materials *Food Hydrocolloids*, 51, 327–337.

Ramakrishnan, S., Ferrando, M. (2014). Influence of emulsification technique and wall composition on physi-cochemical properties and oxidative stability of fish oil microcapsules produced by spray drying. *Food and Bioprocess Technology*, 7, 1959–1972.

Ramakrishnan, S., Ferrando, M., Aceña-Muñoz, L., Mestres, M., De Lamo-Castellví, S., Güell, C. (2012). Fish oil microcapsules from o/w emulsions produced by premix membrane emulsification. *Food and Bioprocess Technology*. doi:10.1007/s11947-012-0950-2.

Ray S., Raychaudhuri, U., Chakraborty, R. (2016). An overview of encapsulation of active compounds used in food products by drying technology. *Food Bioscience*, 13, 76–83.

Reineccius, G. A. (1988). Spray-drying of food flavours. In: Risch, S. J., Reineccius, G. A., Eds., *Flavour Encapsulation*, Washington, DC: American Chemical Society, ACS Symposium Series 370, pp. 55–66.

Rezvankhah, A., Emam-Djomeh, Z., Askari, G. (2019). Encapsulation and delivery of bioactive com-pounds using spray and freeze-drying techniques: A review. *Drying Technology*, 1–24. doi: 10.1080/07373937.2019.1653906.

Roccia, P., Mart'ınez, M. L., Llabot, J. M., Ribotta, P. D. (2014). Influence of spray-drying operating conditions on sunflower oil powder qualities. *Powder Technology*, 254, 307–313.

Rodriguez, J., Martin, M., Ruiz, M. A., Clares, B. (2016). Current encapsulation strategies for bioactive oils: From alimentary to pharmaceutical prospectives. *Food Research International*, 83, 41–59.

Rubilar, M., Morales, E., Contreras, K., Ceballos, C., Acevedo, F., Villarroel, M., Shene, C. (2012). Development of a soup powder enriched with microencapsulated linseed oil as a source of omega-3 fatty acids. *European Journal of Lipid Science and Technology*, 114, 423–433.

Sagiri, S. S., Anis, A., Pal, K. (2015). Review on encapsulation of vegetable oils: Strategies, preparation methods, and applications. *Polymer-Plastics Technology and Engineering*, 55(3), 291–311. doi:10.1080/036 02559.2015.1050521

Saini, R. K., Keum, Y. -S. (2018). Omega-3 and omega-6 polyunsaturated fatty acids: Dietary sources, metabolism, and significance—A review. *Life Sciences*, 203, 255–267.

Sanchez-Reinoso, Z., Gutiérrez, L. -F. (2017). Effects of the emulsion composition on the physical properties and oxidative stability of Sacha Inchi (*Plukenetia volubilis* L.) oil microcapsules produced by spray drying. *Food and Bioprocess Technology*, 10(7), 1354–1366.

Sarabandi, K., Gharehbeglou, P., Jafari, S. M. (2020) Spray-drying encapsulation of protein hydrolysates and bioactive peptides: Opportunities and challenges. *Drying Technology*, 38, 577–595.

Seddighi Pashaki, A., Emam-Djomeh, Z., Askari, G. (2016). *Evaluation of spray drying parameters on physicochemical properties of seedless black Barberry (Berberis vulgaris L.) juice. 20th International Drying Symposium*, Japan.

Serdaroglu, M., Kerimoğlu, B. Ö., Kara, A. (2015). An overview of food emulsions: Description, classification and recent potential applications. *Turkish Journal of Agriculture – Food Science and Technology*, 3(6), 430–438.

Shafiei, G., Ghorbani, M., Hosseini, H., Sadeghi Mahoonak, A., Maghsoudlou, Y., Jafari, S. M. (2020). Estimation of oxidative indices in the raw and roasted hazelnuts by accelerated shelf-life testing. *Journal of Food Science and Technology*, 57, 2433–2442.

Shahidi, F., Zhong, Hy. J. (2020). Methods for Measuring Lipid Oxidation. In: Shahidi, F., ed, *Bailey's Industrial Oil and Fat Products*, 7th Edition; pp. 437–436.

Shamaei, S., Seiiedlou, S. S., Aghbashlo, M., Tsotsas, E., Kharaghani, A. (2017). Microencapsulation of walnut oil by spray drying: Effects of wall material and drying conditions on physicochemical properties of microcapsules. *Innovative Food Science & Emerging Technologies*, 39, 101–112.

Shao, W., Pan, X., Liu, X., Teng, F., Yua, S. (2018). Microencapsulation of algal oil using spray drying technology. *Food Technology and Biotechnology*, 56(1), 65–70, doi: 10.17113/ftb.56.01.18.5452

Shireman, R. (2003). *Essential Fatty Acids. In: Encyclopedia of Food Sciences and Nutrition* 2nd Edition; Elsevier series, 2169–2176, doi: 10.1016/B0-12-227055-X/00424-7

Shishir, M. R. I., Taip, F. S., Aziz, N. A., Talib, R. A., Sarker M. S. H. (2016). Optimization of spray drying parameters for pink guava powder using RSM. *Food Science and Biotechnology*, 25, 1–8.

Silva-James, N. K., Nogueira, R. I., Freitas, S. P. (2018). Blending of pressed vegetable oils from pomegranate seeds and soybean to increase functional lipids consume. *Journal of Analytical and Pharmaceutical Research*, 7(3), 268–269. doi: 10.15406/japlr.2018.07.00237

Simopoulos, A. P. (2002). Omega-3 fatty acids in inflammation and autoimmune diseases. *Journal of the American College of Nutrition*, 21, 495–505.

Simopoulos, A. P. (2006). Evolutionary aspects of diet, the omega-6/omega-3 ratio and genetic variation: Nutritional implications for chronic diseases. *Biomedicine & Pharmacotherapy*, 60, 502–507.

Simopoulos, A. P. (2016). An increase in the omega-6/omega-3 fatty acid ratio increases the risk for obesity. *Nutrients*, 8(3), 128, doi: 10.3390/nu8030128

Soottitantawat, A., Takayama, K., Okamura, K., Muranaka, D., Yoshii, H., Furuta, T., Ohkawara, M., Linko, P. Microencapsulation of l-menthol by spray drying and its release characteristics. *Innovative Food Science & Emerging Technologies*, 2005, 6(2), 163–170.

Soottitantawat, A., Yoshii, H., Furuta, T., Ohkawara, M., Linko, P. (2003).Microencapsulation by spray drying: Influence of emulsion size on the retention of volatile compounds. *Journal of Food Science*, 68, 2256–2262.

Subtil, S. F., Rocha-Selmi, G. A., Thomazini, M., Trindade, M. A., Netto, F. M., Favaro-Trindade, C. S. (2014). Effect of spray drying on the sensory and physical properties of hydrolysed casein using gum arabic as the carrier. *Journal of Food Science and Technology-Mysore*, 51(9), 2014–2021.

Takashige, S., Hermawan Dwi, A., Sultana, A., Shiga, H., Adachi, S., Yoshii, H. (2018). *Encapsulation of krill oil by spray drying. 21st International Drying Symposium*, València, Spain, 11–14 September 2018. doi: 10.4995/ids2018.2018.7323

Takeungwongtrakul, S., Benjakul, S. (2016). Effect of glucose syrup and fish gelatin on physicochemical properties and oxidative stability of spray-drier micro- encapsulated shrimp oil. *Journal of Food Processing and Preservation*, doi:10.1111/jfpp.12876

Takeungwongtrakul, S., Benjakul, S. (2017). Biscuits fortified with micro-encapsulated shrimp oil: Characteristics and storage stability. *Journal of Food Science and Technology*, 54(5), 1126–1136. doi:10.1007/s13197-017-2545-4

Taktak, W., Nasri, R., Lopez-Rubio, A., Hamdi, M., Gomez-Mascaraque, L. G., Ben Amor, N., ... Karra-Chaâbouni, M. (2019). Improved antioxidant activity and oxidative stability of spray dried European eel (*Anguilla anguilla*) oil microcapsules: Effect of emulsification process and eel protein isolate concentration. *Materials Science and Engineering: C*, 104, 109867. doi: 10.1016/j.msec.2019.109867

Tallima, H., El Ridi, R. (2018). Arachidonic acid: Physiological roles and potential health benefits – A review. *Journal of Advanced Research*, 11, 33–41.

Tatar Turan, F., Cengiz, A., Sandıkçı, D., Dervisoglu, M., Kahyaoglu, T. (2016). Influence of an ultrasonic nozzle in spray-drying and storage on the properties of blueberry powder and microcapsules. *Journal of the Science of Food and Agriculture*, 96(12), 4062–4076.

Ton, N. M. N.; Tran, T. T. T., Le, V. V. M. (2016). Microencapsulation of rambutan seed oil by spray-drying using different protein preparations. *International Food Research Journal*, 23(1), 123–128.

Tonon, R. V., Grosso, C. R. F., Hubinger, M. D. (2011). Influence of emulsion composition and inlet air temperature on the microencapsulation of flaxseed oil by spray drying. *Food Research International*, 44(1), 282–289.

Toure, A. Z. Xiaoming, C. -S. Jia, Zhijian, D. (2007). Microencapsulation and oxidative stability of ginger essential oil in maltodextrin/whey protein isolate (MD/WPI). *International Journal of Dairy Science*, 2, 387–392.

Tullberg, C., Vegarud, G., Undeland, I., Scheers, N. (2017). Effects of marine oils, digested with human fluids, on cellular viability and stress protein expression in human intestinal caco-2 cells. *Nutrients*, 9, 11.

Us-Medina, U., Julio, L. M., Segura-Campos, M. R., Ixtaina, V. Y., Tomás, M. C. (2018). Development and characterization of spray-dried chia oil microcapsules using by-products from chia as wall material. *Powder Technology*, 334, 1–8.

Veiga, R. D. S. D., Aparecida Da Silva-Buzanello, R., Corso, M. P., Canan, C. (2019). Essential oils microencapsulated obtained by spray drying: A review. *Journal of Essential Oil Research*, 31, 457–473.

Velasco, J., Dobarganes, C., Holgado, F., Márquez-Ruiz, G. (2009). A follow-up oxidation study in dried microencapsulated oils under the accelerated conditions of the Rancimat test. *Food Research International*, 42(1), 56–62.

Velasco, J., Dobarganes, M. C., Márquez-Ruiz, G. (2000). Application of the accelerated test Rancimat to evaluate oxidative stability of dried microencapsulated oils. *Grasas y Aceites*, 51, 261–267.

Vishnu, K. V., Chatterjee Niladri, S., Ajeeshkumar, K. K., Lekshmi, R. G. K., Tejpal, C. S., Mathew, Suseela, Ravishankar, C. N. (2017). Microencapsulation of sardine oil: Application of vanillic acid grafted chitosan as a bio-functional wall material. *Carbohydrate Polymers*. doi: 10.1016/j.carbpol.2017.06.076

Voucher, A. C. S., Dias, P. C. M., Coimbra, P. T., Costa, I. S. M., Marreto, R. N., Dellamora-Ortiz, G. M., Freitas, O., Ramos, M. F. S. (2019). Microencapsulation of fish oil by casein-pectin complexes and gum arabic microparticles: Oxidative stabilization, *Journal of Microencapsulation*. doi: 10.1080/02652048.2019.1646335.

Wan, P. J. (1995). Accelerated stability methods. In: Warner, K., Michael Eskin, N. A., Eds., *Methods to Assess Quality and Stability of Oils and Fat-containing Foods*, pp. 179–189, Champaign: AOCS Press.

Wang, B., Adhikari, B., Barrow, C. J. (2018). Highly stable spray dried tuna oil powders encapsulated in double shells of whey protein isolate-agar gum and gellan gum complex coacervates. *Powder Technology*. doi: 10.1016/j.powtec.2018.07.084

Wang, Y., Liu, B., Wen, X., Li, M., Wang, K., Ni, Y. (2017). Quality analysis and microencapsulation of chili seed oil by spray drying with starch sodium octenylsuccinate and maltodextrin. *Powder Technology*, 312, 294–298.

Watanabe, Y., Fang, X., Adachi, S., Fukami, H., Matsuno, R. (2004). Oxidation of 6-O-arachidonoyl l-ascorbate microencapsulated with a polysaccharide by spray-drying. *LWT Food Science and Technology*, 37(4), 395–400.

Williams, P. A., Phillips, G. O. (2003). The use of hydrocolloids to improve food texture. In: McKenna, B. M., ed., *Texture in Foods, Volume 1, Chapter 11, Semi-solid Foods*. Boca Raton, FL: CRC Press.

Yu, F., Li, Z., Zhang, T., Wei, Y., Xue, Y., Xue, C. (2017). Influence of encapsulation techniques on the structure, physical properties, and thermal stability of fish oil microcapsules by spray drying. *Food Process Engineering*. doi: 10.1111/jfpe.12576.

Zhang, W. H., Shi, B., Jeff, S. (2007). A theoretical study on autoxidation of unsaturated fatty acids and antioxidant activity of phenolic compounds. *The Journal of the American Leather Chemists Association*, 102, 99–105.

Zhou, D., Zhou, F., Ma, J., Ge, F. (2019). Microcapsulation of *Ganoderma Lucidum* spores oil: Evaluation of its fatty acids composition and enhancement of oxidative stability. *Industrial Crops and Products*, 131, 1–7.

10 Spray Drying Encapsulation of Proteins and Bioactive Peptides

Khashayar Sarabandi
Zahedan University of Medical Sciences, Iran

Pouria Gharehbeglou and Seid Mahdi Jafari
Gorgan University of Agricultural Sciences and Natural Resources, Iran

Zahra Akbarbaglu
University of Tabriz, Iran

CONTENTS

10.1 INTRODUCTION

Over recent years, different studies have been performed with regard to the role of food products on public health. The results of these investigations indicated that there are numerous food-related diseases such as diabetes and cardiovascular diseases, for which lipid oxidation and free radicals are the most important factors contributing to the deterioration of food quality (e.g., unpleasant flavor and shortened shelf-life) (Sarabandi, Mahoonak, & Akbari, 2018a). Additionally, among the most effective factors that contribute to the oxidation of lipids, metal ions are the main pro-oxidants in food emulsions such as beverages, sauces, soups, desserts, and mayonnaise (McClements & Decker, 2000). In addition to food degradation, the oxidation of lipids and generated free radicals can lead to numerous diseases such as atherosclerosis in patients with metabolic syndrome, diabetes, rheumatoid arthritis, Alzheimer's, and cancer (Collins, 2005; Palmieri, Grattagliano, Portincasa, & Palasciano, 2006).

Considering these challenges, it is essential to use bioactive compounds to prevent lipid oxidation and reduce the rate of free radical production so as to increase the shelf-life of foods (especially fat-based types) and inhibit diseases caused by them. Synthetic antioxidants such as butylated hydroxyl anisole (BHA), butylated hydroxyl toluene (BHT), and propyl gallates are among the most important compounds that can be used in the food industry (Shahidi & Zhong, 2015). In addition to antioxidants, various chelators such as EDTA, citric acid, and polyphosphates are used to control prooxidant metal ions. However, the use of any synthetic antioxidant and prevalent chelators has some problems, such as functional disadvantages, regulatory limitations, and consumer concerns about their potential hazards (Sarabandi & Jafari, 2020). Because of these disadvantages, extraction, and the use of antioxidants and natural chelating agents, have received substantial attention (Shahidi & Zhong, 2015).

One group of the most important natural antioxidants are bioactive peptides. These compounds have high immunity as a potential alternative to synthetic antioxidants in stabilizing various food products (Tong, Sasaki, McClements, & Decker, 2000). In addition to antioxidant activity, peptides are known as bioactive compounds that can be used in the production of functional foods (Gibson & Williams, 2005). These peptides are rich in low molecular weight (MW) fractions with high digestibility, nutritional and functional value, and low allergenicity and toxicity (Xie, Huang, Xu, & Jin, 2008). The release of buried and inactive peptides occurs in the primary protein sequence after their enzymatic hydrolysis (You, Zhao, Regenstein, & Ren, 2010). Enzymatic hydrolysis causes significant increases in the biological and antioxidant properties of peptides by producing specific protein fractions; mainly, those containing 2–10 amino acids and with an MW of <6 kDa (Pihlanto-Leppälä, 2000). The most important biological and health-promoting effects of bioactive peptides include the ability to bind minerals (Cross, Huq, Palamara, Perich, & Reynolds, 2005), opiate-like activities (Sienkiewicz-Szłapka et al., 2009), antimicrobial activity (McCann et al., 2006), antithrombotic properties (Shimizu et al., 2009), antihypertensive activity (Jia et al., 2010), chelating of metal ions

(Jamdar et al., 2010), hypocholesterolemic effect (Zhong, Liu, Ma, & Shoemaker, 2007), immuno-modulatory mechanism (Gauthier, Pouliot, & Saint-Sauveur, 2006), anti-allergenic and improved bio-availability (Sarabandi & Jafari, 2020), anticancer (Sarmadi & Ismail, 2010) and antioxidant properties (Akbarbaglu et al., 2019).

In spite of the numerous therapeutic and health-promoting effects that are known for bioactive peptides and proteins, there are some problems with using them. These include bitterness, hydro-phobicity, hygroscopicity, incompatibility, low bioavailability, biostability, and reaction with com-ponents of the food matrix (Udenigwe, 2014). As a result of these challenges, the direct use of bioactive peptides in the formulation of a variety of dietary and beneficial products is problematic (McClements, 2015).

Microencapsulation of bioactive peptides is one of the best techniques to minimize such dis-advantages. During this process, most of the bioactive compounds in the structure of the carrier/wall materials are dispersed or entrapped (Assadpour & Jafari, 2018, 2019; Mohan, McClements, & Udenigwe, 2016). Concerning the type of the delivery system, carriers, and the techniques used, the microencapsulation method can control the bioavailability, biostability, physicochemical, and biological stability of bioactive peptides, while it can also modify their functional properties. Furthermore, the microencapsulation of bioactive compounds makes it possible to use these com-pounds in the various formulations and production of functional foods (McClements, 2015). Among various microencapsulation techniques, the spray drying process is a common, economical, flex-ible, and effective method of converting various liquid food and pharmaceutical formulations into a powder form, which can lead to an increase in their shelf-life and physicochemical and microbial stability (Ray, Raychaudhuri, & Chakraborty, 2016).

However, the different structural and functional properties of bioactive peptides/proteins com-pared with other compounds such as vitamins and phenols can influence the selection of the microencapsulation system. Additionally, the occurrence of thermal and shear stresses during the atomization and drying process results in the accumulation, denaturation, and loss of their biological activity through changes in the structure of the corresponding peptides/proteins (Sarabandi & Jafari, 2020; Sarabandi et al., 2019c). Therefore, the selection of optimum process conditions (e.g., temper-ature, type, composition, and concentration of the carrier) determines the physicochemical proper-ties and stability of the spray dried product (Jafari, Assadpoor, He, & Bhandari, 2008). Considering the health benefits of bioactive peptides and protein-based drugs, as well as the necessity of stabi-lization and microencapsulation of these compounds, this chapter discusses the drying mechanism, components of the system, and the importance, advantages, and applications of the spray drying process for the encapsulation of bioactive peptides and proteins. Descriptions are also given of some of the parameters of the spray drying process and the challenges that affect the biological and functional stability of the dried peptides/proteins, as well as some of the most important challenges relating to the nature and structure of bioactive peptides that make the microencapsulation of these compounds necessary. Correspondingly, the importance of a variety of protein and polysaccharide carrier compounds in maintaining the activity and stabilization of bioactive peptides during the dry-ing and storage process is presented. Finally, the results of some recent studies in the spray drying microencapsulation of proteins and peptides are reported.

10.2 SPRAY DRYING PARAMETERS FOR THE ENCAPSULATION OF BIOACTIVE PEPTIDES AND PROTEINS

In terms of the solidifying methods of food products and pharmaceutical formulations such as encap-sulation of peptides and proteins, freeze and spray drying are among the most important techniques. Freeze drying is an effective process for the microencapsulation of bioactive compounds and various heat-sensitive nutraceuticals such as vitamins, extracts, aromatic compounds, antibiotics, antioxi-dants, probiotics, and other microbial cultures (Rahman, 2007). Other applications of this process include the stabilization of peptides and protein-based drugs (Roy & Gupta, 2004). This process has a number of drawbacks, such as being time-consuming (the process takes from several hours to days), high energy consumption, high process cost (over 50 times of spray drying), changes of pH, the formation of ice crystals, and an increase in the concentration of soluble compounds (e.g., salts or sugars), the stresses resulting from drying and freezing and the high cost of storage, packaging and transportation of freeze dried products (Ajmera & Scherließ, 2014; Jacquot & Pernetti, 2004).

Unlike freeze drying, spray drying is a frequent industrial technique to convert liquid materials into solids (in the form of powders, granules, and agglomerates) to lengthen the shelf-life, and facilitate storage and transportation, and also the handling of food and pharmaceutical products (Mujumdar, 2004). Higher physicochemical and microbial stability, easier usability, transportation, and storage are among the main advantages of dried compounds rather than liquid forms. Even though there has been extensive investigation about the spray drying technique, the energy requirement of this type of dryer is relatively high when compared to other methods. This means a significant portion of water in food products is excreted using heat energy (Shrestha, Howes, Adhikari, & Bhandari, 2007). By contrast, hot air is used in the spray drying process, but most of the applied heat is spent for the evaporation of water from the product; therefore, the temperature does not rise too high and drying is done at a wet-bulb temperature (at least, at the beginning of the drying process when the air temperature is high). Thus, due to the small droplet size (10–200µm), the heat transfer surface for drying increases. For that reason, the drying process is completed in a short time (1–20 seconds) and, because of the high evaporation rate, this can result in the production of a powder with low thermal damage and high quality (Brennan, 2005).

In spray drying, a significant part of drying occurs at a constant drying rate period. For this reason, this drying method is appropriate for heat-sensitive products. In addition, a unique property of this process is its large production capacity, which can produce between 100 g and several tons of powder per hour and the end product can be either free-flowing, agglomerate, or granular powder particles (Bhandari, Patel, & Chen, 2008). Because of these positive impacts, spray drying can be used for the production of powder of all kinds of juices, syrups, concentrates, plant extracts, dairy products, vitamins, emulsions, a variety of bioactive proteins and peptides, and inhaled powders containing drugs (Akbarbaglu et al., 2019; Assadpour & Jafari, 2017, 2019; Bürki, Jeon, Arpagaus, & Betz, 2011; Sarabandi, Mahoonak, & Akbari, 2018a). Nonetheless, the effect of each of spray drying parameters (i.e., inlet temperature, atomization pressure, feed rate, and aspiration rate) on the functional properties and biological activity of bioactive compounds has led to the importance of recognizing these factors and achieving optimum production conditions. Hence, the structure and effective parameters in the spray drying method are briefly discussed in the following sections of this chapter. In addition, the effect of some process factors on the preservation and stability of bioactive compounds (especially peptides and proteins) are described in the following sections. Figure 10.1 shows the most important components of the spray drying system.

10.3 CHALLENGES FOR THE SPRAY DRYING ENCAPSULATION OF PEPTIDES AND PROTEINS

The global trading of functional proteins and peptides has been reported to amount to annual revenues of more than $75 billion (Sun, 2013). Given their potential functional properties and health benefits, bioactive peptides are a perfect choice for the production of functional foods. On the other hand, some problems like solubility, physicochemical instability, reactivity with other nutrients, low bioavailability and biostability, high hygroscopicity, and bitterness reduce the commercialization and applicability of these compounds in food formulations and dietary supplements (Sarabandi & Jafari, 2020). These challenges are described in more detail in the following sections. Several studies have investigated the possibility of using the spray drying encapsulation technique to minimize these problems; however, the structures of proteins and peptides, as well as their susceptibility to spray drying conditions, lead to the decrease or loss of their functional properties and their initial biological activity. The following sections offer descriptions of some of the most important

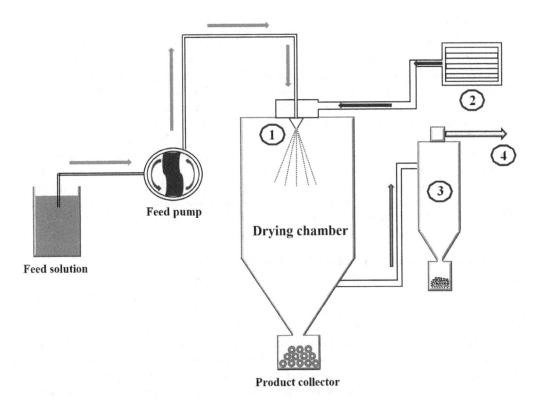

FIGURE 10.1 A schematic presentation of spray dryer system; (1) atomizer, (2) air heater, (3) cyclone separator, and (4) exhaust air. Reprinted from Sarabandi, Gharehbeglou, and Jafari (2019a), with permission.

instabilities depending on the nature of the protein/peptide, the effect of process mechanisms, and the available procedures to reduce these challenges.

10.3.1 STRUCTURE- AND NATURE-DEPENDENT CHALLENGES

While some studies have focused on the ability of peptides to replace synthetic antioxidants and exert health-promoting properties (Akbarbaglu et al., 2019; Jamdar et al., 2010; Maqsoudlou et al., 2019; Khashayar Sarabandi, Mahoonak, & Akbari, 2018a; Sarmadi & Ismail, 2010), several protein-based drugs have a low capability for storage and *in vivo* applications, because of their undesirable physicochemical properties and biological instability (Lafarga & Hayes, 2017; Mohan et al., 2016; Rao, Klaassen Kamdar, & Labuza, 2016b; Sarabandi & Jafari, 2020).

10.3.1.1 Solubility

One of the main purposes of producing bioactive compounds is their use in the formulation of functional food products, most of which (e.g., cheese, butter, margarine, and mayonnaise) are fat-based. Because of the hydrophilic structure of peptides/proteins, they have low solubility in fat-based formulations (Sarabandi & Jafari, 2020). The solubility and reactivity of peptides with other compounds are influenced by their polarity. However, the degree of hydrophilicity or hydrophobicity is also affected by polar or non-polar surface groups. Furthermore, most of these compounds can be adsorbed at the water–air interface as a consequence of their specific surface properties (McClements, 2018). This factor, as well as the design of the colloidal systems, can trigger the instability of these compounds during the spray drying microencapsulation, which causes instability or inability to direct the application of bioactive peptides/proteins in food systems.

10.3.1.2   Chemical Instability

The chemical instability of proteins and peptides can be related to the ionic changes, pH (especially in food systems), temperature, oxygen, and light. These factors play a major role in the stability of bioactive peptides. For example, many proteins (e.g., globular species) undergo structural and denaturation changes at high temperatures (Kim, Decker, & McClements, 2002; Uversky & Goto, 2009). Also, the pH sensitivity of several peptides and proteins results in their incapability to be used in acidic food formulations. These parameters cause physicochemical instability and loss of quality and shelf-life of the products enriched with these compounds. Nevertheless, some factors such as environmental stresses have a substantial impact on the instability of peptides and proteins (McClements, 2018). Complementarily, the presence of reducing and oxidizing agents, hydrogen, and metal ions in the dietary composition are other effective factors that cause the destruction and loss of biological and health-promoting activities of peptides/proteins (Sarabandi, Jafari, Mahoonak, & Mohammadi, 2019b). Therefore, the protection of bioactive compounds against environmental stresses and the composition of food ingredients is very important.

10.3.1.3   Physical Instability and the Effects of Processing

Bioactive peptides and proteins have complex structures compared to many other bioactive compounds such as polyphenols. Due to the aforementioned characteristics, consideration should be given to a number of factors, including their sensitivity to the composition and formulation of the food products, environmental conditions, reactivity with other compounds, and direct use of these compounds in pure form in dietary food formulations (McClements, 2015). In addition to the chemical reactions between peptides and components of the food system, some physical reactions can result in the loss of functional properties and degradation of the quality properties of the foods. One of the most important compounds that react with proteins is the group of metal ions. Different metal ions, including iron, copper, and calcium, react with proteins/peptides, and through gelation and precipitation processes it can lead to the loss of functional and structural properties, as well as the destruction of the stability of the food product (Burey, Bhandari, Howes, & Gidley, 2008). Therefore, designing an appropriate system and separating the peptides from the reactive compounds present in the dietary formulations is essential to prevent physical and chemical reactions.

It is also worth noting that several fortified food products are processed in order to increase their shelf-life. For example, processes of freezing, drying, and pasteurization are used to improve the stability and shelf-life of such products. Nevertheless, these methods also lead to structural changes and the loss of bioactivity of the dietary compounds such as peptides and proteins. For example, forming ice crystals, changing the concentration of soluble substances, pH, phase separation, and protein accumulation are effective parameters in freezing and thawing processes. It should be mentioned that these changes may be influenced by the freezing speed and the size of the ice crystals. Meanwhile, the drying and pasteurization processes of solutions and protein products can lead to the accumulation, precipitation, and denaturation of proteins (Wang, Nema, & Teagarden, 2010).

10.3.1.4   Hygroscopicity

Hygroscopicity is one of the most important challenges related to the nature of bioactive peptides and hydrolysates (Akbarbaglu et al., 2019). Due to the absorption of a high amount of moisture from the environment, these compounds can accelerate destructive physicochemical reactions and microbial spoilage. These reactions result in the loss of functional properties, physical and biological stability, shelf-life, and their applications in various products. Additionally, moisture absorption in the inhalable powders containing pharmaceutical peptides causes loss of flow properties and their usability on account of aggregation and structural changes (Chan et al., 2004). It should be mentioned that packaging permeability, temperature, and the relative humidity (RH) of the environment are the most important external factors that can affect the level of hygroscopicity of bioactive peptides/proteins. The effect of these parameters on the moisture absorption depends on amino acid composition, degree of hydrophobicity, moisture content, presence of reducing sugar, pH, glass transition temperature, and the degree of hydrolysis of peptides and proteins (Rao, Bajaj, Mann, Arora, & Tomar, 2016a; Sarabandi & Jafari, 2020; Subtil et al., 2014). Because of these circumstances, measuring this index plays a key role in assessing the physical and functional stability of bioactive peptides and hydrolysates during storage (Rao, Bajaj, et al., 2016a). Therefore, the microencapsulation of these compounds in carriers with low adhesion and hydrophilicity degree is one of the most important techniques to reduce the impact of hygroscopicity and increase the stability of peptides and hydrolysates (Mohan, Rajendran, He, Bazinet, & Udenigwe, 2015).

10.3.1.5 Glass Transition Temperature

Glass transition temperature (T_g) is one of the determining factors of the stability of spray dried powders. The value of T_g is significantly affected by the amount of moisture absorption. As the temperature of the product rises above T_g, the powders change from glass to rubber form. Under these conditions, the occurrence of hydrophilic surface reactions, particle adhesion, and cake formation lead to particle flux and the loss of the initial structure (Rao, Bajaj, et al., 2016a). T_g can be affected by several factors, with the compositions of the raw material and amount of moisture being the most influential parameters. It is worth noting that the plasticizing effect of water molecules plays a considerable role in reducing the T_g value (Kurozawa, Park, & Hubinger, 2009a).

In the case of proteins, increasing the degree of hydrolysis and diminishing the molecular weight are important in lessening the T_g value and improving the thermos-plasticity properties. Therefore, increasing the T_g value of peptides/proteins is necessary to increase the physicochemical stability of these compounds during storage. One of the best techniques for this purpose is the spray drying microencapsulation of these compounds using carriers with a higher T_g value; e.g., whey proteins, soy protein isolate, and gelatin (Truong, Bhandari, & Howes, 2005).

10.3.1.6 Bitterness

One of the most important goals of identifying and producing bioactive and health-promoting peptides is to use them in the production of a variety of functional foods. For the purposes of illustration, in the case of bioactive peptides derived by enzymatic hydrolysis of various sources, their application is divided according to their characteristics and degree of hydrolysis (DH). In other words, peptides with low, variable, and extensive hydrolysis levels are used to improve nutritional supplements as well as flavorings and functional properties in specific medical diets (Vioque, Clemente, Pedroche, Yust, & Millán, 2001). However, their bitter taste is one of the main challenges in the production of functional products and dietary supplements. The bitterness of these compounds arises from the hydrolysis of peptide bonds and the release of hydrophobic amino acids from the internal structures of proteins (Sarabandi, Mahoonak, Hamishekar, Ghorbani, & Jafari, 2018b). Even so, the bitterness intensity varies in bioactive peptides produced from different protein sources. This difference can be attributed to the alteration in the composition and sequence of amino acids. As an instance, amino acids in hydrolyzed C-terminals such as leucine, phenylalanine, and tryptophan play an important role in intensifying the bitterness of these compounds (Lafarga & Hayes, 2017). Various studies have suggested different methods such as treatment with activated carbon, chromatography, plastein reaction, and hydrolysis with exopeptidases to reduce the bitterness of peptides. On the other hand, the disadvantages of these processes include low yield processes, the loss of essential hydrophobic amino acids, and excessive hydrolysis (Sarabandi & Jafari, 2020). Considering the practical applications of these peptides as additives in food formulations, their incorporation into the carrier matrix to minimize their bitterness is very important (Chaudhry et al., 2008; Ma et al., 2014).

10.3.1.7 Bioavailability and Targeted Delivery

In the previous sections, some of the health characteristics of bioactive peptides and hydrolysates were mentioned and the importance of using these compounds in the production of non-perishable food products to increase consumer health was also declared. However, the data from antioxidant properties of peptides in laboratory conditions cannot be generalized to natural conditions (i.e., *in vivo*) (Sarabandi & Jafari, 2020). This can be attributed to the structural changes and loss of biological activity of peptides in the digestive system (Sarmadi & Ismail, 2010).

It should be considered that the effectiveness and biological activity of these compounds depend on the oral bioavailability and maintaining physicochemical stability under *in vivo* conditions (McClements, 2018). After consuming products enriched with peptides and hydrolysates, they lose biological activity due to acidic conditions and digestive enzymes as a result of their denaturation. In addition, the access of these compounds into the intercellular space in the intestinal lymphatic system is affected by lipid solubility and their permeability to the portal circulation (McClements, 2018; Sarabandi & Jafari, 2020). On the other hand, the biological effects of these compounds in the body require increased bioavailability and biostability during digestion. For that reason, loading these bioactive compounds in the delivery systems via encapsulation techniques such as spray drying for increasing their stability is crucial.

10.3.2 AGGREGATION OF PROTEINS AND PEPTIDES

Numerous fortified foods are processed in order to prolong their shelf-life. Freezing, fermentation, drying, and pasteurization are among the considerable methods in this regard (Wang et al., 2010). Nonetheless, these processes can cause structural changes (aggregation) in proteins and peptides (especially, in their pure form). Protein aggregation is one of the main influential factors in the loss of their biological activity and health-benefiting properties. One of the main areas of this research is the spray drying microencapsulation technique of peptides and proteins. Since shear stress (during atomization) and heat (during drying) are the main mechanisms of the spray drying process, it is necessary to investigate the effect of these factors on the structure and stability of proteins and peptides (Sarabandi & Jafari, 2020).

10.3.3 TECHNICAL CHALLENGES

Among different microencapsulation techniques, spray drying is the most common method for stabilizing formulations and bioactive compounds by converting them from liquid to powder form (Assadpour & Jafari, 2019). But, unlike many other bioactive compounds, peptides and proteins have a complex structure and are particularly sensitive to the process conditions. The basic mechanism of a spray dryer includes shear stress and the conversion of the feed into fine droplets for the duration of atomization, thermal stress, and drying of droplets/particles (Ajmera & Scherließ, 2014). Therefore, one of the most significant challenges in the process of microencapsulation of peptides and proteins is their nature and structural properties. Due to the amphiphilic nature of peptides and proteins, they incorporate in the air–liquid interface for the period of atomization. In these conditions, the hydrophobic areas of the proteins are directed towards the air phase. This process results in structural changes, aggregation and denaturation of proteins, and the intensity of these changes are multiplied by the thermal stress (during drying of droplets/particles) (Lee, Heng, Ng, Chan, & Tan, 2011).

In the beginning, the optimization of spray drying parameters is essential to achieve the highest production efficiency and qualitative properties (Ajmera & Scherließ, 2014). However, performing the microencapsulation process under optimal conditions does not prevent the aggregation, denaturation, and loss of biological activity of the peptides and proteins. The main method to minimize these challenges is to use carrier agents to increase stability, reduce aggregation and denaturation, and maintain physicochemical, functional, and flow behavior of the dried proteins/peptides (Maury, Murphy, Kumar, Mauerer, & Lee, 2005). The mechanism of action of carriers in maintaining protein/peptide activity is mainly through the incorporation of these compounds into a carrier matrix/shell/film formation around the particles, water replacement, and glassy immobilization (Bürki et al., 2011). For that reason, several studies have investigated the effect of various carriers on the preservation of physicochemical properties and the prevention of protein-based drugs, and the stability of bioactive peptides during the spray drying process (Sarabandi & Jafari, 2020).

10.4 SUITABLE WALL MATERIALS FOR THE SPRAY DRYING ENCAPSULATION OF BIOACTIVE PEPTIDES/PROTEINS

In the previous sections, mention was made of the importance of using different carriers to stabilize and maintain the physicochemical and bioactive properties of proteins and peptides during spray drying and storage. Depending on the type and characteristics of the process and its application, specific compounds can be used as wall materials/carriers. At first, considering the use of these compounds in the food formulations, the carriers should be food-grade, biodegradable, non-toxic, and cost-effective (Fathi, Martin, & McClements, 2014). Some of the other important carrier characteristics include: (i) appropriate rheological properties and low viscosity at high concentrations; (ii) ability to incorporate and maintain bioactive compound(s) in the matrix; (iii) high solubility, especially in food-grade solvents used in the food industry; (iv) release of compounds during storage and dissolution; (v) non-reactivity with the microencapsulated compounds; and (vi) the effective protection of compounds under harmful environmental conditions such as temperature, humidity, light, and oxygen (Ray et al., 2016). For these reasons, the commonly used carriers in the spray drying microencapsulation of proteins/peptides can be categorized based on polysaccharides, proteins, and other low-molecular-weight stabilizing compounds. Some of the most important carriers and their functional properties are discussed in the following sections.

10.4.1 Polysaccharide-based Carriers

10.4.1.1 Starch and Its Derivatives

Starch is a complex polysaccharide made up of several glucose molecules as monomeric units joined together by glycosidic bonds. Hydrophilicity, digestibility, and high biodegradability are the main properties of these compounds. It must be pointed out that the lack of emulsifying ability is one of the limiting properties of these compounds. However, the structural modification and physicochemical or enzymatic alterations have been carried out to enhance emulsifying ability

(Jain, Khar, Ahmed, & Diwan, 2008). The most important types of such modified starches include cross-linked, acetylated, hydrolyzed, oxidized, and hydroxypropylated (Wandrey, Bartkowiak, & Harding, 2009). Each group of these starches has its own distinctive characteristics. For example, acetylated starches are highly resistant to enzymatic hydrolysis and hydrolyzed starches shows better stability against heat and oxidation. The ability to form a film is one of the considerable characteristics of oxidized starch. Also, cross-linked types of starch exhibit good stability against shear and thermal stresses (Shishir, Xie, Sun, Zheng, & Chen, 2018). A variety of dextrins and maltodextrins are the most common derivatives of starch. These compounds are derivatives of starch hydrolysis with different degrees of dextrose equivalent (DE). Compounds with DE < 20 are mainly called maltodextrin, whereas those with a higher DE are known as glucose and corn syrups. Dried dextrin products can be called pyrodextrin, brown, or yellow dextrins, and these can be used as carriers for spray drying encapsulation of bioactives such as peptides and proteins (Alvani, Qi, & Tester, 2011).

In contrast, maltodextrins are water-soluble, flavor-free, low-viscose, cost-effective, and digestible compounds with film-forming capability. These properties have led to the widespread use of these compounds as carriers in the microencapsulation of proteins/peptides. In addition to the declared applications, various studies have evaluated the impact of maltodextrin as a carrier to reduce the bitterness of a range of peptides and hydrolysates (Akbarbaglu et al., 2019; Yang et al., 2012). For example, some studies have focused on the encapsulation of peptides from casein (Sarabandi, Peighambardoust, Sadeghi Mahoonak, & Samaei, 2018c), flaxseed (Akbarbaglu et al., 2019), whey (Yang et al., 2012), and chicken meat (Kurozawa et al., 2009a) using a variety of starches and maltodextrin as carriers.

10.4.1.2 Cyclodextrins

Cyclodextrins contain a cone-shaped structure that can be used for the entrapment of a variety of hydrophobic bioactive compounds (into their internal cavity). Depending on the number of glucopyranose units, they are classified into three types of α (6 units), β (7 units), and γ (8 units) (Duchêne & Bochot, 2016). Due to the appropriate structural properties and cost-effectiveness of these carriers, they can be used in the microencapsulation of a diverse range of lipophilic compounds (Shishir et al., 2018). Another advantage of these carriers is their application in improving physicochemical properties, stability, and reducing the bitterness of peptides and hydrolysates. Cyclodextrin as a carrier has been used for spray drying of whey protein hydrolysate (Rukluarh, Kanjanapongkul, Panchan, & Niumnuy, 2019).

10.4.1.3 Cellulose and Its Derivatives

Cellulose is composed of β-D-glucose units attached through glycosidic bonds. Primary cellulose is slightly soluble in water. Consequently, in order to increase its possible uses in the food and pharmaceutical industries, structural modification is done by physical and chemical methods (Đorđević et al., 2016). One of the cellulose derivatives is carboxymethyl cellulose (CMC), which is a water-soluble, biodegradable, and stabilizing agent that can encapsulate a large number of pharmaceutical compounds (Rokhade et al., 2006). Other types of cellulose include methylcellulose (water-soluble and capable of film-forming), cellulose ethers (soluble in water with the ability to cover flavors and fragrances), hydroxypropyl celluloses (water-soluble and suited for film making and as an appropriate wall material for the microencapsulation of different oils), cellulose acetate (increasing the microencapsulation efficacy and targeted release of pharmaceutical compounds), cellulose nano-crystals (acid hydrolysis of cellulose and the cellulose derivatives to increase encapsulation efficiency and stabilizing agent of food and medicinal compounds), and nanofibrillar cellulose (immobilization drugs and stabilization of emulsions) (Shishir et al., 2018). Other applications of sodium carboxymethyl cellulose include microencapsulation by spray drying technique and the protective effect on alkaline phosphatase in pulmonary drug production (Li, Song, & Seville, 2010).

10.4.1.4 Gum Arabic

Gums are food-grade, biocompatible, biodegradable, non-toxic, economical, and efficient compounds, and, because of their functional properties, they are widely used in the food and pharmaceutical industries. One of the most important exudate gums (from Acacia tree) is gum Arabic. This gum is a complex heteropolysaccharide composed mainly of β-d-(1→4) mannopyranosyl, and α-d-(1→6) galactopyranosyl units (Fathi et al., 2014). Owing to its remarkable properties such as availability, flavorlessness, lack of odor, high solubility, emulsifying ability, and low viscosity at high concentrations, this gum can be used in the spray drying microencapsulation of proteins and peptides. Numerous studies have investigated the efficacy of gum Arabic either alone or in combination with maltodextrin, alginate, pectin, and other carriers in the microencapsulation of peptides and hydrolysates (Kurozawa, Park, & Hubinger, 2009b; Murthy et al., 2017; Subtil et al., 2014). For instance, investigations have been made of the microencapsulation of both pink perch hydrolysate (Murthy et al. (2017) and casein peptides (Subtil et al., 2014).

10.4.1.5 Pectin

Pectin is one of the most frequently used plant polysaccharides. The heterogeneous structure of pectin consists of linear units of alpha (1→4) galacturonic acid. Its functional properties vary depending on the degree of esterification. Pectin is a nontoxic and safe dietary fiber that is used as a thickening and gelling agent in various products (Sansone et al., 2011). It is also used as a carrier in the encapsulation of bioactive compounds and delivery systems. Microencapsulation of peptides and hydrolysates and masking the bitter taste of these compounds (Mendanha et al., 2009) are other applications of pectin. In one study, the complexation of nisin with pectin and alginate biopolymers was considered (Amara, Kim, Oulahal, Degraeve, & Gharsallaoui, 2017). The results showed that the use of carriers maintained the structural properties and antibacterial activity of nisin.

10.4.1.6 Chitosan

Chitosan is one of the most common polysaccharides in nature (after cellulose) that is broadly used in the food and pharmaceutical industries. It is a linear polysaccharide composed of randomly distributed β-(1→4)-linked D-glucosamine (deacetylated unit) and N-Acetyl-D-glucosamine (acetylated unit). Non-toxicity, film-forming ability, antibacterial activity, biocompatibility, and biodegradability are among the main advantages and properties of chitosan (Shishir et al., 2018). This biopolymer enhanced the stability and controlled release of nanoliposomes loaded with flaxseed peptides through coating formation (Sarabandi & Jafari, 2020). Moreover, coating with chitosan increases the liposomal membrane resistance in contrast to drying, freezing, and thawing (Sarabandi & Jafari, 2020). Further applications of chitosan include its role as an absorption enhancer, increasing the physicochemical stability and recovery of spray dried calcitonin (Yang et al., 2007). Additionally, chitosan nanoparticles improve the loading capacity, absorption, and bioavailability of drugs in oral, nasal, and skin applications (Wang, Jung, & Zhao, 2017). On the other hand, spray drying of bioactive peptides derived from spirulina was investigated using chitosan as the wall material (Aquino et al., 2020).

10.4.1.7 Sodium Alginate

Alginates are linear, anionic, and hydrophilic polysaccharides that contain different amounts of (1→4)-linked β-d-mannuronic acid and α-l-guluronic acid residues. This polysaccharide is extracted from brown seaweed and produced in the form of sodium and calcium salts. Alginate salts can be widely used for the microencapsulation of bioactive peptides and proteins by spray drying technique, owing to their low cost, non-toxicity, biocompatibility, hydration in hot and cold water, and viscosity (Sarabandi & Jafari, 2020). Applications of sodium alginate in the stabilization of bioactive peptides include its role as a carrier in spray drying of hydrolyzed whey protein (Ma et al., 2014).

10.4.1.8 Stabilizers and Other Carbohydrates

In recent years, the use of mono- and disaccharides as carriers has been particularly important in stabilizing therapeutic peptides and inhalable powders. Among these compounds are trehalose, mannitol, sorbitol, lactose, and sucrose. These compounds biologically stabilize peptides and prevent their aggregation and denaturation through water replacement and glassy immobilization (Bürki et al., 2011). For example, sorbitol and trehalose (Maury et al., 2005), trehalose (Bürki et al., 2011), lactose and mannitol (Grenha, Seijo, & Remunán-López, 2005), trehalose (Adler, Unger, & Lee, 2000), and mannitol (Chan et al., 2004) were used for the spray drying and stabilization of the immunoglobulin G, β-galactosidase, insulin, BSA and calcitonin, respectively.

10.4.2 PROTEIN-BASED WALL MATERIALS

In addition to polysaccharides, some of the proteins and protein-based carriers can also be used to encapsulate a variety of drugs and hydrolysates using the spray drying technique. Due to their film-forming ability and emulsifying activity, proteins result in a decrease in the adhesion of syrups and stabilization of the extracts and different antioxidants during spray drying (Fang & Bhandari, 2012; Sarabandi, Peighambardoust, Sadeghi Mahoonak, & Samaei, 2018c). The stabilizing role of these compounds can be attributed to the preferential migration to the droplet/particle surface during atomization and the formation of shells or protective walls around the particles (Jayasundera, Adhikari, Howes, & Aldred, 2011). The most important proteins used for the spray drying encapsulation of bioactive peptides and proteins are soybean proteins, milk proteins (i.e., caseins and whey proteins), gelatin, cereal proteins (e.g., gluten and rice proteins), and some amino acids (e.g., arginine, glycine, and histidine).

10.4.2.1 Soy Proteins

The main soybean proteins are glycinin and conglycinin. Soy proteins are valuable for food and pharmaceutical applications because of some advantages and functional properties such as biodegradability, availability, nutritional value, emulsifying, water and oil retention, gel, and film formation (Ortiz et al., 2009). Thanks to the above-mentioned characteristics, soy protein isolates can be used for the microencapsulation of different peptides. The film-forming ability, increasing glass transition temperature, reducing the level of hygroscopicity, and bitterness are the main benefits of this compound that make it proper for the microencapsulation of casein hydrolysates (Fávaro-Trindade, Santana, Monterrey-Quintero, Trindade, & Netto, 2010) and soybeans (Wang et al., 2020).

10.4.2.2 Milk Proteins

Caseins (types αs1, αs2, β, and κ-casein) are the main proteins in cow's milk and encompass about 80% of the total protein content in this liquid. Caseins are rich sources of nutrients, functional, and bioactive peptides (López-Fandiño, Otte, & VanCamp, 2006). The emulsifying properties of caseins make these proteins appropriate carriers for the microencapsulation of bioactive peptides. In addition, one of the richest sources of protein comes from the by-products of the dairy industry; i.e., whey proteins (mainly β-lactoglobulin (β-lg) and α-lactalbumin (α-la)). These proteins can be widely used as a drying aid or carrier in spray drying and microencapsulation of various concentrates and adhesive products, including bioactive peptides and proteins (Bazaria & Kumar, 2016; Bhusari, Muzaffar, & Kumar, 2014) for the sake of high nutritional value, cost-effectiveness, digestibility, film-forming ability, and high emulsifying ability. The use of whey protein concentrates (WPC) in combination with alginate decreased the bitterness of whey protein hydrolysates and improved the physicochemical properties of their spray dried particles (Ma et al., 2014).

10.4.2.3 Gelatin

Functional advantages of gelatin include its biodegradability, biocompatibility, high water solubility, film-forming ability, emulsifying power, and high stability (Akhavan Mahdavi, Jafari, Assadpoor, & Dehnad, 2016). Such properties have led to the widespread use of this protein as a carrier for the encapsulation and controlled release of bioactive peptides such as hydrolyzed whey (Gómez-Mascaraque, Miralles, Recio, & López-Rubio, 2016) and casein (Favaro-Trindade, Santana, Monterrey-Quintero, Trindade, & Netto, 2010).

10.4.2.4 Cereal Proteins

Zein is an insoluble, biocompatible, and biodegradable protein that is able to load a variety of bioactive food and pharmaceutical compounds, including peptides (Yang et al., 2017). Wheat proteins (mainly gliadin and glutenin) can form gels and films, and encapsulate, stabilize, and control the release of various compounds (Shishir et al., 2018). Hordenine (a protein naturally found in barley) acts as an effective carrier in the encapsulation of lipophilic compounds under its emulsifying activity and film-forming ability. Additionally, Amaranth protein (*Amaranthus hypochondriacus*) can be used to stabilize bioactive compounds due to its reasonable price, film, and gel-forming ability along with emulsifying and foaming power (Suarez & Añón, 2018).

10.4.2.5 Amino Acids

One of the applications of spray drying is the microencapsulation of pharmaceutical peptides, especially for inhalation medications. But, like other peptides and hydrolysates, these compounds are also susceptible to aggregation, thermal degradation, and loss of their biological activity during spray drying and storage (Sarabandi & Jafari, 2020). Therefore, the preservation of physicochemical properties and their biological activity are essential for pharmaceutical applications (Bilati, Allémann, & Doelker, 2005). In addition, because of some notable properties such as solubilizing, bulking, and buffering capabilities, amino acids may also cause reduction of the aggregation of peptides, and retention of conformational structure and their biological stability (Ajmera & Scherließ,

2014). Glycine (Pikal-Cleland, Cleland, Anchordoquy, & Carpenter, 2002), arginine (Das et al., 2007), and histidine (Hasija, Li, Rahman, & Ausar, 2013) are some examples of the amino acids that can stabilize peptide-based drugs.

10.4.3 OTHER ENCAPSULATING AGENTS

One of the unstable processes of proteins and peptides is their migration to the water–air interface during atomization, which causes denaturation and conformational changes. Aggregation of proteins at the interface results in the loss of their functional and biological properties (Sarabandi & Jafari, 2020). The surface activity of proteins and peptides is affected by the size, composition, and molecular weight of these compounds. This phenomenon is especially important for hydrolysates, because the enzymatic hydrolysis of proteins increases the surface activity and aggregation of these compounds at the interface (Drusch, Hamann, Berger, Serfert, & Schwarz, 2012). The effect of enzymatic hydrolysis to increase the surface activity of peptides can be observed in the foaming capacity of these compounds (Jamdar et al., 2010). The high surface activity of peptides and hydrolysates makes these compounds susceptible to processes such as aggregation and denaturation during atomization and thermal shear stresses. Accordingly, the incorporation of compounds with a higher surface activity like low-molecular-weight surfactants on the interface can be considered (Sarabandi & Jafari, 2020). It should be stated that surfactants mainly reduce the effect of shear stress on the aggregation and denaturation of these compounds by increasing the viscosity of the feed, competing with the protein molecules for incorporation in the interface, sticking to the protein/peptide molecules, and inhibiting or reducing the motion of them (Wang et al., 2010).

10.5 CHARACTERIZATION OF SPRAY DRIED PEPTIDES AND PROTEINS

In the previous Sections (10.1 and 10.2) of this chapter, mention was made of the steps of the spray drying process for encapsulation of bioactive peptides and proteins. The production efficiency, physicochemical and functional properties, flow behavior, physical stability, color properties, sensory and microbial quality, long shelf-life, as well as the preservation of bioactivity of the microencapsulated peptides/proteins are noticeably affected by different parameters during the spray drying process. The most important parameters that affect these properties are inlet and outlet air temperatures, dryer air humidity, feed flow rate, and atomizer speed and pressure. Hence, evaluating variables and achieving optimal process conditions will result in maximum product yield with appropriate qualitative characteristics(Wang & Selomulya, 2020). In addition to process parameters, the concentration, composition, and type of carriers are other factors that influence the spray drying process and the characteristics of the end product (Sarabandi, Sadeghi Mahoonak, Hamishekar, Ghorbani, & Jafari, 2018d). Spray dried compounds can be evaluated in terms of microbial and physical properties such as bulk, particle and tapped densities, flowability, instant characteristics, hygroscopicity, degree of caking, thermo-stability, solubility index, occluded air in tissue, particle size distribution, biological activity, morphology, color, sensory characteristics and dispersibility, wettability, and wettability indexes (Chegini & Ghobadian, 2005). In the following sections, the influence of process parameters of spray drying on the properties of powders containing peptides and proteins is discussed.

10.5.1 PRODUCTION YIELD AND PHYSICAL PROPERTIES

Numerous factors are responsible for the efficiency of the spray drying process as well as the physicochemical properties of peptides containing powders: powder production efficiency (efficiency index of process variables in reducing particle adhesion, conversion, and complete recovery of feed to powder), moisture content and water activity (physical and microbiological stability index, crystallization behavior and glass transition temperature of spray dried peptides), bulk and tapped densities (economic indicator affecting packaging, storage, and transportation), flow behavior (effective indexes on the handling), color indices (demonstrating the effect of process variables on the quality of the product and chemical reactions), solubility, and wettability (influential indexes on the performance characteristics and reconstitution capability of instant powder) (Sarabandi & Jafari, 2020). In various studies, the influence of spray drying process parameters (e.g., inlet and outlet air temperature, type, composition, and carrier concentration, as well as feed and inlet air flow rate) has been investigated. Bürki et al. (2011) investigated the efficiency of powder production as an economic indicator of the process during the spray drying of β-galactosidase (as a protein therapeutic model in the production of respirable powders). Besides, trehalose was added as a stabilizer in this study. Process variables included inlet air temperature (80°C–120°C), nozzle pore size (4–7 µm), and ethanol concentration (0%–20%). Production efficiency varied between 60% and 93%, where the highest values of this index were obtained at the lowest inlet air temperature and pore size and highest ethanol concentration.

However, in another study, the effect of inlet air temperature (110°C, 130°C, and 150°C) on the properties of encapsulated lactate dehydrogenase with trehalose (as the carrier) was studied and the highest production efficiency (78%) was obtained at 150°C (Adler & Lee, 1999). Akbarbaglu et al. (2019) reported that increasing maltodextrin to peptide ratio caused a reduction in moisture content (from 3.5% to 2.5%), water activity (from 0.33 to 0.28), bulk density (from 0.44 to 0.33 g/mL), and angle of repose (from 33°C to 29°C) in the spray dried flaxseed protein hydrolysates. In another study, spray drying microencapsulation resulted in increasing moisture content and water activity (Ortiz et al., 2009). These findings indicate the effect of concentration and type of the carrier on the retention of moisture content and water activity of the compounds. Regarding the density of the powders, contrary to the aforementioned results, the bulk density of the hydrolyzed whey powders after microencapsulation (with maltodextrin and the combination of maltodextrin and β-cyclodextrin) was increased from 0.351 to 0.451 g and 0.478 g/ml, respectively (Yang et al., 2012).

In view of the variations in numerous reports, it can be stated that particle production with different sizes, size distribution, moisture content, porosity, and different morphological characteristics affect the bulk and tapped density of the powders (Sarabandi, Gharehbeglou, & Jafari, 2019a). Solubility is one of the most important functional properties, which has a significant impact on the nature and type of carrier. According to the results reported by Akbarbaglu et al. (2019), increasing the concentration of maltodextrin caused a rise in solubility (from 92% to 96%). On the other hand, the use of soy protein isolates as the carrier reduced the solubility of casein hydrolysates from 100% to 86.7% (Ortiz et al., 2009). These described outcomes demonstrate the influence of type, composition, carrier concentration, and inlet air temperature on the physical properties of spray dried peptides. The results of some other studies are also summarized in Table 10.1.

10.5.2 Physical Stability (Hygroscopicity and T_G)

In addition to accelerating chemical reactions and microbial spoilage, hygroscopicity results in the physical inconsistency of the powders through their aggregation and structural changes (Sarabandi, Sadeghi Mahoonak, Hamishekar, et al., 2018d). On the other hand, aggregation and caking, adhesion, agglomeration, and the structural collapse of particles containing pharmaceutical peptides as a consequence of moisture absorption cause the loss of flow properties and potential use in inhalation applications (Möbus, Siepmann, & Bodmeier, 2012). Furthermore, there is a direct relationship between the degree of hydrolysis and the absorbed moisture content with the glass transition temperature of the peptides; this would mean that increasing the hydrolysis rate and decreasing the molecular weight will cause an increase in the hygroscopicity resulting from the availability of more hydrophilic groups. Additionally, these groups improve the plasticizing properties and decrease the T_g of the compound(s) (Rao, Bajaj, et al., 2016a). Therefore, controlling these two factors through the spray drying microencapsulation within different carriers is necessary. The incorporation of peptides and hydrolysates in the carrier matrix with low adhesion and high T_g stabilizes the peptides (Favaro-Trindade et al., 2010). Several studies have investigated the effect of different spray drying parameters and carriers on the T_g and hygroscopicity of peptides and proteins (Table 10.2).

TABLE 10.1

The effects of encapsulation parameters on the physical and functional properties of spray dried bioactive peptides.

Peptide Type	Carrier Agent	Spray Drying Conditions	Results	Reference
Casein hydrolysates	MD	Carrier to hydrolysate ratio: 60: 40, inlet and outlet air temperatures: 130°C and 70°C, respectively.	The amount of moisture, water activity, and bulk density of powders were not affected by the type of peptide. The solubility of powders was more than 94%.	Sarabandi, Peighambardoust, Sadeghi Mahoonak, and Samaei (2018c)
Flaxseed protein hydrolysates	MD	Carrier to hydrolysate ratios: 1: 1–3: 1 w/w, inlet and outlet air temperatures: 130°C and 75°C, respectively.	Production efficiency, flowability, and solubility improved with increasing carrier concentrations. The amount of moisture and density of the particles decreased.	Akbarbaglu et al. (2019)
Whey protein concentrate hydrolysate	γ-CD	Carrier to hydrolysate ratios: 20: 80–60: 40w/w, inlet and outlet air temperatures: 180°C and 127°C, respectively.	Physical characteristics such as moisture content, water activity, density, and solubility of powders were affected by carrier concentrations.	Rukluarh et al. (2019)
Casein hydrolysates	GA	Carrier to hydrolysate ratios: 70: 30 to 90: 10 w/w, inlet and outlet air temperatures: 140°C and 110°C, respectively.	The physical stability and solubility of peptides were improved by spray drying. Also, the amount of moisture and water activity of the produced powders by the carrier were decreased.	Subtil et al., 2014
Chicken meat protein hydrolysate	MD and GA	Carrier concentrations: 10%–30% w/w, inlet and outlet air temperatures: 180°C and 91°C–102°C, respectively.	Increasing the carrier concentration led to a decrease in moisture content and particle density. Some physical properties such as the amount of moisture and density of the powders decreased with increasing carrier concentration.	Kurozawa et al. (2009a)
Whey protein concentrate hydrolysate	MD and β-CD	Carrier to hydrolysate ratio: 70: 30 w/w, inlet and outlet air temperatures: 200°C and 90°C, respectively.	Spray drying and type of carrier did not affect the solubility of the hydrolyzed.	Yang et al. (2012)
Casein hydrolysates	SPI	Carrier to hydrolysate ratios: 70: 30–80: 20 w/w, inlet and outlet air temperatures: 140°C and 110°C, respectively.	Encapsulation of peptides with soy protein isolates led to an increase in the amount of moisture and water activity of the particles. But using a carrier reduced the solubility.	Molina Ortiz et al. (2009)
Casein hydrolysates	Gelatin and SPI	Carrier mixture to hydrolysate ratios: 70: 30–80: 20 w/w, inlet and outlet air temperatures: 140°C and 110°C, respectively.	Powder properties such as moisture content and solubility were affected by the type and ratio of the carrier.	Favaro-Trindade et al. (2010)

γ-CD: γ-Cyclodextrin; SPI: Soybean protein isolate; GA: Gum Arabic; MD: Maltodextrin; SA: Sodium alginate; β-CD: β-Cyclodextrin.

TABLE 10.2

The effects of encapsulation parameters on the physical stability of spray dried bioactive peptides.

Peptide Type	Carrier Agent	Spray Drying Conditions	Results	Reference
Casein hydrolysates	MD	Carrier to hydrolysate ratio: 60: 40, inlet and outlet air temperatures: 130°C and 70°C, respectively.	The use of maltodextrin significantly reduced the hygroscopicity of casein hydrolysates.	Sarabandi, Peighambardoust, Sadeghi Mahoonak, and Samaei (2018c)
Flaxseed protein hydrolysates	MD	Carrier to hydrolysate ratios: 1: 1–3: 1 w/w, inlet and outlet air temperatures: 130°C and 75°C, respectively.	The hygroscopicity of pure hydrolysates decreased from about 40% to 17% after using the carrier. The value of this index was affected by the carrier concentration.	Akbarbaglu et al. (2019)
Whey protein concentrate hydrolysate	WPC and SA	Carrier mixture (WPC 34: 1 SA) to hydrolysate ratio: 70: 30 w/w, inlet and outlet air temperatures: 200°C and 90°C, respectively.	Hygroscopicity was influenced by microencapsulation. The use of carriers reduced the hygroscopicity value of hydrolyzed from 30% to 20%.	Ma et al. (2014)
Casein hydrolysates	GA	Carrier to hydrolysate ratios: 70: 30 to 90: 10 w/w, inlet and outlet air temperatures: 140°C and 110°C, respectively.	The use of gum Arabic did not affect the hygroscopicity of the sprayed dried hydrolysates.	Subtil et al., 2014
Chicken meat protein hydrolysate	MD and GA	Carrier concentrations: 10-30% w/w, inlet and outlet air temperatures: 180°C and 91°C–102°C, respectively.	Hygroscopicity of capsulated hydrolysates with maltodextrin and gum Arabic decreased from 40% to 16% and 21%, respectively. The value of the glass transition temperature increased from 44°C to 137°C after the use of maltodextrin.	Kurozawa et al. (2009a)
Whey protein concentrate hydrolysate	MD and β-CD	Carrier to hydrolysate ratio: 70: 30 w/w, inlet and outlet air temperatures: 200°C and 90°C, respectively.	The combination of carriers led to a decrease in the hygroscopicity value from about 64% to 36%. Also, the glass transition temperature of peptide increased with the use of carriers.	Yang et al. (2012)
Casein hydrolysates	SPI	Carrier to hydrolysate ratios were 70: 30–80: 20 w/w; inlet and outlet air temperatures were 140°C and 110°C, respectively.	Hygroscopicity of the hydrolyzed particles increased from about 53% to more than 106% after microencapsulation and use of the carrier.	Molina Ortiz et al. (2009)
Casein hydrolysates	Gelatin and SPI	Carrier mixture to hydrolysate ratios: 70: 30–80: 20w/w, inlet and outlet air temperatures: 140°C and 110°C, respectively.	Hygroscopicity of the peptides decreased from about 53% to 28% after microencapsulation.	Favaro-Trindade et al. (2010)
Casein hydrolysates	MD (DE 10-20)	Carrier to hydrolysate ratio: 90: 10 w/w, inlet and outlet air temperatures: 140°C and 110°C, respectively.	The glass transition temperature and hygroscopicity of the peptides increased and decreased, respectively, after microencapsulation.	Rocha, Trindade, Netto, and Favaro-Trindade (2009)

γ-CD: γ-Cyclodextrin; SPI: Soybean protein isolate; GA: Gum Arabic; MD: Maltodextrin; SA: Sodium alginate; β-CD: β-Cyclodextrin.

Some researchers have focused on the influence of moisture content and the relationship of water activity with T_g of peptide-containing powders. As an example, Kurozawa et al. (2009a) evaluated the effect of maltodextrin and gum Arabic on moisture absorption and T_g of spray dried chicken hydrolysates. They reported that the hygroscopicity of microencapsulated powders with maltodextrin and gum Arabic decreased from 40.9% to 15.9% and 21.2%, respectively. Also in the non-carrier hydrolysate, the process of liquefaction occurred in a_w=0.176; by contrast, the use of different ratios of maltodextrin (10%, 20%, and 30% w/w) increased the caking point in the water activities of 0.529, 0.689, and 0.753, respectively. Furthermore, water activities of 0.689, 0.753, and 0.843 were starting points of collapse and fluidization of the particles. Correspondingly, the T_g values versus solid material showed a high correlation with the Gordon-Taylor model.

Kurozawa et al. (2009a) indicated that a higher amount of absorbed moisture could lead to a decrease in the value of T_g and the use of high molecular weight carriers such as maltodextrin and gum Arabic increased the T_g of the particles. Properly speaking, the amount of T_g in the pure peptides increased from 44°C to 137°C after using maltodextrin. Chan et al. (2004) investigated the stability of spray dried salmon calcitonin at different RH values. Mannitol was used as the carrier and the influence of moisture content and degree of crystallization on the structural changes and protein aggregation during storage were scrutinized (Chan et al., 2004). According to their results, the preservation of purified proteins at an RH of > 20% increased the aggregation of the peptides. However, the use of the carrier reduced the aggregation and adhesion of the particles. In another study, the decrease in the hygroscopicity and the T_g of whey protein hydrolysate (Yang et al., 2012) and casein (Rocha et al., 2009) were reported after using different carriers. However, contrary to the above research, the use of soy protein isolates resulted in a duplication of the hygroscopicity of the spray dried casein hydrolysate (Ortiz et al., 2009).

In some studies, the effect of moisture absorption of spray dried proteins on various factors (e.g., modification in the structure of amorphous to crystalline, and morphological) has been investigated. Li, Woo, and Selomulya (2016) researched the effect of different relative humidities (11% to 94%) and milk composition on solubility, protein structure, and the reaction of proteins with lactose and water. Two types of model milk powder were used to compare commercial milk powder. Correspondingly, browning, and the denaturation of proteins, were evaluated. According to their outcomes, the model sample was more stable than commercial powders and due to the presence of minerals in commercial samples, protein denaturation, browning, and loss of solubility were higher than the model sample. X-ray diffraction was used for the evaluation of the effect of moisture absorption (Figure 10.2a) on the deformation of various powders such as model emulsions and commercial spray dried powders. It was found that the greater contents of protein (casein and whey protein) in model emulsion (No. 1) than the treated emulsion (No. 2) led to a delay in its crystallization, due to the reaction with lactose. However, storage of model specimens (amorphous powders) for two months at high RH conditions resulted in moisture absorption and phase change, as a result of lactose crystallization (especially at an RH of 42%) (Figure 10.2b and c). Moreover, differential scanning calorimetry (DSC) evaluation indicated the structural rearrangement of the non-covalent bonds, owing to heat absorption. In the commercial milk powder, a narrow peak was observed in the temperature range of 55°C to 65°C and, according to reported results, fresh milk powder retained its natural (folded) conformation structure. Yet there was no peak observed in the case of the sample maintained at 87% RH for six months, meaning the complete unfolding of the protein structure and its denaturation (Figure 10.2d and e). However, these changes could be observed in the surface structure of the particles (conversion of the flat surface of the particle to the rough surface) after moisture absorption (Figure 10.2f and g).

10.5.3 Thermal, Functional, Chemical, and Structural Stability

As mentioned in the previous Sections (10.3.1–10.3.3), the structure, amphiphilicity, and the surface activity of proteins and peptides lead to the migration and incorporation of these compounds

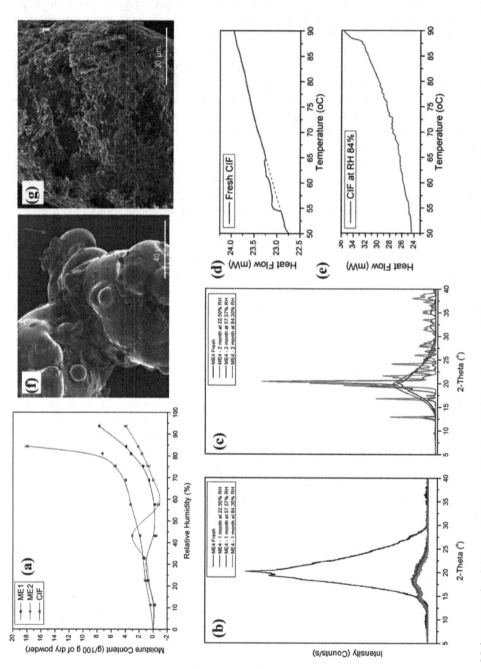

FIGURE 10.2 Moisture sorption isotherm (a), X-ray diffraction (XRD) patterns at a different relative humidity (RH) when samples stored for one month (b) and two months (c); differential scanning calorimetry (DSC) results for fresh commercial infant formula CIF; (d) and CIF stored at 84% RH for six months (e); scanning electron micrographs of fresh CIF (f) and CIF after two months storage (g). ME_1 (model emulsion 1) and ME_2 (model emulsion 2) and CIF at room temperature. Reprinted from Li et al. (2016), with permission.

at the interface, changes in the secondary structure, aggregation, denaturation, and loss of their health-promoting and functional properties (Sarabandi & Jafari, 2020). Therefore, the physical and functional properties, biological activity during the spray drying process and maintenance period, and assessment of the possible changes in chemical structure are the essential factors. In this regard, various investigations have been carried out about the effect of process parameters and stabilizing compounds on such characteristics, maintaining the functional properties of proteins and hydrolysates, and their impact on the structural and conformational changes. For example, the effect of spray and freeze drying processes on the physicochemical and functional properties of rice dreg protein was investigated by Zhao et al. (2013). The findings showed that the spray dried proteins had a better foaming capacity (127%) compared to the freeze-dried samples (118%), although the freeze dried samples had more water and oil storage capacity and thermal stability. These specimens had higher beta-tern structures and less random helix than the spray dried proteins. These changes indicated the presence of ordered conformations and integrate structures in freeze dried proteins (Zhao et al., 2013).

In some other studies, the efficiency of spray and freeze drying processes on maintaining the functional properties of proteins has been investigated and compared. As an illustration, freeze drying has been reported to be a better method to preserve the functional properties of collagen hydrolysates (Zeng, Zhang, Adhikari, & Mujumdar, 2013). Similar results were reported about the effect of spray and freeze drying processes on the functional characteristics of egg white hydrolysates. In this way, the foaming activity of the hydrolysate was improved under freeze drying, but spray drying decreased both foaming and emulsifying abilities (Chen, Chi, & Xu, 2012).

Murthy et al. (2017) investigated the effect of the spray drying process using maltodextrin and gum Arabic (as the carriers) on the physicochemical properties of pink perch meat protein hydrolysate. Based on their obtained results, the emulsifying power of hydrolyzed increased from about 9 to 76 m^2/g after microencapsulation with carriers; however, this process did not affect foaming capacity. Considering the mechanisms and influential parameters of the spray drying process, the cause of such changes can be attributed to the thermal damage, aggregation, and denaturation of proteins during atomization and drying processes.

Changes in the molecular structure of peptides and proteins lead to a loss of bioactivity and bioavailability of these compounds. The most common mechanisms of inactivation are unfolding and aggregation. Taking this into account, research has been conducted on the impact of the spray drying process on the thermal stability and structural and chemical changes in peptides. Fourier transform infrared (FTIR) spectroscopy has been used as a non-destructive and accurate method to evaluate the structural changes and functional groups of peptides during spray drying. As a case in point, the structural properties and functional groups of maltodextrin-based carriers for spray drying of hydrolyzed flaxseed (Figure 10.3) have been studied. The most important structural regions of peptides are the spectra in the frequency at 3296 cm^{-1} (NH stretch) and 1800–800 cm^{-1} (bands that form the amide groups including C=O, NH, and CN). This region includes amide I vibrations (1700–1600 cm^{-1}) and amide II vibrations (1600–1500 cm^{-1}), respectively, related to the stretch of carbonyl bonds (C=O) and the deformation of NH bands and the stretch of CN bonds. Structural changes and major peaks were investigated after the incorporation of peptides into the maltodextrin structure (the carrier). The results of the FTIR spectroscopy for the microencapsulated peptides indicated a combination of the main peaks in pure hydrolyzed and maltodextrin. For example, the peaks at the 3296 cm^{-1} and 2959 cm^{-1} frequencies in the hydrolyzed after microencapsulation shifted to 3352 cm^{-1} and 2931 cm^{-1}, respectively (the distance between the hydrolyzed and maltodextrin frequencies). The peak of the amide I region (stretch C = O) overlapped at 1661 cm^{-1} after microencapsulation of the peptide with the O-H stretch in maltodextrin. After the microencapsulation process, the amide II vibration band without changing the frequency and with less intensity appeared in the spray dried powder at a frequency of 1533 cm^{-1} (not present in the maltodextrin). Also, the peaks at the 1155 cm^{-1} and 1017 cm^{-1} frequencies in maltodextrin appeared after peptide microencapsulation at the 1149cm^{-1} and 1021 cm^{-1} frequencies, respectively. Furthermore, there was no new bond in the spray

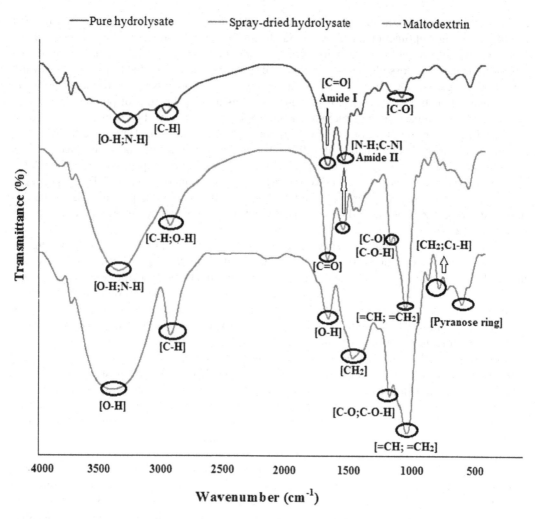

FIGURE 10.3 Fourier transform infrared (FTIR) spectra of the protein hydrolysate, spray dried peptide, and pure maltodextrin. Reprinted from Akbarbaglu et al. (2019), with permission.

dried samples, indicating a uniform distribution of peptide in the maltodextrin matrix without interaction between the core and the wall (Akbarbaglu et al., 2019). Similar consequences were reported after the spray drying microencapsulation of casein hydrolysate (Sarabandi, Sadeghi Mahoonak, Hamishekar, et al., 2018d).

Zhang et al. (2018) investigated the effect of spray and freeze drying processes on the stabilization of trypsin enzyme in trehalose particles. The secondary structure of the enzyme and the conformational changes play an important role in its biological activity. The amide I region in the trypsin enzyme at the frequency of 1600–1720 cm^{-1} and its main constituents include C=O stretch bonds. The main spectra in the amide I region of trypsin include the frequencies 1682 cm^{-1} (beta-helix), 1662 cm^{-1} (alpha helix), and 1638 cm^{-1} (β-sheet). Their results showed that the secondary structure of pure trypsin was intensely altered after spray drying. For instance, the peak severity of alpha-helix and β turns decreased. Furthermore, the position of the β-sheets converted to 1633 cm^{-1}, indicating a structural change. These changes were reported because of dehydration of β-sheets (hydrated carbonyl group) and loss of hydrogen bonds between and within the molecule. In addition, the peak severity of β-turns, β-sheets, and alpha-helix structures in the produced particles by trehalose improved in comparison with the pure spray dried enzyme. These results indicated the maintenance of the structural properties, as well as the biological activity of the enzyme. During the spray drying process, trehalose formed a glassy structure around enzyme molecules by making hydrogen bonds. In addition, the replacement of water molecules at the surface of trypsin molecules with trehalose during the dehydration process resulted in a reduction of motion, maintains the secondary structure and biological activity of the enzyme (Zhang et al., 2018).

In consideration of the specified studies, the importance of maintaining structural properties, and the impact of chemical reactions in functional groups is noticeable. Therefore, process optimization and the use of stabilizing agents are essential to preserve the secondary structure of proteins and prevent any possible aggregation. Maury et al. (2005) performed research about the effect of various sorbitol and trehalose ratios as stabilizing agents on the aggregation and chemical structure of immunoglobulin G (IgG) during the spray drying process. After the regeneration of spray dried pure protein, the aggregation increased from 1% to 17%, meaning that the spray drying process of pure proteins could lead to their aggregation and bioactivity loss. According to this report, the β sheet structure was detected in the spray dried solid; however, the use of sorbitol increased the stability of IgG during spray drying and reduced the aggregation to 0.7% after particle reconstruction.

Schüle, Frieß, Bechtold-Peters, and Garidel (2007) evaluated the influence of spray drying and mannitol as a carrier on the secondary antibody structure (IgG$_1$). The effect of antibody to mannitol ratios between 20: 80 and 100: 0 on protein stability was evaluated. Their results illustrated that high levels of mannitol (80%–50%) led to the instability of IgG$_1$ and increasing the aggregates and there was no structural change in the mannitol/IgG$_1$ solution after FTIR evaluation. Besides, the evaluation of the chemical structure and functional groups of the powders after reconstitution indicated that the secondary structure of the protein was maintained during the spray drying process. Correspondingly, the aggregation of proteins in the solution containing antibodies/mannitol did not indicate any structural changes in the protein. In another study (Chan et al., 2004), the use of mannitol prevented the formation of β-sheets structures as a result of moisture absorption and the aggregation of salmon calcitonin spray dried particles.

10.5.4   BIOLOGICAL AND ANTIOXIDANT STABILITY

Peptide and protein complex structures are sensitive to a variety of unstable reactions during thermal processes, shear stresses, and environmental destructive factors. Among the noticeable chemical reactions are isomerization, deamidation, racemization, and oxidation, while the main physical destabilizing mechanisms include aggregation, denaturation, and precipitation (Bilati et al., 2005). Loss of primary structure, changes in the effective functional groups, and the production of new compounds are the key factors that have substantial impacts on the biological and therapeutic activity of proteins and bioactive peptides (Sarabandi & Jafari, 2020). Thus, microencapsulation of these compounds is essential in the structures of stabilizing compounds and various carriers. Even though thermal and shear stresses, surface activity of peptides, and their incorporation at the water/air interface are important factors in denaturation that can induce the loss of biological activity of these compounds.

Consequently, in various studies, the effect of spray drying parameters on the antioxidant and biological activity of proteins and peptides has been investigated (Sarabandi, Gharehbeglou, & Jafari, 2019a), as summarized in Table 10.3. These studies have focused on parameters and the effects of carriers/stabilizers. Moreover, the effect of process air temperature, the stability of encapsulated lactate dehydrogenase (as the carrier) during the spray drying process, and maintenance have been investigated. The drying process was performed at different temperatures (110, 130, and 150°C), and it was found that increasing the temperature led to a further decrease in the activity of the enzyme. However, the use of higher temperatures, owing to the reduction of moisture content, triggered the stability of the enzymes during storage (Adler & Lee, 1999).

Sarabandi, Peighambardoust, Sadeghi Mahoonak, and Samaei (2018c) investigated the effect of the hydrolysate type (produced with alcalase and pancreatin) on maintaining the antioxidant activity

TABLE 10.3

The effects of encapsulation parameters on the biological and antioxidant stability of spray dried bioactive peptides

Peptide Type	Carrier Agent	Spray Drying Conditions	Results	Reference
Casein hydrolysates	MD	Carrier to hydrolysate ratio: 60: 40, inlet and outlet air temperatures: 130°C and 70°C, respectively.	Depending on the type of antioxidant test, between 77%–99% of the antioxidant and chelating activities of peptides were maintained.	Sarabandi, Peighambardoust, Sadeghi Mahoonak, and Samaei (2018c)
Flaxseed protein hydrolysates	MD	Carrier to hydrolysate ratios: 1: 1–3: 1w/w, inlet and outlet air temperatures: 130°C and 75°C, respectively.	Increasing the carrier concentration led to the maintenance of the antioxidant activity of peptides. Between 69%–97% of the antioxidant activity of hydrolysates was maintained.	Akbarbaglu et al. (2019)
β-Galactosidase	Trehalose	Inlet air temperature: 80°C–120°C, nozzle hole size: 4–7 micrometers, and ethanol concentration: 0%–20%.	Enzyme activity (78%–100%) and its stability during 3 weeks of storage (55%–92%) were affected by process variables.	Bürki et al. (2011)
Trypsin	Trehalose	The core: carrier ratio of 1: 1.	The crystalline structure of trehalose and the hydrogen bonds with trypsin resulted in the preservation of 80% of the enzymatic activity during the process.	Zhang et al. (2018)
Pink perch hydrolysate	MD and GA	Carrier mixture: MD 2: 1 GA, inlet and outlet air temperatures: 160°C and 80°C, respectively.	Inhibitory activity against DPPH radical and chelation of metal ions decreased from 66% to 49% and 26% to 25%, respectively, after microencapsulation with carriers.	Murthy et al. (2017)
Soy milk	-	Inlet air temperature: 200°C–140°C, the feed flow rate: 3–12 mL/min, and feed concentration: 25%–40%.	Inhibitory activity capacities against ABTS and DPPH radicals at 140°C and 200°Cwere approximately 6.8 and 5.3 mg Trolox/g and 9.7 and 7.1 mg Trolox/g, respectively.	Wang et al., 2017

MD: Maltodextrin; GA: Gum Arabic.

of casein hydrolysates (degree of hydrolysis) during the spray drying process, where maltodextrin (1: 1 ratio) was used as a carrier. According to their report, 90.36% to 98.52% of DPPH (1,1-diphenyl-2-picrylhydrazyl) free radical scavenging activity, 77.48% to 91.62% of ABTS (2,2′-azino-bis (3-ethylbenzothiazoline-6-sulfonic acid) diammonium salt) radical, 77.19% to 93–14% of hydroxyl radical, 97.93% to 99.58% of ferrous ion chelating, and 76.66% to 98.35% of copper ions chelating were maintained during the spray drying process. Further, the thermal process had a different influence on the destruction of antioxidant groups. Zhang et al. (2018) used spray drying for the microencapsulation of the enzyme as a protein model. As an exemplification process, trehalose was applied as a carrier for the microencapsulation of trypsin that was performed by spray and spray freeze drying. Pure trypsin particles produced by the spray freeze drying had more enzymatic activity (about 88%) than spray dried types (about 80%). Additionally, the use of trehalose (in the optimum ratio of 1: 1 with the enzyme) led to the maintenance of enzymatic activity (more than 97%)

in the produced particles by both methods. The use of trehalose increased the thermal stability of the spray dried particles. Decreased thermal degradation of trypsin was attributed to the higher thermal stability and crystalline structure of trehalose. Finally, based on the acquired results, the protective effect of trehalose could be related to the formation of hydrogen bonds with trypsin, the production of highly amorphous glass matrices, and the reduction of unfolding and aggregation.

In another study, trehalose (in a ratio of 2: 1) was used as a stabilizer in the spray drying process of β-galactosidase. Depending on the process variables (inlet air temperature, nozzle hole size, and ethanol concentration), between 78% and 100% of enzyme activity was maintained after spray drying. Also, the value of this index reached about 55% to 92% after three weeks of keeping the powders at 70°C. Among the produced samples, the amorphous structure and porosity of the fabricated particles were effective in the stability and maintenance of enzymatic activity (Bürki et al., 2011). Murthy et al. (2017) investigated the effect of the spray drying method using maltodextrin and gum Arabic as carrier agents on the physicochemical properties and antioxidant activity of pink perch meat protein hydrolysate. DPPH radical scavenging activity and metal ion chelating decreased from 66% to 49% and from 26% to 25%, after microencapsulation using maltodextrin and gum Arabic, respectively. Based on their results, this declining trend could be attributed to the low access of antioxidant compounds to radicals. Additionally, alteration in the composition of the particles after the use of maltodextrin and gum Arabic caused a reduction in metal chelating activity.

10.5.5 MORPHOLOGICAL FEATURES

One of the most important factors that affect the stability, usability, and release of the spray dried microencapsulated peptides and proteins is morphology, along with their particle size. The morphological and structural properties of spray dried particles are affected by several parameters such as air temperature, concentration, composition, and type of feed (carrier and core compositions) (Assadpour & Jafari, 2019). Depending on the composition of the feed, the nature of the core (lipophilic or hydrophilic) and the type of the drying process, particles with internal structures (matrix or reservoir types), morphological properties (particles with smooth, dented, wrinkled, or donut-shaped surfaces), and in different sizes can be produced (Sarabandi & Jafari, 2020). Evaluating the morphological properties of the particles can improve the formulation and optimization of process parameters. For example, the production of capsules with a spherical structure and smooth surfaces are important indicators for the production of pharmaceutical peptides (Maury et al., 2005). Furthermore, the manufacture of particles with aerodynamic, morphological, and appropriate size properties is important for the pulmonary delivery in pharmaceutical powders (Möbus et al., 2012).

Considering the importance of the morphological assessment and the corresponding characteristics, several studies have scrutinized the effect of spray drying parameters on the structure of particles. Evaluation of the effect of concentration and carrier type and process air temperature on the morphological characteristics of hydrolyzed flax protein (Akbarbaglu et al., 2019), chicken meat protein hydrolysates (Kurozawa et al., 2009b), casein (Sarabandi, Sadeghi Mahoonak, Hamishekar, et al., 2018d), pink perch protein (Murthy et al., 2017), and WPC (Rukluarh et al., 2019) are among the research published in this area. In the present study, particles of various sizes were observed in a variety of spherical and asymmetrical structures with wrinkled, dented, cracked, or healthy surfaces. Microencapsulation of hydrolyzed casein using soy protein isolate at different wall: core ratios (70: 30 and 80: 20) was performed by spray drying technique. The scanning electron microscopy (SEM) images showed that there were microcapsules with uniform walls and pores in the crust. In addition, the particle size decreased with increasing the ratio of wall material to hydrolyzed casein from 11.32 to 9.18 microns (Ortiz et al., 2009).

Rocha et al. (2009) microencapsulated hydrolyzed casein using maltodextrin as a carrier to reduce the bitterness. They reported that microcapsules had continuous surfaces and hollow structures, and they were matrix-shaped without any apparent porosity. Various parameters affect the formation of the shell and the internal and external structure of the spray dried particles. As an exemplification, the rapid evaporation of the solvent from the droplet surface resulted in the formation of a hard shell of polymer material. As a result of this process, the rate of diffusion and release of solvent from the inner parts decreased concerning the penetration of heat into the capsules. In addition, due to pressure increase inside the shell, the particles/droplets expanded, which caused fast diffusion and the creation of a thinner shell. It is worth noting that, owing to the low diffusion rate of shell material, it causes cracks or pores in the wall of capsules (Ting, Gonda, & Gipps, 1992). On the other hand, it is important to evaluate the effect of various microencapsulation techniques such as spray drying and spray freeze drying on particle morphology. For example, the stabilization of the trypsin enzyme on trehalose particles was performed by these two methods (Zhang et al., 2018). The results showed that the size, morphology, and internal structure of the produced particles differed between the two methods (Figure 10.4a–i). With regards to the effect of type of material, differences in morphology were observed between pure trehalose particles, the combination of trypsin and trehalose, and pure spray dried trypsin (Figure 10.4a–c). For the duration of the spray drying process, pure trehalose particles and trypsin had spherical and wrinkled structures, respectively. Slow migration of dissolved compounds around the droplets during drying and the formation of viscoelastic shells could lead to the formation of wrinkle structures. Trypsin has more surface activity than trehalose. Therefore, trypsin migrates to droplet/air interface in the particles containing trehalose/trypsin, but due to the larger hydrodynamic size, trypsin may have a slower penetration and diffusion.

It can be concluded that in the initial stage of spray drying, the surface of the droplets will be covered more extensively by trypsin. However, the application of capillary force in the shell can lead to the deformation of spherical droplets or semi-dried particles. It is necessary to clarify that spray freeze dried particles had spherical structures with a very high porosity degree. This can be related to maintaining the spherical structure of the particles as a result of their immediate freezing. For

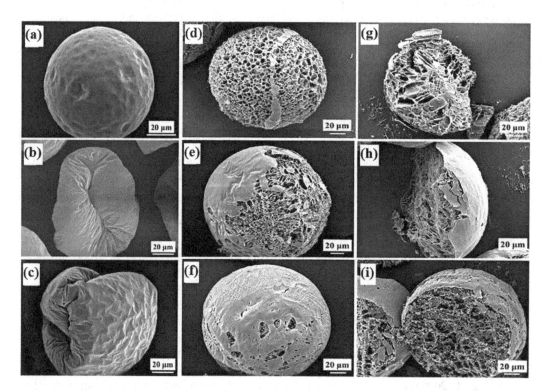

FIGURE 10.4 Scanning electron micrographs of the spray dried (a–c) and spray freeze dried (d–i) of the pure trehalose microparticles (a/d–g), the trypsin/trehalose microparticles at a ratio of 1: 1 (b/e–h), and the pure trypsin microparticles (c/f–i). Reprinted from Zhang et al. (2018), with permission.

the period of this process, a low sublimation rate causes weak capillary forces. In addition, much open porosity could be observed as a result of ice sublimation. The surface of pure dried trehalose particles was completely porous while much of the surface of the particles was covered with trypsin/trehalose, which can be attributed to the formation of a dense layer of trypsin on the surface of the particles (Figure 10.4d–i) (Zhang et al., 2018).

10.5.6 Food Formulations and Bitterness Masking

One of the most important challenges preventing the direct use of bioactive peptides and hydrolysates in food formulations and dietary supplements is their unpleasant/bitter taste. Enzymatic hydrolysis of proteins and production of bioactive peptides results in the release of various amino acids (in particular, hydrophobic types such as alanine, phenylalanine, leucine, isoleucine, methionine, proline, tryptophan, and valine). Some of these compounds, even though creating some functional properties and biological activities, cause a high degree of bitterness in the product (Breternitz, Bolini, & Hubinger, 2017). As a result, the microencapsulation of these compounds is crucial to cover their bitterness (Sarabandi, Sadeghi Mahoonak, Hamishekar, et al., 2018d). The incorporation of peptides in the matrix of different carriers and decreasing the reaction with taste buds can mask their bitterness (Chaudhry et al., 2008). Sensory characteristics of spray dried peptides and the effect of the microencapsulation process on masking the bitterness of peptides in various food products have been analyzed in different studies. For example, Yang et al. (2012) investigated the effect of microencapsulation of whey protein hydrolysate with maltodextrin and the combination of maltodextrin and β-cyclodextrins on the decrease of such compounds. The results of the sensory evaluation showed a decrease of up to eight times in the bitterness intensity of hydrolyzed whey proteins after the microencapsulation process.

In another study, a decrease in the bitterness of hydrolyzed casein was reported after its encapsulation with gelatin and soy protein isolate-based carriers (Favaro-Trindade et al., 2010). Ma et al. (2014) examined the effect of spray and freeze drying on the bitter taste and hygroscopicity of hydrolyzed WPC, where they used sodium alginate and WPC as wall materials. According to their report, the use of WPC in combination with sodium alginate could lead to a decrease in the bitterness of hydrolyzed proteins.

In recent years, a significant amount of work has been carried out on the effectiveness of spray drying microencapsulation on masking the bitterness of hydrolysates in food formulations. Sarabandi, Peighambardoust, Sadeghi Mahoonak, and Samaei (2018c) investigated the spray drying microencapsulation of casein hydrolysates using maltodextrin. Then pure and microencapsulated samples were added to the gummy candy formulation. The obtained results showed a reduction of the bitterness in specimens enriched with the microencapsulated hydrolysates. Murthy et al. (2017) investigated the effect of the spray drying process using maltodextrin and gum Arabic as carriers on the physicochemical and sensory properties of pink perch meat protein hydrolysate. Concerning the results of sensory evaluation of vegetable soup enriched with free hydrolyzed and microencapsulated capsules, spray drying could considerably decrease the bitterness and mask the undesirable taste of the peptides. As an illustration, bitterness was not detected in the samples that were enriched with 4% microencapsulated hydrolysate.

Segura-Campos, Salazar-Vega, Chel-Guerrero, and Betancur-Ancona (2013) studied the enzymatic hydrolysis of chia protein and the production of bioactive peptides with antioxidant and antimicrobial activity. Then, hydrolysate with the highest degree of hydrolysis was selected for addition to white bread and carrot cream. According to the results, the use of hydrolyzed chia protein improved the inhibitory activity of ACE in fortified bread but showed no effect on the antioxidant activity of the product. Furthermore, hydrolysate can be used as a flavoring agent in addition to antioxidant and health-promoting activities. This can be achieved when the assessment of sensory properties of mussel protein hydrolysate as a flavoring in the instant noodles was performed (Breternitz et al., 2017). The same combination of Hi-Cap®100 and maltodextrin (as the carrier) was used for the spray drying encapsulation. The manufactured noodles using two grams of hydrolyzed powders in combination with two grams of seasoning had the highest score for taste, aroma, and overall acceptance (Breternitz et al., 2017).

10.6 CONCLUDING REMARKS

Numerous health benefits of bioactive peptides and proteins, as well as protein-based drugs, have significantly increased the use of these compounds in the food and pharmaceutical industries. Nonetheless, some disadvantages, such as physicochemical instability, low glass transition temperatures (T_g) value, high hygroscopicity and bitterness, reaction with other compounds, and sensitivity to environmental factors, result in a short shelf-life and difficulties in the direct use of these compounds. Hence, spray drying microencapsulation of these compounds with different carriers has been studied extensively. The structure of the complex, the dependence of biological activity on the conformational and chemical structure of peptides, and their surface activity are among the factors that cause the instability of bioactive peptides/proteins during the spray drying process (dehydration stress and shear stress during atomization). In addition, maintaining the physiochemical and biological stability of peptides in the dried powders is imperative. For food and pharmaceutical applications (e.g., inhalable formulations), some properties, such as particle size, morphology, and stability, are particularly important. Further to this, maintaining functional characteristics, flow behavior, and biological activity are other important indicators that should be considered. All the aforementioned parameters are influenced by the variables of the spray drying process, composition and type of feed, and the carriers (wall materials). It should also be emphasized that the use of modified systems and nano spray driers is increasing. Therefore, the design of the process variables and the composition of the formulation depends on the application of the final product.

REFERENCES

Adler, M., Lee, G., 1999. Stability and surface activity of lactate dehydrogenase in spray-dried trehalose. *J. Pharm. Sci.* 88, 199–208.

Adler, M., Unger, M., Lee, G., 2000. Surface composition of spray dried particles of bovine serum albumin/trehalose/surfactant. *Pharm. Res.* 17, 863–870.

Ajmera, A., Scherließ, R., 2014. Stabilisation of proteins via mixtures of amino acids during spray drying. *Int. J. Pharm.* 463, 98–107.

Akbarbaglu, Z., Mahdi Jafari, S., Sarabandi, K., Mohammadi, M., Khakbaz Heshmati, M., Pezeshki, A., 2019. Influence of spray drying encapsulation on the retention of antioxidant properties and microstructure of flaxseed protein hydrolysates. *Colloids Surfaces B Biointerfaces.* doi:10.1016/j.colsurfb.2019.03.038

Akhavan Mahdavi, S., Jafari, S.M., Assadpoor, E., Dehnad, D., 2016. Microencapsulation optimization of natural anthocyanins with maltodextrin, gum Arabic and gelatin. *Int. J. Biol. Macromol.* 85. doi:10.1016/j.ijbiomac.2016.01.011

Alvani, K., Qi, X., Tester, R.F., 2011. Use of carbohydrates, including dextrins, for oral delivery. *Starch-Stärke* 63, 424–431.

Amara, C. Ben, Kim, L., Oulahal, N., Degraeve, P., Gharsallaoui, A., 2017. Using complexation for the microencapsulation of nisin in biopolymer matrices by spray-drying. *Food Chem.* 236, 32–40.

Aquino, R.P., Auriemma, G., Conte, G.M., Esposito, T., Sommella, E., Campiglia, P., Sansone, F., 2020. Development of chitosan/mannitol microparticles as delivery system for the oral administration of a spirulina bioactive peptide extract. *Molecules* 25, 2086.

Assadpour, E., Jafari, S.M., 2017. Spray drying of folic acid within nano-emulsions: optimization by Taguchi approach. *Dry. Technol.* 35, 1152–1160.

Assadpour, E., Jafari, S.M., 2018. A systematic review on nanoencapsulation of food bioactive ingredients and nutraceuticals by various nanocarriers. *Crit. Rev. Food Sci. Nutr.* 59(19), 3129–3151.

Assadpour, E., Jafari, S.M., 2019. Advances in spray-drying encapsulation of food bioactive ingredients: from microcapsules to nanocapsules. *Annu. Rev. Food Sci. Technol.* 10, 103–131.

Bazaria, B., Kumar, P., 2016. Effect of whey protein concentrate as drying aid and drying parameters on physicochemical and functional properties of spray dried beetroot juice concentrate. *Food Biosci.* 14, 21–27.

Bhandari, B.R., Patel, K.C. and Chen, X.D., 2008. Spray drying of food materials-process and product characteristics. *Drying Technol. Food Process*, 4, 113–157.

Bhusari, S.N., Muzaffar, K., Kumar, P., 2014. Effect of carrier agents on physical and microstructural properties of spray dried tamarind pulp powder. *Powder Technol.* 266, 354–364.

Bilati, U., Allémann, E., Doelker, E., 2005. Strategic approaches for overcoming peptide and protein instability within biodegradable nano-and microparticles. *Eur. J. Pharm. Biopharm.* 59, 375–388.

Brennan, J.G., 2005. Evaporation and dehydration, *Food Processing Handbook* WILEY-VCH Verlag GMBH & Co. KGaA, pp. 71–124.

Breternitz, N.R., Bolini, H.M.A., Hubinger, M.D., 2017. Sensory acceptance evaluation of a new food flavoring produced by microencapsulation of a mussel (Perna perna) protein hydrolysate. *LWT-Food Sci. Technol.* 83, 141–149.

Burey, P., Bhandari, B.R., Howes, T., Gidley, M.J., 2008. Hydrocolloid gel particles: formation, characterization, and application. *Crit. Rev. Food Sci. Nutr.* 48, 361–377.

Bürki, K., Jeon, I., Arpagaus, C., Betz, G., 2011. New insights into respirable protein powder preparation using a nano spray dryer. *Int. J. Pharm.* 408, 248–256.

Chan, H.-K., Clark, A.R., Feeley, J.C., Kuo, M.-C., Lehrman, S.R., Pikal-Cleland, K., Miller, D.P., Vehring, R., Lechuga-Ballesteros, D., 2004. Physical stability of salmon calcitonin spray dried powders for inhalation. *J. Pharm. Sci.* 93, 792–804.

Chaudhry, Q., Scotter, M., Blackburn, J., Ross, B., Boxall, A., Castle, L., Aitken, R., Watkins, R., 2008. Applications and implications of nanotechnologies for the food sector. *Food Addit. Contam.* 25, 241–258.

Chegini, G.R., Ghobadian, B., 2005. Effect of spray-drying conditions on physical properties of orange juice powder. *Dry. Technol.* 23, 657–668.

Chen, C., Chi, Y.-J., Xu, W., 2012. Comparisons on the functional properties and antioxidant activity of spray dried and freeze-dried egg white protein hydrolysate. *Food Bioprocess Technol.* 5, 2342–2352.

Collins, A.R., 2005. Antioxidant intervention as a route to cancer prevention. *Eur. J. Cancer* 41, 1923–1930.

Cross, K.J., Huq, N.L., Palamara, J.E., Perich, J.W., Reynolds, E.C., 2005. Physicochemical characterization of casein phosphopeptide-amorphous calcium phosphate nanocomplexes. *J. Biol. Chem.* 280, 15362–15369.

Das, U., Hariprasad, G., Ethayathulla, A.S., Manral, P., Das, T.K., Pasha, S., Mann, A., Ganguli, M., Verma, A.K., Bhat, R., 2007. Inhibition of protein aggregation: supramolecular assemblies of arginine hold the key. *PLoS One* 2, e1176.

Đorđević, V., Paraskevopoulou, A., Mantzouridou, F., Lalou, S., Pantić, M., Bugarski, B., Nedović, V., 2016. Encapsulation technologies for food industry, In: Nedović V., Raspor P., Lević J., Tumbas Šaponjac V., Barbosa-Cánovas G. (Eds) *Emerging and Traditional Technologies for Safe, Healthy and Quality Food.* Springer, pp. 329–382.

Drusch, S., Hamann, S., Berger, A., Serfert, Y., Schwarz, K., 2012. Surface accumulation of milk proteins and milk protein hydrolysates at the air–water interface on a time-scale relevant for spray-drying. *Food Res. Int.* 47, 140–145.

Duchêne, D., Bochot, A., 2016. Thirty years with cyclodextrins. *Int. J. Pharm.* 514, 58–72.

Fang, Z., Bhandari, B., 2012. Comparing the efficiency of protein and maltodextrin on spray drying of bayberry juice. *Food Res. Int.* 48, 478–483.

Fathi, M., Martin, A., McClements, D.J., 2014. Nanoencapsulation of food ingredients using carbohydrate based delivery systems. *Trends Food Sci. Technol.* 39, 18–39.

Fávaro-Trindade, C.S., Santana, A. dos S., Monterrey-Quintero, E.S., Trindade, M.A., Netto, F.M., 2010. The use of spray drying technology to reduce bitter taste of casein hydrolysate. *Food Hydrocoll.* 24, 336–340.

Favaro-Trindade, C.S., Santana, A.S., Monterrey-Quintero, E.S., Trindade, M.A., Netto, F.M., 2010. The use of spray drying technology to reduce bitter taste of casein hydrolysate. *Food Hydrocoll.* 24, 336–340.

Gauthier, S.F., Pouliot, Y., Saint-Sauveur, D., 2006. Immunomodulatory peptides obtained by the enzymatic hydrolysis of whey proteins. *Int. dairy J.* 16, 1315–1323.

Gibson, G.R., Williams, C.M., 2005. *Functional Foods.* IFIS Publishing.

Gómez-Mascaraque, L.G., Miralles, B., Recio, I., López-Rubio, A., 2016. Microencapsulation of a whey protein hydrolysate within micro-hydrogels: Impact on gastrointestinal stability and potential for functional yoghurt development. *J. Funct. Foods* 26, 290–300.

Grenha, A., Seijo, B., Remunán-López, C., 2005. Microencapsulated chitosan nanoparticles for lung protein delivery. *Eur. J. Pharm. Sci.* 25, 427–437.

Hasija, M., Li, L., Rahman, N., Ausar, S.F., 2013. Forced degradation studies: an essential tool for the formulation development of vaccines. *Vaccine Dev. Ther.* 3, 11.

Jacquot, M., Pernetti, M., 2004. Spray coating and drying processes, In: Nedović V., Willaert R. (Eds) *Fundamentals of Cell Immobilisation Biotechnology.* Springer, pp. 343–356.

Jafari, S.M., Assadpoor, E., He, Y., Bhandari, B., 2008. Encapsulation efficiency of food flavours and oils during spray drying. *Dry. Technol.* 26, 816–835.

Jain, A.K., Khar, R.K., Ahmed, F.J., Diwan, P.V., 2008. Effective insulin delivery using starch nanoparticles as a potential trans-nasal mucoadhesive carrier. *Eur. J. Pharm. Biopharm.* 69, 426–435.

Jamdar, S.N., Rajalakshmi, V., Pednekar, M.D., Juan, F., Yardi, V., Sharma, A., 2010. Influence of degree of hydrolysis on functional properties, antioxidant activity and ACE inhibitory activity of peanut protein hydrolysate. *Food Chem.* 121, 178–184.

Jayasundera, M., Adhikari, B., Howes, T., Aldred, P., 2011. Surface protein coverage and its implications on spray-drying of model sugar-rich foods: solubility, powder production and characterisation. *Food Chem.* 128, 1003–1016.

Jia, J., Ma, H., Zhao, W., Wang, Z., Tian, W., Luo, L., He, R., 2010. The use of ultrasound for enzymatic preparation of ACE-inhibitory peptides from wheat germ protein. *Food Chem.* 119, 336–342.

Kim, H.J., Decker, E.A., McClements, D.J., 2002. Impact of protein surface denaturation on droplet flocculation in hexadecane oil-in-water emulsions stabilized by β-lactoglobulin. *J. Agric. Food Chem.* 50, 7131–7137.

Kurozawa, L.E., Park, K.J., Hubinger, M.D., 2009a. Effect of maltodextrin and gum Arabic on water sorption and glass transition temperature of spray dried chicken meat hydrolysate protein. *J. Food Eng.* 91, 287–296.

Kurozawa, L.E., Park, K.J., Hubinger, M.D., 2009b. Effect of carrier agents on the physicochemical properties of a spray dried chicken meat protein hydrolysate. *J. Food Eng.* 94, 326–333.

Lafarga, T., Hayes, M., 2017. Bioactive protein hydrolysates in the functional food ingredient industry: overcoming current challenges. *Food Rev. Int.* 33, 217–246.

Lee, S.H., Heng, D., Ng, W.K., Chan, H.-K., Tan, R.B.H., 2011. Nano spray drying: a novel method for preparing protein nanoparticles for protein therapy. *Int. J. Pharm.* 403, 192–200.

Li, H.-Y., Song, X., Seville, P.C., 2010. The use of sodium carboxymethylcellulose in the preparation of spray dried proteins for pulmonary drug delivery. *Eur. J. Pharm. Sci.* 40, 56–61.

Li, K., Woo, M.W., Selomulya, C., 2016. Effects of composition and relative humidity on the functional and storage properties of spray dried model milk emulsions. *J. Food Eng.* 169, 196–204.

López-Fandiño, R., Otte, J., VanCamp, J., 2006. Physiological, chemical and technological aspects of milk-protein-derived peptides with antihypertensive and ACE-inhibitory activity. *Int. Dairy J.* 16, 1277–1293.

Ma, J.-J., Mao, X.-Y., Wang, Q., Yang, S., Zhang, D., Chen, S.-W., Li, Y.-H., 2014. Effect of spray drying and freeze drying on the immunomodulatory activity, bitter taste and hygroscopicity of hydrolysate derived from whey protein concentrate. *LWT-Food Sci. Technol.* 56, 296–302.

Maqsoudlou, A., Mahoonak, A.S., Mora, L., Mohebodini, H., Toldrá, F., Ghorbani, M., 2019. Peptide identification in alcalase hydrolysated pollen and comparison of its bioactivity with royal jelly. *Food Res. Int.* 116, 905–915.

Maury, M., Murphy, K., Kumar, S., Mauerer, A., Lee, G., 2005. Spray-drying of proteins: effects of sorbitol and trehalose on aggregation and FT-IR amide I spectrum of an immunoglobulin G. *Eur. J. Pharm. Biopharm.* 59, 251–261.

McCann, K.B., Shiell, B.J., Michalski, W.P., Lee, A., Wan, J., Roginski, H., Coventry, M.J., 2006. Isolation and characterisation of a novel antibacterial peptide from bovine αS1-casein. *Int. Dairy J.* 16, 316–323.

McClements, D.J., 2015. Encapsulation, protection, and release of hydrophilic active components: potential and limitations of colloidal delivery systems. *Adv. Colloid Interface Sci.* 219, 27–53.

McClements, D.J., 2018. Encapsulation, protection, and delivery of bioactive proteins and peptides using nanoparticle and microparticle systems: a review. *Adv. Colloid Interface Sci.* 253, 1–22.

McClements, D.J., Decker, E.A., 2000. Lipid oxidation in oil-in-water emulsions: Impact of molecular environment on chemical reactions in heterogeneous food systems. *J. Food Sci.* 65, 1270–1282.

Mendanha, D.V., Ortiz, S.E.M., Favaro-Trindade, C.S., Mauri, A., Monterrey-Quintero, E.S., Thomazini, M., 2009. Microencapsulation of casein hydrolysate by complex coacervation with SPI/pectin. *Food Res. Int.* 42, 1099–1104.

Möbus, K., Siepmann, J., Bodmeier, R., 2012. Zinc–alginate microparticles for controlled pulmonary delivery of proteins prepared by spray-drying. *Eur. J. Pharm. Biopharm.* 81, 121–130.

Mohan, A., McClements, D.J., Udenigwe, C.C., 2016. Encapsulation of bioactive whey peptides in soy lecithin-derived nanoliposomes: influence of peptide molecular weight. *Food Chem.* 213, 143–148.

Mohan, A., Rajendran, S.R.C.K., He, Q.S., Bazinet, L., Udenigwe, C.C., 2015. Encapsulation of food protein hydrolysates and peptides: A review. *RSC Adv.* doi:10.1039/c5ra13419f

Mujumdar, A.S., 2004. Research and development in drying: recent trends and future prospects. *Dry. Technol.* 22, 1–26.

Murthy, L.N., Phadke, G.G., Mohan, C.O., Chandra, M.V., Annamalai, J., Visnuvinayagam, S., Unnikrishnan, P., Ravishankar, C.N., 2017. Characterization of spray dried hydrolyzed proteins from pink perch meat added with maltodextrin and gum Arabic. *J. Aquat. Food Prod. Technol.* 26, 913–928.

Ortiz, S.E.M., Mauri, A., Monterrey-Quintero, E.S., Trindade, M.A., Santana, A.S., Favaro-Trindade, C.S., 2009. Production and properties of casein hydrolysate microencapsulated by spray drying with soybean protein isolate. *LWT-Food Sci. Technol.* 42, 919–923.

Palmieri, V.O., Grattagliano, I., Portincasa, P., Palasciano, G., 2006. Systemic oxidative alterations are associated with visceral adiposity and liver steatosis in patients with metabolic syndrome. *J. Nutr.* 136, 3022–3026.

Pihlanto-Leppälä, A., 2000. Bioactive peptides derived from bovine whey proteins: opioid and ace-inhibitory peptides. *Trends food Sci. Technol.* 11, 347–356.

Pikal-Cleland, K.A., Cleland, J.L., Anchordoquy, T.J., Carpenter, J.F., 2002. Effect of glycine on pH changes and protein stability during freeze–thawing in phosphate buffer systems. *J. Pharm. Sci.* 91, 1969–1979.

Rahman, M.S., 2007. *Handbook of Food Preservation*. CRC Press. Boca Raton, FL 33487–32742.

Rao, P.S., Bajaj, R.K., Mann, B., Arora, S., Tomar, S.K., 2016a. Encapsulation of antioxidant peptide enriched casein hydrolysate using maltodextrin–gum Arabic blend. *J. Food Sci. Technol.* 53, 3834–3843.

Rao, Q., Klaassen Kamdar, A., Labuza, T.P., 2016b. Storage stability of food protein hydrolysates—a review. *Crit. Rev. Food Sci. Nutr.* 56, 1169–1192.

Ray, S., Raychaudhuri, U., Chakraborty, R., 2016. An overview of encapsulation of active compounds used in food products by drying technology. *Food Biosci.* 13. doi:10.1016/j.fbio.2015.12.009

Rocha, G.A., Trindade, M.A., Netto, F.M., Favaro-Trindade, C.S., 2009. Microcapsules of a casein hydrolysate: production, characterization, and application in protein bars. *Food Sci. Technol. Int.* 15, 407–413.

Rokhade, A.P., Agnihotri, S.A., Patil, S.A., Mallikarjuna, N.N., Kulkarni, P.V., Aminabhavi, T.M., 2006. Semi-interpenetrating polymer network microspheres of gelatin and sodium carboxymethyl cellulose for controlled release of ketorolac tromethamine. *Carbohydr. Polym.* 65, 243–252.

Roy, I., Gupta, M.N., 2004. Freeze-drying of proteins: Some emerging concerns. *Biotechnol. Appl. Biochem.* 39, 165–177.

Rukluarh, S., Kanjanapongkul, K., Panchan, N., Niumnuy, C., 2019. Effect of inclusion conditions on characteristics of spray dried whey protein hydrolysate/γ-cyclodextrin complexes. *J. Food Sci. Agric. Technol.* 5, 5–12.

Sansone, F., Mencherini, T., Picerno, P., d'Amore, M., Aquino, R.P., Lauro, M.R., 2011. Maltodextrin/pectin microparticles by spray drying as carrier for nutraceutical extracts. *J. Food Eng.* 105, 468–476.

Sarabandi, K., Gharehbeglou, P., Jafari, S.M., 2019a. Spray-drying encapsulation of protein hydrolysates and bioactive peptides: opportunities and challenges. *Dry. Technol.* 38, 577–595.

Sarabandi, K., Jafari, S.M., 2020. Effect of chitosan coating on the properties of nanoliposomes loaded with flaxseed-peptide fractions: stability during spray-drying. *Food Chem.* 310, 125951.

Sarabandi, K., Jafari, S.M., Mahoonak, A.S., Mohammadi, A., 2019b. Application of gum Arabic and maltodextrin for encapsulation of eggplant peel extract as a natural antioxidant and color source. *Int. J. Biol. Macromol.* 140, 59–68.

Sarabandi, K., Jafari, S.M., Mohammadi, M., Akbarbaglu, Z., Pezeshki, A., Khakbaz Heshmati, M., 2019c. Production of reconstitutable nanoliposomes loaded with flaxseed protein hydrolysates: Stability and characterization. *Food Hydrocoll.* 96, 442–450. doi:10.1016/j.foodhyd.2019.05.047

Sarabandi, K., Peighambardoust, S.H., Sadeghi Mahoonak, A.R., Samaei, S.P., 2018c. Effect of different carriers on microstructure and physical characteristics of spray dried apple juice concentrate. *J. Food Sci. Technol.* doi:10.1007/s13197-018-3235-6

Sarabandi, Khashayar, Mahoonak, A.S., Akbari, M., 2018a. Physicochemical properties and antioxidant stability of microencapsulated marjoram extract prepared by co-crystallization method. *J. Food Process Eng.* doi:10.1111/jfpe.12949

Sarabandi, Khashayar, Mahoonak, A.S., Hamishekar, H., Ghorbani, M., Jafari, S.M., 2018b. Microencapsulation of casein hydrolysates: physicochemical, antioxidant and microstructure properties. *J. Food Eng.* 237, 86–95.

Sarabandi, Khashayar, Sadeghi Mahoonak, A., Hamishekar, H., Ghorbani, M., Jafari, M., 2018d. Microencapsulation of casein hydrolysates: Physicochemical, antioxidant and microstructure properties. *J. Food Eng.* 237, 86–95. doi:10.1016/j.jfoodeng.2018.05.036

Sarmadi, B.H., Ismail, A., 2010. Antioxidative peptides from food proteins: a review. *Peptides* 31, 1949–1956.

Schüle, S., Frieß, W., Bechtold-Peters, K., Garidel, P., 2007. Conformational analysis of protein secondary structure during spray-drying of antibody/mannitol formulations. *Eur. J. Pharm. Biopharm.* 65, 1–9.

Segura-Campos, M.R., Salazar-Vega, I.M., Chel-Guerrero, L.A., Betancur-Ancona, D.A., 2013. Biological potential of chia (*Salvia hispanica* L.) protein hydrolysates and their incorporation into functional foods. *LWT-Food Sci. Technol.* 50, 723–731.

Shahidi, F., Zhong, Y., 2015. Measurement of antioxidant activity. *J. Funct. Foods* 18, 757–781.

Shimizu, M., Sawashita, N., Morimatsu, F., Ichikawa, J., Taguchi, Y., Ijiri, Y., Yamamoto, J., 2009. Antithrombotic papain-hydrolyzed peptides isolated from pork meat. *Thromb. Res.* 123, 753–757.

Shishir, M.R.I., Xie, L., Sun, C., Zheng, X., Chen, W., 2018. Advances in micro and nano-encapsulation of bioactive compounds using biopolymer and lipid-based transporters. *Trends food Sci. Technol.* 78, 34–60.

Shrestha, A.K., Howes, T., Adhikari, B.P., Bhandari, B.R., 2007. Water sorption and glass transition properties of spray dried lactose hydrolysed skim milk powder. *LWT-Food Sci. Technol.* 40, 1593–1600.

Sienkiewicz-Szłapka, E., Jarmołowska, B., Krawczuk, S., Kostyra, E., Kostyra, H., Iwan, M., 2009. Contents of agonistic and antagonistic opioid peptides in different cheese varieties. *Int. Dairy J.* 19, 258–263.

Suarez, S.E., Añón, M.C., 2018. Comparative behaviour of solutions and dispersions of amaranth proteins on their emulsifying properties. *Food Hydrocoll.* 74, 115–123.

Subtil, S.F., Rocha-Selmi, G.A., Thomazini, M., Trindade, M.A., Netto, F.M., Favaro-Trindade, C.S., 2014. Effect of spray drying on the sensory and physical properties of hydrolysed casein using gum Arabic as the carrier. *J. Food Sci. Technol.* 51, 2014–2021.

Sun, L., 2013. Peptide-based drug development. *Mod Chem Appl.* 1, 1–2.

Ting, T.-Y., Gonda, I., Gipps, E.M., 1992. Microparticles of polyvinyl alcohol for nasal delivery. I. Generation by spray-drying and spray-desolvation. *Pharm. Res.* 9, 1330–1335.

Tong, L.M., Sasaki, S., McClements, D.J., Decker, E.A., 2000. Mechanisms of the antioxidant activity of a high molecular weight fraction of whey. *J. Agric. Food Chem.* 48, 1473–1478.

Truong, V., Bhandari, B.R., Howes, T., 2005. Optimization of co-current spray drying process of sugar-rich foods. Part I—Moisture and glass transition temperature profile during drying. *J. Food Eng.* 71, 55–65.

Udenigwe, C.C., 2014. Bioinformatics approaches, prospects and challenges of food bioactive peptide research. *Trends Food Sci. Technol.* 36, 137–143.

Uversky, V.N., Goto, Y., 2009. Acid denaturation and anion-induced folding of globular proteins: multitude of equilibrium partially folded intermediates. *Curr. Protein Pept. Sci.* 10, 447–455.

Vioque, J., Clemente, A., Pedroche, J., Yust, M. Del M., Millán, F., 2001. Obtención y aplicaciones de hidrolizados proteicos. *Grasas y Aceites.* 52, 132–136.

Wandrey, C., Bartkowiak, A., Harding, S.E., 2009. Materials for encapsulation, In: Zuidam N.J., Nedovic, V.A. (Eds.) *Encapsulation Technologies for Food Active Ingredients and Food Processing.* Springer: 31–100.

Wang, H., Tong, X., Yuan, Y., Peng, X., Zhang, Q., Zhang, S., Xie, C., Zhang, X., Yan, S., Xu, J., 2020. Effect of spray-drying and freeze-drying on the properties of soybean hydrolysates. *J. Chem.* 2020, Article ID 9201457,1–8.

Wang, W., Jung, J., Zhao, Y., 2017. Chitosan-cellulose nanocrystal microencapsulation to improve encapsulation efficiency and stability of entrapped fruit anthocyanins. *Carbohydr. Polym.* 157, 1246–1253.

Wang, W., Nema, S., Teagarden, D., 2010. Protein aggregation—pathways and influencing factors. *Int. J. Pharm.* 390, 89–99.

Wang, Y., Selomulya, C., 2020. Spray drying strategy for encapsulation of bioactive peptide powders for food applications. *Adv. Powder Technol.* 31, 409–415.

Xie, Z., Huang, J., Xu, X., Jin, Z., 2008. Antioxidant activity of peptides isolated from alfalfa leaf protein hydrolysate. *Food Chem.* 111, 370–376.

Yang, H., Feng, K., Wen, P., Zong, M.-H., Lou, W.-Y., Wu, H., 2017. Enhancing oxidative stability of encapsulated fish oil by incorporation of ferulic acid into electrospun zein mat. *LWT* 84, 82–90.

Yang, M., Velaga, S., Yamamoto, H., Takeuchi, H., Kawashima, Y., Hovgaard, L., VanDeWeert, M., Frokjaer, S., 2007. Characterisation of salmon calcitonin in spray dried powder for inhalation: effect of chitosan. *Int. J. Pharm.* 331, 176–181.

Yang, S., Mao, X.-Y., Li, F.-F., Zhang, D., Leng, X.-J., Ren, F.-Z., Teng, G.-X., 2012. The improving effect of spray-drying encapsulation process on the bitter taste and stability of whey protein hydrolysate. *Eur. Food Res. Technol.* 235, 91–97.

You, L., Zhao, M., Regenstein, J.M., Ren, J., 2010. Changes in the antioxidant activity of loach (Misgurnus anguillicaudatus) protein hydrolysates during a simulated gastrointestinal digestion. *Food Chem.* 120, 810–816.

Zeng, Q., Zhang, M., Adhikari, B.P., Mujumdar, A.S., 2013. Effect of drying processes on the functional properties of collagen peptides produced from chicken skin. *Dry. Technol.* 31, 1653–1660.

Zhang, S., Lei, H., Gao, X., Xiong, X., Wu, W.D., Wu, Z., Chen, X.D., 2018. Fabrication of uniform enzyme-immobilized carbohydrate microparticles with high enzymatic activity and stability via spray drying and spray freeze drying. *Powder Technol.* 330, 40–49.

Zhao, Q., Xiong, H., Selomulya, C., Chen, X.D., Huang, S., Ruan, X., Zhou, Q., Sun, W., 2013. Effects of spray drying and freeze drying on the properties of protein isolate from rice dreg protein. *Food Bioprocess Technol.* 6, 1759–1769.

Zhong, F., Liu, J., Ma, J., Shoemaker, C.F., 2007. Preparation of hypocholesterol peptides from soy protein and their hypocholesterolemic effect in mice. *Food Res. Int.* 40, 661–667.

11 Spray Drying Encapsulation of Probiotics

Surajit Sarkar
Keventer Agro Limited, India

Arup Nag
Massey University, New Zealand

CONTENTS

11.1 INTRODUCTION

Global demand for fermented milk products is booming due to consumers' increased interest in health-promoting foods. Probiotic cultures are capable of resisting and seeding in the intestinal environment and can be successfully incorporated in fermented milk products as a bioactive ingredient. It is of the utmost importance that in order to extend nutritional and health benefits, probiotics must colonize and maintain the equilibrium between the beneficial and pathogenic organisms in the gastrointestinal tract (GIT). Recently, fermented food products containing probiotic organisms have attracted increased consumer attention as innovative functional food due to their documented healthy image and their application as self-care complementary medicine (Sarkar, 2010, 2013). Kołozyn-Krajewskaa and Dolatowski (2012) declared that probiotic foods have a 60%–70% share of the total functional food market.

Probiotics are live microorganisms that confer health benefits to the host, when administered in sufficient quantities (FAO/WHO, 2001). The definition of probiotics has also been accepted and adopted by the International Scientific Association for Probiotics and Prebiotics (Hill et al., 2014). It has been established that multispecies probiotics are more efficacious than single species preparations (Chapman, Gibson, & Rowland, 2011; Hell, Bernhofer, Stalzer, Kern, & Claassen, 2013), and live microorganisms at a level 10^6 CFU/g are required for exhibiting health benefits (Foligne, Daniel, & Pot, 2013; Pinto et al., 2015). Recent reports revealed that the viability of probiotics may be unnecessary as dead or inactivated probiotic strains, or even their cellular extracts or bacterial cell wall components, can exhibit health beneficial effects (Adams, 2010; Awad et al., 2010; Gueniche et al., 2010; Hoang, Shaw, Pham, & Levine, 2010); however, live cells are more efficacious (Sarkar, 2018).

The optimum balance between beneficial and potentially harmful bacteria in the GIT is important for extending health benefits and can be modulated by dietary interventions such as probiotics. The viability of probiotics in any supplemented food is dependent on their metabolic activity in the specific food matrix, processing parameters, storage conditions, as well as their capability to withstand the GIT environment during gastric transit. The viability of probiotic is influenced by the contents of lactic acid, hydrogen peroxide, molecular oxygen, antimicrobial substances, artificial flavoring and coloring agents (Perricone, Bevilacqua, Altieri, Sinigaglia, & Corbo, 2015), low pH (Fang et al., 2012), enhanced osmotic stress (Bustos & Borquez, 2013; Fu & Chen, 2011), due to fermentation and the detrimental effect of higher processing temperatures (Boza, Barbin, & Scamparini, 2004). In addition to the stability of probiotics during food processing and storage (Anal & Singh, 2007), they must be tolerant to gastric acidity of the stomach and bile in the upper digestive tract (Cook, Tzortzis, Charalampopoulos, & Khutoryanskiy, 2012; Lee, 2014), while they also must reach live to host for exhibiting beneficial roles (Ying et al., 2016).

The proper selection of ingredients intended for use as the protective matrix is important and the application of a combination of two types of ingredients has been proven to be better for higher cell viability than using only one type of coating material (Maciel, Chaves, Grosso, & Gigante, 2014). Encapsulating materials are generally recognized as safe ingredients (Ei-Salam & Ei-Shibiny, 2012); should be food-grade, non-toxic, non-antimicrobial, and render the protection against acid and bile salts (Cook et al., 2012). Diverse materials used for encapsulation are carbohydrates, polysaccharides, plant hydrocolloids, bacterial exopolysaccharides, plant proteins, cellulose, starches, and gums (Anal & Singh, 2007; Yonekura, Sun, Soukoulis, & Fisk, 2014). Sodium alginate is the most commonly used encapsulating material as it confers protection to probiotic microorganisms from heat, pH, oxygen, and other factors during the processing and storage stages (Goh, Heng, & Chan, 2012). Application of alginate induced significantly higher viability of *Lactobacillus acidophilus* PTCC1643 and *Lactobacillus rhamnosus* PTCC1637, due to its preventive effect from acids in simulated gastric juice (Mokarram, Mortazavi, Habibi Najafi, & Shahidi, 2009). Use of milk proteins such as casein (Heidebach, Forst, & Kulozik, 2009) and whey protein (Doherty et al., 2012) for the encapsulation of probiotics may be preferred due to certain functional properties such as bland flavor, high solubility, emulsifying properties, film-forming nature, low viscosity in solutions (Heidebach et al., 2009; Madene, Jacquot, Scher, & Desobry, 2006) and desirable stabilizing effects (Chen, Remondetto, & Subirade, 2006). Significant improvement in survival of *L. bulgaricus* in alginate-milk microspheres may be due to excellent pH tolerance and less acid diffusion into the microspheres resulting from the buffering capacity of milk or formation of denser hydrogel network (Shi et al., 2013). Burgain, Gaiani, Cailliez-Grimal, Jeandel, and Scher (2013) also noted 99% survival of *Lactobacillus rhamnosus* GG due to microencapsulation with micellar casein and denatured whey proteins.

The exploitation of probiotic cultures which are stress-adapted, sonicated, resistant to acid and bile, the inclusion of micronutrient in the food matrix, the adoption of two-step fermentation, the use of oxygen-impermeable packaging materials, and microencapsulation have been suggested for retaining higher viability of probiotics (Sarkar, 2010). Microencapsulation of bacteria in the polymeric and biodegradable matrix is more promising than other techniques for conferring protection to the probiotics from unfavorable environments of processing and GIT transit (Chaikham,

Apichartsrangkoon, George, & Jirarattanarangsri, 2013; Favaro-Trindade, Heinemann, & Pedroso, 2011; Gbassi & Vandamme, 2012; Nedovic, Kalusevic, Manojlovic, Levic, & Bugarski, 2011; Ribeiro et al., 2014; Sousa et al., 2012). Encapsulation of bacteria can deliver a minimum of 10^7 CFU/g probiotics at the time of consumption (Xavier dos Santos et al., 2019) and retaining viability in the digestive tract (Silva, Cezarino, Michelon, & Sato, 2018). An improvement in organoleptic and nutritional qualities of encapsulated foods was noted as encapsulation aids food materials to be unaffected by processing and packaging conditions (Parra-Huertas, 2010). Spray drying is the most economical and extensively adopted technique for the encapsulation of bioactive compounds (Shishir & Chen, 2017; Shishir, Taip, Aziz, Talib, & Sarker, 2016) and the formulation of high value-added products possessing health properties (de Melo Barros et al., 2018).

Global demand for probiotic-containing foods is increasing and now accounts for a major share of the functional food market. Probiotic supplementation has been reported in dairy products (Laroia & Martin, 1991; Penna, Rao-Gurram, & Barbosa-Ca'novas, 2007; Vijayendra & Gupta, 2013a, 2013b), as well as non-dairy foods (Gupta & Abu-Ghannaman, 2012; Kumar, Vijayendra, & Reddy, 2015; Peres, Peres, Hernández-Mendoza, & Malcata, 2012). The global market for spray dried encapsulated probiotics is common for the formulation of functional food products (De & Mauriello, 2016), and is used in dairy products (49%), fruit and vegetable-based products (28%), meat products (13%), and bakery products (10%). Probiotics are stabilized and fortified into powdered foods or nutraceutical formulations and the dried probiotic powders must ensure the retention of viable probiotic bacteria after manufacture and throughout the shelf-life (FAO/WHO, 2006). Probiotic food products should preferably be stored at a temperature of 4°C–5°C (Tripathi & Giri, 2014; Weinbreck, Bodnár, & Marco, 2010) as an elevated temperature of 20°C can induce a significant decline in probiotic viability counts in dried products (Alves, Messaoud, Desobry, Correia Costa, & Rodrigues, 2016; Avila-Reyesa, Garcia-Suareza, Jiménez, Martín-Gonzalez, & Bello-Perez, 2014; Perdana, Fox, Siwei, Boom, & Schutyser, 2014).

Fenster et al. (2019) reported stability of probiotics up to 24 months at ambient temperature and humidity in dietary supplements and other dry food matrices. Transportation and storage of dietary supplements containing probiotics at ambient temperatures and humidity induced greater viability loss in comparison to refrigerated/frozen storage and handling. Sreeja and Prajapati (2013) suggested an excess amount (addition of much higher viable counts at the starting) for compensating viability losses during storage and handling and providing the target dose until the end of shelf-life. An overage is considered safe as a higher dose of probiotics is not unfavorable (Morovic et al., 2017; Zhou et al., 2000) and in few instances, a higher viable population may be beneficial (Ouwehand, Invernici, Furlaneto, & Messora, 2018). Probiotic cultures are expensive ingredients (usually costing > USD 1400/kg) and the need to add up to 10–100-fold overages to compensate for 1–2 log10 decline in viability further adds considerable cost to the final product.

11.2 THE SIGNIFICANCE OF PROBIOTIC VIABILITY

Survival of probiotics is essential to populate the human gut for exerting health benefits (Anselmo, McHugh, Webster, Langer, & Jaklenec, 2016; De & Mauriello, 2016), but there is no agreement among the International Scientific Community on effective probiotic dosages required for extending health effects; however, researchers have suggested minimum dosages of 10^6–10^9 CFU/day (Espirito Santo, Perego, Converti, & Oliveira, 2011) or 10^8–10^9 CFU/day (Bhadoria & Mahapatra, 2011; Ng, Yeung, & Tong, 2011). Minelli and Benini (2008) denoted that for exerting clinical effects, probiotic concentration should be $\geq 10^6$ CFU/ml in the small intestine and $\geq 10^8$ CFU/g in the colon. According to the International Dairy Foods Association, probiotic viability in any probiotic product should be more than 10^7CFU/g up to the date of minimum durability (Divya, Varsha, Nampoothiri, Ismail, & Pandey, 2012). Probiotic viability of $>10^6$–10^7 CFU/g is required up to the end of the shelf-life (typically 6–12 months for dried products) for conferring health benefits to the human host (FAO/WHO, 2003). Effective physicochemical barriers to avoid viability losses under unfavorable processing conditions must be provided to probiotics for their commercial exploitation (Burgain, Gaian, Linder, & Scher, 2011; Jankovic, Sybesma, Phothirath, Ananta, & Mercenier, 2010; Meng, Stanton, Fitzgerald, Daly, & Ross, 2008). A key factor for the successful marketing of probiotics as a functional food is the retention of the probiotic viability during processing, storage, and at the time of consumption (Sarkar et al., 2016).

In order to extend health benefits, probiotics must retain their viability under diverse unfavorable conditions and must reach the target site in a metabolically active form (Vandenplas, Huys, & Daube, 2015; Yarullina et al., 2015). Higher probiotic dosages are more efficacious in preventing antibiotic-associated diarrhea (Gao, Mubasher, Fang, Reifer, & Miller, 2010; Ouwehand et al., 2014), and the recommended minimum daily intake is 5×10^9 CFU/mL (Hayes & Vargas, 2016). Determination of the exact dosage is based on the selected probiotic, probiotic blend, and desired clinical outcome (Italian Health Ministry, 2013). Other reports also confirmed the requirement of higher minimum dosages of $>10^{10}$ CFU for the prevention of atopic dermatitis (Pelucchi et al., 2012) and $> 10^{11}$ for reducing blood pressure (Khalesi, Sun, Buys, & Jayasinghe, 2014). However, Zhang et al. (2016) could not note any significant differences in the beneficial effect of probiotics at dosages $<10^{10}$ CFU or $> 10^{10}$ CFU. The viability of probiotics is paramount for exhibiting health benefits (Table 11.1).

Viability is not always important as both alive and non-viable probiotics are efficacious (Adams, 2010; Sarkar, 2018; Zou, Dong, & Yu, 2009). In some instances, even dead cells provide health effects as not all cell mechanisms are directly related to viability (Dargahi, Johnson, Donkor, Vasiljevic, & Apostolopoulos, 2019), but viable probiotics are more promising (Pelletier, Laure-Boussuge, &

TABLE 11.1
Probiotic Viability for Exhibition of Health Effects

Diseases	Cell Viability (CFU/mL)	References
Irritable Bowel Syndrome	2.5×10^{10}	Roberts, McCahon, Holder, Wilson, and Hobbs (2013)
	3×10^9 to 6×10^9	Lorenzo-Zuniga et al. (2014)
Antibiotic associated diarrhea	$> 5 \times 10^9$ more effective than $<5 \times 10^9$	Johnston, Goldenberg, Vandvik, Sun, and Guyatt (2011)
Clostridium difficile associated diarrhoea	Similar efficacy $>10^{10}$ or $< 10^{10}$	McFarland, 2015
Systolic and diastolic blood pressure	$> 10^{11}$	Khalesi et al. (2014)
Acute gastroenteritis	$> 10^{10}$	Szajewska, Skórka, Ruszczyński, and Gieruszczak-Białek (2013)
Fecal lactobacilli recovery	10^9	Gianotti et al. (2010)
Influenza	5×10^9	Bosch et al. (2012)
Diarrhea	10^{10} more effective than 10^7	Van, Feudtner, Garrison, and Christakis (2002)
Necrotizing enterocolitis	$1.5–3.0 \times 10^9$	Deshpande, Rao, Patole, and Bulsara (2010)
Autism spectrum disorder	$4.5–10^{10}$	Umbrello and Esposito (2016)
Obesity	2×10^{10}	Gomes et al. (2017)
Hypercholesterolemia	3×10^9	Ivey et al. (2015)
Non-alcoholic fatty liver disease	5×10^8	Park et al. (2020)

Donazzolo, 2001) and certain reports revealed that lower concentration is more efficacious than higher concentration (Ritchie & Romanuk, 2012). Health benefits extended by probiotics differ with strains (Amara & Shib, 2015; Fontana, Bermudez-Brito, Plaza-Diaz, Munoz-Quezada, & Gil, 2013; Vemuri et al., 2018), and a single strain of probiotic cultures or their combinations is efficacious in preventing diverse diseases (Table 11.2). Thus, it can be concluded that in order to secure better health benefits, the viability of probiotics is of prime importance but not its concentration.

TABLE 11.2
Health-Promoting Properties of Probiotics

Probiotic Strains	Health Benefits	References
Lactobacillus rhamnosus GG + *Bifidobacterium lactis* Bb12	Promoted innate immune responses to human rotavirus	Sgouras et al. (2004)
Lactobacillus rhamnosus GR-1 + *Lactobacillus reuteri* RC-14	Resolved moderate diarrhea and increased CD4 count in HIV/AIDS patients	Anukam, Osazuwa, Osadolor, Bruce, & Reid, 2008
Lactobacillus helveticus R0052 + *Bifidobacterium longum* R0175	Anxiolytic effects	Messaoudi et al. (2011)
Bifidobacterium infantis 35,624	Alleviate symptoms of irritable bowel syndrome	O'Mahony et al. (2005)
L. reuteri NCIMB 701089	Cardiovascular diseases therapy	Tomaro-Duchesneau et al. (2014)
Lactobacillus paracasei N1115	Protect against the risk of acute upper respiratory tract infections	Pu et al. (2017)
Lactobacillus GG	Treatment of acute diarrhea	Szajewska et al. (2013)
Bifidobacterium bifidum R0071 + *B. infantis* R0033 + *L. helveticus* R0052	Boost infant's immunity by changing the salivary immunoglobulin A and the digestive system	Xiao et al. (2017)
Lactobacillus paracasei subsp. paracasei DC412 + *Lactobacillus acidophilus* NCFB 1748	Induced early innate immune responses and specific immune markers through phagocytosis, polymorphonuclear cell recruitment, and TNF-alpha production	Kourelis et al. (2010)
L. acidophilus DDS-1	Immunomodulatory effect	Vemuri et al. (2018)
Lactobacillus casei strain Shirota	Stimulate immune responses and prevent enterobacterial infections	Matsuzaki (1998)
L. rhamnosus KL53A + *Bifidobacterium breve* PB04	Lower incidence of staphylococcal sepsis, increase in counts of *Lactobacillus* and *Bifidobacterium* counts in the gut microbiota,	Strus et al. (2018)
Bifidobacterium animalis AHC7	Antimicrobial activity against indicator pathogenic organisms	O'Mahony et al., 2009
L. acidophilus CL1285, *L. casei* LBC80R, *L. rhamnosus* CLR2, Bio-K+	Prevent *Clostridium difficile* infection in patients receiving antibiotics.	McFarland, Ship, Auclair, & Millette, 2018
Lactobacillus rhamnosus CNCM I-3690	Protects the intestinal barrier by stimulating both mucus production and cytoprotective response	Martin et al. (2019)
Lactobacillus GG	Alleviate symptoms of *C. difficile* infection	Biller, Katz, Flores, Buie, and Gorbach (1995)

(Continued)

TABLE 11.2
(*Continued*)

Probiotic Strains	Health Benefits	References
Lactobacillus plantarum	Regulate tight-junction proteins to protect against chemical-induced disruption of the epithelial barrier	Karczewski et al. (2010)
Lactobacillus acidophilus DDS-1	Adhesion capacity, the immunomodulatory effect	Vemuri et al. (2018)
Lactobacillus acidophilus CL1285 and *Lactobacillus casei* LBC80R	Antibiotic-associated diarrhea	Gao et al. (2010)
Lactobacillus rhamnosus strain GG (LGG)	Inhibit human gastric cancer cells and colonic cancer cells	Orlando and Russo (2013)
L. reuteri ATCC 55730	Diarrhea	Shornikova, Isolauri, Burkanova, Lukovnikova, and Vesikari (1997)
Lactobacillus rhamnosus LCR35	Reduced rate of necrotizing enterocolitis	Bonsante, Iacobelli, and Gouyon (2013)
Lactobacillus reuteri DSM17938	Prevents diarrhea	Urbanska, Gieruszczak-Bialek, Szymanski, and Szajewska (2016)

11.3   FACTORS AFFECTING THE VIABILITY OF PROBIOTICS

Diverse factors affecting the viability of probiotics may be broadly categorized into three major groups as product environment, food processing and storage, and gastrointestinal environment (Sarkar, 2018). The most critical factors affecting the viability of probiotics during formulation, storage, and gastrointestinal transit are product characteristics, such as food ingredients and additives used, temperature, pH, water activity, oxygen contents, and redox potentials of food matrix (Tripathi & Giri, 2014), technological conditions, such as heat, mechanical stress or osmotic stress (Bustos & Borquez, 2013; Fu & Chen, 2011), packaging conditions (Mortazavian, Mohammadi, & Sohrabvandi, 2012) and storage conditions, such as temperature, relative humidity, and duration of storage (Min, Bunt, Mason, & Hussain, 2019).

Probiotics are now extensively applied as a functional ingredient during functional food formulations. The selection of probiotic cultures should be based on technological and health beneficial aspects as both intrinsic and extrinsic properties influence the viability of probiotics in food. All probiotics and even all strains of a particular probiotic genus are not equally efficient to exhibit positive outcome; therefore, probiotic selection should be done based on the following characteristics:

- Ability to adhere to mucosal and epithelial surfaces;
- Possess antimicrobial spectrum against pathogenic bacteria;
- Must be acid and bile tolerance (Kechagia et al., 2013);
- Must have antibiotic resistance and potential efficacy (Sorokulova, 2008);
- Must be safe for human use (Gueimonde, Sanchez, & de los Reyes-Gavilan, C.G. and Margolles, A., 2013);
- Should be of human origin;
- Should not have pathogenicity and toxicity;
- Should not possess transferable antibiotic resistance genes (Saarela, Mogensen, Fonden, Matto, & Mattila-Sandholm, 2000);
- Suitable for large-scale production (Lacroix & Yildirim, 2007);

- Physiological state of the probiotic organisms added during product formulation;
- Must be stable at storage conditions of probiotic foods;
- Metabolic activities should not be affected by the chemical composition of the food matrix;
- Must have interactive behavior with other starter cultures (Khan, 2014);
- Must have phenotype and genotype stability including plasmid stability;
- Capable of utilizing carbohydrate and protein (Wedajo, 2015);
- Capable of producing exopolysaccharide (Han et al., 2016); and,
- Capable of withstanding processing and environmental factors such as low pH and heat (Fang et al., 2012; Iannitti & Palmieri, 2010).

11.4 AN OVERVIEW OF PROBIOTIC MICROENCAPSULATION

Microencapsulation may be defined as a technology of packaging solids, liquids, or gaseous materials in a sealed capsule, capable of releasing their contents at controlled rates under specific conditions (Doraisamy, Karthikeyan, & Elango, 2018). Encapsulation provides a barrier and protects sensitive bioactive materials such as probiotic microorganisms during processing and GIT transit (Nedovic et al., 2011). The application of encapsulation in pharmaceutical industries, agrochemical, cosmetics (Suganya & Anuradha, 2017), biotherapeutics (Tomaro-Duchesneau et al., 2014), and biotechnology (Farheen, Shaikh, & Shahi, 2017) has been reported. Originally, encapsulation was introduced for biotechnological applications to make production processes by efficiently separating the producer cells from the metabolites, but presently adopted by the food industry for the addition of functional compounds in food products including probiotics (Champagne & Fustier, 2007).

Immobilization and encapsulation are two terms that are often used synonymously. Basically, in the case of immobilization materials are trapped in material within or throughout the matrix, whereas in the case of encapsulation, the core particle is continuously coated around an inner matrix so that it is surrounded by the capsule wall (Gauri & Shiwangi, 2017). Encapsulation may occur naturally when due to the production of bacterial exopolysaccharides, which act as a protective layer, and entrapped cells are less exposed to adverse or unfavorable environmental conditions during production, storage, and GIT transit. All probiotics are not capable of synthesizing exopolysaccharides or produced exopolysaccharides that may not be adequate to encapsulate them completely (Shah, 2000).

Microencapsulation has many advantages over the immobilization technique. Encapsulated bacterial cells are permeable to nutrients, gases, and metabolites for retaining cell viability within the beads (Ding & Shah, 2009; John, Tyagi, Brar, Surampalli, & Prevost, 2011). Microencapsulation is preferred over immobilization due to the formation of relatively smaller sized particles, a larger area for diffusion of nutrients, the easy separation of cells, minimal cell washout (Liu, Ren, & Yao, 2010; Tan, Heng, & Chan, 2011), tolerance to stress factors (Park & Chang, 2000), and controlled release of encapsulated cells (Kailasapathy, 2002; Ye et al., 2018). Microcapsules are protected against adverse reactions such as lipid oxidation of bacterial cell and adverse storage, handling conditions, and environmental factors, such as temperature, oxygen, water, pH changes, and light, and during their passage through GIT or avoid their interaction with food components (Sobel, Versic, & Gaonkar, 2014).

Microencapsulation by spray drying technique was first introduced in 1865 for the drying of eggs. Spray drying was adopted at an industrial scale for producing milk and washing powder in the 1920s and afterward, it was extensively used by diverse processing industries (Silveira, Perrone, Rodrigues Junior, & de, 2013). Lactic acid bacteria were first immobilized in 1975 and later applied for continuous yogurt fermentation (Linko, 1985; Gibbs, Kermasha, Alli, & Mulligan, 1999). Microencapsulation is now successfully employed for encapsulating a wide range of bioactive food ingredients, such as polyphenols, lipids, proteins, vitamins, minerals, enzymes, flavors, and probiotic organisms (Jeyakumari, Zynudheen, & Parvathy, 2016). A schematic diagram of a spray dry encapsulation process along with an image of a Mini Spray Dryer B290 (BÜCHI), available at TECNALIA, is shown in Figure 11.1.

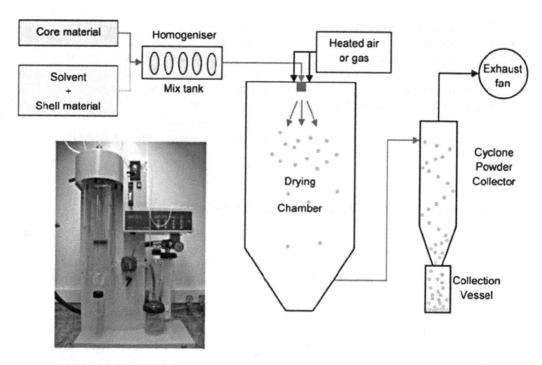

FIGURE 11.1   Schematic diagram of a spray dry encapsulation process and image of a Mini Spray Dryer B290 (BÜCHI), available at TECNALIA (Chavarri, Maranon, & Villaran, 2012).

11.5   THE SPRAY DRYING ENCAPSULATION OF PROBIOTIC CULTURES

Makinen, Berger, Bel-Rhlid, and Ananta (2012) stated that probiotic bacteria can be incorporated into a food product either by incorporating probiotics directly into a carrier like milk or yogurt, followed by fermentation (to yield the final product), or inclusion of dried or encapsulated probiotic microorganisms directly added to the product. Korbekandi, Mortazavian, and Iravani (2011) suggested that oxygen can affect probiotic viability as it is directly toxic to some cells; certain cultures produce toxic peroxides in the presence of oxygen and free radicals produced from the oxidation of components are toxic. Reduction of dissolved oxygen level, application of ruptured or microencapsulated cells, and adding probiotics are considered as the means for maintaining probiotic bacteria required for exhibiting therapeutic effects (Ozen, Pons, & Tur, 2012). The efficiency of encapsulation by the spray drying method is dependent upon the nature of the encapsulating material, the viscosity of the oil-in-water (O/W) emulsions, oil/solid content ratio, and the spray drying parameters including flow rate of gas and liquid, and inlet and outlet temperature of air (Estevinho, Rocha, Santos, & Alves, 2013; Turchiuli, Jimenez Munguia, Hernandez Sanchez, Cortes Ferre, & Dumoulin, 2014). Recommended process parameters for spray drying encapsulation of probiotics have been depicted in Table 11.3. The following factors must be considered during the spray drying of probiotics.

11.5.1   ENCAPSULATING MEDIA

Coated material(s) or core material(s) are those which are being coated; the coating material(s) used are known as packing material, capsule, wall material, film, membrane, carrier, or outer shell (Fang & Bhandari, 2010). The coating protects the active component (core) from environmental stresses, such as acidity, oxygen, and gastric conditions. The adoption of encapsulating technology enables direct and continuous delivery of probiotic bacteria to the gut, where it effectively disintegrates to provide higher viable probiotic bacteria to confer benefits (Gbassi & Vandamme, 2012).

TABLE 11.3
Spray Drying Encapsulation of Probiotics Reported in the Literature So Far

Probiotic Cultures	Encapsulating Materials	Spray Drying Temperature		References
		Inlet	Outlet	
L. acidophilus	20% Maltodextrin +10% skim milk	150°C	80°C	Yudiastuti, Sukarminah, Mardawati, and Kastaman (2019)
L. paracasei NFBC	Gum Arabic, reconstituted skim milk	170°C	95°C	Desmond, Ross, O'Callaghan, Fitzgerald, and Stanton (2002).
B. infantis	Canola oil, caseinate, fructooligosaccharide, dehydrated glucose syrup, or starch	160°C	65°C	Crittenden, Weerakkody, Sanguansri, and Augustin (2006)
Lactobacilli acidophilus	Microcrystalline cellulose, skimmed milk, and lactose,	140°C	-	Lohiya and Avari (2016)
Lactobacilli acidophilus	Chitosan-inulin or chitosan-maltodextrin (1: 15 or 1: 25)	130°C	-	Flores-Belmont, Palou, Lopez-Malo, and Jimenez-Munguia (2015)
Bifidobacterium cells	Gelatinized modified starch	100°C	45°C	O'Riordan, Andrews, Buckle, and Conway (2001)
Lactobacillus paracasei NFBC 338 + *Lactobacillus salivarius* UCC 118	skim milk	-	80°C to 85°C	Gardiner et al. (2000)
L. acidophilus NCDC 016	Maltodextrin, gum Arabic	-	55 ± 2°C	Arepally and Goswami (2019)
Lactobacillus acidophilus + *Lactobacillus rhamnosus*	Maltodextrin	-	-	Anekella and Orsat (2013)
Lactobacillus rhamnosus	Rice starch, inulin	135°C to 155°C	-	Avila-Reyesa et al., 2014
		150°C	-	Behboudi-Jobbehdar, Soukoulis, Yonekura, and Fisk (2013)
Lactobacillus acidophilus La-5 *Bifidobacterium lactis* (Bb-12)	Cellulose acetate phthalate	130°C	75°C	Favaro-Trindade and Grosso (2002)
Bifidobacterium breve R070 *Bifidobacterium longum* R023	Milk, whey protein	-	-	Picot and Lacroix (2004)
L. paracasei NFBC	Gum Arabic, reconstituted skim milk	170°C	95°C to 105°C	Desmond et al. (2002)
Lactobacillus casei UFPEDA	25% (w/v) Maltodextrin	70°C		dos, Finkler, and Finkler (2014)
B. lactis INL1 + *B. lactis* BB12	20% (w/v) Skim Milk	137.5°C	82.5°C	Burns et al. (2017)
Lactobacillus casei subsp. *rhamnosus* TISTR 047		170°C	80°C	Chaipojjana, Phosuksirikul, and Leejeerajumnean (2014)

(Continued)

TABLE 11.3
(*Continued*)

Probiotic Cultures	Encapsulating Materials	Spray Drying Temperature		References
		Inlet	Outlet	
Lactobacillus acidophilus La-5	Sweet Whey or skim milk (30% Total Solids)	180°C	85°C to 95°C	Maciel et al. (2014)
Bifidobacterium animalis spp. *lactis*	Maltodextrin, inulin, oligofructose	140°C	65°C	Paim, Costa, Walter, and Tonon (2016)
L. casei Shirota	Reconstitute skim milk, gum Arabic	119.55°C	-	Gul and Dervisoglu (2020)
Lactobacillus paracasei LMG S-27487	Maltodextrin, inulin	150°C	-	Spigno, Garrido, Guidesi, and Elli (2015)
Bacillus coagulans	Fructooligosaccharide, maltodextrin, Skimmed milk, xanthan gum	110°C	-	Pandey and Vakil (2017)
Lactobacillus acidophilus NCIMB 701748	Whey protein concentrate, D-glucose, Maltodextrin, sodium alginate Hydroxypropyl methylcellulose	134°C	76°C	Yonekura et al. (2014)
	Maltodextrin, skim milk powder, sodium caseinate, whey protein concentrate	134°C	76°C	Soukoulis, Behboudi-Jobbehdar, Yonekura, Parmenter, and Fisk (2014)

Natural or modified polysaccharides, gums, proteins, lipids, and synthetic polymers are generally used as the coating material (de Souza Simoes et al., 2017). The selection of coating materials depends upon the nature of the core material, the encapsulation process, and applied product. Carbohydrates such as alginate, maltodextrin, gums, starch, chitosan, or cellulose derivatives, and proteins are generally used as wall materials for the spray drying encapsulation of probiotics (Kavitake, Kandasamy, Devi, & Shetty, 2018). Pandey and Vakil (2017) denoted that amongst various encapsulation materials like fructooligosaccharide, maltodextrin, skimmed milk, and xanthan gum, the highest protection to *Bacillus coagulans* was conferred by skimmed milk and the lowest by xanthan gum.

Prebiotics are food ingredients that are not digested in the human upper intestinal tract and on arrival in the colon, but are selectively fermented by a limited number of colonic bacteria (van, Coussement, De, Hoebregs, & Smits, 1995). Conjugate application of prebiotics with probiotics results in an improvement in the survivability of probiotics due to the availability of substrate for its fermentation (Sarkar, 2007). Attempts were executed for greater viability of spray dried probiotics with the conjugated application of coating materials with prebiotic (Fritzen-Freire et al., 2012) or fruit juices (Kingwatee et al., 2015). An attempt has been made for the double microencapsulation of probiotics by spray drying (Flores-Belmont et al., 2015). For double microencapsulation, initially spray dried *L. acidophilus* using gelatin-maltodextrin (1: 25) solution was added to chitosan-inulin or chitosan-maltodextrin (1: 25) solution and then again spray dried. When exposed to simulated gastrointestinal conditions, a lower reduction in probiotic viability (3 logs vs. 7 logs) was encountered due to double microencapsulation than simple microencapsulation (Flores-Belmont et al., 2015). However, no significant difference in the viability of spray dried *B. animalis* spp. *lactis* could be noted with the use of jussara juice, along with different combinations of wall materials such as maltodextrin, maltodextrin + inulin, maltodextrin + oligofructose, or maltodextrin + inulin + oligofructose, and encountered a reduction of approximately 1 log cycle (Paim et al., 2016). Higher emulsion viscosity of gum Arabic at higher concentration forms a protective layer of coating on the bacterial cell wall, which hinders water molecule diffusion rate, thereby resulting in lower viability loss during spray drying (Arepally & Goswami, 2019). In an early investigation, Rajam and Anandharamakrishnan (2015) also encountered higher encapsulation efficiency (70.77% to 72.82%) with the application of fructooligosaccharides and whey protein isolate. Coating materials should have the following features:

- They must be food-grade and capable of forming a barrier between the active agent and its surroundings (Zuidam & Nedovic, 2010);
- They should be non-toxic, non-antimicrobial, and capable of resisting acid and bile salts (Cook et al., 2012);
- They enable the release of encapsulated compounds at a specific rate under certain conditions (Shishir, Xie, Sun, Zheng, & Chen, 2018);
- They should be capable of retaining flavors and aromas (Bakry et al., 2016);
- They should have low viscosity at high concentration, good solubility, emulsification, and film-forming and efficient drying properties (Shishir, Taip, Saifullah, Aziz, & Talib, 2017; Sultana et al., 2017);
- They should have good biocompatibility and gel-forming ability (Xu, Gagne-Bourque, Dumont, & Jabaji, 2016);
- They should be biodegradable; and,
- They should have low hygroscopicity and be economic (Silva, Stringheta, Teófilo, Rebouças, & Oliveira, 2013).

11.5.2 Processing Conditions

Outlet air temperature is the most critical drying parameter affecting the viability of spray dried starter cultures and is dependent on the inlet air temperature, airflow rate, product feed rate, medium composition, and atomized droplet size (Boza et al., 2004, Santivarangkna, Kulozik, & Foerst, 2007, 2008). Spray drying of *Bifidobacterium* cells at higher inlet temperatures (>120°C) resulted in higher outlet temperatures (>60°C), causing a significantly reduced viability of encapsulated cells. Chavez and Ledeboer (2007) announced a linear decline in the logarithmic number of probiotics with an elevation in outlet air temperature during the spray drying process from 50°C to 80°C. Gardiner et al. (2000) reported a 97% survival rate of *L. paracasei* NFBC 338 after spray drying at an outlet air temperature of 70°C, but the survival rate reached zero with the increase in temperature to 120°C. Arepally and Goswami (2019) also noted a decline in the viability of *L. acidophilus* NCDC 016 (9.97 to 7.3 log CFU/g) with the increasing inlet air temperature. The decline in viability may be attributed to cellular injuries arising from denaturation of DNA and RNA, dehydration of cytoplasmic membranes, rupture, and the collapse of the cell membrane due to water removal (Behboudi-Jobbehdar et al., 2013).

No significant difference in the viability for *L. acidophilus* La-5 and *B. lactis* Bb-12 could be noted at drying temperature of 110°C, 120°C, and 130°C (Nunes et al., 2018); however, lower viability of *L. acidophilus* (Nunes et al., 2018) and *Saccharomyces cerevisiae* var. *boulardi* (Arslan, Erbas, Tontul, & Topuz, 2015) were reported at 140°C. Nevertheless, in an earlier investigation, spray drying at an inlet temperature of 130°C and outlet temperature of 75°C, no reduction in the viability of *B. lactis* Bb-12 was noted but the viability of *L. acidophilus* La-5 reduced by two log cycle (Favaro-Trindade and Grosso (2002). Results indicate that all probiotic cultures are not equally stable at particular processing conditions during spray drying. Sunny-Ghandi, Powell, Chen, and Adhikari (2012) also encountered an increase in the survival rate of *Lactococcus lactis ssp. cremoris* from 0.1% to 14.7% with the decrease of inlet/outlet temperature (from 200°C to 65°C to 130°C/38°C). Sunny-Roberts and Knorr (2009) established that the survival rate of spray dried *L. rhamnosus* was inversely proportional to the outlet air temperature. Recommended inlet and outlet temperatures were 100°C and 45°C, respectively (O'Riordan et al., 2001).

It has been reported that microorganisms entrapped in the particles are subjected to more heat damage, causing greater loss of viability of bacterial cultures (Santivarangkna et al., 2007) and is influenced by the residence time (Santivarangkna, Higl, & Foerst, 2008). Atomized droplets have a very large surface area due to the formation of large numbers of micrometer-sized droplets (10–200 mm); thus, they require very short drying time during drying (Morgan, Herman, White, & Vesey, 2006; Santivarangkna et al., 2007). Recent reports stated that atomization induced a significant reduction in survival of *Lactococcus lactis* ssp. *cremoris* (Ghandi, Powell, Howes, Chen, & Adhikari, 2012); however, an increment in *Lactobacillus rhamnosus* GG due to reorganization was noted (Guerin et al., 2017). Nazzaro, Orlando, Fratianni, and Coppola (2012) reported that 1–1000 mm sized microcapsules of lactic acid bacteria and probiotics were stable both in the food matrix as well as during GIT transit. The effect of atomization on probiotic viability should also be considered during probiotic selection.

11.5.3 STORAGE CONDITIONS

During two months of storage at 4°C, the viability (CFU/g) of *Lactobacillus paracasei* NFBC 338 declined slightly from 3.2×10^9 to 1×10^9, whereas the counts of *Lactobacillus salivarius* UCC 118 declined from 5.2×10^7 to 9.5×10^6 (Gardiner et al., 2000). Abd-Talib, Mohd-Setapar, Khamis, Nian-Yian, and Aziz (2013) observed a decline in viability (CFU/ml) of *Lactobacillus plantarum* B13 (1.28×10^8 to 2.10×10^6) and *Lactobacillus plantarum* B18 (3.25×10^7 to 2.15×10^7) due to spray drying, which respectively reached a level of 3.00×10^5 and 4.00×10^2 during two weeks storage at 25°C. Spray dried maltodextrin-encapsulated *Lactobacillus casei* had a viable count (CFU/g) of 1.1×10^{10}, which declined to 1.0×10^9 and 3.0×10^8, respectively, after 60 days of storage at −8°C and 4°C (dos et al., 2014). Results indicated that all probiotics are not equally stable during storage and for greater retention of probiotic viability, lower storage temperature should be preferred.

11.6 DESICCATION OF BACTERIAL CELLS DUE TO SPRAY DRYING ENCAPSULATION

Spray drying encapsulation technique can be applied for food ingredients; due to the lower survival rate of probiotics during drying and its subsequent storage, however, its industrial application may be limited. The encapsulated state of bacteria achieved by the spray drying process is equivalent to an inert state that is achieved by desiccation and the process is also referred to as anhydrobiosis. Anhydrobiosis is a phenomenon related to the partial or total desiccation of live organisms, retaining their vital functions after rehydration. In this state, the bacterium stops its metabolism for a temporary period (Garcia, 2011), when the water content in an anhydrobiote is <0.1 g of free water per gram of dry cell weight (Alpert, 2005). Desiccation tolerance is generally defined as the ability of an organism to reach desiccation up to a state of equilibrium with moderately or extremely dry air, followed by regaining its normal functions after rehydration. In a more specific definition, the anhydrobiosis or stabilized state is believed to be achieved when the free water content is reduced to less than 0.1 g for every gm of dry cell mass of the bacterium (Alpert, 2005). Mechanisms involved in desiccation tolerance during drying are balancing dehydration in the tolerant cells due to reversible interactions with other molecules (Rebecchi, Altiero, & Guidetti, 2007), thereby protecting biomolecules and structures to retain their original form after rehydration (Wolkers, Tablin, & Crowe, 2002). Potts et al. (2005) reported that the exact effects and explanation of the post-desiccation behavior of lactic acid bacteria are still unknown. It has been postulated that due to accumulation of Mn^{2+} inside the cells and having a high Mn^{2+}: Fe^{2+} ratio, gram-positive bacteria are more tolerant to desiccation (Daly et al., 2004), due to the protection of proteins from oxidative damage after desiccation (Fredrickson et al., 2008; Poddar et al., 2014). Figure 11.2 shows a schematic diagram for the mechanisms of damage due to desiccation.

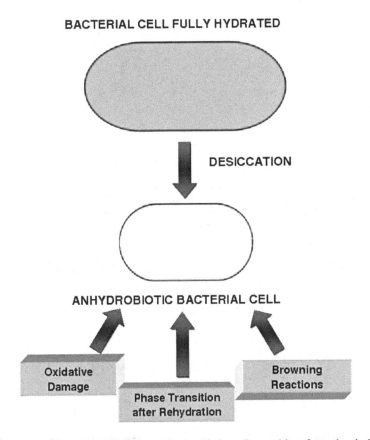

FIGURE 11.2 The mechanisms of damage to the probiotic cells resulting from the desiccation process (Garcia, 2011).

11.7 PROBLEMS ASSOCIATED WITH THE VIABILITY LOSS OF PROBIOTICS DURING SPRAY DRYING

Significant viability loss of probiotics encountered during spray drying and their subsequent storage is attributed to changes in the physical state of membrane lipids or the bacterial cell protein structure (Fritzen-Freire, Prudêncio, Pinto, Munñoz, & Amboni, 2013; Ghandi, Powell, Howes, et al., 2012; Semyonov et al., 2010). Principal mechanisms involved in the viability loss of probiotics are heat stress and dehydration (Janning, & in't Veld, P.H., 1994; Perdana et al., 2013). Inlet temperature, outlet temperature, and residence time of probiotics in the drying chamber determine the extent of heat stress and its effect on the viability of probiotics. During spray drying, probiotic cells are not exposed to a higher inlet temperature for a longer time, but retain for a longer time in the drying chamber until the end of the drying cycle that causes their heat inactivation (Broeckx, Vandenheuve, Claes, Lebeer, & Kiekensa, 2016). This indicates that the outlet temperature is a critical parameter that influences cell viability (Behboudi-Jobbehdar et al., 2013; Favaro-Trindade & Grosso, 2002; Ghandi, Powell, Chen, & Adhikari, 2012). Bielecka and Majkowska, 2000) reported higher survival of yogurt cultures at lower outlet air temperature (60°C–65°C) than those noted at higher outlet air temperature (80°C). Recommended outlet air temperature is 70°C–75°C to ensure satisfactory survival of yogurt cultures in yogurt powder. The viability of spray dried probiotics is also influenced by the strains of probiotics and their storage conditions. Gardiner et al. (2000) reported a sharp decline in survival rates of spray dried *L. paracasei* NFBC 338 and *L. salivarius* UCC 118 with an increment in storage temperature from 4°C to 15°C or 30°C (Figure 11.3).

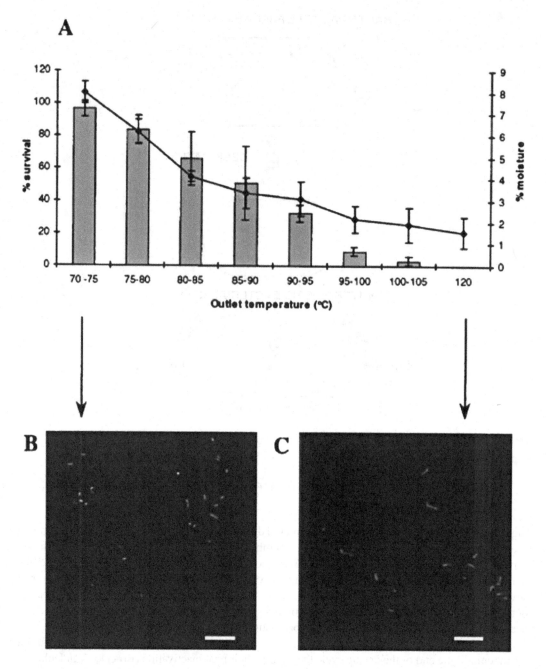

FIGURE 11.3 (A) The survival of *L. paracasei* NFBC 338 during spray drying in 20% (wt/vol) RSM supplemented with 0.5% (wt/vol) yeast extract at different air outlet temperatures (bar graph). The line shows the moisture contents of the resulting powders. The air inlet temperature was maintained at 170°C. (B and C) Confocal micrographs of NFBC 338-containing powders produced at air outlet temperatures of 70°C to 75°C (B) and 120°C (C). The powders were stained with the LIVE/DEAD BacLight viability stain; live cells are green, and dead cells are red. Bars = 10 μm (Gardiner et al., 2000).

11.8 TECHNIQUES FOR THE GREATER SURVIVAL OF PROBIOTICS DURING SPRAY DRYING

Loss of probiotic viability during spray drying is due to heat stress and dehydration (Janning, & in't Veld, P.H., 1994; Perdana et al., 2013), which result in the leakage of the intracellular substances due to the formation of cellular pores (Anekella & Orsat, 2013) and an increase in the cell permeability (De & Mauriello, 2016). Probiotic cells are prone to heat-inactivation at the last stage of spray drying (Peighambardoust, Tafti, & Hesari, 2011) and the extent of heat-inactivation cannot be generalized as it is strain-specific and is dependent on the growth stage (Fu & Chen, 2011). Process optimization of the spray drying is emerging for greater viability of microencapsulated probiotics (Fang & Bhandari, 2010; Kandansamy & Somasundaram, 2012; Krishnaiah, Nithyanandam, & Sarbatly, 2014; Nesterenko, Alric, Silvestre, & Durrieu, 2013). Techniques adopted for greater survival of spray dried probiotics are summarised in Table 11.4.

11.8.1 PROPER SELECTION OF PROBIOTICS

Not all probiotic cultures are equally stable to spray drying conditions, leading to variability in probiotic viability and are influenced by species and strain, thermal and osmotic adaptation as well as growth stages (Champagne, Ross, Saarela, Hansen, & Charalampopoulos, 2011; Corcoran, Ross, Fitzgerald, & Stanton, 2004; Fu & Chen, 2011). Probiotics with greater heat and oxygen tolerance had better survival after spray drying (Dijkstra et al., 2014; Lavari et al., 2015). Diversity in the

TABLE 11.4

Techniques for Greater Survival of Spray Dried Probiotics

Factors Affecting Viability	Recommendations
Selection of probiotics	Greater resistance to heat and oxygen
	Proper media and optimum growth conditions
	Capable of thermal and osmotic adaptation
	Gram-positive bacteria
	Tolerant to harsh drying conditions
Cell harvesting	Harvesting at the stationary phase of growth
	Higher cell concentration ($>1 \times 10^8$ cells/ml)
Processing parameters	Lower inlet temperature (135°C) and outlet temperatures (65°C)
	Adaptation of bacteria to sub-lethal heat treatment
	Drop-in spray pressure
Proper selection of coating materials	Inclusion of prebiotics in the medium
	Coating of spray dried capsules with an additional layer
	Application of optimum concentration of coating materials
Application of protectants	Exclusive use of whey proteins
	Partial replacement of skim milk powder with acacia gum in the growth or drying medium

probiotic viability was noted with different species of the same genera and different strains of the same species during spray drying and subsequent storage (Simpson, Stanton, Fitzgerald, & Ross, 2005), due to variability in the extent of dehydration (Santivarangkna et al., 2007). *Propionibacteria freudenreichii* ITG P20 was found more stable than *Lactobacillus casei* BL23 at higher inlet and outlet temperatures during spray drying (Huang et al., 2016). It has been reported that the viability of bacterial cultures during spray drying is also governed by growth conditions and media. During spray drying, starter cultures accumulate solutes from a highly concentrated extracellular environment (Morgan et al., 2006; Santivarangkna et al., 2007), which results in the stabilization of proteins and the cell membrane from the effects of osmotic stress (Morgan et al., 2006).

Process parameters (e.g., inlet and outlet temperatures and drying time), product parameters (e.g., type and concentrations of carrier medium), and pre-treatments of bacterial cells also affect the viability of probiotics during spray drying (Desmond, Stanton, Fitzgerald, Collins, & Paul Ross, 2001; Lian, Hsiao, & Chou, 2002; O'Riordan et al., 2001). Paez et al. (2012) reported that the survival rate of probiotics is affected not only by spray drying conditions and carrier but also by the strain of probiotics. Gram-positive bacteria are more stable than gram-negative bacteria during the spray drying process. Gram-negative bacteria are surrounded by a thin peptidoglycan cell wall; by contrast, gram-positive bacteria lack an outer membrane but are surrounded by layers of peptidoglycan many times thicker than those found in the gram-negative bacteria (Silhavy et al., 2010). The presence of a thick peptidoglycan layer in gram-positive bacteria (40 or more layers) compared to gram-negative cells (1–5 layers) surrounding the cell membrane has been reported (Donsi, Ferrari, Lenza, & Maresca, 2009; Fu & Chen, 2011; Pispan, Hewitt, & Stapley, 2013).

11.8.2 PROPER CELL HARVESTING

Survival of bacterial cells after drying is also affected by their growth phase, cell concentration, drying medium, and growth media in addition to drying, storage, and packaging conditions (Meng et al., 2008; Morgan et al., 2006). It has been established that microbial cells in the stationary phase are more tolerant to various stresses than log-phase cells (Corcoran, Stanton, Fitzgerald, & Ross, 2005; Michida et al., 2006), due to the greater resistance of stationary phase microorganisms to osmotic and heat stress (Van de Guchte et al., 2002). Corcoran et al. (2004) recorded a greater recovery of *Lactobacillus rhamnosus* after desiccation when harvested at the stationary phase (31%–50%) than those harvested at the early-log phase (14%) and lag phase (2%). Morgan et al. (2006) pointed out that the higher viability of stationary phase cells after desiccation may be due to their greater adaptation toward subsequent drying stresses resulting from starvation of carbon and other nutrients and recommended cell concentration of $>1 \times 10^8$ cells/mL before drying to ensure higher cell viability after desiccation (Morgan et al., 2006).

11.8.3 MANIPULATION OF PROCESSING PARAMETERS

Among the factors affecting the viability of probiotics during and after spray drying are media, stress treatment, drying temperature, and water activity (Daemen & van der, 1982). It has been reported that irreversible alterations in the functional integrity of bacterial membranes and proteins at higher spray drying temperatures lead to the inactivation of probiotics (Bustamante, Villarroel, Rubilar, & Shene, 2015). Thermal effect and loss of bound water from the cytoplasmic membrane induced injury to spray dried bacteria (Santivarangkna, Higl, & Foerst, 2008), and the threshold temperature for the microbial cell damage is within the range of the upper limit of the microbial growth temperature (Foerst & Kulozik, 2011). Scanning electron microscopy of a single dried droplet with skimmed milk at 90°C/5 min and 110°C/4 min revealed fine holes at the surface of *Lactococcus cremoris* but the cell surface remained intact at 70°C/6 min (Fu, Woo, Selomulya, & Chen, 2013).

O'Riordan et al. (2001) encountered a significant decline in the viability of encapsulated bifidobacteria with the application of higher inlet temperature (>120°C) and outlet temperatures (>60°C).

Controlling the inlet/outlet temperature of the spray dryer may minimize viability losses during spray drying. It has been established that the outlet temperature is a more critical parameter influencing its viability of encapsulated bacterial cells. Another way to retain the higher viability of probiotics without manipulating the inlet temperature is by modulating the feed rate. An increase in feed rate induced higher viability *L. plantarum* CIDCA 83114 during spray drying at a constant inlet temperature of 180°C, due to a drop in the outlet temperature from 85 to 70 °C (Golowczyc, Silva, Abraham, De, & Teixeira, 2010). Schuck, Dolivet, Méjean, Hervé, and Jeantet (2013) also reported higher viability (10^{10}) of spray dried *Propionibacteria acidipropionici* using sweet whey with an inlet temperature of 140°C and an outlet temperature of 60°C. According to Romano et al. (2014), the highest survival rates of *L. rhamnosus* GG (56.53%) and *L. rhamnosus* RBM 526 (52.63%) were recorded at an inlet temperature of 135°C and an outlet temperature of 65°C. It has been established the extent of viability loss (log CFU/ml) of *Lactobacillus salivarius* NRRL B-30514 decreased (about 4.60 to 0.56) with the lowering of outlet temperature (98°C–100°C to 70°C–72°C) irrespective of heat adaptation and media employed, least being noted for heat-adapted bacteria (Zhang et al., 2016). An additional advantage obtained due to the adoption of lower outlet temperature is improved storage stability of probiotics (Desmond et al., 2002). To and Etzel (1997) encountered greater delayed lactic acid production by spray dried probiotics such as *Lactococcus cremoris*, *Lactobacillus casei* ssp. *pseudoplantarum*, and *S. thermophiles* with the implementation of higher outlet temperatures. Lag time before acid production by *L. casei* ssp. *pseudoplantarum* was shorter (2 h) at lower outlet temperature (65°C) but higher (10 h) at higher outlet temperature (90°C), indicating greater viability of probiotics at lower outer temperature.

Adaptation of bacteria to sub-lethal heat treatment can improve its thermal tolerance (Desmond et al., 2001; Teixeira, Castro, & Kirby, 1994), and the preheating of probiotics such as *Lactobacillus casei* Nad and *L. plantarum* 8329 at 52°C at 15 min induced their higher survival after spray drying (Paez et al., 2012). However, Lavari et al. (2015) noted no positive effect of pre-treating (52°C at 15 min) on the viability of *L. rhamnosus* 64. Amongst diverse sub-lethal heat treatment (45°C–55°C) investigations, a heat-treatment of 50°C for 15 min was suggested, due to the least viability loss of *Lactobacillus salivarius* NRRL B-30514 (Zhang et al., 2016). Water activity and spray pressure also affect the probiotic viability during spray drying. Monitoring of water activity (0.11 to 0.23) during spray drying decreased the death rate of spray dried probiotic bacteria during storage (Chavez & Ledeboer, 2007). Poddar et al. (2014) reported better survival of spray dried *Lb. paracasei* CRL 431 at a water activity of <0.33. Higher survival of spray dried *Lactobacillus bulgaricus* (Lievense & van't, 1994) and *Lactobacillus acidophilus* (Riveros, Ferrer, & Borquez, 2009) were reported due to a drop in spray pressure.

11.8.4 PROPER SELECTION OF COATING MATERIALS

The concentration of the polymer, its solubility, its rate of solvent removal, and the solubility of organic solvent in water also affect the encapsulation efficiency (Jyothi et al., 2010). Pandey and Vakil (2017) encountered higher viability loss of *Bacillus coagulans* during spray drying (19%–40%) than freeze drying (4%–27.5%). Akhiar (2010) declared that the inclusion of prebiotics in the medium can lower the negative effect of the spray drying process by providing carbon and nitrogen sources enhancing probiotic growth. Conjugated application of spray drying and freeze drying proved to be advantageous for higher probiotic viability. Spray dried *Lactobacillus casei* using alginate matrix containing fructooligosaccharides after subsequently freeze drying had higher cell viability (about 10.98 log CFU/g) and revealed stability during exposure to simulated gastric and intestinal juices and were also capable of releasing viable cells above the therapeutic value (about 8.31 log CFU/g) in the simulated colonic pH (Ivanovska et al., 2012).

Both et al. (2012) recorded higher viability of sodium alginate and maltodextrin-microencapsulated *Lactobacillus brevis* than non-encapsulated bacterial cells. Further, the authors pointed out that amongst different encapsulation materials such as skimmed milk, fructooligosaccharides, maltodextrin, and xanthan gum, the highest and lowest protective abilities were noted for skimmed milk and xanthan gum, respectively. Livney (2010) found milk proteins to be a suitable encapsulating material as these biopolymers have the appropriate physicochemical properties required for encapsulants. Semyonov et al. (2010) suggested an additional layer of coating on spray dried capsules to render protection against the acidic environment of the stomach or bile salts. de, Silva, Fritzen-Freire, Lorenz, and Sant'Anna (2012) noted the greater survival of spray dried capsules of bifidobacteria, previously gelatinized with whey proteins by heating during simulated gastric digestion. The higher concentration of carrier medium during spray drying is detrimental for probiotic viability, due to the formation of larger particles that require longer drying time (Santivarangkna et al., 2007) and have prolonged contact time (Lievense & van't, 1994), which causes their inactivation. A decline in the viability of *Lactobacillus acidophilus* (Espina & Packard, 1979) and Bifidobacteria (Lian et al., 2002) were encountered with the increasing concentration of encapsulating media.

11.8.5 APPLICATION OF PROTECTANTS

The incorporation of protectants into the feed prior to spray drying also protects probiotics (Broeckx et al., 2016) and an additional thermo-protection can be provided with the addition of materials such as whey proteins or by reducing water mobility through the cell membranes and the cell wall that can modulate dehydration upon heating (Ying, Sun, Sanguansri, Weerakkody, & Augustin, 2012). Bustos and Borquez (2013) also reported better viability of *Lactobacillus plantarum* after spray drying with the exclusive inclusion of whey proteins in contrast to gum Arabic or pectin. In an earlier investigation, a 50% replacement of skimmed milk powder with acacia gum in the growth or drying medium induced the better survival of spray dried *Lactobacillus paracasei* (Desmond et al., 2002). Spray drying of *Lactobacilli acidophilus* employing different protecting agents such as microcrystalline cellulose, lactose, and skimmed milk, maximum viability (80%–82%) with skimmed milk and lowest viability (48%) with microcrystalline cellulose was encountered (Lohiya & Avari, 2016). Zheng, Fu, Huang, Jeantet, and Chen (2016) declared that the protective effect of common protectants is due to the provision of physical extracellular protection, thus enhancing the stress tolerance of probiotic cells.

11.9 APPLICATION OF SPRAY DRIED ENCAPSULATED PROBIOTICS IN FOOD PRODUCTS

Cell viability of probiotic function is important for exhibiting health benefits (Jankovic et al., 2010). Conventional methods employing free probiotic cells result in their poor survival in products (de, Faas, Spasojevic, & Sikkema, 2010) due to certain intrinsic factors of the food matrix such as low pH, low water activity, and antibiotic residues (Rokka & Rantamaki, 2010). Encapsulation renders protection to the probiotics (De & Mauriello, 2016; Lawless & Heymann, 2010) and spray dried capsules of *L. paracasei* NFBC and *Lb. plantarum* BM-1 (Desmond et al., 2002; Zhang et al., 2015). Further, Dimitrellou et al. (2016) noted a greater survival rate of spray dried *Lactobacillus casei* ATCC393 in fermented milk during refrigerated storage in contrast to free bacteria. Greater stability of encapsulated cells than free cells may be due to the encapsulation of the cells within the drying medium (De & Mauriello, 2016; Sabikhi, Babu, Thompkinson, & Kapila, 2010), lower water activity in the encapsulated cells (Huang et al., 2017), and possible adaptation to the heat shock during spray drying (Rossi, Zotta, Iacumin, & Reale, 2016).

Crittenden et al. (2006) also reported greater stability of microencapsulated *B. infantis* than non-encapsulated bacteria at room temperature and in the GIT conditions. Spray drying of *Lactobacillus reuteri* (DSM 20016) employing whey containing yeast extract at 55°C and 65°C induced viability loss (CFU/g) from 1.6×10^9 to 2.5×10^7, but encapsulated cells had a higher survival rate (46% vs. 14%) during 5 h exposure to digestive juices in contrast to non-encapsulated cells (Jantzen,

Gopel, & Beermann, 2013). Spray drying encapsulation of *L. acidophillus* La-5 and *B. lactis* Bb-12 revealed better encapsulation efficiency, thermal resistance, gastrointestinal simulation, and storage stability of the former organism than the latter (Nunes et al., 2018). Spray drying encapsulation of *Bifidobacterium breve* R070 and *Bifidobacterium longum* R023 employing milk whey protein had greater tolerance to acidic pH and stability during GIT transit (Picot & Lacroix, 2004). Spray dried *Lactobacillus plantarum* 83,114 and *Lactobacillus kefir* 8321 did not exhibit any alteration in the adhesion capacity to the intestinal Caco-2/TC-7 cells; however, *Lactobacillus kefir* 8348 displayed a significant loss of adhesion capacity (Golowczyc, Silva, Teixeira, De, & Abraham, 2011). Paez et al. (2013) recorded a significantly higher number of Immunoglobulin A (IgA)-producing cells in the small intestines of mice with the administration of spray dried powder of *Lactobacillus acidophilus* A9, *Lactobacillus paracasei* A13, and *Lactobacillus casei* Nad.

Application of spray dried encapsulated probiotics in food products has been depicted in Table 11.5. The inclusion of spray dried *Lactobacillus plantarum* Dad 13 during yogurt manufacture induced a decline in viable cells by 1 log cycle of *L. plantarum* Dad 13, but had comparable microbiological and chemical characteristics to those made from freeze dried preparation (Utami et al., 2016). Spray dried *Lactococcus lactis* Gh1 obtained involving gum Arabic together with *Synsepalum dulcificum* (miracle fruit) having a viable count of about 10^9 CFU/mL was used for yogurt formulation. Higher retention (about 10^7 vs. about 10^5 CFU/mL) of microencapsulated *L. lactis* in yogurt compared to non-microencapsulated cells was encountered (Fazilah et al., 2019).

An attempt was made to incorporate spray dried encapsulated *Lactobacillus paracasei* LMG S-27487 into ice cream. Spray drying induced lowering of viable counts (CFU/ml) from 5.2×10^{10} to 9.5×10^9, accounting for 18% of loss in viability. The ice cream mix containing 5% substitution of the commercial base with the modified base containing spray dried encapsulated *L. paracasei* had a viability of 3.5×10^7, which declined to 3.1×10^7 after two days of storage at $-18°C$ (Spigno et al., 2015). Spray dried *L. rhamnosus* GG employing whey protein isolate or in combination with a physically-modified resistant starch remained stable in apple juice or citrate buffer for more than five weeks at 4°C and were more effective than those held at a higher temperature of 25°C (Ying et al., 2013). Anekella and Orsat (2013) also incorporated spray dried *L. acidophilus* NRRL B-4495 and *L. rhamnosus* NRRL B-442 using maltodextrin in raspberry juice and noted an increase in the survival rate of the probiotics with increasing concentration of the microencapsulating material.

Microencapsulated probiotics can also be incorporated into fermented milk products as dietary supplements, owing to retention of cell viability and enzyme activity in a simulated human

TABLE 11.5
The Application of Spray Dried Encapsulated Probiotics in Functional Foods

Probiotic Cultures	Foods	Application Benefits	References
Lactobacillus casei ATCC393	Fermented milk	The enhanced survival rate of spray dried probiotics during refrigerated storage in contrast to free bacteria.	Dimitrellou et al. (2016)
Lactobacillus plantarum Dad 13	Yogurt	Induced a decline in viable cells by1 log cycle	Utami et al. (2016)
Lactococcus lactis Gh1	Yogurt	Higher retention of probiotics compared to non-microencapsulated cells	Fazilah et al. (2019)
Lactobacillus paracasei LMG S-27487	Ice cream	Retention of probiotic viability up to 9.5 $\times 10^9$	Spigno et al. (2015)
Lactobacillus rhamnosus GG	Apple Juice	Storage stability for >5 weeks at 4°C	Ying et al. (2013)
L. acidophilus NRRL B-4495 + *L. rhamnosus* NRRL B-442	Raspberry Juice	Enhanced survival rate	Anekella and Orsat (2013)

gastrointestinal model (Martoni et al., 2007), as well as effectiveness in delivering viable cells to the colon compared with the non-encapsulated cells (Piano et al., 2012). Microencapsulated probiotic cultures also demonstrated hypercholesterolemic effects (Bhathena et al., 2009) and anti-tumorigenesis properties (McIntosh, Royle, & Playne, 1999) and were efficacious for improvement in glucose tolerance (Lee et al., 2019). Burns et al. (2017) observed no impact of the spray drying process on the viability of *Bifidobacterium animalis* subsp. *lactis* INL1 and *B. animalis* subsp. *lactis* BB12 or protective capacities against acute and chronic inflammation in inflammatory bowel disease.

11.10   CONCLUDING REMARKS

Consumer awareness toward health and the realization of its link with food has resounded the global market for health-promoting foods. The health-promoting properties of probiotics have seen their promotion as a functional additive in the current era of functional foods. Probiotics can maintain the gut microbiota ecosystem, thereby extending health benefits. It has been established that both dead and live probiotic cells are capable of exhibiting health effects, but viable probiotics are more efficacious. Probiotic viability can be retained by providing a physical barrier such as encapsulation. Amongst diverse techniques, spray drying encapsulation is used widely on an industrial scale, as it is the most simple and economical method that can retain great cell viability during functional food formulations.

REFERENCES

Abd-Talib, N., Mohd-Setapar, S.H., Khamis, A.K., Nian-Yian, L. and Aziz, R. 2013. Survival of encapsulated probiotics through spray drying and non-refrigerated storage for animal feeds application. *Agric. Sci.* 4: 78–83.

Adams, C.A. 2010. The probiotic paradox: live and dead cells are biological response modifiers. *Nutr. Res. Rev.* 23: 37–46.

Akhiar, N.S.A.M. 2010. Enhancement of probiotics survival by microencapsulation with alginate and probiotics. *Basic Biotechnol.* 6: 13–18.

Alpert, P. 2005. The limits on frontiers of desiccation-tolerant life. *Integr. Comp. Biol.* 45: 685–695.

Alves, N.N., Messaoud, G.B., Desobry, S., Correia Costa, J.M. and Rodrigues, S. 2016. Effect of drying technique and feed flow rate on bacterial survival and physicochemical properties of a non-dairy fermented probiotic juice powder. *J. Fd. Eng.* 189: 45–54.

Amara, A.A. and Shib, A. 2015. Role of probiotics in health improvement, infection control and disease treatment and management. *Saudi Pharmac. J.* 23: 107–114.

Anal, A.K. and Singh, H. (2007). Recent advances in microencapsulation of probiotics for industrial applications and targeted delivery. *Trends Fd. Sci. Technol.* 18: 240–251.

Anekella, K. and Orsat, V. (2013). Optimization of microencapsulation of probiotics in raspberry juice by spray drying. *LWT – Fd. Sci. Technol.* 50: 17–24.

Anselmo, A.C., McHugh, K.J., Webster, J., Langer, R. and Jaklenec, A. 2016. Layer-by-Layer encapsulation of probiotics for delivery to the microbiome. *Adv. Mater.* 28: 9486–9490.

Anukam, K.C., Osazuwa, E.O., Osadolor, H.B., Bruce, A.W. and Reid, G. 2008. Yogurt containing probiotic *Lactobacillus rhamnosus* GR-1 and *L. reuteri* RC-14 helps resolve moderate diarrhea and increases CD4 count in HIV/AIDS patients. *J Clin Gastroenterol.* 42: 239–243.

Arepally, D. and Goswami, T.K. 2019. Effect of inlet air temperature and gum Arabic concentration on encapsulation of probiotics by spray drying. *LWT – Fd. Sci. Technol.* 99: 583–593.

Arslan, S., Erbas, M., Tontul, I. and Topuz, A. 2015. Microencapsulation of probiotic *Saccharomyces cerevisiae* var. boulardii with different wall materials by spray drying. *LWT – Fd. Sci. Technol.* 63: 685–690.

Avila-Reyesa, S.V., Garcia-Suareza, F.J., Jiménez, M.T., Martín-Gonzalez, M.F.S. and Bello-Perez, L.A. 2014. Protection of L. rhamnosus by spray drying using two prebiotics colloids to enhance the viability. *Carbohyd. Polym.* 102: 423–430.

Awad, H., Mokhtar, H., Imam, S.S., Gad, G.I., Hafez, H. and Aboushady, N. 2010. Comparison between killed and living probiotic usage versus placebo for the prevention of necrotizing enterocolitis and sepsis in neonates. *Pak. J. Biol. Sci.* 13: 253–262.

Bakry, A.M., Abbas, S., Ali, B., Majeed, H., Abouelwafa, M.Y., Mousa, A. and Liang, L. 2016. Microencapsulation of oils: A comprehensive review of benefits, techniques, and applications. *Comp. Rev. Fd. Sci. Fd. Safety* 15: 143–182.

Behboudi-Jobbehdar, S., Soukoulis, C., Yonekura, L. and Fisk, I. 2013. Optimization of spray drying process conditions for the production of maximally viable microencapsulated 1000 L. acidophilus NCIMB 701748. *Dry. Technol.* 31: 1274–1283.

Bhadoria, P.B.S. and Mahapatra, S.C. 2011. Prospects, technological aspects and limitations of probiotics – A Worldwide Review. *Eur. J. Nutr. Fd. Safety.* 1: 23–42.

Bhathena, J., Martoni, C., Kulamarva, A., Urbanska, A.M., Malhotra, M. and Prakash, S. 2009. Orally delivered microencapsulated live probiotic formulation lowers serum lipids in hypercholesterolemic hamsters. *J. Med. Fd* 12: 310–319.

Bielecka, M. and Majkowska, A. 2000. Effect of spray drying temperature of yogurt on the survival of starter cultures, moisture content and sensoric properties of yogurt powder. *Nahrung* 44: 257–260.

Biller, J.A., Katz, A.J., Flores, A.F., Buie, T.M., and Gorbach, S.L. 1995. Treatment of recurrent clostridium difficile colitis with Lactobacillus GG. *J. Pediatr. Gastroenterol. Nutr.* 21: 224–226.

Bonsante, F., Iacobelli, S. and Gouyon, J.B. 2013. Routine probiotic use in very preterm infants: Retrospective comparison of two cohorts. *Am. J. Perinatol.* 30: 41–46.

Bosch, M., Mendez, M., Perez, M., Farran, A., Fuentes, M.C. and Cune, J. 2012. Lactobacillus plantarum CECT7315 and CECT7316 stimulate immunoglobulin production after influenza vaccination in elderly. *Nutrición Hospitalaria* 27: 504–509.

Both, E., Gyenge, L., Bodor, Z., György, E., Lányi, S. and Ábrahám, B. 2012. Intensification of probiotic microorganisms Viability by microencapsulation using Ultrasonic atomizer. *U.P.B. Sci. Bull., Series B.* 74: 27–32.

Boza, Y., Barbin, D. and Scamparini, A.R.P. 2004. Effect of spraydrying on the quality of encapsulated cells of Beijerinckia sp. *Process Biochem.* 39: 1275–1284.

Broeckx, G., Vandenheuve, D., Claes, I.J.J., Lebeer, S. and Kiekensa, F. 2016. Drying techniques of probiotic bacteria as an important step towards the development of novel pharmabiotics. *Int. J. Pharma.* 505: 303–318.

Burgain, J., Gaian, C., Linder, M. and Scher, J. 2011. Encapsulation of probiotic living cells: From laboratory scale to industrial applications. *J. Fd. Engin.* 104: 467–483.

Burgain, J., Gaiani, C., Cailliez-Grimal, C., Jeandel, C. and Scher, J. 2013. Encapsulation of Lactobacillus rhamnosus GG in microparticles: Influence of casein to whey protein ratio on bacterial survival during digestion. *Innov. Fd. Sci. Emerg. Technol.* 19: 233–242.

Burns, P., Alard, J., Hrdy, J., Boutillier, D., Paez, R., Reinheimer, J., Pot, B., Vinderola, G. and Grangette, C. 2017. Spray-drying process preserves the protective capacity of a breast milk derived Bifidobacterium lactis strain on acute and chronic colitis in mice. *Sci. Rep.* 7: 43211, 10 pages.

Bustamante, M., Villarroel, M., Rubilar, M. and Shene, C. 2015. Lactobacillus acidophilus La-05 encapsulated 442 by spray drying: Effect of mucilage and protein from flaxseed (*Linum usitatissimum* L.). *LWT-Fd. Sci. Technol.* 62: 1162–1168.

Bustos, P. and Borquez, R. 2013. Influence of osmotic stress and encapsulating materials on the stability of autochthonous lactobacillus plantarum after spray drying. *Drying Technol.* 31: 57–66.

Chaikham, P., Apichartsrangkoon, A., George, T. and Jirarattanarangsri, W. 2013. Efficacy of polymer coating of probiotic beads suspended in pressurized and pasteurized longan juices on the exposure to simulated gastrointestinal environment. *Int. J. Fd. Sci. Nutr.* 64: 862–869.

Chaipojjana, R., Phosuksirikul, S. and Leejeerajumnean, A. 2014. Survival of four probiotic strains in acid, bile salt and after spray drying. *Int. J. Nutr. Fd. Engin.* 8: 1060–1063.

Champagne, C. and Fustier, P. 2007. Microencapsulation for the improved delivery of bioactive compounds into foods. *Curr. Opin. Biotechnol.* 18: 184–190.

Champagne, C.P., Ross, P.R., Saarela, M., Hansen, K.F. and Charalampopoulos, D. 2011. Recommendations for the viability assessment of probiotics as concentrated cultures and in food matrices. *Int. J. Fd. Microbiol.* 149: 185–193.

Chapman, C.M., Gibson, G.R. and Rowland, I. 2011. Health benefits of probiotics: are mixtures more effective than single strains?. *Europ. J. Nutr.* 50: 1–17.

Chavarri, M., Maranon, I. and Villaran, M.C. 2012. Encapsulation technology to protect probiotic bacteria, Chapter 23, In: *book: Probiotics*, Ed. Rigobelo, E.C., pp. 505, INTECH, Croatia.

Chavez, B.E. and Ledeboer, A.M. 2007. Drying of probiotics: optimization of formulation and process to enhance storage survival. *Drying Technol.* 25: 1193–1201.

Chen, L.Y., Remondetto, G.E. and Subirade, M. 2006. Food protein-based materials as nutraceutical delivery systems. *Trends Fd. Sci. Technol.* 17: 272–283

Cook, M.T., Tzortzis, G., Charalampopoulos, D. and Khutoryanskiy, V.V. (2012). Microencapsulation of probiotics for gastrointestinal delivery. *J. Controlled Release* 162: 56–67.

Corcoran, B.M., Ross, R.P., Fitzgerald, G.F. and Stanton, C. 2004. Comparative survival of probiotic lactobacilli spray dried in the presence of prebiotic substances. *J. Appl. Microbiol.* 96: 1024–1039.

Corcoran, B.M., Stanton, C., Fitzgerald, G.F. and Ross, R.P. 2005. Survival of probiotic lactobacilli in acidic environments is enhanced in the presence of metabolizable sugars. *Appl. Environ. Microbiol.* 71: 3060–3067.

Crittenden, R., Weerakkody, R., Sanguansri, L. and Augustin, M.A. 2006. Synbiotic microcapsules that enhance microbial viability during nonrefrigerated storage and gastrointestinal transit. *Appl. Environ. Microbiol.* 72: 2280–2282.

de Melo Barros, D., de Castro Lima Machado, E., de Moura, D.F., de Oliveira, M.H.M., Rocha, T.A., de Oliveira Ferreira, S.A., de Albuquerque Bento da Fonte, R. and de Souza Bezerra, R. 2018. Potential application of microencapsulation in the food industry. *Int. J. Adv. Res.* 6: 956–976.

Daemen, A.L.H. and van der Stege, H.J. 1982. The destruction of enzymes and bacteria during spray drying of milk and whey. 2. The effect of the drying process. Neth. *Milk Dairy J.* 36: 211–229.

Daly, M.J., Gaidamakova, E.K., Matrosova, V.Y., Vasilenko, A., Zhai, M., Venkateswaran, A., Hess, M., Omelchenko, M.V., Kostandarithes, H.M., Makarova, K.S., Wackett L.P., Fredrickson, J.K. and Ghosal, D. 2004. Accumulation of Mn(II) in Deinococcus radiodurans facilitates gamma-radiation resistance. *Science* 306: 1025–1028.

Dargahi, N., Johnson, J., Donkor, O., Vasiljevic, T. and Apostolopoulos, V. 2019. Immunomodulatory effects of probiotics: Can they be used to treat allergies and autoimmune diseases?. *Maturitas.* 119: 25–38.

De Prisco, A. and Mauriello, G. 2016. Probiotication of foods: a focus on microencapsulation tool. *Trends Fd. Sci. Technol.* 48: 27–39.

Deshpande, G., Rao, S., Patole, S. and Bulsara, M. 2010. Updated meta-analysis of probiotics for preventing necrotizing enterocolitis in preterm neonates. *Pediat.* 125: 921–930.

Desmond, C., Ross, R.P., O'Callaghan, E., Fitzgerald, G. and Stanton, C. 2002. Improved survival of Lactobacillus paracasei NFBC 338 in spray dried powders containing gum acacia. *J. Appl. Microbiol.* 93: 1003–1011.

Desmond, C., Stanton, C., Fitzgerald, G.F., Collins, K. and Paul Ross, R. 2001. Environmental adaptation of probiotic lactobacilli towards improvement of performance during spray drying. *Int. Dairy J.* 11: 801–808.

Dijkstra, A.R., Setyawati, M.C., Bayjanov, J.R., Alkema, W., Van Hijum, S.A.F.T., Bron, P. A. and Hugenholtz, J. 2014. Diversity in robustness of Lactococcus lactis strains during heat stress, oxidative stress, and spray drying stress. *Appl. Environ. Microbiol.* 80: 603–611.

Dimitrellou, D., Kandylis, P., Petrovic, T., Dimitrijevic-Brankovic, S., Levic, S., Nedovic, V. and Kourkoutas, Y. 2016. Survival of spray dried microencapsulated Lactobacillus casei ATCC 393 in simulated gastrointestinal conditions and fermented milk. *LWT – Fd. Sci. Technol.* 71: 169–174.

Ding, W.K. and Shah, N.P. 2009. An improved method of microencapsulation of probiotic bacteria for their stability in acidic and bile conditions during storage. *J. Fd. Sci.* 74: 53–61.

Divya, J.B., Varsha, K.K., Nampoothiri, K.M., Ismail, B. and Pandey, A. 2012. Probiotic fermented foods for health benefits. *Eng. Life Sci.* 12: 377–390.

Doherty, S.B., Auty, M.A., Stanton, C., Ross, R.P., Fitzgerald, G.F. and Brodkorb, A. 2012. Survival of entrapped Lactobacillus rhamnosus GG in whey protein micro-beads during simulated ex vivo gastrointestinal transit. *Int. Dairy J.* 22: 31–43.

Donsi, F., Ferrari, G., Lenza, E. and Maresca, P. 2009. Main factors regulating microbial inactivation by high-pressure homogenization: Operating parameters and scale of operation. *Chem. Engin. Sci.* 64: 520–532.

Doraisamy, K.A., Karthikeyan, N. and Elango, A. 2018. Microencapsulation of probiotics in Functional dairy products development. *Int. J. Adv. Res. Sci. Engn.* 7: 374–387.

dos Santos, R.C.S., Finkler, L. and Finkler, C.L.L. 2014. Microencapsulation of Lactobacillus casei by spray drying. *J. Microencapsul.* 31: 759–767.

Ei-Salam, M.H. and Ei-Shibiny S. 2012. Formation and potential uses of milk proteins as nano delivery vehicles for nutraceuticals: A review. *Int. J. Dairy Technol.* 65: 13–21.

Espina, F. and Packard, V.S. 1979. Survival of Lactobacillus acidophilus in a spray drying process. *J. Fd Protection.* 42: 149–152.

Espirito Santo, A.P., Perego, P., Converti, A. and Oliveira, M.N. 2011. Influence of food matrices on probiotic viability – A review focusing on the fruity bases. *Trends Fd. Sci. Technol.* 22: 377–385.

Estevinho, B.N., Rocha, F., Santos, L. and Alves, A. 2013. Microencapsulation with chitosan by spray drying for industry applications - A review. *Trends Fd. Sci. Technol.* 31: 138–155.

de Castro-Cislaghi, F.P., Silva, C.D.E., Fritzen-Freire, C.B., Lorenz, J.G. and Sant'Anna, E.S. 2012. Bifidobacterium Bb-12 microencapsulated by spray drying with whey: survival under simulated gastrointestinal conditions, tolerance to NaCl, and viability during storage. *J. Fd. Eng.* 113:186–193.

Fang, Y., Kennedy, B., Rivera, T., Han, K.S., Anal, A.K. and Singh, H. 2012. Encapsulation system for protection of probiotics during processing. U.S. Patent 20120263826.

Fang, Z.X. and Bhandari, B. 2010. Encapsulation of polyphenols – A review. *Trends Food Sci. Technol.* 21: 510–523.

FAO/WHO. 2001. *Regulatory and clinical aspects of dairy probiotics. Food and Agriculture Organization of the United Nations*, World Health Organization Cordoba, Argentina.

FAO/WHO. 2003. Standard for fermented milks. Codex standard 243 (pp.1e8). Rome, FAO/WHO.

FAO/WHO. 2006. Probiotics in Food. Health and Nutritional Properties and Guidelines for Evaluation, FAO Food and Nutrition Paper No. 85. World Health Organization and Food and Agriculture Organization of the United Nations, Rome.

Farheen, T., Shaikh, A. and Shahi, S. 2017. A Review on a Process: Microencapsulation. *Int J. Pharm. Res. Hlth. Sci.* 5: 1823–1830.

Favaro-Trindade, C.S. and Grosso, C.R. 2002. Microencapsulation of L. acidophilus (La-05) and B. lactis (Bb-12) and evaluation of their survival at the pH values of the stomach and in bile. *J. Microencap.* 19: 485–494.

Favaro-Trindade, C.S., Heinemann, R.J.B. and Pedroso, D.L. 2011. Developments in probiotic encapsulation. CAB Reviews: Perspectives in agriculture, veterinary science, *Nutr. Nat. Resources* 6: No. 004

Fazilah, N.F., Hamidon, N.H., Ariff, A.B., Khayat, M.E., Wasoh, H. and Halim, M. 2019. Microencapsulation of Lactococcus lactis Gh1 with gum arabic and Synsepalum dulcificum via spray drying for potential inclusion in functional yogurt. *Molecules* 24, 1422, 21 pages.

Fenster, K., Freeburg, B., Hollard, C., Wong, C., Laursen, R.R. and Ouwehand, A.C. 2019. The production and delivery of probiotics: A Review of a practical approach. *Microorganisms.* 7: 83.

Flores-Belmont, I.A., Palou, E., Lopez-Malo, A. and Jimenez-Munguia, M.T. 2015. Simple and double microencapsulation of Lactobacillus acidophilus with chitosan using spray drying. *Int. J. Fd. Stud.* 4: 188–200.

Foerst, P. and Kulozik, U. 2011. Modelling the dynamic inactivation of the probiotic bacterium L. paracasei ssp. paracasei during a low-temperature drying process based on stationary data in concentrated systems. *Fd. Bioprocess. Technol.* 5: 2419–2427.

Foligne, B., Daniel, C. and Pot, B. 2013. Probiotics from research to market: the possibilities, risks and challenges. *Curr. Opin. Microbiol.* 16: 284–292.

Fontana, L., Bermudez-Brito, M., Plaza-Diaz, J., Munoz-Quezada, S. and Gil, A. 2013. Sources, isolation, characterisation and evaluation of probiotics. *Br. J. Nutr.* 109: 35–50.

Fredrickson, J.K., Li, S.W., Gaidamakova, E.K., Matrosova, V.Y., Zhai, M., Sulloway, H.M., Scholten, J.C., Brown, M.G., Balkwill, D.L. and Daly, M.J. 2008. Protein oxidation: key to bacterial desiccation resistance?. *Int. Soc. Microbial. Ecol. J.* 2: 393–403.

Fritzen-Freire, C.B., Prudencio, E.S., Amboni, R.D.M.C., Pinto, S.S., Negrao-Murakami, A.N. and Murakami, F.S. 2012. Microencapsulation of bifidobacteria by spray drying in the presence of probiotics. *Fd. Res. Int.* 45: 306–312.

Fritzen-Freire, C.B., Prudêncio, E.S., Pinto, S.S., Munñoz, I.B. and Amboni, R.D.M.C. 2013. Effect of microencapsulation on survival of Bifidobacterium BB-12 exposed to simulated gastrointestinal conditions and heat treatments. *LWT – Fd. Sci. Technol.* 50: 39–44.

Fu, N., and Chen, X.D. 2011. Towards a maximal cell survival in convective thermal drying processes. *Fd. Res. Int.* 44: 1127–1149.

Fu, N., Woo, M.W., Selomulya, C. and Chen, X.D. 2013. Inactivation of *Lactococcus lactis* ssp. cremoris cells in a droplet during convective drying. *Biochem. Engin. J.* 79: 46–56.

Gao, X.W., Mubasher, M., Fang, C.Y., Reifer, C. and Miller, L.E. 2010. Dose-response efficacy of a proprietary probiotic formula of Lactobacillus acidophilus CL1285 and Lactobacillus casei LBC80R for antibiotic-associated diarrhea and clostridium difficile-associated diarrhea prophylaxis in adult patient. *Am. J. Gastroenterol.* 105: 1636–1641.

Garcia, A.H. 2011. Anhydrobiosis in bacteria: From physiology to applications. *J. Biosci.* 36: 939–950.

Gardiner, G.E., O' Sullivan, E., Kelly, J., Auty, M.A., Fitzgerald, G.F., Collins, J.K., Ross, R. P. and Stanton, C. 2000. Comparative survival rates of human-derived probiotic Lactobacillus paracasei and L. salivarius strains during heat treatment and spray drying. *Appl. Environ. Microbiol.*, 66: 2605–2612.

Gauri A. and Shiwangi, M. 2017. Immobilization and microencapsulation. *J. Adv. Res. Biotech.* 2: 1–4.

Gbassi, G.K. and Vandamme, T. 2012. Probiotic encapsulation technology: From microencapsulation to release into the gut. *Pharm.* 4: 149–163.

Ghandi, A., Powell, I.B., Chen, X.D. and Adhikari, B. 2012. The effect of dryer inlet and outlet air temperatures and protectant solids on the survival of Lactococcus lactis during spray drying. *Dry. Technol.* 30: 1649–1657.

Ghandi, A., Powell, I.B., Howes, T., Chen, X.D. and Adhikari, B. 2012. Effect of shear rate and oxygen stresses on the survival of *Lactococcus lactis* during the atomization and drying stages of spray drying: A laboratory and pilot scale study. *J. Fd. Eng.* 113: 194–200.

Gianotti, L., Morelli, L., Galbiati, F., Rocchetti, S., Coppola, S., Beneduce, A., Gilardini, C., Zonenschain, D., Nespoli, A. and Braga, M. 2010. A randomized double-blind trial on perioperative administration of probiotics in colorectal cancer patients. *World J. Gastroenterol.* 16: 167–175.

Gibbs, B.F., Kermasha, S., Alli, I. and Mulligan, C.N. 1999. Encapsulation in the food industry: A review. *Int. J. Fd. Sci. Nutr.* 50: 213–224.

Goh, C.H. Heng, P.W.S. and Chan. L.W. 2012. Alginates as a useful natural polymer for microencapsulation and therapeutic applications. *Carbohydrate Polymers.* 88: 1–12.

Golowczyc, M.A., Silva, J., Abraham, A.G., De Antoni, G.L. and Teixeira, P. 2010. Preservation of probiotic strains isolated from kefir by spray drying. *Letters Appl. Microbiol.* 50: 7–12.

Golowczyc, M.A., Silva, J., Teixeira, P., De Antoni, G.L., Abraham, A.G. 2011. Cellular injuries of spray dried Lactobacillus spp. isolated from kefir and their impact on probiotic properties. *Int. J. Fd. Microbiol.* 144: 556–560.

Gomes, A.C., de Sousa, R.G.M., Botelho, P.B., Gomes, T.L.N., Prada, P.O. and Mota, J.F. 2017. The additional effects of a probiotic mix on abdominal adiposity and antioxidant Status: A double-blind, randomized trial. *Obesity.* 25: 30–38.

Gueimonde, M., Sanchez, B., de los Reyes-Gavilan, C.G. and Margolles, A. 2013. Antibiotic resistance in probiotic bacteria. *Front. Microbiol.* 4: 202.

Gueniche, A., Bastien, P., Ovigne, J.M., Kermici, M., Courchay, G., Chevalier, V., Breton, L. and Castiel-Higounenc, I. 2010. Bifidobacterium longum lysate, a new ingredient for reactive skin. *Exp. Dermatol.* 19: 1–8.

Guerin, J., Petit, J., Burgain, J., Borges, F., Bhandari, B., Perroud, C., Desobry, S. Scher, J. and Gaiani, C. 2017. Lactobacillus rhamnosus GG encapsulation by spray drying: Milk proteins clotting control to produce innovative matrices. *J. Fd. Eng.* 193: 10–19.

Gul, O. and Dervisoglu, M. 2020. Optimization of spray drying conditions for microencapsulation of *Lactobacillus casei* Shirota using response surface methodology. *Eur. Fd. Sci. Eng.* 1: 1–8.

Gupta, S. and Abu-Ghannaman, N. 2012. Probiotic fermentation of plant based products: Possibilities and opportunities. *Crit. Rev. Fd. Sci. Nutr.* 52: 183–199.

Han, X., Yang, Z., Jing, X., Yu, P., Zhang, Y., Yi, H. and Zhang, L. 2016. Improvement of the texture of yogurt by use of exopolysaccharide producing lactic acid bacteria. *BioMed. Res. Int.* 2016: 7945675, 6 pages.

Hayes, S.R. and Vargas, A.J. 2016. Probiotics for the prevention of pediatric antibiotic associated diarrhea. *Explor. J. Sci. Hlth.* 12: 463–466.

Heidebach, T., Forst, P. and Kulozik, U. 2009. Microencapsulation of probiotic cells by means of rennet-gelation of milk proteins. *Fd. Hydrocoll.* 23: 1670–1677.

Hell, M., Bernhofer, C., Stalzer, P., Kern, J.M. and Claassen, E. 2013. Probiotics in Clostridium difficile infection: reviewing the need for a multi-strain probiotic. *Beneficial Microbes* 4(1): 39–51.

Hill, C., Guarner, F., Reid, G., Gibson, G.R., Merenstein, D.J., Pot, B., Morelli, L., Canani, R.B., Flint, H.J., Salminen, S., Calder, P.C. and Sanders, M.E. 2014. Expert consensus document. The international scientific association for probiotics and prebiotics consensus statement on the scope and appropriate use of the term probiotic. *Nature Rev. Gastroenterol. Hepatol.* 11: 506–514.

Hoang, B.X., Shaw, G., Pham, P. and Levine, S.A. 2010. Lactobacillus rhamnosus cell lysate in the management of resistant childhood atopic eczema. *Inflammation Allergy-Drug Targets.* 9: 192–196.

Huang, S., Cauty, C., Dolivet, A., Le Loir, Y., Chen, X. D., Schuck, P., Jan, G. and Jeantet, R. 2016. Double use of highly concentrated sweet whey to improve the biomass production and viability of spray dried probiotic bacteria. *J. Func. Fd.* 23: 453–463.

Huang, S., Vignolles, M.L., Le Loir, X.D.C.Y., Jan, G., Schuck, P. and Jeantet, R. 2017. Spray drying of probiotics and other food-grade bacteria: A review. *Trends Fd. Sci. Technol.* 63: 1–17 pages.

Iannitti, T. and Palmieri, B. 2010. Therapeutical use of probiotic formulations in clinical practice. *Clin. Nutr.* 29: 701–725.

Italian Health Ministry. 2013. Guidelines on probiotics and prebiotics, Revision, May 2013.

Ivanovska, T.P., Petrushevska-Tozi, L., Kostoska, M.D., Geskovski, N., Grozdanov, A., Stain, C., Stafilov, T. and Mladenovska, K. 2012. Microencapsulation of Lactobacillus casei in Chitosan-Ca-Alginate microparticles using spray drying method. *Macedonian J. Chem. Chemical Engin.* 31: 115–123.

Ivey, K.L., Hodgson, J.M., Kerr, D.A., Thompson, P.L., Stojceski, B. and Prince, R.L. 2015. The effect of yogurt and its probiotics on blood pressure and serum lipid profile: a randomised controlled trial. *Nutr. Metab. Cardiovasc. Dis.* 25: 46–51.

van Loo, J.A.E., Coussement, P., De Leenheer, L., Hoebregs, H. and Smits, G. 1995. On the presence of inulin and oligofructose as natural ingredients in the western diet. *CRC Crit. Rev. Fd. Sci. Nutr.* 35: 525–552.

Jankovic, I., Sybesma, W., Phothirath, P., Ananta, E., Mercenier, A. 2010. Application of probiotics in food products - challenges and new approaches. *Curr. Opin. Biotechnol.* 21: 175–181.

Janning, B. and in't Veld, P.H. 1994. Susceptibility of bacterial strains to desiccation: a simple method to test their stability in microbiological reference materials. *Anal. Chim. Acta.* 286: 469–476.

Jantzen, M., Gopel, A. and Beermann, C. 2013. Direct spray drying and microencapsulation of probiotic *Lactobacillus reuteri* from slurry fermentation with whey. *J. Appl. Microbiol.* 115: 1029–1036.

Jeyakumari, A., Zynudheen, A.A. and Parvathy, U. 2016. Microencapsulation of bioactive food ingredients and controlled release—a review. *MOJ Fd. Process. Technol.* 2: 214–224.

John, R.P., Tyagi, R.D., Brar, S.K., Surampalli, R.Y. and Prevost, D. 2011. Bioencapsulation of microbial cells for targeted agricultural delivery. *Crit. Rev. Biotechnol.* 31: 211–226.

Johnston, B.C., Goldenberg, J.Z., Vandvik, P.O., Sun, X. and Guyatt, G.H. 2011. Probiotics for the prevention of pediatric antibiotic-associated diarrhea. *Cochrane Database Syst Rev.* 11: CD004827, PMID:22071814.

Jyothi, N.V.N., Prasanna, P.M., Sakarkar, S.N., Prabha, K.S., Ramaiah, P.S. and Srawan, G.Y. 2010. Microencapsulation techniques, factors influencing encapsulation efficiency. *J. Microencapsul.* 27: 187–197.

Kailasapathy, K. 2002. Microencapsulation of probiotic bacteria: technology and potential applications. *Curr. Iss. Intest. Microbiol.* 3: 39–48.

Kandansamy, K. and Somasundaram, P.D. 2012. Microencapsulation of colors by spray drying - A review. *Int. J. Fd. Engin.* 8: 1.

Karczewski, J., Troost, F.J., Konings, I., Dekker, J., Kleerebezem, M., Brummer, R.J. and Wells, J.M. 2010. Regulation of human epithelial tight junction proteins by Lactobacillus plantarum in vivo and protective effects on the epithelial barrier. *Am. J. Physiol. Gastrointest. Liver Physiol.* 289: 851–859.

Kavitake, D. Kandasamy, S., Devi, P.B. and Shetty, P.H. 2018. Recent developments on encapsulation of lactic acid bacteria as potential starter culture in fermented foods – A review. *Fd. Biosci.* 21: 34–44.

Kechagia, M., Basoulis, D., Konstantopoulou, S., Dimitriadi, D., Gyftopoulou, K. Skarmoutsou, N. and Fakiri, E.M. 2013. Health benefits of probiotics - A review. *ISRN Nutr.* 2013: 481651, 7 pages.

Khalesi, S., Sun, J., Buys, N. and Jayasinghe, R. 2014. Effect of probiotics on blood pressure: a systematic review and meta-analysis of randomized, controlled trials. *Hypertension.* 64: 897–903.

Khan, S.U. 2014. Probiotics in dairy foods: a review. *Nutr. Fd. Sci.* 44: 71–88.

Kingwatee, N., Apichartsrangkoon, A., Chaikham, P., Worametrachanon, S., Techarung, J. and Pankasemsuk, T. 2015. Spray drying Lactobacillus casei 01 in lychee juice varied carrier materials. *LWT – Fd. Sci. Technol.* 62: 847–853.

Kołozyn-Krajewskaa, D, and Dolatowski, Z. J. 2012. Probiotic meat products and human nutrition. *Process Biochem.* 47: 1761–1772.

Korbekandi, H., Mortazavian, A.M. and Iravani, S. 2011. Technology and stability of probiotic in fermented milks In: *Probiotic and Prebiotic Foods: Technology, Stability and Benefits to the Human Health*, Ed. N. Shah, A.G. Cruz and J.A.F. Faria, pp. 131–169. New York: Nova Science Publishers.

Kourelis, A., Zinonos, I., Kakagianni, M., Christidou, A., Christoglou, N., Yiannaki, E., Testa, T., Kotzamanidis, C., Litopoulou-Tzanetaki, E., Tzanetakis, N. and Yiangou, M. 2010. Validation of the dorsal air pouch model to predict and examine immunostimulatory responses in the gut. *J. Appl. Microbiol.* 108: 274–284.

Krishnaiah, D., Nithyanandam, R. and Sarbatly, R. 2014. A critical review on the spray drying of fruit extract: effect of additives on physicochemical properties. *Crit. Rev. Fd. Sci. Nutr.* 54: 449–473.

Kumar, B.V., Vijayendra, S.V. and Reddy, O.V. 2015. Trends in dairy and non-dairy probiotic products – A review. *J. Fd. Sci. Technol.* 52: 6112–6124.

de Souza Simoes, L., Madalena, D.A., Pinheiro, A.C., Teixeira, J.A., Vicente, A.A. and Ramos, Ó.L. 2017. Micro and nano bio-based delivery systems for food applications: In vitro behavior. *Adv. Colloid Interface Sci.* 243: 23–45.

Lacroix, C. and Yildirim, S. 2007. Fermentation technologies for the production of probiotics with high viability and functionality. *Curr. Opin. Biotechnol.* 18: 176–183.

Laroia. S. and Martin, J.H. 1991. Effect of pH on survival of Bifidobacterium bifidum and Lactobacillus acidophilus in frozen fermented dairy desserts. *Cult. Dairy Prod. J.* 26:13–21.

Lavari, L., Ianniello, R., Páez, R., Zotta, T., Cuatrin, A., Reinheimer, J., Parente, E., Vinderola, G. 2015. Growth
   of Lactobacillus rhamnosus 64 in whey permeate and study of the effect of mild stresses on survival to
   spray drying. *LWT – Fd. Sci. Technol.* 63: 322–330.

Lawless, H.T. and Heymann, H. 2010. *Sensory evaluation of food: Principles and practices* (2nd ed.). Berlin:
   Springer.

Lee, I. 2014. Critical pathogenic steps to high risk Helicobacter pylori gastritis and gastric carcinogenesis.
   World J. Gastroenterol. 20: 6412–6419.

Lee, S., Kirkland, R., Grunewald, Z.I., Sun, Q., Wicker, L. and de La Serre, C.B. 2019. Beneficial effects of
   non-encapsulated or encapsulated probiotic supplementation on microbiota composition, intestinal barrier functions, inflammatory profiles, and glucose tolerance in high fat fed rats. *Nutrients* 11: 1975.

Lian, W.C., Hsiao, H.C. and Chou, C.C. 2002. Survival of bifidobacteria after spray drying. *Int. J. Fd. Microbiol.*
   74: 79–86.

Lievense, L.C. and van't Riet, K. 1994. Convective drying of bacteria. II. Factors influencing survival. *Adv.
   Biochem. Engin./ Biotechnol.* 51: 71–89.

Liu, J., Ren, Y. and Yao, S. 2010. Repeated-batch cultivation of encapsulated Monascus purpureus by polyelectrolyte complex for natural pigment production. *Chin. J. Chem. Engine.* 18: 1013–1017.

Livney, Y.D. 2010. Milk proteins as vehicles for bioactives. *Curr. Opin. Colloid Interface Sci.* 15: 73–83.

Lohiya, G.K. and Avari J.G. 2016. Optimization of feed composition for spray drying of probiotics. *J. Inno.
   Pharm. Biol. Sci.* 3:154–161.

Lorenzo-Zuniga, V., Llop, E., Suarez, C., Alvarez, B., Abreu, L., Espadaler, J. and Serra, J. 2014. I.31,
   a new combination of probiotics, improves irritable bowel syndrome related quality of life. World J.
   Gastroenterol. 20: 8709–8716.

Maciel, G.M., Chaves, K.S., Grosso, C.R.F. and Gigante, M.L. 2014. Microencapsulation of Lactobacillus
   acidophilus La-5 by spray drying using sweet whey and skim milk as encapsulating materials *J. Dairy
   Sci.* 97: 1991–1998.

Madene, A., Jacquot, M., Scher, J. and Desobry, S. 2006. Flavour encapsulation and controlled release – A
   review. *Int. J. Fd. Sci. Technol.* 41: 1–21.

Makinen, K., Berger, B., Bel-Rhlid, R. and Ananta, E. 2012. Science and technology for the mastership of
   probiotic applications in food products. *J. Biotechnol.* 162: 356–365.

Martin, R., Chamignon, C., Mhedbi-Hajri, N., Chain, F., Derrien, M., Escribano-Vázquez, U., Garault, P.,
   Cotillard, A., Pham, H.P., Chervaux, C., Bermúdez-Humarán, L.G., Smokvina, T. and Langella, P. 2019.
   The potential probiotic Lactobacillus rhamnosus CNCM I-3690 strain protects the intestinal barrier by
   stimulating both mucus production and cytoprotective response. *Sci. Rep.* 9: 5398.

Martoni, C., Bhathena, J., Jones, M.L., Urbanska, A.M., Chen, H. and Prakash, S. 2007. Investigation of microencapsulated BSH active Lactobacillus in the simulated human GI tract. *J. Biomedicine Biotechnology.*
   2007: 13684, 9 pages.

Matsuzaki, T. 1998. Immunomodulation by treatment with Lactobacillus casei strain Shirota. *Int. J. Fd.
   Microbiol.* 41: 133–140.

McFarland, L.V. 2015. Deciphering meta-analytic results: a mini-review of probiotics for the prevention of
   paediatric antibiotic-associated diarrhoea and Clostridium difficile infections. *Beneficial Microbes.* 6:
   189–194.

McFarland, L.V., Ship, N., Auclair, J. and Millette, M. 2018. Primary prevention of Clostridium difficile infections with a specific probiotic combining Lactobacillus acidophilus, L. casei, and L. rhamnosus strains:
   assessing the evidence. *J. Hosp. Infect.* 99:443–452.

McIntosh, G.H., Royle, P.J. and Playne, M.J. 1999. A probiotic strain of L. acidophilus reduces DMH-induced
   large intestinal tumors in male Sprague-Dawley rats. *Nutr. Cancer.* 35: 153–159.

Meng, X.C., Stanton, C., Fitzgerald, G.F., Daly, C. and Ross, R.P. 2008. Anhydrobiotics: the challenges of drying probiotic cultures. *Fd. Chem.* 106: 1406–1416.

Messaoudi, M., Lalonde, R., Violle, N., Javelot, H., Desor, D., Nejdi, A., Bisson, J.F., Rougeot, C., Pichelin,
   M., Cazaubiel, M. and Cazaubiel, J.M. 2011. Assessment of psychotropic-like properties of a probiotic
   formulation (Lactobacillus helveticus R0052 and Bifidobacterium longum R0175) in rats and human
   subjects. *Br. J. Nutr.* 105: 755–764.

Michida, H., Tamalampudi, S., Pandiella, S.S., Webb, C., Fukuda, H. and Kondo, A. 2006. Effect of cereal
   extracts and cereal fiber on viability of Lactobacillus plantarum under gastrointestinal tract conditions.
   Biochem. Eng. J. 28: 73–78.

Min, M., Bunt, C.R., Mason, S.L. and Hussain, M.A. 2019. Non-dairy probiotic food products: An emerging group of functional foods. *Crit. Rev. Fd. Sci. Nutr.* 59: 2626–2641.

Minelli, E.B. and Benini, A. 2008. Relationship between number of bacteria and their probiotic effects. *Microbial Eco. Hlth. Dis.* 20: 180–183.

Mokarram, R.R., Mortazavi, S.A., Habibi Najafi, M.B., and Shahidi, F. 2009. The influence of multi stage alginate coating on survivability of potential probiotic bacteria in simulated gastric and intestinal juice. *Fd. Res. Int.* 42: 1040–1045.

Morgan, C.A., Herman, N., White, P.A. and Vesey, G. 2006. Preservation of micro-organisms by drying: A review. *J. Microbiol. Methods,* 66: 183–193.

Morovic, W., Roper, J.M., Smith, A.B., Mukerji, P., Stahl, B., Rae, J.C. and Ouwehand, A.C. 2017. Safety evaluation of HOWARU (R) restore (Lactobacillus acidophilus NCFM, Lactobacillus paracasei LPC-37, Bifidobacterium animalis subsp. lactis Bl-04 and B. lactis Bi-07) for antibiotic resistance, genomic risk factors, and acute toxicity. *Fd. Chem. Toxicol.,* 110: 316–324.

Mortazavian, A.M., Mohammadi, R. and Sohrabvandi, S. 2012. Delivery of probiotic microorganisms into gastrointestinal tract by food products, In: *New Advances in the Basic and Clinical Gastroenterology.* Ed. Tomasz Brzozowski, pp.126. IntechOpen publications, Publisher InTech.

Nazzaro. F., Orlando. P., Fratianni. F. and Coppola, R. 2012. Microencapsulation in food science and biotechnology. *Curr. Opin. Biotech.* 23: 182–186.

Nedovic, V., Kalusevic, A., Manojlovic, V., Levic, S. and Bugarski, B. 2011. An overview of encapsulation technologies for food applications. *Procedia Fd. Sci.* 1: 1806–1815.

Nesterenko, A., Alric, I., Silvestre, F. and Durrieu, V. 2013. Vegetable proteins in microencapsulation: A review of recent interventions and their effectiveness. *Indus. Crops Prod.* 42: 469–479.

Ng, E.W., Yeung, M. and Tong, P.S. 2011. Effects of yogurt starter cultures on the survival of Lactobacillus acidophilus. *Int. J. Fd. Microbiol.* 145: 169–175.

Nunes, G.L., Etchepare, M.A., Cichoski, A.J., Zepka, L.Q., Lopes J.E., Barin, S., de Moraes Flores, E.M., de Bona da Silva, C. and de Menezes, C. R. 2018. Inulin, hi-maize, and trehalose as thermal protectants for increasing viability of Lactobacillus acidophilus encapsulated by spray drying. *LWT Fd. Sci. Technol.* 89: 128–133.

O'Riordan, K., Andrews, D., Buckle, K. and Conway, P. 2001. Evaluation of microencapsulation of a Bifidobacterium strain with starch as an approach to prolonging viability during storage. *J. Appl. Microbiol.* 91: 1059–1066.

O'Mahony, D., Murphy, K.B., Macsharry, J., Boileau, T., Sunvold, G., Reinhart, G., Kiely, B., Shanahan, F. and O'Mahony, L. 2009. Portrait of a canine probiotic Bifidobacterium – From gut to gut. *Vet. Microbiol.* 139: 106–112.

O'Mahony, L., McCarthy, J., Kelly, P., Hurley, G., Luo, F., Chen, K., O'Sullivan, G.C., Kiely, B., Collins, J.K., Shanahan, F. and Quigley E.M. 2005. Lactobacillus and bifidobacterium in irritable bowel syndrome: Symptom responses and relationship to cytokine profiles. *Gastroenterology* 128: 541–551.

Orlando, A. and Russo, F. 2013. Intestinal microbiota, probiotics and human gastrointestinal cancers. *J. Gastrointest. Cancer.* 44:121–131.

Ouwehand, A.C., Dong Lian, C., Weijian, X., Stewart, M., Ni, J., Stewart, T. and Miller L.E. 2014. Probiotics reduce symptoms of antibiotic use in a hospital setting: A randomized dose response study. *Vaccine.*32: 458–463.

Ouwehand, A.C., Invernici, M.M., Furlaneto, F.A.C. and Messora, M.R. 2018. Effectiveness of multistrain versus single-strain probiotics: Current status and recommendations for the future. *J. Clin. Gastroenterol.* 52: Suppl 1, Proceedings from the 9th Probiotics, Prebiotics and New Foods, Nutraceuticals and Botanicals for Nutrition & Human and Microbiota Health Meeting, held in Rome, Italy from September 10 to 12, 2017:S35–S40.

Ozen, A. E., Pons, A. and Tur, J. A. 2012. Worldwide consumption of functional foods: A systematic review. *Nutr. Rev.* 70: 472–481.

de Vos, P., Faas, M.M., Spasojevic, M. and Sikkema, J. 2010. Encapsulation for preservation of functionality and targeted delivery of bioactive food components. *Int. Dairy J.* 20: 292–302.

Paez, R., Lavari, L., Audero, G., Cuatrin, A., Zariztky, N., Reinheimer, J. and Vinderola, G. 2013. Study of the effects of spray drying on the functionality of probiotic lactobacilli. *Int. J. Dairy Technol.* 66:155–161.

Paez, R., Lavari, L., Vinderola, G., Audero, G., Cuatrin, A., Zaritzky, N. and Reinheimer, J. 2012. Effect of heat treatment and spray drying on lactobacilli viability and resistance to simulated gastrointestinal digestion. *Fd. Res. Int.* 48: 748–754.

Paim, D.R.S.F., Costa, S.D.O., Walter, E.H.M., Tonon, R.V. 2016. Microencapsulation of probiotic jussara (Euterpe edulis M.) juice by spray drying. *Int. J. Fd. Microbiol.* 145: 169–175. *Sci. Technol.* 74: 21–25.

Pandey, K.R. and Vakil, B.V. 2017. Encapsulation of probiotic Bacillus coagulans for enhanced shelf life. *J. Appl. Biol. Biotechnol.* 5: 57–65.

Park, E.J., Lee, Y.S., Kim, S.M., Park, G.S., Lee, Y.H., Jeong, D.Y., Kang, J. and Lee, H.J. 2020. Beneficial effects of *Lactobacillus plantarum* strains on non-alcoholic fatty liver disease in high fat/high fructose diet-fed rats. *Nutrients.* 12: 542

Park, J.K. and Chang, H.N. 2000. Microencapsulation of microbial cells. *Biotechnol. Adv.* 18: 303–319.

Parra-Huertas. R.A. 2010. Revisión: Microencapsulación de Alimentos. *Revista Facultad Nacional de Agronomía, Medellín.* 63: 5669–5684.

Peighambardoust, S.H., Tafti, A.G. and Hesari, J. 2011. Application of spray drying for preservation of lactic acid starter cultures: A review. *Trends Fd. Sci. Technol.* 22: 215–224.

Pelletier, X., Laure-Boussuge, S. and Donazzolo, Y. 2001. Hydrogen excretion upon ingestion of dairy products in lactose-intolerant male subjects: Importance of the live flora. *Eur. J. Clin. Nutr.* 55: 509–512.

Pelucchi, C., Chatenoud, L., Turati, F., Galeone, C., Moja, L., Bach, J.F. and La Vecchia, C. 2012. Probiotics supplementation during pregnancy or infancy for the prevention of atopic dermatitis: a meta-analysis. *Epidemiology.* 23: 402–414.

Penna, A.L.B., Rao-Gurram, S. and Barbosa-Ca'novas, G.V. 2007. Effect of milk treatment on acidification, physicochemical characteristics, and probiotic cell counts in low fat yogurt. *Milchwiss.* 62: 48–52.

Perdana, J., Bereschenko, L., Fox, M.B., Kuperus, J.H., Kleerebezem, M., Boom, R.M. and Schutyser, M.A. 2013. Dehydration and thermal inactivation of Lactobacillus plantarum WCFS1: Comparing single droplet drying to spray and freeze drying. *Fd. Res. Int.* 54: 1351–1359.

Perdana, J., Fox, M.B., Siwei, C., Boom, R.M., and Schutyser, M.A.I. 2014. Interactions between formulation and spray drying conditions related to survival of Lactobacillus plantarum WCFS1. *Fd. Res. Int.* 56: 9–17.

Peres, C.M., Peres, C., Hernández-Mendoza, A. and Malcata, F.X. 2012. Review on fermented plant materials as carriers and sources of potentially probiotic LAB – With an emphasis on table olives. *Trends Fd. Sci. Technol.* 26: 31–42.

Perricone, M., Bevilacqua, A., Altieri, C., Sinigaglia, M. and Corbo, M.R. 2015. Challenges for the production of probiotic fruit juices. *Beverages.* 1: 95–103.

Piano, M.D., Carmagnola, S., Ballarè, M., Balzarini, M., Montino, F., Pagliarulo, M., Anderloni, A., Orsello, M., Tari, R., Sforza, F., Mogna, L. and Mogna, G. 2012. Comparison of the kinetics of intestinal colonization by associating 5 probiotic bacteria assumed either in a microencapsulated or in a traditional, uncoated form. *J. Clin. Gastroenterol.* 46: 85–92.

Picot, A. and Lacroix, C. 2004. Encapsulation of bifidobacteria in whey protein-based microcapsules and survival in simulated gastrointestinal conditions and in yogurt. *Int. Dairy J.* 14: 505–515.

Pinto, S.S., Fritzen-Freire, C.B., Benedetti, S., Murakami, F.S., Petrus, J.C.C., Prudêncio, E.S. and Amboni, R.D.M.C. 2015. Potential use of whey concentrate and prebiotics as carrier agents to protect Bifidobacterium-BB-12 microencapsulated by spray drying. *Fd. Res. Int.* 67: 400–408.

Pispan, S., Hewitt, C.J., and Stapley, A.G.F. 2013. Comparison of cell survival rates of E. coli K12 and L. acidophilus undergoing spray drying. *Fd. Bioprod. Process.* 91: 362–369.

Poddar, D., Das, S., Jones, G., Palmer, J., Jameson, G.B., Haverkamp, R.G., Singh, H. 2014. Stability of probiotic Lactobacillus paracasei during storage as affected by the drying method. *Int. Dairy J.* 39: 1–7.

Pu, F., Guo, Y., Li, M., Zhu, H., Wang, S., Shen, X., He, M., Huang, C. and He, F. 2017. Yogurt supplemented with probiotics can protect the healthy elderly from respiratory infections: A randomized controlled open-label trial. *Clin. Interv. Aging.* 12: 1223–1231

Rajam, R. and Anandharamakrishnan, C. 2015. Microencapsulation of Lactobacillus plantarum (MTCC 5422) with fructooligosaccharide as wall material by spray drying. *LWT Fd. Sci. Technol.* 60: 773–780.

Rebecchi, L., Altiero, T. and Guidetti, R. 2007. Anhydrobiosis: the extreme limit of desiccation tolerance. *Invertebr. Surv. J.* 4: 65–81.

Ribeiro, M.C.E., Chaves, K.S., Gebara, C., Infante, F.N.S., Grosso, C.R.F. and Gigante, M.L. 2014. Effect of microencapsulation of Lactobacillus acidophilus LA-5 on physic chemical, sensory and microbiological characteristics of stirred probiotic yogurt. *Fd. Res. Int.* 66: 424–431.

Ritchie, M.L. and Romanuk, T.N. 2012. A meta-analysis of probiotic efficacy for gastrointestinal diseases. *PLoS One.* 7: e34938.

Riveros, B., Ferrer, J. and Borquez, R. 2009. Spray drying of a vaginal probiotic strain of Lactobacillus acidophilus. *Drying Technol..* 27: 123–132.

Roberts, L.M., McCahon, D., Holder, R., Wilson, S. and Hobbs, F.D. 2013. A randomized controlled trial of a probiotic 'functional food' in the management of irritable bowel syndrome. *BMC Gastroenterol.* 13: 1.

Rokka, S. and Rantamaki, P. 2010. Protecting probiotic bacteria by microencapsulation: challenges for industrial applications. *Europ. Fd. Res. Technol.* 231: 1–12.

Romano, A., Blaiotta, G., Di Cerbo, A., Coppola, R., Masi, P. and Aponte, M. 2014. Spray-dried chestnut extract containing *Lactobacillus rhamnosus* cells as novel ingredient for a probiotic chestnut mousse. *J. Appl. Microbiol.* 116: 1632–1641.

Rossi, F., Zotta, T., Iacumin, L. and Reale, A. 2016. Theoretical insight into the heat shock response (HSR) regulation in *Lactobacillus casei* and *L. rhamnosus*. *J. Theoretical Biol.* 402: 21–37.

Saarela, M., Mogensen, G., Fonden, R., Matto, J., and Mattila-Sandholm, T. 2000. Probiotic bacteria: safety, functional and technological properties. *J. Biotechnol.* 84: 197–215.

Sabikhi, L., Babu, R., Thompkinson, D.K. and Kapila, S. 2010. Resistance of microencapsulated lactobacillus acidophilus LA1 to processing treatments and simulated gut conditions. *Fd. Bioprocess Technol.* 3: 586–593.

Santivarangkna, C., Higl, B. and Foerst, P. 2008. Protection mechanisms of sugars during different stages of preparation process of dried lactic acid starter cultures. *Fd. Microbiol.* 25: 429–441.

Santivarangkna, C., Kulozik, U. and Foerst, P. (2007). Alternative drying processes for the industrial preservation of lactic acid starter cultures. *Biotechnology Progress.* 23: 302–315.

Santivarangkna, C., Kulozik, U. and Foerst, P. 2008. Inactivation mechanisms of lactic acid starter cultures preserved by drying processes. *J. Appl. Microbiol.* 105: 1–13.

Sarkar, S. 2007. Potential of prebiotics as functional foods – A Review. *Nutr. Fd. Sci.* 37: 168–177.

Sarkar, S. 2010. Approaches for enhancing the viability of probiotic – a review, *Br. Fd. J.*, 112: 329–349.

Sarkar, S. 2013. Probiotics as functional foods: documented health benefits. *Nutr. Fd. Sci.* 43: 107–115.

Sarkar, S. 2018. Whether viable and dead probiotic are equally efficacious?. *Nutr. Fd. Sci.* 48: 285–300.

Sarkar, S., Sarkar, K., Majhi, R., Sur, A., Basu, S. and Chatterjee, K. 2016. Probiotics: a way of value addition in functional food. *Int. J. Fd. Sci. Nutr. Diet.* 3: 290–293.

Schuck, P., Dolivet, A., Méjean, S., Hervé, C. and Jeantet, R. 2013. Spray drying of dairy bacteria: New opportunities to improve the viability of bacteria powders. *Int. Dairy J.* 31: 12–17.

Semyonov, D., Ramon, O., Kaplun, Z., Levin-Brener, L., Gurevich, N., Shimoni, E., 2010. Microencapsulation of Lactobacillus paracasei by spray freeze drying. *Fd. Res. Int.* 43: 193–202.

Sgouras, D., Maragkoudakis, P., Petraki, K., Martinez-Gonzalez, B., Eriotou, E., Michopoulos, S., Kalantzopoulos, G., Tsakalidou, E. and Mentis, A. 2004. In vitro and in vivo inhibition of Helicobacter pylori by Lactobacillus casei strain Shirota. *Appl. Environ. Microbiol.* 70: 518–526.

Shah, N.P. 2000. Probiotic bacteria: Selective enumeration and survival in dairy foods. *J. Dairy Sci.* 83: 894–907.

Shi, L.E., Li, Z.H., Li, D.T., Xu, M., Chen, H.Y., Zhang, Z.L. and Tang, Z.X. 2013. Encapsulation of probiotics Lactobacillus bulgaricus in alginate-milk microspheres and evaluation of survival in simulated gastrointestinal conditions. *J Fd. Engn.* 117: 99–104.

Shishir, M.R.I. and Chen, W. 2017. Trends of spray drying: A critical review on drying of fruit and vegetable juices. *Trends Fd. Sci. Technol.* 65: 49–67.

Shishir, M.R.I., Taip, F.S., Aziz, N.A., Talib, R.A. and Sarker, M.S.H. 2016. Optimization of spray drying parameters for pink guava powder using RSM. *Fd. Sci. Biotechnol.* 25: 1–8.

Shishir, M.R.I., Taip, F.S., Saifullah, M., Aziz, N.A. and Talib, R.A. 2017. Effect of packaging materials and storage temperature on the retention of physicochemical properties of vacuum packed pink guava powder. *Fd. Pack. Shelf Life.* 12: 83–90.

Shishir, M.R.I., Xie, L., Sun, C., Zheng, X., Chen, W. 2018. Advances in micro and nano-encapsulation of bioactive compounds using biopolymer and lipid-based transporters. *Trends Fd. Sci. Technol.* 78: 34–60.

Shornikova, A.V., Isolauri, E., Burkanova, L., Lukovnikova, S., and Vesikari, T. 1997. A trial in the Korelian Republic of oral rehydration and Lactobacillus GG for treatment of acute diarrhea. *Acta Paediatr.* 86: 460–465.

Silva, K.C.G., Cezarino, E.C., Michelon, M. and Sato, A.C.K. 2018. Symbiotic microencapsulation to enhance Lactobacillus acidophilus survival. *LWT – Fd. Sci. Technol.* 89: 503–509.

Silva, P.I., Stringheta, P.C., Teófilo, R.F., Rebouças, I. and Oliveira, N. 2013. Parameter optimization for spray drying microencapsulation of jaboticaba (Myrciaria jaboticaba) peel extracts using simultaneous analysis of responses. *J. Fd. Engn.* 117: 538–544.

Silveira, A.C.P., Perrone, I.P., Rodrigues Junior, P.H. and de Carvalho, AF 2013. Secagem por Spray: uma revisão. *Rev. Inst. Laticínios Cândido Tostes.* 68: 51–58.

Simpson, P.J., Stanton, C., Fitzgerald, G.F. and Ross, R.P. 2005. Intrinsic tolerance of Bifidobacterium species to heat and oxygen and survival following spray drying and storage. *J. Appl. Microbiol.* 99: 493–501.

Sobel, R., Versic, R., Gaonkar, A. G. 2014. Introduction to Microencapsulation and Controlled Delivery in Foods. In: *Microencapsulation in the Food Industry*. Ed. Gaonkar, A.G., Vasisht, N., Khare, A.R., Sobel, R., Tokyo: Academic Press, pp. 3–12.

Sorokulova, I. 2008. Preclinical testing in the development of probiotics: A regulatory perspective with Bacillus strains as an example. *Clin. Infect. Dis.* 46: 92–95.

Soukoulis, C., Behboudi-Jobbehdar, S., Yonekura, L., Parmenter, C. and Fisk, I. 2014. Impact of milk protein type on the viability and storage stability of microencapsulated *Lactobacillus acidophilus* NCIMB 701748 using spray drying. *Fd. Bioprocess Technol.* 7: 1255–1268.

Sousa, S., Gomes, A.M., Pintado, M.M., Malcata, F.X., Silva, J.P., Sousa, J.M., Costa, P., Amaral, M.H., Rodrigues, D., Rocha-Santos, T.A.P. and Freitas, A.C. 2012. Encapsulation of probiotic strains in plain or cysteine-supplemented alginate improves viability at storage below freezing temperatures. *Engin. Life Sci.* 12: 457–465.

Spigno, G., Garrido, G.D., Guidesi, E. and Elli, M. 2015. Spray-drying encapsulation of probiotics for ice-cream application. *Chem. Engin. Transact.* 43: 49–54.

Sreeja, V. and Prajapati, J.B. 2013. Probiotic formulations: Application and status as pharmaceuticals – A review. *Probiotics Antimicrobiol. Proteins.* 5: 81–91.

Strus, M., Helwich, E., Lauterbach, R., Rzepecka-Węglarz, B., Nowicka, K., Wilińska, M., Szczapa, J., Rudnicka, M., Sławska, H., Szczepański, M., Waśko, A., Mikołajczyk-Cichońska, A., Tomusiak-Plebanek, A. and Heczko, P.B. 2018. Effects of oral probiotic supplementation on gut Lactobacillus and Bifidobacterium populations and the clinical status of low-birth-weight preterm neonates: a multicenter randomized, double-blind, placebo-controlled trial. *Infect. Drug Resist.* 11: 1557–1571.

Suganya, V. and Anuradha, V. 2017. Microencapsulation and nanoencapsulation: A Review. *Int. J. Pharma. Clin. Res.* 9: 233–239.

Sultana, A., Miyamoto, A., Lan Hy, Q., Tanaka, Y., Fushimi, Y. and Yoshii, H. 2017. Microencapsulation of flavors by spray drying using Saccharomyces cerevisiae. *J. Fd. Engin.* 199: 36–41.

Sunny-Roberts, E.O. and Knorr, D. 2009. The protective effect of monosodium glutamate on survival of Lactobacillus rhamnosus GG and Lactobacillus rhamnosus E-97800 strains during spray drying and storage in trehalose-containing powders. *Int. Dairy J.* 19: 209–214.

Szajewska, H., Skórka, A., Ruszczyński, M. and Gieruszczak-Białek, D. 2013. Meta-analysis: Lactobacillus GG for treating acute gastroenteritis in children - Updated analysis of randomised controlled trials. *Aliment. Pharmacol. Ther.* 38: 467–476.

Tan, S.M., Heng, P.W.S. and Chan, L.W. 2011. Development of re-usable yeast-gellan gum micro-bioreactors for potential application in continuous fermentation to produce bio-ethanol. *Pharm.* 3: 731–744.

Teixeira, P., Castro, H. and Kirby, R. 1994. Inducible thermotolerance in Lactobacillus bulgaricus. *Lett. Appl. Microbiol.* 18: 218–221.

To, B.C.S. and Etzel, M.R. 1997. Spray drying, freeze drying, or freezing of three different lactic acid bacteria species. *J. Fd. Sci.* 62: 576–578.

Tomaro-Duchesneau, C., Jones, M.L., Shah, D., Jain, P., Saha, S. and Prakash, S. 2014. Cholesterol assimilation by Lactobacillus probiotic bacteria: an in vitro investigation. *Biomed. Res. Int.* 2014: 380316, 9 pages.

Tripathi, M.K. and Giri, S.K. 2014. Probiotic functional foods: survival of probiotics during processing and storage. *J. Funct. Fds.* 9: 225–241.

Turchiuli, C., Jimenez Munguia, M.T., Hernandez Sanchez, M., Cortes Ferre, H., and Dumoulin, E. 2014. Use of different supports for oil encapsulation in powder by spray drying. *Powder Technol.* 255: 103–108.

Umbrello, G. and Esposito S. 2016. Microbiota and neurologic diseases: potential effects of probiotics. *J. Transl. Med.* 14: 1–11.

Urbanska, M., Gieruszczak-Bialek, D., Szymanski, H. and Szajewska, H., 2016. Effectiveness of *Lactobacillus reuteri* DSM 17938 for the prevention of nosocomial diarrhea in children: a randomized, double-blind, placebo-controlled trial. *Pediatric Infec. Dis. J.* 35: 142–145.

Van Niel, C.W., Feudtner, C., Garrison, M.M. and Christakis, D.A. 2002. Lactobacillus therapy for acute infectious diarrhoea in children: A meta-analysis. *Pediat.* 109: 678–684.

Van de Guchte, M., Serror, P., Chervaux, C., Smokvina, T., Ehrlich, S. D. and Maguin, E. 2002. Stress responses in lactic acid bacteria. *Int. J. General Molecular Microbiol.* 82: 187–216.

Vandenplas, Y., Huys, G. and Daube, G. 2015. Probiotics: an update. *J. Pediatr.* 91: 6–21.

Vemuri, R., Gundamaraju, R., Shastri, M.D., Shukla, S.D., Kalpurath, K., Ball, M., Tristram, S., Shankar, E.M., Ahuja, K. and Eri, R. 2018. Gut microbial changes, interactions, and their implications on human lifecycle: An ageing perspective. *BioMed. Res. Int.* 2018, 4178607, 13 pages.

Vijayendra, S.V.N. and Gupta, R.C. 2013a. Associative growth behavior of dahi and yogurt starter cultures with *Bifidobacterium bifidum* and *Lactobacillus acidophilus* in buffalo skim milk. *Ann. Microbiol.* 63: 461–469.

Vijayendra, S.V.N. and Gupta, R.C. 2013b. Performance evaluation of bulk freeze dried starter cultures of dahi and yogurt along with probiotic strains in standardized milk of cow and buffalo. *J. Fd. Sci. Technol.* 51: 4114–4119.

Wedajo, B. 2015. Lactic Acid Bacteria: Benefits, selection criteria and probiotic potential in fermented food. *J. Prob. Hlth.* 3: 129.

Weinbreck, F., Bodnár, I. and Marco, M.L. 2010. Can encapsulation lengthen the shelf-life of probiotic bacteria in dry products?. *Int. J. Fd. Microbiol.* 136: 364–367.

Wolkers, W.F., Tablin, F. and Crowe, J.H. 2002. From anhydrobiosis to freeze-drying of eukaryotic cells. *Comp. Biochem. Phys.* 131A: 535–543.

Xavier dos Santos, D., Casazza, A.A., Aliakbarian, B., Bedani, R., Saad, S.M.I. and Perego, P. 2019. Improved probiotic survival to in vitro gastrointestinal stress in a mousse containing Lactobacillus acidophilus La-5 microencapsulated with inulin by spray drying. *LWT – Fd. Sci.Technol.*99: 404–410.

Xiao, L., Ding, G., Ding, Y. Deng, C., Ze, X., Chen, L., Zhang, Y., Song, L., Yan, H., Liu, F. and Ben, X. 2017. Effect of probiotics on digestibility and immunity in infants. *Med.* 96: e5953.

Xu, M., Gagne-Bourque, F., Dumont, M. J. and Jabaji, S. 2016. Encapsulation of Lactobacillus casei ATCC 393 cells and evaluation of their survival after freeze drying, storage and under gastrointestinal conditions. *J. Fd. Engn.* 168: 52–59.

Yarullina, D.R., Damshkaln, L.G., Bruslik, N.L., Konovalova, O.A., Ilinskaya, O.N. and Lozinsky, V.I. 2015. Towards effective and stable probiotics. *Int. J. Risk Safety Med.* 27: 65–66.

Ying, D., Sun, J., Sanguansri, L., Weerakkody, R. and Augustin, M.A. 2012. Enhanced survival of spray dried microencapsulated Lactobacillus rhamnosus GG in the presence of glucose. *J. Fd. Engin.* 109: 597–602.

Ying, D.Y., Sanguansri, L., Weerakkody, R. Bull, M., Singh, T.K. and Augustin, M.A. 2016. Effect of encapsulant matrix on stability of microencapsulated probiotics. *J. Func. Fds.* 25: 447–458.

Ying, D.Y., Schwanderc, S., Weerakkodya, R., Sanguansria, L., Gantenbein-Demarchic, C. and Augustina, M.A. 2013. Microencapsulated Lactobacillus rhamnosus GG in whey protein and resistant starch matrices: Probiotic survival in fruit juice. *J. Funct. Fds.* 5, 98–105.

Yonekura, L., Sun, H., Soukoulis, C. and Fisk, I. 2014. Microencapsulation of *Lactobacillus acidophilus* NCIMB 701748 in matrices containing soluble fibre by spray drying: Technological characterization, storage stability and survival after *in vitro* digestion. *J. Funct. Foods.* 6: 205–214

Yudiastuti, S., Sukarminah, E., Mardawati, E. and Kastaman, R. 2019. *Evaluation study of Lactobacillus acidophilus drying. IOP Conference Series: Earth and Environmental Science*, Vol. 250: 012016, Purwokerto, Indonesia.

Zhang, G.Q., Hu, H.J., Liu, C.Y., Zhang, Q., Shakya, S. and Li, Z.Y. 2016. Probiotics for prevention of atopy and food hypersensitivity in early childhood. A PRISMA-Compliant Systematic Review and Meta-Analysis of Randomized Controlled Trials. *Med.* 95: e2562.

Zhang, Y.J., Li, S., Gan, R.Y., Zhou, T., Xu, D.P. and Li, H.B. 2015. Impacts of gut bacteria on human health and diseases. *Int. J. Mol. Sci.* 16: 7493–7519.

Zheng, X., Fu, N., Huang, S., Jeantet, R. and Chen, X.D. 2016. Exploring the protective effects of calcium-containing carrier against drying-induced cellular injuries of probiotics using single droplet drying technique. *Fd. Res. Int.* 90: 226–234.

Zhou, J.S., Shu, Q., Rutherfurd, K.J., Prasad, J., Gopal, P.K. and Gill, H. 2000. Acute oral toxicity and bacterial translocation studies on potentially probiotic strains of lactic acid bacteria. *Fd. Chem. Toxicol.* 38: 153–161.

Zou, J., Dong, J. and Yu, X. 2009. Meta-analysis: Lactobacillus containing quadruple therapy versus standard triple first-line therapy for *Helicobacter pylori* eradication. *Helicobacter.* 14: 97–107.

Zuidam, N.J. and Nedovic, V.A. 2010. *Encapsulation Technologies for Active Food Ingredients and Food Processing.* Springer, Verlag New York.

12 Nano Spray Drying for the Encapsulation of Bioactive Ingredients

Arlete Maria Lima Marques

INL – International Iberian Nanotechnology Laboratory, Portugal;
CEB – Centre of Biological Engineering, University of Minho, Portugal

Ana Isabel Bourbon, Lorenzo Pastrana, and Miguel Angelo Cerqueira

INL – International Iberian Nanotechnology Laboratory, Portugal

Jose Antonio Couto Teixeira

CEB – Centre of Biological Engineering, University of Minho, Portugal

CONTENTS

12.1 INTRODUCTION

Spray drying technology consists of a fast one-step process that can dry various types of solutions (e.g., aqueous or organic), and colloidal systems (emulsions, dispersions, and suspensions) to obtain dried materials that can be in the form of either particles or capsules. This technology has been used for several years in the food, pharmaceutical, and chemical industries where it is seen as a simple, secure, fast, cost-effective, and efficient method for manufacturing dried powders. Additionally, due to the low moisture content, the spray dried structures can be stored for a long time. Despite the high temperature applied during the spray drying process, the short exposure makes this technique usable for drying even heat-sensitive materials (e.g., enzymes, probiotics, polyphenols, natural food colorants, omega-3 oils, essential oils) without causing negative effects such as chemical degradation (Arpagaus, 2011;Nedovic, Kalusevic, Manojlovic, Levic, & Bugarski, 2011; Arpagaus, John, Collenberg, & Rutti, 2017; Assadpour & Jafari, 2019; Li, Anton, Arpagaus, Belleteix, & Vandamme, 2010; Marques et al., 2019).

Recently, spray drying equipment has been modified to make it able to produce smaller size structures, from microscale to nanoscale. In addition, this change allowed easy equipment handling and higher recovery of the dried particle. The first nano spray drier was launched by Büchi under the name of the 'Nano spray Dryer (NSD) B-90'. More recently, this was superseded by the 'B-90 HP' model, which can produce micro and submicron particles using a low amount of sample. This equipment also provides a high-yield drying process (when compared with conventional cyclone traps in regular spray dryers), and, therefore, decreases the research and development costs. Similar to the conventional spray drying process, nano spray drying technology is a continuous and one-step process that goes from the liquid solution to the dried structures (Arpagaus, Collenberg, Rütti, Assadpour, & Jafari, 2018; Büchi Labortechnik, 2015; Harsha et al., 2015).

In this chapter, the application of nano spray drying technology for the encapsulation of bioactive ingredients is discussed and compared with other spray driers. There is also some presentation and discussion of the published studies using the nano spray drying technology in both 'closed-' and 'open-loop', as well as various materials that are used in the nano spray drying of bioactive ingredients for various applications.

12.2 NANO SPRAY DRYING TECHNIQUE

In terms of the process, a nano spray dryer operates under the same process as that of a conventional spray dryer; i.e., drying a liquid solution to dry micro- and nanoparticles by spraying the solution into a hot drying medium (Heng, Lee, Ng, & Tan, 2011). This process has the advantage of being a simple one-step process that can convert different liquid formulations, like nanoemulsions, nanosuspensions, organic, and aqueous solutions, into dry powders with sizes ranged from nanometers to micrometers (Arpagaus, 2019b; Arpagaus et al., 2018, 2017). Nano spray drying was developed as a response to the increasing demand from pharmaceutical, food, and cosmetic industries for smaller particles using a faster production process with a more efficient powder recovery and a better collection approach (Heng et al., 2011).

Büchi Labortechnik AG Co. has been working on spray drying technology for several years. Firstly, they launched the 'Mini Spray Dryer (MSD) B-290', which was capable of producing microparticles with a cyclone separator to collect particles. Then, in 2009, they launched the 'NSD B-90', which allows the production of submicron particles. Such submicron particles are produced based on the vibrating mesh spray technology, followed by a gentle flow of laminar drying gas and an electrostatic collector for the efficient collection of the dried materials (Arpagaus et al., 2017; Assadpour & Jafari, 2019; Büchi Labortechnik, 2015, 2020a). In 2017, Büchi Labortechnik AG launched its most recent product, the 'NSD B-90 HP', which has a 2nd generation spray head that increases the surface contact area allowing the production of smaller particles with submicron size. Using this equipment, it is also possible to obtain particles from a small sample volume and with higher productivity (>90%) (Büchi Labortechnik, 2017c).

There are several differences between the traditional spray dryers and the nano spray dryers that are currently available in the market. Table 12.1 compiles the main differences between the equipment 'MSD B-290', 'NSD B-90', and the most recent version, the 'NSD B-90 HP' (Büchi Labortechnik, 2017c, 2015, 2020a).

All of the spray dryers have the same working bases, and different companies manufacture this equipment. In this chapter, the NSD was compared with the MSD produced by Büchi Labortechnik AG, in an attempt to compare equipment manufactured by the same company. The 'MSD B-290' (Figure 12.1) contains an electric heater that is responsible for heating the inlet gas (1), an atomizer that breaks the feed solution into tiny droplets forming a spray (2) that enters in contact with the turbulent hot gas in co-current, for a very short time, which evaporates the solvent and dries

TABLE 12.1

Main Differences Between Various Spray Drying Equipment; 'MSD B-290', 'NSD B-90', and 'NSD B-90 HP' (Arpagaus, 2019b; Arpagaus et al., 2017; Büchi Labortechnik, 2015, 2017c, 2017b, 2020a, 2016, 2017d)

Characteristics	Mini Spray Dryer B-290**	Nano Spray Dryer B-90**	Nano Spray Dryer B-90 HP**
Minimum sample volume (mL)	30	2	2
Maximum sample viscosity (mPa.s)	<300	<10	<5
Spray frequency (kHz)		60	80–140
Airflow (m³/h)	35 (turbulent)	10 (laminar)	10 (laminar)
Mean residence time (s)	1.0–1.5	1–4	1–4
Spray gas (L/min)	3.33–13.33	80–160	80–160
Spray generation device	Two-fluid nozzle (optional: ultrasonic nozzle and three –fluid nozzle)	Vibrating mesh	Innovative nebulizers
Nozzle diameter	0.7; 1.4; 2 mm (Nozzle tip diameter)	4.0; 5.5; 7.0 μm (spray caps)	Small, medium, large (nebulizers)
Maximum temperature (°C)	220	120	120
Heating capacity (W)	2300	1400	1400
Water evaporation capacity* (L/h)	1.0 *	0.2 *	0.2 *
Mean droplet size (μm)	5–30	8–21	3–15
Particles size (μm)	2–60	0.3–5	0.2–5
Collector	Cyclone and glass vessel	Electrostatic particles collector	Electrostatic particles collector
Yield (%)	>70	>90	>90

* Different values if using organic solvents.
** The equipment MSD B-290', 'NSD B-90', and 'NSD B-90 HP' are produced by the company Büchi Labortechnik AG.

the particles in the glass cylinder (3). The dried particles are sedimented in the cyclone (4) and are collected at the bottom. The gas passes through the outlet filter (5) to filter the very fine particles that were not collected in the cyclone, and, finally, the gas is aspirated by the pump gas through the system (6) (Büchi Labortechnik, 2020a; Ousset et al., 2018).

The mini spray dryer equipped with a two-fluid nozzle contains a nozzle tip that has a hole with a diameter of 0.7 mm (and a nozzle cap). This two-fluid nozzle allows the mixing of the liquid and the gas to produce a stable cone that will be dried in the cylinder. The spray is obtained by passing the solution and the gas through the nozzle's ruby stone, which has a precise opening and sharp edges, to precisely control the spray cone. Droplet size can be controlled by using the other two available nozzle tips with diameter sizes of 1.4 mm and 2.0 mm which are also suitable to spray more viscous solutions or dispersions. The nozzle cap is available in two different diameters: 1.4 mm and 1.5 mm. The smaller nozzle cap is more appropriate for minimizing the operation costs when nitrogen is used as the spray gas, because with this nozzle the gas consumption is lower. The larger nozzle cap (1.5 mm in diameter) is more appropriate when air is used as the spray gas, since this design is more robust regarding concentric alignment to form a vertical and uniform spray cone (Büchi Labortechnik, 2020a).

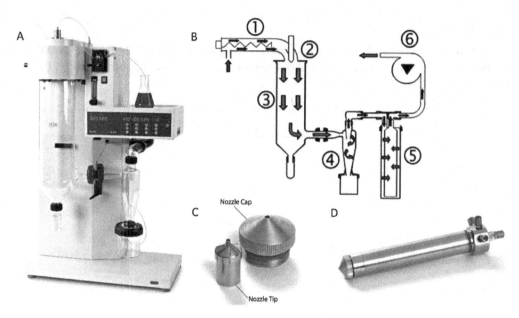

FIGURE 12.1 'MSD B-290' equipment from Büchi Labortechnik AG (A) and equipment scheme and components (B). 'MSD B-290' spray head (D) and nozzle cap for solution atomization (C) (adapted from Büchi Labortechnik, 2016, 2020a, 2020b).

Despite being widely used for obtaining dry particles, the mini spray dryer has some drawbacks such as not being capable of producing submicron particles and having a low efficiency of sedimentation and collection of particles by cyclone separators. The equipment is unable to collect particles with a size of <2 μm because of their low weight and high kinetic energy (Heng et al., 2011; Lee, Heng, Ng, Chan, & Tan, 2011). Another problem is the deposition of the particles in the drying chamber due to its turbulent gas flow.

The Mini Spray Dryer has been used to dry different structures, solutions, emulsions, gels, and even suspensions. It has also been used for the encapsulation of different compounds. Today, however, one of the most recent trends in the pharmaceutical and food industry is the use of structures at the nanoscale. The nanoencapsulation of bioactive compounds presents unique advantages such as the increase of encapsulation efficiency, the stabilization of the structures that are less predisposed to aggregate and sediment, the promotion of the controlled release of the nanoencapsulated bioactive compounds, and an increase in the absorption in the gastrointestinal tract (GIT). Due to their subcellular size, nanostructures possess a high surface area per mass unit, which can increase their efficiency during the delivery of bioactive compounds (Cerqueira et al., 2014, 2016; Costa et al., 2018; Davidov-Pardo, Joye, & McClements, 2015; Marques et al., 2019). Following this trend, spray drying technology was improved in order to be able to produce nanostructures from the feed sample. It is known that conventional spray drying equipment is unable to produce droplets small enough to obtain nanostructures. Conventional atomizers need to be improved in order to produce very small particles, and the drying gas needs to have a laminar flow to have a controlled spray, and also an efficient collector for the collection of even small particles (Assadpour & Jafari, 2019; Li et al., 2010; Oliveira, Guimarães, Cerize, Tunussi, & Poço, 2013).

The new 'NSD B-90' (Figure 12.2) presents a novel vibrating mesh spray technology capable of producing fine droplets. These small droplets are dried by gentle flow laminar drying gas in a glass drying column. The length of the glass drying column is possible to change according to the operation needs. The obtained dried particles can range between 300 nm and 5 μm, being also able to produce nanoparticles (<100 nm) when the drying solutions are nanosuspensions or nanoemulsions.

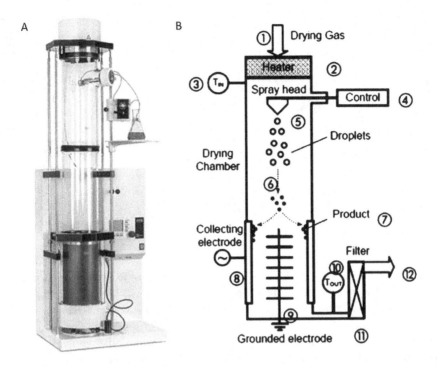

FIGURE 12.2 'NSD B-90' equipment from Büchi Labortechnik AG (A) and 'NSD B-90' drying process and components (B) (Büchi Labortechnik, 2015, 2020c).

Further, it has an electrostatic collector that attracts and collects the produced dried particles with high efficiency (Arpagaus, 2011; Arpagaus et al., 2017; Büchi Labortechnik, 2015; Bürki, Jeon, Arpagaus, & Betz, 2011; Heng et al., 2011).

The 'NSD B-90' process (Figure 12.2) starts with the drying gas (1) that is heated in the electrical heater (2) at the selected temperature (maximum 120°C) on the top of the glass drying chamber (6), where a temperature sensor (3) is installed to control the gas inlet temperature. Then, the solution feed is pumped by a peristaltic pump to the spray head (5) which uses piezo technology to generate microdroplets that are dried when in contact with the co-current laminar flow drying gas inside the spray cylinder (6). To efficiently collect the produced particles, this equipment uses an electric field that is generated by high voltage created between the collecting electrode (8) and the High Voltage (HV)-electrode (9) to attract particles. The HV-electrode (9) deflects the particles while the collecting electrode attracts them, resulting in their deposition in the collecting electrode (8). After the drying process and particle collection (7), the drying gas passes through the outlet temperature sensor (10) and into the outlet filter (11) (Arpagaus, 2019b; Arpagaus et al., 2018; Büchi Labortechnik, 2017b; Heng et al., 2011).

The particle production technology in 'NSD B-90' is based on Piezo technology, which generates precisely-controlled microdroplets from the feed solution. The spray head has a piezoelectric actuator driven by an electronic circuit with a thin stainless steel membrane. The membrane has controlled-sized holes that consequently and precisely control the generated liquid droplet size. The solution passes through the holes due to the piezoelectric actuator that drives at a fixed frequency (60 kHz). The piezoelectric crystal contracts and expands, which results in the membrane vibration and an upward and downward movement. The upward movement fills all the membrane holes and the downward ejects the droplets (Figure 12.3). Membranes are available with a spray mesh of 4.0, 5.5, or 7.0 μm, allowing the production of the droplets with different sizes (Arpagaus et al., 2017; Büchi Labortechnik, 2015). The 'NSD B-90 HP' was created and designed to produce particles

A

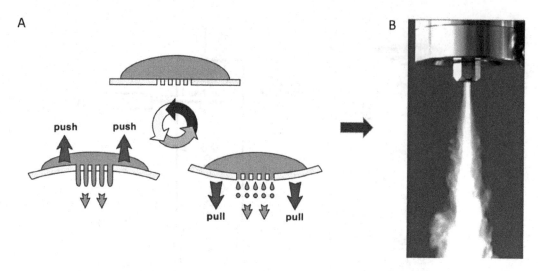

B

FIGURE 12.3 'NSD B-90' from Büchi spray head components; Piezo technology mechanism to produce precisely controlled size droplets (A) and spray head from equipment 'NSD B-90', producing a spray of dried particles (B). Adapted from Büchi Labortechnik (2015) and Büchi Labortechnik (2017b).

with a range of sizes from submicron to the micrometer. Like the other spray dryer models, this nano spray drier can also dry and obtain particles from different sources such as nanosuspensions and even nanoemulsions; the difference is in the case of viscous solutions, the viscosity of which should be lower than 5 cps. It has an efficient and faster particle production capacity and also higher recovery yields have advantages when compared with older versions of the equipment (Arpagaus, 2019b, 2019a; Büchi Labortechnik, 2017c).

The difference between the 'NSD B-90' and the 'NSD B-90 HP' is mostly in the head design and drying technology. The new equipment is equipped with new nebulizers that enhance the handling of the spray head and perform with higher throughput for particle production. The equipment is also equipped with a new 'Auto Stop' that will shut down the spray process when the nebulizer has no solution to spray; this stop mechanism will protect the material as well as increasing the nebulizer lifetime (Büchi Labortechnik, 2017c, 2015, 2017b).

The drying process of this new 'NSD B-90 HP' is practically the same as the equipment previously described (i.e., the 'NSD B-90'). The method includes the inlet drying gas on top of the drying column that introduces gas with a controlled laminar flow rate, passing through the heater to reach the desired temperature (max. 120°C) into the drying column. The equipment is equipped with a relative pressure sensor that when there is a pressure loss the sensor will end the drying process. In order to retain possibly harmful particles, there is an outlet filter that is fixed on the drying gas outlet which works as an exhaust. In 'closed-loop' mode, the outlet filter is also an essential part of the process to clean the recycled gas from unwanted residues and impurities (Büchi Labortechnik, 2017c).

The 'NSD B-90 HP' has the 2nd generation of spray heads, which has three nebulizers with different hole sizes for droplet generation by vibrating mesh technology, resulting in droplets with different sizes and therefore particles with a range of sizes. For the fluid to reach the spray head, a peristaltic pump with a controlled speed, makes the liquid sample circulate in plastic tubes reaching the nebulizer for continuous and efficient atomization and then circulates back to the solution reservoir, making the spray generation a continuous and rapid process. This process is continuously repeated until solutions are all dried into particles.

Nebulizers are available in three different hole sizes: small, medium, and large (Figure 12.4). Due to the Piezo technology, the nebulizers produce precisely-controlled droplets from the liquid solution, as described before (Figure 12.3A). Like the spray caps in the case of 'NSD B-90', the nebulizer in the new equipment contains a piezoelectric actuator (Büchi Labortechnik, 2017c).

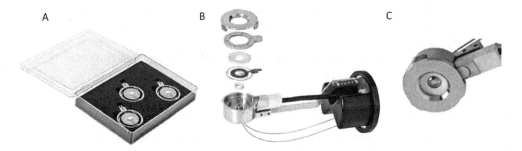

FIGURE 12.4 'NSD B-90 HP' spray head; nebulizers (A), spray head components (B), and spray head after assembling all the components (C). Adapted from Büchi Labortechnik, 2017c, 2020c (Reprinted from Arpagaus, 2019b, with permission from Elsevier).

The 'NSD B-90 HP' has a higher throughput in particles productions, due to the higher spray frequency (from 80 kHz to 140 kHz), compared with the 1st generation head of the 'NSD B-90' (which has a fixed spray frequency of 60 kHz) (Arpagaus, 2019b, 2019a; Büchi Labortechnik, 2017c, 2017b).

The 'NSD B-90 HP' is available in two different set-ups that have different applications and accessories that were created to allow handling and drying organic-based samples safely. Therefore, there are two possible modes of operation for the 'NSD B-90 HP', including the 'Basic' model that works in an 'open-loop' and 'Advanced' mode that is more indicated for samples containing organic solvents and works in 'closed-loop' with two more accessories connected to the equipment (i.e., the 'Inert-Loop B-295' and the 'Dehumidifier B-296 Nano').

12.2.1 'NANO SPRAY DRYER B-90 HP BASIC' (OPEN-CIRCUIT (WITH OXYGEN))

The 'NSD B-90 HP Basic' is suitable for drying aqueous solutions, nanoemulsions, or nanosuspensions with a water content >90% (Table 12.2), in an 'open-loop'. This set-up requires a drying gas with controlled pressure that can come from the compressed air or by using an aspirator with an inlet filter in blowing mode, which works as a pump to establish a constant gas flow. The compressed air used in the equipment must be dried and free from impurities and oil. In an 'open-loop' configuration and when the air humidity is too high, the use of a dehumidifier (e.g., 'Dehumidifier B-296') is possible.

12.2.2 'NANO SPRAY DRYER B-90 HP ADVANCED' (CLOSED-CIRCUIT (WITHOUT OXYGEN))

The 'NSD B-90 HP Advanced' allows the drying of organic solvent-based solutions in a 'closed-loop'. It is suitable for solutions with a mixture of organic solvents and water (20%–90% of water) and organic solutions (<80% water) (Table 12.2). The drying gas used in this mode is a combination of N_2 and CO_2. This operation mode of 'NSD B-90 HP' is combined with one more accessory, the

TABLE 12.2
Nano Spray Drying of Different Sample Solutions. Adapted from Büchi Labortechnik (2017i)

Drying Solutions	Operation Mode
Aqueous solutions (>90% water)	'Open-loop' or 'closed-loop'
Mixtures (20–90% water)	'Closed-loop'
Organic solutions (<80% water)	'Closed-loop'

'Inert Loop B-295' that allows the safe use of organic solvents in the nano spray drying process. The 'closed-loop' configuration also has the possibility of adding a dehumidifier system, as in the 'open-loop'. The dehumidifier is strongly recommended when mixtures of organic solvent and water are used. The dehumidifier condensates the water and prevents water from entering the 'Inert Loop B-295'.

12.2.3 ATOMIZATION SYSTEM AND NOZZLE SIZE

During atomization, the frequency and power of spraying should be adjusted for each solution and solvent, so as to improve spray quality and throughput. The nebulizer size (small, medium, or large) has a direct influence on the size of the dried droplets; usually, when a smaller mesh size is used, the generated droplets are smaller, and consequently, the produced particles have smaller sizes. By contrast, when a large nebulizer is used, the droplets are bigger. However, droplet size depends not just on the nebulizer that is used but also on the feed solution properties, its concentrations, viscosity, and surface tension. The tendency is to have a higher particle size with higher viscosity and concentration of the solution used (Büchi Labortechnik, 2017c, 2017j). When compared with the droplets obtained from water solutions, the droplets generated from solutions with organic solvents are relatively small, which is explained by their lower viscosity, surface tension, and density (Arpagaus, 2012; Arpagaus et al., 2018, 2017; Marques et al., 2019). The viscosity of the solution and the size of the particles in solutions should be checked before processing, because the vibrating mesh technology has a risk of clogging if the product is overconcentrated (e.g., high viscosity) or if the particle size of nanoemulsions or nanosuspensions in the feed solution is bigger than about one-tenth of the nebulizer hole diameter. This can result in discontinuous droplet generation or even no droplet generation at all. The maximum liquid viscosity for obtaining favorable droplets should be <5 mPa.s (Arpagaus, 2019b; Büchi Labortechnik, 2017c; Oliveira et al., 2013).

12.2.4 DRYING GAS

After particles are sprayed, they are dried with a heated gas and the dried particles are obtained. This gas is heated on top of the vertical cylinder to a selected temperature that can reach a maximum of 120°C. The 'NSD B-90 HP' has the advantage of drying particles with a laminar flow gas (controlled by the user from 80 to 160 L/min), that allows for a more precise and controlled spray, without the accumulation of particles in the drying chamber inner walls. This allows for a more efficient drying process with high yields on the particle recovery. In particular, this equipment is made of modular glass assembly, which allows the easy modification of the drying chamber length; having a short mode (around 0.3 m of length) or the long one (around 0.8 m of length) (Arpagaus, 2019b). Another advantage of the modular assembly is that it helps the cleaning process after using the nano spray dryer (Arpagaus, 2019b; Büchi Labortechnik, 2017c).

Although during the nano spray drying process, high temperatures are used to dry the solution, this technique is suitable to dry heat-sensitive compounds, since the time that they are exposed to the high temperature is so short that the degradation of the compounds can be negligible (Arpagaus, 2011; Arpagaus et al., 2017; Assadpour & Jafari, 2019).

12.2.5 RECOVERY OF SAMPLES

The nano spray dryer is equipped with a highly efficient electrostatic particle collector. The collector mechanism is based on electrostatic charging that attracts the particles and separates them from the drying gas, which improves particle recovery. During the nano spray dryer process, the mechanism to attract the particles from the drying gas to the collector consists of the generation of an electric field with a voltage of 15 kV between the cylinder (anode) and the star-shaped counter electrode (cathode). Then the negatively charged particles are attracted and collected in the internal

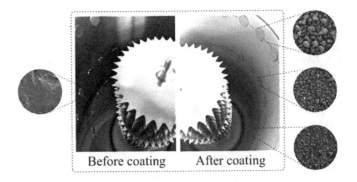

FIGURE 12.5 'NSD B-90' collector before particles and coating production and after drying PLGA particles. Reprinted from Baghdan et al. (2019), with permission from Elsevier.

wall of the cylinder (Arpagaus, 2012; Arpagaus et al., 2017; Lee et al., 2011). This particle collecting method is very efficient and has the advantage of being able to attract particles with submicron sizes, which results in great separation efficiency and consequently a high yield, even with small sample quantities.

After the drying process, the particles are gently collected from the internal wall of the collecting cylinder using a particle scrapper. This process causes no damage in the particle wall, but should be very gentle to avoid losses of powder and guarantee an ideal yield (Bürki et al., 2011; Feng et al., 2011; Heng et al., 2011; Lee et al., 2011). The particles can be then stored in airtight vials and a controlled and dry atmosphere (usually a desiccator with silica gel at room temperature), in order to avoid moisture absorption (Arpagaus, 2019b; Arpagaus et al., 2017).

One interesting characteristic of the particle deposition in the cylinder length is the segregation effect over the cylinder length. Normally, particles with a big diameter are collected earlier in the collector than the small particles; this happens because of the larger surface charge and a larger electrostatic force acting on those particles. This mechanism results in the deposition of the bigger particles on the top part of the cylinder and the small ones at the bottom part of the cylinder. Baghdan et al. (2019) used the 'NSD B-90' to produce poly(lactic-co-glycolic acid) (PLGA) particles to encapsulate the antibiotic norfloxacin, and to produce nanocoatings for dental implants. Titanium discs were used as a model material for dental implants and were placed in the collector cylinder in different positions (top, middle, and bottom) with approximately 6 cm of distance between them. Results showed that the spherical particles with smooth surface particles with a size range between 400 and 600 nm were produced, while the particles with smaller size were formed at the bottom of the cylinder and the bigger particles on the top, as shown in Figure 12.5 (Baghdan et al., 2019; Li et al., 2010).

12.3 MATERIALS USED FOR ENCAPSULATION USING THE NANO SPRAY DRYER

Over the past few years, the encapsulation of bioactive compounds has been one of the main topics of research. In the field of bioactive nanoencapsulation, different types of nanostructures (size below <1 μm) have been developed. Among the examples of these nanostructures are nanoparticles, nanocapsules, nanoemulsions, nanofibers, solid lipid nanoparticles, and nanohydrogels. The main objective is to develop a structure that can encapsulate a higher amount of the specific bioactive compound(s) and to be able to protect and release them at the specific sites.

The encapsulation process using a nano spray dryer is accomplished by dissolving, emulsifying, or dispersing the core substance in a solution of the carrier material and then spraying the mixture into the hot drying chamber. The encapsulated material may be present in either liquid or solid form.

It is possible, through the nano spray dryer, to encapsulate active compounds and produce nanoparticles in a single step, within a continuous and scalable process (Abdel-Mageed et al., 2019).

To choose the right wall/carrier material, it is very important to target high encapsulation efficiency and capsule stability (Anandharamakrishnan & Ishwarya, 2015). Wall materials can be selected from a wide variety of natural (Abdel-Mageed et al., 2019; Bourbon, Barbosa-Pereira, Vicente, Cerqueira, & Pastrana, 2020; Kyriakoudi & Tsimidou, 2018) and synthetic polymers (Arpagaus et al., 2018; Draheim, Crécy, Hansen, Collnot, & Lehr, 2015). When considering food applications, one of the important considerations is to use materials that are food grade and approved by the regulatory agencies (e.g., FDA or EFSA) (Nedovic et al., 2011).

Several papers have been published on the evaluation of the use of the drying technique for the encapsulation of the bioactive compounds and their process conditions to obtain homogeneous particles. Some examples are presented in Table 12.3.

Particles obtained by the nano spray dryer are highly influenced by different parameters such as the solution composition (e.g., concentration, ionic strength, viscosity, and solvent) and processing parameters (inlet temperature, flow air drying, spray percentage, frequency, mesh size, and open or closed cycle). The conjugation of these variables results in a final product with distinct properties such as size, homogeneity, roughness, and porosity, which will influence the final application of the dried product. Several reviews have been published with actual information, reporting the effect of using different wall materials on the final product properties (Arpagaus et al., 2017; Li et al., 2010; Shehata & Ibrahim, 2019).

The selection of the wall material for nanoencapsulation is usually related to the compatibility of this material with the encapsulated bioactive compound, the release properties, efficiency capacity, storage stability, and water solubility (Arpagaus, 2019b). Several properties have been highlighted on the selection of the wall materials, such as the grade (depending on if it is for food or pharmaceutical purposes), presenting emulsifying properties, low viscosity with a high level of solids, and low hygroscopicity. In the applications, the most used materials to produce stable particles using 'NSD B-90' are polysaccharides and proteins, but lipids also are being used frequently.

12.3.1 Polysaccharides

Biodegradable polymers such as polysaccharides are frequently used as wall materials to encapsulate bioactive compounds. These biopolymers present numerous unique properties (e.g., thermal resistance, non-toxic, and food-grade) and can protect active compounds and release them in different environmental conditions. Most of the recent research using 'NSD B-90' includes polysaccharides (e.g., pectin, carboxymethyl cellulose (CMC), and carrageenan), which are being used to produce nanoparticles and nanocapsules. Wang, Hu, Zhou, Xue, and Luo (2016) tested the application of five types of polysaccharides to coat nanostructured lipid carriers with oleic acid with different concentrations using the 'NSD B-90', so as to obtain an ultra-fine powder. These authors observed that, depending on the polysaccharide used, nanoparticles with different properties could be obtained. They observed that after the drying process, particles that were coated using carrageenan exhibited the best effect on producing small, spherical, and well-separated powders, followed by pectin, alginate, CMC, and gum Arabic (Wang, Hu, et al., 2016). Other authors have also used a nano spray dryer to dry polysaccharide-based systems prepared previously. Hu, Wang, Zhou, Xue, and Luo (2016) fabricated polyelectrolyte complex nanoparticles using chitosan, gallic acid, and gum Arabic, and dried these structures with 'NSD B-90'. To form well-separated dried nanoparticles, these authors added polyethylene glycol (PEG) and observed that spherical, homogeneous, and smooth powders of nanoparticles were obtained. The use of an appropriate amount of PEG improved the structural properties and uniformity of the powders during the spray drying (Hu, Wang, et al., 2016).

New polymers are also being tested using this innovative technique (nano spray drying) to obtain micro- and nanoparticles that can be able to be applied in different areas (e.g., glycol chitosan and

TABLE 12.3

The Available Literature on the use of Nano Spray Drying Technology Using Different Wall Materials

Wall Material	Material Features	Applications		References
		Goal	Nano Spray Dryer	
α-amylase	The hydrolytic enzyme that catalyzes the hydrolysis of starch	Production of α-amylase nanopowder using nano spray dryer with full enzyme activity and maximum possible yield	α-amylase particles were obtained with an average size of 600 nm, a production yield of 94%, using a 7 mm spray cap. Sucrose was selected as the enzyme stabilizer (0.15% w/v). The drying flow rate was 100 L/min with an inlet temperature of 80°C.	Abdel-Mageed et al. (2019)
Maltodextrin	Polysaccharide with high water solubility, low viscosity, and mild flavor	Encapsulate aqueous saffron extracts in maltodextrin using nano spray dryer	Saffron extracts were encapsulated in maltodextrin using a size 1.5 μm (mesh of 4 μm) and 4.2 μm (mesh of 7 μm). The best inlet temperature was 100°C with a product yield between 71% and 87%.	Kyriakoudi and Tsimidou (2018)
Lactoferrin-glycomacropeptide	Milk protein with bioactive properties	Evaluate the influence of the drying process of protein-based nanohydrogels in their physical–chemical properties	Increasing the drying temperature (80°C, 100°C, and 120°C) of 'NSD B-90' resulted in increasing protein degradation. The nozzle with a higher size (7 μm) promoted larger particles.	(Bourbon et al. (2020)
Gelatin/sodium alginate	Polymeric materials for encapsulation and controlled drug delivery; biocompatible polymers and possess mucoadhesive properties	Encapsulate metformin hydrochloride (MET)	MET was encapsulated in gelatin/sodium alginate particles using a 7 mm spray cap, a flow rate of 3.5 L/min, an inlet temperature of 120°C, and a frequency of 60 KHz.	Shehata and Ibrahim (2019)

(Continued)

TABLE 12.3
(Continued)

Wall Material	Material Features	Applications		References
		Goal	Nano Spray Dryer	
Carbopol®	Vinyl polymer used as gelling/swelling agent	Encapsulate Vildagliptin and metformin	Ildagliptin and metformin hydrochloride-loaded Carbopol nanoparticles (VMCN) were prepared using a spray cap of 5.5 µm, an inlet temperature of 80 °C, and drying air of 100–110 L/min.	Harsha (2015)
2,3,6- triacetyl-β-cyclodextrin	Sustained-release carrier for several hydrophilic drugs	Encapsulate sepiapterin	'Closed-loop' with nitrogen flow 20 mL.min⁻¹, and an inlet temperature of 55°C using an 80% spray.	Kuplennik and Sosnik (2019)
Sodium caseinate and pectin	Natural Biopolymers and their interaction under acidic pH condition have been successfully exploited as a novel approach to prepare and stabilize solid lipid nanoparticles (SLNs)	Encapsulate curcumin	Curcumin-loaded SLNs were dried using an inlet temperature of 100°C, a flow rate at 120 L/min, and a mesh size of 7 µm.	Xue, Wang, Hu, Zhou, and Luo (2018)
Lactoferrin	A milk protein with bioactive properties	Dry nanoemulsions with omega-3 fatty acids	Nanoemulsions were dried at 80°C, with a 100% spray, a drying flow of 100–120 L/min, and a mesh size of 7 µm.	Nunes et al. (2020)
Zein-caseinate-pectin complex	A protein obtained from corn	Encapsulate Eugenol	Eugenol-loaded complex nanoparticles were dried with a mesh cap of 5.0 µm, an airflow rate of 120 L/min, and an inlet temperature of 100°C.	Veneranda et al. (2018)
Poly (lactic-co-glycolic acid) (PLGA)	Biodegradable and biocompatible copolymer	Prepare poly (PLGA) nano- and microparticles	'Closed-loop', using a spray mesh of 4 µm, using a temperature range of 50°C–90°C with a spray rate at 70%–100%	Draheim et al. (2015)

their derivate polymers). The selection of a suitable polysaccharide combined with the optimized process parameters can result in particles with unique properties. Several works reported that in the spray drying process, the properties of the initial feed material and the operating conditions, for example, nozzle size, flow rate, and inlet and outlet temperature, strongly affect the characteristics and production yield of the nanoparticles/nanocapsules (Ngan et al., 2014). Harsha et al. (2017) tested different conditions, namely the wall material concentration (1.35%–2% alginate), the inlet temperature (80°C,100°C, and 120°C) and feed flow (20, 25, and 30 mL/min), and observed that the obtained powder presented a morphology close to a shriveled particle. This behavior was explained by the shrinking of the core when the droplets are dried at high temperatures. The authors also observed that the increase in the concentration of the solid feed could lead to higher particle size (Harsha et al., 2017).

12.3.2 PROTEINS

Peptides and proteins are highly-sensitive compounds to physicochemical stresses during processing and storage. One of the strategies is the preparation of these compounds as solid forms, because better stability can be achieved in the solid rather than in the liquid state (Haggag & Faheem, 2015). Aiming an increasing stability of proteins, a drying process is usually applied, or stabilizing compounds are used. Spray drying technologies, such as nano spray drying, are currently being used for drying sensitive compounds such as proteins. The low time of residence and the low droplet temperature (relatively) do not result in the significant degradation of these compounds during the drying process. Bourbon et al. (2020) dried protein nanohydrogels using 'NSD B-90' and compared them with the nanohydrogels dried by freeze-drying. These authors observed that the nano spray drying technique did not affect the denaturation, structure, and morphology of protein-based nanohydrogels (Bourbon et al., 2020). Previous studies of various research groups showed that the extent of protein degradation during spray drying was strongly dependent on the process conditions, as well as the formulation; inlet temperature, spray rate, and the addition of excipients were reported to be of major importance (Abdel-Mageed et al., 2019; Arpagaus et al., 2017; Lee et al., 2011; Li et al., 2010). Process parameters and formulation also have an impact on the particle morphology and the aerodynamic properties of the powder (Arpagaus, 2019b; Arpagaus et al., 2018; Bourbon et al., 2020; Lee et al., 2011; Li et al., 2010; Veneranda et al., 2018). Most of the works using protein samples dried by the nano spray dryer mentioned that spray-dried protein samples are a fragile powder that can be easily deformed during the recovering process (Zhou, Wang, Hu, & Luo, 2016).

12.3.3 LIPIDS

Some of the bioactive compounds that are needed to be encapsulated are lipophilic. One of the strategies to protect such lipophilic compounds against oxidation, improve their dispersion in water, and increase their bioavailability is to encapsulate them in a nanodispersion (e.g., nanoemulsion, solid-lipid nanoparticles, and micelles). These isotropic dispersed systems of two non-miscible liquids, normally consisting of an oily system dispersed in an aqueous system, or an aqueous system dispersed in an oil system, forming droplets or oil phases of nanometric sizes (Bourbon, Gonçalves, Vicente, & Pinheiro, 2018). Only a few studies (Büchi Labortechnik, 2017e; Nunes et al., 2020; Prasad Reddy, Padma Ishwarya, & Anandharamakrishnan, 2019; Wang, Soyama, & Luo, 2016) have been reported for the use of 'NSD B-90' for drying nanoemulsions on a homogeneous powder. Such a limitation can be explained by the effect of the drying processes in the lipid structure, which can easily induce the severe aggregation, agglomeration, or melting of lipid, and, consequently, changing their structure. Recently, Nunes et al. (2020) reported the use of this technique to dry bio-based nanoemulsions containing encapsulated omega-3. These authors observed that nanoemulsions were successfully dried by nano spray drying, resulting in particles with a spherical morphology at the submicron size (Nunes et al., 2020). In addition, other authors tested the use of gum Arabic and lecithin (food grade natural emulsifiers) to develop eugenol oil nanoemulsions and dried them using this technique. As a result, an ultrafine spherical powder with a size distribution lower than 500 nm was obtained, and the dried powders exhibited excellent re-dispersibility in water and maintained their physicochemical properties after rehydration (Hu, Gerhard, Upadhyaya, Venkitanarayanan, & Luo, 2016). Some authors have also reported the use of 'NSD B-90' to produce solid phospholipid nanoparticles (Brinkmann-Trettenes, Barnert, & Bauer-Brandl, 2014). Solid Lipid Nanoparticles (SLNs) are nanoscale delivery systems for highly lipophilic compounds with limited bioavailability. These structures offer a high loading capacity in their lipid core, when compared with other colloidal delivery systems. Different studies demonstrated that SLNs could be successfully dried using nano spray drying technology (Arpagaus et al., 2018). Xue, Wang, Hu, Zhou, and Luo (2017) produced submicron SLNs by homogenizing melted stearic acid directly in an aqueous phase containing sodium caseinate and pectin. After nano spray drying, the obtained dry SLN powders could be effectively re-dispersed in water without any variation in size, shape, and morphology. In another study, Xue et al. (2018) encapsulated curcumin in SLNs and produced nano spray dried powders with uniform distribution. They mentioned the nano spray dryer as a method of manufacturing nanoparticles for oral administration that is solvent-free and that promises to encapsulate heat-sensitive lipophilic drugs that are stable in SLNs (Xue et al., 2018).

12.3.4 OTHERS

Biopolymers based on lactic acid and glycolic acid and their copolymers have attracted considerable interest in various applications, including carrier substances for the production of bioactive compounds with excellent biocompatibility, adjustable degradation rate, and non-toxicity to humans (Arpagaus, 2019a). Several works are reporting the use of poly(lactic-co-glycolic acid) (PLGA), a copolymer of polylactic acid (PLA) and polyglycolic acid (PGA) to form nanoparticles using spray drying techniques (Chauhan et al., 2019; Draheim et al., 2015; Maghrebi, Joyce, Jambhrunkar, Thomas, & Prestidge, 2020). To prepare nanoparticles using these biopolymers, authors reported the use of organic solvents such as dichloromethane (DCM), chloroform (CFM), ethyl acetate, or ethyl formate, which implies that 'NSD B-90' works in 'closed-loop' (used for organic solvents).

12.4 DRIED STRUCTURES (OBTAINED PARTICLES FROM NANO SPRAY DRYER)

Tables 12.4 and 12.5 report on some of the published works using the nano spray drying technology for the manufacture of micro- and nanostructures that are used in the food, pharmaceutical, and chemical industries. Nano spray dryer technologies are used for different purposes, an encapsulation methodology as reported in Table 12.4 or to dry a material solution, nanoemulsion, and nanosuspension and convert it into a powder (Table 12.5). There are just a few published works using the most recent 'NSD B-90 HP' and most of the available works are from the developer company, Büchi Labortechnik AG. On the other hand, several works are available using the 'NSD B-90' for various applications and using different materials and conditions.

Pérez-Masiá et al. (2015) compared two different methods (the 'NSD B-90' and electrospraying (ES)) to encapsulate folic acid. They also tested whey protein concentrate (WPC) and a commercial resistant starch as wall materials. The electrospraying method allowed the use of higher polymer concentrations (20% w/v), while in the case of nano spray drying the polymer concentration was only 0.4% (w/v), to avoid the blockage of the head and the nozzle membrane. Similarly, guar gum (0.5 wt%) and span 20 (5 wt%) were added to the solution, to achieve better spraying during the electrospray process. Folic acid was added to obtain a dispersion with a concentration of 1.5% in relation to the amount of polymer. Both methodologies and wall materials could produce capsules loaded with folic acid. Regarding the morphology of the particles, the results showed that it was

TABLE 12.4

The published literature on the use of nano spray drying technology for the encapsulation of bioactive compounds.

Wall material	Concentration	Equipment version	Nozzle diameter/ Nebulizer	Gas flow (L/min)	Inlet temp. (°C)	Encapsulated bioactive compound	Particles diameter	Drying yield (%)	Encapsulation efficiency (%)	References
Arabic gum, polyvinyl alcohol, maltodextrin, whey protein, and modified starch	0.1, 1, and 10 wt%	Nano Spray Dryer B-90 ('closed loop')	4.0 μm	100	100	Vitamin E acetate nanoemulsions	625–801 nm	70–90	N/A	Li et al. (2010)
Bovine serum albumin (BSA) and Surfactant - polyoxyethylene, sorbitan monoleate 80	0.5, 1, and 2% (BSA), 0, 0.05, and 0.5% (Tween 80)	Nano Spray Dryer B-90 ('closed loop')	4.0 μm 5.5 μm 7 μm	90 120 150	80 100 120	N/A	460 ± 10 nm (mean size)	72 ± 4	N/A	Lee et al. (2011)
B-galactosidase with Trehalose	5% (protein to trehalose ratio 1:2)	Nano Spray Dryer B-90 ('open loop')	4.0 μm 5.5 μm 7 μm	100–110	80 100 120	N/A	1–5 μm	60–90	N/A	Bürki et al. (2011)
Arabic gum, cashew nut gum, sodium alginate, and carboxymethyl cellulose	0.1% 0.5% 1%	Nano Spray Dryer B-90 ('open loop')	4 μm 7 μm	130	120	Vitamin B12	698 nm–34.23 μm	NA	N/A	Oliveira et al. (2013)
Chitosan	10% 20%	Nano Spray Dryer B-90 ('open loop')	4 μm 7 μm	110–120	120	N/A	206 ± 20 nm to 487 ± 15 nm	13–85	N/A	Demir and Degim (2013)
Polylactide-co-glycolide acid (PLGA)	0.21%	Nano Spray Dryer B-90 ('closed loop')	4.0 μm 5.5 μm 7.0 μm	112–118	60 and 80	Clozapine and Risperidone	248.48 ± 11.71 nm to 392.83 ± 24.92 nm	48–64	90.4–94.7	Panda et al. (2014)

(Continued)

TABLE 12.4
(Continued)

Wall material	Concentration	Equipment version	Nozzle diameter/ Nebulizer	Gas flow (L/min)	Inlet temp. (°C)	Encapsulated bioactive compound	Particles diameter	Drying yield (%)	Encapsulation efficiency (%)	References
Alginate and pectin	0.10%, 0.25% or 0.50%	Nano Spray Dryer B-90 ('open loop')	4.0 μm 5.5 μm 7.0 μm	100	90	Gentamicin sulphate	310–1003 nm	82–92	70.8–83.5	Cicco et al. (2014)
Whey protein concentrate (WPC) Resistant starch	0.4%	Nano Spray Dryer B-90 ('open loop')	0.7 μm	140	90	Folic acid (vitamin B$_9$)	50 nm–4.5 μm	N/A	83.9 ± 7.8 (WPC) 52.5 ± 7.6 (Resistant starch)	Pérez-Masiá et al. (2015)
Crosslinked double layer solid lipid Nanoparticles (Pectin and Sodium caseinate and Compritol ATO 888)	Sodium caseinate – 1.2 mg/mL	Nano Spray Dryer B-90 ('open loop')	5.5 μm	120	100	Curcumin	500 nm–1 μm	N/A	40–90	Wang, Ma, et al. (2016)
Sodium caseinate and pectin (nanoemulsions)	0.25% 0.5%	Nano Spray Dryer B-90	5.5 μm	120	100	Peppermint oil Vitamin D3 and E Vitamin B3, B6, B12, C and BCAA	1–2 μm	N/A	N/A	Wang, Soyama, and Luo (2016)
Egg yolk low density lipoprotein (LDL)/ pectin and LDL/ carboxymethylcellulose (Crosslinked LDL/ pectin and LDL/CMC nanogels)	0.5 mg/ml (Crosslinked nanogels concentration)	Nano Spray Dryer B-90 ('open loop')	5.5 μm	120	100	Curcumin	<500 nm–>1 μm	N/A	N/A	Zhou et al. (2018)

(Continued)

Wall material	Concentration	Equipment version	Nozzle diameter/ Nebulizer	Gas flow (L/min)	Inlet temp. (°C)	Encapsulated bioactive compound	Particles diameter	Drying yield (%)	Encapsulation efficiency (%)	References
Mannitol and poly(lactic-*coglycolic* acid) (PLGA)	0% 20% 33% 50% and 100% (mannitol solution:PLGA nanosuspensio ratio)	Nano Spray Dryer B-90	7.0 μm	140	80	N/A	1.1 μm–7.2 μm	N/A	N/A	Torge et al. (2017)
Sodium carboxymethyl cellulose (CMC)	0.2% (S-Iso:CMC ratio between 3:1 and 1:3)	Nano Spray Dryer B-90	5.5 μm	100	60 and 80	soy isoflavone extract	303 nm–1857 nm	61.2–87.1	78-89	Gaudio, Sansone, et al. (2017)
Hydromellose acetate succinate with different acetate and succinate substitution levels (HPMCAS)	0.15–0.50%	Nano Spray Dryer B-90	4.0 μm 5.5 μm 7.0 μm	100	60	Soy isoflavones extract (Daidzein and genistein)	555 nm–830 nm	59.8–90.2	62-86	Gaudio, Russo, et al. (2017)
Curcumin	0.1%	Nano Spray Dryer B-90 HP ('closed loop')	Small	120–150	65–75	NA	0.367–1.29 μm	N/A	N/A	Büchi Labortechnik AG (2017a)
Maltodextrin and maltodextrin sunflower oil nanoemulsion	1% and 20% (maltodextrin solutions) and 8% (maltodextrin in nanoemulsion)	Nano Spray Dryer B-90 HP ('open loop')	Small and Large	100, 120 and 150	100 and 120	N/A	0.548–5.57 μm	41–90	N/A	Büchi Labortechnik AG (2017e)
Poly(lactic-*coglycolic* acid) (PLGA)	0.1%	Nano Spray Dryer B-90 HP ('closed loop')	Medium	140–160	55	N/A	156 nm–2 μm	N/A	N/A	Büchi Labortechnik AG (2017f)

(Continued)

TABLE 12.4 (Continued)

Wall material	Concentration	Equipment version	Nozzle diameter/ Nebulizer	Gas flow (L/min)	Inlet temp. (°C)	Encapsulated bioactive compound	Particles diameter	Drying yield (%)	Encapsulation efficiency (%)	References
Silicon dioxide (nanosuspension)	1%	Nano Spray Dryer B-90 HP ('open loop')	Large	140	120	N/A	0.372–4.02 µm	>60	N/A	Büchi Labortechnik AG (2017g)
Kollidon 30® (polyvinylpyrrolidone polymer)	2–5%	Nano Spray Dryer B-90 HP ('closed loop') and ('open loop')	Small Large	120–150	70–100	N/A	0.345–7.87 µm	51–88	N/A	Büchi Labortechnik AG (2017h)
Bovine Serum Albumine (BSA)	.1% 1% 10%	Nano Spray Dryer B-90 HP ('open loop')	Small Medium Large	150	100	N/A	0.133–6.34 µm	>60	N/A	Büchi Labortechnik AG (2017j)
Bovine serum albumin-dextran and pectin (Solid lipid-polymer hybrid nanoparticles with different solid lipids (glycerids and saturated fatty acid))	Bovine serum albumin-dextran – 1 mg/mL Pectin – 5 mg/ mL	Nano Spray Dryer B-90	5.5 µm	120	100	Curcumin	211–713 nm	N/A	NA	Wang et al. (2018)
Maltodextrin	10%	Nano Spray Dryer B-90 ('open loop')	4.0 µm 7.0 µm	100	100	Saffron hydrophilic apocarotenoids (crocins and picrocrocin)	1.5–4.2 µm	71–87	54–82	Kyriakoudi and Tsimidou (2018)
Whey protein isolate (WPI) (emulsion with Tween 80)	0.4 % (WPI)	Nano Spray Dryer B-90 ('open loop')	7 µm	90–110	90	Roasted coffe bean oil	315.429 ± 1.193 nm	N/A	N/A	Prasad Reddy, Padma Ishwarya, and Anandharamakrishnan (2019)

(Continued)

Wall material	Concentration	Equipment version	Nozzle diameter/ Nebulizer	Gas flow (L/min)	Inlet temp. (°C)	Encapsulated bioactive compound	Particles diameter	Drying yield (%)	Encapsulation efficiency (%)	References
Sucrose, polyethylene glycol 1000; Polyoxyethylene sorbitan monooleate 80; Pluronic F127	0.05–0.2%	Nano Spray Dryer B-90 ('open loop')	4 μm 7 μm	95–105	80	α-amylase	337–1057 nm	75–94	N/A	Abdel-Mageed et al. (2019)
β-cyclodextrin	5 mM	Nano Spray Dryer B-90	4 μm	100	100	Hydroxytyrosol	0.4–3.4 μm	53	84.4 ± 3.2	Malapert et al. (2019)
Dimethylaminoethyl methacrylate, butyl methacrylate and methyl methacrylate (DBM) based copolymer	0.05%	Nano Spray Dryer B-90 HP	4 μm	128–135	120	α-galactosylceramide	263	N/A	N/A	Gonzatti et al. (2019)
Gelatin and sodium alginate	1%	Nano spray dryer B-90	7.0 μm	110	120	Metformin hydrochloride	650 nm -850 nm	81.1 -93.5	90.5–95.3	Shehata and Ibrahim (2019)
Poly(lactic-co- glycolic acid) (PLGA)	PLGA 2.5 % NFX PLGA 5 % NFX PLGA 10 % NFX	Nano Spray Dryer B-90 ('closed loop')	4 μm	100	70	Norfloxacin (NFX)	400–600 nm	N/A	0-73–10.33 μg/cm²	Baghdan et al. (2019)
Lactoferrin (LF) and Glycomacropeptide (GMP) nanohydrogels	LF – 2.5 μM GMP–8.33 μM Mixture molar ratio of 1:7 (LF:GMP)	Nano Spray Dryer B-90 ('open loop')	4 and 7 μm	100 and 110	80, 100, and 120	NA	N/A	N/A	N/A	Bourbon et al. (2020)
Bovine serum albumin (BSA)	0.25%	Nano Spray Dryer B-90 ('closed loop')	4 μm	101	100	Rutin	316 ± 210 nm	N/A	32	Pedrozo et al. (2020)
Bovine lactoferrin nanoemulsionjs	2%	Nano Spray Dryer B-90 ('open loop')	7 μm	100–120	80	Omega -3 fatty acids	<500 nm	N/A	>99	Nunes et al. (2020)
Dextrin	10 mg/mL	Nano Spray Dryer B-90 HP ('open loop')	Large	10	120	Amphotericin B	347.5 ± 36.1 nm	51.5	14.38 ± 2.79	Silva-Carvalho et al. (2020)

TABLE 12.5

The Published Literature on the use of Nano Spray Drying Technology to Produce a Stable Powder by Drying Solutions, Nanoemulsions, Nanosuspensions, and Nanohydrogels

Solution	Concentration	Equipment Version	Nozzle Diameter/ nebulizer	Gas Flow (L/Min)	Inlet Temp (°C)	Particles Diameter	Drying Yield (%)	References
Bovine serum albumin (BSA) and surfactant - polyoxyethylene, and sorbitan monooleate 80	0.5, 1 and 2% (BSA) 0, 0.05 and 0.5% (Tween 80)	'NSD B-90' ('closed-loop')	4.0 µm 5.5 µm 7 µm	90 120 150	80 100 120	460 ± 10 nm (mean size)	72	Lee et al. (2011)
B-galactosidase with trehalose	5% (protein to trehalose ratio 1:2)	'NSD B-90' ('open-loop')	4.0 µm 5.5 µm 7 µm	100–110	80 100 120	1–5 µm	60–90	Bürki et al. (2011)
Chitosan	10% 20%	'NSD B-90' ('open-loop')	4 µm 7 µm	110–120	120	206 ± 20 nm to 487 ± 15 nm	13–85	Demir and Degim (2013)
Mannitol andpoly(lactic-coglycolic acid) (PLGA)	0% 20% 33% 50% and 100% (mannitol solution: PLGA nanosuspension ratio)	'NSD B-90'	7.0 µm	140	80	1.1 µm–7.2 µm	N/A	Torge, Grützmacher, Mücklich, and Schneider (2017)
Curcumin	0.1%	'NSD B-90' HP' ('closed-loop')	Small	120–150	65–75	0.367–1.29 µm	N/A	Büchi Labortechnik (2017a)
Maltodextrin and maltodextrin sunflower oil nanoemulsion	1% and 20% (maltodextrin solutions) and 8% (maltodextrin in nanoemulsion)	'NSD B-90 HP' ('open-loop')	Small and Large	100, 120 and 150	100 and 120	0.548–5.57 µm	41–90	Büchi Labortechnik (2017e)
Poly(lactic-coglycolic acid) (PLGA)	0.1%	'NSD B-90 HP' ('closed-loop')	Medium	140–160	55	156 nm–2 µm	N/A	Büchi Labortechnik (2017f)
Silicon dioxide (nanosuspension)	1%	'NSD B-90 HP' ('open-loop')	Large	140	120	0.372–4.02 µm	>60	Büchi Labortechnik (2017g)
Kollidon 30 ® (polyvinylpyrrolidone polymer)	2–5%	'NSD B-90 HP' ('closed-loop') and ('open-loop')	Small Large	120–150	70–100	0.345–7.87 µm	51–88	Büchi Labortechnik (2017h)

(Continued)

Solution	Concentration	Equipment Version	Nozzle Diameter/ nebulizer	Gas Flow (L/Min)	Inlet Temp (°C)	Particles Diameter	Drying Yield (%)	References
Bovine saerum albumine (BSA)	0.1% 1% 10%	'NSD B-90 HP' ('open-loop')	Small Medium Large	150	100	0.133–6.34 μm	> 60	Büchi Labortechnik (2017j)
Lactoferrin (LF) and glycomacro-peptide (GMP) nanohydrogels	LF – 2.5 μM GMP – 8.33 μM Mixture molar ratio of 1: 7 (LF: GMP)	'NSD B-90' ('open-loop')	4 μm and 7 μm	100 and 110	80, 100 and 120	N/A	N/A	Bourbon et al. (2020)

N/A: information not available.

possible to obtain spherical capsules for both cases; in the case of resistant starch, however, some agglomerations were formed (Figure 12.6). Regarding their sizes, submicron and micron capsules were obtained in both cases, but nano spray drying led to bigger and broader size distribution. Electrospraying led to smaller particles with enhanced control over the size of the capsules and nar-rower size distribution. The authors also determined the encapsulation efficiency and concluded that there were no significant differences between the nano spray drying and the electrospray when using WPC, reaching values of 83.9% and 80.8% for the nano spray drying and electrospray, respectively. In the case of resistant starch, the encapsulation efficiency was 52.5% for the nano spray drying method and 44.0% for the electrospraying method. Both materials and encapsulation techniques led to improved folic acid stability, although better stability was obtained for WPC wall material (Pérez-Masiá et al., 2015).

Gaudio, Sansone, et al. (2017b) used the nano spray drying technology to encapsulate soy iso-flavone extract in sodium carboxymethyl cellulose (CMC) nanoparticles. They obtained stable nanoparticles with narrow size distribution and an average size of 650 nm. Soy isoflavone extracts were encapsulated, with an efficiency ranging between 78% and 89%. In this work, the bioavailabil-ity of soy isoflavone extract in the nanoparticles was also evaluated, where the permeation studies

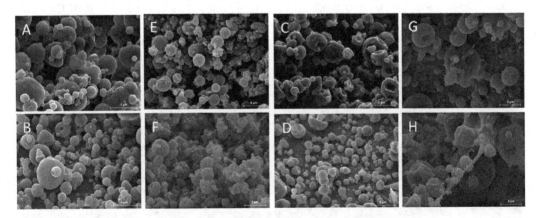

FIGURE 12.6 Nano spray drying (NSD) and electrospray (ES) produced particles: A) WPC by NSD; B) WPC/Folic acid by NSD; E) WPC by ES; F) WPC/Folic acid by ES; C) Resistant starch (RS by NSD); D) RS/Folic acid by NSD; G) RS by ES; H)RS/Folic acid by ES. (Reprinted from Pérez-Masiá et al., 2015, with permission from Elsevier.)

were conducted using Franz-type vertical diffusion cells. Due to the low solubility, the free soy iso-flavone extract permeation value was very low (8.5 $\mu g/cm^2$ at 180 min); however, nanoencapsulated soy isoflavone extracts had an enhanced permeation rate, up to 4.5 times higher than pure material and twice the permeation observed for the materials produced using the mini spray dryer. The presence of CMC wall material was able to improve isoflavone extract affinity with aqueous media and stabilize the extract. These results suggest that isoflavones extract-loaded nanoparticles obtained using nano spray drying have great potential to enhance the bioavailability of the encapsulated material (Gaudio, Sansone, et al., 2017b).

Nunes et al. (2020) produced lactoferrin-based nanoemulsions to encapsulate omega-3 fatty acids and dried the nanoemulsions using the nano spray dryer. In this work, the authors used two different drying processes, including spray drying using the 'NSD B-90' and freeze drying. Nano spray drying process resulted in smooth and defined particles, while the particles obtained using freeze drying contained an amorphous structure (Figure 12.7). Obtained dried nanoemulsions were evaluated by circular dichroism (CD) to detect changes in the protein structure, and the results showed that the nano spray dried process caused a loss of the regular secondary structure of the protein, possibly due to the drying temperature. The results obtained by CD showed that the α-helix structure decreased (due to the drying process) from 8% in the nanoemulsions to 2% in the case of the freeze drying process, and to 0% in the case of the nano spray drying process. On the other hand, the β-sheet was increased in the case of the nano spray drying process (51%), but it was maintained for the freeze drying process (46%). Random coil increased from 46% in the native lactoferrin to 51% with the nanoemulsion process, but remained stable with the drying processes (Nunes et al., 2020).

Silva-Carvalho et al. (2020) have also used the nano spray drying and freeze drying processes to produce dried dextrin-amphotericin complex for Leishmaniasis treatment. Results showed that it was possible to obtain dried nanocomplexes. Obtained particles from 'NSD B-90' had a bigger mean diameter (about 347.5 nm), when compared with the particles obtained from the freeze drying process (about 214.2 nm). The drying yield for the freeze drying process (86.8%) was also higher, compared with the nano spray drying process (51.5%). However, the drug content was similar in both methodologies; about 14.38% and 14.68% for nano spray dryer and freeze dryer, respectively. Nonetheless, the association efficiency determined for the nano spray dried (about 33%) material was slightly inferior to those obtained by freeze drying (about 60%). The cytotoxicity experiments (tested in different cell types, and the free amphotericin B) showed to be cytotoxic, depending on the concentration for cell viability, especially for BMMΦ cells; at the same concentrations, however, nanocomplexes dried from both methodologies were significantly less toxic against BMMΦ cells

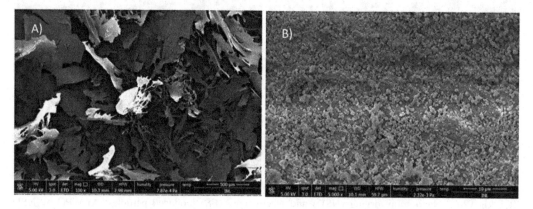

FIGURE 12.7 Scanning electron micrographs of dried lactoferrin and omega-3 fatty acids nanoemulsions by: A) Freeze drying, and B) Nano spray drying. Adapted from Nunes et al. (2020), with permission from The Royal Society of Chemistry.

(CC$_{50}$ of 5.77 and 6.26 μM for freeze dryer and nano spray dryer nanocomplex, respectively). *In vitro* leishmanicidal activity of nanocomplexes was determined during 24 h exposure of *L. infantum* and *L. amazonensis* promastigotes or BMMΦ infected with *L. infantum* amastigotes to different concentrations of the produced materials. Results showed that nanocomplexes led to a decrease in *L. amazonensis* and *L. infantum* viability, although slightly higher concentrations of nanocomplexes were needed when compared with the free drug. This work shows that dried nanocomplexes obtained by either freeze drying or nano spray drying are approximately 2.47- and 1.98-fold more selective than the free drug (Silva-Carvalho et al., 2020).

Kyriakoudi and Tsimidou (2018) encapsulated saffron hydrophilic apocarotenoids extracts (crocins and picrocrocin) using maltodextrin as wall material using an 'NSD B-90'. They tested three different cores: wall material ratios (i.e., 1: 5, 1: 10, and 1: 20 w/w) and evaluated their morphology using SEM. The results showed that the dried particles produced with 4 μm mesh size presented a spherical and smooth surface morphology; however, the particles produced with the 7 μm mesh size resulted in a mixture of smooth and wrinkled surface particles. The nanoparticles were characterized in terms of stability and were evaluated under thermal and *in vitro* gastrointestinal conditions, in the presence or absence of caffeic acid (as a phenolic antioxidant). Stability results showed that the nanoencapsulation of crocin and picrocrocin reinforced their stability, the core: wall ratio had more impact on the stability than the mesh size used during the nano spray drying process, and the presence of caffeic acid improved the stability of the particles. Under *in vitro* gastrointestinal conditions, it was observed that the nanoencapsulation had a positive effect on the bioaccessibility of crocin and picrocrocin when compared with their free form. The bioaccessibility of the picrocrocin increased from 71.9% (free form) to 80.7% (nanoencapsulated form), and for the crocin was increased from 60% (in the free form) to 70% (nanoencapsulated form). The presence of caffeic acid, also improved the bioaccessibility of saffron extracts (Kyriakoudi & Tsimidou, 2018).

Baghdan et al. (2019) used the 'NSD B-90' to produce antibacterial nanocoatings for dental implants. Nanocoatings were produced using PLGA-based particles loaded with norfloxacin (NFX) as an antibiotic with different concentration ratios. Titanium discs were used as a model material for dental implants. To evaluate the efficiency of the encapsulated antibiotic, they performed bacterial viability assays and obtained a reduction of 99.83% and 95.42% with the nanocoatings of PLGA 5% NFX from the top and the bottom position, respectively. Particles without the addition of NFX did not affect bacterial viability. Furthermore, the biocompatibility test using L929 cells were made *in vitro* for assessing the *in vitro* cytotoxicity. This study was carried out by counting the number of cells adhered to the surface of titanium discs with and without nanocoatings (nanocoatings with and without NFX antibiotic) after 24 and 96 h of incubation. After 24 h of incubation, there was no significant difference in the cell count between the nanocoated and the uncoated titanium discs. It was observed that at 96 h of incubation there was a considerable increase in the cell count for all samples. Uncoated and nanocoated without NFX had comparable results, which indicates that nanocoating had no toxic effect on cell proliferation. Although, titanium discs nanocoated with PLGA 5% NFX from the top position of the collector had a significantly lower cell proliferation (Baghdan et al., 2019).

As seen in Table 12.5, some studies used the nano spray drying technology to dry solutions, emulsions, and nanosuspensions in order to obtain a stable powder. Some works were performed by the company Büchi Labortechnik AG. In one of the studies, they tested the 'NSD B-90 HP', the most recent nano spray drying technology in a 'closed-loop' mode using the 'Inert Loop B-295' to obtain curcumin submicron particles. In this test, they used two different formulations; in one of the formulations, curcumin at a concentration of 0.1% (w/v) was dissolved in ethanol and in the other formulation it was dissolved in a mixture of ethanol and acetone (1: 1). The solution of curcumin with ethanol was dried at a temperature of 65°C, with a gas flow of 120 L.min^{-1}. Curcumin solubilized in the mixture of ethanol and acetone was dried at 75°C and with 150 L/min of gas flow rate. Both formulations were dried using the small nebulizer, with an 80% spray rate and with a feed rate of 10%. Obtained particles presented a spherical morphology in both cases and sizes ranged between 0.367–1.29 µm and 0.428–0.974 µm for ethanol and ethanol–acetone mixture solutions, respectively. This work shows that the solvents used did not influence particle size and morphology, but had an influence on process throughput; they observed an improvement of the spray solution throughput from 14.8 mL/h using ethanol to 20.4 mL/h using the ethanol–acetone mixture (Büchi Labortechnik, 2017a). Büchi Labortechnik AG company also studied the possibility of using malto-dextrin as a suitable encapsulation agent for the nano spray dryer. In this study, the equipment 'NSD B-90 HP' in 'open-loop' mode was used NSD and the effect of small and large nebulizers in the final particles was evaluated. The solutions were dried at 100°C, with a gas flow rate of 120 L/min for maltodextrin solution (1% w/v) and 150 L/min for a concentration of 20% (w/v). It was con-cluded that high concentrations were possible to dry using the small nebulizer, resulting in spherical particles with a range size of 0.867 to 5.57 µm for the large nebulizer and of 0.548 to 4.0 µm for the small one. The most significant difference in this study was the throughput that was 3–4 times smaller for the concentration of 20% compared with the 1% when using the small nebulizer (Büchi Labortechnik, 2017e). In another study, Büchi Labortechnik AG company evaluated the possibility to produce protein submicron particles, using bovine serum albumin (BSA) as a model. They tested different concentrations of BSA (i.e., 10%, 1%, and 0.1%), with the addition of 0.05% Tween 80. The solutions were dried in the new 'NSD B-90 HP' in an 'open-loop' at 100°C and with an airflow rate of 100 L/min. Three sizes for the nebulizers, including small, medium, and large, were tested. The results showed that particle size and size distribution were increasing with nebulizer diameter and solution concentration; the particle size ranged between 0.133 and 6.34 µm. The pH and the drying temperature had a minor influence on the mean particle size and size distribution (Büchi Labortechnik, 2017j).

12.5 CONCLUDING REMARKS

In the micro- and nanoencapsulation fields, one of the main challenges is the drying process of the nanostructures or nanosystems, despite the methodologies available (e.g., freeze-drying and spray drying). Thus, it is not easy to maintain and control the size, structure, and morphology of the mate-rials using these technologies. The nano spray dryer is a good alternative to these methodologies. The nano spray drying technology is a suitable technique for drying bio-based nanostructures loaded with bioactive compounds. It is also very helpful for the production of submicron and nanoparticles and capsules from solutions or suspensions. Several works have shown it is possible to obtain ultra-fine powders at the micro- and nanoscales using different types of biopolymers that can load bioac-tive compounds, opening the opportunity for several applications in the food and pharmaceutical industries. It is also possible to dry bioactive compounds without using any wall material, resulting in submicron particles of these compounds. One of the drawbacks of the nano spray drying meth-odology is its low productivity. Therefore, despite the improvements performed in the 'NSD B-90', which allowed productivity increase for industrialization purposes, it will be essential to further increase the productivity of this drying technology.

ACKNOWLEDGMENTS

Arlete M. Marques (SFRH/BD/132911/2017) is recipient of a fellowship from Fundação para a Ciência e Tecnologia (FCT, Portugal). This work was supported by the Portuguese Foundation for Science and Technology (FCT) under the scope of the strategic funding of UIDB/04469/2020 unit and BioTecNorte operation (NORTE-01-0145-FEDER-000004) funded by the European Regional Development Fund under the scope of Norte2020 - Programa Operacional Regional do Norte. The authors would like to thank the H2020 MSCA-RISE project FODIAC - Food for Diabetes and Cognition (reference number 778388).

REFERENCES

Abdel-Mageed, Heidi M., Shahinaze A. Fouad, Mahmoud H. Teaima, Azza M. Abdel-Aty, Afaf S. Fahmy, Dalia S. Shaker, and Saleh A. Mohamed. 2019. "Optimization of Nano Spray Drying Parameters for Production of α-Amylase Nanopowder for Biotheraputic Applications Using Factorial Design." *Drying Technology* 37 (16): 2152–2160. https://doi.org/10.1080/07373937.2019.1565576.

Anandharamakrishnan, C., and S. Ishwarya Padma. 2015. "Selection of Wall Material for Encapsulation by Spray Drying." In *Spray Drying Techniques for Food Ingredient Encapsulation*, edited by C. Anandharamakrishnan and S.P. Ishwarya, 77–100. New Jersey, USA: Wiley Online Books. https://doi.org/doi:10.1002/9781118863985.ch4.

Arpagaus, Cordin. 2011. "Nano Spray Dryer B-90: Literature Review and Applications." *Büchi Information Bulletin*,63: 8.

Arpagaus, Cordin. 2012. "A Novel Laboratory-Scale Spray Dryer to Produce Nanoparticles." *Drying Technology* 30 (10): 1113–1121. https://doi.org/10.1080/07373937.2012.686949.

Arpagaus, Cordin. 2019a. "PLA/PLGA Nanoparticles Prepared by Nano Spray Drying." *Journal of Pharmaceutical Investigation* 49 (4): 405–426. https://doi.org/10.1007/s40005-019-00441-3.

Arpagaus, Cordin. 2019b. "Production of Food Bioactive-Loaded Nanoparticles by Nano Spray Drying." In *Nanoencapsulation of Food Ingredients by Specialized Equipment*, edited by Seid Mahdi Jafari, Volume 3, 151–211. Cambridge, MA: Academic Press (Elsevier Inc.). https://doi.org/10.1016/B978-0-12-815,671-1.00004-4.

Arpagaus, Cordin, Andreas Collenberg, David Rütti, Elham Assadpour, and Seid Mahdi Jafari. 2018. "Nano Spray Drying for Encapsulation of Pharmaceuticals." *International Journal of Pharmaceutics* 546: 194–214.

Arpagaus, Cordin, Philipp John, Andreas Collenberg, and David Rutti. 2017. "Nanocapsules Formation by Nano Spray Drying." In *Nanoencapsulation Technologies for the Food and Nutraceutical Industries*, edited by Seid Mahdi Jafari, 346–401. Cambridge, MA: Academic Press.

Assadpour, Elham, and Seid Mahdi Jafari. 2019. "Advances in Spray-Drying Encapsulation of Food Bioactive Ingredients: From Microcapsules to Nanocapsules." *Annual Review of Food Science and Technology* 10 (1): 103–131. https://doi.org/10.1146/annurev-food-032818-121,641.

Baghdan, Elias, Michael Raschpichler, Walaa Lutfi, Shashank Reddy Pinnapireddy, Marcel Pourasghar, Jens Schäfer, Marc Schneider, and Udo Bakowsky. 2019. "Nano Spray Dried Antibacterial Coatings for Dental Implants." *European Journal of Pharmaceutics and Biopharmaceutics* 139 (March): 59–67.

Bourbon, Ana I., Letricia Barbosa-Pereira, António A. Vicente, Miguel A. Cerqueira, and Lorenzo Pastrana. 2020. "Dehydration of Protein Lactoferrin-Glycomacropeptide Nanohydrogels." *Food Hydrocolloids* 101 (August 2019): 105550. https://doi.org/10.1016/j.foodhyd.2019.105550.

Bourbon, Ana I., Raquel F. S. Gonçalves, António A. Vicente, and Ana C. Pinheiro. 2018. "Characterization of Particle Properties in Nanoemulsions." In *Nanoemulsions - Formulation, Applications, and Characterization*, edited by Seid Mahdi Jafari and B. T. David Julian - Nanoemulsions McClements, 519–546. Cambridge, MA: Academic Press. https://doi.org/10.1016/B978-0-12-811,838-2.00016-3.

Brinkmann-Trettenes, Ulla, Sabine Barnert, and Annette Bauer-Brandl. 2014. "Single Step Bottom-up Process to Generate Solid Phospholipid Nano-Particles." *Pharmaceutical Development and Technology* 19 (3): 326–332. https://doi.org/10.3109/10837450.2013.778875.

Büchi Labortechnik, A. G.. 2015. "Nano Spray Dryer B-90 - Operation Manual." *Büchi* Version B.

Büchi Labortechnik, A. G. 2016. "Mini Spray Dryer B-290: Technical Data Sheet." 1–9. *Büchi Labortechnik AG*. 1st of March 2020. http://static1.buchi.com/sites/default/files/downloads/B-290_Data_Sheet_en_D.pdf?83b925aae302a8e76f002d1ce679fa06904d1039.

Büchi Labortechnik, A. G. 2017a. "Curcumin Sub-Micrometer Particles by Nano Spray Drying." *Application Note No. 276/2017*, no. 276.

Büchi Labortechnik, A. G. 2017b. "Nano Spray Dryer B-90 HP." *Büchi Labortechnik AG*. 1st of March 2020. https://www.buchi.com/en/products/spray-drying-and-encapsulation/nano-spray-dryer-b-90-hp.

Büchi Labortechnik, A. G. 2017c. "Nano Spray Dryer B-90 HP - Operation Manual." *Büchi Labortechnik AG*. 1st of March 2020. https://static1.buchi.com/sites/default/files/downloads/093261_B-90_Operation_Manual_en_0.pdf?6184640abe255f77115287bb82b78c058b6217b0.

Büchi Labortechnik, A. G. 2017d. "Nano Spray Dryer B-90 HP Technical Data Sheet."

Büchi Labortechnik, A. G. 2017e. "Nano Spray Drying Maltodextrin from Solution and Emulsion." *Application Note No. 275/2017*, no. 275.

Büchi Labortechnik, A. G. 2017f. "PLGA Sub-Micron Particles by Nano Spray Drying." *Application Note No. 273/2017*, no. 273.

Büchi Labortechnik, A. G. 2017g. "Preparation of SiO$_2$ Agglomerated Particles." *Application Note No. 277/2017*, no. 277: 2–5.

Büchi Labortechnik, A. G. 2017h. "Production of Kollidon® 30 Particles Nano." *Application Note No. 274/2017*, no. 274.

Büchi Labortechnik, A. G. 2017i. "Spray Drying & Encapsulation Solutions Particle Formation for Lab Scale." *Büchi Labortechnik AG.*

Büchi Labortechnik, A. G.. 2017j. "Sub-Micron Bovine Serum Albumin Particles." *Application Note No. 272/2017*, no. 272.

Büchi Labortechnik, A. G. 2020a. "B-290 Mini Spray Dryer Operation Manual." *Büchi Labortechnik AG.*

Büchi Labortechnik, A. G. 2020b. "Büchi - Mini Spray Dryer B-290." *Mini Spray Dryer B-290 The World Leading R&D Solution for Spray Drying.* 21st April 2020. https://www.buchi.com/en/products/spray-drying-and-encapsulation/mini-spray-dryer-b-290.

Büchi Labortechnik, A. G. 2020c. "Nano Spray Dryer B-90 HP - Small Particles, Small Samples, High Yields." *Nano Spray Dryer B-90 HP.* 15th April 2020. https://www.buchi.com/en/products/spray-drying-and-encapsulation/nano-spray-dryer-b-90-hp.

Bürki, K., I. Jeon, C. Arpagaus, and G. Betz. 2011. "New Insights into Respirable Protein Powder Preparation Using a Nano Spray Dryer." *International Journal of Pharmaceutics* 408: 248–256. https://doi.org/10.1016/j.ijpharm.2011.02.012.

Cerqueira, Miguel A., Ana C. Pinheiro, Hélder D. Silva, Philippe E. Ramos, Maria A. Azevedo, María L. Flores-López, Melissa C. Rivera, Ana I. Bourbon, Óscar L. Ramos, and António A. Vicente. 2014. "Design of Bio-Nanosystems for Oral Delivery of Functional Compounds." *Food Engineering Reviews* 6 (1): 1–19. https://doi.org/10.1007/s12393-013-9074-3.

Cerqueira, Miguel A., Ana Cristina Braga Pinheiro, Cátia Vanessa Saldanha Carmo, Catarina Maria Martins Duarte, Maria G. Carneiro-da-Cunha, and António Augusto Vicente. 2016. "Nanostructures Biobased Systems for Nutrient and Bioactive Compounds Delivery." In *Nutrient Delivery*, edited by Alexandru Mihai Grumezescu, 43–85. Cambridge, MA: Academic Press (Elsevier Inc.) https://doi.org/10.1016/B978-0-12-804,304-2/00021-4.

Chauhan, Rajat, Rayeanne Balgemann, Christopher Greb, Betty M. Nunn, Shunichiro Ueda, Hidetaka Noma, Kevin McDonald, Henry J. Kaplan, Shigeo Tamiya, and Martin G. O'Toole. 2019. "Production of Dasatinib Encapsulated Spray-Dried Poly (Lactic-Co-Glycolic Acid) Particles." *Journal of Drug Delivery Science and Technology* 53 (October): 101204. https://doi.org/10.1016/J.JDDST.2019.101204.

Cicco, Felicetta De, Amalia Porta, Francesca Sansone, Rita P. Aquino, and Pasquale Del Gaudio. 2014. "Nano spray Technology for an in Situ Gelling Nanoparticulate Powder as a Wound Dressing." *International Journal of Pharmaceutics* 473: 30–37.

Costa, Maria José, Philippe Emmanuel Ramos, Pablo Fuciños, José António Teixeira, Lorenzo M. Pastrana, and Miguel A. Cerqueira. 2018. "Development of Bio-Based Nanostructured Systems by Electrohydrodynamic Processes." In *Nanotechnology Applications in the Food Industry*, edited by V. Ravishankar Rai and Jamuna A. Bai. 3–20, Boca Raton: Taylor and Francis Group. ISBN: 9780429950216.

Davidov-Pardo, Gabriel, Iris J. Joye, and David Julian McClements. 2015. "Food-Grade Protein-Based Nanoparticles and Microparticles for Bioactive Delivery: Fabrication, Characterization, and Utilization." In *Protein and Peptide Nanoparticles for Drug Delivery*, edited by B. T. Rossen - Advances in Protein Chemistry and Structural Biology Donev, 98: 293–325. Cambridge, MA: Academic Press (Elsevier Inc.). https://doi.org/10.1016/bs.apcsb.2014.11.004.

Demir, Gulen Melike, and Ismail Tuncer Degim. 2013. "Preparation of Chitosan Nanoparticles by Nano Spray Drying Technology." *Fabad Journal of Pharmaceutical Sciences* 38 (3): 127–133.

Draheim, Christina, Francois De Crécy, Steffi Hansen, Eva Maria Collnot, and Claus Michael Lehr. 2015. "A Design of Experiment Study of Nanoprecipitation and Nano Spray Drying as Processes to Prepare PLGA Nano- and Microparticles with Defined Sizes and Size Distributions." *Pharmaceutical Research* 32 (8): 2609–2624. https://doi.org/10.1007/s11095-015-1647-9.

Feng, A. L., M. A. Boraey, M. A. Gwin, P. R. Finlay, P. J. Kuehl, and R. Vehring. 2011. "Mechanistic Models Facilitate Efficient Development of Leucine Containing Microparticles for Pulmonary Drug Delivery." *International Journal of Pharmaceutics* 409 (1–2): 156–163. https://doi.org/10.1016/j.ijpharm.2011.02.049.

Gaudio, Pasquale Del, Paola Russo, Rosalia Rodriguez Dorado, Francesca Sansone, Teresa Mencherini, Franco Gasparri, and Rita Patrizia Aquino. 2017. "Submicrometric Hypromellose Acetate Succinate Particles as Carrier for Soy Isoflavones Extract with Improved Skin Penetration Performance." *Carbohydrate Polymers* 165: 22–29. https://doi.org/10.1016/j.carbpol.2017.02.025.

Gaudio, Pasquale Del, Francesca Sansone, Teresa Mencherini, Felicetta De Cicco, Paola Russo, and Rita Patrizia Aquino. 2017. "Nano spray Drying as a Novel Tool to Improve Technological Properties of Soy Isoflavone Extracts." *Planta Medica* 83: 426–433.

Gonzatti, Michelangelo Bauwelz, Maria Eduarda Perrud Sousa, Ariane Simões Tunissi, Renato Arruda Mortara, Adriano Marim de Oliveira, Natália Neto Pereira Cerize, and Alexandre de Castro Keller. 2019. "Nano Spray Dryer for Vectorizing α-Galactosylceramide in Polymeric Nanoparticles: A Single Step Process to Enhance Invariant Natural Killer *T Lymphocyte* Responses." *International Journal of Pharmaceutics* 565: 123–132. https://doi.org/10.1016/j.ijpharm.2019.05.013.

Haggag, Yusuf A., and Ahmed M. Faheem. 2015. "Evaluation of Nano Spray Drying as a Method for Drying and Formulation of Therapeutic Peptides and Proteins." *Frontiers in Pharmacology* 6: 140. https://doi.org/10.3389/fphar.2015.00140.

Harsha, Sree N. 2015. "In Vitro and In Vivo Evaluation of Nanoparticles Prepared by Nano Spray Drying for Stomach Mucoadhesive Drug Delivery." *Drying Technology* 33 (10): 1199–1209. https://doi.org/10.1080/07373937.2014.995305.

Harsha, Sree N., Bandar E. Al-Dhubiab, Anroop B. Nair, Mahesh Attimarad, Katharigatta N. Venugopala, and S. A. Kedarnath. 2017. "Pharmacokinetics and Tissue Distribution of Microspheres Prepared by Spray Drying Technique: Targeted Drug Delivery." *Biomedical Research* 28 (8): 3387–3396.

Harsha, Sree N., Bander E. Aldhubiab, Anroop B. Nair, Ibrahim Abdulrahman Alhaider, Mahesh Attimarad, Katharigatta N. Venugopala, Saminathan Srinivasan, Nagesh Gangadhar, and Afzal Haq Asif. 2015. "Nanoparticle Formulation by Buchi B-90 Nano Spray Dryer for Oral Mucoadhesion." *Drug Design, Development and Therapy* 9: 273–282. https://doi.org/10.2147/DDDT.S66654.

Heng, Desmond, Sie Huey Lee, Wai Kiong Ng, and Reginald B. H. Tan. 2011. "The Nano Spray Dryer B-90." *Expert Opinion on Drug Delivery* 8 (7): 965–972. https://doi.org/10.1517/17425247.2011.588206.

Hu, Qiaobin, Hannah Gerhard, Indu Upadhyaya, Kumar Venkitanarayanan, and Yangchao Luo. 2016. "Antimicrobial Eugenol Nanoemulsion Prepared by Gum Arabic and Lecithin and Evaluation of Drying Technologies." *International Journal of Biological Macromolecules* 87 (June): 130–140. https://doi.org/10.1016/J.IJBIOMAC.2016.02.051.

Hu, Qiaobin, Taoran Wang, Mingyong Zhou, Jingyi Xue, and Yangchao Luo. 2016. "Formation of Redispersible Polyelectrolyte Complex Nanoparticles from Gallic Acid-Chitosan Conjugate and Gum Arabic." *International Journal of Biological Macromolecules* 92 (November): 812–819. https://doi.org/10.1016/J.IJBIOMAC.2016.07.089.

Kuplennik, Nataliya, and Alejandro Sosnik. 2019. "Enhanced Nanoencapsulation of Sepiapterin within PEG-PCL Nanoparticles by Complexation with Triacetyl-Beta Cyclodextrin." *Molecules* 24 (15): 1–19.

Kyriakoudi, Anastasia, and Maria Z. Tsimidou. 2018. "Properties of Encapsulated Saffron Extracts in Maltodextrin Using the Büchi B-90 Nano Spray-Dryer." *Food Chemistry* 266 (January): 458–465. https://doi.org/10.1016/j.foodchem.2018.06.038.

Lee, Sie Huey, Desmond Heng, Wai Kiong Ng, Hak Kim Chan, and Reginald B. H. Tan. 2011. "Nano Spray Drying: A Novel Method for Preparing Protein Nanoparticles for Protein Therapy." *International Journal of Pharmaceutics* 403 (1–2): 192–200. https://doi.org/10.1016/j.ijpharm.2010.10.012.

Li, Xiang, Nicolas Anton, Cordin Arpagaus, Fabrice Belleteix, and Thierry F. Vandamme. 2010. "Nanoparticles by Spray Drying Using Innovative New Technology: The Büchi Nano Spray Dryer B-90." *Journal of Controlled Release* 147(2):304–310. https://doi.org/10.1016/j.jconrel.2010.07.113.

Maghrebi, Sajedeh, Paul Joyce, Manasi Jambhrunkar, Nicky Thomas, and Clive A. Prestidge. 2020. "Poly(Lactic-Co-Glycolic) Acid–Lipid Hybrid Microparticles Enhance the Intracellular Uptake and Antibacterial Activity of Rifampicin." *ACS Applied Materials & Interfaces* 12 (7): 8030–8039. https://doi.org/10.1021/acsami.9b22991.

Malapert, Aurélia, Emmanuelle Reboul, Mallorie Tourbin, Olivier Dangles, Alain Thiery, Fabio Ziarelli, and Valérie Tomao. 2019. "Characterization of Hydroxytyrosol-β-Cyclodextrin Complexes in Solution and in the Solid State, a Potential Bioactive Ingredient Aurélia." *LWT - Food Science and Technology* 102: 317–323.

Marques, Arlete M., M. Alexandra Azevedo, José A. Teixeira, Lorenzo M. Pastrana, Catarina Gonçalves, and Miguel A. Cerqueira. 2019. "Engineered Nanostructures for Enrichment and Fortification of Foods." In *Food Applications of Nanotechnology*, edited by Gustavo Molina, Inamuddin Franciele Maria Pelissari, and Abdullah Mohamed Asiri, 61–86. Boca Raton, FL: CRC Press.

Nedovic, Viktor, Ana Kalusevic, Verica Manojlovic, Steva Levic, and Branko Bugarski. 2011. "An Overview of Encapsulation Technologies for Food Applications." *Procedia Food Science* 1: 1806–1815. https://doi.org/10.1016/j.profoo.2011.09.265.

Ngan, La Thi Kim, San-Lang Wang, Đinh Minh Hiep, Phung Minh Luong, Nguyen Tan Vui, Tran Minh Đinh, and Nguyen Anh Dzung. 2014. "Preparation of Chitosan Nanoparticles by Spray Drying, and Their Antibacterial Activity." *Research on Chemical Intermediates* 40 (6): 2165–2175. https://doi.org/10.1007/s11164-014-1594-9.

Nunes, Rafaela, Beatrz D'Avó Pereira, Miguel Cerqueira, Pedro Miguel Silva, Lorenzo Pastrana, A. A. Vicente, Joana Martins, and Ana Isabel Bourbon. 2020. "Lactoferrin-Based Nanoemulsions to Improve the Physical and Chemical Stability of Omega-3 Fatty Acids." *Food & Function* 11 (3): 1966–1981. https://doi.org/10.1039/c9fo02307k.

Oliveira, Adriano M., Kleber L. Guimarães, Natália N. P. Cerize, Ariane S. Tunussi, and João G. R. Poço. 2013. "Nano Spray Drying as an Innovative Technology for Encapsulating Hydrophilic Active Pharmaceutical Ingredients (API)." *Nanomedicine & Nanotechnology* 4 (6):186. https://doi.org/10.4172/2157-7439.1000186.

Ousset, Aymeric, Joke Meeus, Florent Robin, Martin Alexander Schubert, Pascal Somville, and Kalliopi Dodou. 2018. "Comparison of a Novel Miniaturized Screening Device with Büchi B290 Mini Spray-Dryer for the Development of Spray-Dried Solutions Dispersions (SDSDs)." *Processes* 6: 129. https://doi.org/10.3390/pr6080129.

Panda, Apoorva, Jairam Meena, Rajesh Katara, and Dipak K. Majumdar. 2014. "Formulation and Characterization of Clozapine and Risperidone Co-Entrapped Spray-Dried PLGA Nanoparticles." *Pharmaceutical Development and Technology* 7450: 1–11. https://doi.org/10.3109/10837450.2014.965324.

Pedrozo, Regiellen Cristina, Emilli Antônio, Najeh Maissar Khalil, and Rubiana Mara Mainardes. 2020. "Bovine Serum Albumin-Based Nanoparticles Containing the Flavonoid Rutin Produced by Nano Spray Drying." *Brazilian Journal of Pharmaceutical Sciences* 56: 1–8. https://doi.org/10.1590/s2175-97,902,019,000,317,692.

Pérez-Masiá, Rocío, Rubén López-Nicolás, Maria Jesús Periago, Gaspar Ros, Jose M. Lagaron, and Amparo López-Rubio. 2015. "Encapsulation of Folic Acid in Food Hydrocolloids through Nano spray Drying and Electrospraying for Nutraceutical Applications." *Food Chemistry* 168: 124–133. https://doi.org/10.1016/j.foodchem.2014.07.051.

Prasad Reddy, M. N., S. Padma Ishwarya, and C. Anandharamakrishnan. 2019. "Nanoencapsulation of Roasted Coffee Bean Oil in Whey Protein Wall System through Nano spray Drying." *Journal of Food Processing and Preservation* 43 (3): 1–8. https://doi.org/10.1111/jfpp.13893.

Shehata, Tamer M., and Mahmoud Mokhtar Ibrahim. 2019. "BÜCHI Nano Spray Dryer B-90: A Promising Technology for the Production of Metformin Hydrochloride-Loaded Alginate-Gelatin Nanoparticles." *Drug Development and Industrial Pharmacy* 45 (12): 1907–1914. https://doi.org/10.1080/03639045.2019.1680992.

Silva-Carvalho, R., J. Fidalgo, K. R. Melo, M. F. Queiroz, S. Leal, H. A. Rocha, T. Cruz, and P. Parpot. 2020. "Development of Dextrin-Amphotericin B Formulations for the Treatment of Leishmaniasis." *International Journal of Biological Macromolecules* 153. https://doi.org/10.1016/j.ijbiomac.2020.03.019.

Torge, Afra, Philipp Grützmacher, Frank Mücklich, and Marc Schneider. 2017. "The Influence of Mannitol on Morphology and Disintegration of Spray-Dried Nano-Embedded Microparticles." *European Journal of Pharmaceutical Sciences* 104: 171–179. https://doi.org/10.1016/j.ejps.2017.04.003.

Veneranda, Marco, Qiaobin Hu, Taoran Wang, Yangchao Luo, Kepa Castro, and Juan Manuel Madariaga. 2018. "Formation and Characterization of Zein-Caseinate-Pectin Complex Nanoparticles for Encapsulation of Eugenol." *LWT - Food Science and Technology* 89 (March): 596–603. https://doi.org/10.1016/J.LWT.2017.11.040.

Wang, Taoran, Minkyung Bae, Ji-young Lee, and Yangchao Luo. 2018. "Solid Lipid–Polymer Hybrid Nanoparticles Prepared with Natural Biomaterials: A New Platform for Oral Delivery of Lipophilic Bioactives." *Food Hydrocolloids* 84: 581–592. https://doi.org/10.1016/j.foodhyd.2018.06.041.

Wang, Taoran, Qiaobin Hu, Mingyong Zhou, Jingyi Xue, and Yangchao Luo. 2016. "Preparation of Ultra-Fine Powders from Polysaccharide-Coated Solid Lipid Nanoparticles and Nanostructured Lipid Carriers by Innovative Nano Spray Drying Technology." *International Journal of Pharmaceutics* 511 (1): 219–222. https://doi.org/10.1016/j.ijpharm.2016.07.005.

Wang, Taoran, Xiaoyu Ma, Yu Lei, and Yangchao Luo. 2016. "Colloids and Surfaces B: Biointerfaces Solid Lipid Nanoparticles Coated with Cross-Linked Polymeric Double Layer for Oral Delivery of Curcumin." *Colloids and Surface B: Biointerfaces* 148: 1–11. https://doi.org/10.1016/j.colsurfb.2016.08.047.

Wang, Taoran, Shin Soyama, and Yangchao Luo. 2016. "Development of a Novel Functional Drink from All Natural Ingredients Using Nanotechnology." *LWT – Food Science and Technology*. https://doi. org/10.1016/j.lwt.2016.06.050.

Xue, Jingyi, Taoran Wang, Qiaobin Hu, Mingyon Zhou, and Yangchao Luo. 2018. "Insight into Natural Biopolymer-Emulsified Solid Lipid Nanoparticles for Encapsulation of Curcumin: Effect of Loading Methods." *Food Hydrocolloids* 79 (June): 110–116. https://doi.org/10.1016/J.FOODHYD.2017.12.018.

Xue, Jingyi, Taoran Wang, Qiaobin Hu, Mingyong Zhou, and Yangchao Luo. 2017. "A Novel and Organic Solvent-Free Preparation of Solid Lipid Nanoparticles Using Natural Biopolymers as Emulsifier and Stabilizer." *International Journal of Pharmaceutics* 531 (1): 59–66. https://doi.org/10.1016/j. ijpharm.2017.08.066.

Zhou, Mingyong, Keiona Khen, Taoran Wang, Qiaobin Hu, Jingyi Xue, and Yangchao Luo. 2018. "Chemical Crosslinking Improves the Gastrointestinal Stability and Enhances Nu- Trient Delivery Potentials of Egg Yolk LDL/Polysaccharide Nanogels." *Food Chemistry* 239: 840–847.

Zhou, Mingyong, Taoran Wang, Qiaobin Hu, and Yangchao Luo. 2016. "Low Density Lipoprotein/Pectin Complex Nanogels as Potential Oral Delivery Vehicles for Curcumin." *Food Hydrocolloids* 57: 20–29. https://doi.org/10.1016/J.FOODHYD.2016.01.010.

13 Advances in Spray Drying Technology in the Scope of Bioactive Encapsulation

Yongchao Zhu
The University of Auckland, New Zealand

Siew Young Quek
The University of Auckland, New Zealand;
Riddet Institute, New Zealand

CONTENTS

13.1 INTRODUCTION

Spray drying is acknowledged as a widely used technology in the food industry and is also one of the oldest encapsulation methods. In general, particles or powders are produced by spray drying the liquid jets with hot air in a drying chamber. The drying process occurs immediately when the liquid droplets come into contact with the dry airflow. In the context of bioactive encapsulation, the targeted bioactive compounds are usually dissolved in an aqueous medium or prepared as emulsion before spray drying, depending on their solubility in solvent or water. The drying process can be optimized by modifying the spray drying operation parameters and feed composition. These parameters include air flow rate, inlet air humidity, inlet and outlet air temperatures, feed solution parameters (feed rate, feed concentration, feed formulation, rheological properties, and thermodynamic properties), and the spray dryer specification. These parameters are crucial for the spray drying microencapsulation of bioactive compounds and they are closely associated with the quality of the spray dried particles produced.

The use of spray drying techniques possesses several advantages, such as equipment availability, low processing cost, and suitability for a wide range of encapsulating materials. The application of spray drying techniques effectively decreases the overall weight and volume by transforming the liquid feed into solid or powder format, and, hence, final products are more easily and convenient for storage and transportation compared with liquid or gel. The technique can be used to control the particle size and moisture content of powders to meet a special requirement. Furthermore, spray drying can be operated as a continuous process with a relatively short drying time. The encapsulation process helps to offer a protective effect for the bioactive compounds to retain their bioactivity during storage or delivery. Thus, the encapsulation of bioactive compounds by spray drying has been widely applied to protect functional bioactive ingredients for the improvement of their physicochemical properties and stability.

The selection of suitable wall materials is critically important so as to ensure the stability of the encapsulated bioactive compounds. First, wall materials should have adequate stability under different external environments (e.g. pH, temperature, humidity) (Quek, Chen, & Shi, 2015). In addition, there must be desirable compatibility between wall materials and core materials to produce high-quality products (Grumezescu & Oprea, 2017). To achieve a wider application, the water solubility of wall material is very important since most of the powder products need to be re-dissolved in water upon application or consumption. The rheological properties of wall materials also play an important role in spray drying encapsulation. For some lipophilic bioactive ingredients, wall materials that have a certain degree of emulsification capability are desirable for ensuring better encapsulation (Turchiuli, Munguia, Sanchez, Ferre, & Dumoulin, 2014). Furthermore, the controlled-release properties of wall materials enable bioactive ingredients to be released at a specific site, to achieve a higher functional effect (e.g., in the small intestine to enable greater absorption). Ideally, wall materials that are used for the encapsulation of bioactive ingredients should be neutral in taste, so they have no effect on the sensory profile of the final product. They should be food-grade, available, and cost-effective for large-scale production in the food industry. Taking the above factors into consideration, common wall materials used for bioactive encapsulation include polysaccharides, proteins, and their derivatives, which are discussed in other chapters of this book.

Since the development of functional foods has attracted interest from consumers as a current food trend, the production of encapsulated bioactive ingredients by spray drying has inevitably received widespread attention from the research community and food industry. Several bioactive compounds have been encapsulated using the spray drying technique; e.g., including vitamins, minerals, functional lipids, probiotics, amino acids, peptides and proteins, phytochemicals, and polyphenols (Rezvankhah, Emam-Djomeh, & Askari, 2020). The physicochemical properties of the spray dried particles containing bioactive compounds have been widely reported. These include moisture content, water activity, water-solubility, hygroscopicity, morphology, particle size, density, and the stability and retention of encapsulated bioactive compounds (Huang et al., 2017; Rezvankhah et al., 2020; Ziaee et al., 2019). Their control release properties and digestibility have also been studied both *in vitro* and *in vivo* (Marante, Viegas, Duarte, Macedo, & Fonte, 2020; Poozesh & Bilgili, 2019; Saifullah, Shishir, Ferdowsi, Rahman, & Van, 2019). Current advances in spray drying technology for the encapsulation of bioactive compounds focus, in particular, on the modification of atomization technique, which results in particle size decrease (down to nano-size level), and the exploration of the spray drying mechanism to provide a better understanding of particle formation. The following sections of this chapter introduce these technical advances and discuss their applications in the encapsulation of bioactive compounds.

13.2 MICROFLUIDIC-JET SPRAY DRYING

13.2.1 BACKGROUND

Spray dryers are available in several various designs that differ in terms of size and shape, and the atomization technique applied (Sagis, 2015). One of the most important parts of a spray dryer, the atomizer, is used to achieve the atomization of the feed solution. During atomization, large liquid bulk is disintegrated into countless fine droplets that are dried in the drying chamber. Three types of atomization techniques have been widely applied in spray dryers used in the food or pharmaceutical industries: rotary atomization (also known as centrifugal atomization), pressure atomization, and pneumatic atomization (Santos et al., 2017). Rotary atomization is achieved by the centrifugal force generated by the high-speed rotation of a dispersing disc. After the liquid is added to the high-speed rotating dispersing disc, it is spread in a thin film on the surface of the dispersing disc by centrifugal force, and is finally disrupted into droplets (Huang, Kumar, & Mujumdar, 2006). In the case of pressure atomization, the feed solutions are sprayed from a nozzle at a certain size and under high pressure. After spraying from the nozzle, the liquid forms a thin film at the edge of the nozzle, which then converts into many fine droplets. The droplet size produced is therefore related to the nozzle diameter (Hede, Bach, & Jensen, 2008). Finally, the principle of pneumatic atomization is to use high-speed airflow to turn the liquid film into fine droplets. The most common pneumatic atomizer is the two-fluid nozzle atomizer, where the term "two-fluid" refers to the feedstock liquid and the compressed gas. When the compressed air is sprayed from the atomizer at an ultra-high speed, it encounters the liquid phase with a low flow rate. There is a large relative speed difference between the two fluids, which causes friction and shear force between the gas and liquid, causing the liquid to form tiny droplets (Santos et al., 2017).

Microfluidic-jet spray drying, also known as monodisperse droplet spray drying, is a newly emerging spray drying technique. One significant difference between microfluidic-jet spray drying and conventional spray drying is the type of atomizer used in the spray dryer. The primary characteristic of spray droplets produced by the atomization of the conventional spray drying techniques is polydisperse. In contrast, microfluidic-jet spray drying produces monodisperse droplets as a result of its specially designed nozzles. The differences in the spray droplet size between the polydisperse and monodisperse droplets would lead to differences in the particle properties, such as morphology, particle size, water content, encapsulation efficiency, and storage stability.

For conventional spray drying, the atomization process produced polydisperse spray droplets, resulting in a complex droplet trajectory in the drying chamber. These polydisperse droplets collide among themselves or with the dryer wall. This may lead to the agglomeration of particles that could have a negative influence on the final yield of dried powders. Moreover, each individual droplet may follow a different drying path inside the drying chamber and subject to different temperature–time and humidity-time profiles (Wu et al., 2011). This leads to differences in residence time for the spray droplets and complexity in the mathematical modeling of the drying process. The size distribution of the droplets produced by the conventional spray drying process is normally diverse, in the range of 1 to 600 μm for the rotary atomizers, 10 to 800 μm for the pressure nozzles, and 5 to 300 μm for the two-fluid nozzles (Wu, Patel, Rogers, & Chen, 2007).

There is an increased demand for particle size uniformity for the specific applications, such as in high-efficacy drug microcapsules, nutraceuticals, functional food ingredients, and catalysts (Lu et al., 2020; Wu et al., 2007; Zhang et al., 2017). In the pharmaceutical industry, highly spherical particles are desirable for the achievement of a better delivery effect (Shekunov, Chattopadhyay, Tong, & Chow, 2007). Spherical and uniform particles are ideal microcapsules for colloidal drug delivery and dry formulation microcapsules due to their better cell uptake (Jindal, 2017). Spherical

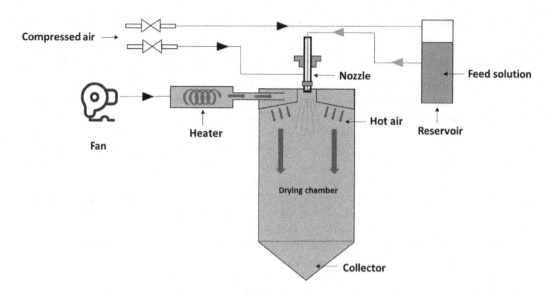

FIGURE 13.1 Schematic diagram of the microfluidic-jet spray dryer.

particles in nano-size and micro-size have been extensively applied as signal reporters or probes for the detection of biomolecules in DNA assay, immunoassay, and cell bio-imaging (Rogers, Fang, Lin, Selomulya, & Chen, 2012).

The rationale for the invention of the microfluidic-jet spray drying is to overcome the drawbacks of polydisperse particles produced by the conventional spray dryers, and to produce particles of a uniform size and spherical shape. A typical schematic diagram of the microfluidic-jet spray dryer is shown in Figure 13.1. Compared to polydisperse particles, the monodisperse spray dried particles have a very narrow size distribution and possess predictable trajectories in the chambers. Over the course of the past decade or so, a series of studies on microfluidic-jet spray drying were reported by Patel and Chen (Liu, Chen, & Selomulya, 2015; Patel & Chen, 2007; Patel, Chen, & Lin, 2006). They developed a microfluidic-jet spray dryer by introducing an ink-jet device to a conventional lab-scale spray dryer. The ink-jet device is a type of monodisperse droplet generator (MDG), which can generate a single stream of droplets for drying (Liu et al., 2015). Through the adoption of this innovative technique, spray dried particles with uniform characteristics can be produced.

13.2.2 Design of Nozzle

The atomization of microfluidic-jet spray drying is based on a combined action of two forces, a mechanical force and a vibration force (Wu et al., 2007). The mechanical force is similar to the force of spray used in conventional spray drying, which forces the feed liquid to pass through a nozzle to form a continuous jet in the chamber. The pneumatic nozzle is typically used in the microfluidic-jet spray drying to generate the mechanical force required for spraying. The vibration force is produced by a piezoceramics transducer (PZT) equipped on the nozzle (Figure 13.2). By transmitting the vibration frequency signal from the piezoelectric ceramic, the piezoelectric element applies periodical disturbance to the liquid jet and breaks it up into the monodisperse droplets (Wu et al., 2011). Previously, the PZT-driven generator has been widely utilized in ink-jet printing applications, and it has then been adapted into the atomization system of microfluidic-jet spray drying. The formation of monodisperse droplets is dependent on the feed properties, flow rate of liquid, and disturbance frequency from the piezoceramics transducer. The liquid flow rate can be adjusted through controlling the pressure applied to the liquid reservoir, with the intention of producing a laminar flow. The droplet size of the liquid jet can be controlled by the nozzle size, the liquid flow rate, and the vibration frequency (Wu et al., 2007).

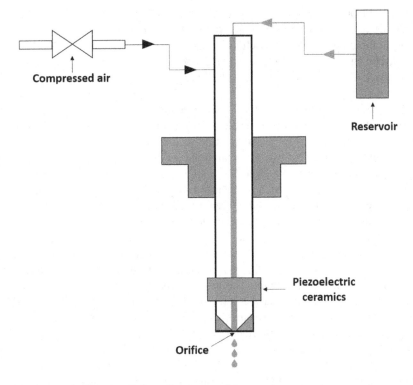

FIGURE 13.2 Schematic diagram of piezoelectric materials transducer nozzle.

Generally, the reduction of nozzle size can decrease the droplet size of feed liquid. The droplet size of liquid produced by the PZT driven nozzle is in the range of 20 to 200 µm (Wu et al., 2007).

13.2.3 USE OF MICROFLUIDIC-JET SPRAY DRYING FOR BIOACTIVE ENCAPSULATION

13.2.3.1 Advantages

The application of microfluidic-jet spray drying can reduce the complexity of spray droplets, decreasing the possibility of powder agglomeration, and hence improving the flowability of the dried particles. Since this drying method manufactures droplets in uniform size, it is possible to simulate the drying process of the droplets and the particle formation by establishing mathematical models. Furthermore, the spherical and broadly uniform particles produced by this drying method may also enhance the encapsulation efficiency for better delivery of bioactive compounds, and this is an advantage in the pharmaceutical industry as mentioned before. This feature enables the easier analysis of the drying process and helps develop an applicable drying kinetics model for upscaling operations. Compared to the conventional spray drying process, the process optimization of mono-disperse spray drying would be relatively easy, because the monodisperse droplets have a higher reproducibility during drying. Furthermore, the spray dried particles are expected to be more efficiently collectible due to the improved flowability of the dried products.

13.2.3.2 Delivery of Bioactive Compounds

The application of the microfluidic-jet spray dryer for bioactives encapsulation has been relatively recent. There have been studies on the encapsulation of both lipophilic and hydrophilic bioactives, with lipophilic compounds received more attention. The lipophilic bioactives studied were Coenzyme Q10, vitamin E, canola oil, fish oil, curcumin, and astaxanthin while the hydrophilic bioactives included fermented noni juice and epigallocatechin gallate (EGCG). These studies mostly involved a single bioactive compound as core material such as curcumin (Liu, Chen, Cheng, & Selomulya, 2016) and astaxanthin (Dai et al., 2013). A few studies have reported the effect of co-encapsulating multiple core materials such as coenzyme Q10 and vitamin E (Fu, You, Quek, Wu, & Chen, 2020; Huang, Quek, Fu, Wu, & Chen, 2019). Table 13.1 shows a summary of previous researches in bioactive encapsulation by microfluidic-jet spray drying.

Microfluidic-jet spray drying was found to be useful for the encapsulation of thermo-sensitive bioactives. Dai et al. (2013) reported high bioactive retention in the monodisperse astaxanthin-containing microparticles. In a later study, Li et al. (2017) reported that stable microcapsules with high oil loading (80%) and relatively high encapsulation efficiency (about 92.6%) could be obtained by microfluidic-jet spray drying. This work indicated the feasibility of encapsulating high oil content emulsions by a microfluidic-jet spray dryer. Some studies demonstrated the efficiency of the mono-disperse droplet spray drying technique in encapsulating hydrophilic bioactives that are sensitive to heat (Fu, Zhou, et al., 2011). These authors reported that lactose could be used to protect EGCG from heat deterioration. During the drying of droplets, the wet-bulb temperature of the droplets is kept at a low temperature during the constant drying rate period. As the drying proceeds, the droplets enter the falling drying rate period. In this stage, the surface temperature of the droplet rises rapidly and finally forms a shell to enclose the bioactive materials. Since the time of the falling drying rate period is very short, the thermal influences towards the bioactive materials are limited.

The microfluidic-jet spray drying has been reported to offer a good encapsulation efficiency for bioactive compounds. Most of the microcapsules produced from the microfluidic-jet spray drying were found to have an encapsulation efficiency of more than 90%. Using this method, Huang et al. (2019) achieved high microencapsulation efficiencies (96.01%–99.79%) of Coenzyme Q_{10} (CoQ_{10}) and vitamin E, which could be associated with high emulsion stability. In this case, it is important to have feed emulsions with good emulsion stabilities (i.e., emulsions with small droplet size and narrow size distribution) to avoid emulsion creaming that would affect the final quality of the spray dried particles. In the study by Huang et al. (2019), the high encapsulation efficiency could also be correlated with the appearance of microcapsules. The absence of apparent pores or cracks on the surface of dried microcapsules can prevent possible oil leakage. In another study, Liu et al. (2016), the encapsulation efficiency of curcumin-loaded microcapsules was between 95.0% and 97.5%.

Li et al. (2017) employed sodium caseinate-lactose complex conjugates (via the Maillard reaction) to produce canola oil-containing microcapsules by microfluidic-jet spray drying. This study also explored the effect of pH variation on the stability of emulsion containing the sodium caseinate-lactose conjugates. The results showed that the interactions between sodium caseinate and lactose at pH 11 enhanced the adsorption of the conjugates onto the oil droplets, and, hence, promoted a better oil entrapment (about 95.2%) by the cross-linked complex.

The effect of solid content on the properties of microcapsules produced by microfluidic-jet spray drying was also investigated. In the study by Huang et al. (2019), the feed emulsions (10% and 30% solid content) were spray dried at 160°C and 190°C, respectively. The spray dried particles produced from the feed emulsion containing 30% w/w solid content showed a significant difference compared with those containing 10% w/w solid content. The former spray dried particles had larger particle size (172–175 μm vs. 116 μm), higher microencapsulation efficiency (98%–99% vs. 96%), lower tapped density (0.532–0.599 g/cm³ vs. 0.406–0.502 g/cm³), and better flow properties (Carr's index of 22.06%–29.19% vs. 22.70%–40.64%). It was suggested that the quality of microcapsules could be improved by increasing the solid content. From the SEM micrographs, the microcapsules produced from emulsions with higher solid content (30% w/w) showed larger particle size and better spherical shape than those from the lower solids content (10% w/w).

Huang et al. (2019) also investigated the effect of drying temperature on the microcapsule properties. Their results showed that microcapsules obtained at 190°C possessed better retention of CoQ_{10} and vitamin E, higher antioxidant capacities, and better color retention than those dried at 160°C. On the other hand, the use of relatively low drying temperature for bioactives encapsulation by micro-fluidic jet spray drying has been reported. The inlet temperatures of 110°C and 150°C have

TABLE 13.1

Previous Studies on Bioactive Encapsulation of by Microfluidic-jet Spray Drying

Type of Bioactives	Bioactives	Wall Material	Drying Temperature (Inlet/Outlet Temperature °C)	Encapsulation Efficiency (%)	Particle Morphology	Authors
Lipophilic bioactives	Coenzyme Q10, vitamin E	Whey protein isolate, soluble corn fiber, maltodextrin	190/90	85–99	Microparticles with uniform size and shape	Fu et al. (2020)
	Coenzyme Q10, vitamin E	OSA starch	160/70 190/90	96.01–99.79	Monodisperse, wrinkled, irregular surfaces	Huang et al. (2019)
	canola oil	Sodium caseinate, lactose	180/80	73.1–95.2	Uniform size, relatively smooth surfaces	Li, Woo, Patel, and Selomulya (2017)
	Whole milk	Carrageenan	200/88	NA	Non-porous surfaces, macro-porous in the internal cross-sections	Foerster, Liu, Gengenbach, Woo, & Selomulya, 2017
	Fish oil (50% DHA)	Whey protein isolate	160/NA	13.1–93.2	Raisin-like particle	Wang, Liu, Chen, and Selomulya (2016)
	Curcumin	Whey protein isolate	150/NA 110/NA	95.0–97.5	Uniform size, pot-like shapes	Liu et al., 2016
	Astaxanthin	Lactose, Sodium caseinate, Whey protein concentrate	140/30	NA	Very smooth surfaces, similar appearance	Dai et al. (2013)
Hydrophilic bioactives	Fermented Noni juice	Maltodextrins, gum Arabic	170/90	NA	Uniform particle size, semi-spherical shapes	Zhang et al. (2020)
	Epigallocatechin gallate (EGCG)	Lactose	95/NA 130/NA	NA	Biconcave or spherical morphology with corrugated surface	Fu et al. (2011)

NA: data not available.

been applied in the encapsulation of curcumin using whey protein isolate as wall material (Liu et al., 2016). It was noted that the inlet temperature of 110°C was sufficient to produce dried microcapsules. This indicated that the microfluidic-jet spray drying could be applied at a relatively low inlet temperature. After spray drying, the microcapsules containing curcumin could be easily re-dispersed into a water solution, demonstrating a good water solubility of these microcapsules.

Morphological results showed consistently that the microfluidic-jet spray drying produced uniform microparticles or microcapsules. These particles presented a uniform size and morphology. Some of the particles presented a pot-like morphology with round or slightly wrinkled surfaces (Dai et al., 2013; Liu et al., 2016), while others presented a raisin-like shape with wrinkled and irregular surfaces (Wang, Liu, et al., 2016; Zhang, Khoo, Chen, & Quek, 2020). The spray dried particles were formed by solvent evaporation and solutes diffusion caused by heat and mass transfers. The differences of surface characteristics in particles may be caused by the difference in wall materials used for encapsulation.

13.2.4 CHALLENGES FACING MICROFLUIDIC-JET SPRAY DRYING

Microfluidic-jet spray drying presents several significant advantages, as discussed above; nevertheless, there are some challenges associated with the application of this relatively new technique. The successful application of this technique depends on the stable performance of the PZT driven nozzle, which produces the monodisperse spray droplets in the drying chamber. One common issue encountered is the blockage of the nozzle during spray drying (Van, Houben, & Koldeweij, 2013). An orifice with a small diameter is prone to blockage, especially for feed liquid containing suspended particles having a diameter larger than the critical diameter (generally, 5% of the orifice diameter). During the spray drying process, the blockage issue may be more severe due to the drying out of the feed liquid at the orifice. To solve this, some measures can be taken to reduce the possibility of blockage, such as filtering the concentrated feed and applying a flow of low-temperature air around the nozzles (Wu et al., 2007). Another issue of the microfluidic-jet spray drying is related to the production scale (Patel & Chen, 2007). The current scale of production is limited compared with the conventional spray dryers, due to the small capacity of the PZT driven nozzle. At present, the monodisperse droplet generators can produce a few milligrams of powder per hour. One possible solution to enhance the output of the PZT driven nozzle is by introducing multiple PZT driven nozzles in a single assembly that work in parallel. For instance, several hundred ink-jet devices can be grouped in a single mechanical-hydraulic plate to produce multiple streams of uniform droplets. This multiple-nozzle design is expected to achieve production capacities up to a few kilograms of powder per hour, and, hence, greatly enhance the production throughput (Wu et al., 2007).

13.3 NANO SPRAY DRYING

13.3.1 BACKGROUND

Nanoencapsulation is now increasingly applied in the pharmaceuticals, materials engineering, and functional foods areas (Saifullah et al., 2019). Nanoencapsulation is defined as the encapsulation of the bioactive compounds within nanocarriers with dimensions between 100 and 1000 nm. Nanoparticles possess a smaller size and larger surface area than common spray dried particles, which may improve the dissolution rate, absorption rates, and bioavailability (Heng, Lee, Ng, & Tan, 2011). Nano spray drying technology can produce sub-micron and nano-scale particles from nanoemulsion or nanosolutions. Compared with micro-sized spray dried particles, the production of nanosized particles requires a special spray drying set-up (Assadpour & Jafari, 2019). This is because the atomizers in conventional spray dryers cannot manage to generate tiny droplets suitable for conversion into nanoparticles. Thus, specially designed atomizers are required for the production of tiny droplets in nano spray drying (Arpagaus, Collenberg, Rütti, Assadpour, & Jafari, 2018). The technical characteristics of nano spray technology include gentle drying airflow, vibrating mesh used to produce tiny droplets, and electrostatic deposition technology used to collect nanoparticles.

The nano spray drying technique has a good application for delivering bioactive compounds. The demand for nanoparticles has increased over these years so there is a need to produce nanoparticles with narrow size distribution and good encapsulation capacity (Mohammadi, Shekaari, &

Zafarani-Moattar, 2020; Patra et al., 2018). For the encapsulation of bioactive food ingredients using a nano spray dryer, different process parameters could be optimized to achieve the desired powder properties, including yield, encapsulation efficiency, particle size, release profile, and powder morphology (Arpagaus et al., 2018). For instance, in order to achieve smaller-sized nanoparticles, a smaller mesh with a low solid concentration feed and surface tension should be applied. Other parameters to consider are inlet and outlet temperatures of the drying gas, the vibration frequency of the ultrasonic nozzle feed flow rate, drying gas type, humidity, and flow rate. It should be noted that some formulation parameters can also affect the final nanoparticle properties, including feed composition, viscosity, solvent type, surface tension, solids concentration, and bioactive content (Assadpour & Jafari, 2019).

Compared to the conventional spray drying techniques, one major challenge of nano spray drying is the collection of nanoparticles, owing to their very small size. Nanoparticles are prone to follow the gas flow; therefore, it is very difficult for it to be separated and collected by cyclone separators. The cyclone separators in conventional spray dryers are suitable for the collection of particles less than 2 mm. The nano spray dryer is equipped with a special collector for particle collecting. Although some nano spray drying processes can achieve a yield of 76 to 96%, sometimes only 50% to 70% may be achieved in the actual production (Oliveira, Guimarães, Cerize, Tunussi, & Poço, 2013). The reasons accounted for this lower yield include the possible deposition of particles on the spray cap or the drying wall, the clogging of the nozzle, and the loss of particles during the manual collection process. Therefore, yield improvement of nano spray drying is still an ongoing important research target.

13.3.2 OPERATION PRINCIPLE OF NANO SPRAY DRYER

Figure 13.3 shows a schematic diagram of the nano spray dryer and its working principles. The nano spray dryer has an ultrasonic atomizer equipped with a vibrating mesh technology, which is capable of producing nano-size droplets with a narrow particle size distribution. The vibration mesh spray technology utilizes a small spray cap with a thin membrane that is equipped with an array of

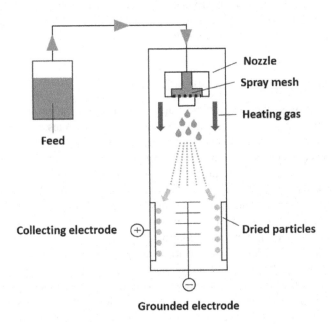

FIGURE 13.3   A schematic of nano spray dryer.

tiny micron-sized holes (4.0, 5.5, or 7.0 μm) (Arpagaus, 2012). The vibration of the membrane is controlled by an actuator, which generates ultrasonic frequency. Through the vibration, tiny droplets were produced in a size range of 3 to 15 μm with a narrow droplet size distribution (Schmid, Arpagaus, & Friess, 2009). The spray droplets are then dried by a heating gas flow which moves with the atomized droplets co-currently in the drying chamber. The gas flow must be controlled as a laminar range to prevent any turbulence inside the drying chamber; otherwise, the spray dried particle may be carried by the cyclone, leading to a low yield. The collection of the spray dried nanoparticles is achieved by an electrostatic field acting on small particles (Feng et al., 2011). A high-voltage electric field (see as Figure 13.3) is set up at the bottom of the drying chamber (Assadpour & Jafari, 2019). It is made up of two electrodes, including a grounded electrode at the center (central anode) and a collecting electrode (surrounding cathode) at the sidewalls of the drying chamber. When the nanoparticles travel down to the bottom, the dried nanoparticles are charged through the central anode and moved to the surrounding cathode for collection (Arpagaus et al., 2018). The resultant dried nanoparticles can be gently scraped from the collector. The electrostatic particle collector can collect nanoparticles with a small batch, from 30 to 500 mg (Arpagaus, 2012; Lee, Heng, Ng, Chan, & Tan, 2011). This small production is suitable for the encapsulation of new pharmaceuticals or high-value bioactive compounds, which are available in small amounts for formulation design (Xue, Wang, Hu, Zhou, & Luo, 2018).

Currently, the most commonly used nano spray dryer in literature is the Nano Spray Dryer B-90, a laboratory benchtop nano spray dryer designed by the Buchi company (Switzerland). There are several publications on the encapsulation of bioactives using this spray dryer. The minimal particle size range of the spray dried nanoparticles reported is around 350–500 nm (Arpagaus, 2012). The morphology of the spray dried particles varies; smooth spherical particles, shrink-wrinkled, donut-shaped, and granules, depending on the wall materials used and the process parameters. The outlet temperature for the Nano Spray Dryer B-90 reported ranges between 28°C and 59°C for aqueous solutions. These low outlet temperatures are favored for the encapsulation of heat-sensitive bioactive components. The feed rates of the Nano Spray Dryer B-90 are from 3 to 25 mL/h using a 4.0 mm spray cap, and 50 mL/h using a 5.5 mm spray cap (Lee et al., 2011). Feed rates are related to the hole size of the spray cap membrane, feed composition, inlet temperature, and spray rate (Bürki, Jeon, Arpagaus, & Betz, 2011; Schmid, Arpagaus, & Friess, 2011). This spray dryer allows a small sample quantity from 5 to 20 mL to be spray dried, making it suitable for the delivering of high-value pharmaceuticals and bioactive compounds.

13.3.3 Bioactive Encapsulation by Nano Spray Drying

Most of the encapsulation studies utilized the nano spray dryer for the purposes of drug delivery in the pharmaceutical field. Nano spray drying can produce dried particles with smaller particle sizes than that of the conventional spray dryers, and, therefore, enhancing bioavailability and release of bioactive components (Arpagaus et al., 2018). The drug-loaded nanoparticles have a higher surface-volume ratio and a higher possibility of targeted release. These features are considered to be ideal in drug encapsulation and delivery. Compared to drug delivery, there are relatively limited studies on bioactives encapsulation for food application using nano spray drying. Table 13.2 shows the previous studies in the last decade or so involving nano spray drying of food materials.

Bürki et al. (2011) studied the nanoencapsulation of β-galactosidase by a nano spray dryer for the development of respirable powders. β-galactosidase and trehalose were employed as a core protein and stabilizer, respectively. The effect of inlet temperature, hole size of the spray cap membrane, and ethanol content on the properties of spray dried particles were investigated. These included particle morphology, enzyme activity, particle size, span, yield, and storage stability. The optimal conditions of nano spray drying were as follows; inlet temperature of 80°C, spray capsize of 4 um, and ethanol content of 0% in the feed solution for spray drying. A high yield of 90% was obtained

TABLE 13.2

Previous Studies on Functional Food Ingredients Encapsulation by Nano Spray Drying (The Years 2010 To 2020)

Bioactives	Wall Materials	Inlet Temperature /°C	Spray Capsize	Particle Morphology	Authors
Lactoferrin, Glycomacropeptide	NA	80, 100 and 120	4.0, 7.0	Particles with different sizes	Bourbon, Barbosa-Pereira, Vicente, Cerqueira, and Pastrana (2020)
Roasted coffee bean oil	Whey protein	90	7.0	Nanosized, smooth, and spherical particles (0.2 ~ 0.4 µm)	Busolo, Torres-Giner, Prieto, and Lagaron (2019)
Egg yolk LDL/ polysaccharide	Pectin, carboxymethyl cellulose, EDC, NHS	100	5.5	Agglomerated particles and large distribution (> 1 µm); smooth surface and homogenous distribution (< 1 µm)	Zhou et al. (2018)
Curcumin	Sodium caseinate, pectin, stearic acid	100	7.0	Aggregation and donut-like shape (with ethanol); spherical particles with uniform distribution (without ethanol)	Xue et al. (2018)
Saffron extracts	Maltodextrin	100	4.0, 7.0	Spherical particles (1.5 ~ 4.2 µm)	Kyriakoudi and Tsimidou (2018)
Eugenol	Zein, sodium caseinate, pectin	100	5.0	Spherical shape and uniform size distribution	Veneranda et al. (2018)
Eugenol oil	gum Arabic, lecithin	100	4.0	Ultrafine spherical powders (< 0.5 µm)	Hu, Gerhard, Upadhyaya, Venkitanarayanan, and Luo (2016)
NA	Soy lecithin, sodium caseinate, pectin	100	5.5	Spherical and large aggregates	Wang et al. (2016)
Folic acid	Whey protein concentrate, resistant starch, guar gum	90	0.7	Spherical submicron and micron particles	Pérez-Masiá et al. (2015)
Resveratrol	Poly (ε-caprolactone), sodium deoxycholate, trehalose	55	7.0	Spherical particles with irregular surfaces	Dimer, Ortiz, Pohlmann, and Guterres (2015)
Vitamin B12	gum Arabic, cashew nut gum, sodium alginate, sodium carboxymethyl cellulose, Eudragit RS100	120	4.0, 7.0	Spherical shapes	Oliveira et al. (2013)

(Continued)

TABLE 13.2
(*Continued*)

Bioactives	Wall Materials	Inlet Temperature /°C	Spray Capsize	Particle Morphology	Authors
Leucine	Trehalose	75	7.0	Smooth spheres, hollow particles with thin shells	Feng et al. (2011)
Bovine serum albumin	Tween 80	80, 100, 120	4.0, 5.5, 7.0	Spherical and smooth, wrinkled, donut-shaped, granules	Lee et al. (2011)
β-Galactosidase	Trehalose	80, 100, 120	4.0, 5.5, 7.0	Spherical and smooth surface, a mixture of smooth and shrivelled spheres	Bürki et al. (2011)
Vitamin E acetate	gum Arabic, whey protein, polyvinyl alcohol, modified starch, maltodextrin	100	4.0	Sphere-shaped particles	Li, Anton, Arpagaus, Belleteix, and Vandamme (2010)

Abbreviations: NA: not applicable; LDL: low-density lipoprotein; EDC: 1-ethyl-3-(3-dimethyl-aminopropyl-1-carbodi-imide); NHS: N-hydroxysuccinimide.

under the aforementioned conditions for the production of particles with respirable size. The results showed the inlet temperature and spray capsize affected enzyme activity. The full enzyme activity of β-galactosidase was maintained on the optimal parameters of spray drying. The morphology of particles was affected by different cap size. The particle size of spray dried particles was influenced by the hole size of the spray cap membrane and the ethanol content. The smallest size of the cap membrane produced a monodisperse particle size distribution and the largest yield of particles. The higher recovery of powders was obtained at the smaller size of the cap, the lower inlet temperatures, and the higher ethanol concentrations. The larger spray capsize and the absence of ethanol gave higher storage stability of the protein.

Dimer et al. (2015) delivered resveratrol-loaded microcapsules by nano spray drying to improve rare disorder pulmonary arterial hypertension. The spray dried powders were used as an inhaler for the pulmonary treatment. The formulations of particles contained resveratrol (therapeutic agent), poly-ε-caprolactone (wall matrix), sodium deoxycholate (for preventing agglomeration), and treha-lose (drying adjuvant). The resultant spray dried particles showed a particle diameter of 2.32 μm, a high yield of 80%, and a low moisture content of 2.0%. The particles had good aerodynamic proper-ties and were suitable for drug deposition on the bronchial and alveolar regions of the lungs due to the low density, the high flowability of particles, the spherical shape, and the irregular surface on particles. The results provided a new approach to formulate a control-delivery system for pulmonary administration by using nano spray drying.

Nano spray drying has been also applied to develop solid lipid nanoparticles (SLNs) for bioactives delivery. A solid lipid nanoparticle is a type of nano-level colloidal carrier, which is composed of lipids, surfactants, and active ingredients. It combines the advantages of liposomes, nanoemulsions, and polymer nanoparticles. Solid lipid nanoparticles have excellent encapsulation efficiency and controlled release properties. The SLNs can be used as an alternative delivery system to traditional colloidal systems, including emulsions and liposomes. Wang, Hu, et al. (2016) utilized layer-by-layer (LbL) technique and nano spray drying technology to develop solid lipid nanoparticles (SLNs). This research aims to improve the re-dispersibility of nanoparticles after drying. The surfactants used in this study were natural ingredients instead of synthetic surfactants. Solid lipid nanoparticles were prepared by solvent diffusion and hot homogenization through three layers of coating. Soy lecithin, sodium caseinate, and pectin was used as coatings for the first, second, and third layers, respectively. The formulations of solid lipid nanoparticles were optimized by surface response methodology (Box–Behnken design). The morphology of the resultant nanoparticles showed a well-separated, spherical shape, and smooth surface. The shape and dimension of the nanoparticles were mainly determined by the external layer (i.e., pectin). The optimal parameters for producing separated and spherical particles were at pH 5 and treated by heating at 80°C for 30 min. The particle size of nanoparticles was about 317.8 nm. The particle size and morphology of nanoparticles remain the same after their re-dispersion in water. The LbL-coated SLNs with excellent water re-dispersity can be used for lipophilic bioactives delivery.

Moreover, Xue et al. (2018) investigated the encapsulation of curcumin into solid lipid nanoparticles (SLNs) through nano spray drying. This encapsulation may improve the bioavailability of lipophilic bioactives, such as curcumin. The solid lipid nanoparticles were prepared without synthetic surfactant. Sodium caseinate (NaCas), pectin, and stearic acid were applied as emulsifiers and stabilizers. Four different loading approaches were used for exploring the optimal encapsulation efficiency, which differed with regard to the use of ethanol, the addition sequence of curcumin and solid lipids, the deprotonation of NaCas at pH 12, and pectin adsorption at pH 4. The properties of curcumin-loaded SLNs were analyzed. The researchers found that the addition of ethanol as a solvent of curcumin had a negative effect on the particle size, stability, and re-dispersibility of nanoparticles. The addition of ethanol could cause a large size and aggregation of solid lipid nanoparticles. The use of deprotonated NaCas as a solvent of curcumin was better than the use of organic solvent. Furthermore, the addition of melted lipid and pectin at pH 12 could produce nanoparticles with a uniform and small size, which is favored for gastrointestinal stability and re-dispersal. The encapsulation of curcumin into SLNs improved the antioxidant capacity. This study gave important information on the production of curcumin containing nanoparticles by nano spray drying.

The nano spray drying technique has been compared with other drying methods for bioactives encapsulation. Oliveira et al. (2013) reported a study on the encapsulation of vitamin B12 by using a nano spray dryer and a conventional spray drying system. The encapsulating matrices in this work include gum Arabic, cashew nut gum, sodium alginate, sodium carboxymethyl cellulose, and Eudragit RS100. The study assessed the effect of the drying method on particle size distribution, specific surface area, and morphological characteristics. The encapsulating efficiency and the release kinetics of spray dried particles were evaluated by *in vitro* digestion experiments that simulated the gastric and intestinal phases. The results showed that a decrease in particle size and an increase of the surface area can enhance the release rate kinetics. By comparison with conventional spray drying equipment (Mini Spray Drier Büchi B190), a nano spray dryer with a vibrating mesh technology can improve the release kinetics through a reduction in particle size.

On the other hand, Pérez-Masiá et al. (2015) reported an encapsulation study of folic acid by two encapsulation techniques, nano spray drying, and electrospraying. Whey protein concentrate (WPC) and a commercial resistant starch were used as wall matrices. It was observed that spray dried particles from the two techniques had different size ranges, including nano-, submicro-, and micro-sizes. Electrospraying produced particles with smaller particle sizes and narrow size distribution. However, there was no significant difference in the encapsulation efficiency between nano spray drying and electrospraying. In another study, Hu et al. (2016) used a nano spray dryer to coat eugenol oil containing nanoemulsions stabilized by gum Arabic and lecithin. The resultant particles were compared with the particles of the same formulation produced by freeze drying. This study also evaluated the effect of different drying techniques (spray drying and freeze drying) on the morphology and re-dispersibility of the dried powders. There was a significant difference between the dried particles produced by the two technologies. The particles produced by the nano spray dryer were spherical with a smooth surface, while the freeze-dried particles presented as aggregated flakes with a rough surface. The addition of ethanol improved the uniformity of spray dried powders, but did not show a significant effect on the morphological characteristics of the freeze-dried particles. The diameter of nano spray dried powders was between 200–500 nm, which was much smaller than that of the dried particles (10 ~ 25 µm) from conventional spray dryers.

The particle size of nano spray dried powders were associated with the wall materials applied for encapsulation. Li et al. (2010) developed nanocrystals containing vitamin E acetate by different polymeric wall materials: gum Arabic, whey protein, modified starch, and maltodextrin. The droplet sizes of nano-emulsion were compared before and after spray drying by re-dispersing the spray dried powder. Before drying, the lipid nano-droplets existed in particle sizes lower than 100 nm: gum Arabic (95.49 nm), modified starch (91.49 nm), maltodextrin (83.49 nm), and whey protein (88.54 nm). After spray drying, the re-dispersed emulsion from the spray dried powder showed no significant degradation or size increase. It indicated that the integrity of nano-structured systems can be well maintained during the spray drying process. It was noted that the particle size distribution of the re-dispersed emulsion showed a significant increase (from 0.117 to 0.511) when using gum Arabic as wall materials. The researchers thought this may be due to droplet flocculation and destruction caused by the surface-active functionalities of gum Arabic. In addition, relatively high yields of spray dried nanoparticles were obtained, and the amounts of powders ranged from 50 mg to 500 mg. The particle sizes of the spray dried powders were lower than 1 µm, and the lowest particle size (350 nm) of particles was obtained from the samples encapsulated by gum Arabic at 0.1 wt% solid concentration. It was reported that the particle size was mainly dependent on the wall material properties and the feed solution concentration.

13.4   UNDERSTANDING THE MECHANISM OF PARTICLE FORMATION

13.4.1   BACKGROUND

Although spray drying technology has wide applications in the delivery of bioactive materials, the basic mechanism for the formation of dry particles is yet to be fully understood. The spray drying process is a process in which liquid droplets are transformed into dry particles under heat transfer. The particle formation process largely determines the final properties of the particles (i.e., particle morphology, particle size, density, and porosity) (Boel et al., 2020). Understanding the particle formation process helps with the optimization of the properties of dry particles so that specific dry particles can be produced to meet specific production needs. In order to gain deeper insights into the drying mechanism of droplets as affected by various processing parameters (including those from drying and wall and core materials) perspectives, establishing a technique to investigate the process of droplet formation is highly required. For this, single droplet drying has been established as a useful method to study and observe droplet formation.

The single droplet drying technique (SDD) is employed to study the drying history of droplets and the formation process of the particles during spray drying. It can monitor the transition process from liquid droplet to dry particle under convective drying conditions that simulate the process of spray drying (Fu, Woo, & Chen, 2012). The SDD system was initially used by Ranz and Marshall in 1952 to study the drying mechanisms of pure solvent droplet or droplets with dissolved solids (Ranz & Marshall, 1952). This technique was further developed to study the morphological changes of droplets and to investigate the correlations between particle properties and drying conditions (Chen, 2004; Lin & Chen, 2002; Lin & Chen, 2004). Although the properties of the feed solution and the final powder products can be investigated based on the spray drying system, it is difficult to track the drying mechanism, history, and drying behavior of the droplets. There is a significant change in the moisture content of the individual droplets under heat transfer during the drying process.

The histories on moisture content changes and temperature effect have significant influences on the physicochemical properties, such as surface morphology, density, and bioactive retention (Fu et al., 2012). Therefore, securing an understanding of the drying history of the droplet is significant to give insight into the final properties of the spray dried particles. In an SDD experiment, an individual droplet is dried under controlled air conditions that mimic the drying temperature, airflow velocity, and the ambient humidity of the real spray drying conditions. The SDD system can provide a fundamental study on the drying kinetics of particular droplets with different wall matrix/matrices and core materials, and, subsequently, the change in particle properties (Zhang et al., 2019). The optimization of spray drying conditions through the SDD technique can also help the spray drying industry to save processing time and cost on feeding solution (Fu et al., 2012). One can view the SDD system as a scaled-down process of spray drying in industry, which utilizes the drying process in a controlled manner to study the drying mechanisms and particle formation.

13.4.2 Principle of Single Droplet Drying

A reliable single droplet drying system is required to ensure the accuracy of the measurement. The basic principles for a reliable SDD system are as follows: (i) the droplet size of the single droplet should be controllable and reproducible; (ii) the drying parameters should be controlled to simulate spray drying (temperature, air velocity, and humidity); and (iii) the mass, temperature, and diameter changes of the droplet can be screened accurately (Fu et al., 2012).

The SDD system with the glass-filament system (Figure 13.4) is now a preferable application in studies of droplet changes during the drying process. In this glass-filament system, a single droplet is produced through a glass capillary and is then suspended on a glass filament in a drying chamber. The droplet is processed under controlled temperature, moisture, and airflow rate. The morphological changes of the droplet are monitored by a video camera and can be observed through the camera images with proper magnification. Furthermore, the mass change and temperature change of the droplet during drying are measured by a glass-filament balance and a pair of inserted thermocouples (Fu et al., 2012). The accurate measurement of the droplet size, weight, and temperature can be used to predict the drying rate and drying process of droplets by the Reaction Engineering Approach (REA) and the master activation-energy curve, respectively (Fu et al., 2012; Fu, Woo, Lin, Zhou, & Chen, 2011; Lin & Chen, 2007; Zhang, Fu, Quek, Zhang, & Chen, 2019).

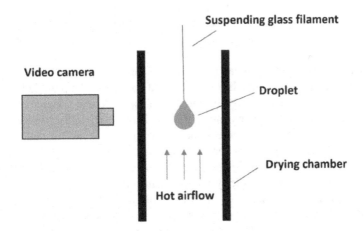

FIGURE 13.4   A schematic diagram of single droplet drying.

13.4.3   Use of SDD to Understand the Spray Drying Encapsulation of Bioactive Compounds

Since the SDD system allows the investigation of droplet drying kinetics in a better-controlled condition, SDD studies have been conducted to investigate the effect of parameter changes on droplet formation when encapsulating some bioactive materials. One important parameter is the effect of the type of wall material(s) applied in the spray drying encapsulation. The wall material has a significant influence on the drying phenomena of the feed droplet. Wang, Che, Fu, Chen, and Selomulya (2015) reported the application of SDD to study the encapsulation of docosahexaenoic acid (DHA) by octenyl succinic anhydride modified (OSA) starch and whey protein concentrate (WPC). The surface formation of the emulsion droplets containing DHA during the drying process and the resulting morphology characteristics of the particle surface, as well as the dissolution behavior of dried and semi-dried particles produced from emulsions, were studied. The DHA-containing spray dried particles encapsulated by OSA-starch showed a better wettability than the particles encapsulated by WPC. The WPC-based emulsion droplet had a higher rate of shell formation than the OSA-starch-based emulsion at the same solid content. The OSA-starch presented a good protective effect on core oil, while the WPC showed a weak protective capacity to encapsulating DHA oil. This study indicated that the SDD system can help to illustrate the surface formation of droplets containing DHA during the drying process, and to study the surface composition and morphology of the dried particles. In a more recent study, Zhang, Quek, et al. (2019) used the SDD technique to study the structure formation of microencapsulated fermented noni juice (FNJ) powder containing maltodextrin (M1, 10–13 DE; M3, 17–20 DE) and gum acacia (GA). This study reported the drying behaviors of FNJ containing different wall materials (M1, M3, and GA) with a solid content of 21% and 28% in 0–300 s at 90°C. It showed that all droplets had consistent and uniform shrinkage at the initial drying stage (0–30 s). Over the course of the drying process, a thin film was formed and became thicker at the drying time of 60 s onwards, which could be observed from the surface morphology of droplets. At the drying time of 90 s, there was an obvious difference between the semi-dried GA particles (bowl-like) and the MD particles (pear-like). The difference in wall properties may be the reason accounted for the differences in particle shape. Between the two wall materials investigated, GA possesses high molecular mass protein while MD does not contain any protein.

Lallbeeharry et al. (2014) studied how the addition of surfactant, including lecithin (1% w/w) and nonionic Tween 80 (0.1% w/w), could affect the drying behavior of milk and particle formation by SDD. Their findings showed that the addition of surfactants improved particle wettability during shell formation. The wetting time of spray dried particles decreased from 35 s to less than 15 s. At the same total loading of 1% w/w in total solids, milk particles added with Tween 80 showed a higher wettability (<5 s) than those with lecithin (>30 s). This study helped elucidate droplet drying behavior in the early shell formation stage and how surface composition influences the powder functionalities.

The solid content of feed emulsion can result in differences in particle formation during drying. A number of studies have investigated the effect of solid content on the particle formation by SDD. Huang et al. (2019) applied SDD to study the particle formation of emulsions containing Coenzyme Q10 and vitamin E during convective drying (at 70°C and 90°C). Their results showed that emulsion droplets with 30% w/w solids content were associated with earlier crust formation (30 s faster) and less particle shrinkage during drying compared with those with 10% w/w solid content. In a recent study, Zhang, Quek, et al. (2019) also investigated the particle formation associated with different initial solid contents, but in a hydrophilic system involving droplets containing fermented noni juice and maltodextrin. The results showed that the droplet of 28% solid content had a faster formation rate of the crust than the droplet of 21% solid content. These two studies imply that the time of surface crust formation is highly related to the initial solid content of the droplet.

Zheng, Fu, Huang, Jeantet, and Chen (2016) utilized the SDD technique to investigate the effects of calcium cation on the history of cellular injuries and cell inactivation of *Lactobacillus rhamnosus*

GG (LGG) during droplet drying. The addition of 1 mM $CaCl_2$ into the lactose carrier improved cell viability, decreased cellular injuries, and increased regrowth capability during the drying progress. At the early drying stage up to 120 s, the LGG viability was unchanged with the carrier of lactose and calcium. The viable cell content decreased as the drying progressed. After drying time of 135 s, LGG cells with the addition of calcium cation showed better survival than the cells having pure sugar as a carrier. This phenomenon may be due to the positive effect of calcium cation on stabilizing sub-cellular structures.

The SDD technique is applied not only to monitor the drying behavior of a single droplet under controlled air conditions, but also to investigate the re-solubility of dried particles. Zhang, Quek, et al. (2019) studied the effect of wall material types and different initial solid content on the solubility of the particles containing fermented noni juice. The particles containing maltodextrin showed better solubility than that of gum acacia, which indicated that maltodextrin was a better wall material than gum acacia in terms of powder solubility. Moreover, the feed solutions containing maltodextrin with a total solid content of 21% and 28% were employed to investigate their effect on the particle solubility by the single droplet drying.

At the drying period from 30 to 180 s, there was no significant difference in the rehydration behaviors between the two different particles. From the 240 s of drying onwards, the semi-dried particles (21%) were thoroughly dissolved in the single droplet within 30 s, while the semi-dried particle (28%) were completely dissolved in 40 s. The results indicated that the increase of maltodextrin concentration may decrease the solubility of particles. In a more recent study, Reaction Engineering Approach (REA) mathematical modeling has been applied to give insight into the drying kinetics and particle formation process (Zhang, Khoo, Swedlund, et al., 2020).

13.5 CONCLUDING REMARKS

The spray drying technique is being improved by researchers so as to produce spray dried particles with high stability and functionality. Some new techniques in spray drying have been applied to study the encapsulation of bioactive materials, including microfluidic-jet spray drying and nano spray drying, which have been covered in this chapter. These techniques presented some groundbreaking changes compared with the traditional spray drying techniques, by producing more controllable particle size and morphology or exploring the in-depth mechanism of dried particle formation. These new spray drying techniques have been applied for bioactive encapsulation with the primary aims of improving the overall encapsulation efficiency, bioactive stability, and control-release properties. It is undeniable that those new techniques still have a number of shortcomings when applied in the commercial production of spray dried particles; e.g., lower yield, higher cost, and longer processing time. However, they offer researchers and manufacturers access to new technological developments in bioactive encapsulation. For example, these new spray drying techniques can be well collaborated with the conventional spray drying for optimization of the production parameters, since the spray dried particles have a more controllable particle size and morphology, while the correlation between particle properties and processing parameters can be more effectively established through a mathematical model using computational fluid dynamics. Therefore, it helps promote the development of spray drying as a promising bioactive encapsulation technique for food and pharmaceutical applications. In addition to the aforementioned emerging spray drying techniques, the use of SDD has contributed to a better understanding of the particle formation process and the effect of drying parameters on the final properties of particles. Moreover, the relation to the type of wall matrices applied and physicochemical properties of the feed solution could be further explored.

Computer simulation methods have also been used to optimize the encapsulation of bioactives by spray drying. In recent years, mathematical modeling has been increasingly influential in spray drying encapsulation. One of the most used methods is computational fluid dynamics (CFD), which can model the fluid flow to predict the flow distribution, mass and heat conversion, and particle separation processes in the spray drying process. The CFD modeling can be used to effectively design or modify the design of the spray dryer and the associated drying parameters to save research time and cost in the spray drying operation. However, the use of the CFD method is still limited in the spray drying encapsulation of bioactives. It will be useful to model the process of bioactive encapsulation so we can predict the stability of bioactive compounds concerning the initial material properties, the particle properties, and processing conditions. In the future, computer simulation could be a huge boost to promote the development of spray drying technology in the digital age.

There are still challenges facing the spray drying techniques for the encapsulation of bioactive materials, despite the developments in recent years. Firstly, the use of drying air temperature during the drying process may lead to the degradation of certain thermos-sensitive products and loss of nutrients. Thus, drying at a lower temperature is desirable to protect the bioactivities of the targeted compounds. Another issue of spray drying is caused by the deposition of particles on the walls of the drying chamber. This reduces the throughput of spray dried particles and influences product quality. The optimization of the dryer operating conditions is required to minimize the wall deposition and improve particle quality. Moreover, the wide droplet size distribution and irregular droplet drying trajectories in conventional spray drying may lead to the difficulty to simulate the particle formation process for the optimization of particle production. More technological development in the spray drying technology and studies on the drying mechanisms of spray droplets are necessary to elucidate the particle formation. In addition, the blockage of atomizers of the viscous feed solutions is another issue that also requires further technological development. Furthermore, the upscaling of new spray drying technologies, such as nano spray drying and microfluidic-jet spray drying, will require further research. Overall, there are still gaps for the improvement of the quality of spray dried powders, to decrease the manufacturing cost, and implement the scaling up process for new spray drying technologies.

REFERENCES

Arpagaus, C. (2012). A novel laboratory-scale spray dryer to produce nanoparticles. *Drying Technology*, *30*(10), 1113–1121.

Arpagaus, C., Collenberg, A., Rütti, D., Assadpour, E., & Jafari, S. M. (2018). Nano spray drying for encapsulation of pharmaceuticals. *International Journal of Pharmaceutics*, *546*(1–2), 194–214.

Assadpour, E., & Jafari, S. M. (2019). Advances in spray-drying encapsulation of food bioactive ingredients: From microcapsules to nanocapsules. *Annual Review of Food Science and Technology*, *10*, 103–131.

Boel, E., Koekoekx, R., Dedroog, S., Babkin, I., Vetrano, M. R., Clasen, C., & Van den Mooter, G. (2020). Unraveling Particle Formation: From Single Droplet Drying to Spray Drying and Electrospraying. *Pharmaceutics*, *12*(7), 625.

Bourbon, A. I., Barbosa-Pereira, L., Vicente, A. A., Cerqueira, M. A., & Pastrana, L. (2020). Dehydration of protein lactoferrin-glycomacropeptide nanohydrogels. *Food Hydrocolloids*, *101*, 105550.

Bürki, K., Jeon, I., Arpagaus, C., & Betz, G. (2011). New insights into respirable protein powder preparation using a nano spray dryer. *International Journal of Pharmaceutics*, *408*(1–2), 248–256.

Busolo, M., Torres-Giner, S., Prieto, C., & Lagaron, J. (2019). Electrospraying assisted by pressurized gas as an innovative high-throughput process for the microencapsulation and stabilization of docosahexaenoic acid-enriched fish oil in zein prolamine. *Innovative Food Science & Emerging Technologies*, 51, 12–19.

Chen, X. D. (2004). Heat-mass transfer and structure formation during drying of single food droplets. *Drying Technology*, *22*(1–2), 179–190.

Dai, Q., You, X., Che, L., Yu, F., Selomulya, C., & Chen, X. D. (2013). An investigation in microencapsulating astaxanthin using a monodisperse droplet spray dryer. *Drying Technology*, *31*(13–14), 1562–1569.

Dimer, F. A., Ortiz, M., Pohlmann, A. R., & Guterres, S. S. (2015). Inhalable resveratrol microparticles produced by vibrational atomization spray drying for treating pulmonary arterial hypertension. *Journal of Drug Delivery Science and Technology*, *29*, 152–158.

Feng, A., Boraey, M., Gwin, M., Finlay, P., Kuehl, P., & Vehring, R. (2011). Mechanistic models facilitate efficient development of leucine containing microparticles for pulmonary drug delivery. *International Journal of Pharmaceutics*, *409*(1–2), 156–163.

Foerster, M., Liu, C., Gengenbach, T., Woo, M. W., & Selomulya, C. (2017). Reduction of surface fat formation on spray dried milk powders through emulsion stabilization with λ-carrageenan. *Food Hydrocolloids*, *70*, 163–180.

Fu, N., Woo, M. W., & Chen, X. D. (2012). Single droplet drying technique to study drying kinetics measurement and particle functionality: A review. *Drying Technology*, *30*(15), 1771–1785.

Fu, N., Zhou, Z., Jones, T.B., Tan, T.T., Wu, W.D., Lin, S.X., Chen, X.D. & Chan, P.P. (2011). Production of monodisperse epigallocatechin gallate (EGCG) microparticles by spray drying for high antioxidant activity retention. *International Journal of Pharmaceutics*, *413*(1–2), 155–166.

Fu, N., Woo, M. W., Lin, S. X. Q., Zhou, Z., & Chen, X. D. (2011). Reaction Engineering Approach (REA) to model the drying kinetics of droplets with different initial sizes—Experiments and analyses. *Chemical Engineering Science*, *66*(8), 1738–1747.

Fu, N., You, Y. -J., Quek, S. Y., Wu, W. D., & Chen, X. D. (2020). Interplaying Effects of Wall and Core Materials on the Property and Functionality of Microparticles for Co-Encapsulation of Vitamin E with Coenzyme Q 10. *Food and Bioprocess Technology*, 13, 705–721.

Grumezescu, A., & Oprea, A. E. (2017). *Nanotechnology applications in food: flavor, stability, nutrition and safety*. Cambridge, MA: Academic Press.

Hede, P. D., Bach, P., & Jensen, A. D. (2008). Two-fluid spray atomisation and pneumatic nozzles for fluid bed coating/agglomeration purposes: A review. *Chemical Engineering Science*, *63*(14), 3821–3842.

Heng, D., Lee, S. H., Ng, W. K., & Tan, R. B. (2011). The nano spray dryer B-90. *Expert Opinion on Drug Delivery*, *8*(7), 965–972.

Hu, Q., Gerhard, H., Upadhyaya, I., Venkitanarayanan, K., & Luo, Y. (2016). Antimicrobial eugenol nanoemulsion prepared by gum arabic and lecithin and evaluation of drying technologies. *International Journal of Biological Macromolecules*, *87*, 130–140.

Huang, E., Quek, S. Y., Fu, N., Wu, W. D., & Chen, X. D. (2019). Co-encapsulation of coenzyme Q10 and vitamin E: A study of microcapsule formation and its relation to structure and functionalities using single droplet drying and microfulidic-jet spray drying. *Journal of Food Engineering*, *247*, 45–55.

Huang, L. X., Kumar, K., & Mujumdar, A. (2006). A comparative study of a spray dryer with rotary disc atomizer and pressure nozzle using computational fluid dynamic simulations. *Chemical Engineering and Processing: Process Intensification*, *45*(6), 461–470.

Huang, S., Vignolles, M. L., Chen, X. D., Le Loir, Y., Jan, G., Schuck, P., & Jeantet, R. (2017). Spray drying of probiotics and other food-grade bacteria: A review. *Trends in Food Science & Technology*, *63*, 1–17.

Jindal, A. B. (2017). The effect of particle shape on cellular interaction and drug delivery applications of micro- and nanoparticles. *International Journal of Pharmaceutics*, *532*(1), 450–465.

Kyriakoudi, A., & Tsimidou, M. Z. (2018). Properties of encapsulated saffron extracts in maltodextrin using the Büchi B-90 nano spray-dryer. *Food Chemistry*, *266*, 458–465.

Lallbeeharry, P., Tian, Y., Fu, N., Wu, W. D., Woo, M. W., Selomulya, C., & Chen, X. D. (2014). Effects of ionic and nonionic surfactants on milk shell wettability during co-spray-drying of whole milk particles. *Journal of Dairy Science*, *97*(9), 5303–5314.

Lee, S. H., Heng, D., Ng, W. K., Chan, H. K., & Tan, R. B. (2011). Nano spray drying: A novel method for preparing protein nanoparticles for protein therapy. *International Journal of Pharmaceutics*, *403*(1–2), 192–200.

Li, K., Woo, M. W., Patel, H., & Selomulya, C. (2017). Enhancing the stability of protein-polysaccharides emulsions via Maillard reaction for better oil encapsulation in spray dried powders by pH adjustment. *Food Hydrocolloids*, *69*, 121–131.

Li, X., Anton, N., Arpagaus, C., Belleteix, F., & Vandamme, T. F. (2010). Nanoparticles by spray drying using innovative new technology: The Büchi nano spray dryer B-90. *Journal of Controlled Release*, *147*(2), 304–310.

Lin, S., & Chen, X. D. (2004). Changes in milk droplet diameter during drying under constant drying conditions investigated using the glass-filament method. *Food and Bioproducts Processing*, *82*(3), 213–218.

Lin, S. X. Q., & Chen, X. D. (2002). Improving the glass-filament method for accurate measurement of drying kinetics of liquid droplets. *Chemical Engineering Research and Design*, *80*(4), 401–410.

Lin, S. X. Q., & Chen, X. D. (2007). The reaction engineering approach to modelling the cream and whey protein concentrate droplet drying. *Chemical Engineering and Processing: Process Intensification*, *46*(5), 437–443.

Liu, W., Chen, X., & Selomulya, C. (2015). On the spray drying of uniform functional microparticles. *Particuology*, *22*, 1–12.

Liu, W., Chen, X. D., Cheng, Z., & Selomulya, C. (2016). On enhancing the solubility of curcumin by microencapsulation in whey protein isolate via spray drying. *Journal of Food Engineering*, *169*, 189–195.

Lu, W., Wang, S., Lin, R., Yang, X., Cheng, Z., & Liu, W. (2020). Unveiling the importance of process parameters on droplet shrinkage and crystallization behaviors of easily crystalline material during spray drying. *Drying Technology*, 1–11.

Marante, T., Viegas, C., Duarte, I., Macedo, A. S., & Fonte, P. (2020). An overview on spray-drying of protein-loaded polymeric nanoparticles for dry powder inhalation. *Pharmaceutics*, *12*(11), 1032.

Mohammadi, B., Shekaari, H., & Zafarani-Moattar, M. T. (2020). Synthesis of nanoencapsulated vitamin E in phase change material (PCM) shell as thermo-sensitive drug delivery purpose. *Journal of Molecular Liquids*, *320*, 114429.

Oliveira, A., Guimarães, K., Cerize, N., Tunussi, A., & Poço, J. (2013). Nano spray drying as an innovative technology for encapsulating hydrophilic active pharmaceutical ingredients (API). *Journal of Nanomedicine and Nanotechnology*, *4*(2), 1–6.

Patel, K. C., & Chen, X. D. (2007). Production of spherical and uniform-sized particles using a laboratory ink-jet spray dryer. *Asia-Pacific Journal of Chemical Engineering*, *2*(5), 415–430.

Patel, K., Chen, X., & Lin, S. (2006). Development of a laboratory ink-jet spray dryer. *Chemeca 2006: Knowledge and Innovation*, 730.

Patra, J. K., Das, G., Fraceto, L. F., Campos, E. V. R., del Pilar Rodriguez-Torres, M., Acosta-Torres, L. S., … Sharma, S. (2018). Nano based drug delivery systems: recent developments and future prospects. *Journal of Nanobiotechnology*, *16*(1), 71.

Pérez-Masiá, R., López-Nicolás, R., Periago, M. J., Ros, G., Lagaron, J. M., & López-Rubio, A. (2015). Encapsulation of folic acid in food hydrocolloids through nanospray drying and electrospraying for nutraceutical applications. *Food Chemistry*, *168*, 124–133.

Poozesh, S., & Bilgili, E. (2019). Scale-up of pharmaceutical spray drying using scale-up rules: A review. *International Journal of Pharmaceutics*, *562*, 271–292.

Quek, S. Y., Chen, Q., & Shi, J. (2015). Microencapsulation of food ingredients for functional foods. *Functional Food Ingredients and Nutraceuticals: Processing Technologies*, *13*, 267.

Ranz, W., & Marshall, W. R. (1952). Evaporation from drops. *Chemical Engineering and Processing*, *48*(3), 141–146.

Rezvankhah, A., Emam-Djomeh, Z., & Askari, G. (2020). Encapsulation and delivery of bioactive compounds using spray and freeze-drying techniques: A review. *Drying Technology*, *38*(1–2), 235–258.

Rogers, S., Fang, Y., Lin, S. X. Q., Selomulya, C., & Chen, X. D. (2012). A monodisperse spray dryer for milk powder: Modelling the formation of insoluble material. *Chemical Engineering Science*, *71*, 75–84.

Sagis, L. M. (2015). *Microencapsulation and microspheres for food applications*. Cambridge, MA: Academic Press.

Saifullah, M., Shishir, M. R. I., Ferdowsi, R., Rahman, M. R. T., & Van Vuong, Q. (2019). Micro and nano encapsulation, retention and controlled release of flavor and aroma compounds: A critical review. *Trends in Food Science & Technology*, *86*, 230–251.

Santos, D., Maurício, A. C., Sencadas, V., Santos, J. D., Fernandes, M. H., & Gomes, P. S. (2017). Spray drying: An overview. In Musumeci, T., & Pignatello (Eds), *Biomaterials-Physics and Chemistry-New Edition*. London, UK: IntechOpen, 9–31.

Schmid, K., Arpagaus, C., & Friess, W. (2009). Evaluation of a vibrating mesh spray dryer for preparation of submicron particles. *Respiratory Drug Delivery Europe*, *30*, 323–326.

Schmid, K., Arpagaus, C., & Friess, W. (2011). Evaluation of the Nano Spray Dryer B-90 for pharmaceutical applications. *Pharmaceutical Development and Technology*, *16*(4), 287–294.

Shekunov, B. Y., Chattopadhyay, P., Tong, H. H., & Chow, A. H. (2007). Particle size analysis in pharmaceutics: principles, methods and applications. *Pharmaceutical Research*, *24*(2), 203–227.

Turchiuli, C., Munguia, M. J., Sanchez, M. H., Ferre, H. C., & Dumoulin, E. (2014). Use of different supports for oil encapsulation in powder by spray drying. *Powder Technology*, *255*, 103–108.

Van Deventer, H., Houben, R., & Koldeweij, R. (2013). New atomization nozzle for spray drying. *Drying Technology*, *31*(8), 891–897.

Veneranda, M., Hu, Q., Wang, T., Luo, Y., Castro, K., & Madariaga, J. M. (2018). Formation and characterization of zein–caseinate–pectin complex nanoparticles for encapsulation of eugenol. *LWT*, *89*, 596–603.

Wang, T., Hu, Q., Zhou, M., Xia, Y., Nieh, M.-P., & Luo, Y. (2016). Development of "all natural" layer-by-layer redispersible solid lipid nanoparticles by nano spray drying technology. *European Journal of Pharmaceutics and Biopharmaceutics*, *107*, 273–285.

Wang, Y., Che, L., Fu, N., Chen, X. D., & Selomulya, C. (2015). Surface formation phenomena of DHA-containing emulsion during convective droplet drying. *Journal of Food Engineering*, *150*, 50–61.

Wang, Y., Liu, W., Chen, X. D., & Selomulya, C. (2016). Micro-encapsulation and stabilization of DHA containing fish oil in protein-based emulsion through mono-disperse droplet spray dryer. *Journal of Food Engineering*, *175*, 74–84.

Wu, W. D., Amelia, R., Hao, N., Selomulya, C., Zhao, D., Chiu, Y. L., & Chen, X. D. (2011). Assembly of uniform photoluminescent microcomposites using a novel micro-fluidic-jet-spray-dryer. *AIChE Journal*, *57*(10), 2726–2737.

Wu, W. D., Patel, K. C., Rogers, S., & Chen, X. D. (2007). Monodisperse droplet generators as potential atomizers for spray drying technology. *Drying Technology*, *25*(12), 1907–1916.

Xue, J., Wang, T., Hu, Q., Zhou, M., & Luo, Y. (2018). Insight into natural biopolymer-emulsified solid lipid nanoparticles for encapsulation of curcumin: Effect of loading methods. *Food Hydrocolloids*, *79*, 110–116.

Zhang, A., Li, X.Y., Zhang, S., Yu, Z., Gao, X., Wei, X., Wu, Z., Wu, W.D. & Chen, X.D. (2017). Spray-drying-assisted reassembly of uniform and large micro-sized MIL-101 microparticles with controllable morphologies for benzene adsorption. *Journal of Colloid and Interface Science*, *506*, 1–9.

Zhang, C., Quek, S. Y., Fu, N., Liu, B., Kilmartin, P. A., & Chen, X. D. (2019). A study on the structure formation and properties of noni juice microencapsulated with maltodextrin and gum acacia using single droplet drying. *Food Hydrocolloids*, *88*, 199–209.

Zhang, C., Fu, N., Quek, S. Y., Zhang, J., & Chen, X. D. (2019). Exploring the drying behaviors of microencapsulated noni juice using reaction engineering approach (REA) mathematical modelling. *Journal of Food Engineering*, *248*, 53–61.

Zhang, C., Khoo, S. L. A., Chen, X. D., & Quek, S. Y. (2020). Microencapsulation of fermented noni juice via microfulidic-jet spray drying: Evaluation of powder properties and functionalities. *Powder Technology*, *361*, 995–1005.

Zhang, C., Khoo, S. L. A., Swedlund, P., Ogawa, Y., Shan, Y., & Quek, S. Y. (2020). Fabrication of spray-dried microcapsules| Containing noni juice using blends of maltodextrin and gum acacia: Physicochemical properties of powders and bioaccessibility of bioactives during in vitro digestion. *Foods*, *9*(9), 1316.

Zheng, X., Fu, N., Huang, S., Jeantet, R., & Chen, X. D. (2016). Exploring the protective effects of calcium-containing carrier against drying-induced cellular injuries of probiotics using single droplet drying technique. *Food Research International*, *90*, 226–234.

Zhou, M., Khen, K., Wang, T., Hu, Q., Xue, J., & Luo, Y. (2018). Chemical crosslinking improves the gastrointestinal stability and enhances nutrient delivery potentials of egg yolk LDL/polysaccharide nanogels. *Food Chemistry*, *239*, 840–847.

Ziaee, A., Albadarin, A. B., Padrela, L., Femmer, T., O'Reilly, E., & Walker, G. (2019). Spray drying of pharmaceuticals and biopharmaceuticals: Critical parameters and experimental process optimization approaches. *European Journal of Pharmaceutical Sciences*, *127*, 300–318.

14 The Characterization and Analysis of Spray Dried Encapsulated Bioactives

Paola Pittia and Marco Faieta
University of Teramo, Italy

CONTENTS

14.1 INTRODUCTION

Spray drying, also known as "atomization," is a technology that is widely used to encapsulate food ingredients as well as bioactives and functional components. Drying, through water evaporation and the decrease in water activity as well as by the structuring of the matrix component, increases the stability compared to the original suspension and can sustain functional compounds for long storage periods (Guterres, Beck, & Pohlmann, 2009). During the drying process, the entrapment of the active compounds occurs due to the rapid evaporation of the dissolving or dispersing agent, which, in the majority of the cases, is water. In general, the type of encapsulation achieved by atomization is a matrix type, i.e., the core component is finely dispersed in the dispersing/carrier agent that acts as structuring and physical barrier agent (Moran, Yin, Cadwallader, & Padua, 2014).

Encapsulation by spray drying leads to the formation of a very low-moisture-to-dry powder made of particles in the amorphous or crystalline state, providing optimal properties of solubility, protection against degradation by chemical agents and physical factors, stabilization and the controlled/modulated release of bioactive compounds (Ray, Raychaudhuri, & Chakraborty, 2016). The process conditions and the natural, physical, physicochemical and chemical properties of the carrier and core components determine the degree of encapsulation of the bioactive (payload, or encapsulation efficiency), physical properties, technological functionality, and stability of the final powder which needs to be determined to maximize its applicability and storability.

The evaluation of the qualitative and technological properties of the encapsulated powders is important in understanding and optimizing their applicability and maximizing their usage in food and non-food products. This chapter reviews the process efficiency indices as well as the main qualitative (physical, structural) and functional properties of spray dried encapsulates.

14.2 PROCESSING EFFICIENCY

14.2.1 ENCAPSULATION YIELD AND PAYLOAD

In general, the efficiency of the encapsulation process refers to the ratio of the amount of core material encapsulated or entrapped within a shell or matrix compared with the initial ones. However, depending on the nature of the functional component, it could be computed in different ways and referred to different indices as follows:

- Encapsulation efficiency (Eq 14.1)

$$EE(\%) = \frac{(W_t - W_s)}{W_t} \cdot 100 \tag{14.1}$$

- Encapsulation yield (Eq 14.2)

$$EY(\%) = (W_t / W_i) \cdot 100 \tag{14.2}$$

- Payload (Eq 14.3)

$$PY(\%) = (W_t / W_m) \cdot 100 \tag{14.3}$$

where W_t and W_s correspond to the total mass of the functional component and to that on the surface of the encapsulated powder, respectively, while W_i and W_m refer to the mass of the initial/added amount of functional component and the amount of the encapsulated powder, respectively.

In the case of oils, lipid substances and flavor compounds encapsulated as the core component, drying process, and formulation do not determine a full entrapment since they could remain partly adsorbed onto the dried particle surface. In such cases, the encapsulation efficiency also has to consider the fraction that remains in the outer layer of the powder particle (Bhandari, D'Arcy, & Padukka, 1999). In order to estimate the content of the non-encapsulated oil/flavor compound remained onto the particle surface, analysis are carried out on the solvent (in general organic solvent, e.g. hexan) recovered after washing the powder. The total encapsulated oil/flavour is then determined only after the microencapsulates are dissolved or disrupted so to allow the full release of the encapsulated component (Anandharamakrishnan & Padma Ishwarya, 2015).

Encapsulation yield and payload are indices related to the ratio of the entrapped bioactive on its initial content or the mass of the encapsulated powder and both useful for quality control. They are also parameters relevant to the bioavailability of the functional ingredient as at the same quantity; high payloads correspond to a higher release and related effects allowing the use of reduced amounts of powder for a specific purpose.

The payload of microencapsulated particles can range from <1% to approximately 99%, depending on the particle morphology, encapsulation material, and process (Moran et al., 2014). The atomization techniques will generally yield lower payloads (10% to 30%) than other encapsulation techniques and higher loadings can be achieved through the adjustment of the core/shell ratio in the spray drying solution. However, at increasing concentrations of the core material, a decrease of the structuring and protective matrix material occurs. This increases the risk of higher amounts of functional ingredient on or near the surface of the microsphere, which, in turn, can result in a faster core degradation or a release of it into the surrounding system or environment (Oxley, 2014).

The payload could be used as an index of the ease of a bioactive compound to be incorporated in a carrier under specific process conditions and/or affinity between the carrier and core material as well as the film- and structure-forming ability of the carrier material during the drying process. Under the same process conditions, high molecular weight carbohydrates present a higher ability to structure and retain bioactives during spray drying than low molecular weight ones (Bhandari, Datta, & Howes, 1997; Reineccius & Risch, 1995; Roos & Karel, 1991).

The estimation of the process efficiency indices generally requires chemical analyses to be carried out on the encapsulated powder in order to quantify the amount of the core component either fully or partly entrapped in the powder as well as the component that remains adsorbed onto the surface of the powder particles (e.g., oils). In the case of bioactive compounds that are ultraviolet-(UV-) absorbing species (e.g., vitamins and polyphenolic compounds) or that have absorption maximum in the UV-visible wavelengths, the total content in dried microencapsulates can be evaluated by spectrophotometry or a high-performance liquid chromatography (HPLC) technique combined with the UV-detector (Anandharamakrishnan & Padma Ishwarya, 2015). Gas chromatography (GC) techniques are also widely used to determine the content of both volatile and non-volatile components (Reineccius, 2005).

The encapsulation efficiency of probiotic microcapsules is based on the estimation of the viability of probiotic cells before and after spray drying by the use of conventional microbiological analyses (e.g., standard plate count method using the de Man, Rogosa and Sharpe (MRS) agar medium). The method requires the release of probiotic cells from the microencapsulates into the growth medium that could be achieved only by a proper dissolution of the wall material.

14.2.2 MOISTURE AND WATER ACTIVITY

The water content of spray dried powders and encapsulates is a key parameter to evaluate the process efficiency and estimate its stability due to the relationship between moisture and water activity (via the specific sorption isotherms) and the effect on the Tg (glass transition temperature) of the product (via the Gordon-Taylor equation). Moisture can be estimated based on different principles, including, among others, gravimetry, refractometry, titrimetry (Karl Fischer titration) and the electrical conductance method. The gravimetric methods imply the determination of the moisture based on the weight loss after oven drying at 105°C until a constant weight is obtained. In the case of thermo-sensitive compounds, sugar-based carrier materials, or volatile core components (e.g., aroma, essential oils), the use of a vacuum oven is recommended (set at 70°C and operating for 24 h; AOAC, 2006). Alternative and less time-consuming methods are those that use an infrared moisture analyzer at 130°C (Quispe-Condori, Saldaña, & Temelli, 2011) or the Near-Infrared Reflectance (NIR) instrument, both of which require a preliminary calibration with the oven drying method for accurate estimation of moisture content.

Water activity (a_w) is a physicochemical parameter that measures the state of water and its availability or "freedom" in a food system under thermodynamic equilibrium. It is computed as the ratio between the partial vapor pressure of water in the food and the standard state partial vapor pressure of water at the same temperature and under equilibrium conditions; its value (dimensionless, $0 < a_w < 1$)

depends on the specific composition of the food system and, in particular, the presence of polar and hydrophilic compounds able to bound water. Water activity is a successful factor which is commonly used in correlation with food safety and quality (Labuza & Altunakar, 2020). In particular, in food manufacturing, it finds its main application with regard to the microbiological status; as a general rule, at decreasing water activity (i.e., lower availability of free water) a reduction or inhibition of microbial growth and spoilage occurs; it could also affect the rate of other degradative reactions in food (e.g., lipid oxidation, enzymatic reactions, bioactive compounds degradation) even if for these phenomena also other factors seems to be involved in their occurrence (Roos & Drusch, 2015). A food stability map that reports the rate of some degradative reactions (i.e., lipid oxidation, enzymatic and non-enzymatic reactions, microbial growth) as a function of water activity of food has been developed by Labuza, Tannembaum, and Karel (1970).

The success of spray drying in the encapsulation of bioactives and functional ingredients is, thus, highly related to the low moisture content and water activity of the end product (aw <0.15–0.10) and the corresponding enhancement of the storage stability of final products that widens their applicability as ingredients in foods and non-food sectors.

The high microbial stability related to the low water activity of dried encapsulates is, however, opposite to the high risk of lipid oxidation in the case of lipids or lipophilic core compounds such as polyunsaturated fatty acids. High inlet and outlet air temperatures during spray drying could cause a ballooning effect which could favor the mass transfer of the core to the surface of the microencapsulate, leading to autoxidation of the unencapsulated surface oil (Drusch & Berg, 2008). Water activity could be evaluated by either a resistive electrolytic, a capacitance, or a dew point hygrometer. Care must be taken in the case of dried encapsulates with core volatile compounds (flavor, essential oils) as they can interfere with the measurements.

14.3   CHARACTERIZATION OF THE POWDERS

14.3.1   Particle Characterization

Particle characterization is a key element in spray drying encapsulation and underlines the design and development of products with the expected properties, stability, and functionality. It includes the description of primary physical and microstructural properties of food powders and, among others, the most important are particle morphology, particle size and distribution, shape, density, Tg, degree of crystallinity, and color.

14.3.1.1   Physical Properties

In dried encapsulated powders, physical characterization allows the achievement of a detailed understanding of several properties useful to evaluate the impact of the drying process conditions on the macroscopic (e.g., color) and structural (shape and morphology) properties as well as to estimate some technological functionalities (crystallinity, the controlled release behavior) and stability (e.g., glass transition temperature).

14.3.1.1.1   Morphology and Structural Properties

Spray drying process variables and initial feed composition determine the surface morphology and structural properties of dried encapsulates which are indeed crucial for their applications. Although, filled homogenous spheres may form during the spray drying process, many other grain/ microparticle morphologies have been observed, such as doughnut-like or deflated balloons (Lintingre, Lequeux, Talini, & Tsapis, 2016). For the given drying conditions, particles may inflate, distort, shrivel, or fracture depending upon the rheological properties, porosity, or non-porosity of the particle skin or crust formed (Walton, 2000).

In addition to the nature and concentration of the carrier and core components of the feed dispersion, a series of process parameters, such as residence time of particles within the drying chamber,

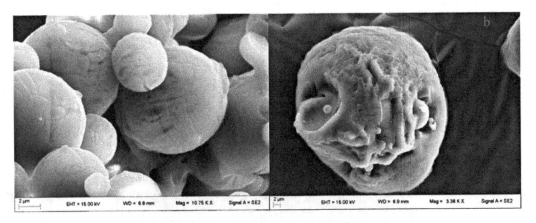

FIGURE 14.1 Microstructure of spray dried trehalose (a) and trehalose + limonene 5000 ppm (b) obtained from initial systems at 40% (w/w) saccharide concentration (Pittia and Faieta, unpublished data).

method and conditions of atomization, type of spray/air contact, drying air temperature, and feed parameters such as concentration, temperature, and degree of feed aeration, may affect the morphology. Too high and too low drying temperatures can cause non-uniform solvent evaporation rate and determine the collapse and internal bubble formation which, in turn, lead to hollowed, shriveled, blistered or, broadly, irregularly shaped powder particles, as shown in Figure 14.1 and Figure 14.2.

The general morphology, particle size, and distribution of encapsulates could be evaluated by a wide array of analytical methods; the most commonly used are the microscopy techniques (Champion, Katare, & Mitragotri, 2007) such as optical light microscopy, electron microscopy (SEM, TEM, ESEM), fluorescence microscopy (Confocal Laser Scanning Microscopy, CLSM), dynamic light scattering (DLS), X-ray photoelectron spectroscopy (XPS), and atomic force microscopy (AFM) (Moran et al., 2014). Depending on the selected microscopic techniques employed, different particle size and morphology details can be qualitatively and/or quantitatively evaluated. Optical microscopy is a highly accessible, low-cost, simple technique that also allows the evaluation of color properties, but it could be used for the morphological characterization for particles with a minimum average diameter of 1–2 μm.

TEM and SEM are high-resolution techniques that allow the study of structural properties of particles in the order of 0.1–1000 μm and 0.02–1000 μm, respectively. High-resolution, 3-D images can be obtained by AFM (particle size range: 0.02–1.2 μm) and CLSM (particle size range 0.1–1000 μm) that, however, require expensive equipment. For a large number of spherical encapsulates, and if the refractive indices of the cell wall and bioactives are available, laser light scanning could represent an interesting technique. Morphological evaluations of dried particles and powders require an adequate selection and, if necessary, a combination of the microscopy techniques, based on the information of interest, the method of sample/specimen preparation, and the suitability of the sample to be evaluated through the use of the specific methodology.

The effect of drying conditions on morphology is still under investigation due to the complexity of the thermal conditions in the drying chamber and their impact on the initial droplet. Simulation approaches such as those applied in the 'single droplet drying' technique have been developed and applied; results obtained contribute to understand the variation related to the effect of the drying parameters on the likely morphology of the dried particles, but results are, in general, difficult to be transferred on the pilot and industrial scale (Walton, 2000).

14.3.1.1.2 Particle Size and Distribution

Particle size and its distribution are among the most important physical characteristics of microencapsulated powders, especially with regard to handling and storage, as they have an impact

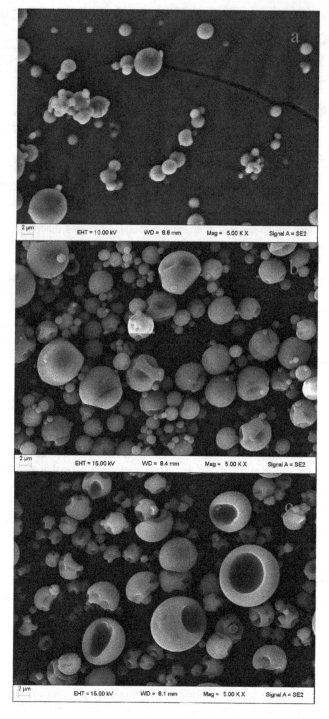

FIGURE 14.2 Microstructure of spray dried 100% trehalose solution (a), 50% trehalose +50% maltodextrin solution (b) and 20% trehalose +80% maltodextrin solution (c). All solutions had initial solid concentration equal to 20% (w/v). Maltodextrin DE 8.5-10 (Faieta and Ppittia, unpublished data).

on flowability, compaction, and/or segregation (Barbosa-Canovas, Ortega-Rivas, Juliano, & Yan, 2005a; Moran et al., 2014). These properties are directly related to their dissolution and/or to the controlled release of the encapsulated core bioactives while, at the macroscopic level, they could affect color, taste, and appearance. As a general rule, a decrease in particle size to the micron scale, as a result of spray drying encapsulation, results in the improved solubility and bioavailability of bioactives, due to the increase in surface area.

Depending on the spray drying process used, particles of different sizes can be obtained. In particular, the average size of powders produced from bench and pilot spray dryer can range from 5–40 μm, while for commercial spray dryers (two stages), it is about 200–400 μm (Woo & Bhandari, 2013). The rheological properties of the feed solution affect the particle size during spray drying, as the higher the viscosity of the initial liquid system, the larger the droplets and particles will be formed during atomization and, subsequently, during the spray drying process (Jinapong, Suphantharika, & Jamnong, 2008).

Moreover, each droplet will have a different thermal story in the drying chamber and the various phenomena occurring during the evaporation process and with different kinetics may lead to a wide heterogeneity of particle size. This highlights the importance of the evaluation of the size distribution of powder particles as the peculiar qualitative property of the final product.

The determination of the average particle size and distribution could be performed by different methods depending on the information of interest and a knowledge of the materials used. As seen for morphology, microscopy methods are widely used to determine particle size, including optical, SEM, TEM, and CSLM. Standard upright optical microscopes could be used for particles with a size equal to or higher than 0.2 μm due to light wavelength restrictions. TEM is capable of resolving structures with smaller dimensions than optical microscopy (nano- to micro-) and could also be used to quantify the shell thickness of (Chiu et al., 2005; Zhang, Law, & Lian, 2010). AFM, with a resolution of approximately 3 nm, provides 3-D images of samples, with particles ranging in size from nano- to centi-meter scale and it is widely used. In recent times, new techniques such as X-ray micro-computed tomography (CT) are increasingly used to image complex 3-D structures with a spatial resolution in the micrometer range (Law, 2007).

Digital images or video images of samples and specimens could be taken from microscopic analyses and could be used for quantitative analysis using image analysis software.

When encapsulates are not spherical, the shape could be defined by a circularity shape parameter, called "Roundness (RN)," which corresponds to a minimum value of unity for a circle and it is defined by the following equation:

$$RN = P^2 / 4\pi A \tag{14.4}$$

where P is the perimeter of single encapsulates and A is their cross-sectional area.

In the alternative, different spherical equivalent diameters could be used depending on the main physical property of the dried particle, such as the hydrodynamic diameter (i.e., the particle diffuses like a spherical particle of that size) or a Stokes diameter (i.e., the particle sediments like a spherical particle of that size) (O'Hagan, Hasapidis, Coder, Helsing, & Pokrajac, 2005).

Visual observation of the particles by microscopy techniques (e.g., SEM, TEM, optical) are generally used to determine particle size despite the various limitations and the need for a large number of specimens and images to be evaluated in order to obtain a reliable average particle size. Thus, to complete the physical characterization of dried encapsulates and their functionality, in addition to size, particle distribution is also performed and the instrumental techniques available could be categorized using the following four methodologies: sieving, microscope counting techniques, sedimentation, and stream scanning. A thorough description of the methods, instruments, and modeling approaches for particle size distribution analysis, along with the *pros* and *cons* of their usage, is reported in various book chapters and reviews (Barbosa-Canovas et al., 2005a; Dodds, 2013).

A sieve analysis is performed by equipment made of a series of screens having different-sized spacings; it allows for the estimation of the particle size distribution by computing the weight fraction of the initial sample present in each of the sieves after standard shaking conditions. The main limitation of this method is the fact that it requires a relatively high amount of sample material.

In the food industry, laser light scattering instruments are an alternative for determining the particle size distribution of samples in the order of micrograms. They determine particle size based on the light scattering intensity, whose entity varies according to the scattering angle, size of particles, refractive indices of particles, and the medium according to the Mie theory of light scattering. Limitations in its use are the difficulty to evaluate particles with large size distribution and the need to have a difference between the particle refractive index and the dispersion medium. Moreover, particles should be spherical, a characteristic that could be difficult to achieve in the case of spray dried particles. Thus, for a better characterization, laser light scattering measurements are sometimes combined with others obtained with different techniques, like the microscopy methods (e.g., OM, SEM, and TEM).

The particle size distribution of particles could be determined based on a number of different units (number, surface, mass or volume, size) and the results are generally graphically described in plots representing the particle size data of the powder (*x*-axis) *vs.* the particle amount or size frequency (*y*-axis). To describe the distribution trend, different average or mean sizes, i.e., measures of central tendency, can be defined, such as the mode, the median, and the mean (Barbosa-Canovas et al., 2005a).

14.3.1.1.3 Density

Density is a physical property that is relevant for dried encapsulates as it complements other particle properties such as the bulk powder structure and the particle size. As a general definition, the density of a particle is defined as its total mass divided by its total volume. Depending on the total volume measured, it can be determined in different ways; in the case of dried particles, for example, the *apparent* particle density, or the *effective* (or aerodynamic) particle density are generally determined. The *apparent* particle density is defined as the mass of a particle divided by its volume, excluding only the open pores, and is evaluated by gas or liquid displacement methods and tools such as the air (other gas, e.g., helium) or liquid pycnometry. In the latter cases, special solvents that do not dissolve, react or penetrate the particulate food solid have to be used (e.g., toluene) (Barbosa-Canovas, Ortega-Rivas, Juliano, & Yan, 2005b; Regiane Victória de, Borges, Botrel, & de, 2014).

Effective or bulk particle density is based on the average density within an aerodynamic envelope around it, including any open or closed pores in the specific volume. In the case of coarse solids, mercury porosimeter is the preferred method to measure the volume of the open pores, but the equipment is very expensive. Alternative methods include the bed voidage method, the bed pressure drop method, and the sand displacement method (Barbosa-Canovas et al., 2005a). More simple methods are based on the evaluation of the weight of the powder gently loaded in a tared graduated cylinder followed by the calculation of the bulk density (ρ_d) according to the relationship: mass/volume (Jinapong et al., 2008). In addition, the bulk tapped density could also be calculated; for this purpose, a certain amount of powder is initially freely loaded into a glass graduated cylinder and the samples repeatedly tapped manually by lifting and dropping the cylinder under its own weight at a vertical distance until a negligible difference in volume between succeeding measurements is observed. Given the mass (m) and the apparent (tapped) volume (V) of the powder, the powder bulk density is computed as m/V (gcm^{-3}) (Barbosa-Canovas et al., 2005a; Goula & Adamopoulos, 2008).

From the ratio of particle density (ρp) to bulk tapped density (ρb), the porosity index (ε_b) can be calculated using the following equation:

$$\varepsilon_b = \left(1 - \rho_b / \rho_p\right) \cdot 100$$

$$(14.5)$$

where ρ_b is the bulk tapped density of microencapsulated powder and ρ_p is the particle density of the microencapsulated powder (Krokida & Maroulis, 1997).

14.3.1.1.4   Mechanical Strength

The mechanical strength of the encapsulated dried powders is an important qualitative criterion as it could influence the stability and release patterns of the products. The evaluation of these properties as a function of shell thickness and composition allows an efficient design of a stable delivery system (Gharsallaoui, Roudaut, Chambin, Voilley, & Saurel, 2007). Various methods could be used to assess mechanical shell properties, which could be performed either on macroscopic surfaces or on the microcapsule itself. Bubble or droplet tensiometry methods, which evaluate the dilatational elastic moduli of the shell, or surface shear methods, that probe the surface shear storage and loss moduli, are applied on macroscopic surfaces, mimicking the surface of encapsulates.

Mechanical properties of encapsulating particles can be evaluated by both indirect and direct methods. The former ones determine the resistance to fluid shear force (Peirone, Delaney, Kwiecin, Fletch, & Chang, 1998), to agitation in presence of glass beads for a certain time (Leblond, Tessier, & Hallé, 1996), or to bubble disengagement in a bubble column (Lu, Gray, & Thompson, 1992). They evaluate, in general, the damages caused by physical stress, and the results depend not only on the specific mechanical resistance of the encapsulated particle, but also on other factors (e.g., hydrodynamics of the process equipment). Alternative methods include osmotic pressure methods, or thermal expansion methods (Zhang et al., 2010).

Direct methods are based on compression methodologies like those performed on a layer of encapsulates or a single particle between two glass plates (Ohtsubo, Tsuda, & Tsuji, 1991) or by using uniaxial compression. In all cases, a series of limitations of their applicability exist related to the particle size of dried encapsulates, and the information obtained could be not fully reliable. As an alternative and to overcome size limitations, the micro-particulation technique has been developed and has been used to evaluate the mechanical strength of dried encapsulates of different size and composition (Zhang et al., 2010; Zhang, Saunders, & Thomas, 1999). This method is based on the compression of a single encapsulation between a probe and a glass plate while the probe is connected to a force transducer, to which the data acquisition system is linked.

AFM probe can also be used to determine the mechanical strength and elastic properties of dried particles with the limitation of the expensive instrumentation required (Sagis, 2015; Zhang et al., 2010).

14.3.1.1.5   Glass Transition Temperature and Degree of Crystallinity

The Tg and the degree of crystallinity of wall materials are physical properties that characterize the matrix structure of dried encapsulates and influence their qualitative and technological functionalities (e.g., flowability, dispersibility) as well as their stability. The knowledge of these properties contributes to the design of encapsulates with specific functionalities and applicability under specific environmental and/or system conditions.

The glass transition is defined as the transformation of a supercooled liquid (rubbery) to a highly viscous, solid-like amorphous, or glassy state (Roos, 2020). This occurs over a temperature range as the molecules reduce their translational mobility, become immobile, and exhibit only rotational motion and vibrations. Tg is, thus, the temperature at which the phase transition occurs and its value is governed by both the molecular complexity of the component or matrix and the water content (Slade, Levine, & Reid, 1991). In encapsulation technology by spray drying, the Tg concept is of main relevance as the majority of the carrier/wall materials (e.g., high and low molecular weight carbohydrates, hydrocolloids, proteins) exhibit a Tg, and the rapid water removal occurring during drying determines the transition from a liquid or super-cooled (rubbery) state to a glassy or amorphous solid one. An amorphous solid is characterized by a microstructure consisting of short-range order and regions of high and low densities with a higher entropy than the corresponding crystals. From a thermodynamic point of view, this condition corresponds to a non-equilibrium state; thus, its structure can undergo crystallization or structural relaxation to achieve an equilibrium condition (Yu, 2001) that could be triggered by either the increase of temperature above the Tg or the addition of plasticizers, in particular water, that decrease the Tg of the system below the process or environmental temperature. In an amorphous state, molecules are also disordered, more open and porous, and possess more sites for external interactions with other molecules, and this leads to a higher hygroscopicity.

Furthermore, in the amorphous state, translational mobility is hindered and only the rotational and vibrational ones occur; hence, encapsulates are recognized as a stable system, able to protect the core components to a large extent. The degradation of dried particles can occur very slowly and is related to the diffusion of small molecules (e.g., volatile aroma compounds, oxygen) through the porosity of the amorphous system led by diffusive phenomena. Above the Tg, the material is in the rubbery, plastic state, and, consequently, molecules diffusion can occur at a rate that depends on the difference between the actual temperature and the Tg. Under these conditions, several physical and structural processes are triggered (e.g., crystallization, caking, sticking) that can impair the qualitative properties of the encapsulated powder as well as decrease the quality and functionality of the core component. Thus, the information about the Tg can help in determining the storage conditions and shelf-life stability of an encapsulate.

The Tg of a pure compound (also referred to as Tg_{dry}) depends on its nature and molecular weight and, as a general rule, as higher is the molecular weight and complexity, as higher is the Tg. Moreover, due to the plasticizing effect of water, Tg decreases at a higher water content of the system with a trend described by the Gordon-Taylor equation (Roos, 2020; Roos & Drusch, 2015). Besides water, also other small compounds (e.g., small saccharides, polyols) can act as a plasticizer and induce similar effects. In spray drying systems, for instance, the addition of a different amount of trehalose to maltodextrin to encapsulate *Spirulina platensis* extracts determined a decrease of the Tg_{dry} of the pure maltodextrin from ca. 190°C to 181°C (20% w/w) and 132°C (50% w/w) (Faieta, Corradini, Michele, Ludescher, & Pittia, 2020).

To increase the stability of encapsulates, the use of carrier materials with high Tg_{dry} values is recommended either as a single component or in the mix with others (e.g., maltodextrin, starch, trehalose) to limit the risk to reach the moisture conditions that would decrease the Tg of the dried encapsulated powder as low as the ambient temperature at which stickiness and aggregation phenomena would occur or even trigger the glass to rubbery state transition with consequent release of the core component.

To determine the Tg of wall materials and dried encapsulates, different instrumental techniques can be used. At Tg many of the physical properties of the material change dramatically, including the increase in heat capacity, and/or kinetically- hindered and diffusion-limited reactions and processes, like the release of the volatile compound. Thanks to this, thermo-analytical techniques are used and among them, the most used ones are the Differential Scanning Calorimetry (DSC) and Thermo-Gravimetric Analysis (TGA).

By DSC, *Tg* can be determined by the measurement of the temperature at which a sudden increase in heat capacity occurs (Bhandari & Howes, 1999). Moreover, when a sample is heated at a constant increasing temperature rate, DSC provides besides the Tg, thermograms from which additional information (temperature, enthalpy) on first-order, heat-induced processes like melting and solidification of the components of the system can be retrieved. Thermograms can be also exploited to determine crystallization profiles, i.e., the temperature at which the crystals start to form and the degree of crystallization (Saadi et al., 2012). DSC analysis allows also to evaluate the chemical interactions between the core and carrier material that could be evidenced by variations in the Tg or the enthalpy and/or peak temperature of exothermic and endothermic reactions.

Dynamic Mechanical Analysis (DMA) and Thermo-Mechanical Analysis (TMA) are also used to determine Tg; both measure changes occurring in the mechanical and physical properties (expansion, rheological parameters) of the materials while they are heated (or cooled).

To evaluate the degree of crystallinity of dried encapsulates X-ray diffraction (XRD) and nuclear magnetic resonance (NMR) are techniques commonly used for this purpose. Results from these instrumental techniques can complement those obtained from DSC and provide an overview of the physical properties and stability of the encapsulated powders.

14.3.1.1.6 Color

Color is a sensory perceived property derived from the reflected or transmitted spectrum of light interacting with the human eye. The color of an encapsulated powder is determined by its physical properties; it has its main importance as quality attribute as it could affect acceptability by the consumers of the compound itself or of the product in which the encapsulated powder is added.

The colorimetric properties of an encapsulated powder depend on various factors, including the initial color of the core (bioactive, functional ingredient) and carrier materials due to the naturally present pigments; the core/carrier weight ratio in the colloidal or dispersion system before spray drying; the thermal degradation of the pigments and/or the formation of new colored compounds by heat-induced reactions (e.g., Maillard reaction, lipid oxidation) during drying.

In general, when colored functional components are taken into account, the nature and concentration of the carrier affect the final color of the powder due to the entrapment effect occurring during the spray drying. For example, a higher concentration of maltodextrin determined a decrease of the colorimetric parameters of the initial persimmon pulp feed solution and the final product significantly as maltodextrin is usually a bright white compound in its natural form (Du et al., 2014). Moreover, different carrier compounds can affect the entrapment degree of colored active ingredients and, in turn, affect the final color of the powders as observed in some studies (Comunian et al., 2011; Otálora, Carriazo, Osorio, & Nazareno, 2018; Tupuna et al., 2018; Xu et al., 2018).

While sensory tests can be performed on the encapsulated powders by trained panelists to obtain a general visual appearance, objective instrumental evaluation of color properties of powders is in general carried out by spectrophotometers, spectrocolorimeters, and tristimulus colorimeters taking into account that, due to the opaque nature of powder, only reflectance measurements can be used. Under this condition, instruments measure and quantify the fraction of an incident standard light that is reflected from the powder surface, and how reflection occurs is determined by the physical and chemical properties of the sample. In particular, the presence of chromophores causes the specific absorption or scattering of certain wavelengths of light, leaving the remaining of the incident light to travel onwards to the 'observer'. Moreover, the physical properties of the system could also prevent light from interacting with the chromophores or cause it to be trapped inside the sample (Joshi & Brimelow, 2002).

A tristimulus colorimeter comprises a source of illumination (standard illuminant or observer), a combination of three filters to modify the energy distribution of the incident/reflected light, and one photoelectric detector that converts the reflected light into an electrical output. The filters correspond to the three primary colors in the spectrum (red, green, and blue) and could be combined to match most colors. The computer or the interface connected to the colorimeter performs data conversion between CIE and other color scale systems or between different standard white light sources (Clydesdale & Ahmed, 1978; Hunter & Harold, 1987; Joshi & Brimelow, 2002). The most used color scale systems are the XYZ, the HunterLab Y, x, y, and the CIELAB L*a*b* that represent colors in a three-dimensional space. The CIELAB colorimetric space, in particular, includes two-color coordinates, a* (red/green chromaticity) and b* (yellow/blue chromaticity), as well as a psychometric index of lightness, L* (lightness, luminosity), and the corresponding colorimetric parameters like the hue angle (h°), chroma (C*).

Hue angle (h°) is a qualitative attribute of color, used to define the difference of a certain color with reference to grey color with the same lightness. This attribute is related to the differences in absorbance at different wavelengths and could be computed by the following equation.

$$h^\circ = \tan^{-1}\left(b^* / a^*\right) \tag{14.6}$$

Angle values of 0° or 360° correspond to red hue, whilst angles of 90°, 180°, and 270° represent yellow, green, and blue hues, respectively.

Chroma (C*), is a quantitative attribute of colorfulness or intensity of the color and is calculated by the following equation.

$$C^* = \sqrt{a^{*2} + b^{*2}} \tag{14.7}$$

Other chromatic indices could be also used and obtained from the main colorimetric parameters like the Whiteness index (WI) (Equation 14.8) and the Yellowing index (YI) (Equation 14.9):

$$WI = 100 - \left[\left(100 - L^*\right)^2 + \left(a^*\right)^2 + \left(b^*\right)^2\right]^{1/2} \tag{14.8}$$

$$YI = 142.86 \cdot b^* / L^* \tag{14.9}$$

Measurements made on a tristimulus colorimeter are comparative and the use of calibrated standards is required.

Color changes due to processing, formulation, composition, or storage can be determined as the modulus of the distance vector between the initial color values and the actual color coordinates (Pathare, Opara, & Al-Said, 2013) and various indices can be obtained, such as the total color change (ΔE), which takes into account the sum of the changes of the three colorimetric parameters (L*, a*, b*).

In contrast to tristimulus colorimeters, reflectance spectrocolorimeters measure the whole spectrum of visible light reflected from a sample (range: 380–700 nm) via an integrating sphere and a diffraction grating; the results are expressed as the ratio between the reflected light from the sample and that from a known reference standard. A spectrophotometer collects all surface-reflected light across the visible wavelength range at the desired wavelength intervals (e.g., 10 nm or 20 nm) and is graphically analyzed; the result is reported as a reflectance curve that is what the perceived color should be (Joshi & Brimelow, 2002).

Currently, there is good availability of both bench-top and portable models facilitating both lab analysis and near-line studies. In recent times, new advancements have been made in the colorimetric tools to evaluate color in food matrices with the intention of sorting out some limits of the spectrophotometric and colorimetric instruments as in the case of non-uniform color samples. The technology of calibrated color imaging analysis of a food is based on the use of digital cameras that collect images in digital format without film processing. The digital images can be easily processed, duplicated, modified, or transmitted and the camera *RGB* signals are then correlated with a series of color patches in a standardized reference chart (Hutchings, Luo, & Ji, 2002).

As aforementioned, the colorimetric data obtained will reflect the physical and compositional properties of the dried encapsulates, and among the various factors that could influence the results the size, which should be as uniform as possible to avoid scattering effects, the surface properties of the particles, as well as the presence of superficial translucent compounds onto the outer layer of the particles (e.g. not encapsulated oil), are included. Water content can also influence the amount of light reflected by a sample. In general, a dried material appears opaque, without or with a limited chromaticity due to the presence of air in the intercellular and intracellular spaces which causes the diffusion of light inside a material at interphases of different refraction indexes (Saarela, Heikkinen, Fabritius, Haapala, & Myllylä, 2008). When moisture content increases (e.g., due to moisture uptake during storage or processing), the samples recover transparency and chromaticity (Buera, Farroni, & Agudelo-Laverde, 2015).

For quality control purposes, the effect of the spray drying on the natural pigments present in the infeed system could be evaluated as the index of the heat damage induced by the process conditions. To this aim, the evaluation of the differences between the color of the dispersion before spray drying and that of the dissolved encapsulated powder at a similar solute concentration could be performed via both colorimetric methods or spectrophotometric analysis in the visible region (Faieta et al., 2020; Lao & Giusti, 2017; Zhang et al., 2020).

14.4 TECHNOLOGICAL FUNCTIONALITY OF POWDERS

In this section, a series of post-drying properties of encapsulated powders will be briefly reviewed, taking into account their potential usage and application in food and non-food products.

14.4.1 Flowability

The flowability of a powder is an important property in handling and processing operations, such as transportation, conveying, mixing, formulation, compression, packaging, and storage (Ortega-Rivas, 2005). Flowability depends on the morphological properties, particle size and distribution, surface and particle chemical composition. The presence of water has a significantly influence on flowability and caking properties. In general, at higher moisture content a more cohesive matrix can originate, decreasing the flowability of the powder. Furthermore, the majority of food powders, including dried encapsulates, present a critical value of relative humidity (at a given temperature) above which the powder will cake, and this impairs the powder flowability.

Several empirical tests are available to assess flowability, including those based on the measurement of angles of repose (Teunou, Vasseur, & Krawczyk, 1995), Hausner ratio, Carr indices (Peleg, 1977), flow through a defined opening, and compression testing (Carr, 1965), which could be useful in quality control, where a change in a measured value may be indicative of a change in the flow behavior of a given material (Fitzpatrick, 2005). More commonly flowability is measured by the Jenike shear test, which is considered the most accurate method (Fitzpatrick, 2005). This method allows producing failure properties that can be used for quantitative design purposes and for qualitative comparisons of powders (Teunou, Fitzpatrick, & Synnott, 1999) by the flow index value that allow the classification of the powders as hardened, very cohesive, cohesive, easy flow, free-flowing. Thanks to recent instrumental advances in physical techniques, shear and flow cells as an accessory for rheometers have been recently designed and developed, which allow for accurate powder analysis under relative humidity (RH) and temperature-controlled conditions.

14.4.2 Reconstitution and Dissolution Properties

Reconstitution represents an important property of dried encapsulates related to the mechanism of release of the encapsulated/core ingredient, and thus, its functionality and/or bioaccessibility upon consumption.

Reconstitution or dissolution of food powder develops via four steps occurring in sequence but, depending on the composition and physical properties of the particles, may also overlap: (i) wetting of powder particles; (ii) particles sinking into solution; (iii) particles dispersing evenly in solution; and (iv) solid particles dissolving completely. Based on this processing mechanism, different reconstitution properties of a food powder could be evidenced, in particular, wettability, sinkability, dispersibility, and solubility, respectively.

The overall dissolution of a food powder or a dried encapsulate depends on both the nature of the carrier/matrix component and its affinity for the solvent which, in the case of foods, is mostly water or oil, as well as their physical properties. The solubility and wetting behaviors depend on the particle size, structural properties (e.g., amorphous vs. crystalline state, presence of pores), and composition of encapsulates. Moreover, during spray drying, the formation of insoluble materials in the particle could occur and the rate of formation depends largely on the temperature and moisture content profiles of the droplet in the drying chamber (Chen & Patel, 2008).

To evaluate the reconstitution properties of encapsulates, the solubility index, along with the wettability, is generally evaluated to obtain information about the powder's behavior after manufacturing or during reconstitution in the aqueous phase. These evaluations are, in general, performed with rather simple laboratory methods. In particular, solubility testing methods are based on the determination of the ratio of the weight of dissolved dried sample with regard to that of the initial powder, after dissolution, stirring, and centrifugation to remove the undissolved fraction (e.g., Cano-Chauca, Stringheta, Ramos, & Cal-Vidal, 2005; Sarabandi, Peighambardoust, Sadeghi Mahoonak, & Samaei, 2018).

Wettability is similarly evaluated with even more empirical approaches and based on the rate or time required by a certain amount of powder particles to submerge in distilled water at a specific temperature (Fuchs et al., 2006; Sarabandi et al., 2018). In recent times, advanced techniques such as laser diffraction have been used to evaluate wettability with higher reliability of the results (Murrieta-Pazos et al., 2012).

14.4.3 Water Sorption Isotherms

In general, a water (ads-, de-)sorption isotherm is the graphical representation of the thermodynamic equilibrium between moisture content and the water activity at a given temperature and pressure of a compound, ingredient, or food and it is used to describe the water sorption behavior and estimate the water-matrix affinity (Iglesias & Chirife, 1982; Labuza & Altunakar, 2020). Depending on the initial sample, adsorption isotherms are determined from the dried material while the desorption one is determined from the fresh, moisty, and/or moistened ones.

Each encapsulated powder, depending on its physical and chemical properties, is characterized by specific sorption isotherms whose determination is a key element to estimate general effects on the mass transfer of active components in the particles, to evaluate solute diffusivity of matrix materials and changes of the quality of encapsulates during storage and transportation (Zhang et al., 2010. Moreover, desorption isotherms could also be helpful in the design and development of the drying process parameters to determine the drying behavior of the infeed system (Anandharamakrishnan & Padma Ishwarya, 2015).

Sorption isotherms could be obtained by both static and dynamic methodologies. In static methods, samples in a hermetic and closed environment are kept in contact with air at a constant RH and temperature until equilibrium between the sample and air is reached (Caballero-Cerón, Guerrero-Beltrán, Mújica-Paz, Torres, & Welti-Chanes, 2015). In this group, hygrometric, manometric, and gravimetric methods are included (Iglesias & Chirife, 1982).

In dynamic methods, samples are placed on a microbalance and an airstream at controlled RH and constant temperature passes continuously over them (Caballero-Cerón et al., 2015). These methods reduce the determination time (from days to hours) as constant exposure to airflow and a small amount of sample make it possible to measure points at high a_w levels without the risk of any microbial growth or physicochemical changes in the powder (Domian, Brynda-Kopytowska, Cieśla, & Górska, 2018; Maher, Auty, Roos, Zychowski, & Fenelon, 2015; Rahman & Al-Belushi, 2006; Spackman & Schmidt, 2010). For this method, one sample is needed for the complete sorption isotherm, weight is automatically recorded, and sample handling is reduced (Rahman & Al-Belushi, 2006; Yu, Martin, & Schmidt, 2008).

Different mathematical models (e.g., BET,[1] GAB,[2] Langmuir, Peleg equation) could be applied to the water sorption isotherm data to determine the main physicochemical parameters of the water-matrix interactions occurring in encapsulates like the monolayer moisture value, an index that is related to the stability of many foods systems. By combining the data on the Tg as a function of water content (i.e., the iso-viscosity curve) with those of the water adsorption isotherm where the moisture content is reported as a function of the Relative Humidity (or a_w), it is possible to obtain a graph that allows a determination of the water-related critical conditions (i.e., critical water content and corresponding critical water activity) for stability, corresponding to those at which the Tg coincides with the storage temperature (Haque & Roos, 2004). For instance, in a study carried out on a microencapsulated paprika oleoresin made of soy protein isolates and maltodextrins the critical a_w value for all studied systems was close to 0.6, and the critical water content was around $11\%_{db}$ (Porras-Saavedra et al., 2019).

14.4.4 HYGROSCOPICITY

Hygroscopicity is another index of the water-dried encapsulates interactions and could be simply estimated based on the moisture uptake of a sample of the encapsulated powder (e.g., 1 g) under a specific water activity or Relative Humidity (e.g., 75%) at 25°C at a fixed time or until a constant weight is obtained (Fritzen-Freire et al., 2012).

14.4.5 MASS TRANSFER PROPERTIES

Mass transfer is an intrinsic key element of dried encapsulates associated with both the intrinsic nature and the functionality of the dried particles. It is expected that the core component remains entrapped and immobile in the dried particle matrix and its mass transfer hindered until one or more desired changes induced by chemical and physical factors trigger its mobility and diffusion, initially in the carrier matrix and then in the specific environment (e.g., food, human mouth, stomach, gut). Various physical, structural, and mechanical properties could affect, favor, or hinder the mass transfer of the core, the encapsulated component of the dry powder particles, and several methods are used to characterize the dynamic mass transfer process of encapsulates.

In dried encapsulates, different trends of the core component release and mass transfer could be found depending on the mechanism of encapsulation, the nature of the carrier, and the physical properties of the powder. A burst release is typical of highly soluble carriers or broken particles. The prompt release might be triggered by changes in the environmental conditions, e.g., pH change, the addition of enzymes, and an increase in the level of water content.

In general, methods measure the release of the bioactive or functional component from encapsulates dispersed in an aqueous liquid media at fixed temperatures by applying agitation to assure the good mixing and suspension of the dispersed encapsulates (Zhang et al., 2010). The variation of the bioactive/active component concentration in the liquid phase as a function of time allows to determine the release kinetics and corresponding diffusion coefficients. Different instrumental (e.g., GC–MS,[3] HPLC,[4] spectrophotometry), analytical, or microbiological methods are used depending on the nature of the component of interest. Gas chromatographic techniques also allow to determine the release kinetics of volatile flavour compounds in the vapor/gas phase from both liquid and solid systems (G Lian & Astill, 2002).

Overall, these methods could be applied when a medium-to-low rate of dissolution of the carrier encapsulant is estimated; as a progressively increasing trend of the bioactive/functional ingredient concentration in the dispersing solvent will occur, data could be properly modeled and the kinetics of the release determined. If the carrier is too soluble in the solvent (e.g., pure water or dilute solutions), like low- to medium-molecular-weight carbohydrates, the risk of these approaches is to observe an excessively quick release, detected as a steep change of concentration of the bioactive that does not allow to determine its kinetic behavior..

In this case, alternative "dissolving" systems mimicking the real conditions (e.g., food systems) in terms of water content and availability (e.g., water activity) could be developed. They could better control or modulate the water mass transfer to the carrier of the microcapsule and induce a controlled release, depending on the effects that the moisture can induce on the physical properties (in particular, Tg), the water activity, and the related effects on the mobility of the system. In the case of core colored components, the dissolving release systems could be made of transparent gels or white powdered systems (e.g., starch, flours) preliminarily equilibrated at the desired equilibrium relative humidity (ERH[5]%), in which the dried encapsulated powder is properly mixed and the release evaluated by colorimetric measurements (da Silva Felipe et al., 2013).

For encapsulated volatile aroma compounds, the kinetics of the release could also be performed by evaluating the mass transfer in the air or headspace of a closed container (vial, bottle, etc.) where the encapsulated sample is inserted (Lopes et al., 2019). Flavor release from the mouth and into the nasal headspace during eating could be evaluated by some advanced instrumental techniques such as the atmospheric pressure chemical ionization or proton-transfer-reaction mass spectroscopy (APCI-MS or PTR-MS) that could measure the real-time release of volatile compounds. Analyses are carried out on the exhaled air from the nose gently sucked into the mass spectrometer, where the concentration of the volatile compounds is continuously detected as protonated ions. This method also offers the possibility to relate the real-time release profile to sensory properties despite the need for the complex mathematical models to obtain the mass transfer properties of encapsulates (Guoping Lian, Malone, Homan, & Norton, 2004).

Temperature is an important factor for the mass transfer processes and needs to be properly defined when release properties are investigated as it affects, both directly and indirectly, the kinetics of the mass transfer. For the release and stability studies of encapsulated materials, two or more temperatures can be selected, as in the Accelerated Shelf-Life Test (ASLT) approaches, and results could be modeled to obtain additional kinetic parameters. However, for ASLT, the selection of the temperatures is crucial as the selection of temperatures, both above and below the Tg of the system, can lead to the evaluation of the mass transfer or release kinetics from systems characterized by a different physical state, i.e., one related to the amorphous state (when the temperature is < Tg) and the other to the rubbery system (when the temperature is > Tg).

Mass transfer properties and kinetic parameters of encapsulated powders could be determined by fitting experimental data with reliable mathematical models that allow the prediction of the release kinetics. The zero-order kinetic release is determined when over-saturated amounts of bioactive are present in the core or with micro-capsules where the thin shell is the rate-limiting step. First- or multiple-order release is common for the matrix type of encapsulates and release due to other interactions (Zhang et al., 2010).

14.5 STABILITY

The stability of spray dried micro- and nano-encapsulates is related to the changes in the physical and structural, physicochemical, and chemical properties that can cause, in turn, qualitative functionality changes such as the retention ability or barrier efficiency of the encapsulates triggering undesired release of the core component, flowability, and density, and/or chemical degradation of the core components due to oxidative or other chemical reactions.

Food powders are highly hygroscopic and susceptible to absorbing moisture. A difference between the ERH % of dried encapsulates (usually <15%) and that of the atmosphere surrounding the product (i.e., the ambient, environment, or in a package) represents the driving force for moisture sorption. While diffusivity of water vapor into a bulk powder is very low, it occurs very quickly when an intimate contact between the dried particles and the air is favored (i.e., during manipulation or mixing). Moisture uptake affects the physical and thermal properties of the powder and, in particular, the Tg and, depending on the decrease of the Tg, this can cause undesired changes (like stickiness and caking) to occur and thus, the flowability, or even induce the glass-to-rubbery state transition.

Stickiness occurs easily in encapsulates containing whey, lactose, protein hydrolysate, high fat-containing milk, and sugar-rich products. It happens when individual particles or solids of a free-flowing powder stick to one another to ultimately form a mass of solids (Roudaut, 2020). It can be described in terms of cohesion (particle-particle stickiness) and adhesion (particle–wall surface stickiness) phenomena (Papadakis & Bahu, 1992). In amorphous dried encapsulates, adhesion and cohesion exist when the particle surface has a viscosity $<10^8$ Pa·s due to water and/or thermal plasticization determine a decrease of the Tg in the outer layer of the particle (Boonyai, Bhandari, & Howes, 2004). The occurrence of powder stickiness is characterized by the determination of its sticky point, defined as the temperature at which the power needed to stir the powder in a tube increases sharply. The mechanisms and driving forces that lead dried particles to stick along with the testing methods developed to measure the stickiness of powders at different moisture and temperature conditions which can provide the conditions where the stickiness can occur are described and reviewed by some authors (Bhandari, 2013; Boonyai et al., 2004; Roudaut, 2020).

Caking is a deleterious phenomenon in dried powders as it determines the formation of aggregates in rock-hard lumps of variable size in which the particles are unable of independent translations and under extreme conditions, a solidification of the whole powder mass can occur (Peleg, 1983). Caking is known to occur during drying or storage and the key parameter of the phenomenon are again related to a critical hydration level reached by water sorption, accidental wetting, or moisture condensation. This causes the particle's surface to become plasticized with a lower viscosity (as in the case of the stickiness), favoring the merging with the neighboring particles through an interparticle liquid bridging (Roudaut, 2020). During the drying process, the dried particles can cake if the temperature is > Tg; to avoid this undesirable event, the product is cooled immediately to an appropriate temperature before packaging (Bhandari, Bansal, Zhang, & Schuck, 2013). In addition to moisture adsorption or accidental wetting, additional causes for caking are the melting of lipids or amorphous sugars at the particle surface due to elevated temperature, the release of absorbed water from the crystallization of amorphous sugars, chemical reactions that produce liquids, and, finally, water immigration from excessive liquid ingredients in the powder (Barbosa-Canovas et al., 2005a; Peleg, 1983; Roos & Drusch, 2015).

Different instrumental methods can be used to determine the critical conditions of caking in dried particles and encapsulates (Aguilera, del, & Karel, 1995; Bhandari, 2007).

The crystallization of powders is another undesired change that could occur in matrices that contain small saccharides (e.g., sugars, lactose) and which is caused by the increase of the molecular mobility (both rotational and translational) occurring when the dried particle has a Tg < the ambient temperature caused by hydration or temperature increase. During crystallization, water is excluded from the crystals and this further plasticizes the system, inducing a "crystallization front" within the system. The water diffusion to the amorphous surrounding matrix can also favor further chemical degradations (Roudaut, 2020).

To achieve an adequate estimate of the stability of dried encapsulates, what is required is the combination of the information on the physical properties and composition of the dried particles and those relative to the environment (storage, formulation, processing) that could trigger the undesired degradative reactions, processes, or phenomena, including the release of the encapsulated bioactive components.

At the laboratory level, the estimation of dried encapsulated powders stability, in general, implies the exposure of samples to stressful environmental conditions such as high temperatures, relative humidity, oxygen concentrations by adopting accelerated shelf-life test approaches.

14.6 CONCLUSIONS

In this chapter, the main process efficiency indices, as well as the main qualitative (physical, structural) and techno-functional properties of spray dried encapsulates, along with the main methodologies and instrumental techniques available for their determination or estimation, have been reviewed. It is highlighted that for some qualitative properties, the evaluations are still based on simple methodologies and, thus, the need for new methodologies or instrumental techniques that could better determine the attributes and functionality of dried encapsulates. The development of advanced instrumental techniques in the future will contribute to enhance the knowledge about the encapsulation of bioactives and functional ingredients by spray drying and the corresponding characteristics of the final products. This, however, highlights the need for specialized trained staff with the proper skills and competencies will perform the analyses and interpret the results.

NOTES

1 Brunauer–Emmett–Teller
2 Guggenheim-Anderson-Boer
3 Gas chromatography–mass spectrometry
4 High Performance Liquid Chromatography
5 Equilibrium Relative Humidity

REFERENCES

Aguilera, JoséM, JoséM del Valle, and Marcus Karel. 1995. "Caking Phenomena in Amorphous Food Powders." *Trends in Food Science & Technology*. doi: 10.1016/S0924-2244(00)89023-8

Anandharamakrishnan, C., and S. Padma Ishwarya. 2015. *Spray Drying Techniques for Food Ingredient Encapsulation. Spray Drying Techniques for Food Ingredient Encapsulation*. Chichester, UK: John Wiley & Sons, Inc. doi: 10.1002/9781118863985.

AOAC. 2006. *Official Methods of Analysis of the Association of Official Analytical Chemists*, 18th ed. Gaithersburg, MD: Association of Official Analytical Chemists.

Barbosa-Canovas, Gustavo V., Enrique Ortega-Rivas, Pablo Juliano, and Hong Yan. 2005a. *Food Powders: Physical Properties, Processing, and Functionality. Food Engineering Series*. New York, NY: Kluwer Academic/Plenum Publishers.

Barbosa-Canovas, Gustavo V., Enrique Ortega-Rivas, Pablo Juliano, and Hong Yan. 2005b. "Particle Properties." In *Food Powders: Physical Properties, Processing, and Functionality*, edited by Gustavo V. Barbosa-Canovas, Enrique Ortega-Rivas, Pablo Juliano, and Hong Yan, 19–54. New York, NY: Kluwer Academic/Plenum Publishers.

Bhandari, Bhesh. 2007. "Stickiness and Caking in Food Preservation." In *Handbook of Food Preservation*, edited by Mohammad Shafiur Rahman, 2nd ed., 387–402. New York, NY: Taylor & Francis.

Bhandari, Bhesh. 2013. "Introduction to Food Powders." In *Handbook of Food Powders: Processes and Properties*, edited by Bhesh Bhandari, Nidhi Bansal, Min Zhang, and Pierre Schuck, 1st ed., 1–25. Cambridge, UK: Woodhead Publishing. doi: 10.1533/9780857098672.1.

Bhandari, Bhesh, Nidhi Bansal, Min Zhang, and Pierre Schuck. 2013. *Handbook of Food Powders: Processes and Properties*.Cambridge, UK: Woodhead Publishing Limited.

Bhandari, Bhesh, Bruce R. D'Arcy, and Indra Padukka. 1999. "Encapsulation of Lemon Oil by Paste Method Using β-Cyclodextrin: Encapsulation Efficiency and Profile of Oil Volatiles." *Journal of Agricultural and Food Chemistry*. doi: 10.1021/jf9902503.

Bhandari, Bhesh, Nivedita Datta, and Tony Howes. 1997. "Problems Associated with Spray Drying of Sugar-Rich Foods." *Drying Technology*. doi: 10.1080/07373939708917253.

Bhandari, Bhesh, and T. Howes. 1999. "Implication of Glass Transition for the Drying and Stability of Dried Foods." *Journal of Food Engineering*. doi: 10.1016/S0260-8774(99)00039-4.

Boonyai, Pilairuk, Bhesh Bhandari, and Tony Howes. 2004. "Stickiness Measurement Techniques for Food Powders: A Review." *Powder Technology* 145(1). Elsevier: 34–46.

Buera, M. P., A. E. Farroni, and L. M. Agudelo-Laverde. 2015. "Water and Food Appearance." In *Water Stress in Biological, Chemical, Pharmaceutical and Food Systems. Food Engineering Series*, edited by Gutierrez-Lopez Gustavo F., Alamilla-Beltran Liliana, M. Pilar Buera, Welti-Chanes Jorge, Parada-Arias Efren, and Gustavo V. Barbosa-Cánovas, 27–39. New York: Springer-Verlag. doi: 10.1007/978-1-4939-2578-0_3.

da Silva Felipe C., Carolina Rodrigues da Fonseca, Severino Matias de Alencar, Marcelo Thomazini, Julio C de Carvalho Balieiro, Paola Pittia, and Carmen Sílvia Favaro-Trindade. 2013. "Assessment of Production Efficiency, Physicochemical Properties and Storage Stability of Spray dried Propolis, A Natural Food

Additive, Using Gum Arabic and OSA Starch-Based Carrier Systems." *Food and Bioproducts Processing* 91 (1). Amsterdam,NL: Elsevier: 28–36.

Caballero-Cerón, C., Guerrero-Beltrán, J. A., Mújica-Paz, H., Torres, J. A. and Welti-Chanes, J. 2015. "Moisture Sorption Isotherms of Foods: Experimental Methodology, Mathematical Analysis, and Practical Applications." In *Water Stress in Biological, Chemical, Pharmaceutical and Food Systems. Food Engineering Series*, edited by F. Gutierrez-Lopez Gustavo, Alamilla-Beltran Liliana, M. Pilar Buera, Welti-Chanes Jorge, Parada-Arias Efren, and Gustavo V. Barbosa-Cánovas, 187–214. New York, NY: Springer Science+Business Media LLC New York. doi: 10.1007/978-1-4939-2578-0_15.

Cano-Chauca, Milton, P. C. Stringheta, A. M. Ramos, and J. Cal-Vidal. 2005. "Effect of the Carriers on the Microstructure of Mango Powder Obtained by Spray Drying and Its Functional Characterization." *Innovative Food Science and Emerging Technologies*. doi: 10.1016/j.ifset.2005.05.003.

Carr, Ralph L. 1965. "Evaluating Flow Properties of Solids." *Chemical Engineering* 18: 163–168.

Champion, Julie A., Yogesh K. Katare, and Samir Mitragotri. 2007. "Making Polymeric Micro- and Nanoparticles of Complex Shapes." *Proceedings of the National Academy of Sciences of the United States of America*. doi: 10.1073/pnas.0705326104.

Chen, Xiao Dong, and Kamlesh C. Patel. 2008. "Manufacturing Better Quality Food Powders from Spray Drying and Subsequent Treatments." *Drying Technology*. doi: 10.1080/07373930802330904.

Chiu, Gigi N. C., Sheela A. Abraham, Ludger M. Ickenstein, Rebecca Ng, Göran Karlsson, Katarina Edwards, Ellen K. Wasan, and Marcel B. Bally. 2005. "Encapsulation of Doxorubicin into Thermosensitive Liposomes via Complexation with the Transition Metal Manganese." *Journal of Controlled Release* 104 (2). Elsevier: 271–288.

Clydesdale, Fergus M., and E. M. Ahmed. 1978. "Colorimetry – Methodology and Applications." *C R C Critical Reviews in Food Science and Nutrition*. doi: 10.1080/10408397809527252.

Comunian, Talita A., Ednelí S. Monterrey-Quintero, Marcelo Thomazini, Julio C.C. Balieiro, Pierpaolo Piccone, Paola Pittia, and Carmen S. Favaro-Trindade. 2011. "Assessment of Production Efficiency, Physicochemical Properties and Storage Stability of Spray dried Chlorophyllide, a Natural Food Colorant, Using Gum Arabic, Maltodextrin and Soy Protein Isolate-Based Carrier Systems." *International Journal of Food Science and Technology*. doi: 10.1111/j.1365-2621.2011.02617.x.

Dodds, John. 2013. "Techniques to Analyse Particle Size of Food Powders." In *Handbook of Food Powders: Processes and Properties*, edited by Bhesh Bhandari, Nidhi Bansal, Min Zhang, and Pierre Schuck, 1st ed., 309–338. Cambridge, UK: Woodhead Publishing. doi: 10.1533/9780857098672.2.309.

Domian, Ewa, Anna Brynda-Kopytowska, Jolanta Cieśla, and Agata Górska. 2018. "Effect of Carbohydrate Type on the DVS Isotherm-Induced Phase Transitions in Spray dried Fat-Filled Pea Protein-Based Powders." *Journal of Food Engineering* 222. Elsevier: 115–125.

Drusch, S., and S. Berg. 2008. "Extractable Oil in Microcapsules Prepared by Spray drying: Localisation, Determination and Impact on Oxidative Stability." *Food Chemistry*. doi: 10.1016/j.foodchem.2007.12.016.

Du, Jing, Zhen Zhen Ge, Ze Xu, Bo Zou, Ying Zhang, and Chun Mei Li. 2014. "Comparison of the Efficiency of Five Different Drying Carriers on the Spray Drying of Persimmon Pulp Powders." *Drying Technology*. doi: 10.1080/07373937.2014.886259.

Faieta, Marco, Maria G. Corradini, Alessandro Di Michele, Richard D. Ludescher, and Paola Pittia. 2020. "Effect of Encapsulation Process on Technological Functionality and Stability of Spirulina Platensis Extract." *Food Biophysics*. doi: 10.1007/s11483-019-09602-1.

Fitzpatrick, John. 2005. "Food Powder Flowability." In *Encapsulated and Powdered Foods*, edited by Charles Onwulata, 247–260. Boca Raton, FL: CRC Press. doi: 10.1201/9781420028300.ch10.

Fritzen-Freire, Carlise B., Elane S. Prudêncio, Renata D. M. C. Amboni, Stephanie S Pinto, Aureanna N Negrão-Murakami, and Fabio S Murakami. 2012. "Microencapsulation of Bifidobacteria by Spray Drying in the Presence of Prebiotics." *Food Research International* 45 (1). Elsevier: 306–312.

Fuchs, M., C. Turchiuli, M. Bohin, M. E. Cuvelier, C. Ordonnaud, M. N. Peyrat-Maillard, and E. Dumoulin. 2006. "Encapsulation of Oil in Powder Using Spray Drying and Fluidised Bed Agglomeration." *Journal of Food Engineering*. doi: 10.1016/j.jfoodeng.2005.03.047.

Gharsallaoui, Adem, Gaëlle Roudaut, Odile Chambin, Andrée Voilley, and Rémi Saurel. 2007. "Applications of Spray drying in Microencapsulation of Food Ingredients: An Overview." *Food Research International*. doi: 10.1016/j.foodres.2007.07.004.

Goula, Athanasia M., and Konstantinos G. Adamopoulos. 2008. "Effect of Maltodextrin Addition during Spray Drying of Tomato Pulp in Dehumidified Air: II. Powder Properties." *Drying Technology*. doi: 10.1080/07373930802046377.

Guterres, Silvia Stanisçuaski, Ruy Carlos Ruver Beck, and Adriana Raffin Pohlmann. 2009. "Spray dry-ing Technique to Prepare Innovative Nanoparticulated Formulations for Drug Administration: A Brief Overview." *Brazilian Journal of Physics* 39 (1A). SciELO Brasil: 205–209.

Haque, M. K., and Y. H. Roos. 2004. "Water Sorption and Plasticization Behavior of Spray-dried Lactose/ Protein Mixtures." *Journal of Food Science* 69 (8). Wiley Online Library: E384–E391.

Hunter, Richard S., and Richard W. Harold. 1987. *The Measurement of Appearance*. 2nd ed. New York, NY: John Wiley & Sons.

Hutchings, J., R. Luo, and W. Ji. 2002. "Calibrated Color Imaging Analysis of Food." In *Color in Food: Improving Quality*, edited by Douglas B MacDougall, 352–366. Cambridge, UK: Woodhead Publishing. doi: 10.1533/9781855736672.2.352.

Iglesias, H A, and J Chirife. 1982. "Handbook of Food Isotherms." New York, NY: Academic Press. doi: 10.1016/B978-0-12-370,380-4.X5001-4.

Jinapong, Nakarin, Manop Suphantharika, and Pimon Jamnong. 2008. "Production of Instant Soymilk Powders by Ultrafiltration, Spray Drying and Fluidized Bed Agglomeration." *Journal of Food Engineering*. doi: 10.1016/j.jfoodeng.2007.04.032.

Joshi, P., and C.J.B. Brimelow. 2002. "Color Measurement of Foods by Color Reflectance." In *Color in Food: Improving Quality*, edited by Douglas B MacDougall, 80–114. Cambridge, UK: Woodhead Publishing. doi: 10.1533/9781855736672.2.80.

Krokida, M. K., and Z. B. Maroulis. 1997. "Effect of Drying Method on Shrinkage and Porosity." *Drying Technology*. doi: 10.1080/07373939708917369.

Labuza, Theodore P, and Bilge Altunakar. 2020. "Water Activity Prediction and Moisture Sorption Isotherms." In *Water Activity in Foods: Fundamentals and Applications*, edited by Gustavo V. Barbosa-Cánovas, Anthony J. Fontana Jr., Shelly J. Schmidt, and Theodore P. Labuza, 2nd ed., 161–205. Chicago, IL: Wiley Online Library.

Labuza, Theodore P., S Tannembaum, and Marcus Karel. 1970. "Water Content and Stability of Low Moisture and Intermediate Moisture Foods." *Food Technology* 24: 543–550.

Lao, Fei, and M. Monica Giusti. 2017. "The Effect of Pigment Matrix, Temperature and Amount of Carrier on the Yield and Final Color Properties of Spray Dried Purple Corn (*Zea mays* L.) Cob Anthocyanin Powders." *Food Chemistry* 227. Elsevier: 376–382.

Law, Ning Geng Daniel. 2007. "Stabilisation and Targeted Delivery of Enzyme Nattokinase by Encapsulation." *Minerva Biotecnologica* 19 (1).

Leblond, François A., Josée Tessier, and Jean Pierre Hallé. 1996. "Quantitative Method for the Evaluation of Biomicrocapsule Resistance to Mechanical Stress." *Biomaterials*. doi: 10.1016/0142-9612(96)00027-0.

Lian, G., and C. Astill. 2002. "Computer Simulation of the Hydrodynamics of Teabag Infusion." *Food and Bioproducts Processing*. doi: 10.1205/096030802760309179.

Lian, Guoping, Mark E. Malone, Jenny E. Homan, and Ian T. Norton. 2004. "A Mathematical Model of Volatile Release in Mouth from the Dispersion of Gelled Emulsion Particles." *Journal of Controlled Release*. doi: 10.1016/j.jconrel.2004.04.017.

Lintingre, E., F. Lequeux, L. Talini, and N. Tsapis. 2016. "Control of Particle Morphology in the Spray Drying of Colloidal Suspensions." *Soft Matter*. doi: 10.1039/c6sm01314g.

Lopes, Sofia, Catherine Afonso, Isabel Fernandes, Maria-Filomena Barreiro, Patrícia Costa, and Alírio E. Rodrigues. 2019. "Chitosan-Cellulose Particles as Delivery Vehicles for Limonene Fragrance." *Industrial Crops and Products* 139. Elsevier: 111407.

Lu, George Z., Murray R. Gray, and B. G. Thompson. 1992. "Physical Modeling of Animal Cell Damage by Hydrodynamic Forces in Suspension Cultures." *Biotechnology and Bioengineering*. doi: 10.1002/bit.260401018.

Maher, P G, M A E Auty, Y H Roos, L M Zychowski, and M A Fenelon. 2015. "Microstructure and Lactose Crystallization Properties in Spray Dried Nanoemulsions." *Food Structure*. doi: 10.1016/j.foostr.2014.10.001.

Moran, Linda L., Yun Yin, Keith R. Cadwallader, and Graciela W. Padua. 2014. "Testing Tools and Physical, Chemical, and Microbiological Characterization of Microencapsulated Systems." In *Microencapsulation in the Food Industry*, edited by Anilkumar G Gaonkar, Niraj Vasisht, Atul Ramesh Khare, and Robert Sobel, 323–352. San Diego, CA: Academic Press. doi: 10.1016/B978-0-12-404,568-2.00026-1.

Murrieta-Pazos, I., C. Gaiani, L. Galet, R. Calvet, B. Cuq, and J. Scher. 2012. "Food Powders: Surface and Form Characterization Revisited." *Journal of Food Engineering*. doi: 10.1016/j.jfoodeng.2012.03.002.

O'Hagan, P, K Hasapidis, A Coder, H Helsing, and G Pokrajac. 2005. "Particle Size Analysis of Food Powders." In *Encapsulated and Powdered Foods*, edited by Charles Onwulata, 215–246. Boca Raton, FL: CRC Press.

Ohtsubo, Toshiro, Shigenori Tsuda, and Kozo Tsuji. 1991. "A Study of the Physical Strength of Fenitrothion Microcapsules." *Polymer*. doi: 10.1016/0032-3861(91)90080-3.

Ortega-Rivas, Enrique. 2005. "Handling and Processing of Food Powders and Particulates." In *Encapsulated and Powdered Foods*, edited by Charles Onwulata, 75–144. Boca Raton, FL: CRC Press. doi: 10.1201/9781420028300.pt2.

Otálora, María Carolina, José G. Carriazo, Coralia Osorio, and Mónica Azucena Nazareno. 2018. "Encapsulation of Cactus (*Opuntia megacantha*) Betaxanthins by Ionic Gelation and Spray Drying: A Comparative Study." *Food Research International* 111. Elsevier: 423–430.

Oxley, James. 2014. "Chapter 4 - Overview of Microencapsulation Process Technologies." In *Microencapsulation in the Food Industry*, edited by Anilkumar G Gaonkar, Niraj Vasisht, Atul Ramesh Khare, and Robert Sobel, 35–46. San Diego, CA: Academic Press. doi: 10.1016/B978-0-12-404,568-2.00004-2.

Papadakis, Spyridon E, and Richard E Bahu. 1992. "The Sticky Issues of Drying." *Drying Technology* 10 (4). Taylor & Francis: 817–837.

Pathare, Pankaj B, Umezuruike Linus Opara, and Fahad Al-Julanda Al-Said. 2013. "Color Measurement and Analysis in Fresh and Processed Foods: A Review." *Food and Bioprocess Technology* 6 (1). Dordrecht, NL: Springer: 36–60.

Peirone, M. A., K. Delaney, J. Kwiecin, A. Fletch, and P. L. Chang. 1998. "Delivery of Recombinant Gene Product to Canines with Nonautologous Microencapsulated Cells." *Human Gene Therapy*. doi: 10.1089/hum.1998.9.2-195.

Peleg, M. 1977. "Flowability of Food Powders and Methods for Its Evaluation—A Review." *Journal of Food Process Engineering* 1 (4). Wiley Online Library: 303–328.

Peleg, M. 1983. "Physical Characteristics of Food Powders." In *Physical Porperties of Foods*, edited by M. Peleg and E. B. Bagley, 293–323. Westport, CT: AVI Publishing Co.

Porras-Saavedra, Josefina, Leonardo Cristian Favre, Liliana Alamilla-Beltrán, María Florencia Mazzobre, Gustavo Fidel Gutiérrez-López, and María del Pilar Buera. 2019. "Thermal Transitions and Enthalpic Relaxations as Related to the Stability of Microencapsulated Paprika Powders." *Journal of Food Engineering* 245. Amsterdam, NL: Elsevier: 88–95.

Quispe-Condori, Sócrates, Marleny D.A. Saldaña, and Feral Temelli. 2011. "Microencapsulation of Flax Oil with Zein Using Spray and Freeze Drying." *LWT - Food Science and Technology*. doi: 10.1016/j.lwt.2011.01.005.

Rahman, Mohammad Shafiur, and Rashid Hamed Al-Belushi. 2006. "Dynamic Isopiestic Method (DIM): Measuring Moisture Sorption Isotherm of Freeze-Dried Garlic Powder and Other Potential Uses of DIM." *International Journal of Food Properties*. doi: 10.1080/10942910600596134.

Ray, Sohini, Utpal Raychaudhuri, and Runu Chakraborty. 2016. "An Overview of Encapsulation of Active Compounds Used in Food Products by Drying Technology." *Food Bioscience*. doi: 10.1016/j.fbio.2015.12.009.

Fernandes Regiane Victória de Barros, Soraia Vilela Borges, Diego Alvarenga Botrel, and Cassiano Rodrigues de Oliveira. 2014. "Physical and Chemical Properties of Encapsulated Rosemary Essential Oil by Spray Drying Using Whey Protein-Inulin Blends as Carriers." *International Journal of Food Science and Technology*. doi: 10.1111/ijfs.12449.

Reineccius, Gary. 2005. "Quality Control." In *Flavor Chemistry and Technology*, edited by Gary Reineccius, 2nd ed., 450–460. Boca Raton, FL: CRC Press. doi: 10.1201/9780203485347.

Reineccius, Gary, and Sara J Risch. 1995. "Encapsulation and Controlled Release of Food Ingredients." In *ACS Symposium Series. 206th National Meeting of the American Chemical Society*, Chicago, IL. Vol. 590. Washington, DC: American Chemical Society. doi: 10.1021/bk-1995-0590.

Roos, Yrjo H. 2020. "Water Activity and Glass Transition." In *Water Activity in Foods*, edited by Gustavo V. Barbosa-Cánovas, Anthony J. Fontana, Shelly J. Schmidt, and Theodore P. Labuza, 2nd ed., 27–44. Chicago, IL: Wiley Blackwell/IFT Press. doi: 10.1002/9781118765982.ch3.

Roos, Yrjo H, and Stephan Drusch. 2015. *Phase Transitions in Foods*. Kidlington, UK: Academic Press.

Roos, Yrjo H, and Marcus Karel. 1991. "Water and Molecular Weight Effects on Glass Transitions in Amorphous Carbohydrates and Carbohydrate Solutions." *Journal of Food Science* 56 (6). Wiley Online Library: 1676–1681.

Roudaut, Gaëlle. 2020. "Water Activity and Physical Stability." In *Water Activity in Foods*, edited by Gustavo V. Barbosa-Cánovas, Anthony J. Fontana, Shelly J. Schmidt, and Theodore P. Labuza, 2nd ed. Chicago, IL: Wiley Blackwell/IFT Press. doi: 10.1002/9781118765982.ch10.

Saadi, Sami, Abdul Azis Ariffin, Hasanah Mohd Ghazali, Mat Sahri Miskandar, Huey Chern Boo, and Sabo Mohammed Abdulkarim. 2012. "Application of Differential Scanning Calorimetry (DSC), HPLC and

PNMR for Interpretation Primary Crystallisation Caused by Combined Low and High Melting TAGs." *Food Chemistry.* doi: 10.1016/j.foodchem.2011.10.095.

Saarela, J. M.S., S. M. Heikkinen, T. E.J. Fabritius, A. T. Haapala, and R. A. Myllylä. 2008. "Refractive Index Matching Improves Optical Object Detection in Paper." *Measurement Science and Technology.* doi: 10.1088/0957-0233/19/5/055710.

Sagis, Leonard M.C. 2015. "Determination of Mechanical Properties of Microcapsules." In *Microencapsulation and Microspheres for Food Applications,* edited by Leonard M. C. Sagis, 195–205. London, UK: Academic Press. doi:10.1016/B978-0-12-800,350-3.00010-8.

Sarabandi, Kh, S. H. Peighambardoust, A. R. Sadeghi Mahoonak, and S. P. Samaei. 2018. "Effect of Different Carriers on Microstructure and Physical Characteristics of Spray Dried Apple Juice Concentrate." *Journal of Food Science and Technology.* doi: 10.1007/s13197-018-3235-6.

Slade, Louise, Harry Levine, and David S Reid. 1991. "Beyond Water Activity: Recent Advances Based on an Alternative Approach to the Assessment of Food Quality and Safety." *Critical Reviews in Food Science & Nutrition* 30 (2–3). Abingdon-on-Thames, UK: Taylor & Francis: 115–360.

Spackman, Christy C W, and Shelly J Schmidt. 2010. "Characterising the Physical State and Textural Stability of Sugar Gum Pastes." *Food Chemistry* 119 (2). Amsterdam, NL: Elsevier: 490–499.

Teunou, E., J. J. Fitzpatrick, and E. C. Synnott. 1999. "Characterization of Food Powder Flowability." *Journal of Food Engineering.* doi: 10.1016/S0260-8774(98)00140-X.

Teunou, E, J. Vasseur, and M. Krawczyk. 1995. "Measurement and Interpretation of Bulk Solids Angle of Repose for Industrial Process Design." *Powder Handling and Processing* 7 (3). Clausthal-Zellerfeld, FR Germany: Trans Tech Publications, 1989-: 219–228.

Tupuna, Diego Santiago, Karina Paese, Silvia Stanisçuaski Guterres, André Jablonski, Simone Hickmann Flôres, and Alessandro de Oliveira Rios. 2018. "Encapsulation Efficiency and Thermal Stability of Norbixin Microencapsulated by Spray drying Using Different Combinations of Wall Materials." *Industrial Crops and Products* 111. Elsevier: 846–855.

Walton, D E. 2000. "The Morphology of Spray dried Particles a Qualitative View." *Drying Technology* 18 (9). Abingdon-on-Thames, UK: Taylor & Francis: 1943–1986.

Woo, Meng Wai, and Bhesh Bhandari. 2013. "Spray Drying for Food Powder Production." In *Handbook of Food Powders: Processes and Properties,* edited by Bhesh Bhandari, Nidhi Bansal, Min Zhang, and Pierre Schuck, 1st ed., 29–56. Cambridge, UK: Woodhead Publishing.

Xu, Duoxia, Yang Xu, Guorong Liu, Zhanqun Hou, Yinghao Yuan, Shaojia Wang, Yanping Cao, and Baoguo Sun. 2018. "Effect of Carrier Agents on the Physical Properties and Morphology of Spray dried Monascus Pigment Powder." *LWT* 98. Amsterdam, NL: Elsevier: 299–305.

Yu, Lian. 2001. "Amorphous Pharmaceutical Solids: Preparation, Characterization and Stabilization." *Advanced Drug Delivery Reviews.* doi: 10.1016/S0169-409X(01)00098-9.

Yu, X., S. E. Martin, and Shelly J. Schmidt. 2008. "Exploring the Problem of Mold Growth and the Efficacy of Various Mold Inhibitor Methods during Moisture Sorption Isotherm Measurements." *Journal of Food Science* 73 (2). Wiley Online Library: E69–E81.

Zhang, Z, Daniel Law, and Guoping Lian. 2010. "Characterization Methods of Encapsulates." In *Encapsulation Technologies for Active Food Ingredients and Food Processing,* edited by N. J. Zuidam and Viktor Nedovic, 101–125. Dordrecht, NL: Springer.

Zhang, Z., R. Saunders, and C. R. Thomas. 1999. "Mechanical Strength of Single Microcapsules Determined by a Novel Micromanipulation Technique." *Journal of Microencapsulation.* doi: 10.1080/026520499289365.

Zhang, Zhi-Hong, Huadong Peng, Meng Wai Woo, Xin-An Zeng, Margaret Brennan, and Charles S. Brennan. 2020. "Preparation and Characterization of Whey Protein Isolate-Chlorophyll Microcapsules by Spray Drying: Effect of WPI Ratios on the Physicochemical and Antioxidant Properties." *Journal of Food Engineering* 267. Elsevier: 109729.

15 Application of Spray Dried Encapsulated Bioactives in Food Products

Amir Pouya Ghandehari Yazdi
Zar Research, and Industrial Development Group, Iran

Zahra Beig Mohammadi
Islamic Azad University, Iran

Khadijeh Khoshtinat
National Nutrition and Food Research Institute,
Shahid Beheshti University of Medical Science, Iran

Leila Kamali Rousta
Zar Research, and Industrial Development Group, Iran

Seid Mahdi Jafari
Gorgan University of Agricultural Sciences and Natural Resources,
Iran

CONTENTS

15.1 INTRODUCTION

Food bioactive compounds are phytochemicals possessing several functional activities, such as antioxidant, anti-inflammatory, anti-carcinogenic, and antimicrobial effects alongside biodiversity. Some examples of these compounds include carotenoids, phenolic acids, anthocyanins, and flavonoids. The application of bioactive components has encountered some limitations due to their instability (physically, chemically, and/or enzymatically) during food processing, storage (oxygen, light, temperature, pH, and interaction with other food ingredients), and digestion, as well as losing activity partially or even completely after isolation. Encapsulation of these compounds for food applications should overcome some challenges that are associated with their composition and the specific interactions with the materials used for their encapsulation. Then, it is necessary to consider the effectiveness, toxicity, and migration of the components to the food matrix, as well as delivery systems issues of the encapsulated bioactive components (Giaconia et al., 2020).

In order to increase the bioavailability, solubility, and applicability of bioactive components, encapsulation is one of the best solutions, although it may face some challenges such as extra cost, production complexity, and a lack of technical knowledge (Dias, Ferreira, & Barreiro, 2015; Zuidam & Shimoni, 2010). Encapsulation is a technique in which one substance (active component) is entrapped into another one (wall material), and particles in the nanometer (nanoencapsulation), micrometer (microencapsulation), or millimeter scales are produced (Burgain, Gaiani, Linder, & Scher, 2011; Lakkis, 2007). For this purpose, several techniques are used; spray drying, freeze drying, molecular inclusion, extrusion processes, liposomes, coacervation, polymeric micelles, supercritical fluids, nanostructured lipid matrices, and solvent evaporation (Aguiar, Costa, Rocha, Estevinho, & Santos, 2017; Franco et al., 2017).

Encapsulation by spray drying is the most widely-used technique on an industrial scale; it is also considered as a low-cost encapsulation method. It is one of the most conventional approaches to encapsulate bioactive agents in which the target components are dissolved, emulsified, or dispersed in an aqueous medium of one or more wall materials. Subsequently, the blend is atomized and sprayed within a chamber with relatively high temperatures. During the spray drying process, forming a thin layer at the surface of the droplets and increasing the concentration of ingredients in the drying droplets will occur. The atomized droplet size will be determined by drying time and particle size. The carrier materials should protect the bioactive materials and they should show high solubility in water, possess suitable film-forming and emulsifying properties, and be cost-effective. Molecular weight, glass transition, crystallinity, and diffusibility are other influential criteria. Natural gums (gum Arabic, alginates, and carrageenan), proteins (dairy proteins, soy proteins, and gelatin), carbohydrates (maltodextrins and cellulose derivatives), and/or lipids (waxes and emulsifiers) are used as wall/carrier materials. Traditional spray dried carriers usually release their content instantly after mixing with water, depending on the porosity of the particles (Dias et al., 2015; Zuidam & Shimoni, 2010).

By spray drying encapsulation, active components can be enclosed within a protective outer layer. Due to the inherent film-forming properties of spray drying, the wall component dries out at a much faster rate than the medium (usually water; in which the core material is suspended) to compose the feed solution. Therefore, the wall material can form a coating around the droplet containing the core. A number of influencing parameters, such as the humidity of the gas, gas flow, inlet and outlet temperature, feed rate, type of atomization, the geometry of the chamber, viscosity, dry content, and the type of solvents and materials, should be considered when encapsulating bioactive materials by spray drying. Generally, the variables which influence the effectiveness of encapsulation by spray drying could be classified into two categories: (A) properties of the food emulsion (including the core and the wall material); and (B) the conditions of the spray drying process (Anandharamakrishnan & Ishwarya, 2015b, 2015c).

The achievement of spray drying on an industrial scale is obviously because of its commercial usage and improvement as an encapsulation method, availability, economic viability, reproducibility, easy scale-up, and flexibility of processing. Assadpour and Mahdi Jafari (2019) listed the main advances of the spray drying method for the encapsulation of food bioactive components, including control over particle size, shape, and morphology (amorphous/crystal form, porosity), the conversion of various feeds directly into dry powders via a one-step semi-continuous procedure, simplicity of process, easy operation, capability to scale up, possibility of highly sticky feeds encapsulation with surfactants and drying aids, low cost, fast operation process, energy-efficiency, closed- or open-cycle design for aqueous or organic solvents, low risk of degradation of the heat-sensitive substances, the possibility of high viscous feeds encapsulation by means of preheating, the design of particles with controlled release properties, suitability for hydrophobic alongside hydrophilic food ingredients, producing powders with an extended shelf-life by high efficiency encapsulation and possibility of optimizing the process by various DOE (design of experiment) techniques.

Wall material(s) should have some ideal properties, including low hygroscopic behavior, low viscosity, the ability to stabilize emulsions containing core ingredients, and form a thin layer around to protect them, without any odd taste or flavor, and at a reasonable economic price (Franco et al., 2017). Gum Arabic, whey protein, modified starches, maltodextrin, sodium caseinate, gelatin, and chitosan are examples of some ordinary wall materials implemented for the encapsulation of food components (Gharsallaoui, Roudaut, Chambin, Voilley, & Saurel, 2007).

Different antioxidant compounds such as caffeic acid, chlorogenic acid, and rosmarinic acid with sodium carboxymethyl cellulose have shown a great encapsulation efficiency, the quality of being stable, and the protection of their antioxidant activity on an industrial scale (Aguiar et al., 2017) as well as gallic acid encapsulated by bacterial exopolysaccharides such as a fucose-rich saccharide, which is known as 'FucoPol' (Lourenco et al., 2017).

In this chapter, following an overview of the bioactive ingredients encapsulated by spray drying, there will be a discussion of the application of spray dried encapsulated ingredients in different food products and the production of functional foods.

15.2 AN OVERVIEW OF DIFFERENT INGREDIENTS ENCAPSULATED BY SPRAY DRYING

In the manufacture of functional and healthy foods, natural antioxidants and antimicrobial ingredients have been regarded as good choices as green additives. Polyphenols, essential oils (EOs), organic acids, and peptides are among examples of these kinds of additives.

EOs that have been used as flavorings, aromas, antimicrobial, and antioxidant agents in food products are highly susceptible to changes caused by external factors, such as light, oxygen, and temperature, and their volatility is also a concerning factor (Botrel, de, & Borges, 2015). The most frequently used EOs are monoterpenes (i.e., pinene, limonene, γ-terpinene, and ß-carbohyllene), monoterpenoids (i.e., citronellai, thymol, carvacol, carvone, and linalool), phenylpropanoids (i.e., cinnamaldehyde, eugenol, vanillin, and safrole), and others (i.e., allyl-isothiocyanat and allicin). To apply EOs as antimicrobial agents, two main factors including their different kinds and the target food should be considered. For example, in meat products, coriander, eugenol, oregano, clove, and thyme oils can show high antimicrobial properties. The antimicrobial property of EOs will be reduced because of the interaction of fat with EOs in fish that contains high levels of fat compared with fish with a lower level of fat. That is why the application of EOs as an antimicrobial agent is successful in vegetables and fruits with low-fat content as most of the vegetables are consumed unprocessed and EOs can be an appropriate choice to improve their shelf-life (Spinelli, Conte, Lecce, Incoronato, & Del, 2015).

Antimicrobial peptides (AMPs) have important functionalities, such as antibacterial, antiviral, antifungal, and inflammatory properties. Organic acids, traditional natural antimicrobial agents, are derived from animals and plants. They decrease the pH of foods, and thereby prevent the growth of pH-sensitive bacteria such as *Listeria monocytogenes, Salmonella spp., Clostridium perfringens, Escherichia coli*, and *Campylobacter*. The anionic part of the organic acids is accumulated into bacteria and causes a disorder in essential functions of bacteria, with no drawback in terms of the sensory properties of foods (Bahrami, Delshadi, Assadpour, Jafari, & Williams, 2020). The direct addition of antimicrobial EOs may cause some undesirable interposition in food products, such as intense taste and interaction with food constituents (Zanetti et al., 2018).

15.3   APPLICATION OF SPRAY DRIED ENCAPSULATED MATERIALS IN DIFFERENT FOOD PRODUCTS

In this part, the application of spray dried encapsulated bioactive components in various food products will be covered. The products that were reviewed for this purpose include dairy products, cereals, bakery, confectionery/chocolate, beverages, and meat and meat-based products, as well as active packaging materials. Summary tables demonstrate the results of the corresponding research, and the main factors, namely core and wall materials, the operation conditions of spray drying such as inlet and outlet air temperatures, atomizing nozzle diameter, airflow rate, and feed flow rate.

15.3.1   DAIRY PRODUCTS

Dairy products are known to perform as excellent carriers of microencapsulated bioactive compounds due to their high consumption rate and high nutritional value. In some products such as cheese, some nutrients can be washed away into the whey. Therefore, fortification has been suggested for the enhanced recovery of these materials. Experiments have been carried out to understand the fortification of different dairy products with vitamins (e.g., A, C, D, and E), minerals (e.g., iron, zinc, and calcium), and probiotics (Banville, Vuillemard, & Lacroix, 2000; Pinto et al., 2012). However, fortification with vitamins and probiotics appears to be more sensitive (than minerals) to heat, light, humidity, oxidation, and pH. Minerals, on the other hand, may express some problems with bioavailability and interactions with other compounds. In addition, they could act as incompatible factors on color, odor, taste, and other sensory properties of the final product (da, Fernandes, Barros, Fernandes, & Barreiro, 2019; Kha, Nguyen, Roach, & Stathopoulos, 2015).

Microencapsulation can be mostly used for probiotics because of their relatively large dimensions that are between 1–5 μm (Assadpour & Mahdi Jafari, 2019). Spray drying is commonly employed as a microencapsulation technique for bioactives in the food industry because it saves money and sounds flexible. Microencapsulation has shown efficiency for the processing of more stable fortified dairy products against adverse environmental changes. This technology has been successfully implemented to protect probiotics from heat and pH effects (Arslan, Erbas, Tontul, & Topuz, 2015). In this section, the encapsulated ingredients by spray drying technology, which have been used in dairy products are considered (Table 15.1).

TABLE 15.1
Application of Spray Dried Encapsulated Bioactives in Dairy Products

Product	Core Material	Wall Material	Spray Drying Conditions	Results	References
Cheddar cheese	*Lactobacillus paracasei* NFBC 338	Reconstituted skim milk	Inlet temperature: 175°C, outlet temperature: 68°C	The number of *Lb. paracasei* were observed 10^9 CFU/g respectively after the spray drying and cheese ripening after three months.	Gardiner et al. (2002)
Milk	Iron and Ascorbic acid	Polyglycerol monostearate (PGM)	Wall: core: 5:1, 10;1, 15:1, and 20:1	The encapsulation efficiency increased to 94.2% and thiobarbituric acid (TBA) was significantly decreased	Lee, Ahn, Lee, and Kwak (2004)

(Continued)

Product	Core Material	Wall Material	Spray Drying Conditions	Results	References
Yogurt	Betacyanin	Maltodextrin	Inlet temperature: 120°C, outlet temperature: 60°C, Feed flow rate: 0.4 L/h, Wall: core: 3:1, 4:1, and 5:1	Changing the ratio of the wall: core did not have a significant effect on the efficiency of microencapsulation.	Azeredo, Santos, Souza, Mendes, and Andrade (2007)
Yogurt	Fish oil	Barley protein	Inlet temperature: 150°C, outlet temperature: 60°C	The peroxide value was significantly decreased during five weeks of storage, compared with the free samples.	Wang, Tian, and Chen (2011)
Frozen Yogurt	*Bifidobacterium BB-12*	Reconstituted skim milk 10 and 20% inulin10 and 20%	Inlet temperature: 150°C, outlet temperature: 55°C, feed flow rate: 6 mL/min, airflow speed: 35 m³/h	The microbial population of encapsulated probiotics in inulin was more than in reconstituted skim milk.	Pinto et al. (2012)
Ricotta cream	*Bifidobacterium BB-12*	Reconstituted skim milk and inulin, reconstituted skim milk, and oligofructose-enriched inulin	Inlet temperature: 150°C, outlet temperature: 55°C, feed flow rate: 6 mL/min, airflow speed: 35 m³/h	The reconstituted skim milk and inulin had the highest effect on the protection of the microorganisms during storage.	Pinto et al. (2012)
Yogurt	β-carotene	Maltodextrin	Inlet temperature: 170°C, outlet temperature: 95°C, feed flow rate: 7.5 mL/min	The encapsulation efficiency was 37.7%. The preservation of β-carotene was observed during storage in low pH.	Donhowe, Flores, Kerr, Wicker, and Kong (2014)
Ice cream	Pomegranate peel extract	Different types of maltodextrin	Inlet temperature: 160°C, outlet temperature: 62°C–68°C, feed flow rate: 8 mL/min, air flow rate: 600 L/h, wall: core: 1: 1 and 3: 1	Encapsulation improved the stability of phenolic compounds. The addition of microcapsules of pomegranate peel extracts (MPPE) into ice cream improved the antioxidant and α-glucosidase inhibitory activities of samples compared to the control sample (without MPPE). About 75% of the panelists admitted the sensory properties of enriched samples.	Çam, İçyer, and Erdoğan (2014)

(Continued)

TABLE 15.1
(*Continued*)

Product	Core Material	Wall Material	Spray Drying Conditions	Results	References
Ice cream	*Lactobacillus paracasei*	Inulin, maltodextrin	Inlet temperature: 150°C, outlet temperature: 61.8°C–65°C, feed flow rate: 6 mL/min	The combination of 46% commercial skim milk powder, 24% anhydrous glucose, 28% inulin, and 2% sodium alginate was selected as the wall material for encapsulation of *Lactobacillus paracasei*.	Spigno, Garrido, Guidesi, and Elli (2015)
Milk and Yogurt	Gas oil	Mixture of whey protein and Arabic gam at a ratio of: 7:3	Inlet temperature: 154°C, outlet temperature: 80°C, feed flow rate: 970 mL/h, air flow speed: 4.3 m/s	In the products with encapsulated gac oil the peroxide value was more stable than the free form.	Kha et al. (2015)
Swiss cheese	Swiss cheese bioaroma	Modified corn starch, maltodextrin, mixture of both	Inlet temperature: 175°C, feed flow rate: 2.97×10^{-7} m³/s, air flow rate: 5.8×10^{-4} m³/s, air pressure: 5 bar	The effect of modified starch on the organic acids retention was greater than that of maltodextrin	(da Costa et al. 2015)
Fermented milk	*Lactobacillus casei* ATCC 393	Skim milk	Inlet temperature: 170°C, outlet temperature: 80°C–85°C	Improvement of survival rate for *L. casei* by approximately $10^7 \log_{10}$ CFU/g.	Dimitrellou et al. (2016)
Probiotic milk chocolate	*Lactobacillus plantarum* HM47	A mixture of maltodextrin and *Moringa oleifera* gum, and Tender coconut water	Inlet temperature: 120°C, outlet temperature: 93°C, nozzle diameter: 0.7 mm	Improvement of the survival rate *L. plantarum* HM47 approximately $10^8 \log_{10}$ CFU/g, as well as decreasing the *Enterobacteriaceae* population.	Nambiar, Sellamuthu, and Perumal (2018)
Milk	*Curcumin*	Skim milk	Inlet temperature: 162.8°C, outlet temperature: 88.8°C	The stability of curcumin antioxidant activity during six months of storage was observed and only 28% of that was decreased.	Neves, Desobry-Banon, Perron, Desobry, and Petit (2019)
Yogurt	*Lactococcus lactis*	A mixture of gum Arabic and *Synsepalum dulcificum*	Inlet temperature: 130°C, outlet temperature: 60°C, feed flow rate: 35.25 mL/min, airflow speed: 4.3 m/s	The miracle fruit pulp (as an encapsulating agent for the growth of *L. lactis* during storage) was affected more than other parts.	Fazilah et al. (2019)

(*Continued*)

Product	Core Material	Wall Material	Spray Drying Conditions	Results	References
Yogurt	*Spirulina platensis*	Maltodextrin, maltodextrin crosslinked with citric acid	Inlet temperature: 170°C, outlet temperature: 95°C, feed flow rate: 6 mL/min, wall: core ratio: 1: 1	Spirulina encapsulated in maltodextrin crosslinked with citric acid had better nutritional profile, color, and antioxidant activity than the others.	da Silva et al. (2019)
Ice cream	Pistachio peel extract	Maltodextrin	Inlet temperature: 150°C, feed flow rate: 7 mL/min, wall: core ratio: 2: 1 w/w	By increasing the number of microcapsules of pistachio hull extract (MPHE), total phenolic content, and antioxidant properties improved. The incorporation of MPHE had no significant effect on the overall liking of ice cream.	Ghandehari Yazdi, Barzegar, Ahmadi Gavlighi, Sahari, and Mohammadian (2020)

15.3.2 MILK

Today, nutritionists recommend adding iron to milk and milk-based products in order to tackle iron deficiency anemia. Fortification by iron in food processing is very difficult, because of the ability of iron to change the color and also the possibility of sedimentation, oxidation, and a metallic taste in food products. Polyacylglycerol monostearate (PGMS) was used as a wall material for the spray dried microencapsulation of ascorbic acid and ferric ammonium sulfate simultaneously in fortified milk (Table 15.1). Microencapsulated samples with encapsulation efficiencies between 80.7 and 94.2% for different rates of wall/core were produced and added to commercial whole milk which was then stored for 12 days (4°C). Thiobarbituric acid (TBA) was used to monitor the effect of encapsulation on milk fat oxidation. Results showed that TBA was significantly lower for the encapsulated samples than in the unencapsulated group after 12 days of storage. In addition, the release rate of vitamin C at the end of storage was 9.2%, which suggested that PGMS was a good carrier for the encapsulation of vitamin C for fortifying milk and dairy products. The panelist also did not recognize any difference between microencapsulated samples and the control (Lee et al., 2004).

The cucurbit *Momordica cochinchinensis* or gac fruit is full of carotenoids (β-carotene and lycopene), α-tocopherol (vitamin E), polyphenol compounds, and polyunsaturated fatty acids. Because of the large number of double bonds in the structure of carotenoids and unsaturated fatty acids, the gac oil is easily vulnerable to oxidation and isomerization. Therefore, it is crucial to use an effective method to stop biodegradation (Kha, Nguyen, Roach, Parks, & Stathopoulos, 2013). Accordingly, the encapsulated gac fruit oil in whey protein and gum Arabic by spray drying was used for milk fortification. After the pasteurization of raw milk (63°C, 20 s), the encapsulated gac oil powder was added and stirred. The pasteurized milk after thermal treatment was then cooled and kept at 4°C for 30 days. Peroxide value, color content, β-carotene, and lycopene content were determined during the storage and were compared with the milk before pasteurization as the control sample. The results showed that the color of fortified milk was the same as the control and the peroxide value increased slightly from 3.8 to 4.3 meq/kg in fortified milk during four weeks of storage. The level of the carotenoids rather decreased from 61.3 to 52.9 µg/g in fortified pasteurized milk during storage. The lycopene content showed more reduction than carotenoids, but was more stable than the free state. The authors recommended encapsulated gac oil as a rich source of β- carotene and lycopene for the fortification of pasteurized milk (Kha et al., 2015).

A spray drying technique was successfully implemented to produce microparticles of *Lactobacillus casei ATCC 393* loaded into skim milk (as wall material) to produce fermented milk. The results showed that a high level of microencapsulated *L. casei* (23.5%) was remained compared with free cells (14.7%) after three weeks. The microencapsulated *L. casei* approximately remained at $10^7 \log_{10}$ CFU/g at the end of storage, which is more than the minimum requirement to have a probiotic effect ($10^6 \log_{10}$ CFU/g). The authors stated that they did not recognize any difference between the milk fermented by microencapsulated *L. casei* and the commercial fermented milk. However, both samples provided higher values in overall acceptance than the control sample, as well as samples produced by free *L. casei* cells. These results confirm the improvement of the survival rate of *L. casei* by microencapsulation (Dimitrellou et al., 2016).

Microencapsulated *Lactobacillus plantarum HM47* in maltodextrin and *Moringa oleifera* gum was incorporated into chocolate milk. The probiotic chocolate milk was stored for 180 days at 25°C. The results showed that at the end of the storage time, more than $10^8 \log_{10}$ CFU/g microorganisms survived, which is more than the amount prescribed for probiotics. Clinical studies have shown that consuming probiotic powder and chocolate with probiotic milk decreases the *Enterobacteriaceae* population simultaneously as it increases the lactic acid bacteria population. In addition, a safety assessment of *L. plantarum HM47* showed that no adverse effects were observed with regard to food intake, blood cell components, and vital organs of treated mice. The authors stated that the probiotic milk chocolate could be a potential carrier for the probiotic by keeping the bacterial cells for humans (Nambiar et al., 2018).

Curcumin is a natural polyphenol that possesses significant antioxidant and anticancer features, but it is sensitive to light, oxygen, and heat and also demonstrates very low solubility in water (Sahu, Kasoju, & Bora, 2008). Due to the structural and physicochemical features of curcumin, milk is a good option for encapsulating it (Livney, 2010). The skim milk powders with curcumin were spray dried. The results showed that the wettability rate of the encapsulated milk powder with curcumin was lower than that of milk powder alone. This problem was greatly eased by changing the spray conditions. Milk powder containing curcumin indicated no noticeable difference in other properties except for color, which was a little yellower than milk powder alone. The stability of curcumin antioxidant activity during storage was observed. This indicates that skim milk was a good carrier for curcumin encapsulation. During storage, the increase in temperature decreased the content of curcumin. In the case of the samples stored at 25°C for six months, for instance, only 28% of curcumin content decreased which represents the protective effect of encapsulation by skim milk (Neves et al., 2019).

15.3.2.1 Yogurt

Fermented foods and beverages vary in cultural preferences and traditions in different areas where they are produced. Fermentation processes to preserve raw materials such as milk and fruit are created by organic acids or alcohol and supply the desired taste, flavor, and texture to food (Swain, Anandharaj, Ray, & Parveen Rani, 2014). Yogurt is one of the products fermented out of milk that has been produced and consumed all over the world as a dairy product for thousands of years. Due to its consumption by people of all ages, yogurt can be used as a good vehicle for bioactive enrichment through the use of various bioactive ingredients such as bio pigments, antioxidants, EOs, vitamins, and minerals (da Silva et al., 2019).

Red beetroot is one of the main vegetables regarding production and consumption. This vegetable contains mineral elements such as iron, calcium, phosphorus, and magnesium. Betalains as water-soluble pigments can create a broad range of colors for betaxanthins from yellow to orange and for betacyanins from purple to violet. In addition, betalains that are used in the food industry has increased due to its antioxidant activity. Betalains, like other pigments, are often sensitive to external factors and their stability depends on pH. Their degradation is simply influenced by temperature, light, oxygen, and moisture (Azeredo et al., 2007; Janiszewska & Wlodarczyk, 2013; Ravichandran et al., 2014). To solve this problem, betacyanin was microencapsulated in different ratios of maltodextrin (Table 15.1) by the spray drying method (Azeredo et al., 2007; Janiszewska & Wlodarczyk, 2013). Microencapsulated betacyanin was added to yogurt to produce colored yogurt that can be especially attractive to children and then compared with synthetic pigments during processing and storage. The results showed the synthetic colorant and microencapsulated betacyanin in yogurt were equivalent to those of the standard. In addition, betacyanin retention while microencapsulation was nearly 90%, independent of the maltodextrin-betacyanin ratio. The panelists did not distinguish the samples containing betacyanin from the blank in terms of flavor. The authors stated that the encapsulated betacyanin remained more in the same condition than in the free samples. The betacyanin content further decreased in the presence of light so that betacyanin degradation in the amber jars and translucent jars was 30% and 57%, respectively after six months of storage (Azeredo et al., 2007).

Carotenoids are one of the highly used pigment groups in nature and responsible for yellow, orange, or red colors. β-Carotene has the most intense activity and efficiency of vitamin A conversion among other carotenoids of provitamin A. Also, β-carotene has antioxidant features and can act as a radical lipid protection compound due to its unique conjugate structure double links. Low β-carotene absorption in the body has led to the food fortification by β-carotene (Grune et al., 2010; Haskell, 2012). Carotenoids are exposed to oxidation because of exposure to oxygen, light, and heat which can cause discoloration, loss of antioxidant activity, and vitamins. Microencapsulation aids in maintaining the stability of carotenoids and releasing them under certain conditions (Donhowe et al., 2014; Rutz, Borges, Zambiazi, Rosa, & Silva, 2016). Spray drying of β-carotene for preservation, release, and bioavailability in yogurt was achieved by maltodextrin. The spray dried microencapsulated β-carotene powder with 37.7% encapsulation efficiency had lower moisture content, particle size, and higher surface β-carotene content in comparison with the commercial water-dispersible β-carotene. The pH of the food affected the release of encapsulated ingredients. The pH of the yogurt in this research was about 4.3. Under a more acidic pH condition, the transfer of the β-carotene to the oil phase is facilitated. Solid and semi-solid food matrices may inhibit the release of β-carotene in the oil phase of the oil–water compounds. The authors stated that the spray drying of β-carotene in maltodextrin in yogurt was approved as an acceptable method for preservation as well as maintaining bioavailability (Donhowe et al., 2014).

Barley protein was applied as a wall material to encapsulate fish oil by spray drying. The results showed the condition with high encapsulation efficiency occurred at the equivalent ratio of oil/protein (15%), and an inlet temperature of 150°C. The microencapsulated fish oil was incorporated into a yogurt product that was stored at 4°C after stirring and pasteurization. The results of oxidative stability by peroxide value were determined below 3 meq/kg oil after five weeks of storage. These findings confirmed that the barley enriched with hordein protein microcapsules possessed an exceptional ability for the protection of the oil from oxidation, resulting in its efficient making for the use in food products (Wang et al., 2011).

For fortification of yogurt, the spray dried microencapsulated gac fruit oil in whey protein and gum Arabic was added. The raw milk was heated until 90°C and was quickly cooled to 45°C. When the starter was incorporated, encapsulated gac oil powder was added and stirred. After incubation, the fortified yogurt was stored at 4°C for 30 days. The amount of β-carotene and lycopene content in the fortified yogurt decreased slightly, but there was no significant difference between them during four weeks of storage. The results showed the levels of the peroxide value increased to some extent from 2 to 2.6 meq/kg in the case of the fortified yogurt during storage and was more stable than the control samples. The authors stated that the encapsulated gac oil could be a good source of β-carotene and lycopene for the fortification of yogurt (Kha et al., 2015).

The survival of different free and microencapsulated probiotics such as *Lactobacillus plantarum* (Brinques & Ayub, 2011), *Bifidobacterium* BB-12 (Pinto et al., 2012), and *Lactococcus lactis* Gh1 (Fazilah et al., 2019) in yogurt was investigated successfully. *Bifidobacterium* BB-12 microcapsules produced with reconstituted skim milk 10 and 20% w/v and inulin 10 and 20% w/v were added to the frozen yogurt samples. The results showed that the number of free cells in frozen yogurt decreased to about 34% after 90 days while the number of microencapsulated *Bifidobacterium* remained constant. The frozen yogurt with microcapsules produced by inulin showed more apparent viscosity and larger hysteresis than the frozen yogurt with microcapsules produced by skim milk. Frozen yogurt samples containing encapsulated probiotics in inulin showed an increase in the microbial population at the end of the storage period. Finally, the results demonstrated that inulin could be a good carrier for probiotic frozen yogurt (Pinto et al., 2012).

Microencapsulated *Lactococcus lactis* Gh1 in gum Arabic and different parts of the miracle fruit (*Synsepalum dulcificum*) was produced and added into yogurt and stored at 4°C for three weeks. The initial counts of the microencapsulated bacteria were at least $10^9 \log_{10}$ CFU/mL. The highest viability (85%) and the highest encapsulation efficiency (99.27%) were observed in the powders containing *L. lactis* in combination with *S. dulcificum* seed. After the incubation and fermentation of yogurt by probiotic at zero time, the highest and the lowest viability of *L. lactis* was observed for yogurt fortified with miracle fruit seeds and yogurt with free cells respectively. After three weeks of storage, the yogurt containing free cells had the minimum cell viability while the highest was observed in the yogurt fortified with miracle fruit pulp. The results showed that the miracle fruit pulp could successfully be used as an encapsulating agent for the growth of *L. lactis* during the storage of a yogurt product (Fazilah et al., 2019).

To produce a novel functional yogurt formulation, *Spirulina platensis* was spray dried by maltodextrin and maltodextrin cross-linked with citric acid and finally added into yogurt. *S. platensis* or *Arthrospira platensis* is a microalga that contains large amounts of protein (55%–70%), essential amino acids, essential fatty acids, and various vitamins, such as B1, B2, B12, E, and provitamin A (Soni, Sudhakar, & Rana, 2017). The *S. platensis* was successfully microencapsulated in maltodextrin and maltodextrin crosslinked with citric acid with 66% and 77% yield, respectively. The results during seven days of storage showed the functional yogurt with spirulina encapsulated in maltodextrin cross-linked with citric acid had a better nutritional pattern, color, and antioxidant value than the others. Despite the nutritional value and beneficial effects of *S. platensis* on health, its fish-like odor and intense green color have not been widely accepted by consumers in various products. In addition, based on the results of the sensory evaluation, when microcapsule forms were used, the undesirable fishy odor of microalgae was not understood and the strong green color in the yogurt was also weakened (da Silva et al., 2019).

15.3.2.2   Cheese

During the cheese-making process, many of the nutrients are released through the whey so the fortification of cheese with bioactive compounds has been considered in recent years. Different materials such as omega 3, enzymes, polyphenols, flavors, and vitamins have been added to different cheese types, to either enrich its nutritional quality or improve the ripening process (Rashidinejad, Birch, Sun-Waterhouse, & Everett, 2014). Due to the sensitivity of bioactive elements under environmental conditions and other compounds in food products, microencapsulation has been proposed to improve the formulation of cheese (Castro, Tornadijo, Fresno, & Sandoval, 2015).

Spray dried probiotic *Lactobacillus paracasei* NFBC 338 in skim milk powder was produced for Cheddar cheese manufactured by Gardiner et al. (2002). The high survival rate (84.5%) of *Lb. paracasei* contained 1×10^9 CFU/g cells was observed after spray drying and remained stable during seven weeks of storage at 4, 15, and 30°C. The control cheese was manufactured using pasteurized milk and 1.5% (w/v) starter culture only and probiotic cheese was a control sample with the addition of 0.1% (w/v) microencapsulated powder of *Lb. paracasei*. On the first day of production, the number of bacteria was determined 2×10^7 CFU/g. After three months of ripening, the cheese showed an increase in the number of bacteria to 7.7×10^7 CFU/g. The results showed that both cheeses were similar in chemical and sensory properties and scored high for flavor and texture, which was consistent with the degree of commercial cheddar in Ireland. The results demonstrated that microencapsulated *Lb. paracasei* was a useful tool for dairy products such as Cheddar cheese (Gardiner et al., 2002).

The *Bifidobacterium* BB-12 was added to ricotta cream through microencapsulation by the spray drying of reconstituted skim milk with inulin and reconstituted skim milk with oligofructose-enriched inulin, in order to evaluate the biocheese. Ricotta, which is an Italian whey cheese processed from sheep, cow, goat, or Italian water buffalo milk whey, is a leftover product from the manufacturing of other cheeses (Castro et al., 2015). To make the ricotta cream, the ricotta cheese was produced from the acid coagulation of cheese whey by adding lactic acid at 90°C. Protein precipitation occurred rapidly; after a short time, it was completely precipitated and drained. The ricotta cheese, butter, and skim milk were heated to 100°C and then cooled at 45°C. Both free and microencapsulated bacteria were added to ricotta cream and stored at 5°C for 60 days. The survival rate of *Bifidobacterium* BB-12 decreased during the storage period. The high protection of the microorganisms achieved in this study by reconstituted skim milk and inulin. The authors stated that prebiotics such as inulin had a bifidogenic effect and could be an effective wall material for the protection of probiotic bacteria in ricotta cream. According to the panelists, the biocheese with microencapsulated *Bifidobacterium* BB-12 in reconstituted skim milk and inulin showed more overall acceptability than other samples (Pinto et al., 2012).

Microencapsulated Swiss cheese bioaroma was produced by the spray drying method. This was produced via the fermentation of whey permeate (bioaroma) by *Propionibacterium freudenreichii*. Different values of inlet temperature and various combinations of modified corn starch and malto-dextrin (as a wall material) were used. The high retention of organic acids especially propionic acid (79.31%) and acetic acid (53.14%) was obtained at 175°C inlet temperature and 50% modified starch concentration. As the results revealed, lowest moisture content and water activity were observed in this condition. In addition, the effect of modified starch on organic acid retention was greater than that of maltodextrin. The low emulsifying capacity and the low volatile compound retention of maltodextrin can be one of the important reasons for the reduction of the amount of organic acid retention by increasing the amount of maltodextrin (da Costa et al., 2015).

15.3.2.3 Ice Cream

Ice cream is a frozen dairy dessert and is one of the most famous leisure foods in the globe (Fazaeli, Emam-Djomeh, & Yarmand, 2016). Ice cream formulations usually consist of milk, sweeteners, stabilizers, emulsifiers, flavorings, and coloring agents. The caloric value of ice cream is high, but it is not usually rich in dietary fiber or natural antioxidants such as vitamin C, natural pigments, and phenolic compounds (Erkaya, Dağdemir, & Şengül, 2012). According to the attractiveness of this dessert among consumers, the low production process, and storage temperature, and the ability to stabilize ingredients, it is considered an appropriate carrier for functional ingredients (Gabbi, Bajwa, & Goraya, 2018). So far, many functional ice cream formulations have been created with different components such as pomegranate peel extract (Çam et al., 2014) and pistachio green peel extract (Ghandehari Yazdi et al., 2020). Improvement of ice cream with these components is a good way to increase the nutritional value and functional properties of ice cream. On the other hand, it may have a negative effect on their texture, overrun, color, sensory characteristics, and melting properties of ice cream. The melting rate depends on different factors such as the formulations, ingredients used, the amount of air that penetrated the ice crystals, the rheological characteristic, and the network of fat globules formed during freezing (Marshall, 2003). Also, it has been shown that the texture characteristics of ice cream are dependent on the ice phase volume, size of ice crystals, fat destabilization, overrun, and the rheological attributes of the mixture.

The incorporation of probiotic bacteria (*Lactobacillus paracasei*) into ice cream has been studied (Spigno et al., 2015). In free form, probiotics can be provided by mixing an acidified/fermented milk base, while in such a form, probiotics can also be added before the whipping-freezing stage. Encapsulation facilitates the consumption of probiotics in the manufacturing process and also prevents damage to living cells during the process (Soukoulis, Fisk, & Bohn, 2014). In this study, probiotic cells were microencapsulated in two different formulations (F1 and F2) by using a spray dryer. The F1 consisted of 46% commercial skim milk powder, 24% anhydrous glucose, 28% maltodextrin (Maltrin® 40), and 2% sodium alginate. The second formulation (F2) was the same as F1 (Fibruline® instant) except for maltodextrin, which was replaced by inulin. The results indicated that encapsulation was solely viable with F2, but with a lower process efficiency (51%), higher aw (0.42), and 82% cell mortality. The results were related to the type of maltodextrin and species of microorganisms. In this research, 18% of mortality was observed after two days of storage at −18°C. These authors suggested that stability should be evaluated after more than two days (Spigno et al., 2015).

The addition of microcapsules of pomegranate peel extract to ice cream has been reported by Çam et al. (2014). They used maltodextrin as a carrier for pomegranate peel extracts by the spray drying method. Pomegranate peel extract is a good source of phenolic compounds, especially, punicalagins and ellagic acids (Çam et al., 2014). Antimicrobial, antioxidant, and antidiabetic properties of pomegranate peel extracts have been demonstrated (Al-Zoreky, 2009; Gautam & Sharma, 2012). The present phenolic compounds in pomegranate peel extract were able to inhibit the enzymes related to carbohydrate digestion; i.e., α-amylase and α-glucosidase. In this study, the optimal inlet temperature and core: wall ratio for encapsulation of pomegranate peel extracts were 160°C and 1: 1 or 3: 1 w/w, respectively. The researchers indicated that microcapsules containing phenolic compounds of pomegranate peel were stable at 4°C for 90 days. They showed that the addition of these microcapsules into ice cream improved the antioxidant and α-glucosidase inhibitory activities of ice cream compared with the control sample (without microcapsules of pomegranate peel extracts). From a sensory point of view, for 75% of the panelists, the sensory characteristics of the fortified ice cream were acceptable. The researchers suggested that microcapsules containing the extract could improve the functional properties of ice cream (Çam et al., 2014).

Ghandehari Yazdi et al. (2020) used the microcapsules of pistachio green peel extract to produce the functional ice cream. In this study, maltodextrin (as a wall) and spray drying method were used to encapsulate phenolic compounds of pistachio green peel. Pistachio green peel extract is rich in phenolic compounds and its antioxidant and antimicrobial activities have been proven (Rajaei, Barzegar, Mobarez, Sahari, & Esfahani, 2010). Researchers found that the most important phenolic compounds in pistachio green peel extracts were gallic acid and phloroglucinol (Barreca et al., 2016; Ghandahari Yazdi, Barzegar, Sahari, & Ahmadi Gavlighi, 2019). Ghandehari Yazdi et al. (2020) indicated that the incorporation of microcapsules of pistachio green peel extract (2%) into ice cream increased the total phenolic content and antioxidant activity about 2 and 3.7 folds, respectively, in comparison with the ice cream without microcapsules of pistachio green peel extracts. The researchers indicated that the addition of microcapsules had no remarkable effect on the overall liking of the ice cream. Also, they claimed that the enrichment of ice cream with microcapsules of pistachio green peel extract improved the melting resistance and first dripping times. The researchers showed that microcapsules of pistachio green peel extract had the potential to be implemented as a functional element in the ice cream section (Ghandehari Yazdi et al., 2020).

15.3.3 Cereals and Bakery Products

After dairy products, one of the major types of products in the functional food market is bakery products, which constitute roughly 20% of the market (Tolve et al., 2016). Due to the simple manufacturing process, the popularity of bakery products, and their market share, they can be used as one of the best carriers for the delivery of bioactive compounds. It should be noted that such an enrichment does not adversely affect the physicochemical properties of bakery products.

15.3.3.1 Bread

Bread is one of the main foods in the world, due to its nutritional properties. It is a good reservoir of complex carbohydrates, proteins, minerals, and vitamins (Roseli, 2008). Due to the high consumption of bread and its simple production process, several studies have been carried out to enrich it. Various compounds have been used in the formulation of functional bread: whole grains, fibers, vitamins, minerals, omega-3, soy, and probiotics (Hamaker, B. R., 2008; Roseli, 2008). The addition of these materials into a food matrix positions a substantial challenge because of the alterations in the physicochemical and sensory properties of bread. Encapsulation of bioactives can be an effective way to prevent these alterations. In several studies, the application of spray dried encapsulated bioactive compounds in bread has been evaluated (Table 15.2). In fortification, the factors that affect the quality of bread should be considered; e.g., loaf volume, specific volume, oven spring, bread firmness, and sensory properties (Hall & Tulbek, 2008).

The use of omega-3 fatty acids in the formulation of functional food and pharmaceuticals has been recommended for the promotion of consumer health. Fish oil is a great reservoir of omega-3 fatty acids. However, using it in food formulations is a big challenge since it is sensitive to oxidation and its individual distinctive flavor and odor. In addition, Frankel, Satué-Gracia, Meyer, and German (2002) revealed that iron increased the oxidation of fish oil and this is important, because bread contains iron. For these reasons, the encapsulation of fish oil is a good way to solve these problems. Yep, Li, Mann, Bode, and Sinclair (2002) reported that the bread containing microencapsulated fish oil was available in some countries. Also, Davidov-Pardo et al. (2008) evaluated the encapsulation of fish oil by spray drying (wall materials: methylcellulose, (WPI), and calcium-gelatin casein) and polymerization with transglutaminase (wall material: soybean protein isolate). The obtained microcapsules were added to the bread and its rheological and sensory properties were investigated. Results revealed that among the wall materials, soybean protein isolate and methylcellulose had the least change in the rheological and sensory properties of bread and dough, while they also had a lower oxidative rate than other wall materials. The spray drying method and methylcellulose as a wall material were recommended for the encapsulation of fish oil compared to the polymerization with the enzyme (Davidov-Pardo et al., 2008). In a similar study, Gallardo et al. (2013) evaluated the effect of microencapsulated linseed oil incorporated into bread. Microencapsulation of linseed oil was done by spray drying using various compounds (gum Arabic, maltodextrin, methylcellulose, and WPI) as wall materials. The highest encapsulation efficiencies and protection from oxidation were reported in the case of the microcapsules made of gum Arabic alone, and a combination of gum Arabic, maltodextrin, and WPI. Due to the efficiency of encapsulation and the simple formulation of microcapsules made of linseed oil and gum Arabic, they were chosen to enrich the bread. The results of this study were contrary to the expectations of the authors, because the amount of alpha-linolenic acid decreased significantly. This significant reduction could be a consequence of the bread production steps, such as adding water, dough kneading, incubation at 80% relative humidity (RH), or baking at 220°C, which could increase the oxidation rate of fatty acids (Gallardo et al., 2013). Similar results have been reported in fortifying bread with ingredients such as commercial micro-encapsulated algae, fish oils, and flax oil by Serna-Saldivar, Zorrilla, De La, Stagnitti, and Abril (2006). Gallardo et al. (2013) suggested that in order to prevent the decrease of alpha-linolenic acid, critical steps in the process of bread manufacture should be identified and the process optimized.

Considering the health benefits of probiotics, the probability of incorporating them into bread was evaluated by Altamirano-Fortoul et al. (2012). *Lactobacillus acidophilus* was microencapsulated in the watery dispersion of WPI, inulin, pectin, fresh agave sap, and carboxymethylcellulose (CMC) by spray drying and added to different starch suspensions. Probiotic coatings were used as follows: S1 (5% starch containing 1% microcapsules), S2 (5% starch containing 1% microcapsules and 5% starch solution), and S3 (5% starch, 2% microcapsules, and 5% starch). Before baking, the probiotic coatings were sprayed on the surface of the bread. After baking and storage time (24 h), a remarkable survival of *Lactobacillus acidophilus* was observed in the product. However, the highest

TABLE 15.2

Application of Spray Dried Encapsulated Bioactives in Bakery Products

Product	Core Material	Wall Material	Spray Drying Conditions	Results	References
Bread	Fish oil	Methylcellulose, WPI, calcium–gelatin casein	Inlet temperature: 160°C, outlet temperature: 78°C, air pressure: 0.4 bar, air flow rate: 20 mL/min, wall: core: 2: 1	Among the wall materials, methylcellulose had the least change in the rheological and sensory properties of bread and dough. The highest encapsulation efficiency was obtained by WPI (87.2%) and methylcellulose (84.1%), respectively.	Davidov-Pardo, Roccia, Salgado, León, and Pedroza-Islas (2008)
Bread	*Lactobacillus acidophilus*	A mixture of WPI, pectin, inulin, fresh agave sap, and carboxymethylcellulose, at a ratio of 45.67: 18.72: 22.83: 1.06: 11.72	Inlet temperature: 130°C, outlet temperature: 65°C, air pressure: 2 bars, air flow rate: 15 mL/min,	After baking the noteworthy survival of *Lactobacillus acidophilus* was observed in the product.	(Altamirano-Fortoul, Moreno-Terrazas, Quezada-Gallo, and Rosell (2012)
Bread	Curcumin	A mixture of starch and gelatin	Inlet temperature: 190°C, feed flow rate: 70 mL/min, drying airflow: 70 m3/h, wall: core: 30: 1	Encapsulation improved the solubility and thermal resistance of curcumin. At concentrations higher than 0.035% microcapsules of curcumin had a conservation effect.	Wang et al. (2012)
Bread	L-5-methyltetrahydrofolic acid with or without sodium ascorbate	Skim milk powder	Inlet temperature: 150°C, outlet temperature: 95°C, nozzle diameter, 0.5 mm, wall: sodium ascorbate: methyltetrahydrofolic: 99.9870: 1: 0.0130 g/L	After spray drying the recoveries of L-5-methyltetrahydrofolic acid were higher than 95%. In simulated gastric fluid (10 min at 37°C), microcapsules were released from the coating. Encapsulation improved stability during storage.	Tomiuk et al. (2012)

(Continued)

Food product	Core	Wall material	Process conditions	Results	Reference
Cake	Lycopene	Modified starch	Inlet temperature: 180°C, outlet temperature: 98°C, nozzle diameter, 0.5 mm, feed flow rate: 10 mL/min, core concentration: 5%, 10%, and 15%	The encapsulation efficiency was in the range of 21%–29%. Encapsulation improved the stability of lycopene.	Rocha, Fávaro-Trindade, and Grosso (2012)
Pasta	Garcinia cowa fruit extract	WPI	Inlet temperature: 145°C and 155°C, outlet temperature: 90 and 105°C, air pressure: 50 psi, feed flow rate: 6 mL/s, wall: core: 1:1 and 1.5:1	In the optimum conditions (wall: core ratio 1.5:1; 90°C), the recovery of total hydroxycitric acid and free hydroxycitric acid was 89.13% and 94.49%, respectively. In these conditions, antioxidant properties and sensory characteristics of samples containing microcapsules were greater than the control sample.	Pillai, Prabhasankar, Jena, and Anandharamakrishnan (2012)
Bread	L-5-methyltetrahydrof olic acid with or without sodium ascorbate	Modified starch	Inlet temperature: 135°C, outlet temperature: 90–95°C, nozzle diameter: 0.5 mm, wall: sodium ascorbate: core: 9: 0: 1, 99: 0: 1, 99: 1: 0.1, 99: 1: 0.01	Encapsulation improved the stability of L-5-methyltetrahydrofolic acid during storage. The highest recovery of L-5-methyltetrahydrofolic acid was 120% (wall: sodium ascorbate: core: 99: 1: 0.01).	Liu, Green, Wong, and Kitts (2013)
Bread	Linseed oil	gum Arabic, a combination of Arabic gam and maltodextrin, the combination of gum Arabic, maltodextrin, and WPI, a combination of maltodextrin and methylcellulose	Inlet temperature: 175°C, outlet temperature: 75°C, air pressure: 2.8 bar, air flow rate: 15 mL/min, gum Arabic (112 g): oil ratio: 4, gum Arabic (72 g) and maltodextrin (56 g): oil ratio: 3.9, gum Arabic (22 g), maltodextrin (85 g), and WPI (22 g): oil ratio: 4, and maltodextrin (10 g), and methylcellulose (20 g): oil ratio: 2	Microcapsules prepared of gum Arabic (100%) and mixtures off gum Arabic, maltodextrin, and WPI presented the greatest microencapsulation efficiencies (higher than 90%) and conservation from oxidation. The appearance of the fortified breads was similar to the control sample (without microcapsules). After preparation, the amount of α-linolenic acid decreased.	Gallardo et al. (2013)

(Continued)

TABLE 15.2
(Continued)

Product	Core Material	Wall Material	Spray Drying Conditions	Results	References
Bread	Garcinia fruit extract	WPI, maltodextrin, a mixture of both at a ratio of 1 : 1	Inlet temperature: 140°C, outlet temperature: 95°C, air pressure: 0.3447 Mpa, feed flow rate: 2.1 mL/s, wall: core: 1: 1	Microcapsules prepared of maltodextrin showed the highest free (86%) and net (90%) recovery. The amount of (−)-hydroxycitric acid in the bread was sufficient to claim the functional products.	Ezhilarasi, Indrani, Jena, and Anandharamakrishnan (2014)
Biscuit	Garden cress seed oil	Whey protein concentrate	Inlet temperature: 180°C, outlet temperature: 90°C, feed flow rate: 5.0 L/h, core: wall: 0.4. Air pressure: 300 kPa, oil: protein ratio: 0.4	Encapsulation improved the stability of the alpha-linolenic acid, as well as the shelf-life of biscuits under different storage conditions. The sensorial attributes of enriched biscuits were acceptable.	Umesha, Manohar, Indiramma, Akshitha, and Naidu (2015)
Noodles	L-5-methyltetrahydrofolate with Sodium ascorbate	Modified starch	Inlet temperatures: 135°C, outlet temperatures: 90°C–95°C, nozzle diameter: 0.5 mm, L-5-methyltetrahydrofolate: starch: sodium ascorbate: 0.1: 91: 9	Encapsulation improved the stability of methyltetrahydrofolate.	Liu, Green, and Kitts (2015)
Bread	Green tea polyphenols	β-cyclodextrin, Maltodextrin, a mixture of both	Inlet temperature: 120°C, outlet temperature: 60°C, air pressure: 165.47 KPa, wall: core: 1 : 1	Maltodextrin was chosen as the best wall because of encapsulation efficiency and antioxidant properties. The physicochemical properties of enriched bread were similar to the control sample (without green tea extract). The overall liking of enriched samples was lower than the control.	Pasrija, Ezhilarasi, Indrani, and Anandharamakrishnan (2015)

Biscuit	Ascorbic acid	gum Arabic	Inlet temperature: 150°C, outlet temperature: 75°C, feed flow rate: 8 mL/min, nozzle diameter: 0.7 mm, atomization gas flow: 10 L/min, aspiration: 35 m3/h wall: core: 4: 1	The encapsulation efficiencies were about 100%. Encapsulation improved the stability of ascorbic acid during the preparation of biscuits.	Alvim, Stein, Koury, Dantas, and Cruz (2016)
Cookie	Fish oil	Maltodextrin, fish gelatin	Inlet temperature: 160°C, outlet temperature: 80°C, feed flow rate: 5 mL/min, air flow rate: 0.5 L/h, nozzle diameter: 3 mm, air pressure: 4 bar, wall: core: 2:1	The encapsulation protected the fish oil from oxidation in the formulation of cookies. Sensory properties of enriched cookies with fish oils microcapsules (prepared with maltodextrin) were acceptable. Cookies containing fish microcapsules had a softer texture than controls (without fish oil).	Jeyakumari, Janarthanan, Chouksey, and Venkateshwarlu (2016)
Biscuit	Shrimp oil	A mixture of sodium caseinate, fish gelatin, and glucose syrup at a ratio of: 1: 1: 4	Inlet temperature: 180°C, outlet temperature: 90°C, feed flow rate: 8.08 mL/min, air flow rate: 4.3 m/s, air pressure: 40.61 psi, nozzle diameter: 0.5 mm	The incorporation of shrimp oil microcapsules decreased the thickness while increased the spread ratio and hardness of biscuits. Biscuit enrichment with shrimp oil microcapsules was acceptable up to 6%. The encapsulation protected the shrimp oil from oxidation in the formulation of cookies.	Takeungwongtrakul & Benjakul, 2017

(Continued)

TABLE 15.2
(Continued)

Product	Core Material	Wall Material	Spray Drying Conditions	Results	References
Biscuit	Artemide black rice	Maltodextrins, gum Arabic, a mixture of both at a ratio of 50: 50	Inlet temperature: 150°C, feed flow rate: 7 mL/min, air flow rate: 40 m³/h, nozzle diameter: 0.7 mm	The highest resistance to thermal degradation was obtained with a mixture of maltodextrin and gum Arabic (50:50). With the addition of microcapsules, the amount of phenolic compounds and antioxidant properties increased compared to the control sample.	Papillo et al. (2018)
Cake	*Bifidobacterium bifidum*, *Lactobacillus acidophilus*, *Saccharomyces boulardii*	A mixture of gum Arabic and β cyclodextrin at a ratio of 9: 1	Inlet temperature: 120°C, outlet temperature: 50°C, air pressure: 0.3 bar, Feed flow rate: 16.5 mL/min, aspiration rate: 38 m³/h, core: wall ratio: 1:20	*L. acidophilus* and *S. boulardii* were suggested for use in bakery products compared to *B. bifidum*, due to thermal resistance. Microcapsules did not have good stability and the shelf-life of these cakes should not be more than 30 days.	Arslan-Tontul, Erbas, and Gorgulu (2019)
Biscuit	Cocoa hulls extract	Maltodextrin, gum Arabic, a mixture of both	Inlet temperature: 150°C, outlet temperature: 80°C, feed flow rate: 7 mL/min, air flow rate: 40 m³/h, maltodextrin: gum Arabic: dry extract: 80: 0: 20, 64: 16: 20, 40: 40: 20, 16: 64: 20, and 0: 80: 20	Maltodextrin was selected as the best wall material and the optimal ratio of the wall to the core was 80 to 20. Antioxidant properties and stability of phenolic compounds improved by encapsulation.	Papillo et al. (2019)

reduction was found in S3. Although this approach affected the physicochemical attributes of the bread, the sensory evaluation showed that the product had a good overall liking. The authors suggested that S2 and S3 could be used to produce such a functional bread product.

Curcumin is one of the natural colorants being applied in the food industry that has several pharmacological properties such as antimicrobial, antidiabetic, and anticancer, and anti-inflammatory potentials (Rafiee, Nejatian, Daeihamed, & Jafari, 2019; Sampathu, Lakshminarayanan, Sowbhagya, Krishnamurthy, & Asha, 2000). Antimicrobial effects of curcumin against microorganisms such as *Bacillus subtilis*, *Escherichia coli*, and *Staphylococcus aureus* has been reported by Egan et al. (2004). However, free curcumin cannot dissolve in water and is vulnerable to decomposition during the process because of the presence of oxygen, light, and high temperatures (Liang, Meng, & Lei, 2007). Curcumin was also microencapsulated in a mixture of starch and gelatin by spray drying. The microcapsules were incorporated into bread and their preservative properties were investigated (Wang et al., 2012). The results showed that the thermal resistance and solubility of microcapsules were higher than free curcumin. The results also showed that at concentrations greater than 0.035%, the curcumin microcapsules showed preservation properties in the manufactured bread product. In conclusion, the authors suggested that curcumin microcapsules could be used as an appropriate food preservative (Wang et al., 2012). Vitaglione et al. (2012) too indicated that the encapsulation of curcumin could improve its bioavailability within enriched bread.

Folate deficiency is a well-known risk factor that contributes to anemia and poor pregnancy consequences. To prevent this disease, food products are enriched with folic acid, due to its stability and high bioavailability (Food and Drug Administration, 1996). Folic acid is an artificial form of folate. The enrichment of foods with folic acid can positively decrease the risk of neural tube defects. The absorption of vitamin B_{12} may decline by fortifying food products with folic acid, because the hematologic signs of vitamin B_{12} deficiency will be masked by folic acid (Weir & Scott, 1999). L-methyltetrahydrofolate can be regarded as a good substitute for folic acid to fortify food products, which, unlike folic acid, does not involve a vitamin B_{12} deficiency. Tomiuk et al. (2012) microencapsulated L-5-methyltetrahydrofolic acid (with or without sodium ascorbate) by spray drying using skim milk powder as wall material and then added to bread. Encapsulation recoveries of L-5-methyltetrahydrofolic in microencapsulated ingredients were higher than 95%. Skim milk powder was performed as a protective coating and bulking agent. Therefore, the dispersion of L-5-methyltetrahydrofolic acid microcapsules was better than their free form throughout the bread dry premix. The immediate recovery results after baking showed that the microencapsulation had improved the stability of L-5-methyltetrahydrofolic in baking processes in comparison with its free form. The sodium ascorbate acted as an antioxidant and reduced the loss of L-5-methyltetrahydrofolic acid. Storage stability results indicated that the concentration of L-5-methyltetrahydrofolic acid in microcapsules with and without sodium ascorbate after seven days of storage decreased about 9.3 and 42.3%, respectively. Overall, the researchers determined that encapsulated L-5-methyltetrahydrofolic could be successfully added to bread, although at the same time its bioavailability should also be considered (Tomiuk et al., 2012). Similar results were reported by Liu et al. (2013), who encapsulated L-5-methyltetrahydrofolic acid in modified starch with or without sodium ascorbate by spray drying and then incorporated them into bread. The authors concluded that the encapsulation with a high core: wall ratio enhanced the stability of L-5-methyltetrahydrofolic during the baking and the storage of bread. Also, they found that sodium ascorbate improved the recovery of methyltetrahydrofolic acid (Liu et al., 2013).

Hydroxycitric acid is naturally present in rinds of *Garcinia cowa*. It is sensitive to high temperature and is transformed into lactone during the thermal process. Garcinia fruit extract, rich in (−)-hydroxycitric acid, was microencapsulated in three various wall materials (WPI, maltodextrin, and a mixture of WPI and maltodextrin) by spray drying and then incorporated to bread. The results revealed that the recovery percent of (−)-hydroxycitric acid within maltodextrin was higher than other walls. Additionally, bread with maltodextrin encapsulates had considerable physicochemical and sensory features and maintained higher amounts of (−)-hydroxycitric acid during the process.

Finally, it was recommended that spray drying could be considered as an appropriate technique for the encapsulation of (−)-hydroxycitric acid and its retention in the bakery process (Ezhilarasi et al., 2014).

Pasrija et al. (2015) investigated the effect of spray dried microencapsulated green tea polyphenols on bread quality characteristics. Antioxidant and anti-carcinogenic properties of green tea polyphenols have been proven. Catechins, epicatechins, epigallocatechin, epicatechin gallate, epigallocatechin gallate, and gallic acid were detected as the main phenolic compounds of green tea extract (Rashidinejad, Birch, & Everett, 2016; Yang & Koo, 1997). Green tea polyphenols were microencapsulated using maltodextrin, β-cyclodextrin, and a combination of them. The results revealed that among all wall materials, maltodextrin possessed higher encapsulation efficiency and antioxidant levels. Researchers found that the physicochemical properties of bread containing green tea extract and its microcapsules were similar to the control sample (which did not contain green tea extract). However, the findings of the sensory evaluation indicated that the overall liking of enriched samples was lower than the control. After baking, the amount of phenolic compounds decreased and the amount of these compounds in treatments containing green tea extract and its microcapsules was 202.71 and 204.31, respectively.

15.3.3.2 Pasta

Pasta is one of the most interesting foods consumed globally due to its low price, easy cooking, and terrific taste. Pasta is usually made from durum wheat semolina, which is full of calories but empty of dietary fibers, minerals, and vitamins (Kamali Rousta, Ghandehari Yazdi, & Amini, 2020). In general, the production process of pasta is simple. For the preparation of pasta, semolina and water are mixed continuously in the chamber and the mixture is laminated or extruded through dies that form the desired shape. After that, it is dried in the dryer to achieve a moisture content of 8%–9% (Bustos, Perez, & Leon, 2015).

Due to the limited raw materials in the production of pasta, the simple production process, and its consumption by all ages and sections of society, it can be used as a good carrier for food enrichment by using various ingredients. So far, several studies have been carried out on the enrichment of pasta with various ingredients, such as *garcinia cowa* fruit extract (Pillai et al., 2012) and L-5-methyltetrahydrofolate (Liu et al., 2015).

Enrichment with various compounds improves the nutritional aspect and functional features of pasta. However, based on the findings of studies, such an enrichment may affect the sensory, texture, cooking (optimal cooking time and cooking loss), and color characteristics of pasta (Kamali Rousta et al., 2020). Van Boekel et al. (2010) and Oliviero and Fogliano (2016) reported that the alteration of color upon fortification with vegetables and herbs is a positive feature while changes in the taste of the products must be evaluated, due to the taste of non-conventional ingredients. Also, Marchetti et al. (2018) indicated that the fortification by functional ingredients can lead to weakness of the gluten-starch network and adversely influence starch gelatinization. As mentioned earlier, it is better to encapsulate these compounds in order to prevent the unpleasant taste, destruction, and interaction of functional ingredients with other compounds during the production process.

Pillai et al. (2012) reported the manufacture of a pasta product containing microcapsules of *garcinia cowa* fruit extract. This fruit is a good source of hydroxycitric acid and its antioxidant and antimutagenic properties have been proven (Jena, Jayaprakasha, & Sakariah, 2002; Negi, Jayaprakasha, & Jena, 2010). In this study, WPI (as a wall material) and the spray drying technique were used to encapsulate *arcinia cowa* fruit extract. The effect of the wall: core ratio (1: 1 and 1.5: 1) and outlet temperatures (90 and 105°C) on the encapsulation efficiency of *garcinia cowa* fruit extract was evaluated. The conditions for conducting this study can be seen in Table 15.2. These researchers showed that by increasing the outlet temperature and wall material content, the moisture of spray drying products decreased. This study showed that wall: core ratio and outlet temperature were effective factors for the encapsulation efficiency and in the optimum conditions (wall: core ratio 1.5: 1; 90°C); the recovery of total hydroxycitric acid and free hydroxycitric acid was 89.13% and 94.49%, respectively. Antioxidant activity and sensory properties of pasta samples containing microcapsules of *garcinia cowa* fruit extract (wall: core ratio of 1.5: 1; 90°C) were higher than the control sample (without microcapsules). They observed that the addition of microcapsules (produced in optimal conditions) into pasta decreased the cooking loss. The authors suggested that a high amount of WPI led to the strengthening of the protein-starch network (Pillai et al., 2012).

Liu et al. (2015) incorporated L-5-methyltetrahydrofolate microcapsules into noodles and studied their stability. In this study, for the encapsulation of L-5-methyltetrahydrofolate, sodium ascorbate (as a stabilizer), modified starch (as a wall material), and spray drying techniques were used. The results revealed that the stability of L-5-methyltetrahydrofolate increased due to the spray drying encapsulation. The rate of decrease of L-5-methyltetrahydrofolate in cooked noodles in free form and encapsulated was 68% and 28%, respectively (based on the initial amount in the flour). Accordingly, the authors suggested that the obtained microcapsules could be used in the production of such noodle products and could enhance dietary folate intake since pasta is one of the main dishes of the Asian population.

15.3.3.3 Cakes

Cakes are one of the most favorite bakery products worldwide, which are usually consumed as a snack/dessert. Cakes usually have a sweet taste and their formulation includes flour, sugar, eggs, fat, and flavor agents (Pasukamonset et al., 2018). Several studies have been carried out on cake enrichment with various bioactive compounds (Lu, Lee, Mau, & Lin, 2010; Pasukamonset et al., 2018), although limited experiments have been carried on the application of spray dried encapsulated ingredients in cakes (Arslan-Tontul et al., 2019; Rocha et al., 2012).

Lycopene is used in food formulations for various purposes such as improving the health benefits and adding as a colorant (Goula & Adamopoulos, 2012). Lycopene has a great potential for isomerization and oxidation during the manufacturing process and storage, because of the high number of its conjugated double bonds. Encapsulation is a good way to increase the stability of these compounds. Rocha et al. (2012) microencapsulated lycopene by spray drying using modified starch as the wall material. They reported that the efficiency of encapsulation was variable (21%–29%) and the stability of lycopene was improved by encapsulation. The researchers observed that by increasing the quantity of core from 5% to 15% the encapsulation efficiency decreased by about 38%. In all treatments, by increasing the temperature of storage there was a reduction in the level of lycopene retention. Therefore, the retention of these compounds depends on the temperature. Authors found that with the addition of lycopene microcapsules into the cake, a* value (which represents redness) increased significantly (Rocha et al., 2012).

Bifidobacterium bifidum, *Lactobacillus acidophilus*, and *Saccharomyces boulardii* were microencapsulated in the mixture of gum Arabic and β-cyclodextrin through the spray drying method. After baking, the microcapsules were added to various kinds of cakes. In addition, microcapsules were injected into the center of the cakes before baking (T). The survival of probiotics was investigated during the storage period of 90 days. The results indicated that *Saccharomyces boulardii* and *Bifidobacterium bifidum* were not detected in T. Storage stability analysis showed that the microcapsules did not have favorable stability, while the shelf-life of these cakes did not exceed 30 days under refrigerated conditions. Since the thermal resistance of *L. acidophilus* and *S. boulardii* was higher than *B. bifidum*, the authors recommended that *L. acidophilus* and *S. boulardii* could be used in the formulation of these products (Arslan-Tontul et al., 2019).

15.3.3.4 Biscuits

The main ingredients for the production of biscuits are flour, sugar, and oil. Biscuit is popular due to its features such as variety in shape and taste, high shelf-life (six months or longer), affordability, suitability for different age groups, and easy access (Ahmad & Ahmed, 2014; Gandhi et al., 2001; Mamat & Hill, 2018). In terms of nutritional value, biscuits have a high amount of carbohydrates, a relatively small content of plant proteins, and a limited number of micronutrients. Thus, the final product is poor in terms of nutritional composition and can be improved by supplementing with functional compounds (Ahmad & Ahmed, 2014).

As can be seen in Table 15.2, several experiments have been implemented on the application of spray dried encapsulated bioactive compounds in biscuits. For example, Umesha et al. (2015) encapsulated Garden cress (*Lepidium sativum*) seed oil, which is a rich source of alpha-linolenic acid, within whey protein concentrate by spray drying with an encapsulation efficiency of 64.8% and a particle size of 15.4 μm. The obtained microcapsules (20 g/100 g) and Garden cress seed oil (5 g/100 g) were added to the biscuits. The authors found that the alpha-linolenic acid content of samples containing microcapsules and Garden cress seed oil were 1.02 and 1.05 g/100 g, respectively. As expected, encapsulation increased the stability of the alpha-linolenic acid, as well as the shelf-life of biscuits under different storage conditions (90% RH at 38°C for three months, 30%–40% RH at 38°C–40°C for four months, and 65% RH at 27°C for five months), compared with Garden cress seed oil-enriched samples. These researchers found that the sensorial characteristic of biscuit enriched with Garden cress seed oil microcapsules was acceptable. Evaluation of the physical properties of biscuits showed that enrichment with Garden cress seed oil microcapsules increased the thickness, hardness, a*, and b* while the density, spread ratio, weight, and L* decreased it. These researchers declared that the microcapsules containing Garden cress seed oil could boost the nutritional value of biscuits (Umesha et al., 2015).

Jeyakumari et al. (2016) microencapsulated fish oil in maltodextrin or fish gelatin, then embedded them into cookies. The efficiency of encapsulation with maltodextrin was higher than fish gelatin. In this study, the use of fish oil microcapsules to enrich cookies was compared to other forms (fish oil as such, fish oil-in-water emulsion). The results showed that the encapsulation process decreased the oil oxidation compared to other treatments. Physicochemical studies have shown that cookies containing fish oil microcapsules had lower brightness and hardness than other samples. Concerning the sensory evaluations, it appeared that fish oil microcapsules containing maltodextrin were comparable with the control samples. Eventually, they concluded that fish oil microcapsules could be successfully embedded into the cookies and improve the nutritional intake of omega-3 fatty acids. These authors declared that more studies were needed to discover the bioavailability of omega-3 fatty acids in these products (Jeyakumari et al., 2016).

In another similar study, encapsulation of shrimp oil within the mixture of sodium caseinate, fish gelatin, and glucose syrup by spray drying was studied by Takeungwongtrakul and Benjakul (2017). Lipids from the hepatopancreas are considered an inexpensive source of valuable polyunsaturated fatty acids and astaxanthin. The main fatty acids of shrimp are linoleic acid and oleic acid, respectively (Takeungwongtrakul & Benjakul, 2017). In addition, Takeungwongtrakul, Benjakul, and Aran (2012) reported that shrimp oil contains other fatty acids namely eicosapentaenoic acid and docosahexaenoic acid (Takeungwongtrakul et al., 2012). Notwithstanding the high nutritional value, the use of shrimp oil in the formulation of bakery products is challenging, due to the susceptibility of these compounds to oxidation. The results revealed that by increasing the concentration of shrimp oil microcapsules, the spread ratio, hardness, redness, and yellowness increased, while the thickness decreased. These researchers realized that the samples should be stored in the dark because of the acceleration of oxidation by light. Results of the sensory analysis revealed that the addition of shrimp oil microcapsules (up to 6%) had no profound effect on the overall acceptability of the product. They reported that the obtained microcapsules could be used to enrich the nutritional quality of biscuits (up to 6%) (Takeungwongtrakul & Benjakul, 2017).

Alvim et al. (2016) successfully encapsulated ascorbic acid in gum Arabic by spray drying. The average diameter of microcapsules and encapsulation efficiency was about 9.3 μm and 100%, respectively. This research demonstrated that gum Arabic was appropriate for the protection of ascorbic acid from degradation in the cooking process. The amount of ascorbic acid loss in treatments containing free ascorbic acid and ascorbic acid microcapsules was 28% and 11%, respectively. The researchers reported that enrichment with ascorbic acid microcapsules had a significant effect on the physicochemical features of biscuits. The water activity and hardness of the samples containing microcapsules were higher than the control samples (without ascorbic acid). After cooking, they found dark spots on the surface of biscuits containing ascorbic acid in its free form while these dark spots were not seen in samples containing microcapsules. The formation of dark spots could be a consequence of thermal degradation or oxidation of ascorbic acid. The authors concluded that the spray drying technique could be used to obtain microcapsules as potential conservation tools while using susceptible ingredients in the bakery industry (Alvim et al., 2016).

The health benefits of pigmented rice have been proven for humans (Samyor, Das, & Deka, 2017). Italian black rice (*Oryza sativa L.*, var. Artemide) is known as a rich reservoir of phenolic compounds, especially anthocyanins (Bordiga et al., 2014). As mentioned earlier, the use of phenolic compounds in food formulations is challenging. Papillo et al. (2018) microencapsulated phenolic compounds from Artemide extract by spray drying using maltodextrin, gum Arabic, and a mixture of them as wall materials. The encapsulated powders attained were added into biscuits in order to investigate the effect of baking on the stability of the phenolic compounds. The authors revealed that the drying temperature had no effect on the amount of total phenolic content. In contrast, they observed that the drying and microencapsulation process had a major effect on the total anthocyanins, condensed tannins, and antioxidant activity. The stability analysis during storage at −20°C for 30 days showed that the combination of maltodextrin and gum Arabic (50: 50 w/w) was better in preserving bioactive compounds than using them alone and the term of storage did not affect the amount of total phenolic content. The amount of total phenolic content, anthocyanins, and antioxidant activity in the baked biscuit were 61%–65%, 31%–34%, and 69%–73% lower than the initial product before baking, respectively. Overall, they proposed that encapsulated Artemide black rice extract could be applied in the bakery industry (Papillo et al., 2018).

Similarly, the encapsulation of cocoa hulls extract and its incorporation into a biscuit product was studied by Papillo et al. (2019). Cocoa hull is a by-product of cocoa production that is rich in phenolic compounds (Dias et al., 2015). Microencapsulation of phenolic compounds from cocoa hulls extract was done by spray drying using maltodextrin, gum Arabic, and a mixture of them as wall materials. The stability investigation during storage at −20°C for 90 days indicated that encapsulation improved the stability of phenolic compounds of cocoa hulls extract and their antioxidant activity. After baking, the maximum quantity of phenolic compounds and antioxidant activity were observed in the treatment consisting of 20% extract and 80% maltodextrin. The authors claimed that more studies are required to improve the properties of microcapsules (such as drying conditions) and their effect on the physicochemical properties of biscuits (Papillo et al., 2019).

15.3.4 CONFECTIONERY PRODUCTS

In confectionery products, encapsulated sweeteners, sugar and sugar derivatives, flavors, colorants, acidulants, and foaming agents can keep the tone of color, prevention of sugar inversion, flavor loss inhibition, moisture sorption, and depigmentation control of the product. They also help maintain brightly colored candies and lead to a blasting release of some specified flavors.

15.3.4.1 Chewing Gum

An extended period of controlled release, unaltered flavor ingredients, releasing of flavor during chewing are among the advantages of chewing gum containing flavors dried by spray drying. In comparison with the free form of sweeteners, the encapsulated high-intensity sweeteners will have controlled sweetness intensities and slower releasing during chewing (Anandharamakrishnan & Ishwarya, 2015a). There are some patents on the encapsulation of aspartame by spray drying in order to increase its stability in chewing gum. Bahoshy and Klose (1978), for example, used gum Arabic and dextrin as wall materials. The aspartame stability was almost twice using encapsulation by spray drying. In an experiment conducted by Cea, Posta, and Glass (1983), researchers showed that microencapsulation by atomization could double the stability of aspartame when it was encapsulated by zein and shellac. Schobel and Yang (1989) used hydrophobic polymers and hydrophobic plasticizers to encapsulate sweeteners. The hydrophobic polymers had film-forming capabilities such as ethylcellulose and phthalic acid esters. The plasticized hydrophobic polymer, which caused the release of sweetness from chewing gum, was significantly extended when aspartame was encapsulated by it. In order to granulate sweetener, hydroxypropylmethycelluse (HPMC) was dispersed in ethanol, after adding water and passing through a standard screen dried at about 60°C.

15.3.4.2 Chocolate

Chocolate is very popular among consumers and its main ingredient is cocoa butter and is available in different forms, textures, and colors. Chocolate can be considered a functional food because of its high flavonoid content. Today, functional chocolates, such as those containing probiotics and prebiotics, are available in the market. Accordingly, chocolate is a good carrier for bioactive compounds. Some of the recent studies on the encapsulation of bioactive compounds by spray drying and their incorporation into chocolate are listed in Table 15.3.

The incorporation of spray dried encapsulated *Lactobacillus plantarum* 564 and *Lactobacillus plantarum* 299v in skim milk (Table 15.3) into a dark chocolate product was investigated. Dark chocolate with 75% cacao was produced and encapsulated bacteria were added to the chocolate after tempering while the temperature was lower than 40°C. During the first two months of storage at room temperature, the cell numbers of both spray dried probiotic bacteria were more than 8 log units. After 360 days of storage, the cell numbers of probiotic bacteria decreased slightly and, finally, they counted more than 5 Log units. According to these results, both probiotic bacteria remained more stable during the first 90 days of storage. In addition, there was no significant difference in chemical composition and sensory evaluation between the control samples and probiotic chocolate, which indicates the lack of effect of probiotic bacteria on the chemical and sensory characteristics of dark chocolate. Sensory evaluation of panelists showed that the control sample and dark probiotic chocolate showed very good sensory quality after 60 and 180 days of storage, especially in taste and texture. Due to the high viability of bacterial cells and acceptable sensory properties, the authors stated that the encapsulated probiotics *Lb. plantarum 564* and *Lb. Plantarum 299v* could be used successfully to produce dark probiotic chocolate and this could be used as a functional product that provides probiotics to the consumers (Mirković et al., 2018).

In another study with the aim of reducing cholesterol levels in individuals, phytosterols were added into several dark chocolates that contained 64%, 72%, and 85% cocoa powder (Tolve et al., 2018). The direct use of phytosterols in food formulations is challenging, because they are susceptible to degradation in the process and storage. Other limitations of direct application of phytosterols in the formulation of products are as follows: unpleasant taste, insolubility in water, low solubility in fats and oils, and interaction with ingredients (Alvim, Souza, Koury, Jurt, & Dantas, 2013). Consequently, the encapsulation of these compounds is an effective way to increase their physical stability. In this study, microencapsulation of phytosterols was performed by spray drying using WPI as the wall material and then they were incorporated into dark chocolates at different levels of 0%, 5%, 10%, and 15%.

The particle size of the microcapsules is important and should be less than 20 μm because the particle size has a significant effect on sensorial attributes (Zawistowski, 2010). The results showed that the particle size distribution of the prepared chocolates was detected to be lower than 30 μm. These researchers reported that by increasing the concentration of cocoa, the antioxidant properties increased, which may be a result of the increase in the concentration of phenolic compounds. The results showed that chocolate containing 85% cocoa was suitable for the production of the functional product. Sensory evaluation showed that the produced chocolate was acceptable. These authors recommended that the daily consumption of 10 g of chocolates enriched with 15% of phytosterols microcapsules could supply about 25%–30% of 0.8 g/day, the least amount needed to experience a 10% decrease in the LDL cholesterol.

The cinnamon oleoresin is rich in antioxidants and usually is extracted by ethanol from *Cinnamomum burmannii*. The high viscosity and sticky texture of cinnamon oleoresin have

TABLE 15.3

Application of Spray Dried Encapsulated Bioactives in Confectionery Products

Product	Core Material	Wall Material	Spray Drying Conditions	Results	References
Chewing gum	Aspartame	A mixture of gum Arabic and dextrin	Inlet temperature: 190°C–218°C, outlet temperature: 93°C–121°C	Encapsulation of aspartame and applied it in chewing gum resulted in reducing its decomposition rate.	Bahoshy and Klose (1978)
Chewing gum	Aspartame	Cellulose, cellulose derivatives, arabinogalactan, gum Arabic, polyolefins, waxes, vinyl polymers, gelatin, and zein	Inlet temperature 37°C–51.5°C	The shelf stability of products containing the encapsulated aspartame is almost doubled. It is suggested to apply to chewing gum formulations.	Cea et al. (1983)
Chewing gum	Aspartame	Hydrophobic polymer, plasticizer	Inlet temperature 40°C–80°C	The encapsulated aspartame was protected from deterioration due to its moisture and provides a controlled release in chewed during consumption.	Schobel and Yang (1989)
Dark chocolate	Phytosterols	WPI	Inlet temperature: 155°C, outlet temperature: 90°C and 105°C, air pressure: 50 psi, flow rate: 0.5 l/h, spray flow: 600 L/h, orifice diameter: 0.7 mm	The content of cocoa and the concentration of microcapsules and their interaction had a significant effect on antioxidant properties. Eating of 10 g of chocolates (enriched with 15% of microcapsules) could provide about 25%–30% of 0.8 g/day, the minimum content necessary to identify a decrease of 10% of the LDL cholesterol.	Tolve et al. (2018)
Dark chocolate	*Lactobacillus plantarum*	Skim milk	Inlet temperature: 170°C, outlet temperature: 80°C	Both encapsulated probiotic bacteria remained more than 8 log units stable during the first 60 days of storage and no significant effect on taste and texture	Mirković et al., 2018

(Continued)

TABLE 15.3 (*Continued*)

Product	Core Material	Wall Material	Spray Drying Conditions	Results	References
Gummy candy	Eggplant peel extract	Maltodextrin, gum Arabic, and their combination: 1: 1	Inlet temperature: 140°C and 170°C, outlet temperature: 75°C, air pressure: 4.5 bar, feed flow rate: 15 mL/ min, feed temperature: 25°C	Wall material and inlet temperature influenced the physicochemical of the powder. The highest amount of phenolic compounds and antioxidant properties were obtained at 170°C by using maltodextrin. The overall liking and color of the gummy candy improved with the addition of obtained microcapsules.	Sarabandi, Jafari, Mahoonak, and Mohammadi (2019)
Dark chocolate	Cinnamon oleoresin	A mixture of gum Arabic and maltodextrin with a ratio of 3: 1	Inlet temperature: 100.2°C, feed flow rate: 15–20 mL/ min, exhaust temperature: 74°C	The encapsulated cinnamon oleoresin showed the highest remained phenol content and antioxidant activities during the thermal process with no significant effect on taste and texture compared with the control sample.	Praseptiangga, Invicta, and Khasanah (2019)

prevented its use in many products. In addition, the presence of highly volatile and heat-sensitive compounds has limited cinnamon oleoresin consumption. To solve these problems, microencapsulation of cinnamon oleoresin was studied with a mixture of gum Arabic and maltodextrin (3:1) by spray drying (Table 15.3). These microcapsules were then added to the dark chocolate bar in varying amounts (from 4% to 8%) during the refining-conching process before the tempering process. The study of chemical composition showed the highest phenol content and antioxidant activities and the lowest moisture content were in the sample with 8% microencapsulated cinnamon oleoresin dark chocolate bar. The panelists did not distinguish between control and the microencapsulated cinnamon oleoresin samples. In addition, in comparison with control samples containing cinnamon, it was observed that in samples containing cinnamon, a higher amount of phenolic and antioxidant compounds remained during the thermal process, which indicates the protective effect of their encapsulation (Praseptiangga et al., 2019).

In another work, eggplant peel extract was microencapsulated in maltodextrin and gum Arabic by spray drying (Sarabandi et al., 2019). Eggplant peel is a waste product and a cheap source of different natural colors and antioxidants such as delphinidin 3-rutinoside and delphinidin 3-(p-coumaroyl rutinoside)-5-glucoside (Azuma et al., 2008). Due to the susceptibility of these compounds to decomposition during the process and storage, encapsulation is a good way to enhance their stability. This study showed that inlet temperature and carrier type were effective on the physicochemical attributes of powder. Within the optimum conditions (i.e., producing powders by maltodextrin at an inlet temperature of 170°C), the total phenolic content, 2,2-diphenyl-1-picrylhydrazyl (DPPH), 2,2′-azino-bis(3-ethylbenzothiazoline-6-sulfonic acid) (ABTS), Trolox equivalent antioxidant capacity (TEAC), hydroxyl radicals scavenging activity, and reducing power were 5.2 mg/g, 73.4%, 90.5%, 2. 5 mM, 79.1%, and 1.2 Abs700, respectively. The microcapsules obtained under optimal conditions were used in the gummy candy. The sensory analysis showed that the incorporation of encapsulated eggplant extract into the formulation of gummy candy at the level of 1.5% enhanced its color and overall liking (Sarabandi et al., 2019).

15.3.5 BEVERAGES

15.3.5.1   Fruit and Vegetable Juices

Fruit and vegetable juices are famous to contain excellent sources of bioactive compounds, namely phenolic compounds and vitamins. The inclusion of fruits and vegetables in the people's daily diet is one of the signs of a healthy diet and has become a principal part of a healthy lifestyle. The effects of the consumption of juices on human health were investigated by many experiments (Drewnowski & Rehm, 2015; Trude et al., 2015; Zheng et al., 2017). Juice can be divided into freshly squeezed, concentrate, juice drinks, nectars, smoothies, and fruit and vegetable juices with added ingredients (Caswell, 2009).

Due to the simple production process and their consumption by all ages of people, fruit and vegetable juices can be used as a reliable vehicle for the enrichment of various ingredients (Shishir & Chen, 2017). To date, several studies have been completed on the enrichment of juices with various functional ingredients. Non-dairy food products such as juices have been studied as suitable carriers for probiotics targeting the consumers who are suffering from lactose intolerance and milk protein allergy (Rivera-Espinoza & Gallardo-Navarro, 2010). As a result of their nutritional value and acidic nature, fruits and vegetables are terrific media for the fermentation of lactic acid bacteria (Chaudhary, 2019; De, Maresca, Ongeng, & Mauriello, 2015).

Even though enrichment with different compounds improves nutritional value and functional properties, bioactive compounds and probiotics are affected by oxygen, heat, pH levels, and moisture. In addition, changes in flavor profile with the loss of fresh compounds, undesirable flavor, solubility and insolubility in water, interaction with other compounds, and packaging considerations have led to protective methods. Maintaining the viability of probiotics in fruit and vegetable juices seems challenging due to the effects of an acidic environment, low pH, and low protein content compared with dairy products (Ying et al., 2013). Furthermore, probiotics should be released into the large intestine (colon) after passing through the stomach and small intestine. Studies showed probiotics and bioactive compounds can be preserved in food products more actively when encapsulated (Anekella & Orsat, 2013; Antunes et al., 2013).

Probiotics are generally loaded into microcapsules in different methods. Spray drying is the most widely applied microencapsulation method for probiotics in food products. Antunes et al. (2013) used spray dried encapsulates of *Bifidobacterium animalis subsp. lactis BB-12* in cellulose acetate phthalate and maltodextrin for acerola nectar. They stated that the encapsulation of *Bifidobacterium* could increase their survival; the survival rate of free cells was 5.9 log CFU/mL while the level of spray dried bacteria was 8 log CFU/mL after one month.

Anekella and Orsat (2013) investigated the microencapsulation of *Lactobacillus rhamnosus NRRL B-4495* and *Lactobacillus acidophilus NRRL B-442* in raspberry juice by spray drying for producing non-dairy probiotic juice. The anthocyanin in berries with probiotics can be a good combination for the improvement of immune and digestive systems. Moreover, dietary fibers in raspberries can act as prebiotics, and, correspondingly, they can improve the performance of probiotics (Chiou & Langrish, 2007). Having optimized the influence of various circumstances related to microencapsulation, they observed that the maximum obtained survival rate of probiotics (81.17%) took place at 100°C inlet temperature and an inlet feed rate of 40 mL/min.

Microencapsulated *Lactobacillus rhamnosus GG* in WPI, physically-modified resistant starch, and their combination was added to apple juice. The pH of the juice was 3.5 and all samples were kept for five weeks at 4°C or 25°C (Ying et al., 2013). All microencapsulated *L. rhamnosus GG* formulations containing WPI alone or WPI in combination with resistant starch provided better protection to *L. rhamnosus GG* in apple juice compared with the formulation containing resistant starch alone. The initial counts of all microencapsulated bacteria powders were between 8.3 and 8.8 $\log_{10}$ CFU/g after spray drying (Table 15.4). Spray drying led to a loss in probiotic viability of less than 1 $\log_{10}$ CFU/g. After five weeks of storage at 25°C, results showed *L. rhamnosus* GG increased in apple juice to the high level of 9–10 $\log_{10}$ CFU/100 mL juice. As the results showed, the encapsulation of *L. rhamnosus* GG in WPI could lead to the preservation of probiotics in low pH beverages.

TABLE 15.4

Application of Spray Dried Encapsulated Bioactives in Juices And Drinks

Product	Core Material	Wall Material	Spray Drying Conditions	Results	References
Acerola nectar	*Bifidobacterium animalis subsp. lactis BB-12*	A mixture of cellulose acetate phthalate and maltodextrin	Inlet temperature: 110°C, outlet temperature: 75°C, feed flow rate: 5 mL/min	Increasing the survival rate of bacteria by about 8 $\log_{10}$ CFU/mL after one month.	Antunes et al. (2013)
Raspberry juice	*Lactobacillus rhamnosus NRRL B-4495 and Lactobacillus acidophilus NRRL B-442*	Maltodextrin	Inlet temperature: 100, 115, and 130°C, Outlet temperature: 69°C–97°C, feed flow rate: 10, 15, 20 mL/min, Wall: core ratio: 1: 1, 1: 1.5, and 1: 2	The maximum survival of probiotics was obtained approximately 81% at 100°C inlet temperature and an inlet feed rate of 40 mL/min.	Anekella and Orsat (2013)
Apple juice	*Lactobacillus rhamnosus GG*	WPI (WPI), resistant starch (RS), WPI/RS: 4:1, 1: 1, 1:4	Inlet temperature: 100, 115, and 130°C, outlet temperature: 69°C–91°C, feed flow rate: 5 mL/min	Encapsulation of *L. rhamnosus GG* in WPI alone or combination with resistant starch provided better protection to *L. rhamnosus* GG in apple juice compared to the formulation containing resistant starch alone.	Ying et al. (2013)
Instant probiotic oat beverages powder	*E. coli K12, L. acidophilus*, oat drink	Maltodextrin	Inlet temperature: 100°C, outlet temperature: 60°C–100°C	The survival of bacteria during spray drying was depended on their growth stage and outlet temperature.	Pispan et al. (2013)

(Continued)

Product	Core Material	Wall Material	Spray Drying Conditions	Results	References
Synbiotic carrot juice	*Lactobacillus casei*	A mixture of alginate, chitosan, and CaCl$_2$	Inlet temperature: 120°C, outlet temperature: 60°C, feed flow rate: 6 mL/min	Increasing the viability of the *L.casei* approximately remained 8.1 log$_{10}$ CFU/g during incubation after 3 months at 4°C in carrot juice, compared to free cells almost 5 log$_{10}$ CFU/g.	Petreska-Ivanovska et al. (2014)
Kiwifruit milk powders	Kiwifruit, Milk, Zinc	Maltodextrin	Inlet temperature: 150°C, feed flow rate: 10 mL/min, air flow rate: 100 L/h, nozzle diameter: 0.7 mm	The retention of almost 90% of vitamin C in powders after 30 at 4°C.	Sun-Waterhouse and Waterhouse (2015)
Probiotic jussara juice powders	*Bifidobacterium spp. Lactis,* Jussara juice	Maltodextrin, inulin, and oligofructose	Inlet temperature: 140°C, outlet temperature: 65°C, feed flow rate: 0.3 L/h, air flow rate: 4.3 m/s, nozzle diameter: 0.5 mm	The inulin cause to increase the survival and thermal stability of bifidobacteria.	Paim, Costa, Walter, and Tonon (2016)
Black chokeberry (*Aroniamelanocarpa*) juice and wine powders	Black chokeberry juice and wine	Maltodextrin, hydroxypropyl-β-cyclodextrin (HP-CD), a combination of maltodextrin and gum Arabic (MGA)	Inlet temperature: 140°C, outlet temperature: 70°C, nozzle diameter: 0.7 mm, drying air flow rate: 75%	The HP-CD had the highest encapsulation efficiency for both juice and wine. Also, the storage stability of wine microcapsules polyphenol content was much higher than the juice at the same condition.	Wilkowska, Ambroziak, Adamiec, and Czyżowska (2016)
Probiotic passion fruit juice powders	*B. animalis ssp. lactis BB-12,* Passion fruit	Maltodextrin (PFM), Inulin (PFI), and a combination of both (PFMI)	Inlet temperature: 130°C; outlet temperature: 45°C; feed flow rate: 6 mL/min; air flow rate: 35 m³/h	The encapsulated bacteria in PFI and PFMI, showed lower reductions of survival rates after spray drying in comparison with the encapsulated bacteria in PFM.	Dias et al. (2017)

(Continued)

TABLE 15.4 (*Continued*)

Product	Core Material	Wall Material	Spray Drying Conditions	Results	References
Carrot juice powders	Carrot Juice	Maltodextrin, gum Arabic, WPI, Mixture of maltodextrin and gum Arabic at a ratio of 1:1; 2:1; 3:1	Inlet temperature: 160°C; outlet temperature: 90°C; feed flow rate: 0.4×10^{-6} m³/s; air flow rate:0.055 m³/s; wall: core ratio: 1:1	High carotenoid retention, higher stability of powders, and lower water activity were observed for gum Arabic samples compared to the other samples.	Janiszewska-Turak et al. (2017)
Asian pear juice powders	Asian pear juice	Maltodextrin	Inlet temperature: 130°C–170°C; outlet temperature: 82°C–88°C; feed flow rate: 6 mL/ min; airflow rate: 35 m³/h	The higher inlet temperature and increasing maltodextrin concentration cause a decrease in vitamin C content and phenol content.	Lee, Chang, Na, and Han (2017)
Tamarillo juice powders	Tamarillo juice	Maltodextrin, gum Arabic, succinic anhydride modified, waxy maize starch, resistant maltodextrin	Inlet temperature: 150°C; outlet temperature: 70°C; airflow rate: 900 m³/h	The encapsulated samples in gam Arabic and OSA1 had the highest encapsulation efficiency while the OSA1 and OSA2 encapsulate samples showed the highest storage stability.	Ramakrishnan, Adzahan, Yusof, and Muhammad (2018)
Goldenberry juice powders	Goldenberry juice	Cellobiose, maltodextrin, modified starch, inulin, alginate, gum Arabic	Inlet temperature: 140°C; outlet temperature: 70°C; feed flow rate: 10 mL/ min; airflow rate: 473 L/h; nozzle diameter: 1.5 mm	The cellobiose samples showed high carotenoid concentration and high encapsulation efficiency.	Etzbach, Pfeiffer, Weber, and Schieber (2018)

Microencapsulated *Lactobacillus casei* in chitosan-Ca-alginate and fructooligosaccharide as a prebiotic was tested to produce synbiotic carrot juice. The initial cell concentration of free and encapsulate *L. casei* of 7.4 $\log_{10}$ CFU/g was applied to ferment functional carrot juice. The survival rate of fermented samples after 24 h of fermentation was 10.46 $\log_{10}$ CFU/g and 9.6 $\log_{10}$ CFU/g for free and encapsulated *L. casei*, respectively. In contrast to the results of the fermentation process, the viability of *L. casei* in the best formulation approximately stayed 8.1 $\log_{10}$ CFU/g during incubation after three months at 4°C in carrot juice while that of free cells was almost 5 $\log_{10}$ CFU/g. This could certify the role of encapsulation in the survival of probiotics (Petreska-Ivanovska et al., 2014).

Fruits and vegetables must sustain prolonged storage periods in order to maintain their nutritional value and enable market supply throughout the year. For this reason, drying these products while maintaining their nutritional value and quality is very important. There have been several studies on the spray drying microencapsulation of fruit and vegetable juices alone or in combination with probiotics. A novel formulation of spray dried kiwifruit and milk powders fortified with zinc was produced by Sun-Waterhouse and Waterhouse (2015). Green or gold kiwifruit puree-skim milk mass with a ratio of 15: 85, zinc citrate, and maltodextrin were spray dried (Table 15.4) to produce pre-prepared powders with high levels of residual phenolic compounds and vitamin C.

Instant reconstitution of the powders in water with dissolution efficiency of less than 30 s and water activity of less than 0.3 were produced. The vitamin C content of fruity–milk powders was retained (88%–90%) after 30 days of storage at 4°C (Sun-Waterhouse & Waterhouse, 2015).

The probiotic microorganism *Bifidobacterium spp. Lactis* was microencapsulated by spray drying using several types of wall materials and added to Jussara juice, subsequently (Paim et al., 2016). Jussara is a fruit from the *Euterpe edulis M. palm*, originally from the Brazilian Atlantic forest with high anthocyanin and phenolic content (Borges et al., 2011). Maltodextrin, inulin and oligofructose, and maltodextrin (1: 1: 2), inulin and maltodextrin (1: 1), and maltodextrin and oligofructose (1: 1) were used as wall materials. Inlet and outlet air temperatures were 140°C and 65°C, respectively. The initial count in the feed juice was around 10 $\log_{10}$ CFU/g before spray drying on a dry basis, which corresponds to approximately 11.5 $\log_{10}$ CFU/g. In this case, Jussara juice and inulin increased the survival of bacterial cells. The high cell survival and good protection of the microorganisms achieved in this study may be related to the wall material ratio of 1: 5. In addition, the authors observed that inulin had a positive effect on the thermal stability of bifidobacteria during the spray drying process (Paim et al., 2016).

Dias et al. (2017) produced a powdered probiotic Yellow passion fruit *(Passiflora edulis flavicarpa)* juice to prepare a new non-dairy probiotic beverage. This fruit is a good source of vitamins, mainly A and C, and minerals (Dias et al., 2015). Passion fruit juice and *Bifidobacterium animalis* ssp. lactis BB-12 was microencapsulated by spray drying using maltodextrin and/or inulin as wall materials. The conditions for conducting this study can be seen in Table 15.4. The results showed that the encapsulation efficiencies were significantly affected by the wall materials. Remarkable reductions in probiotics were observed in the samples of passion fruit juice with *B. lactis* produced only by maltodextrin (PFM). In contrast, for the samples prepared only by inulin (PFI) and a combination of maltodextrin and inulin (PFMI), the reductions in survival rates were lower after the spray drying process. The initial count of *B. lactis* was 9.31 $\log_{10}$ CFU/mL. According to the results, the survived bacteria in PFM, PFMI, and PFI before and after spray drying was 8.88, 8.88, and 9.04 $\log_{10}$ CFU/m versus 7.03, 7.51, and 7.84 $\log_{10}$ CFU/m, respectively. Observed results in this study may also be attributed to the low outlet temperature (45°C) that probably decreased the amount of thermal damage of bacterial cells. The storage at a lower temperature (4°C) showed a better survival rate of bacteria. While being stored at 25°C, the samples made with inulin offered better protection for the encapsulated bacteria and after 30 days, PFI samples presented an increasing number of viable *B. Lactis* (8.41 $\log_{10}$ CFU/m) (Dias et al., 2017).

Janiszewska-Turak et al. (2017) investigated the impact of maltodextrin, gum Arabic, mixtures of maltodextrin, and gum Arabic (1: 1, 2: 1, and 3: 1) and WPI as carriers on properties of microencapsulated pure carrot juice (Table 15.4). The results showed that the water activity of microencapsulated powders increased slowly with the augmentation of the maltodextrin ratio. Also, the lowest carotene content was observed in maltodextrin samples (423 mg/kg). Carotene content in the juice before microencapsulation was about 860 mg/kg. WPI microcapsules showed the highest carotene concentration (573 mg/kg), followed by gum Arabic and maltodextrin. The outputs gained in this research clarify that the powders with gum Arabic resulted in a more useful approach for the microencapsulation of carrot juice. More carotenoids retention, higher stability of powders, and lower water activity were observed for gum Arabic samples in comparison with the maltodextrin samples. Although the carrot powders with WPI had highly similar properties with gum Arabic, these samples may act as a lumping agent during storage. The authors suggested the mixture of maltodextrin and gum Arabic could be used for microencapsulated carotenoids when stored at room temperature.

Whey protein/maltodextrin (WM) and soy protein/maltodextrin (SM) were used as wall materials for spray drying of grape juice with great potential anthocyanins and synthetic food color replacer. SM samples resulted in a higher encapsulation efficiency than WM formula, while the anthocyanin retention in WM (77.9%–94%) samples was higher than for WM samples (58.5%–71.6%). Reconstituted microencapsulated grape juice in water showed similar color parameters to fresh juice, with better results for powders prepared with SM than WM. The authors also stated that SM samples microcapsules were more spherical and more soluble. They also presented low water content and agglomeration property, meaning that they can be used for functional and healthy food products (Moser, Souza, & Nicoletti Telis, 2017).

Maltodextrin, at three levels 15%, 20%, and 25%, and different inlet air temperatures (130°C–170°C) were applied to produce encapsulated Asian pear juice powders (Table 15.4) by Lee et al. (2017). These researchers investigated the effects of wall concentrations and inlet air temperatures on color, bioactive compounds, and other characteristics of encapsulated Asian pear juice powder. The results indicated higher inlet air temperature and augmentation of maltodextrin concentration lead to a reduction in particle size, vitamin C content, and phenol content. Based on the research findings, it was determined that Asian pear juice powder could be manufactured by spray drying at 170°C with 15% (w/v) maltodextrin and would be ready to use in different foods and nutraceutical products (Lee et al., 2017).

Tamarillo is a good source of anthocyanins and carotenoids. Several wall materials such as maltodextrin, n-octenyl succinic anhydride modified starch from waxy maize for high load encapsulation (OSA1), low viscosity gum Arabic alternative (OSA2), resistant maltodextrin, and gum Arabic have been used for the encapsulation of tamarillo juice by spray drying method (Ramakrishnan et al., 2018). The authors observed the powders prepared with gam Arabic and OSA1 had the highest encapsulation efficiency for bioactive compounds while the microcapsules from OSA1 and OSA2 showed the highest storage stability. The tamarillo powders stored in the dark at 4°C showed greater anthocyanin and carotenoid content compared to the presence of light at 25°C.

The goldenberry (*Physalis peruviana L.*) juice containing high bioactive components including carotenoids encapsulated with different wall materials by spray drying method (Etzbach et al., 2018). Maltodextrin, modified starch, inulin, alginate, gum Arabic, and cellobiose were compared. It was seen that cellobiose samples showed better results such as high carotenoid concentration after spray drying, a great encapsulation efficiency (77.2%), and slow carotenoid degradation, compared to the other wall material that were used. In addition, cellobiose protected the carotenoids from degradation processes through light exposure, high temperature, and oxygen. Therefore, it can be offered as a potential wall material for the encapsulation of fruit juices (Etzbach et al., 2018).

15.3.5.2 Drinks

Today, health-promoting food beverages and drinks are designed as functional products that help to meet the need for nutritious materials in order to lead a healthy life (Otles & Cagindi, 2012). Lactose intolerance, milk allergy, and cholesterol level are major causes of dairy-based probiotic products inadequacy, and, with an increase in veganism, there is also a growing demand for dairy-free drinks. The addition of fruit and vegetable juices with probiotic bacteria was discussed in the previous section. In this section, we look at other beverages containing spray dried encapsulated bioactive materials.

Pispan, Hewitt, and Stapley (2013) prepared the probiotic-based powders for instant reconstitution cereal beverages. Maltodextrin was used as a carrier and the outlet air temperature was determined to be between 60 and 100°C. Cell culture harvesting times were determined at mid-log and early stationary phase for *E.coli K12* as 8 and 13 Log_{10} CFU/mL and *L. acidophilus* as 6 and 10 hours, respectively. The findings revealed that the survival of *E. coli K12* and *L. acidophilus NCIMB70225* during spray drying depended on outlet air temperature and bacteria growth stage. The cells harvested from the early stationary phase had a greater rate of retention after spray drying in comparison with the mid-log phase. The results confirmed the effect of the growth phase on the heat tolerance of the bacteria. The survival rates of the mid-log phase and early stationary phase cells are reduced at a drying temperature of 60–80°C and 90–100°C, respectively. The early-stationary phase cells also showed more retention at the beginning of spray drying and storage of probiotic-based instant oat powders. Finally, the authors stated that *L. acidophilus* encapsulated with whey protein and maltose could be more interesting for the market.

Black chokeberry (*Aronia melanocarpa*) juice and wine with high polyphenol compounds were spray dried using different wall materials by Wilkowska et al. (2016). Antioxidant stability in microencapsulated juice and wine by maltodextrin, hydroxypropyl-β-cyclodextrin, and combination of maltodextrin and gum Arabic was studied within a year under different storage conditions such as temperature (8 and 25°C) and light (light or darkness). During the process of wine production, remarkable changes could occur in the concentration and type of polyphenols in comparison with the fresh fruit. The results showed the storage stability of juice polyphenol was much lower in comparison with that of similar wine polyphenol microcapsules stored at 25°C, due to the different quality and quantity of their phenolic compounds. The authors also claimed that the type of wall materials had a significant effect on the encapsulation efficiency of both juice (34.9%–54.2%) and wine (47.5%–59.1%). The HP-CD encapsulated sample had the lowest water content, the highest encapsulation efficiency for juice and wine, the highest total polyphenols, and total anthocyanins retention, and the lowest loss of antioxidant activity under all test conditions. The authors stated that with the increased demand for nonalcoholic drinks and foods, favorable spray dried microencapsulated wine polyphenols could provide healthy components in nonalcoholic products (Wilkowska et al., 2016).

15.3.6 MEAT PRODUCTS

Meat, as a food matrix, is highly susceptible to oxidation because of the interaction between its anti- and pro-oxidant ingredients (Banerjee, Verma, & Siddiqui, 2017). Oxidation of lipids will reduce the meat product acceptability as the texture, color, and nutrients deteriorate (Falowo, Fayemi, & Muchenje, 2014). New and healthier meat products can be produced by food technologists if undesired substances are decreased and the desired components are increased (Hygreeva, Pandey, & Radhakrishna, 2014). The recent trend in the use of natural antioxidants, such as agro-food industry by-products, has been developed for resistant meat products during their shelf-life (Fernandes et al., 2017; Fernandes, Trindade, Lorenzo, & De, 2018). The limitation of natural antioxidant applications is associated with their nature and the process such as exposure to oxygen, light, pH, and temperature (Anandharamakrishnan & Ishwarya, 2015a). On the other hand, the spoilage of meat depends on microorganisms, lipid content, and autolytic enzymes (Ramachandraiah, Choi, & Hong, 2018). Munekata et al. (2017) showed the positive feature of natural antioxidants in Spanish salchichon (a type of pork sausage) enriched with ω-3 fatty acids encapsulated with a stabilizer (konjac glucomannan).

Recently, high saturated fats or added nitrite salts have been considered as the main possible health threats of meat products which concern the consumers (Ramachandraiah et al., 2018). Animal products are the major source of saturated fat with high consumption in wealthy countries. This has caused a high incidence of chronic diseases such as cardiovascular disease (Walker, Rhubart-Berg, McKenzie, Kelling, & Lawrence, 2005). For this reason, innovation is centralized on new and healthier additives to prevent the negative effects of the consumption of processed and red meat (Toldrá & Reig, 2011). Notably, there are a few studies on the application of spray dried microcapsules in meat products, mainly for preservation purposes.

With inoculating thermotolerant lactic acid bacteria encapsulated by spray drying (LAB) in cooked meat batters and storing in a refrigerator (4°C), it was shown that the encapsulation by spray drying could efficiently protect these bacteria during the process of emulsified cooked meat products (Table 15.4). Increasing water activity, total moisture, and fat release, decreasing luminosity and redness without any change in yellowness, and a difference in pH and acidity were found in encapsulated LAB in comparison with free cells. Due to inoculation type, hardness and springiness expressed no change while a decrease in cohesiveness was observed in encapsulated ones. With an improvement in some physicochemical properties and texture profile and enhancing the initial LAB simultaneously with reducing Enterobacteria, the researchers proposed them as bioprotective cultures for improving the microbial safety of cooked meat products (Pérez-Chabela, Lara-Labastida, Rodriguez-Huezo, & Totosaus, 2013).

Fish, especially European sea bass, with white flesh and low-fat content, have low antioxidant content. On the other hand, propolis, a red or brown gummy substance accumulated by honeybees from tree buds, shows great antioxidant activity because of the high content of polyphenols. However, its strong and unpleasant taste and odor limited its application in the food industry. In order to cover its strong odor in sea bass fish burgers and improve its sensory properties, the formulation of fish burgers with spray-dried propolis was optimized by Spinelli et al. (2015) (Table 15.5). Gum Arabic as wall material was used in different ratios. Since an alcohol-free powder of microencapsulated propolis was able to cover its odor, it can be a promoting potential source of phenolics and antioxidants in food industries (Spinelli et al., 2015).

TABLE 15.5
Application of Spray Dried Encapsulated Bioactives in Meat Products

Product	Core Material	Wall Material	Spray Drying Conditions	Results	References
Meat batters	Thermotolerant lactic acid bacteria	Gum Arabic	Inlet temperature: 100°C, outlet temperature: 60°C, air flow rate: 1.5 L/h	The encapsulation of thermotolerant lactic acid bacteria can decrease luminosity, redness, and cohesiveness, with no change in the yellowness of the meat batter. It has an effective manner to protect them, and apply in cooked emulsified meat products.	Pérez-Chabela et al. (2013)
Fish burger	Propolis	Gum Arabic and GGG	Inlet temperature: 120°C, Outlet temperature: 88°C, aspiration rate of: 100%, pump flow rate: 25%	It is feasible to retain a higher quantity of propolis and cover the smell using the capsule (core: wall: 1: 20). Due to the sensory characteristics, the appropriate amount of microcapsules to use in the formula was 5%. The concentration of phenolic compounds and antioxidant properties increased by 3 and 4 times, respectively, compared to the control.	Spinelli et al. (2015)

(Continued)

Product	Core Material	Wall Material	Spray Drying Conditions	Results	References
Pork sausages	Jabuticaba, aqueous extract	Maltodextrin	Inlet temperature: 150°C, air flow rate: 40 mL/min, air flow rate: 30 mL/min, nozzle diameter: 1.5 mm	Microencapsulated jabuticaba extract (MJE) at 2% and 4% can reduce lipid oxidation. 4% MJE had negative effects on color, texture, and overall acceptance, but 2% MJE did not show such an effect.	Baldin et al. (2016)
Meat burger	Propolis	Capsule	Inlet temperature: 150°C, outlet temperature: 100°C	The efficiency of encapsulation was 76.78%. The concentration of phenolic compounds was 24.61 mg Gallic Acid Equivalent.g^{-1}. The strong antioxidant activity of the microcapsules in burger meat was approved. The acceptance rate of the enriched product was 63.80%.	dos Reis et al. (2017)
Pork sausage	Fish oil	A mixture of maltodextrin (13%), gum Arabic (1%), and caseinate (6%)	Inlet temperature: 180°C, outlet temperature: 80°C, aspirator feed rate: 75 L/h	The extract of beer residue, chestnut leaves, and peanut skin could be used as natural antioxidants. With the addition of natural antioxidants, hexanal content was reduced while the content of free fatty acids was increased. The addition of natural antioxidants had a significant effect on the physicochemical properties of the product.	Munekata et al. (2017)
Chicken sausage	Gac extract	Maltodextrin	Inlet temperature: 140, 160, and 180°C, feed flow rate: 12–14 mL/min, air flow rate: 600 L/h, Air pressure: 4 bar	Encapsulated gac powder extract in chicken sausage could reduce cooking yield, improve the redness and yellowness, and maintain sausage characteristics during storage for a week. It could also inhibit microorganisms and lipid oxidation.	Chanshotikul and Hemung (2019)

(Continued)

TABLE 15.5 (*Continued*)

Product	Core Material	Wall Material	Spray Drying Conditions	Results	References
Meat model systems	Fish oil	Maltodextrin, a mixture of maltodextrin and chitosan	Inlet temperature:180°C, outlet temperature: 85°C–90°C, feed flow rate: 1 L/h, aspirator rate: 80%, nozzle diameter: 0.5 mm	The enrichment in EPA and DHA depends on the type of microcapsules and the type of meat model system, with fish oil microcapsules of lecithin + chitosan-maltodextrin added to dry-cured meat model system having the best results.	Solomando, Antequera, and Pérez-Palacios (2020)

Baldin et al. (2016) added microencapsulated jabuticaba extract (MJE) to pork sausage at both 2% and 4% concentrations and observed a noticeable reduction of lipid oxidation during a 15-day storage. Adding 4% MJE had a negative effect on the color, texture, and overall acceptability of the product whereas 2% MJE did not negatively affect such properties. Furthermore, microencapsulated propolis co-product extract (MPC) had a satisfactory oxidation inhibition in meat burgers. The sensory evaluation proved the ideal color, appearance, and texture of burgers containing MPC (dos et al., 2017).

Munekata et al. (2017) showed Spanish salchichon enriched with encapsulated ω-3 fatty acids containing konjac glucomannan (stabilizer agent) has good antioxidant activity on oxidation of lipids, proteins, and volatile compounds. Natural antioxidants were phenolic compounds of beer residue extract (BRE), chestnut leaves extract (CLE), and peanut skin extract (PSE). The mono-layered microcapsules of fish oil were prepared with maltodextrin, gum Arabic, and caseinate and were homogenized and dried using a spray dryer. The authors concluded that natural antioxidants in salchichón cut down the formation of carbonyls in meat proteins and volatile compounds. They suggested that BRE, CLE, and PSE are some excellent sources of phenolic compounds, mainly catechin, gallic acid, and protocatechuic acid (Montesano, Rocchetti, Putnik, & Lucini, 2018).

Chanshotikul and Hemung (2019) had a survey on the encapsulation of gac powder extract and its application in low-nitrite chicken sausage. The ethanolic extract using the microwave-assisted technique of vacuum-dried gac powder (aril and pulp) was encapsulated with maltodextrin through an optimized spray drying method. The encapsulated gac powder extract (EGPE) was used in chicken sausage prepared at a low-nitrite level. During the storage at 10°C for a week, EGPE could enhance the redness and yellowness of sausage and inhibit microorganisms that were indicated by a reduction of total plate count. The lipid oxidation of sausage was also retarded. However, a slight reduction in cooking yield was observed. Based on these results, the researchers suggested that EGPE could be a functional ingredient, a natural colorant, an antimicrobial agent, and an antioxidant in chicken sausage at low nitrate levels.

Solomando et al. (2020) examined the biological availability of EPA (eicosatetraenoic acid) and DHA (docosahexaenoic acid) in various types of fish oil microcapsules as a neat and delivered in two-meat model systems (cooked and dry-cured). To encapsulate these ingredients that have a high susceptibility to oxidation, the wall materials were lecithin-maltodextrin (MO) and lecithin + chitosan-maltodextrin (MU). The highest quantity of these ω-3-polyunsaturated fatty acids was found in the case of the encapsulated samples by lecithin + chitosan-maltodextrin added to the dry-cured model. *In vitro* digestion results confirmed that the highest release percentage of EPA and DHA was obtained in a cooked model containing lecithin-maltodextrin as the wall materials. They concluded that the distribution of the fat particles in a fine paste of the cooked meat model system may cause the release of fat, and, hence, the bioaccessibility of EPA and DHA. Therefore, the importance of these bioactive compounds, their bioaccessibility, and microcapsules characteristics should be considered, concurrently with the determination of EPA and DHA quantity in the enriched food (Solomando et al., 2020).

15.4   ACTIVE PACKAGING

Improvement of food packaging is very important to enhance shelf-life, facilitate handling, prevent storage or transportation physicochemical damages, and decrease food-borne disease hazards (Pilevar, Bahrami, Beikzadeh, Hosseini, & Jafari, 2019). Two types of advanced packagings are known as 'intelligent/smart' and 'active' packagings. In intelligent packaging, due to the presence of some indicators in packaging materials, consumers can notice the food characteristics change over time. Active packaging has great advantages such as controlling the growth of the harmful microorganism during food storage by adding antimicrobials (natural or synthetic) to the package (Khaneghah, Hashemi, & Limbo, 2018). EOs, enzymes, organic acids, antimicrobial peptides, and some biopolymers (e.g., chitosan) are among the most usable natural antimicrobial agents (Hosseinnejad & Jafari, 2016). To compensate for the decrease of antimicrobial activity over time, they should be used in high quantities, which is not economical and might have a negative effect on the sensory properties of foods. Recently, via the loading of antimicrobial ingredients into carriers by encapsulation, natural antimicrobial agents can have better functional properties such as controlled release, high adsorption and solubility, and increased bioavailability (Assadpour & Mahdi Jafari, 2019).

Although food manufacturers should produce insect-free food packages, during transportation, storage in warehouses, or in retails, insects can enter packaged foods such as cereal grains, cereal-derived products, milk powders, cheese, dried fruits, nuts, and dried and smoked meats (Licciardello, 2018). In order to prevent *Sitophilus oryzae* (L.) contamination in brown rice, Lee et al. (2017) made an anti-insect pest-repellent sachet containing EOs, with anti-insect pest activities. EOs were obtained from garlic, ginger, black pepper, onion, and fennel as well as major compounds of garlic (allyl disulfide, AD; allyl mercaptan, AM) and onion (AD and AM) and were measured against *Sitophilus oryzae* (L.). The results showed that garlic EO, onion EO, AD, and AM had strong fumigant insecticidal activities. As AM showed the highest acetylcholinesterase (AChE) inhibition rate, its microcapsules were produced by spray drying procedure, using rice flour as a wall material, with high efficiency (80.02%). Rice flour has a good film-forming feature and is low-cost as well as eco-friendly. Finally, a sachet was produced composed of rice flour microcapsules (RAM) containing garlic EO, onion EO, AD, and AM. RAM revealed a remarkable repelling effect within 48 h, with no undesirable effect on the sensory properties before and after cooking brown rice. The release profile of RAM sachet was expected to extend over 20 months during the distribution period of brown rice.

15.5   CONCLUDING REMARKS

The biological compounds in plants, mainly EOs and natural antioxidants, can be used in healthy and high-quality food products. They may possess strong off-flavors/odors and low water solubility, while they are also generally unstable and sensitive to degradation under the usual process and storage conditions. In food systems, encapsulation makes the protection, delivery, and controlled release of bioactive compounds possible. Spray drying is particularly one of the most common and economic micrometric encapsulation techniques. Although the advances in spray drying are opening new research opportunities for its improvement, there are still several challenges to overcome. Furthermore, basic problems concerning the outcome of encapsulation, at both microscopic and nanoscopic levels, must be resolved to remove consumers' doubts and to consider the green food processing aspects with transparency in terms of safety issues. The findings discussed in this chapter showed that spray drying is a successful method for the encapsulation of bioactives such as plant extracts (or their isolated compounds), probiotic bacteria, shrimp and fish oil, and various vitamins for their subsequent incorporation into various functional foods. Additionally, it appears that among the various wall/coating materials used in the spray drying encapsulation of bioactives, maltodextrin and gum Arabic are the most applicable materials.

REFERENCES

Aguiar, J., Costa, R., Rocha, F., Estevinho, B. N., & Santos, L. (2017). Design of microparticles containing natural antioxidants: Preparation, characterization and controlled release studies. *Powder Technology*, *313*, 287–292.

Ahmad, S., & Ahmed, M. (2014). A review on biscuit, a largest consumed processed product in India, its fortification and nutritional improvement. *International Journal of Science Invention Today*, *3*(2), 169–186.

Altamirano-Fortoul, R., Moreno-Terrazas, R., Quezada-Gallo, A., & Rosell, C. M. (2012). Viability of some probiotic coatings in bread and its effect on the crust mechanical properties. *Food Hydrocolloids*, 29(1), 166–174.

Alvim, I. D., Souza, F. D. S. D., Koury, I. P., Jurt, T., & Dantas, F. B. H. (2013). Use of the spray chilling method to deliver hydrophobic components: physical characterization of microparticles. *Food Science and Technology*, *33*, 34–39.

Alvim, I. D., Stein, M. A., Koury, I. P., Dantas, F. B. H., & Cruz, C. L. D. C. V. (2016). Comparison between the spray drying and spray chilling microparticles contain ascorbic acid in a baked product application. *LWT-Food Science and Technology*, *65*, 689–694.

Al-Zoreky, N. (2009). Antimicrobial activity of pomegranate (*Punica granatum L.*) fruit peels. *International Journal of Food Microbiology*, *134*(3), 244–248.

Anandharamakrishnan, C., & Ishwarya, S. P. (2015a). Encapsulation of bioactive ingredients by spray drying, In *Spray drying Techniques for Food Ingredient Encapsulation* (pp. 156–179). Chicago, USA: John Wiley and Sons Ltd.

Anandharamakrishnan, C., & Ishwarya, S. P. (2015b). Spray drying for encapsulation, In *Spray drying Techniques for Food Ingredient Encapsulation* (pp. 65–76). Chicago, USA: John Wiley and Sons Ltd.

Anandharamakrishnan, C., & Ishwarya, S. P. (2015c). Industrial relevance and commercial applications of spray dried active food encapsulates, *In Spray Drying Techniques for Food Ingredient Encapsulation* (pp. 275–284). Chicago, USA: John Wiley and Sons Ltd.

Anekella, K., & Orsat, V. (2013). Optimization of microencapsulation of probiotics in raspberry juice by spray drying. *LWT - Food Science and Technology*, *50*(1), 17–24.

Antunes, A. E. C., Liserre, A. M., Coelho, A. L. A., Menezes, C. R., Moreno, I., Yotsuyanagi, K., & Azambuja, N. C. (2013). Acerola nectar with added microencapsulated probiotic. *LWT - Food Science and Technology*, *54*(1), 125–131.

Arslan, S., Erbas, M., Tontul, I., & Topuz, A. (2015). Microencapsulation of probiotic *Saccharomyces cerevisiae var. boulardii* with different wall materials by spray drying. *LWT - Food Science and Technology*, *63*, 685–690.

Arslan-Tontul, S., Erbas, M., & Gorgulu, A. (2019). The Use of probiotic-loaded single-and double-layered microcapsules in cake production. *Probiotics and Antimicrobial Proteins*, *11*(3), 840–849.

Assadpour, E., & Mahdi Jafari, S. (2019). A systematic review on nanoencapsulation of food bioactive ingredients and nutraceuticals by various nanocarriers. *Critical Reviews in Food Science and Nutrition*, *59*(19), 3129–3151.

Azeredo, H. M., Santos, A. N., Souza, A. C., Mendes, K. C., & Andrade, M. I. R. (2007). Betacyanin stability during processing and storage of a microencapsulated red beetroot extract. *American Journal of Food Technology*, *2*(4), 307–312.

Bahoshy, B. J., & Klose, R. E. (1978). L-aspartyl-L-phenylalanine methyl ester. *Patent no US4122195*.

Bahrami, A., Delshadi, R., Assadpour, E., Jafari, S. M., & Williams, L. (2020). Antimicrobial-loaded nanocarriers for food packaging applications. *Advances in Colloid and Interface Science*, 102, 140.

Baldin, J. C., Michelin, E. C., Polizer, Y. J., Rodrigues, I., de Godoy, S. H. S., Fregonesi, R. P., ... & Trindade, M. A. (2016). Microencapsulated jabuticaba (Myrciaria cauliflora) extract added to fresh sausage as natural dye with antioxidant and antimicrobial activity. *Meat Science*, *118*, 15–21.

Banerjee, R., Verma, A. K., & Siddiqui, M. W. (2017). Potential Applications of Natural Antioxidants in Meat and Meat Products. In *Natural Antioxidants* (pp. 115–160). New York, USA: Apple Academic Press.

Banville, C., Vuillemard, J. C., & Lacroix, C. (2000). Comparison of different methods for fortifying Cheddar cheese with vitamin D. *International Dairy Journal*, 10(5–6), 375–382.

Barreca, D., Laganà, G., Leuzzi, U., Smeriglio, A., Trombetta, D., & Bellocco, E. (2016). Evaluation of the nutraceutical, antioxidant and cytoprotective properties of ripe pistachio (*Pistacia vera L.*, variety Bronte) hulls. *Food Chemistry*, *196*, 493–502.

Bordiga, M., Gomez-Alonso, S., Locatelli, M., Travaglia, F., Coïsson, J. D., Hermosin-Gutierrez, I., & Arlorio, M. (2014). Phenolics characterization and antioxidant activity of six different pigmented *Oryza sativa* L. cultivars grown in Piedmont (Italy). *Food Research International*, 65, 282–290.

Borges, G. D. S. C., Vieira, F. G. K., Copetti, C., Gonzaga, L. V., Zambiazi, R. C., Mancini Filho, J., & Fett, R. (2011). Chemical characterization, bioactive compounds, and antioxidant capacity of jussara (*Euterpe edulis*) fruit from the Atlantic Forest in southern Brazil. *Food Research International, 44*(7), 2128–2133.

Botrel, D. A., de Barros Fernandes, R. V., & Borges, S. V. (2015). Microencapsulation of EOs using spray drying technology. In *Microencapsulation and Microspheres for Food Applications* (pp. 235–251). London, England: Academic Press.

Brinques, GB., Ayub, M. A. Z. (2011). Effect of microencapsulation on survival of *Lactobacillus plantarum* in simulated gastrointestinal conditions, refrigeration, and yogurt, *Journal of Food Engineering* 103, 123–128.

Burgain, J., Gaiani, C., Linder, M., & Scher, J. (2011). Encapsulation of probiotic living cells: From laboratory scale to industrial applications. *Journal of Food Engineering, 104*(4), 467–483.

Bustos, M., Perez, G., & Leon, A. (2015). Structure and quality of pasta enriched with functional ingredients. *RSC Advances, 5*(39), 30780–30792.

Çam, M., İçyer, N. C., & Erdoğan, F. (2014). Pomegranate peel phenolics: microencapsulation, storage stability and potential ingredient for functional food development. *LWT - Food Science and Technology, 55*(1), 117–123.

Castro, J. M., Tornadijo, M. E., Fresno, J. M., & Sandoval, H. (2015). Biocheese: a food probiotic carrier. *BioMed Research International, 2015*, 1–11.

Caswell, H. (2009). The role of fruit juice in the diet: an overview. *Nutrition Bulletin, 34*(3), 273–288.

Cea, T., Posta, J. D., & Glass, M. (1983). Encapsulated APM and method of preparation. *U.S. Patent No. 4,384,004*. Washington, DC: U.S. Patent and Trademark Office.

Chanshotikul, N., & Hemung, B. O. (2019). Encapsulation of Gac Powder Extract and Its Application in Low-Nitrite Chicken Sausage. *International Journal of Food Engineering, 5*(2), 146–151.

Chaudhary, A. (2019). Probiotic Fruit and Vegetable Juices: Approach Towards a Healthy Gut. *International Journal of Current Microbiology and Applied Sciences*, 8(6), 1265–1279.

Chiou, D., & Langrish, T. A. G. (2007). Development and characterisation of novel nutraceuticals with spray drying technology. *Journal of Food Engineering*, 82, 84–91.

Davidov-Pardo, G., Roccia, P., Salgado, D., León, A., & Pedroza-Islas, R. (2008). Utilization of different wall materials to microencapsulate fish oil. *American Journal of Food Technology, 3*(6), 384–393.

De Prisco, A., Maresca, D., Ongeng, D., & Mauriello, G. (2015). Microencapsulation by vibrating technology of the probiotic strain *Lactobacillus reuteri* DSM 17938 to enhance its survival in foods and in gastrointestinal environment. *LWT - Food Science and Technology, 61*(2), 452–462.

Dias, C. O., de Almeida, J. D. S. O., Pinto, S. S., de Oliveira Santana, F. C., Verruck, S., Müller, C. M. O., ... & Amboni, R. D. D. M. C. (2017). Development and physico-chemical characterization of microencapsulated bifidobacteria in passion fruit juice: A functional non-dairy product for probiotic delivery. *Food Bioscience, 24*, 26–36.

Dias, M. I., Ferreira, I. C., & Barreiro, M. F. (2015). Microencapsulation of bioactives for food applications. *Food & Function, 6*(4), 1035–1052.

Dimitrellou, D., Kandylis, P., Petrovic, T., Dimitrellou, S., Levi, S., Nedovi, V., Kourkoutas, Y. (2016). Survival of spray dried microencapsulated *Lactobacillus casei* ATCC 393 in simulated gastrointestinal conditions and fermented milk. *LWT - Food Science and Technology*, 71: 169–174.

Donhowe, E. G., Flores, F. P., Kerr, W. L., Wicker, L., & Kong, F. (2014). Characterization and in vitro bioavailability of β-carotene: Effects of microencapsulation method and food matrix. *LWT-Food Science and Technology, 57*(1), 42–48.

dos Reis, A. S., Diedrich, C., de Moura, C., Pereira, D., de Flório Almeida, J., da Silva, L. D., ... & Carpes, S. T. (2017). Physico-chemical characteristics of microencapsulated propolis co-product extract and its effect on storage stability of burger meat during storage at −15°C. *LWT - Food Science and Technology, 76*, 306–313.

Drewnowski, A., & Rehm, C. D. (2015). Socioeconomic gradient in consumption of whole fruit and 100% fruit juice among US children and adults. *Nutrition Journal, 14*(1), 3.

Egan, M. E., Pearson, M., Weiner, S. A., Rajendran, V., Rubin, D., Glöckner-Pagel, J., ... & Caplan, M. J. (2004). Curcumin, a major constituent of turmeric, corrects cystic fibrosis defects. *Science, 304*(5670), 600–602.

Erkaya, T., Dağdemir, E., & şengül, M. (2012). Influence of Cape gooseberry (*Physalis peruviana* L.) addition on the chemical and sensory characteristics and mineral concentrations of ice cream. *Food Research International, 45*(1), 331–335.

Etzbach, L., Pfeiffer, A., Weber, F., & Schieber, A. (2018). Characterization of carotenoid profiles in golden-berry (*Physalis peruviana* L.) fruits at various ripening stages and in different plant tissues by HPLC-DAD-APCI-MSn. *Food Chemistry, 245*, 508–517.

Ezhilarasi, P. N., Indrani, D., Jena, B. S., & Anandharamakrishnan, C. (2014). Microencapsulation of Garcinia fruit extract by spray drying and its effect on bread quality. *Journal of the Science of Food and Agriculture, 94*(6), 1116–1123.

Falowo, A. B., Fayemi, P. O., & Muchenje, V. (2014). Natural antioxidants against lipid–protein oxidative deterioration in meat and meat products: A review. *Food Research International, 64*, 171–181.

Fazaeli, M., Emam-Djomeh, Z., & Yarmand, M. (2016). Optimization of Spray Drying Conditions for Product ion of Ice Cream Mix Powder Flavored With Black Mulberry Juice. *Journal of Agricultural Science and Technology, 18*, 1557–1570.

Fazilah, N.F., Khayat, ME., Hamidon, N. H., Ariff, A.B., Wasoh, H., Murni Halim, M. (2019). Microencapsulation of *Lactococcus lactis* Gh1 with gum Arabic and *Synsepalum dulcificum* via spray drying for potential inclusion in functional yogurt, *Molecules* 24, 1422.

Fernandes, R. D. P. P., Trindade, M. A., Lorenzo, J. M., & De Melo, M. P. (2018). Assessment of the stability of sheep sausages with the addition of different concentrations of *Origanum vulgare* extract during storage. *Meat Science, 137*, 244–257.

Fernandes, R. D. P. P., Trindade, M. A., Tonin, F. G., Pugine, S. M. P., Lima, C. G. D., Lorenzo, J. M., & De Melo, M. P. (2017). Evaluation of oxidative stability of lamb burger with *Origanum vulgare* extract. *Food Chemistry, 233*, 101–109.

Food and Drug Administration. (1996). Food standards: Amendment of standards of identity for enriched grain products to require addition of folic acid; final rule (21 CFR Parts 136, 137, and 139). *Federal Register, 61*, 8781–8797.

Franco, D., Antequera, T., de Pinho, S. C., Jiménez, E., Pérez-Palacios, T., Fávaro-Trindade, C. S., & Lorenzo, J. M. (2017). The use of microencapsulation by spray drying and its aplication in meat products. *Strategies for Obtaining Healthier Foods*. New York, USA: Nova Science Publishers.

Frankel, E. N., Satué-Gracia, T., Meyer, A. S., & German, J. B. (2002). Oxidative stability of fish and algae oils containing long-chain polyunsaturated fatty acids in bulk and in oil-in-water emulsions. *Journal of Agricultural and Food Chemistry, 50*(7), 2094–2099.

Gabbi, D. K., Bajwa, U., & Goraya, R. K. (2018). Physicochemical, melting and sensory properties of ice cream incorporating processed ginger (*Zingiber officinale*). *International Journal of Dairy Technology, 71*(1), 190–197.

Gallardo, G., Guida, L., Martinez, V., López, M. C., Bernhardt, D., Blasco, R., … & Hermida, L. G. (2013). Microencapsulation of linseed oil by spray drying for functional food application. *Food Research International, 52*(2), 473–482.

Gandhi, A. P., Kotwaliwale, N., Kawalkar, J., Srivastav, D. C., Parihar, V. S., & Nadh, P. R. (2001). Effect of incorporation of defatted soyflour on the quality of sweet biscuits. *Journal of Food Science and Technology, 38*, 502–503.

Gardiner, G. E., Bouchier, P., O'Sullivan, E., Kelly, J., Collins, J. K., Fitzgerald, G., … & Stanton, C. (2002). A spray-dried culture for probiotic Cheddar cheese manufacture. *International Dairy Journal, 12*(9), 749–756.

Gautam, R., & Sharma, S. (2012). Effect of *Punica granatum* Linn.(peel) on blood glucose level in normal and alloxan-induced diabetic rats. *Research Journal of Pharmacy and Technology, 5*(2), 226–227.

Ghandahari Yazdi, A. P., Barzegar, M., Sahari, M. A., & Ahmadi Gavlighi, H. (2019). Optimization of the enzyme-assisted aqueous extraction of phenolic compounds from pistachio green hull. *Food Science & Nutrition, 7*(1), 356–366.

Ghandehari Yazdi, A. P., Barzegar, M., Ahmadi Gavlighi, H., Sahari, M. A., & Mohammadian, A. H. (2020). Physicochemical properties and organoleptic aspects of ice cream enriched with microencapsulated pistachio peel extract. *International Journal of Dairy Technology*. doi: 10.1111/1471-0307.12698.

Gharsallaoui, A., Roudaut, G., Chambin, O., Voilley, A., & Saurel, R. (2007). Applications of spray drying in microencapsulation of food ingredients: An overview. *Food Research International, 40*(9), 1107–1121.

Giaconia, M. A., dos Passos Ramos, S., Pereira, C. F., Lemes, A. C., De Rosso, V. V., & Braga, A. R. C. (2020). Overcoming restrictions of bioactive compounds biological effects in food using nanometer-sized structures. *Food Hydrocolloids, 105*, 939.

Goula, A. M., & Adamopoulos, K. G. (2012). A new technique for spray-dried encapsulation of lycopene. *Drying Technology, 30*(6), 641–652.

Grune, T., Lietz, G., Palou, A., Ross, A. C., Stahl, W., Tang, G., … & Biesalski, H. K. (2010). β-Carotene is an important vitamin A source for humans. *The Journal of Nutrition, 140*(12), 2268S–2285S.

Hall, C., & Tulbek, M.C. (2008). Omega-3-enriched bread. In B. R. Hamaker (Ed.), *Technology of Functional Cereal Products* (pp. 388–404). Boca Raton, Boston, New York, Washington, D.C, USA: CRC Press.

Hamaker, B. R. (Ed.). (2008). *Technology of functional cereal products*. Boca Raton, Boston, New York, Washington, D.C, USA: CRC Press.

Haskell, M. J. (2012). The challenge to reach nutritional adequacy for vitamin A: β-carotene bioavailability and conversion—evidence in humans. *The American Journal of Clinical Nutrition, 96*(5), 1193S–1203S.

Hosseinnejad, M., & Jafari, S. M. (2016). Evaluation of different factors affecting antimicrobial properties of chitosan. *International Journal of Biological Macromolecules, 85*, 467–475.

Hygreeva, D., Pandey, M. C., & Radhakrishna, K. (2014). Potential applications of plant based derivatives as fat replacers, antioxidants and antimicrobials in fresh and processed meat products. *Meat Science, 98*(1), 47–57.

da Costa, J. M. G., Silva, E. K., Hijo, A. A. C. T., Azevedo, V. M., Malta, M. R., Alves, J. G. L. F., & Borges, S. V. (2015). Microencapsulation of Swiss cheese bioaroma by spray drying: Process optimization and characterization of particles. *Powder Technology, 274*, 296–304.

Janiszewska, E., & Wlodarczyk, J. (2013). Influence of spray drying conditions on beetroot pigments retention after microencapsulation process. *Acta Agrophysica, 20*(2), 343–356.

Janiszewska-Turak, E., Dellarosa, N., Tylewicz, U., Laghi, L., Romani, S., Dalla Rosa, M., & Witrowa-Rajchert, D. (2017). The influence of carrier material on some physical and structural properties of carrot juice microcapsules. *Food Chemistry, 236*, 134–141.

Jena, B. S., Jayaprakasha, G. K., & Sakariah, K. K. (2002). Organic acids from leaves, fruits, and rinds of Garcinia cowa. *Journal of Agricultural and Food Chemistry, 50*(12), 3431–3434.

Jeyakumari, A., Janarthanan, G., Chouksey, M. K., & Venkateshwarlu, G. (2016). Effect of fish oil encapsulates incorporation on the physico-chemical and sensory properties of cookies. *Journal of Food Science and Technology, 53*(1), 856–863.

Kamali Rousta, L., Ghandehari Yazdi, A. P., & Amini, M. (2020). Optimization of athletic pasta formulation by D-optimal mixture design. *Food Science & Nutrition*. doi: 10.1002/fsn3.1764.

Kha, T. C., Nguyen, M. H., Roach, P. D., Parks, S. E., & Stathopoulos, C. (2013). Gac fruit: nutrient and phytochemical composition, and options for processing. *Food Reviews International, 29*(1), 92–106.

Kha, T. C., Nguyen, M. H., Roach, P. D., & Stathopoulos, C. (2015). A storage study of encapsulated gac (*Momordica cochinchinensis*) oil powder and its fortification into foods. *Food and Bioproducts Processing, 96*, 113–125.

Khaneghah, A. M., Hashemi, S. M. B., & Limbo, S. (2018). Antimicrobial agents and packaging systems in antimicrobial active food packaging: An overview of approaches and interactions. *Food and Bioproducts Processing, 111*, 1–19.

Lakkis, J. M. (Ed.). (2007). *Encapsulation and Controlled Release Technologies in Food Systems*. Oxford, UK: Blackwell Pub.

Lee, J. B., Ahn, J., Lee, J., & Kwak, H. S. (2004). L-ascorbic acid microencapsulated with polyacylglycerol monostearate for milk fortification. *Bioscience, Biotechnology, and Biochemistry, 68*(3), 495–500.

Lee, S. H., Chang, Y., Na, J. H., & Han, J. (2017). Development of anti-insect multilayered films for brown rice packaging that prevent Plodia interpunctella infestation. *Journal of Stored Products Research, 72*, 153–160.

Liang, J. L., Meng, Y. Z., & Lei, C. G. (2007). Study on antiseptic effects of curcumin. *Science and Technology of Cereals, Oils and Foods, 2*, 73–79.

Licciardello, F. (2018). Development of insect-repellent food packaging materials. In *Reference Module in Food Science* (pp. 1–11). Amsterdam, Netherlands: Elsevier.

Liu, Y., Green, T. J., & Kitts, D. D. (2015). Stability of microencapsulated L-5-methyltetrahydrofolate in fortified noodles. *Food Chemistry, 171*, 206–211.

Liu, Y., Green, T. J., Wong, P., & Kitts, D. D. (2013). Microencapsulation of L-5-methyltetrahydrofolic acid with ascorbate improves stability in baked bread products. *Journal of Agricultural and Food Chemistry, 61*(1), 247–254.

Livney, Y. D. (2010). Milk proteins as vehicles for bioactives. *Current Opinion in Colloid & Interface Science, 15*(1–2), 73–83.

Lourenco, S. C., Torres, C. A., Nunes, D., Duarte, P., Freitas, F., Reis, M. A., … & Alves, V. D. (2017). Using a bacterial fucose-rich polysaccharide as encapsulation material of bioactive compounds. *International Journal of Biological Macromolecules, 104*, 1099–1106.

Lu, T. M., Lee, C. C., Mau, J. L., & Lin, S. D. (2010). Quality and antioxidant property of green tea sponge cake. *Food Chemistry, 119*(3), 1090–1095.

Mamat, H., & Hill, S. E. (2018). Structural and functional properties of major ingredients of biscuit. *International Food Research Journal, 25*(2), 462–471.

Marchetti, N., Bonetti, G., Brandolini, V., Cavazzini, A., Maietti, A., Meca, G., & Mañes, J. (2018). Stinging nettle (*Urtica dioica* L.) as a functional food additive in egg pasta: Enrichment and bioaccessibility of Lutein and β-carotene. *Journal of Functional Foods, 47*, 547–553.

Marshall, R. (2003). *Ice Cream* Marshall RT, Goff HD, Hartel RW, New York, USA: Kluwer Academic.

Mirković, M., Seratlić, S., Kilcawley, K., Mannion, D., Mirković, N., & Radulović, Z. (2018). The sensory quality and volatile profile of dark chocolate enriched with encapsulated probiotic Lactobacillus plantarum bacteria. *Sensors, 18*(8), 2570.

Montesano, D., Rocchetti, G., Putnik, P., & Lucini, L. (2018). Bioactive profile of pumpkin: an overview on terpenoids and their health-promoting properties. *Current Opinion in Food Science, 22*, 81–87.

Moser, P., Souza, R. T. D., & Nicoletti Telis, V. R. (2017). Spray drying of grape juice from hybrid cv. BRS Violeta: microencapsulation of anthocyanins using protein/maltodextrin blends as drying aids. *Journal of Food Processing and Preservation, 41*(1), e12852.

Munekata, P. E. S., Domínguez, R., Franco, D., Bermúdez, R., Trindade, M. A., & Lorenzo, J. M. (2017). Effect of natural antioxidants in Spanish salchichón elaborated with encapsulated n-3 long chain fatty acids in konjac glucomannan matrix. *Meat Science, 124*, 54–60.

Nambiar, R. B., Sellamuthu, P. S., & Perumal, A. B. (2018). Development of milk chocolate supplemented with microencapsulated Lactobacillus plantarum HM47 and to determine the safety in a Swiss albino mice model. *Food Control, 94*, 300–306.

Negi, P. S., Jayaprakasha, G. K., & Jena, B. S. (2010). Evaluation of antioxidant and antimutagenic activities of the extracts from the fruit rinds of Garcinia cowa. *International Journal of Food Properties, 13*(6), 1256–1265.

Neves, M. I. L., Desobry-Banon, S., Perron, I. T.., Desobry, S., & Petit, J. (2019). Encapsulation of curcumin in milk powders by spray drying: Physicochemistry, rehydration properties, and stability during storage, *Powder Technology, 345*, 601–607.

Oliviero, T., & Fogliano, V. (2016). Food design strategies to increase vegetable intake: The case of vegetable enriched pasta. *Trends in Food Science & Technology, 51*, 58–64.

Otles, S., & Cagindi, O. (2012). Safety considerations of nutraceuticals and functional foods. In *Novel Technologies in Food Science* (pp. 121–136). New York, USA: Springer.

Paim, D. R., Costa, S. D., Walter, E. H., & Tonon, R. V. (2016). Microencapsulation of probiotic jussara (*Euterpe edulis* M.) juice by spray drying. *LWT - Food Science & Technology, 74*, 21–25.

Papillo, V. A., Locatelli, M., Travaglia, F., Bordiga, M., Garino, C., Arlorio, M., & Coïsson, J. D. (2018). Spray-dried polyphenolic extract from Italian black rice (*Oryza sativa* L., var. Artemide) as new ingredient for bakery products. *Food Chemistry, 269*, 603–609.

Papillo, V. A., Locatelli, M., Travaglia, F., Bordiga, M., Garino, C., Coïsson, J. D., & Arlorio, M. (2019). Cocoa hulls polyphenols stabilized by microencapsulation as functional ingredient for bakery applications. *Food Research International, 115*, 511–518.

Pasrija, D., Ezhilarasi, P. N., Indrani, D., & Anandharamakrishnan, C. (2015). Microencapsulation of green tea polyphenols and its effect on incorporated bread quality. *LWT - Food Science and Technology, 64*(1), 289–296.

Pasukamonset, P., Pumalee, T., Sanguansuk, N., Chumyen, C., Wongvasu, P., Adisakwattana, S., & Ngamukote, S. (2018). Physicochemical, antioxidant and sensory characteristics of sponge cakes fortified with *Clitoria ternatea* extract. *Journal of Food Science and Technology, 55*(8), 2881–2889.

Pérez-Chabela, M. L., Lara-Labastida, R., Rodriguez-Huezo, E., & Totosaus, A. (2013). Effect of spray drying encapsulation of thermotolerant lactic acid bacteria on meat batters properties. *Food and Bioprocess Technology, 6*(6), 1505–1515.

Petreska-Ivanovska, T., Petrushevska-Tozi, L., Grozdanov, A., Petkovska, R., Hadjieva, J., Popovski, E., … & Mladenovska, K. (2014). From optimization of synbiotic microparticles prepared by spray drying to development of new functional carrot juice. *Chemical Industry and Chemical Engineering Quarterly, 20*(4), 549–564.

Pilevar, Z., Bahrami, A., Beikzadeh, S., Hosseini, H., & Jafari, S. M. (2019). Migration of styrene monomer from polystyrene packaging materials into foods: Characterization and safety evaluation. *Trends in Food Science & Technology*. doi: 10.1016/j.tifs.2019.07.020.

Pillai, D. S., Prabhasankar, P., Jena, B. S., & Anandharamakrishnan, C. (2012). Microencapsulation of Garcinia cowa fruit extract and effect of its use on pasta process and quality. *International Journal of Food Properties, 15*(3), 590–604.

Pinto, S. S., Fritzen-Freire, C. B., Muñoz, I. B., Barreto, P. L., Prudêncio, E. S., & Amboni, R. D. (2012). Effects of the addition of microencapsulated Bifidobacterium BB-12 on the properties of frozen yogurt. *Journal of Food Engineering, 111*(4), 563–569.

Pispan, S., Hewitt, C. J., & Stapley, A. G. F. (2013). Comparison of cell survival rates of E. coli K12 and L. acidophilus undergoing spray drying. *Food and Bioproducts Processing, 91*(4), 362–369.

Praseptiangga, D., Invicta, S. E., & Khasanah, L. U. (2019). Sensory and physicochemical characteristics of dark chocolate bar with addition of cinnamon (*Cinnamomum burmannii*) bark oleoresin microcapsule. *Journal of Food Science and Technology, 56*(9), 4323–4332.

Rafiee, Z., Nejatian, M., Daeihamed, M., & Jafari, S. M. (2019). Application of different nanocarriers for encapsulation of curcumin. *Critical Reviews in Food Science and Nutrition, 59*(21), 3468–3497.

Rajaei, A., Barzegar, M., Mobarez, A. M., Sahari, M. A., & Esfahani, Z. H. (2010). Antioxidant, anti-microbial and antimutagenicity activities of pistachio (*Pistachia vera*) green hull extract. *Food and Chemical Toxicology, 48*(1), 107–112.

Ramachandraiah, K., Choi, M. J., & Hong, G. P. (2018). Micro-and nano-scaled materials for strategy-based applications in innovative livestock products: A review. *Trends in Food Science & Technology, 71*, 25–35.

Ramakrishnan, Y., Adzahan, N. M., Yusof, Y. A., & Muhammad, K. (2018). Effect of wall materials on the spray drying efficiency, powder properties and stability of bioactive compounds in tamarillo juice microencapsulation. *Powder Technology, 328*, 406–414.

Rashidinejad, A., Birch, E. J., & Everett, D. W. (2016). Antioxidant activity and recovery of green tea catechins in full-fat cheese following gastrointestinal simulated digestion. *Journal of Food Composition and Analysis, 48*, 13–24.

Rashidinejad, A., Birch, E. J., Sun-Waterhouse, D., & Everett, D. W. (2014). Delivery of green tea catechin and epigallocatechin gallate in liposomes incorporated into low-fat hard cheese. *Food Chemistry, 156*, 176–183.

Ravichandran, K., Palaniraj, R., Saw, N. M. M. T., Gabr, A. M., Ahmed, A. R., Knorr, D., & Smetanska, I. (2014). Effects of different encapsulation agents and drying process on stability of betalains extract. *Journal of Food Science and Technology, 51*(9), 2216–2221.

Rivera-Espinoza, Y., & Gallardo-Navarro, Y. (2010). Non-dairy probiotic products. *Food Microbiology, 27*(1), 1–11.

Rocha, G. A., Fávaro-Trindade, C. S., & Grosso, C. R. F. (2012). Microencapsulation of lycopene by spray drying: characterization, stability and application of microcapsules. *Food and Bioproducts Processing, 90*(1), 37–42.

Roseli, C. M. (2008). Vitamin and mineral fortification of bread. In B. R. Hamaker (Ed.), *Technology of functional cereal products* (pp. 336–361). Boca Raton, Boston, New York, Washington, D.C, USA: CRC Press.

Rutz, J. K., Borges, C. D., Zambiazi, R. C., Rosa, C. G., Silva, M. M. (2016). Elaboration of microparticles of carotenoids from natural and synthetic sources for applications in food, *Food Chemistry, 202*, 324–333.

da Silva, S. C., Fernandes, I. P., Barros, L., Fernandes, Â., …& Barreiro, M. F. (2019). Spray-dried *Spirulina platensis* as an effective ingredient to improve yogurt formulations: Testing different encapsulating solutions, *Journal of Functional Foods, 60*, 103427.

Sahu, A., Kasoju, N., & Bora, U. (2008). Fluorescence study of the curcumin– casein micelle complexation and its application as a drug nanocarrier to cancer cells. *Biomacromolecules, 9*(10), 2905–2912.

Sampathu, S. R., Lakshminarayanan, S., Sowbhagya, H. B., Krishnamurthy, N. A., & Asha, M. R. (2000, February). *Use of curcumin as a natural yellow colourant in ice cream. In Proceedings of the National Seminar on Natural Colouring Agents.* Lucknow, India.

Samyor, D., Das, A. B., & Deka, S. C. (2017). Pigmented rice a potential source of bioactive compounds: A review. *International Journal of Food Science & Technology, 52*(5), 1073–1081.

Sarabandi, K., Jafari, S. M., Mahoonak, A. S., & Mohammadi, A. (2019). Application of gum Arabic and maltodextrin for encapsulation of eggplant peel extract as a natural antioxidant and color source. *International Journal of Biological Macromolecules, 140*, 59–68.

Schobel, A. M., & Yang, R. K. (1989). Encapsulated sweetener composition for use with chewing gum and edible products. *U.S. Patent No. 4,824,681*. Washington, DC: U.S. Patent and Trademark Office.

Serna-Saldivar, S. O., Zorrilla, R., De La Parra, C., Stagnitti, G., & Abril, R. (2006). Effect of DHA containing oils and powders on baking performance and quality of white pan bread. *Plant Foods for Human Nutrition, 61*(3), 121–129.

Shishir, M. R. I., & Chen, W. (2017). Trends of spray drying: A critical review on drying of fruit and vegetable juices. *Trends in Food Science & Technology, 65*, 49–67.

Solomando, J. C., Antequera, T., & Pérez-Palacios, T. (2020). Study on fish oil microcapsules as neat and added to meat model systems: Enrichment and bioaccesibility of EPA and DHA. *LWT-Food Sciencew and Technology, 120,* 108946.

Soni, R. A., Sudhakar, K., & Rana, R. S. (2017). Spirulina–From growth to nutritional product: A review. *Trends in Food Science & Technology, 69,* 157–171.

Soukoulis, C., Fisk, I. D., & Bohn, T. (2014). Ice cream as a vehicle for incorporating health-promoting ingredients: Conceptualization and overview of quality and storage stability. *Comprehensive Reviews in Food Science and Food Safety, 13*(4), 627–655.

Spigno, G., Garrido, G., Guidesi, E., & Elli, M. (2015). Spray drying encapsulation of probiotics for ice-cream application. *Chemical Engineering Transactions, 43,* 49–54.

Spinelli, S., Conte, A., Lecce, L., Incoronato, A. L., & Del Nobile, M. A. (2015). Microencapsulated propolis to enhance the antioxidant properties of fresh fish burgers. *Journal of Food Process Engineering, 38*(6), 527–535.

Sun-Waterhouse, D., & Waterhouse, G. I. (2015). Spray drying of green or gold kiwifruit juice–milk mixtures; novel formulations and processes to retain natural fruit colour and antioxidants. *Food and Bioprocess Technology, 8*(1), 191–207.

Swain, M. R., Anandharaj, M., Ray, R. C., & Parveen Rani, R. (2014). Fermented fruits and vegetables of Asia: a potential source of probiotics. *Biotechnology Research International, 2014,* 1–19.

Takeungwongtrakul, S., & Benjakul, S. (2017). Biscuits fortified with micro-encapsulated shrimp oil: characteristics and storage stability. *Journal of Food Science and Technology, 54*(5), 1126–1136.

Takeungwongtrakul, S., Benjakul, S., & Aran, H. (2012). Lipids from cephalothorax and hepatopancreas of Pacific white shrimp (*Litopenaeus vannamei*): Compositions and deterioration as affected by iced storage. *Food Chemistry, 134*(4), 2066–2074.

Toldrá, F., & Reig, M. (2011). Innovations for healthier processed meats. *Trends in Food Science & Technology, 22*(9), 517–522.

Tolve, R., Condelli, N., Caruso, M. C., Barletta, D., Favati, F., & Galgano, F. (2018). Fortification of dark chocolate with microencapsulated phytosterols: Chemical and sensory evaluation. *Food & Function, 9*(2), 1265–1273.

Tolve, R., Galgano, F., Caruso, M. C., Tchuenbou-Magaia, F. L., Condelli, N., Favati, F., & Zhang, Z. (2016). Encapsulation of health-promoting ingredients: applications in foodstuffs. *International Journal of Food Sciences and Nutrition, 67*(8), 888–918.

Tomiuk, S., Liu, Y., Green, T. J., King, M. J., Finglas, P. M., & Kitts, D. D. (2012). Studies on the retention of microencapsulated L-5-methyltetrahydrofolic acid in baked bread using skim milk powder. *Food Chemistry, 133*(2), 249–255.

Trude, A. C., Kharmats, A., Jock, B., Liu, D., Lee, K., Martins, P. A., … & Gittelsohn, J. (2015). Patterns of food consumption are associated with obesity, self-reported diabetes and cardiovascular disease in five American Indian communities. *Ecology of Food and Nutrition, 54*(5), 437–454.

Umesha, S. S., Manohar, R. S., Indiramma, A. R., Akshitha, S., & Naidu, K. A. (2015). Enrichment of biscuits with microencapsulated omega-3 fatty acid (Alpha-linolenic acid) rich Garden cress (*Lepidium sativum*) seed oil: Physical, sensory and storage quality characteristics of biscuits. *LWT - Food Science and Technology, 62*(1), 654–661.

Van Boekel, M., Fogliano, V., Pellegrini, N., Stanton, C., Scholz, G., Lalljie, S., … & Eisenbrand, G. (2010). A review on the beneficial aspects of food processing. *Molecular Nutrition & Food Research, 54*(9), 1215–1247.

Vitaglione, P., Barone Lumaga, R., Ferracane, R., Radetsky, I., Mennella, I., Schettino, R., … & Fogliano, V. (2012). Curcumin bioavailability from enriched bread: The effect of microencapsulated ingredients. *Journal of Agricultural and Food Chemistry, 60*(13), 3357–3366.

Walker, P., Rhubart-Berg, P., McKenzie, S., Kelling, K., & Lawrence, R. S. (2005). Public health implications of meat production and consumption. *Public Health Nutrition, 8*(4), 348–356.

Wang, R., Tian, Z., & Chen, L. (2011). A novel process for microencapsulation of fish oil with barley protein. *Food Research International, 44*(9), 2735–2741.

Wang, Y. F., Shao, J. J., Zhou, C. H., Zhang, D. L., Bie, X. M., Lv, F. X., … & Lu, Z. X. (2012). Food preservation effects of curcumin microcapsules. *Food Control, 27*(1), 113–117.

Weir, D. G., & Scott, J. M. (1999). Brain function in the elderly: role of vitamin B_{12} and folate. *British Medical Bulletin, 55*(3), 669–682.

Wilkowska, A., Ambroziak, W., Adamiec, J., & Czyżowska, A. (2016). Preservation of antioxidant activity and polyphenols in chokeberry juice and wine with the use of microencapsulation. *Journal of Food Processing and Preservation, 41*(3), e12924.

Yang, T. T. C., & Koo, M. W. L. (1997). Hypocholesterolemic effects of Chinese tea. *Pharmacological Research*, 35(6), 505–512.

Yep, Y. L., Li, D., Mann, N. J., Bode, O., & Sinclair, A. J. (2002). Bread enriched with microencapsulated tuna oil increases plasma docosahexaenoic acid and total omega-3 fatty acids in humans. *Asia Pacific Journal of Clinical Nutrition*, 11(4), 285–291.

Ying, D., Schwander, S., Weerakkody, R., Sanguansri, L., Gantenbein-Demarchi, C., & Augustin, M. A. (2013). Microencapsulated Lactobacillus rhamnosus GG in whey protein and resistant starch matrices: Probiotic survival in fruit juice. *Journal of Functional Foods*, 5(1), 98–105.

Zanetti, M., Carniel, T. K., Dalcanton, F., dos Anjos, R. S., Riella, H. G., de Araujo, P. H., ... & Fiori, M. A. (2018). Use of encapsulated natural compounds as antimicrobial additives in food packaging: A brief review. *Trends in Food Science & Technology*, 81, 51–60.

Zawistowski, J. (2010). 17 Tangible health benefits of phytosterol functional foods. *Functional Food Product Development*, 362.

Zheng, J., Zhou, Y., Li, S., Zhang, P., Zhou, T., Xu, D. P., & Li, H. B. (2017). Effects and mechanisms of fruit and vegetable juices on cardiovascular diseases. *International Journal of Molecular Sciences*, 18(3), 555.

Zuidam, N. J., & Shimoni, E. (2010). Overview of microencapsulates for use in food products or processes and methods to make them. In *Encapsulation Technologies for Active Food Ingredients and Food Processing* (pp. 3–29). New York, USA: Springer.

16 Application of Spray Dried Encapsulated Bioactives in Pharmaceuticals

Paul Joyce, Hayley B. Schultz, Tahlia R. Meola,
Ruba Almasri, and Clive A.Prestidge
University of South Australia, Australia

CONTENTS

16.1 INTRODUCTION

It is well established that solid dosage forms (administered as tablets, capsules, or powders) are favored over liquid dosage forms by the pharmaceutical industry and its consumers (Fonteyne et al., 2015). This is largely attributed to the possible enhancements in short- and long-term stability, the ease of manufacturing, shipping, and handling, and the ability to introduce higher dosing accuracy (Desai, Wang, Wen, Li, & Timmins, 2013; Hardy & Cook, 2003; Joyce et al., 2019). Despite this, the physical and chemical nature of several bioactive molecules (e.g., poor aqueous solubility and susceptibility to the environment- or enzyme-mediated degradation) commonly require these drugs to be reconstituted within a liquid formulation to overcome such key limitations (Maghrebi, Prestidge,

& Joyce, 2019). For example, the overwhelming majority (up to 70%) of the newly-identified drug candidates are poorly water-soluble, which limits their therapeutic potential when administered to the body (Pouton, 2006).

A common approach used within the pharmaceutical industry to overcome the solubility challenges of such compounds is to formulate them within liquid lipid-based systems, to enable the drug to be delivered in a molecular/amorphous state (Joyce et al., 2019; Porter, Trevaskis, & Charman, 2007). However, as for most liquid dosage forms, these liquid lipid-based formulations suffer from poor physicochemical stability, which limits their translatability and commercial potential as pharmaceutical products. Subsequently, solidification techniques are required to transform precursor liquid dosage forms into dry powders, which can be achieved by either adsorbing the liquid drug solution/dispersion onto solid excipients/carriers, or through the utilization of specific drying techniques (Hardy & Cook, 2003). In recent decades, spray drying has gained increasing attention as an industry leader for the reconstitution of dry powders, in large part due to the reduced economic and resource burden of this drying approach and the ability to readily modulate the properties of the fundamental formulations for an optimized drug delivery performance (Davis & Walker, 2018). That is, spray drying is capable of engineering various forms of amorphous solid drug dispersions, microparticles, nanoparticles, or solid self-emulsifying drug delivery systems (Figure 16.1), where the final form of the dry powder can be controlled by the precursor formulation properties and/or the spray drying process. The process of spray drying can be simplified into four key steps: (1) heating of the drying gas; (2) atomization of the liquid feed into small droplets; (3) spraying of atomized droplets through a continuous flow of the heated gas; and (4) separation of the manufactured particles from the gas stream for powder collection (Arpagaus, Collenberg, Rütti, Assadpour, & Jafari, 2018). In this chapter, we discuss the advantages and challenges of this simple and continuous drying technique over alternate pharmaceutical drying processes, while attributing specific focus to established and foreseen applications of spray drying, which are ultimately designed to enhance the performance of drug delivery vehicles. The overall benefits of spray drying, which are highly

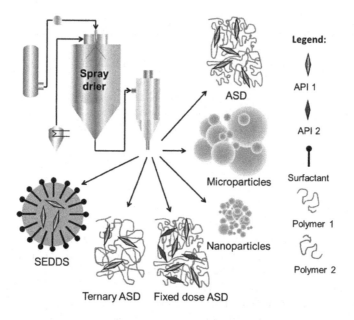

FIGURE 16.1 A schematic representation of the various final solid dosage forms that can be fabricated through the process of spray drying; amorphous solid dispersions (ASD), fixed-dose and ternary ASD, nano- and micro-particles, and solid self-emulsifying drug delivery systems (solid SEDDS). Reproduced from Davis and Walker (2018), with permission from Elsevier.

dependent on the desired administration approach and the therapeutic compound of interest, demonstrate the vast potential to translate into future dry powder technologies that are both clinically- and commercially-relevant in the pharmaceutical industry.

16.2 THE ADVANTAGES OF SPRAY DRYING FOR ENCAPSULATING BIOACTIVES IN THE PHARMACEUTICAL INDUSTRY

Several functional approaches exist for transforming liquid pharmaceutical excipients/formulations into solid dosage forms; e.g., spray drying (Ré, 2006), freeze drying (Lane, Brennan, & Corrigan, 2005; Singh, Vuddanda, Singh, & Srivastava, 2013), hot-melt extrusion (Guns et al., 2011), solvent evaporation (Atuah, Walter, Merkle, & Alpar, 2010; Homayouni, Sadeghi, Nokhodchi, Varshosaz, & Afrasiabi Garekani, 2015; Hong et al., 2016), fluid bed granulation (Burggraeve, Monteyne, Vervaet, Remon, & De, 2013), electrospraying (Nguyen, Clasen, & Van, 2016), precipitation (Hahn, Kim, & Shimobouji, 2007), and co-precipitation (Rashidinejad, Loveday, Jameson, Hindmarsh, & Singh, 2019). It is well established that the quality of the dried product is greatly influenced by the type and conditions of the drying process, along with the composition of excipients, since each drying technique introduces varying stresses on the formulation. Therefore, by using the same raw materials, formulations with contrasting physical and chemical properties can be attained when different drying methods are employed (Walters et al., 2014). Consequently, it is important to take into consideration the impact of the drying technique on the physicochemical characteristics and performance of the formulation. Spray drying provides numerous well-documented advantages over other drying techniques that will be discussed herein. Specific focus will be attributed to comparing and contrasting the advantages of spray drying for the other most commonly employed drying techniques in pharmaceutical manufacturing; i.e., freeze drying and hot-melt extrusion.

16.2.1 BENEFITS AFFORDED BY THE SPRAY DRYING PROCESS

Spray drying is considered a single-step continuous process that is not only simpler and more cost-effective than other drying techniques but also permits high throughput potential (Chiou & Langrish, 2007; Patel, Patel, & Chakraborty, 2014). This is mainly due to the reproducibility of this drying process, which, in turn, allows scaling up to any production size. Spray drying is an adaptable drying process, whereby several parameters can be manipulated (e.g., the spray dryer settings or composition of the formulation itself), in order to optimize the quality characteristics of the manufacture powders and/or their subsequent performance (Elversson & Millqvist-Fureby, 2005; Ré, 2006; R. Vehring, 2008). Hence, spray drying can be used in a wide range of applications, including drying heat-sensitive materials such as bioactive compounds/materials without the risk of degradation (Saluja et al., 2010), modifying the release of the drug (Atuah et al., 2010; Bhalekar, Madgulkar, Gunjal, & Bagal, 2013; Nunes et al., 2018), taste masking (Chranioti, Chanioti, & Tzia, 2016), and protecting the drug from the gastrointestinal (GI) environment (Dimitrellou et al., 2016; Lane et al., 2005; Nunes et al., 2018). Furthermore, the spray drying process can produce free-flowing powders consisting of single particles (Singh et al., 2013) or controlled agglomerates to be reconstituted for local administration, pressed into tablets, filled into capsules, or incorporated into implants and dressings (Ré, 2006). This remarkable versatility makes the technique preferred to a growing number of manufacturers in the pharmaceutical industry.

Comparisons between spray drying and other drying techniques reveal that the composition, structure of the matrix, processing time, drying temperature, and particle morphology are the critical factors governing the stability and performance of the generated formulations. With regard to the encapsulation of bioactives and their pharmaceutical delivery, spray drying is much more than a simple drying technique and offers unique opportunities that cannot be delivered by other drying techniques. This is evident in terms of its: (i) fast and gentle drying process; (ii) potential for one-step, high-throughput production; (iii) less costly manufacturing; (iv) high encapsulation efficiency; and (v) versatility in producing a wide range of products ranging from fine particles for pulmonary or parenteral delivery (Vidgrén, Vidgrén, & Paronen, 1987) to significantly larger granules for oral drug delivery (Davis & Walker, 2018). Despite the presence of numerous publications on the advantages of spray drying over other drying techniques, the full potential of spray drying of bioactive pharmaceuticals has yet to be fully exploited.

16.2.2 SPRAY DRYING VERSUS FREEZE DRYING

From a manufacturing standpoint, spray drying affords a multitude of advantages over freeze drying (or lyophilization), specifically with regard to the production costs, energy consumption, and quality control. High production costs associated with long processing time, high energy consumption, and poor quality of the resultant product due to long-term exposure to stress (heat and pressure), have limited the scope of applications of several conventional drying methods (Walters et al., 2014). Freeze drying is one such example, where the high costs and batch format have hindered its use in large-scale production (Horaczek & Viernstein, 2004). Despite the wide application of freeze drying in various fields, this approach is considered a time- and energy-consuming process that can take days or weeks to complete the drying cycle, especially if the formulation is not optimized and for large batches (Tang & Pikal, 2004). In contrast, spray drying provides less costly and ultrafast manufacturing compared to freeze drying (Ré, 2006).

For freeze drying, in addition to the costs associated with manufacturing solid dosage forms, several other industrial perspectives must be considered, which are typically built into the development of a new formulation. In particular, a patient-focussed design must be utilized to consider the impact on patient compliance (Joyce et al., 2019). Spray drying has been utilized to enhance palatability and mask the bitter taste of multiple drugs (Bora, Borude, & Bhise, 2008; Chranioti et al., 2016). In a study by Chranioti et al. (2016), spray drying encapsulation of steviol glycosides with maltodextrin and inulin presented a superior approach for masking the bitter aftertaste of steviol glycosides, when compared to both freeze drying and oven drying. The enhanced taste-masking ability of spray drying was stipulated to be linked to the superior physicochemical properties of the spray dried formulation, compared to the other drying techniques. That is, particle morphology observations revealed that spray drying generated spherical particles, where a complete maltodextrin/inulin coating was present, whereas porous flakes or blocks with some degree of crystallinity were produced by both freeze and oven drying.

Further to the manufacturing advantages afforded by spray drying, several studies have reported enhancements in the physical properties in the generated spray dried particles with uniform particle size distribution and enhanced flow properties, which introduces simpler and less costly handling when compared to freeze drying (Chranioti et al., 2016; Febriyenti et al., 2014; Lane et al., 2005; Singh et al., 2013). This is mainly attributed to the inability of particles to segregate during the freeze drying process, which triggers the formation of large flakes with poor dispersibility. To avoid this, the addition of a cryoprotectant is critical to improve particle redispersibility and mitigate the destabilizing effects of the freeze/thaw process (Yasmin, Tan, Bremmell, & Prestidge, 2014), which ultimately changes the composition and physicochemical properties of the formulation. A study by Singh et al. (2013) revealed that the performance of solid self-microemulsifying drug delivery systems (SMEDDS) of valsartan was superior in terms of flowability, compressibility, emulsification efficiency, and slightly enhanced bioavailability (about 5%) for spray dried solid SMEDDS, compared to freeze dried solid SMEDDS. Spray drying generated spherical porous particles with a uniform particle size distribution, whereas voluminous flakes were produced from freeze drying with different sizes and shapes. It was hypothesized that the uniform submicron-sized spray dried particles promoted the enhanced drug dissolution and solubilization in the gastrointestinal tract (GIT), which subsequently led to improved systemic drug absorption.

A major advantage of spray drying, when compared to freeze drying, is its application to thermolabile biological products (e.g., lipids, proteins, and nucleic acids, polyphenols, carotenoids, and vitamins), where the short residence time and reduced exposure to thermal stress limit oxidation and degradation of the biological molecules (Ré, 2006). For example, microencapsulation of plasmid DNA has been investigated for vaccine development, as a strategy to protect DNA from degradation, provide sustained release and enhance *in vivo* efficacy (Atuah et al., 2010; Lane et al., 2005). Typically, this requires encapsulation within a polymeric scaffold, which can be achieved via spray drying or freeze drying. Importantly, Lane et al. (2005) demonstrated the ability to increase DNA encapsulation efficiency within spray dried polymeric particles (> 90% encapsulation efficiency) compared to freeze drying (30% encapsulation efficiency). In doing so, the spray dried formulation achieved an 8-fold increase in the level of biological activity of encapsulated DNA extract when compared to the freeze dried formulation. Moreover, spray drying produces uniform particles with size ranges below 10 μm, which has important implications for vaccine development since this introduces the ability for the formulation to be successfully administered via either the pulmonary/mucosal (Husband, 1993; Yuki & Kiyono, 2009) or parenteral routes (Johansen, Men, Merkle, & Gander, 2000). On this basis, the spray drying technique proves to be an optimum, rapid, and efficient method for the preparation of delivery systems for DNA vaccination, as well as alternate thermolabile bioactive compounds.

16.2.3 SPRAY DRYING VERSUS HOT-MELT EXTRUSION

Hot-melt extrusion is the process of applying heat and pressure to a solid substrate, typically a polymer, to melt and force it through an orifice in a continuous process (Maniruzzaman, Boateng, Snowden, & Douroumis, 2012). For its application in the pharmaceutical industry, all materials (e.g., drug, polymer, and additional excipients) must be soluble or miscible and stable under the processing conditions (i.e., high temperature and high shear stress), to ensure the drug is successfully loaded within the solid carrier (Li, Hui Zhou, & Ping Xu, 2019). As a result of the high heat and stress exposed to the drug formulation, only a limited number of the active pharmaceutical entities can retain their biological activity during the extrusion process. Furthermore, biologics, such as proteins and peptides, which are highly heat-sensitive, cannot be formulated with hot-melt extrusion, which significantly limits the application of this drying approach, when compared to spray drying (Nelson, 2015).

One key study demonstrating degradation caused by hot-melt extrusion was performed by Hengsawas Surasarang et al. (2017), where the amorphous solid dispersions of albendazole were synthesized via spray drying and hot-melt extrusion. Despite initial processability and thermal degradation studies, high heat along with the long processing time caused 97.4% degradation of albendazole during the hot-melt extrusion with or without a polymer (Kollidon®). In contrast, the short residence time during the spray drying process in the combination with acid and polymer was shown to produce a stable amorphous solid dispersion, which ultimately displayed an 8-fold enhancement in the extent of dissolution and was stable for 1 year (displaying <5% degradation). These findings were in agreement with a stability study performed on fixed-dose combination systems comprising hydrochlorothiazide and ramipril (Kelleher et al., 2018). In this study, all spray dried formulations remained amorphous for 60 days, while the formulations generated by hot-melt extrusion with one of two polymers (*i.e.* Kollidon VA 64® or Soluplus®) both partially recrystallized, with the drying process triggering a high rate of degradation, which required the addition of a plasticizer. Despite the crystalline drug content, hot-melt extrusion samples achieved slightly higher dissolution than spray dried formulations, which was attributed to the surface roughness of the compressed discs generated from Wood's apparatus and not the solid-state variations. In contrast, Haser et al. (2017) observed a higher rate of recrystallization for a spray dried amorphous solid dispersion of naproxen (21.6%) compared to a corresponding naproxen-povidone K25 formulation prepared by melt extrusion. This was only exhibited at a high drug load of 60%, but not at 30%, and was attributed to the high surface area of spray dried particles, which in turn, provides a high degree of nucleation sites.

While hot-melt extrusion suffers from fundamental limitations in the development of pharmaceuticals, studies have shown that these limitations can be overcome by combining spray drying and extrusion to produce stable solid dosage forms. For example, Guns et al. (2011) examined the influence of combining spray drying with hot-melt extrusion on the mixing capacity and phase behavior of solid dispersions of miconazole and a graft copolymer. Despite no added benefit in the kinetic miscibility of the drug, spray drying led to a remarkable decrease in the torque experienced by the extruder, which enabled successful extrusion to occur at a lower temperature. Thus, the inclusion of spray drying as an additional step to generate an amorphous polymer before hot-melt extrusion reduced the degrading effect on the active ingredient, thus producing a final formulation with greater drug loading and activity. However, it should be noted that the combination of two drying techniques is a highly time-, energy-, and cost-consuming process and is more likely to suffer from poor commercial translation and uptake.

16.3   ADMINISTRATION OF ROUTE-DEPENDENT APPLICATIONS OF SPRAY DRIED BIOACTIVES

Spray dried products are free-flowing drug formulations that can be administered via multiple dosage routes. The administration route selected is typically dependent on the drug, therapeutic application and/or disease state. In the context of spray dried powders, the pharmaceutical industry typically utilizes one of four administration routes: oral, pulmonary, systemic or dermal delivery (Figure 16.2). This section will discuss, in detail, the various administration route-dependent applications of spray dried pharmaceutical powder formulations, their subsequent mechanisms of action, and the benefits afforded by the spray drying process. An overview of the key studies and insights derived for applying spray dried pharmaceutical formulations are summarized in Table 16.1.

16.3.1   ORAL DELIVERY

Oral delivery is the mainstay of pharmaceutical formulations, with 54% of drugs approved in 2019 by the U.S. Food and Drug Administration (2019) intended for administration via a tablet or capsule form. It is the most preferred route of drug administration and provides a high degree of patient compliance owing to convenience, non-invasiveness, and low production costs for the

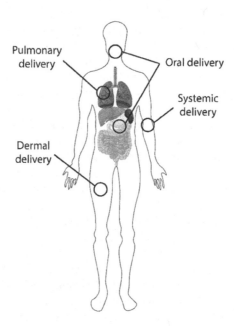

FIGURE 16.2   A schematic representation of the various drug administration routes that can be applied for spray dried bioactive formulations.

TABLE 16.1

An overview of key case studies demonstrating the application of spray dried pharmaceutical powders via various administration routes

Drug/ Bioactive	Formulation Excipients	Application(s)	Key Outcome Relating to Formulation Performance	Reference
Oral delivery				
Griseofulvin	Poloxamer 407	Treatment of fungal infections	1.8-fold enhancement in oral bioavailability compared to the pure drug, when dosed to rats.	Wong, Kellaway, and Murdan (2006)
Apigenin	Poloxamer 407	Model poorly water-soluble bioactive, with demonstrated anti-inflammatory and antioxidant properties	2.4-fold enhancement in oral bioavailability, compared to the pure bioactive, when dosed orally to beagle dogs.	Altamimi et al. (2018)
Allisartan isoproxil	Mannitol, sodium lauryl sulfate	Model poorly water-soluble drug, used for the treatment of hypertension	4.7-fold enhancement in oral bioavailability, compared to the pure drug, when dosed orally to rats.	Hou et al. (2017)
Ziprasidone	Hypromellose acetate succinate	Treatment of psychotic conditions	2.3-fold enhancement in oral bioavailability, compared to the commercial formulation, when dosed orally to rats.	Thombre, Caldwell, Friesen, McCray, and Sutton (2012)
Itraconazole	Hydroxymethyl propylcellulose	Treatment of fungal infections	1.8-fold increase in oral bioavailability and reduction of food effect, when compared to the commercial formulation and dosed orally to rats.	Mou, Chen, Wan, Xu, and Yang (2011)
Itraconazole	Maltodextrin and lipid emulsion	Treatment of fungal infections	A 1.15-fold increase in oral bioavailability, compared to the unformulated drug, when dosed orally to rats.	Rao and Aghav (2014)
Malotilate	Carboxymethyl cellulose and lipid emulsion	Treatment of liver disease	1.7-fold enhancement in oral bioavailability, compared to the precursor liquid lipid emulsion, when dosed orally to rats.	Zhang, Gao, Qian, Liu, and Zu (2011)
Cinnarizine	PLGA and medium-chain triglycerides	A model poorly water-soluble drug that serves as an antihistamine	2-fold improvement in oral bioavailability compared to two liquid lipid systems, when dosed orally to rats.	Joyce et al. (2017)

(Continued)

TABLE 16.1 (*Continued*)

Drug/ Bioactive	Formulation Excipients	Application(s)	Key Outcome Relating to Formulation Performance	Reference
Blonanserin	Smectite clay particles and medium-chain triglycerides	A model poorly water-soluble drug used for the treatment of psychotic conditions	A decrease in oral bioavailability for spray dried particles, compared to liquid lipid emulsions, due to the interference of clay particles with drug release.	Dening et al. (2018)
Simvastatin	Porous silica and medium-chain triglycerides	A model poorly water-soluble drug used as cholesterol and lipid-lowering agent	2.9- to 6.1-fold increase in oral bioavailability, compared to the pure drug, when dosed orally to rats. Bioavailability enhancement was dependent on silica geometry.	Meola, Abuhelwa, Joyce, Clifton, and Prestidge (2020)
Celecoxib	Porous silica and medium-chain triglycerides	A model poorly water-soluble drug used as an anti-inflammatory	1.2- and 1.6-fold improvement in oral bioavailability compared to the commercial formulation and liquid lipid system, when dosed orally in rats.	Tan, Simovic, Davey, Rades, and Prestidge (2009)
Nifedipine	Eudragit® L	Treatment of angina and lowering of blood pressure	11.9-fold improvement in oral bioavailability, compared to the pure drug, due to the controlled drug release within the intestinal environment.	Choi et al. (2011)
Inactivated influenza virus	Trehalose and Eudragit	Influenza virus vaccine	10-fold higher response in antigen-specific immunoglobulins than naive mice controls, due to the gastro-protective mechanism of Eudragit.	Shastri, Kim, Quan, D'Souza, and Kang (2012)
Tumor cell lysate	β-cyclodextrin, ethylcellulose, trehalose, and HPMCAS	Breast cancer vaccine	Mice vaccinated with spray dried tumor cell lysate developed significantly smaller tumors, compared to an unvaccinated control group, but no direct comparison was made with the non-spray dried lysate.	Chablani et al. (2012)

(Continued)

Drug/ Bioactive	Formulation Excipients	Application(s)	Key Outcome Relating to Formulation Performance	Reference
Pulmonary delivery				
siRNA	Polyplexes are formed with transferrin-polyethyleneimine (Tf-PEI) and melittin-polyethyleneimine (Mel-PEI), then spray dried with trehalose	Targeting transcription factor GATA3 for the treatment of asthma	siRNA was successfully delivered to TH2 cells, which promoted *in vivo* gene silencing and high transfection efficacies since the spray dried particles were of optimal size for dry powder inhalation.	Keil, Baldassi, and Merkel (2020)
Salbutamol sulphate	Lactose, β-cyclodextrin, starch or sodium carboxy-methylcellulose (NaCMC)	A model drug used for the treatment of asthma	β-cyclodextrin, starch, and NaCMC revealed optimal deep lung deposition, compared to lactose particles. NaCMC demonstrated a sustained drug release mechanism and further increased the aerosolization performance.	Xu, Guo, Xu, Li, and Seville (2014)
Paclitaxel	PEGylated liposomes, spray dried in an organic solvent	Treatment of lung cancer	Excellent aerosolization performance and redistribution of precursor liposomes during in vitro characterization.	Meenach, Anderson, Zach Hilt, McGarry, and Mansour (2013)
Sildenafil citrate	NaCMC, sodium alginate, and sodium hyaluronate	Treatment of pulmonary arterial hypertension	The pharmacokinetic evaluation revealed a > 17-fold and > 4-fold increase in the lung and plasma area-under-the-curve, respectively, when the spray dried microparticles were inhaled, compared to the orally dosed pure drug.	Shahin et al. (2019)
Sodium alendronate	Ammonium bicarbonate in hydroalcoholic solutions	A model drug used in the treatment of bone diseases	Inhalation delivery of spray dried microparticles triggered a 3.5-fold increase in bioavailability, compared to the orally dosed pure drug.	Cruz et al. (2011)
Budesonide	Chitosan	Treatment of asthma and chronic obstructive pulmonary disease	Spray dried chitosan particles revealed excellent aerosolization properties, allowing for deep lung deposition and extended *in vitro* release up to 4 hours.	Naikwade and Bajaj (2009)

(Continued)

TABLE 16.1 (*Continued*)

Drug/ Bioactive	Formulation Excipients	Application(s)	Key Outcome Relating to Formulation Performance	Reference
Dermal & transdermal delivery				
Dexamethasone	Polycaprolactone	Treatment of dermatological diseases, including psoriasis and acne	Significantly greater accumulation and controlled release of dexamethasone when spray dried polymeric particles were combined within conventional skin cream, compared to the pure drug.	Beber et al. (2014)
Soy isoflavones	HPMCAS	Model bioactive used as an anti-aging agent	Using nano spray drying to produce submicron particles enhanced the skin penetration of soy isoflavones 4.5-fold compared to the unformulated extract.	Del et al. (2017))
Amoxicillin	Chitosan	Treatment of wound infections	2-fold enhancement in the anti-bacterial efficacy against *Staphlyococcus aureus* compared to pure amoxicillin.	Ngan et al. (2014)
Gentamicine	Alginate and pectin	Treatment of wound infections	Spray dried particles gelled within 15 min of skin contact and subsequently released the anti-bacterial agent over the course of 3–6 days, considerably increasing the anti-bacterial efficacy compared to the pure drug.	De, Porta, Sansone, Aquino, and Del (2014)
Systemic delivery				
Etanidazole	PLGA	Used for radiosensitizing in the treatment of cancer	*In vitro* drug release demonstrated a sustained release mechanism over the course of a 6 h period, which was predicted to reduce the overall frequency of dosing.	Wang and Wang (2002)

(*Continued*)

Drug/ Bioactive	Formulation Excipients	Application(s)	Key Outcome Relating to Formulation Performance	Reference
siRNA	Loaded in primary human serum albumin nanoparticles that were subsequently spray dried with PLGA	Used as a model gene targeting agent	Nano-in-nano formulation approach demonstrated the ability to protect thermolabile bioactives and exerted a unique drug delivery mechanism that could be used for systemic or pulmonary delivery.	Amsalem et al. (2017)
Carboplatin	Gelatin	Treatment of cancer	24 hours after the intravenous injection to albino mice, the drug concentration in the lungs was significantly higher than in any other tissues including the liver, spleen, or blood.	Harsha et al. (2014)
Salbutamol	Albumin	Treatment of asthma	*In vivo* pharmacokinetics revealed that the albumin microspheres increased the concentration of salbutamol >14-fold, compared to the pure drug, due to the tissue targeting potential of the spray dried formulation.	Harsha et al. (2017)
Erythropoietin (EPO)	Hyaluronic acid and sodium tetrathionate	Model protein drug	*In vitro* release studies revealed bi-phasic release kinetics, with burst release over the first three days, followed by a 9-day prolonged-release period in media simulating plasma. Elevated EPO levels were observed *in vivo* for seven days.	Hahn et al. (2007)
Paclitaxel	Poly(3-hydroxybutyrate)	Treatment of cancer	Paclitaxel-loaded microparticles revealed a significantly enhanced *in vivo* anti-tumor activity, compared to the pure drug, in intraperitoneally transplanted mice tumor models.	Bonartsev et al. (2017)

(Continued)

TABLE 16.1 (*Continued*)

Drug/ Bioactive	Formulation Excipients	Application(s)	Key Outcome Relating to Formulation Performance	Reference
Risperidone	PLGA	Treatment of psychosis	PLGA nanoparticles were successfully fabricated with average diameters of 250 nm, through nano spray drying, and triggered a sustained release mechanism over 10 days, *in vitro*.	Panda, Meena, Katara, and Majumdar (2016)
Rifampicin	PLGA, medium-chain triglycerides, and mannitol	Treatment of intracellular *Staphlyococcus aureus* infections	Spray dried hybrid formulations triggered a 7.2-fold increase in macrophage uptake and a 2.5-log reduction in intracellular bacterial counts, compared to the pure drug.	Maghrebi, Joyce, Jambhrunkar, Thomas, and Prestidge (2020)

manufacturers (Colombo, Sonvico, Colombo, & Bettini, 2009). However, the high number of poorly water-soluble drugs emerging from the drug discovery pipeline highlights the need for bio-enabling techniques, such as spray drying, which can encapsulate bioactives and modify their biopharmaceutical performance.

16.3.1.1   Enhancing oral Bioavailability

Originally, considered as a primary dehydration process, the spray drying of pharmaceuticals is now a means for developing novel pharmaceutical carriers for the encapsulation of poorly water-soluble drugs that suffer from poor oral bioavailability. The process of spray drying introduces the ability to formulate poorly water-soluble drugs within micro- or nano-particles with a high surface area, which promotes drug dissolution, while the use of solubility-enhancing excipients can promote drug solubilization in the GIT. Specifically, spray drying can be used to fabricate amorphous solid dispersions, solid dispersions, microparticles, nanoparticles, and solid self-emulsifying systems, which will be discussed in the subsequent section (Paudel, Worku, Meeus, Guns, & Van den, 2013; Singh & Van den, 2016; Sollohub & Cal, 2010; Reinhard Vehring, 2008). Additional to the solidification benefits, spray drying enables the control of particle size, morphology, porosity, and drug solid-state (crystalline or amorphous), which can play a substantial role in influencing oral pharmacokinetics and bioavailability (Salama, 2020). The ability to control particle size, morphology, and porosity of spray dried dry emulsion powders composed of entirely the same excipients, but in varying concentrations and by employing different processing parameters, is highlighted in Figure 16.3.

Polymeric excipients are commonly chosen as solid carrier materials for spray drying due to their amphiphilic nature, which allows for the incorporation of poorly water-soluble compounds within the hydrophobic regions, while the hydrophilic regions provide high colloidal stability in aqueous environments. However, the selection of the excipients in a formulation is highly dependent on the individual API and the following examples reflect the need for polymer variety. A study executed by Wong et al. (2006) demonstrated the ability for spray drying to increase solubility and bioavailability of the anti-fungal, griseofulvin, when spray dried alone, or with 0.05% w/v Poloxamer 407 (Pluronic F127), from organic dichloromethane solutions. Drug crystallinity did not

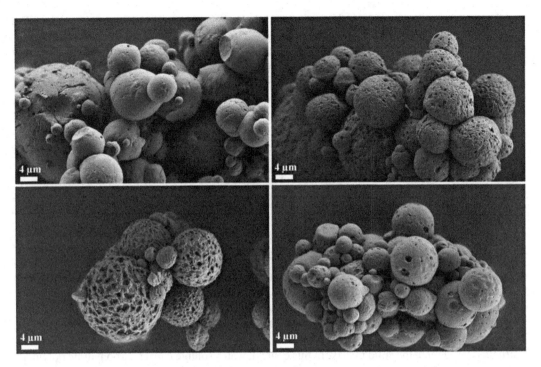

FIGURE 16.3 A comparison of the particle sizes and morphologies of four sets of simvastatin-loaded dry emulsions, comprised of various excipients concentrations and fabricated using various spray drying parameters. The primary excipients in each dry emulsion are: 1-oleoyl-rac-glycerol, Miglyol 812, Tween 20, hydroxypropyl methylcellulose (HPMC), and mannitol. Adapted from Pohlen, Lavrič, Prestidge, and Dreu (2020), with permission from Springer Nature.

differ between formulations; however, *in vitro* dissolution studies revealed a significant increase in drug release from the griseofulvin/Poloxamer 407 and griseofulvin-alone particles compared to the pure drug compound, signifying that the spray drying process itself enhanced the dissolution properties. Interestingly, after the oral administration to rats, no significant difference in bioavailability was observed between the spray dried griseofulvin and the pure drug, while a 1.8-fold improvement was evident for the griseofulvin/Poloxamer 407 particles. Although the addition of Poloxamer 407 increased the particle size (mean diameter of 11.2 μm, compared to 8.5 μm for griseofulvin-only particles), the results demonstrated that the inclusion of a hydrophilic surfactant was important in enhancing the wettability, and thus the bioavailability of a spray dried formulation. Similarly, Altamimi et al. (2018) observed a 2.4-fold increase in apigenin exposure when spray dried with Poloxamer 407, compared to the pure drug after oral administration to beagle dogs, owing to a combination of increased wettability and amorphization of the drug.

Spray drying can also meet the emerging demands for nanostructured therapeutics by generating nanosized particles that facilitate dissolution and enhance bioavailability, owing to their increased surface area (Labortechnik, 2009). Hou et al. (2017) used a combinatorial approach of ball milling followed by spray drying to produce spray dried nanocrystals using allisartan isoproxil as a model drug. The spray dried formulation containing a drug, mannitol, and sodium lauryl sulfate displayed an equivalent particle size to the original nanocrystals, and oral administration to rats revealed a 4.7-fold enhancement in bioavailability compared to the unformulated drug. In an alternate study, solid nanocrystalline dispersions of ziprasidone were successfully fabricated by Thombre et al. (2012), through spray drying with hypromellose acetate succinate (HPMCAS) as the matrix polymer. Drug domains between 50 to 100 nm in size were successfully encapsulated within the spherical and slightly rough particles, indicating the presence of drug crystals at or near the particle surface. The oral administration to beagle dogs demonstrated a 2.3-fold improvement in the drug bioavailability (compared to the commercial formulation) and, importantly, the successful removal of the fed/fasted variation associated with the drug. Furthermore, Mou et al. (2011) investigated the feasibility of facile acid–base neutralization to generate an itraconazole nanosuspension prior to spray drying using a HPMC and mannitol matrix. Particles appeared spherical and although particle aggregation was observed in the solid state, the dispersibility in an aqueous environment was not impacted. In comparison to the commercial formulation (Sporanox®), the dried nanosuspension exhibited a 1.8-fold greater oral bioavailability in the fasted-state, whilst removing the influence of food and reducing inter-individual variability in bioavailability. Thus, the wealth of studies performed highlight the ability for drug encapsulation via spray drying to enhance the biopharmaceutical performance of various poorly water-soluble compounds.

16.3.1.2   Solidification of Lipid-based Formulations

The solidification of lipid-based formulations, such as lipid emulsions and self-emulsifying drug delivery systems (SEDDS), affords a panoply of advantages compared to the precursor liquid-state formulation (Dening, Rao, Thomas, & Prestidge, 2016a; Joyce et al., 2019). Liquid lipid emulsions can be readily converted to dry emulsions by dehydrating the aqueous phase of an oil-in-water emulsion. To achieve this, a solid carrier is typically dispersed within the aqueous phase of the emulsion, which, upon dehydration, forms a three-dimensional particulate network whereby the oil phase is distributed throughout the solid matrix. In this context, water-soluble carrier carbohydrate polymers, such as mannitol, lactose, and dextrins, are most commonly utilized due to their ability to protect the oil phase during spray drying, as well as their ability to form powders of high flowability and low adhesiveness. An example of employing carbohydrate polymers for the manufacture of dried emulsions was performed by Rao and Aghav (2014), who prepared itraconazole-loaded dry emulsion via spray drying the liquid precursor with maltodextrin. The dry emulsion consisted of well-separated spherical particles with a smooth surface. While *in vitro* studies revealed no significant difference in itraconazole performance between the liquid and dry emulsions, the oral bioavailability of itraconazole was improved by 115% for the spray dried emulsion, compared to the unformulated drug. In a similar study, the influence of carboxymethyl cellulose as a carrier for a malotilate encapsulated emulsion was investigated by Zhang et al. (2011), who revealed that a 1.7-fold enhancement in oral bioavailability for the dry emulsion, compared to the precursor liquid emulsion. Subsequently, such findings demonstrated that spray drying oil-in-water emulsions with carbohydrate polymers, to form dry emulsions, was an effective formulation strategy to improve drug performance, when compared to both the pure drug and a liquid emulsion. While the exact mechanism for promoting enhanced bioavailability varies, depending on the drug, lipid, and solid carrier type, the spray drying process typically increases the surface area of lipid within the system, which promotes enhanced digestibility by lipase enzymes that further promote the solubilization of the drug in the GIT (Joyce, Dening, Gustafsson, & Prestidge, 2017).

Other insoluble organic materials, such as synthetic polymers, have also shown the ability to form dry emulsions when spray dried with lipid-in-water emulsions. Importantly, polymeric excipients may offer prolonged and/or controlled drug release for an extended period due to their slower rate of matrix erosion, compared to carbohydrate polymers (Joyce, Schultz, Meola, & Prestidge, 2020). Moreover, the spray drying of Pickering emulsions (solid particle-stabilized emulsions) has been shown to produce particles with unique nanostructures that are capable of enhancing the *in vivo* performance of a wide range of poorly water-soluble drugs (Dening, Rao, Thomas, & Prestidge, 2016b; Meola, Schultz, Peressin, & Prestidge, 2020; Simovic et al., 2009; Tan et al., 2009). Joyce, Yasmin, et al. (2017) fabricated polymer-lipid hybrid solid microparticles by spray drying dispersed medium-chain triglyceride droplets stabilized by poly(lactic-co-glycolic) acid (PLGA) nanoparticles for the delivery of poorly water-soluble weak base, cinnarizine. By utilizing synthetic polymeric nanoparticles, such as PLGA nanoparticles, cinnarizine could be loaded into both the polymer and lipid phases, which resulted in dual release kinetics where rapid drug release was observed from the lipid droplets in the acidic environment, followed by a slow gradual release from the polymeric micelles. Additionally, a 2-fold enhancement in the drug bioavailability was observed after the oral administration to rats, compared to two alternative lipid-based formulations, due to the prolonged-release kinetics and the ability for PLGA to serve as a precipitator inhibitor of cinnarizine (Joyce & Prestidge, 2018).

While organic materials have been explored to a greater degree for the fabrication of dry emulsions, inorganic materials (e.g., porous silica particles) are also suitable solid carrier systems that promote enhanced drug solubilization and oral absorption. Tan et al. (2009) were the first to develop solid nanostructured silica-lipid hybrid (SLH) microcapsules via spray drying a silica nanoparticle-stabilized Pickering emulsion, composed of medium-chain triglycerides stabilized by phospholipid emulsifiers, to enhance the pharmacokinetic properties of celecoxib. As shown by Figure 16.4, the nanostructure, particle size, and surface morphology can be readily manipulated by altering the silica nanoparticle size and charge of the stabilizing emulsifier, among other variables. Additionally, cross-sectional morphology revealed that the internal nanoporous matrix, with pore sizes ranging from 100 to 500 nm, is controlled by the self-assembly of silica nanoparticles at the lipid droplet surface and inner oil core during the water removal process, and, thus, is also mediated by key physicochemical properties of the silica particles and emulsion droplets. The oral administration to rats demonstrated a 1.3- and 1.6-fold improvement in the bioavailability for the celecoxib SLH, compared to the precursor oil-in-water emulsion and celecoxib suspension, respectively. Similarly, the bioavailability of indomethacin SLH was enhanced 1.4- and 1.7-fold in comparison to the precursor emulsion and conventional drug suspension (Simovic et al., 2009). Furthermore, SLH particles have shown the ability to mimic the pharmaceutical food effect, thereby reducing the fasted-fed state variability associated with some poorly water-soluble drugs. For example, Dening et al. (2016b) spray dried ziprasidone-loaded SLH in an attempt to reduce the influence of food on drug solubilization. SLH formulations exhibited up to a 43-fold increase in the fasted-state solubilization compared to a pure drug, and no significant difference was observed between the fed and fasted state, highlighting the food-mimicking ability. Various mechanisms can explain the improved performance offered by spray dried SLH: (i) preservation of the drug in a pre-solubilized state; (ii) high surface area owing to the nanostructured porous matrix; and (iii) enhanced lipase accessibility to facilitate lipid digestion (Joyce, Barnes, Boyd, & Prestidge, 2016; Joyce, Tan, Whitby, & Prestidge, 2014;). The commercial potential of a spray dried SLH formulation was demonstrated in a phase I study evaluating the safety and pharmacokinetic profile of spray dried ibuprofen encapsulated SLH whereby the formulation was safe, well-tolerated and enhanced bioavailability 1.95-fold compared to Nurofen®, after oral administration to healthy male subjects (Tan, Eskandar, Rao, & Prestidge, 2014).

An alternate category of solid lipid-based systems to dry emulsions are solid SEDDS, which are isotropic mixtures of oil, surfactant, and solvents in the form of a powder (Joyce et al., 2019). Upon redispersion in aqueous media, such as the GIT, solid SEDDS spontaneously emulsify to produce a fine oil-in-water emulsion, which further serves to solubilize the poorly soluble drug compounds (Tan, Rao, & Prestidge, 2013). The solidification of SEDDS via a spray drying approach arises from the primary objective of: (i) solidifying liquid-SEDDS excipients that self-emulsify *in vivo*; and (ii) stabilization of dispersed SEDDS that re-emulsify *in vivo* (Joyce et al., 2019). However, the opportunities for improved drug performance, presented by forming solid SEDDS include: (i) prolonged gastric transit through the development of floating and mucoadhesive systems; (ii) enhanced GI lipolysis, promoting an increase in intestinal solubilization; (iii) potential to incorporate precipitation inhibitors to prolong drug solubilization, and ultimately; and (iv) improved absorption across the intestinal epithelium (Figure 16.5) (Joyce et al., 2019).

Similar to dry emulsions, the drug delivery performance of solid SEDDS relies heavily on the type of the solid carrier material used. The impact of the carrier type on drug performance within solid SEDDS was investigated by Kang, Oh, Oh, Yong, and Choi (2012), who prepared

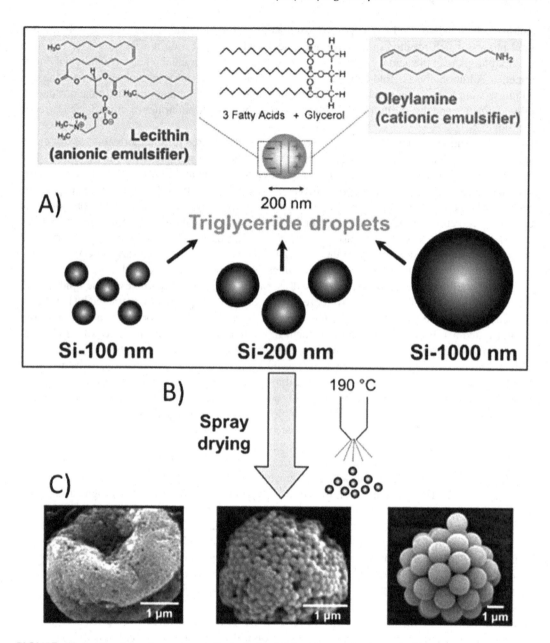

FIGURE 16.4 A schematic representation of the fabrication mechanism of silica-lipid hybrid (SLH) particles. (A) Silica nanoparticles of varying sizes can be combined with oil-in-water emulsion droplets of varying charges, to form silica-stabilized Pickering emulsions; (B) The Pickering emulsions are spray dried; and (C) forming nanostructured microparticles in the size range of 1–5 μm composed of highly porous three-dimensional networks, where lipid droplets are dispersed throughout a silica nanoparticle matrix. Reproduced from Tan, Colliat-Dangus, Whitby, and Prestidge (2014), with permission from American Chemical Society.

solid SEDDS for flurbiprofen using silicon dioxide, magnesium stearate, polyvinyl alcohol (PVA), sodium carboxymethylcellulose (Na-CMC), and hydroxypropyl-β-cyclodextrin (HP-β-CD). Whilst no comparisons were made to the precursor liquid SEDDS, *in vivo* studies revealed superior flurbiprofen bioavailability, relative to the pure drug, when SEDDS were spray dried with hydrophobic carriers (i.e., silicon dioxide and magnesium stearate), compared to spray drying with the

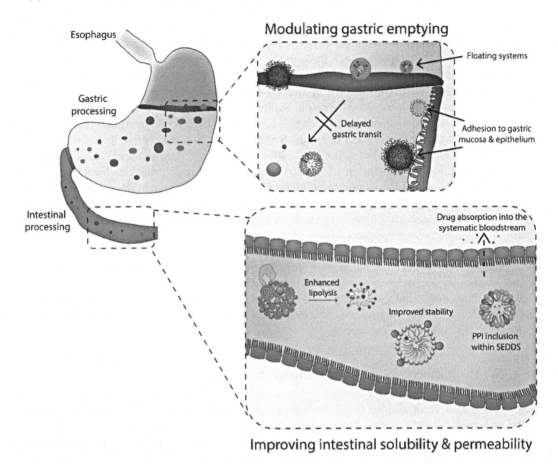

FIGURE 16.5 A schematic illustration of the potential benefits obtained by creating solid self-emulsifying drug delivery systems (SEDDS) through the spray drying isotropic mixtures of lipids with solid carrier materials. Reproduced from Joyce et al. (2019), with permission from Elsevier.

hydrophilic carriers (*i.e.* PVA, Na-CMC and HP-β-CD). It was stipulated that the hydrophilic carriers retarded the emulsification of lipid species when dispersed within the aqueous phase, which subsequently inhibited drug solubilization and absorption *in vivo*. While the impact of carrier type nanostructure on the drug solubilization and absorption mechanism is not well understood, it emphasizes the importance of solid carrier selection in spray drying of liquid lipid-based systems for optimizing oral delivery performance. Moreover, the critical need for future studies that elucidate the physicochemical parameters of solid carriers that lead to enhanced solubilization and absorption is highlighted.

16.3.1.3 Enteric Coating
Enteric coatings are utilized in oral drug delivery to protect the encapsulated drug from the acidic gastric environment, by preventing the release of the drug until it reaches the small intestine. Polymeric materials such as Eudragit®, cellulose acetate phthalate, and hydroxyl propyl methyl cellulose phthalate/acetate succinate (HPMCP or HPMCAS) are commonly utilized due to being unionized and insoluble at low pH, and ionized and soluble in the higher pH of the GIT (Chablani, Tawde, & D'Souza, 2012). Nevertheless, Eudragit® appears to be the predominant enteric coating utilized in spray drying for the oral delivery of drugs.

Simonoska Crcarevska, Glavas Dodov, and Goracinova (2008) coated budesonide-loaded chi-tosan-alginate microparticles with a Eudragit® S100 solution to fabricate spherical particles with a smooth surface and collapsed center, through the process of spray drying. The Eudragit enteric coating permitted only 20% drug release in acidic media (simulating the gastric environment), prior to releasing 100% of the encapsulated drug in pH 7.4 media, and, thus, providing the potential for the efficient release of budesonide in the colon. Similarly, an enterically-coated nifedipine elixir was spray dried with an organic solution of Eudragit® L to form spherical microparticles that prevented drug release in an acidic media (i.e. pH 2.0 and 6.8) by up to 80%, compared to the uncoated elixir during *in vitro* dissolution studies (Choi et al., 2011). Pharmacokinetic studies revealed that after the oral administration to rats, the enteric-coated elixir significantly enhanced the bioavailability of nifedipine 1.6-fold and 11.9-fold compared to the uncoated elixir and pure drug, respectively. In a similar study, Li et al. (2009) attempted to reduce the burst release associated with nifedipine encap-sulated gelatin microcapsules, by coating particles with Eudragit® L100–55 via spray drying. The Eudragit-coated particles were successful at hindering *in vitro* drug dissolution in acidic media by up to 50% compared to uncoated gelatin microcapsules. In comparison to the unformulated powder, nifedipine exposure was enhanced about 40-fold for the coated particles after oral administration to rats, confirming the delivery system could maintain the effective plasma concentrations for a prolonged time period. Enteric coating via spray drying has been utilized in numerous additional cases, specifically for the encapsulation and delivery of sensitive biologics which will be discussed in greater detail in the following section.

16.3.1.4 The Microencapsulation of Bioactive Proteins and Peptides

The solidification of proteins and peptides provides superior stability, compared to when in the liquid state, while the use of an additional carrier material introduces the ability to protect the sensitive bio-active molecules from the harsh environmental and enzymatic conditions of the GIT. Lyophilization has been the most common approach for the production of solid protein powders; nevertheless, since a secondary process is generally required to break up the lyophilizate to achieve a flowable powder, solidification via spray drying is deemed more desirable from a commercial viewpoint (Costantino & Pikal, 2004; Sollohub & Cal, 2010).

The oral delivery of proteins and peptides is limited, primarily due to their susceptibility to hydro-lysis and degradation at gastric pH levels, and by proteolytic enzymes in the small intestine, leading to an associated poor bioavailability (Sinha & Trehan, 2003; Truong-Le, Lovalenti, & Abdul-Fattah, 2015). To date, numerous proteins and peptides have been encapsulated in microparticulate systems via spray drying with careful selection of excipients to overcome these challenges. Studies on oral vaccines have reported that the microparticles with a diameter of less than 10 µm demonstrate good adjuvant effects, which result in boosting the immune response to produce antibodies and prolong immunity (Hori, Onishi, & Machida, 2005; Tabata, Inoue, & Ikada, 1996). Such particle size is eas-ily obtained utilizing a spray drying solidification process.

Although the drying process for the spray drying technique utilizes high temperatures, the ther-mal denaturation is usually negligible, and any denaturation that occurs may be attributed to dehy-dration, rather than exposure to the drying process itself. Therefore, the use of excipients, such as sugars or polyols that can replace hydrogen-bonded water, are critical for the preservation of spray dried proteins and peptides. Although intended for respiratory use, Andya et al. (1999) investigated the stabilizing effect of trehalose, lactose, and mannitol on the stability of a spray dried immuno-globulin G (IgG) monoclonal antibody and found trehalose was most effective in stabilizing the pro-tein. The superior stabilizing effect of trehalose was confirmed by Maury, Murphy, Kumar, Mauerer, and Lee (2005), who explored the effect of trehalose and sorbitol on the stability and aggregation of a spray dried IgG solution. Trehalose was again found to be a more effective stabilizer; however, both sugars significantly reduced the particle aggregation after spray drying owing to the water replacement mechanism.

To avoid degradation within the GIT, enteric polymer coatings can be used as solid carriers for oral spray dried proteins (Truong-Le et al., 2015). Importantly, polymeric excipients offer additional advantages for the delivery of proteins, such as stimulating a systemic immune response, prolonged residence of antigen to the immune system, and removal of the need for vaccine adjuvants since the microparticle can act as an adjuvant itself (Chablani, Tawde, & D'Souza, 2012; O'Hagan & Singh, 2003). Shastri et al. (2012) formulated an influenza vaccine by spray drying inactivated influenza virus with Eudragit and trehalose, as gastroprotective and stabilizing agents, respectively. The oral administration of the microparticulate powder, with particle sizes of 1–6 μm, to mice, revealed the protection from the gastric environment prior to absorption, which induced a 10-fold higher response in antigen-specific immunoglobulins than naive mice controls. Thus, it may provide increased levels of protection against influenza. The ability to promote a vaccine response with solid protein systems was also shown by Pastor et al. (2013), who encapsulated a cholera vaccine within an enterically-coated cellulose-alginate polymer matrix via spray drying to produce monodispersed smooth spherical microparticles with an average particles size of 6 μm. *In vivo* studies revealed that the solid microencapsulated system induced an equivalent immune response to a liquid vaccination. The potential for the clinical translation was further highlighted by the high stability of the spray dried vaccine, where under the accelerated conditions (40°C), the vaccine remained stable for six months (Pastor, Esquisabel, Marquínez, Talavera, & Pedraz, 2014), indicating its suitability as a cold-chain free oral vaccine.

Furthermore, it has been reported that electrostatic attractions between the negatively charged mucus linings of the GIT and positively charged particles (+30 to +50 mV) result in a decreased penetration toward epithelia, which, in turn, may lessen the uptake of the particles (Shakweh, Besnard, Nicolas, & Fattal, 2005). Chablani, Tawde, and D'Souza (2012) investigated a FITC-albumin microparticle-based spray dried system comprising of specific excipients and carriers to tailor particle size, charge, and hydrophobicity to maximize the uptake via M-cells in a single vaccine. Particles were doughnut-shaped and homogenous with a mean diameter of 1.5 μm. The use of enteric polymers, Eudragit®, and HPMCAS imparted sufficient hydrophobicity and the addition of chitosan glycol induced a positive charge of +15.7 mV, facilitating the penetration across the mucus layer and uptake by M-cells, as confirmed by fluorescent imaging. In a subsequent study, the authors incorporated a tumor cell lysate into an aqueous polymer matrix, prior to spray drying to form an enterically-protected prophylactic breast cancer vaccine (Chablani, Tawde, Akalkotkar, et al., 2012). The microparticles obtained a particle size distribution between 1 and 4 μm, which was ideal for M cell uptake in the small intestine, associated with mimicking the dimensions of a pathogen. Orally-vaccinated mice developed significantly smaller tumors, compared to an unvaccinated control group, indicating that the spray dried vaccine was effective in generating protective immunity.

An alternative to the reconstitution of spray dried powders for the oral delivery of proteins and antigens, orally disintegrating films are of interest as the buccal cavity provides a large surface area for the rapid disintegration, release, and absorption of therapeutic entities. Gala et al. (2017) produced spray dried microparticles (mean particle size of 3.65 μm) containing a live attenuated measles vaccine which were incorporated into an orally disintegrating film and administered to juvenile pigs via the buccal route. A significant increase in antibody presentation and co-stimulatory markers were observed, partly attributed to the irregularly shaped and rough particle morphology which stimulates biodistribution and uptake by macrophages. The results suggest that the orally disintegrating vaccine is a viable formulation for non-invasive immunization that may increase patient compliance.

16.3.2 Pulmonary Delivery

Drug delivery via the inhalation administration route has received increasing attention in recent decades, specifically for the treatment of chronic lung diseases and infections, such as cystic fibrosis, tuberculosis, and chronic obstructive pulmonary disease (Smola, Vandamme, &

Sokolowski, 2008). In order to successfully deliver therapeutics to the deep lung tissue, they must be formulated with a carrier system composed of particles large enough to avoid direct exhalation during inhalation and small enough to avoid mucociliary clearance in the upper airways (Garbuzenko, Mainelis, Taratula, & Minko, 2014). Studies have shown that the optimal size for the deposition of drug carriers into the lower airways is about 2–5 µm (Lee, Loo, Traini, & Young, 2015). The most commonly employed approach to achieve drug-loaded particles of this size for the inhalation therapy is via nebulization, where a liquid drug dispersion/solution is aerosolized into fine droplets in the optimal size range for lung deposition (O'Callaghan & Barry, 1997). However, this administration approach is severely limited by the solubility and dissolution rate of the drug in the nebulizing media, as well as other factors associated with liquid drug dispersions/solutions (e.g., stability and shelf-life) (Respaud, Vecellio, Diot, & Heuzé-Vourc'h, 2015). Subsequently, a specific focus has been attributed to overcoming these limitations by developing solid particulate systems capable of delivering drugs into the lower airways of the respiratory tract via inhalation.

Due to the size constraints required for efficacious inhalation and drug deposition, spray drying has emerged as an ideal development approach for dry powder inhalation, where 2–5 µm particles, which either encapsulate the drug or are composed purely of solid amorphous drug molecules, can be readily synthesized by controlling key drying conditions, such as nozzle size, flow rate, and inlet/outlet temperature (Cabral-Marques & Almeida, 2009). Furthermore, spray drying introduces the ability to modulate additional characteristics associated with: (i) particle morphology, shape, and surface chemistry; (ii) the aerosolization performance (e.g., particle cohesiveness and flowability); and (iii) the drug release/absorption profiles following inhalation (Amaro, Tajber, Corrigan, & Healy, 2011; Minne, Boireau, Horta, & Vanbever, 2008). Due to these advantages, spray drying is considered a more suitable technique for producing inhalable dry powders over other conventional methods, such as jet milling and micronization, since these approaches provide little control over particle size, shape, and morphology, as well as fundamental properties for effective aerosolization (Seville, Li, & Learoyd, 2007; Vidgrén et al., 1987).

Advances in nanotechnology-based engineering approaches have gained additional potential for the fabrication of effective inhalable dry powders through the process of spray drying. Several emerging studies have demonstrated the ability to synthesize a 'nanoparticle-in-microparticle' system for an optimal deep lung deposition and uptake into epithelial lung cells (Jensen et al., 2010; Keil et al., 2020; Simon, Amaro, Cabral, Healy, & de, 2016). That is, a dispersion of drug-confining nanoparticles is spray dried to form microparticles of optimal particle size for inhalation (2–5 µm), which are subsequently composed of a three-dimensional matrix of countless primary nanoparticles (Joyce, Whitby, & Prestidge, 2015). In doing so, the particles are large enough to avoid exhalation and subsequently undergo redispersion/disintegration into the primary particles upon deposition in the lower airways (Imperiale & Sosnik, 2013; Muralidharan, Malapit, Mallory, Hayes, & Mansour, 2015). This approach to inhalation administration allows for optimal uptake into epithelial lung cells and other key target cells within the lung, such as phagocytic cells (Lee, Johnson, Robbins, & Bridson, 2013; Maghrebi et al., 2020).

Of key interest with respect to the nanoparticle-in-microparticle inhalation approach is the ability to effectively deliver protein and gene therapeutics to the lung (Kaye, Purewal, & Alpar, 2009; Rudolph, Lausier, Naundorf, Müller, & Rosenecker, 2000). Biological molecules, such as proteins and nucleic acids, are extremely sensitive to the external environment, due to their ability to be readily hydrolyzed by changes in pH and the presence of enzymes in the lung. Subsequently, they must be delivered within a carrier system that protects the sensitive cargo until its release at the target site. Keil et al. (2020) have performed a wealth of research dedicated to delivering siRNA to activated T-helper cells subtype 2 (TH2) in the lung, to target transcription factor GATA3 for the treatment of asthma. To achieve this, polyplexes approximately 100 nm in diameter were synthesized by combining conjugates of transferrin-polyethyleneimine (Tf-PEI) and melittin-polyethyleneimine (Mel-PEI) with siRNA, through electrostatic-mediated interactions (Liu, Nguyen, Steele, Merkel, & Kissel, 2009). The peptide-polymer conjugates were selected based on the ability for transferrin to promote uptake into TH2 cells via transferrin receptor-mediated uptake and for melittin to promote the endosomal release, once internalized by TH2 cells (Kim, Nadithe, Elsayed, & Merkel, 2013). The polyplexes were spray dried with trehalose (10 wt%), which served as a cryoprotectant to control the aerodynamic properties of the formed microparticles, thereby promoting the fabrication of microparticles ideal for inhalation (Keil et al., 2019). In proof-of-concept studies, siRNA was successfully delivered to TH2 cells, which promoted *in vivo* gene silencing and high transfection efficacies were demonstrated after particle redispersion of the dry powder formulation (Kandil, Feldmann, Xie, & Merkel, 2019; Keil et al., 2019). The polyplex microparticles were of optimal size for dry powder inhalation, which, upon deposition within the lung, were capable of redispersing into the primary particles for enhanced uptake and endosomal escape, through transferrin and melittin receptor-mediated processes, within activated T-cells. The mechanism of synthesis, inhalation approach, and targeted uptake of this spray dried formulation into TH2 cells is represented in Figure 16.6. The initial findings indicate that this formulation approach is a highly promising gene therapy for the treatment of asthma, with continual development and optimization of the formulation expecting to facilitate the translation from bench to bedside (Keil et al., 2020). Furthermore, the work performed is a prominent example for the potential application of spray drying in developing nanoparticle-in-microparticle type formulations for optimal inhalation therapy.

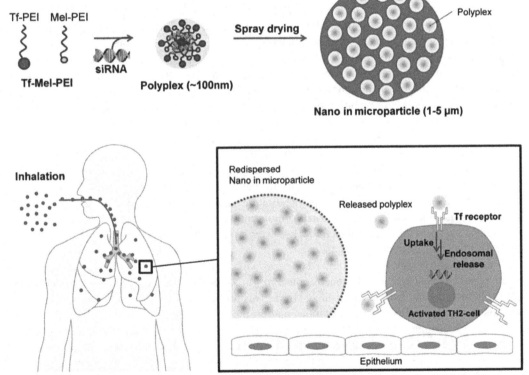

FIGURE 16.6 A schematic representation of the synthesis of nanoparticle-in-microparticles composed of transferrin-polyethyleneimine (Tf-PEI) and melittin-polyethyleneimine (Mel-PEI) siRNA polyplexes. Reproduced from Keil et al. (2020), with permission from John Wiley & Sons.

16.3.3   Dermal and Transdermal Delivery

Particulate drug delivery systems, such as liposomes (Elsayed, Abdallah, Naggar, & Khalafallah, 2007; Honeywell-Nguyen & Bouwstra, 2005), solid lipid nanoparticles (SLN) (Passerini et al., 2009; Pople & Singh, 2006; Santander-Ortega et al., 2010; Teskač & Kristl, 2010), and polymeric micro- and nanoparticles (Lboutounne, Chaulet, Ploton, Falson, & Pirot, 2002; Marchiori et al., 2010; Ourique et al., 2011; Pohlmann et al., 2013), have received increasing interest as carriers for topical drug delivery. Particulate formulations offer several advantages, such as limiting direct contact of the drug with the skin and improving the stability of the encapsulated molecule (Weiss-Angeli et al., 2010). These micro- and nanocarriers can modulate drug penetration of the skin by either improving or limiting its permeation, depending on the intended use of the encapsulated bioactive. For example, nanocarriers may be fabricated to improve the transdermal delivery of the drugs, such as vaccines (Kohli & Alpar, 2004; Mahe et al., 2009) and hormones (Malik, Tondwal, Venkatesh, & Misra, 2008), to enable them to reach systemic circulation (Weiss-Angeli et al., 2010). In contrast, they may be fabricated to limit drug penetration by improving uptake within a certain layer of the skin and reducing systemic absorption for drugs such as psoralen (Doppalapudi, Jain, Chopra, & Khan, 2017), anti-acne agents (Patel & Prabhu, 2020), and some sunscreens. Prow et al. (2011) reported that a ≤ 10 nm particle size was required to reach the dermis, while nanoparticles >10 nm in diameter were unlikely to penetrate through the stratum corneum into the viable human skin and tended to accumulate in the hair follicle openings. Furthermore, particulate formulations could allow a sustained release of the entrapped ingredient to the skin, which may be important for the bioactives that are irritating at high concentrations on the skin over a prolonged period, or to decrease systemic absorption.

Particulate formulations that have been prepared for enhanced dermal and transdermal delivery are most often fabricated using methods such as high-pressure homogenization, precipitation methods, and freeze drying, with very few fabricated using spray drying. However, spray drying has been used to develop some promising skin drug delivery technologies. For example, Beber et al. (2014) encapsulated the anti-inflammatory dexamethasone in submicron polymeric particles via spray drying, with a mean particle size of 0.975 μm. The particles were incorporated into oil-in-water emulsion creams for local administration to the skin. The creams showed controlled drug release kinetics and a significant increase in retention of drug in the epidermis. Del, Russo, et al. (2017) produced several particulate formulations to improve the performance of soy isoflavones extracts. One formulation was prepared using HPMCAS to encapsulate the extracts via nano spray drying, which produced submicron particles (approximately 550 nm) that enhanced skin penetration (Del et al., 2017). An alternate study was employed to investigate the encapsulation of soy isoflavone extract within carboxymethyl cellulose to enhance its affinity for aqueous media (Del, Russo, et al., 2017), which demonstrated a 4.5-fold increase in the permeation through the biological membranes compared to the unformulated extract. Both spray dried formulations were able to increase the bioavailability of soy isoflavones and have potential as topical delivery systems for skin cancer treatments or as anti-aging agents within cosmetic products.

Where the skin barrier is compromised (e.g., wounded, diseased skin or aged skin), there may be potential for an enhanced particle penetration (Prow et al., 2011). Thus, many of the reported methods for fabricating spray dried encapsulated bioactives for dermal delivery is intended for the use on wounds, which typically contain antibiotics and/or anti-inflammatories. Conti, Giunchedi, Genta, and Conte (2000) manufactured spray dried microparticles composed of ampicillin (encapsulated in chitosan) to form powders that demonstrated topical wound-healing properties in albino rats, as evidenced by rapid cicatrization of skin tissue. Similarly, Ngan et al. (2014) prepared amoxicillin-loaded chitosan nanoparticles by nano spray drying, achieving particles in the range of 95–336 nm, which displayed 2-fold greater antibacterial activity against *Staphylococcus aureus* compared to pure amoxicillin. Maged, Mahmoud, and Ghorab (2017) investigated the influence of increasing weight ratio of econazole (antifungal) to cyclodextrin (hydroxypropyl beta-cyclodextrin), on drug release and long-term stability of spray dried nanoparticles. Drug release increased with an increase in drug content, even though it negatively influenced long-term stability by promoting the drug recrystallization over 6 months. De Cicco et al. (2014) were able to encapsulate the antibacterial gentamicin in spray dried alginate-pectin nanopowders, that gelled within 15 min of being administered locally to a wound. The particles had a diameter ranging from 310 to 405 nm, a high yield of 82%–92%, and a high drug loading of 25%. The gelling particles were tested for drug release using Franz cells, demonstrating sustained drug release over 3–6 days and displaying greater antimicrobial activity than pure gentamicin.

The skin barrier can be compromised by microneedles, iontophoresis, electrophoresis, or ultrasound technologies to enhance particle delivery (Marwah, Garg, Goyal, & Rath, 2016). Tawde, Chablani, Akalkotkar, and D'Souza (2016) developed a spray dried microparticulate dermal vaccine for ovarian cancer, using a whole-cell lysate of a murine ovarian cancer cell line. The particles contained methacrylic copolymer Eudragit® FS 30 D and hydroxyl propyl methylcellulose acetate succinate, known for their sustained and controlled release properties. The particles had a diameter of 1.58 ± 0.62 µm (mimicking the size of pathogenic species), and, thus, they could enhance the probability of being phagocytosed by dendritic cells and initiating an immune response. The vaccine was administered to C57BL/6 mice transdermally using a microneedle device, to penetrate the stratum corneum, as well as orally via gavage, in combination with interleukins. The mice were challenged with live tumor cells, and after 15 weeks the vaccinated mice displayed 9-fold greater tumor suppression than the control mice.

While the literature surrounding spray drying to generate particulate formulations for dermal and transdermal skin delivery is limited, there is an opportunity to apply spray drying to a range of particulate formulation strategies and excipients to enhance drug delivery through the skin. Specifically, harnessing nano spray drying to fabricate particles as small as 100 nm and employing strategies such as microneedles to penetrate the skin barrier offers significant opportunities for spray drying as a fabrication method for dermal and transdermal drug delivery formulations.

16.3.4 SYSTEMIC DELIVERY

Spray drying is a key fabrication method for the systemic administration (e.g. intravenous, subcutaneous) of particulate formulations (Schwendeman, Shah, Bailey, & Schwendeman, 2014). The key consideration of spray dried particulate formulations for the systemic route is particle size. Since the smallest capillaries in the body are approximately 5 to 6 µm in diameter, particles should be fabricated significantly smaller than 5 µm, without forming aggregates, to avoid causing an embolism (Singh & Lillard, 2009). Furthermore, the formulation must pass freely through a needle without clogging, which could occur with one single oversized particle (Schwendeman et al., 2014). A typical intravenous needle gauge is approximately 25 G, which has an inner diameter of 260 µm. This emphasizes the importance of optimized spray drying parameters to ensure a narrow particle size distribution and limited agglomeration. Particulate formulations are often utilized to achieve bioactive protection, site-specific delivery, or controlled release/depot delivery.

16.3.4.1 Bioactive Protection

By encapsulating a bioactive within a particulate carrier system, the active can be protected from the *in vivo* environment, or the body can be protected from the active. Some bioactives may undergo rapid enzymatic degradation or have a short half-life (e.g., peptides, proteins, or nucleic acids); thus, they require large or multiple doses to achieve an adequate exposure (Ge et al., 2002; Wan & Yang, 2016). By encapsulating the sensitive active, and controlling the release, the active agent is protected from degradation and is released over time to achieve adequate exposure with a dose reduction (Bowey & Neufeld, 2010). Alternatively, harmful actives (i.e., cytotoxic agents) can be encapsulated to ensure the body is not exposed to the action until it reaches its target site, reducing side-effects (Twaites, De, & Alexander, 2005). For example, the anticancer drug, etanidazole, possesses a short *in vivo* half-life (about 2–3 h in dogs) that needs to be overcome. Wang and Wang (2002) developed the spray dried PLGA microspheres containing etanidazole for injectable delivery to sustain the release of the drug over six hours, limiting its exposure and degradation. In doing so, the concentration and frequency of doses could be reduced, which is expected to lessen the toxic and adverse effects of etanidazole chemotherapy. To protect the siRNA from acidic environments, such as within phagosomes, Amsalem et al. (2017) utilized a unique fabrication method to encapsulate siRNA. Briefly, siRNA-loaded human serum albumin (HSA) nanoparticles of about 100 nm in diameter were first synthesized through the process of cross-linking. The primary HSA nanoparticles were then encapsulated within PLGA nanoparticles (with or without PEG) through nano spray drying, by employing a Büchi Nano Spray Dryer B-90 fitted with a vibrating-mesh perforated head (Amsalem et al., 2017), since this technique affords the ability to create spray dried particles <1 μm in diameter. Nano spray drying resulted in the formation of 'nano-in-nano' spherical particles with an average diameter of 580–770 nm, where the inner morphology was hypothesized to consist of many encapsulated siRNA-loaded HSA nanoparticles embedded within a PLGA polymeric fiber matrix. Importantly, the structural integrity and the gene silencing activity of the encapsulated siRNA were preserved, due to the use of a low temperature during nano spray drying (50°C) and the coating of HSA nanoparticles with a polymeric shell. Since these particles protect the integrity of siRNA from extracellular media and exert controlled drug release mechanisms, it is stipulated that such an approach will be suitable for systemic or pulmonary siRNA delivery.

16.3.4.2   Site-specific Delivery

Injectable particulate formulations can be fabricated to target specific organs and enhance cellular uptake, through the careful fabrication of particles with a specific particle size (Arpagaus et al., 2018). Site-specific drug delivery can be important to prevent the possible undesired side effects, especially for the delivery of cytotoxic chemotherapeutics. Harsha et al. (2014) fabricated 5–15 μm gelatin microspheres, containing the anticancer agent carboplatin, using nano spray drying to treat lung cancer. 24 hours after the intravenous injection to albino mice, the drug concentration in the lungs was significantly higher than in any other tissues, including liver, spleen, or blood. The particles accumulated in the lung, since the size of the injected microspheres (11.3 μm) was larger than the diameter of the capillaries in the lung (about 6 μm) (Harsha, Chandramouli, & Rani, 2009), and demonstrated its potential for lung-specific targeted delivery of therapeutic agents. Similarly, Harsha et al. (2017) encapsulated salbutamol in 8 μm-sized albumin microspheres for an intravenous injection to target the lung, as an alternative to inhalation-based delivery, through the use of nano spray drying. *In vivo* pharmacokinetics revealed that the albumin microspheres increased the concentration of salbutamol >14-fold, compared to the pure drug, due to the tissue targeting potential of the spray dried formulation.

Using a different approach, polymeric particles intended for vaccine delivery can be formulated to mimic the size of the pathogens to increase particulate uptake by antigen-presenting cells such as dendritic cells and macrophages (O'Hagan, Singh, & Ulmer, 2006). The optimal size of the microparticles for efficient uptake is within the 1–3 μm range, with cationic particles being most effective (O'Hagan et al., 2006). Alternatively, nanoparticles (< 200 nm) can access a greater range of target sites, as they allow for the efficient uptake and accumulation by a variety of cell types, especially endothelium in inflammatory sites, epithelium, tumors, or microcapillaries where they can extravasate through (Panyam & Labhasetwar, 2003; Panyam, Sahoo, Prabha, Bargar, & Labhasetwar, 2003).

16.3.4.3   Sustained-release and Depot Delivery

Drug release from the spray dried particles can be systematically controlled by the encapsulation process and the excipients used to encapsulate the bioactive; e.g., drug release from the spray dried microparticles by diffusion of water or drug into or out of the particle, swelling behavior, and/or surface/bulk erosion of the particle (Figure 16.7) (Lengyel et al., 2019). Subsequently, particulate

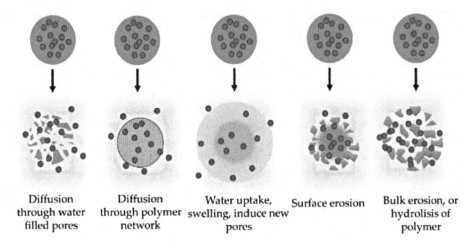

| Diffusion through water filled pores | Diffusion through polymer network | Water uptake, swelling, induce new pores | Surface erosion | Bulk erosion, or hydrolisis of polymer |

FIGURE 16.7   The release behavior of the encapsulated bioactives from spray dried particle-based systems. Reproduced from Lengyel, Antal, Laki, and Antal (2019), with permission from MDPI.

systems can be designed to achieve sustained release over many days, or as depot delivery systems to achieve drug release over many months. Depot delivery systems are often used for the gradual release of bioactives such as vaccines (B. Baras et al., 2000, B. Baras, Benoit, Poulain-Godefroy, et al., 2000), contraceptives, or anticancer agents (Anton et al., 2017) that require chronic exposure over an extended period of time. Many *in vitro* studies document the optimization of spray dried microparticulate carriers (i.e., size and excipient ratios) to achieve a desired sustained release profiles (Baras, Benoit, & Gillard, 2000; Meeus et al., 2015); nonetheless, few document their performance *in vivo*.

An injectable hyaluronic acid micro-hydrogel was successfully developed as a controlled release formulation for the hormone, erythropoietin, and other protein actives by Hahn et al. (2007). The micro-hydrogels were prepared by spray drying a diluted gel precursor solution to achieve a particulate formulation with a mean particle size of 2.3 μm and a 95% recovery of erythropoietin (Hahn et al., 2007). An *in vivo* pharmacokinetic study in Sprague–Dawley rats demonstrated the ability of the injectable micro-hydrogel to maintain erythropoietin concentrations in plasma above 0.1 ng/mL (minimum therapeutic concentration) up to 7 days, compared to 2 days for the injected aqueous erythropoietin formulation.

Bonartsev et al. (2017) demonstrated significant antitumor activity of spray dried paclitaxel loaded microparticles both *in vitro* on murine hepatoma cells and *in vivo* by intraperitoneal administration against transplanted tumor models in mice. The particles were prepared using poly(hydroxyalkanoates), specifically poly(3-hydroxybutyrate), that possess desirable biodegradability and biocompatibility properties. The sustained release of paclitaxel over two months suggested that the biodegradable particles would be suitable for the application in long-term anticancer therapy.

An injectable microparticulate risperidone formulation consisting of PLGA, RISPERDAL® CONSTA®, was approved by the US FDA in 2003 as the first atypical long-acting antipsychotic medication (Panda et al., 2016). Panda et al. (2016) aimed to co-encapsulate a second antipsychotic drug, clozapine, within a similar formulation utilizing nano spray drying and PLGA. The particles were successfully fabricated with a diameter of 250 nm, and were shown to be non-cytotoxic, and sustained drug release up to 10 days *in vitro* (Panda et al., 2016). This nano spray drying process proved to be a suitable strategy for developing polymeric nanopowders with dual-drug delivery properties for the intravenous treatment of schizophrenia. In general, spray drying has demonstrated its broad application to systemically administered bioactives, enabling a range of particle properties such as protection of the bioactive, site-specific delivery, controlled release, and depot delivery.

16.4 CHALLENGES ASSOCIATED WITH SPRAY DRYING BIOACTIVES IN THE PHARMACEUTICAL INDUSTRY

Despite the widespread advantages of spray drying for the encapsulation of bioactives in the pharmaceutical industry, several limitations and challenges exist that have prevented the rapid and frequent translation of spray dried formulations into the clinical and commercial setting. Two important considerations that are fundamental to the success of all pharmaceutical formulations are drug loading and yield. Drug loading for all particulate systems is controlled by the chemistry and structure of both the drug and formulation excipients. For oral administered systems, it is integral that drugs are within their molecular/amorphous forms to allow for efficient absorption across the intestinal epithelium. Thus, the drug loading is controlled by the solubility of the drug within the formulation excipients (Almasri et al., 2020). This is a major challenge for poorly water-soluble bioactive compounds, since loading typically ranges between 1% and 20% for lipid- and polymer-based spray dried systems (Porter et al., 2007; Schultz, Joyce, Thomas, & Prestidge, 2020). Therefore, this approach is typically not feasible or practical for high-dose drugs, and additional formulation design is required to overcome such challenges.

Unlike other drying techniques, such as freeze drying and hot-melt extrusion, which are capable of producing high yields, the spray drying yield is largely dependent on work scale, with yields being lower on smaller-scale operations since the fraction of material lost exists as a greater component of the production volume (Sosnik & Seremeta, 2015). That is, laboratory-based spray drying typically leads to low yields (i.e., about 10%–70%), due to a large proponent of the drying material accumulating on the glass walls of the drying chamber or within the filter (Lee, Heng, Ng, Chan, & Tan, 2011). Not only does this lead to inefficient manufacturing, but the scale-dependent yields also introduce difficulties in scaling spray drying production to an industrial scale. While industrial-scale spray dryers are significantly more efficient and produce greater yields, it can be difficult to predict how the scale-up process leads to changes in powder properties (e.g., particle size, porosity, morphology, flowability) and the behavior of encapsulated bioactives (e.g., loading, solubility, release) (Gaspar, Vicente, Neves, & Authelin, 2014). In this sense, considerable work is required to understand key correlations related to scaling from the lab- to industry-scale spray drying and how this influences overall formulation performance. By improving this understanding, it is expected that the translation from the lab to the clinic and industry will become more economically and commercially viable.

16.5 CONCLUDING REMARKS

Spray drying has transformed from a simple drying technique into a more sophisticated technology for overcoming the biopharmaceutical challenges associated with problematic therapeutics. The strategy provides several advantages over other popular drying techniques, owing to its simple, one-step, high-throughput, and gentle process, allowing the manufacture of uniform spherical particles or agglomerates. This technique can be used in conjunction with a range of solid carriers, and the drying parameters can be fine-tuned to achieve the desired particulate formulation, with the required biopharmaceutical properties. Spray dried formulations have established their potential for drug delivery via a range of administration routes, whereby they exhibit specific advantages over conventional therapies. The translation of such formulations to clinical settings has been limited; however, with the continual investigation and further *in vivo* testing, in combination with the advancement of the new spray drying technologies (e.g., nano spray drying, spray freeze drying, superheated steam spray drying, and two-stage horizontal spray drying), the clinical application of spray dried pharmaceutical formulations holds great potential. The significant focus should continue to be attributed to overcoming the fundamental limitations that prevent clinical and commercial translation of spray dried pharmaceuticals, specifically the poor drug loading and low yields, as well as overcoming the challenges of scaling -up from laboratory-scale to industrial-scale production.

ACKNOWLEDGMENTS

The Australian Research Council Centre of Excellence in Convergent Bio-Nano Science and Technology (ARC CE140100036) is gratefully acknowledged for research funding and support. The Australian Government Research Training Program is acknowledged for the PhD Scholarship of Tahlia R. Meola.

REFERENCES

Administration, U. S. F. D. (2019). *New Drug Therapy Approvals 2019*. January 26 2020. Retrieved from https://www.fda.gov/media/134493/download

Almasri, R., Joyce, P., Schultz, H. B., Thomas, N., Bremmell, K. E., & Prestidge, C. A. (2020). Porous nanostructure, lipid composition, and degree of drug supersaturation modulate in vitro fenofibrate solubilization in silica-lipid hybrids. *Pharmaceutics*, *12*, 687. doi: 10.3390/pharmaceutics12070687

Altamimi, M. A., Elzayat, E. M., Alshehri, S. M., Mohsin, K., Ibrahim, M. A., Al Meanazel, O. T.,… Alsarra, I. A. (2018). Utilizing spray drying technique to improve oral bioavailability of apigenin. *Advanced Powder Technology*, *29*(7), 1676–1684. Retrieved from http://www.sciencedirect.com/science/article/pii/S0921883118301328, https://www.sciencedirect.com/science/article/abs/pii/S0921883118301328?via%3Dihub. doi: 10.1016/j.apt.2018.04.002

Amaro, M. I., Tajber, L., Corrigan, O. I., & Healy, A. M. (2011). Optimisation of spray drying process conditions for sugar nanoporous microparticles (NPMPs) intended for inhalation. *International Journal of Pharmaceutics*, *421*(1), 99–109.

Amsalem, O., Nassar, T., Benhamron, S., Lazarovici, P., Benita, S., & Yavin, E. (2017). Solid nano-in-nanoparticles for potential delivery of siRNA. *Journal of Controlled Release*, *257*, 144–155. Retrieved from http://www.sciencedirect.com/science/article/pii/S0168365916303224, https://www.sciencedirect.com/science/article/abs/pii/S0168365916303224?via%3Dihub. doi: 10.1016/j.jconrel.2016.05.043

Andya, J. D., Maa, Y.-F., Costantino, H. R., Nguyen, P.-A., Dasovich, N., Sweeney, T. D.,… Shire, S. J. (1999). The effect of formulation excipients on protein stability and aerosol performance of spray dried powders of a recombinant humanized anti-IgE monoclonal antibody1. *Pharmaceutical Research*, *16*(3), 350–358.

Anton, P. B., Anton, L. Z., Sergey, G. Y., Irina, I. Z., Vera, L. M., Tatiana, K. M.,… Helena, M. T. (2017). New poly(3-hydroxybutyrate) microparticles with paclitaxel sustained release for intraperitoneal administration.. *Anti-Cancer Agents in Medicinal Chemistry*, *17*(3), 434–441. Retrieved from http://www.eurekaselect.com/node/141730/article, http://www.eurekaselect.com/141730/article. doi: 10.2174/1871520615666160504095433

Arpagaus, C., Collenberg, A., Rütti, D., Assadpour, E., & Jafari, S. M. (2018). Nano spray drying for encapsulation of pharmaceuticals. *International Journal of Pharmaceutics*, *546*(1–2), 194–214.

Atuah, K. N., Walter, E., Merkle, H. P., & Alpar, H. O. (2010). Encapsulation of plasmid DNA in PLGA-stearylamine microspheres: A comparison of solvent evaporation and spray-drying methods. *Journal of Microencapsulation*, *20*(3), 387–399. doi: 10.3109/02652040309178077

Baras, B., Benoit, M. A., & Gillard, J. (2000). Parameters influencing the antigen release from spray dried poly(dl-lactide) microparticles. *International Journal of Pharmaceutics*, *200*(1), 133–145. Retrieved from http://www.sciencedirect.com/science/article/pii/S037851730000363X, https://www.sciencedirect.com/science/article/pii/S037851730000363X?via%3Dihub. doi: 10.1016/S0378-5173(00)00363-X

Baras, B. t., Benoit, M.-A., Poulain-Godefroy, O., Schacht, A.-M., Capron, A., Gillard, J., & Riveau, G. (2000). Vaccine properties of antigens entrapped in microparticles produced by spray-drying technique and using various polyester polymers. *Vaccine*, *18*(15), 1495–1505.

Beber, T. C., Andrade, D. F., Kann, B., Fontana, M. C., Coradini, K., Windbergs, M., & Beck, R. C. R. (2014). Submicron polymeric particles prepared by vibrational spray-drying: Semisolid formulation and skin penetration/permeation studies. *European Journal of Pharmaceutics and Biopharmaceutics*, *88*(3), 602–613. Retrieved from http://www.sciencedirect.com/science/article/pii/S0939641114002331, https://www.sciencedirect.com/science/article/abs/pii/S0939641114002331?via%3Dihub. doi: 10.1016/j.ejpb.2014.07.008

Bhalekar, M., Madgulkar, A., Gunjal, S., & Bagal, A. (2013). Formulation and optimisation of sustained release spray-dried microspheres of glipizide using natural polysaccharide. *PDA Journal of Pharmaceutical Science and Technology*, *67*(2), 146–154. Retrieved from https://journal.pda.org/content/pdajpst/67/2/146.full.pdf. doi: 10.5731/pdajpst.2013.00909

Bonartsev, A. P., Zernov, A. L., Yakovlev, S. G., Zharkova, I. I., Myshkina, V. L., Mahina, T. K.,... Borisova, J. A. (2017). New Poly (3-hydroxybutyrate) microparticles with paclitaxel sustained release for intraperitoneal administration. *Anti-cancer Agents in Medicinal Chemistry, 17*(3), 434–441.

Bora, D., Borude, P., & Bhise, K. (2008). Taste masking by spray-drying technique. *AAPS PharmSciTech, 9*(4), 1159–1164. Retrieved from https://pubmed.ncbi.nlm.nih.gov/19016332, https://www.ncbi.nlm.nih.gov/pmc/articles/PMC2628278/. doi: 10.1208/s12249-008-9154-5

Bowey, K., & Neufeld, R. J. (2010). Systemic and Mucosal Delivery of Drugs within Polymeric Microparticles Produced by Spray Drying. *BioDrugs, 24*(6), 359–377. doi: 10.2165/11539070-000000000-00000

Burggraeve, A., Monteyne, T., Vervaet, C., Remon, J. P., & De Beer, T. (2013). Process analytical tools for monitoring, understanding, and control of pharmaceutical fluidized bed granulation: a review. *European Journal of Pharmaceutics and Biopharmaceutics, 83*(1), 2–15.

Cabral-Marques, H., & Almeida, R. (2009). Optimisation of spray-drying process variables for dry powder inhalation (DPI) formulations of corticosteroid/cyclodextrin inclusion complexes. *European Journal of Pharmaceutics and Biopharmaceutics, 73*(1), 121–129.

Chablani, L., Tawde, S. A., Akalkotkar, A., D'Souza, C., Selvaraj, P., & D'Souza, M. J. (2012). Formulation and evaluation of a particulate oral breast cancer vaccine. *Journal of Pharmaceutical Sciences, 101*(10), 3661–3671. Retrieved from http://www.sciencedirect.com/science/article/pii/S0022354915314064, https://jpharmsci.org/article/S0022–3549(15)31406–4/fulltext. doi: 10.1002/jps.23275

Chablani, L., Tawde, S. A., & D'Souza, M. J. (2012). Spray-dried microparticles: a potential vehicle for oral delivery of vaccines. *Journal of Microencapsulation, 29*(4), 388–397. Retrieved from https://doi.org/1 0.3109/02652048.2011.651503, https://www.tandfonline.com/doi/full/10.3109/02652048.2011.651503. doi: 10.3109/02652048.2011.651503

Chiou, D., & Langrish, T. A. G. (2007). Development and characterisation of novel nutraceuticals with spray drying technology. *Journal of Food Engineering, 82*(1), 84–91. Retrieved from http://www.sciencedirect.com/science/article/pii/S0260877407000684. doi: 10.1016/j.jfoodeng.2007.01.021

Choi, J.-Y., Jin, S.-E., Park, Y., Lee, H.-J., Park, Y., Maeng, H.-J., & Kim, C.-K. (2011). Development of coated nifedipine dry elixir as a long acting oral delivery with bioavailability enhancement. *Archives of Pharmacal Research, 34*(10), 1711–1717. Retrieved from doi: 10.1007/s12272-011-1015-1.

Chranioti, C., Chanioti, S., & Tzia, C. (2016). Comparison of spray, freeze and oven drying as a means of reducing bitter aftertaste of steviol glycosides (derived from *Stevia rebaudiana* Bertoni plant)--Evaluation of the final products. *Food Chemistry, 190*, 1151–1158. Retrieved from https://www.ncbi.nlm.nih.gov/pubmed/26213089. doi: 10.1016/j.foodchem.2015.06.083

Colombo, P., Sonvico, F., Colombo, G., & Bettini, R. (2009). Novel Platforms for Oral Drug Delivery. *Pharmaceutical Research, 26*(3), 601–611. Retrieved from https://doi.org/10.1007/s11095-008-9803-0, https://link.springer.com/article/10.1007%2Fs11095–008–9803-0. doi: 10.1007/s11095-008-9803-0

Conti, B., Giunchedi, P., Genta, I., & Conte, U. (2000). The preparation and in vivo evaluation of the wound-healing properties of chitosan microspheres. *STP Pharma Sciences, 10*(1), 101–104.

Costantino, H. R., & Pikal, M. J. (2004). *Excipients for use in lyophilized pharmaceutical peptide, protein* (Vol. 2). United States of America: American Association of Pharmaceutical Sciences.

Cruz, L., Fattal, E., Tasso, L., Freitas, G. C., Carregaro, A. B., Guterres, S. S.,... Tsapis, N. (2011). Formulation and in vivo evaluation of sodium alendronate spray dried microparticles intended for lung delivery. *Journal of Controlled Release, 152*(3), 370–375. Retrieved from http://www.sciencedirect.com/science/article/pii/S0168365911001350. doi: 10.1016/j.jconrel.2011.02.030

Davis, M., & Walker, G. (2018). Recent strategies in spray drying for the enhanced bioavailability of poorly water-soluble drugs. *Journal of Controlled Release, 269*, 110–127. Retrieved from http://www.sciencedirect.com/science/article/pii/S0168365917309768. doi:10.1016/j.jconrel.2017.11.005

De Cicco, F., Porta, A., Sansone, F., Aquino, R. P., & Del Gaudio, P. (2014). Nanospray technology for an in situ gelling nanoparticulate powder as a wound dressing. *International Journal of Pharmaceutics, 473*(1), 30–37. Retrieved from http://www.sciencedirect.com/science/article/pii/S0378517314004785, https://www.sciencedirect.com/science/article/pii/S0378517314004785?via%3Dihub. doi: 10.1016/j.ijpharm.2014.06.049

Del Gaudio, P., Russo, P., Rodriguez Dorado, R., Sansone, F., Mencherini, T., Gasparri, F., & Aquino, R. P. (2017). Submicrometric hypromellose acetate succinate particles as carrier for soy isoflavones extract with improved skin penetration performance. *Carbohydrate Polymers*, *165*, 22–29. Retrieved from http:// www.sciencedirect.com/science/article/pii/S0144861717301455, https://www.sciencedirect.com/science/article/abs/pii/S0144861717301455?via%3Dihub. doi: 10.1016/j.carbpol.2017.02.025

Del Gaudio, P., Sansone, F., Mencherini, T., De Cicco, F., Russo, P., & Aquino, R. P. (2017). Nanospray Drying as a Novel Tool to Improve Technological Properties of Soy Isoflavone Extracts. *Planta Med*, *83*(05), 426–433. Retrieved from https://www.thieme-connect.de/products/ejournals/abstract/10.105 5/s-0042-110,179. doi:10.1055/s-0042-110179

Dening, T. J., Rao, S., Thomas, N., & Prestidge, C. A. (2016a). Novel nanostructured solid materials for modulating oral drug delivery from solid-state lipid-based drug delivery systems. *The AAPS Journal*, *18*(1), 23–40. Retrieved from https://doi.org/10.1208/s12248-015-9824-7, https://www.ncbi.nlm.nih.gov/pmc/articles/PMC4706287/pdf/12248_2015_Article_9824.pdf. doi: 10.1208/s12248-015-9824-7

Dening, T. J., Rao, S., Thomas, N., & Prestidge, C. A. (2016b). Silica encapsulated lipid-based drug delivery systems for reducing the fed/fasted variations of ziprasidone in vitro. *European Journal of Pharmaceutics and Biopharmaceutics*, *101*, 33–42. Retrieved from http://www.sciencedirect.com/science/article/pii/S0939641116000217. doi: 10.1016/j.ejpb.2016.01.010

Dening, T. J., Thomas, N., Rao, S., van Looveren, C., Cuyckens, F., Holm, R., & Prestidge, C. A. (2018). Montmorillonite and Laponite clay materials for the solidification of lipid-based formulations for the basic drug Blonanserin: in vitro and in vivo investigations. *Molecular Pharmaceutics*, *15*(9), 4148–4160.

Desai, D., Wang, J., Wen, H., Li, X., & Timmins, P. (2013). Formulation design, challenges, and development considerations for fixed dose combination (FDC) of oral solid dosage forms. *Pharmaceutical Development and Technology*, *18*(6), 1265–1276. Retrieved from https://doi.org/10.3109/10837450.20 12.660699. doi: 10.3109/10837450.2012.660699

Dimitrellou, D., Kandylis, P., Petrović, T., Dimitrijević-Branković, S., Lević, S., Nedović, V., & Kourkoutas, Y. (2016). Survival of spray dried microencapsulated Lactobacillus casei ATCC 393 in simulated gastrointestinal conditions and fermented milk. *LWT - Food Science and Technology*, *71*, 169–174. Retrieved from http://www.sciencedirect.com/science/article/pii/S002364381630144X. doi: 10.1016/j.lwt.2016.03.007

Doppalapudi, S., Jain, A., Chopra, D. K., & Khan, W. (2017). Psoralen loaded liposomal nanocarriers for improved skin penetration and efficacy of topical PUVA in psoriasis. *European Journal of Pharmaceutical Sciences*, *96*, 515–529. Retrieved from http://www.sciencedirect.com/science/article/pii/S092809871630464X, https://www.sciencedirect.com/science/article/abs/pii/S092809871630464X?via%3Dihub. doi: 10.1016/j.ejps.2016.10.025

Elsayed, M. M., Abdallah, O. Y., Naggar, V. F., & Khalafallah, N. M. (2007). Lipid vesicles for skin delivery of drugs: reviewing three decades of research. *International Journal of Pharmaceutics*, *332*(1–2), 1–16. Retrieved from https://www.sciencedirect.com/science/article/pii/S0378517306010623?via%3Dihub.

Elversson, J., & Millqvist-Fureby, A. (2005). Particle size and density in spray drying-effects of carbohydrate properties. *J Pharm Sci*, *94*(9), 2049–2060. Retrieved from https://www.ncbi.nlm.nih.gov/pubmed/16052553. doi: 10.1002/jps.20418

Febriyenti, F., Mohtar, N., Mohamed, N., Hamdan, M. R., Najib, S., Salleh, M.,… Baie, B. (2014). Comparison of freeze drying and spray drying methods of haruan extract. *International Journal of Drug Delivery*, *6*, 286–291.

Fonteyne, M., Vercruysse, J., De Leersnyder, F., Van Snick, B., Vervaet, C., Remon, J. P., & De Beer, T. (2015). Process Analytical Technology for continuous manufacturing of solid-dosage forms. *TrAC Trends in Analytical Chemistry*, *67*, 159–166. Retrieved from http://www.sciencedirect.com/science/article/pii/S0165993615000424. doi: 10.1016/j.trac.2015.01.011

Gala, R. P., Popescu, C., Knipp, G. T., McCain, R. R., Ubale, R. V., Addo, R.,… D'Souza, M. J. (2017). Physicochemical and preclinical evaluation of a novel buccal measles vaccine. *AAPS PharmSciTech*, *18*(2), 283–292. Retrieved from https://doi.org/10.1208/s12249-016-0566-3, https://link.springer.com/article/10.1208/s12249-016-0566-3, https://link.springer.com/article/10.1208%2Fs12249–016–0566-3. doi: 10.1208/s12249-016-0566-3

Garbuzenko, O. B., Mainelis, G., Taratula, O., & Minko, T. (2014). Inhalation treatment of lung cancer: the influence of composition, size and shape of nanocarriers on their lung accumulation and retention. *Cancer Biology & Medicine*, *11*(1), 44.

Gaspar, F., Vicente, J., Neves, F., & Authelin, J.-R. (2014). Spray drying: scale-up and manufacturing. In *Amorphous solid dispersions* (pp. 261–302): Springer.

Ge, H., Hu, Y., Jiang, X., Cheng, D., Yuan, Y., Bi, H., & Yang, C. (2002). Preparation, characterization, and drug release behaviors of drug nimodipine-loaded poly (ε-caprolactone)-poly (ethylene oxide)-poly (ε-caprolactone) amphiphilic triblock copolymer micelles. *Journal of Pharmaceutical Sciences*, *91*(6), 1463–1473.

Guns, S., Dereymaker, A., Kayaert, P., Mathot, V., Martens, J. A., & Van den Mooter, G. (2011). Comparison between hot-melt extrusion and spray-drying for manufacturing solid dispersions of the graft copolymer of ethylene glycol and vinylalcohol. *Pharm Res*, *28*(3), 673–682. Retrieved from https://www.ncbi.nlm. nih.gov/pubmed/21104299. doi: 10.1007/s11095-010-0324-2

Hahn, S. K., Kim, J. S., & Shimobouji, T. (2007). Injectable hyaluronic acid microhydrogels for controlled release formulation of erythropoietin. *Journal of Biomedical Materials Research Part A*, *80*(4), 916–924.

Hardy, I. J., & Cook, W. G. (2003). Predictive and correlative techniques for the design, optimisation and manufacture of solid dosage forms. *Journal of Pharmacy and Pharmacology*, *55*(1), 3–18. Retrieved from https://onlinelibrary.wiley.com/doi/abs/10.1111/j.2042-7158.2003.tb02428.x. doi: 10.1111/j.2042-7158.2003.tb02428.x

Harsha, S., Al-Dhubiab, B. E., Nair, A. B., Attimarad, M., Venugopala, K. N., & Kedarnath, S. A. (2017). Pharmacokinetics and tissue distribution of microspheres prepared by spray drying technique: Targeted drug delivery. *Biomedical Research*, 28, 3387–3396

Harsha, S., Al-Khars, M., Al-Hassan, M., Kumar, N. P., Nair, A. B., Attimarad, M., & Al-Dhubiab, B. E. (2014). Pharmacokinetics and tissue distribution of spray dried carboplatin microspheres: lung targeting via intravenous route. *Archives of Pharmacal Research*, *37*(3), 352–360.

Harsha, S., Chandramouli, R., & Rani, S. (2009). Ofloxacin targeting to lungs by way of microspheres. *International Journal of Pharmaceutics*, *380*(1–2), 127–132. Retrieved from https://www.sciencedirect. com/science/article/pii/S0378517309004864?via%3Dihub.

Haser, A., Cao, T., Lubach, J., Listro, T., Acquarulo, L., & Zhang, F. (2017). Melt extrusion vs. spray drying: The effect of processing methods on crystalline content of naproxen-povidone formulations. *European Journal of Pharmaceutical Sciences*, *102*, 115–125. Retrieved from http://www.sciencedirect.com/science/article/pii/S092809871730115X. doi: 10.1016/j.ejps.2017.02.038

Hengsawas Surasarang, S., Keen, J. M., Huang, S., Zhang, F., McGinity, J. W., & Williams, R. O. (2017). Hot melt extrusion versus spray drying: hot melt extrusion degrades albendazole. *Drug Development and Industrial Pharmacy*, *43*(5), 797–811. Retrieved from https://doi.org/10.1080/03639045.2016.1220577. doi: 10.1080/03639045.2016.1220577

Homayouni, A., Sadeghi, F., Nokhodchi, A., Varshosaz, J., & Afrasiabi Garekani, H. (2015). Preparation and characterization of celecoxib dispersions in Soluplus(®): comparison of spray drying and conventional methods. *Iranian Journal of Pharmaceutical Research*, *14*(1), 35–50. Retrieved from https://pubmed. ncbi.nlm.nih.gov/25561910, https://www.ncbi.nlm.nih.gov/pmc/articles/PMC4277617/, https://www. ncbi.nlm.nih.gov/pmc/articles/PMC4277617/pdf/ijpr-14-035.pdf.

Honeywell-Nguyen, P. L., & Bouwstra, J. A. (2005). Vesicles as a tool for transdermal and dermal delivery. *Drug Discovery Today: Technologies*, *2*(1), 67–74. Retrieved from https://www.sciencedirect.com/science/article/abs/pii/S1740674905000089?via%3Dihub.

Hong, S., Shen, S., Tan, D. C., Ng, W. K., Liu, X., Chia, L. S.,… Gokhale, R. (2016). High drug load, stable, manufacturable and bioavailable fenofibrate formulations in mesoporous silica: a comparison of spray drying versus solvent impregnation methods. *Drug Deliv*, *23*(1), 316–327. Retrieved from https://www. ncbi.nlm.nih.gov/pubmed/24853963. doi: 10.3109/10717544.2014.913323

Horaczek, A., & Viernstein, H. (2004). Comparison of three commonly used drying technologies with respect to activity and longevity of aerial conidia of Beauveria brongniartii and Metarhizium anisopliae. *Biological Control*, *31*(1), 65–71. Retrieved from http://www.sciencedirect.com/science/article/pii/S104996440400088X. doi: 10.1016/j.biocontrol.2004.04.016

Hori, M., Onishi, H., & Machida, Y. (2005). Evaluation of Eudragit-coated chitosan microparticles as an oral immune delivery system. *International Journal of Pharmaceutics*, *297*(1), 223–234. Retrieved from http://www.sciencedirect.com/science/article/pii/S0378517305002437, https://www. sciencedirect.com/science/article/pii/S0378517305002437?via%3Dihub. doi: 10.1016/j.ijpharm.2005. 04.008

Hou, Y., Shao, J., Fu, Q., Li, J., Sun, J., & He, Z. (2017). Spray-dried nanocrystals for a highly hydrophobic drug: Increased drug loading, enhanced redispersity, and improved oral bioavailability. *International Journal of Pharmaceutics*, *516*(1), 372–379. Retrieved from http://www.sciencedirect.com/science/article/pii/S037851731631106, https://www.sciencedirect.com/science/article/pii/S0378517316311061?via%3Dihub. doi: 10.1016/j.ijpharm.2016.11.0431

Husband, A. J. (1993). Novel vaccination strategies for the control of mucosal infection. *Vaccine*, *11*(2), 107–112. Retrieved from http://www.sciencedirect.com/science/article/pii/0264410X9390003G. doi: 10.1016/0264-410X(93)90003-G

Imperiale, J. C., & Sosnik, A. (2013). Nanoparticle-in-Microparticle delivery systems (NiMDS): production, administration routes and clinical potential. *Journal of Biomaterials and Tissue Engineering*, *3*(1), 22–38.

Jensen, D. M. K., Cun, D., Maltesen, M. J., Frokjaer, S., Nielsen, H. M., & Foged, C. (2010). Spray drying of siRNA-containing PLGA nanoparticles intended for inhalation. *Journal of Controlled Release*, *142*(1), 138–145.

Johansen, P., Men, Y., Merkle, H. P., & Gander, B. (2000). Revisiting PLA/PLGA microspheres: an analysis of their potential in parenteral vaccination. *European Journal of Pharmaceutics and Biopharmaceutics*, *50*(1), 129–146. Retrieved from http://www.sciencedirect.com/science/article/pii/S0939641100000795. doi: 10.1016/S0939-6411(00)00079-5

Joyce, P., Barnes, T. J., Boyd, B. J., & Prestidge, C. A. (2016). Porous nanostructure controls kinetics, disposition and self-assembly structure of lipid digestion products. *RSC Advances*, *6*(82), 78385–78395. Retrieved from http://dx.doi.org/10.1039/C6RA16028J.

Joyce, P., Dening, T. J., Gustafsson, H., & Prestidge, C. A. (2017). Modulating the lipase-mediated bioactivity of particle-lipid conjugates through changes in nanostructure and surface chemistry. *European Journal of Lipid Science & Technology*, *119*(12), 1700213–1700222.

Joyce, P., Dening, T. J., Meola, T. R., Schultz, H. B., Holm, R., Thomas, N., & Prestidge, C. A. (2019). Solidification to improve the biopharmaceutical performance of SEDDS: Opportunities and challenges. *Advanced Drug Delivery Reviews*, *142*, 102–117. Retrieved from http://www.sciencedirect.com/science/article/pii/S0169409X18303016. doi: 10.1016/j.addr.2018.11.006

Joyce, P., & Prestidge, C. A. (2018). Synergistic effect of PLGA nanoparticles and submicron triglyceride droplets in enhancing the intestinal solubilisation of a lipophilic weak base. *European Journal of Pharmaceutical Sciences*, *118*, 40–48.

Joyce, P., Schultz, H. B., Meola, T. R., & Prestidge, C. A. (2020). Polymer lipid hybrid (PLH) formulations: a synergistic approach to oral delivery of challenging therapeutics. In R. Shegokar (Ed.), *Delivery of Drugs* (pp. 1–27). Amsterdam, Oxford and Cambridge, MA: Elsevier.

Joyce, P., Tan, A., Whitby, C. P., & Prestidge, C. A. (2014). The role of porous nanostructure in controlling lipase-mediated digestion of lipid loaded into silica particles. *Langmuir*, *30*(10), 2779–2788. Retrieved from https://doi.org/10.1021/la500094b. doi: 10.1021/la500094b

Joyce, P., Whitby, C. P., & Prestidge, C. A. (2015). Bioactive hybrid particles from poly(d,l-lactide-co-glycolide) nanoparticle stabilized lipid droplets. *ACS Applied Materials & Interfaces*, *7*(31), 17460–17470. Retrieved from http://dx.doi.org/10.1021/acsami.5b05068.

Joyce, P., Whitby, C. P., & Prestidge, C. A. (2016). Nanostructuring biomaterials with specific activities towards digestive enzymes for controlled gastrointestinal absorption of lipophilic bioactive molecules. *Advances in Colloid and Interface Science*, *237*, 52–75. Retrieved from http://www.sciencedirect.com/science/article/pii/S0001868616301658. doi: 10.1016/j.cis.2016.10.003

Joyce, P., Yasmin, R., Bhatt, A., Boyd, B. J., Pham, A., & Prestidge, C. A. (2017). Comparison across three hybrid lipid-based drug delivery systems for improving the oral absorption of the poorly water-soluble weak base cinnarizine. *Molecular Pharmaceutics*, *14*(11), 4008–4018. Retrieved from https://doi.org/10.1021/acs.molpharmaceut.7b00676, https://pubs.acs.org/doi/10.1021/acs.molpharmaceut.7b00676. doi: 10.1021/acs.molpharmaceut.7b00676

Kandil, R., Feldmann, D., Xie, Y., & Merkel, O. M. (2019). Evaluating the regulation of cytokine levels after siRNA treatment in antigen-specific target cell populations via intracellular staining. In *Nanotechnology for Nucleic Acid Delivery* (pp. 323–331). New York, NY.: Springer.

Kang, J. H., Oh, D. H., Oh, Y.-K., Yong, C. S., & Choi, H.-G. (2012). Effects of solid carriers on the crystalline properties, dissolution and bioavailability of flurbiprofen in solid self-nanoemulsifying drug delivery system (solid SNEDDS). *European Journal of Pharmaceutics and Biopharmaceutics*, *80*(2), 289–297. Retrieved from http://www.sciencedirect.com/science/article/pii/S0939641111003237https://www.sciencedirect.com/science/article/abs/pii/S0939641111003237?via%3Dihub. doi: 10.1016/j.ejpb.2011.11.005

Kaye, R. S., Purewal, T. S., & Alpar, H. O. (2009). Simultaneously manufactured nano-in-micro (SIMANIM) particles for dry-powder modified-release delivery of antibodies. *Journal of Pharmaceutical Sciences*, *98*(11), 4055–4068.

Keil, T. W., Feldmann, D. P., Costabile, G., Zhong, Q., da Rocha, S., & Merkel, O. M. (2019). Characterization of spray dried powders with nucleic acid-containing PEI nanoparticles. *European Journal of Pharmaceutics and Biopharmaceutics*, *143*, 61–69.

Keil, T. W. M., Baldassi, D., & Merkel, O. M. (2020). T-cell targeted pulmonary siRNA delivery for the treatment of asthma. *WIREs Nanomedicine and Nanobiotechnology*, e1634. Retrieved from https://doi.org/10.1002/wnan.1634. doi: 10.1002/wnan.1634

Kelleher, J. F., Gilvary, G. C., Madi, A. M., Jones, D. S., Li, S., Tian, Y.,… Healy, A. M. (2018). A comparative study between hot-melt extrusion and spray-drying for the manufacture of anti-hypertension compatible monolithic fixed-dose combination products. *International Journal of Pharmaceutics*, *545*(1), 183–196. Retrieved from http://www.sciencedirect.com/science/article/pii/S0378517318303016https://www.sciencedirect.com/science/article/pii/S0378517318303016?via%3Dihub. doi: 10.1016/j.ijpharm.2018.05.008

Kim, N., Nadithe, V., Elsayed, M., & Merkel, O. (2013). Tracking and treating activated T cells. *Journal of Drug Delivery Science and Technology*, *23*(1), 17–21.

Kohli, A., & Alpar, H. (2004). Potential use of nanoparticles for transcutaneous vaccine delivery: effect of particle size and charge. *International Journal of Pharmaceutics*, *275*(1–2), 13–17.

Labortechnik, B. (2009). *Switzerland Brochure Nano Spray Dryer B-90*. In. Flawil, Switzerland: Büchi.

Lane, M. E., Brennan, F. S., & Corrigan, O. I. (2005). Comparison of post-emulsification freeze drying or spray drying processes for the microencapsulation of plasmid DNA. *Journal of Pharmacy and Pharmacology*, *57*(7), 831–838. Retrieved from https://onlinelibrary.wiley.com/doi/abs/10.1211/0022357056406. doi: 10.1211/0022357056406

Lboutounne, H., Chaulet, J.-F., Ploton, C., Falson, F., & Pirot, F. (2002). Sustained ex vivo skin antiseptic activity of chlorhexidine in poly(ε-caprolactone) nanocapsule encapsulated form and as a digluconate. *Journal of Controlled Release*, *82*(2), 319–334. Retrieved from http://www.sciencedirect.com/science/article/pii/S0168365902001426, https://www.sciencedirect.com/science/article/abs/pii/S0168365902001426?via%3Dihub. doi: 10.1016/S0168-3659(02)00142-6

Lee, S. H., Heng, D., Ng, W. K., Chan, H.-K., & Tan, R. B. (2011). Nano spray drying: a novel method for preparing protein nanoparticles for protein therapy. *International Journal of Pharmaceutics*, *403*(1–2), 192–200.

Lee, W.-H., Loo, C.-Y., Traini, D., & Young, P. M. (2015). Inhalation of nanoparticle-based drug for lung cancer treatment: advantages and challenges. *Asian Journal of Pharmaceutical Sciences*, *10*(6), 481–489.

Lee, Y.-S., Johnson, P. J., Robbins, P. T., & Bridson, R. H. (2013). Production of nanoparticles-in-microparticles by a double emulsion method: A comprehensive study. *European Journal of Pharmaceutics and Biopharmaceutics*, *83*(2), 168–173.

Lengyel, M. K.-S. N., Antal, V., Laki, A.J., Antal, I. (2019). Microparticles, Microspheres, and Microcapsules for Advanced Drug Delivery. *Scientia Pharmaceutica*, *87*, 20.

Li, D. X., Kim, J. O., Oh, D. H., Lee, W. S., Hong, M. J., Kang, J. Y.,… Choi, H.-G. (2009). Development of nifedipine-loaded coated gelatin microcapsule as a long acting oral delivery. *Archives of Pharmacal Research*, *32*(1), 127–132. Retrieved from https://doi.org/10.1007/s12272-009-1126-0, https://link.springer.com/article/10.1007%2Fs12272-009-1126-0. doi: 10.1007/s12272-009-1126-0

Li, L., Hui Zhou, C., & Ping Xu, Z. (2019). Chapter 14 - self-nanoemulsifying drug-delivery system and solidified self-nanoemulsifying drug-delivery system. In S. S. Mohapatra, S. Ranjan, N. Dasgupta, R. K. Mishra, & S. Thomas (Eds.), *Nanocarriers for Drug Delivery* (pp. 421–449). Amsterdam, Oxford, Cambridge, MA: Elsevier.

Liu, Y., Nguyen, J., Steele, T., Merkel, O., & Kissel, T. (2009). A new synthesis method and degradation of hyper-branched polyethylenimine grafted polycaprolactone block mono-methoxyl poly (ethylene glycol) copolymers (hy-PEI-g-PCL-b-mPEG) as potential DNA delivery vectors. *Polymer*, *50*(16), 3895–3904.

Maged, A., Mahmoud, A., & Ghorab, M. (2017). Hydroxypropyl-beta-cyclodextrin as cryoprotectant in nanoparticles prepared by nano spray drying technique. *J Pharm Sci Emerg Drugs*, *5*, *1*, 2.

Maghrebi, S., Joyce, P., Jambhrunkar, M., Thomas, N., & Prestidge, C. A. (2020). PLGA-Lipid Hybrid (PLH) Microparticles Enhance the Intracellular Uptake and Anti-Bacterial Activity of Rifampicin. *ACS Applied Materials & Interfaces*, *12*(7), 8030–8039.

Maghrebi, S., Prestidge, C. A., & Joyce, P. (2019). An update on polymer-lipid hybrid systems for improving oral drug delivery. *Expert Opinion on Drug Delivery*, *16*(5), 507–524.

Mahe, B., Vogt, A., Liard, C., Duffy, D., Abadie, V., Bonduelle, O.,… Blume-Peytavi, U. (2009). Nanoparticle-based targeting of vaccine compounds to skin antigen-presenting cells by hair follicles and their transport in mice. *Journal of Investigative Dermatology*, *129*(5), 1156–1164. Retrieved from https://www.jidonline.org/article/S0022-202X(15)34322-0/pdf.

Malik, R., Tondwal, S., Venkatesh, K., & Misra, A. (2008). Nanoscaffold matrices for size-controlled, pulsatile transdermal testosterone delivery: nanosize effects on the time dimension. *Nanotechnology*, *19*(43), 435101.

Maniruzzaman, M., Boateng, J. S., Snowden, M. J., & Douroumis, D. (2012). A Review of Hot-Melt Extrusion: Process Technology to Pharmaceutical Products. *ISRN Pharmaceutics*, *2012*, 436,763. Retrieved from https://doi.org/10.5402/2012/436763. doi: 10.5402/2012/436763

Marchiori, M., Lubini, G., Dalla Nora, G., Friedrich, R., Fontana, M., Ourique, A.,… Tedesco, S. (2010). Hydrogel containing dexamethasone-loaded nanocapsules for cutaneous administration: preparation, characterization, and in vitro drug release study. *Drug Development and Industrial Pharmacy*, *36*(8), 962–971. Retrieved from https://www.tandfonline.com/doi/full/10.3109/03639041003598960.

Marwah, H., Garg, T., Goyal, A. K., & Rath, G. (2016). Permeation enhancer strategies in transdermal drug delivery. *Drug Delivery*, *23*(2), 564–578. Retrieved from https://doi.org/10.3109/10717544.2014.935532, https://www.tandfonline.com/doi/pdf/10.3109/10717544.2014.935532?needAccess=true. doi: 10.3109/10717544.2014.935532

Maury, M., Murphy, K., Kumar, S., Mauerer, A., & Lee, G. (2005). Spray-drying of proteins: effects of sorbitol and trehalose on aggregation and FT-IR amide I spectrum of an immunoglobulin G. *European Journal of Pharmaceutics and Biopharmaceutics*, *59*(2), 251–261. Retrieved from http://www.sciencedirect.com/science/article/pii/S0939641104002103, https://www.sciencedirect.com/science/article/abs/pii/S0939641104002103?via%3Dihub. doi: 10.1016/j.ejpb.2004.07.010

Meenach, S. A., Anderson, K. W., Zach Hilt, J., McGarry, R. C., & Mansour, H. M. (2013). Characterization and aerosol dispersion performance of advanced spray dried chemotherapeutic PEGylated phospholipid particles for dry powder inhalation delivery in lung cancer. *European Journal of Pharmaceutical Sciences*, *49*(4), 699–711. Retrieved from http://www.sciencedirect.com/science/article/pii/S0928098713001917. doi: 10.1016/j.ejps.2013.05.012

Meeus, J., Lenaerts, M., Scurr, D. J., Amssoms, K., Davies, M. C., Roberts, C. J., & Van den Mooter, G. (2015). The influence of spray-drying parameters on phase behavior, drug distribution, and in vitro release of injectable microspheres for sustained release. *Journal of Pharmaceutical Sciences*, *104*(4), 1451–1460. Retrieved from http://www.sciencedirect.com/science/article/pii/S0022354915301684, https://onlinelibrary.wiley.com/doi/abs/10.1002/jps.24361?sid=nlm%3Apubmed. doi: 10.1002/jps.24361

Meola, T. R., Abuhelwa, A., Joyce, P., Clifton, P. M., & Prestidge, C. A. (2020). A safety, tolerability and pharmacokinetic study of a novel simvastatin silica-lipid hybrid formulation in healthy male subjects. *Drug Delivery and Translational Research*, (in press), 1–12. doi: 10.1007/s13346-020-00853-x.

Meola, T. R., Schultz, H. B., Peressin, K., & Prestidge, C. A. (2020). Enhancing the oral bioavailability of simvastatin with silica-lipid hybrid particles: The effect of supersaturation and silica geometry. *European Journal of Pharmaceutical Sciences*(Accepted: In Press). 150, 105357.

Minne, A., Boireau, H., Horta, M. J., & Vanbever, R. (2008). Optimization of the aerosolization properties of an inhalation dry powder based on selection of excipients. *European Journal of Pharmaceutics and Biopharmaceutics*, *70*(3), 839–844.

Mou, D., Chen, H., Wan, J., Xu, H., & Yang, X. (2011). Potent dried drug nanosuspensions for oral bioavailability enhancement of poorly soluble drugs with pH-dependent solubility. *International Journal of Pharmaceutics*, *413*(1), 237–244. Retrieved from http://www.sciencedirect.com/science/article/pii/S0378517311003619, https://www.sciencedirect.com/science/article/pii/S0378517311003619?via%3Dihub. doi: 10.1016/j.ijpharm.2011.04.034

Muralidharan, P., Malapit, M., Mallory, E., Hayes, D., & Mansour, H. M. (2015). Inhalable nanoparticulate powders for respiratory delivery. *Nanomedicine: Nanotechnology, Biology and Medicine*, *11*(5), 1189–1199. Retrieved from http://www.sciencedirect.com/science/article/pii/S1549963415000313. doi: 10.1016/j.nano.2015.01.007

Naikwade, S., & Bajaj, A. (2009). Preparation and in vitro evaluation of budesonide spray dried microparticles for pulmonary delivery. *Scientia Pharmaceutica*, *77*(2), 419–442.

Nelson, K. D. (2015). 6 - Absorbable, drug-loaded, extruded fiber for implantation. In T. Blair (Ed.), *Biomedical Textiles for Orthopaedic and Surgical Applications* (pp. 119–143). Cambridge, Kidlington, UK: Woodhead Publishing.

Ngan, L. T. K., Wang, S.-L., Hiep, Đ. M., Luong, P. M., Vui, N. T., Đinh, T. M., & Dzung, N. A. (2014). Preparation of chitosan nanoparticles by spray drying, and their antibacterial activity. *Research on Chemical Intermediates*, *40*(6), 2165–2175. Retrieved from https://doi.org/10.1007/s11164-014-1594-9, https://link.springer.com/article/10.1007%2Fs11164–014–1594-9. doi: 10.1007/s11164-014-1594-9

Nguyen, D. N., Clasen, C., & Van den Mooter, G. (2016). Pharmaceutical applications of electrospraying. *Journal of Pharmaceutical Sciences*, *105*(9), 2601–2620.

Nunes, G. L., Etchepare, M. D. A., Cichoski, A. J., Zepka, L. Q., Jacob Lopes, E., Barin, J. S.,… de Menezes, C. R. (2018). Inulin, hi-maize, and trehalose as thermal protectants for increasing viability of Lactobacillus acidophilus encapsulated by spray drying. *LWT*, *89*, 128–133. Retrieved from http://www.sciencedirect.com/science/article/pii/S0023643817307703. doi: 10.1016/j.lwt.2017.10.032

O'Callaghan, C., & Barry, P. W. (1997). The science of nebulised drug delivery. *Thorax*, *52*(Suppl 2), S31.

O'Hagan, D. T., & Singh, M. (2003). Microparticles as vaccine adjuvants and delivery systems. *Expert Review of Vaccines*, *2*(2), 269–283. Retrieved from https://doi.org/10.1586/14760584.2.2.269, https://www.tandfonline.com/doi/abs/10.1586/14760584.2.2.269. doi: 10.1586/14760584.2.2.269

O'Hagan, D. T., Singh, M., & Ulmer, J. B. (2006). Microparticle-based technologies for vaccines. *Methods*, *40*(1), 10–19. Retrieved from http://www.sciencedirect.com/science/article/pii/S1046202306001575, https://www.sciencedirect.com/science/article/pii/S1046202306001575?via%3Dihub. doi: 10.1016/j.ymeth.2006.05.017

Ourique, A. F., Melero, A., Silva, C. D. B. D., Schaefer, U. F., Pohlmann, A. R., Guterres, S. S.,… Beck, R. C. R. (2011). Improved photostability and reduced skin permeation of tretinoin: Development of a semisolid nanomedicine. *European Journal of Pharmaceutics and Biopharmaceutics*, *79*(1), 95–101. Retrieved from http://www.sciencedirect.com/science/article/pii/S093964111100107X, https://www.sciencedirect.com/science/article/abs/pii/S093964111100107X?via%3Dihub. doi: 10.1016/j.ejpb.2011.03.008

Panda, A., Meena, J., Katara, R., & Majumdar, D. K. (2016). Formulation and characterization of clozapine and risperidone co-entrapped spray dried PLGA nanoparticles. *Pharmaceutical Development and Technology*, *21*(1), 43–53. Retrieved from https://doi.org/10.3109/10837450.2014.965324, https://www.tandfonline.com/doi/full/10.3109/10837450.2014.965324. doi: 10.3109/10837450.2014.965324

Panyam, J., & Labhasetwar, V. (2003). Biodegradable nanoparticles for drug and gene delivery to cells and tissue. *Advanced Drug Delivery Reviews*, *55*(3), 329–347.

Panyam, J., Sahoo, S. K., Prabha, S., Bargar, T., & Labhasetwar, V. (2003). Fluorescence and electron microscopy probes for cellular and tissue uptake of poly (D, L-lactide-co-glycolide) nanoparticles. *International Journal of Pharmaceutics*, *262*(1–2), 1–11.

Passerini, N., Gavini, E., Albertini, B., Rassu, G., Di Sabatino, M., Sanna, V.,… Rodriguez, L. (2009). Evaluation of solid lipid microparticles produced by spray congealing for topical application of econazole nitrate. *Journal of Pharmacy and Pharmacology*, *61*(5), 559–567. Retrieved from https://onlinelibrary.wiley.com/doi/abs/10.1211/jpp.61.05.0003. doi: 10.1211/jpp.61.05.0003

Pastor, M., Esquisabel, A., Marquínez, I., Talavera, A., & Pedraz, J. L. (2014). Cellulose acetate phthalate microparticles containing Vibrio cholerae: steps toward an oral cholera vaccine. *Journal of Drug Targeting*, *22*(6), 478–487. Retrieved from https://doi.org/10.3109/1061186X.2014.888071, https://www.tandfonline.com/doi/full/10.3109/1061186X.2014.888071. doi: 10.3109/1061186X.2014.888071

Pastor, M., Esquisabel, A., Talavera, A., Año, G., Fernández, S., Cedré, B.,… Pedraz, J. L. (2013). An approach to a cold chain free oral cholera vaccine: in vitro and in vivo characterization of Vibrio cholerae gastro-resistant microparticles. *International Journal of Pharmaceutics*, *448*(1), 247–258. Retrieved from http://www.sciencedirect.com/science/article/pii/S0378517313002123, https://www.sciencedirect.com/science/article/pii/S0378517313002123?via%3Dihub. doi: 10.1016/j.ijpharm.2013.02.057

Patel, B.B., Patel, J. K., & Chakraborty, S. (2014). Review of patents and application of spray drying in pharmaceutical, food and flavor industry. *Recent Patents on Drug Delivery & Formulation*, *8*(1), 63–78. Retrieved from https://www.ingentaconnect.com/content/ben/ddf/2014/00000008/00000001/art00007.

Patel, R., & Prabhu, P. (2020). Nanocarriers as versatile delivery systems for effective management of acne. *International Journal of Pharmaceutics*, *119*, 140.

Paudel, A., Worku, Z. A., Meeus, J., Guns, S., & Van den Mooter, G. (2013). Manufacturing of solid dispersions of poorly water soluble drugs by spray drying: Formulation and process considerations. *International Journal of Pharmaceutics*, *453*(1), 253–284. Retrieved from http://www.sciencedirect.com/science/article/pii/S0378517312006904, https://www.sciencedirect.com/science/article/pii/S0378517312006904?via%3Dihub. doi: 10.1016/j.ijpharm.2012.07.015

Pohlen, M., Lavrič, Z., Prestidge, C., & Dreu, R. (2020). Preparation, physicochemical characterisation and doe optimisation of a spray-dried dry emulsion platform for delivery of a poorly soluble drug, simvastatin. *AAPS PharmSciTech*, *21*(4), 119. Retrieved from https://doi.org/10.1208/s12249-020-01651-x. doi: 10.1208/s12249-020-01651-x

Pohlmann, A. R., Fonseca, F. N., Paese, K., Detoni, C. B., Coradini, K., Beck, R. C. R., & Guterres, S. S. (2013). Poly(ε-caprolactone) microcapsules and nanocapsules in drug delivery. *Expert Opinion on Drug Delivery*, *10*(5), 623–638. Retrieved from https://doi.org/10.1517/17425247.2013.769956, https://www.tandfonline.com/doi/full/10.1517/17425247.2013.769956. doi: 10.1517/17425247.2013.769956

Pople, P. V., & Singh, K. K. (2006). Development and evaluation of topical formulation containing solid lipid nanoparticles of vitamin A. *AAPS PharmSciTech*, *7*(4), E63–E69. Retrieved from https://doi.org/10.1208/pt070491, https://link.springer.com/article/10.1208%2Fpt070491. doi: 10.1208/pt070491

Porter, C. J. H., Trevaskis, N.L., Charman, W.N. (2007). Lipids and lipid-based formulations: optimising the oral delivery of lipophilic drugs. *National Review of Drug Discovery*, *6*(3), 231–248.

Pouton, C. W. (2006). Formulation of poorly water-soluble drugs for oral administration: Physicochemical and physiological issues and the lipid formulation classification system. *European Journal of Pharmaceutical Sciences*, *29*(3–4), 278–287. Retrieved from http://www.sciencedirect.com/science/article/pii/S0928098706001151. doi: 10.1016/j.ejps.2006.04.016

Prow, T. W., Grice, J. E., Lin, L. L., Faye, R., Butler, M., Becker, W.,… Roberts, M. S. (2011). Nanoparticles and microparticles for skin drug delivery. *Advanced Drug Delivery Reviews*, *63*(6), 470–491. Retrieved from http://www.sciencedirect.com/science/article/pii/S0169409X11000160, https://www.sciencedirect.com/science/article/abs/pii/S0169409X11000160?via%3Dihub. doi: 10.1016/j.addr.2011.01.012

Rao, M. R. P., & Aghav, S. S. (2014). Spray-Dried Redispersible Emulsion to Improve Oral Bioavailability of Itraconazole. *Journal of Surfactants and Detergents*, *17*(4), 807–817. Retrieved from https://doi.org/10.1007/s11743-013-1538-1, https://aocs.onlinelibrary.wiley.com/doi/abs/10.1007/s11743-013-1538-1. doi: 10.1007/s11743-013-1538-1

Rashidinejad, A., Loveday, S. M., Jameson, G. B., Hindmarsh, J. P., & Singh, H. (2019). Rutin-casein co-precipitates as potential delivery vehicles for flavonoid rutin. *Food Hydrocolloids*, *96*, 451–462. Retrieved from http://www.sciencedirect.com/science/article/pii/S0268005X18324871. doi: 10.1016/j.foodhyd.2019.05.032

Ré, M.-I. (2006). Formulating drug delivery systems by spray drying. *Drying Technology*, *24*(4), 433–446. Retrieved from https://doi.org/10.1080/07373930600611877. doi: 10.1080/07373930600611877

Respaud, R., Vecellio, L., Diot, P., & Heuzé-Vourc'h, N. (2015). Nebulization as a delivery method for mAbs in respiratory diseases. *Expert Opinion on Drug Delivery*, *12*(6), 1027–1039.

Rudolph, C., Lausier, J., Naundorf, S., Müller, R. H., & Rosenecker, J. (2000). In vivo gene delivery to the lung using polyethylenimine and fractured polyamidoamine dendrimers. *The Journal of Gene Medicine*, *2*(4), 269–278.

Salama, A. H. (2020). Spray drying as an advantageous strategy for enhancing pharmaceuticals bioavailability. *Drug Delivery and Translational Research*, *10*(1), 1–12. Retrieved from https://doi.org/10.1007/s13346-019-00648-9, https://link.springer.com/article/10.1007%2Fs13346–019–00648-9. doi: 10.1007/s13346-019-00648-9

Saluja, V., Amorij, J. P., Kapteyn, J. C., de Boer, A. H., Frijlink, H. W., & Hinrichs, W. L. J. (2010). A comparison between spray drying and spray freeze drying to produce an influenza subunit vaccine powder for inhalation. *Journal of Controlled Release*, *144*(2), 127–133. Retrieved from http://www.sciencedirect.com/science/article/pii/S0168365910001586. doi: 10.1016/j.jconrel.2010.02.025

Santander-Ortega, M. J., Stauner, T., Loretz, B., Ortega-Vinuesa, J. L., Bastos-González, D., Wenz, G.,… Lehr, C. M. (2010). Nanoparticles made from novel starch derivatives for transdermal drug delivery. *Journal of Controlled Release*, *141*(1), 85–92. Retrieved from http://www.sciencedirect.com/science/article/pii/S0168365909005768, https://www.sciencedirect.com/science/article/abs/pii/S0168365909005768?via%3Dihub. doi: 10.1016/j.jconrel.2009.08.012

Schultz, H. B., Joyce, P., Thomas, N., & Prestidge, C. A. (2020). Supersaturated-silica lipid hybrids improve in vitro solubilization of abiraterone acetate. *Pharmaceutical Research*, *37*(4), 77–77.

Schwendeman, S. P., Shah, R. B., Bailey, B. A., & Schwendeman, A. S. (2014). Injectable controlled release depots for large molecules. *Journal of Controlled Release*, *190*, 240–253. Retrieved from https://www.ncbi.nlm.nih.gov/pmc/articles/PMC4261190/pdf/nihms604679.pdf.

Seville, P. C., Li, H.-y., & Learoyd, T. P. (2007). Spray-dried powders for pulmonary drug delivery. *Critical Reviews™ in Therapeutic Drug Carrier Systems*, *24*(4), 307–360.

Shahin, H. I., Vinjamuri, B. P., Mahmoud, A. A., Shamma, R. N., Mansour, S. M., Ammar, H. O.,… Chablani, L. (2019). Design and evaluation of novel inhalable sildenafil citrate spray dried microparticles for pulmonary arterial hypertension. *Journal of Controlled Release, 302*, 126–139. Retrieved from http://www.sciencedirect.com/science/article/pii/S0168365919301890. doi: 10.1016/j.jconrel.2019.03.029

Shakweh, M., Besnard, M., Nicolas, V., & Fattal, E. (2005). Poly (lactide-co-glycolide) particles of different physicochemical properties and their uptake by peyer's patches in mice. *European Journal of Pharmaceutics and Biopharmaceutics, 61*(1), 1–13. Retrieved from http://www.sciencedirect.com/science/article/pii/S0939641105001426, https://www.sciencedirect.com/science/article/abs/pii/S0939641105001426?via%3Dihub. doi: 10.1016/j.ejpb.2005.04.006

Shastri, P. N., Kim, M. C., Quan, F. S., D'Souza, M. J., & Kang, S. M. (2012). Immunogenicity and protection of oral influenza vaccines formulated into microparticles. *Journal of Pharmaceutical Sciences, 101*(10), 3623–3635. Retrieved from http://www.sciencedirect.com/science/article/pii/S0022354915313654, https://www.ncbi.nlm.nih.gov/pmc/articles/PMC5558794/pdf/nihms890733.pdf. doi: 10.1002/jps.23220

Simon, A., Amaro, M. I., Cabral, L. M., Healy, A. M., & de Sousa, V. P. (2016). Development of a novel dry powder inhalation formulation for the delivery of rivastigmine hydrogen tartrate. *International Journal of Pharmaceutics, 501*(1), 124–138. Retrieved from http://www.sciencedirect.com/science/article/pii/S0378517316300667. doi: 10.1016/j.ijpharm.2016.01.066

Simonoska Crcarevska, M., Glavas Dodov, M., & Goracinova, K. (2008). Chitosan coated Ca–alginate microparticles loaded with budesonide for delivery to the inflamed colonic mucosa. *European Journal of Pharmaceutics and Biopharmaceutics, 68*(3), 565–578. Retrieved from http://www.sciencedirect.com/science/article/pii/S0939641107002123, https://www.sciencedirect.com/science/article/abs/pii/S0939641107002123?via%3Dihub. doi: 10.1016/j.ejpb.2007.06.007

Simovic, S., Heard, P., Hui, H., Song, Y., Peddie, F., Davey, A. K.,… Prestidge, C. A. (2009). Dry Hybrid Lipid–Silica Microcapsules Engineered from Submicron Lipid Droplets and Nanoparticles as a Novel Delivery System for Poorly Soluble Drugs. *Molecular Pharmaceutics, 6*(3), 861–872. Retrieved from https://doi.org/10.1021/mp900063t, https://pubs.acs.org/doi/10.1021/mp900063t. doi: 10.1021/mp900063t

Singh, A., & Van den Mooter, G. (2016). Spray drying formulation of amorphous solid dispersions. *Advanced Drug Delivery Reviews, 100*, 27–50. Retrieved from http://www.sciencedirect.com/science/article/pii/S0169409X15300041, https://www.sciencedirect.com/science/article/abs/pii/S0169409X15300041?via%3Dihub. doi: 10.1016/j.addr.2015.12.010

Singh, R., & Lillard, J. W. (2009). Nanoparticle-based targeted drug delivery. *Experimental and Molecular Pathology, 86*(3), 215–223. Retrieved from http://www.sciencedirect.com/science/article/pii/S001448000800141X, https://www.ncbi.nlm.nih.gov/pmc/articles/PMC3249419/pdf/nihms87089.pdf. doi: 10.1016/j.yexmp.2008.12.004

Singh, S. K., Vuddanda, P. R., Singh, S., & Srivastava, A. K. (2013). A comparison between use of spray and freeze drying techniques for preparation of solid self-microemulsifying formulation of valsartan and in vitro and in vivo evaluation. *Biomed Res Int, 2013*, 909045. Retrieved from https://www.ncbi.nlm.nih.gov/pubmed/23971048. doi: 10.1155/2013/909045

Sinha, V. R., & Trehan, A. (2003). Biodegradable microspheres for protein delivery. *Journal of Controlled Release, 90*(3), 261–280. Retrieved from http://www.sciencedirect.com/science/article/pii/S0168365903001949, https://www.sciencedirect.com/science/article/abs/pii/S0168365903001949?via%3Dihub. doi: 10.1016/S0168-3659(03)00194-9

Smola, M., Vandamme, T., & Sokolowski, A. (2008). Nanocarriers as pulmonary drug delivery systems to treat and to diagnose respiratory and non respiratory diseases. *International Journal of Nanomedicine, 3*(1), 1.

Sollohub, K., & Cal, K. (2010). Spray drying technique: II. Current applications in pharmaceutical technology. *Journal of Pharmaceutical Sciences, 99*(2), 587–597. Retrieved from http://www.sciencedirect.com/science/article/pii/S0022354916304129, https://jpharmsci.org/article/S0022-3549(16)30412-9/fulltext. doi: 10.1002/jps.21963

Sosnik, A., & Seremeta, K. P. (2015). Advantages and challenges of the spray-drying technology for the production of pure drug particles and drug-loaded polymeric carriers. *Advances in Colloid and Interface Science, 223*, 40–54. Retrieved from http://www.sciencedirect.com/science/article/pii/S0001868615000767. doi: 10.1016/j.cis.2015.05.003

Tabata, Y., Inoue, Y., & Ikada, Y. (1996). Size effect on systemic and mucosal immune responses induced by oral administration of biodegradable microspheres. *Vaccine, 14*(17), 1677–1685. Retrieved from http://www.sciencedirect.com/science/article/pii/S0264410X96001491, https://www.sciencedirect.com/science/article/pii/S0264410X96001491?via%3Dihub. doi: 10.1016/S0264-410X(96)00149-1

Tan, A., Colliat-Dangus, P., Whitby, C. P., & Prestidge, C. A. (2014). Controlling the enzymatic digestion of lipids using hybrid nanostructured materials. *ACS Applied Materials & Interfaces*, 6(17), 15,363–15,371. Retrieved from http://dx.doi.org/10.1021/am5038577.

Tan, A., Eskandar, N. G., Rao, S., & Prestidge, C. A. (2014). First in man bioavailability and tolerability studies of a silica–lipid hybrid (Lipoceramic) formulation: a Phase I study with ibuprofen. *Drug Delivery and Translational Research*, 4(3), 212–221. Retrieved from https://doi.org/10.1007/s13346-013-0172-9, https://link.springer.com/article/10.1007%2Fs13346–013–0172-9. doi: 10.1007/s13346-013-0172-9

Tan, A., Rao, S., & Prestidge, C. A. (2013). Transforming lipid-based oral drug delivery systems into solid dosage forms: An overview of solid carriers, physicochemical properties, and biopharmaceutical performance. *Pharmaceutical Research*, 30(12), 2993–3017. Retrieved from https://doi.org/10.1007/s11095-013-1107-3, https://link.springer.com/article/10.1007%2Fs11095–013–1107-3. doi: 10.1007/s11095-013-1107-3

Tan, A., Simovic, S., Davey, A. K., Rades, T., & Prestidge, C. A. (2009). Silica-lipid hybrid (SLH) microcapsules: A novel oral delivery system for poorly soluble drugs. *Journal of Controlled Release*, 134(1), 62–70. Retrieved from http://www.sciencedirect.com/science/article/pii/S0168365908006779, https://www.sciencedirect.com/science/article/abs/pii/S0168365908006779?via%3Dihub. doi: 10.1016/j.jconrel.2008.10.014

Tang, X., & Pikal, M. J. (2004). Design of freeze-drying processes for pharmaceuticals: Practical advice. *Pharmaceutical Research*, 21(2), 191–200. Rewtrieved from https://doi.org/10.1023/B:PHAM.0000016234.73023.75. doi: 10.1023/B:PHAM.0000016234.73023.75

Tawde, S. A., Chablani, L., Akalkotkar, A., & D'Souza, M. J. (2016). Evaluation of microparticulate ovarian cancer vaccine via transdermal route of delivery. *Journal of Controlled Release*, 235, 147–154. Retrieved from https://www.sciencedirect.com/science/article/abs/pii/S0168365916303480?via%3Dihub.

Teskač, K., & Kristl, J. (2010). The evidence for solid lipid nanoparticles mediated cell uptake of resveratrol. *International Journal of Pharmaceutics*, 390(1), 61–69. Retrieved from http://www.sciencedirect.com/science/article/pii/S0378517309007200, https://www.sciencedirect.com/science/article/pii/S0378517309007200?via%3Dihub. doi: 10.1016/j.ijpharm.2009.10.011

Thombre, A. G., Caldwell, W. B., Friesen, D. T., McCray, S. B., & Sutton, S. C. (2012). Solid nanocrystalline dispersions of ziprasidone with enhanced bioavailability in the fasted state. *Molecular Pharmaceutics*, 9(12), 3526–3534. Retrieved from https://doi.org/10.1021/mp3003607, https://pubs.acs.org/doi/pdf/10.1021/mp3003607. doi: 10.1021/mp3003607

Truong-Le, V., Lovalenti, P. M., & Abdul-Fattah, A. M. (2015). Stabilization challenges and formulation strategies associated with oral biologic drug delivery systems. *Advanced Drug Delivery Reviews*, 93, 95–108. Retrievedw from http://www.sciencedirect.com/science/article/pii/S0169409X15001830, https://www.sciencedirect.com/science/article/abs/pii/S0169409X15001830?via%3Dihub. doi: 10.1016/j.addr.2015.08.001

Twaites, B., De Las Heras Alarcón, C., & Alexander, C. (2005). Synthetic polymers as drugs and therapeutics. *Journal of Materials Chemistry*, 15(4), 441–455.

Vehring, R. (2008). Pharmaceutical particle engineering via spray drying. *Pharmaceutical Research*, 25(5), 999–1022. Retrieved from https://www.ncbi.nlm.nih.gov/pubmed/18040761. doi: 10.1007/s11095-007-9475-1

Vidgrén, M. T., Vidgrén, P. A., & Paronen, T. P. (1987). Comparison of physical and inhalation properties of spray dried and mechanically micronized disodium cromoglycate. *International Journal of Pharmaceutics*, 35(1), 139–144. Retrieved from http://www.sciencedirect.com/science/article/pii/0378517387900822, https://www.sciencedirect.com/science/article/pii/0378517387900822?via%3Dihub. doi: 10.1016/0378-5173(87)90082-2

Walters, R. H., Bhatnagar, B., Tchessalov, S., Izutsu, K.-I., Tsumoto, K., & Ohtake, S. (2014). Next generation drying technologies for pharmaceutical applications. *Journal of Pharmaceutical Sciences*, 103(9), 2673–2695. Retrieved from http://www.sciencedirect.com/science/article/pii/S0022354915304330. doi: https://doi.org/10.1002/jps.23998

Wan, F., & Yang, M. (2016). Design of PLGA-based depot delivery systems for biopharmaceuticals prepared by spray drying. *International Journal of Pharmaceutics*, 498(1–2), 82–95. Retrieved from https://www.sciencedirect.com/science/article/pii/S0378517315304269?via%3Dihub.

Wang, F., & Wang, C.-H. (2002). Effects of fabrication conditions on the characteristics of etanidazole spray dried microspheres. *Journal of Microencapsulation*, 19(4), 495–510. Retrieved from https://www.tandfonline.com/doi/abs/10.1080/02652040210140483.

Weiss-Angeli, V., Bourgeois, S., Pelletier, J., Guterres, S. S., Fessi, H., & Bolzinger, M.-A. (2010). Development of an original method to study drug release from polymeric nanocapsules in the skin. *Journal of Pharmacy and Pharmacology*, 62(1), 35–45. Retrieved from https://onlinelibrary.wiley.com/doi/abs/10.1211/jpp.62.01.0003. doi: 10.1211/jpp.62.01.0003

Wong, S. M., Kellaway, I. W., & Murdan, S. (2006). Enhancement of the dissolution rate and oral absorption of a poorly water soluble drug by formation of surfactant-containing microparticles. *International Journal of Pharmaceutics*, *317*(1), 61–68. Retrieved from http://www.sciencedirect.com/science/article/pii/S0378517306001980, https://www.sciencedirect.com/science/article/pii/S0378517306001980?via%3Dihub. doi: 10.1016/j.ijpharm.2006.03.001

Xu, E.-Y., Guo, J., Xu, Y., Li, H.-Y., & Seville, P. C. (2014). Influence of excipients on spray dried powders for inhalation. *Powder Technology*, *256*, 217–223. Retrieved from http://www.sciencedirect.com/science/article/pii/S0032591014001491. doi: 10.1016/j.powtec.2014.02.033

Yasmin, R., Tan, A., Bremmell, K. E., & Prestidge, C. A. (2014). Lyophilized silica lipid hybrid (SLH) carriers for poorly water-soluble drugs: physicochemical and in vitro pharmaceutical investigations. *Journal of Pharmaceutical Sciences*, *103*(9), 2950–2959.

Yuki, Y., & Kiyono, H. (2009). Mucosal vaccines: novel advances in technology and delivery. *Expert Review of Vaccines*, *8*(8), 1083–1097. Retrieved from https://doi.org/10.1586/erv.09.61. doi: 10.1586/erv.09.61

Zhang, J., Gao, Y., Qian, S., Liu, X., & Zu, H. (2011). Physicochemical and pharmacokinetic characterization of a spray dried malotilate emulsion. *International Journal of Pharmaceutics*, *414*(1), 186–192. Retrieved from http://www.sciencedirect.com/science/article/pii/S0378517311004625, https://www.sciencedirect.com/science/article/pii/S0378517311004625?via%3Dihub. doi: 10.1016/j.ijpharm.2011.05.032

Index

Page numbers in **bold** indicate tables, page numbers in *italics* indicate figures and page numbers followed by n indicate notes.